Compressible Flow

Compressible Flow

STEFAN SCHREIER
Associate Professor of Mechanical Engineering
University of Connecticut

1807 175 YEARS OF PUBLISHING 1982

A WILEY-INTERSCIENCE PUBLICATION

JOHN WILEY & SONS

New York · Chichester · Brisbane · Toronto · Singapore

Library of Congress Cataloging in Publication Data:

Schreier, Stefan

Compressible flow.

"A Wiley-Interscience publication."

Includes bibliographical references and index.

1. Gas dynamics. I. Title.

QA930.S383 533'.2 80-20607

ISBN 0-471-05691-X

Printed in the United States of America

10 9 8 7 6 5 4 3 2 1

Preface

The present work had its origin in the one-year graduate course in compressible flow which the author has taught for a number of years at the University of Connecticut. The work is somewhat broader in scope than most works in the field. In addition to the conventional material on perfect gas flow, the book contains chapters on the Navier–Stokes equations, transonic flow, laminar boundary layers, turbulent boundary layers, real gas effects and computational methods.

In the basic theory the approach has been to derive the equations in their most general form first and then to apply the results to particular cases. Thus, the momentum, energy, and continuity equations are first derived for three-dimensional unsteady flow and the results are then reduced to the particular cases of quasi-one-dimensional steady flow, steady two- and three-dimensional flow, unsteady one-dimensional flow, and normal and oblique shocks. In order to make the derivations more readily comprehensible, they are presented in more detail than usual.

Every attempt has been made to discuss the physical significance of the derivations as opposed to a merely formal presentation of the mathematics. Where philosophical or logical questions arise, these are brought out and discussed.

The derivation of the Navier–Stokes equation is presented in some detail in a manner which, it is hoped, will enable the reader to readily grasp the significance of this important equation. In the chapters on transonic flow, laminar boundary layers, and turbulent boundary layers, the most important theories are discussed, at some length, in preference to presenting a cursory overview of as many works as possible. Emphasis is placed on presenting experimental results as well as theoretical results as checks on the validity of the theories. In Chapter 9 the properties of real, as opposed to ideal gases are discussed in detail, as are flows where real gas effects are significant. Finally, a chapter on numerical methods is presented, since numerical analysis has become an indispensable part of the solution of problems in compressible flow.

The present work will appeal mainly to graduate students and workers in the field, but may also be used by advanced undergraduates. A background in vector analysis, ordinary and partial differential equations is desirable as well as a knowledge of elementary fluid mechanics and thermodynamics.

STEFAN SCHREIER

Storrs, Connecticut
March 1981

Contents

Compressible
Flow

1 The Basic Equations

The Nature of the Problem

A compressible fluid differs from an incompressible one in that the density ρ of the fluid is not constant. As a result, an additional unknown is introduced. In incompressible flow it is found that only two natural laws are required for the solution of problems: the law of the conservation of mass and the law of the conservation of momentum. These laws are usually expressed as the continuity equation

$$\nabla \mathbf{V} = 0 \tag{1}$$

and the momentum equation

$$\rho \frac{D\mathbf{V}}{Dt} = -\nabla p \tag{2}$$

If ρ is known, there are only two unknowns in incompressible flow: the pressure p and the velocity $\mathbf{V}$. Equations (1) and (2) are, therefore, sufficient to render the problem determinate.

In compressible flow, the addition of another unknown, ρ, requires the introduction of some other relation. Such a relation exists in the law of the conservation of energy or the first law of thermodynamics. For closed systems this relation is usually written

$$đQ = dE + đW \tag{3}$$

where $đQ$ represents the heat added to the system, $đW$ the work done by the system, and dE the change in the energy of the system. The bar through the $đ$ in $đW$ and $đQ$ indicates that these quantities are not exact differentials—in other words, Q and W are not properties of the system. The quantity E, in most compressible flow problems, consists of the internal energy E_t and the kinetic energy $\frac{1}{2}mV^2$. Thus, if the heat added and the work done are presumed given, the use of equation (3) adds one new unknown to the system of equations, the internal energy E_t. The internal energy, however, is a state property; that is, an intrinsic property of the gas independent of its history. Pressure, temperature and density, for example, are state properties. Heat and work are not. From elementary thermodynamics we know that for a homogeneous gas, any two state properties are sufficient to determine all the rest. Hence, the internal energy may be related to the pressure and density through an equation of state of the form,

$$E_t = E_t(p, \rho) \tag{4}$$

Thus, through the introduction of equations (3) and (4), the system of equations is once again rendered determinate; that is, there are now four equations and four unknowns: p, ρ, E_t, and $\mathbf{V}$.

The equations cited above apply only to the simplest types of compressible flow problems—those in which the fluid is homogeneous and inviscid. The presence of viscosity and chemical reactions, for example, introduces additional unknowns and renders the problem more complicated. It may thus be seen that problems in compressible flow tend to be more complicated than those in incompressible flow.

Inviscid Compressible Flow—Derivation of the Governing Equations

The equations governing the inviscid compressible flow of a homogeneous fluid were first derived by Euler [1]. Euler considered all the properties of the fluid to be continuous functions of time and space. Consider, for example, the coordinate system shown in Figure 1 and assume a fluid to be flowing through it.

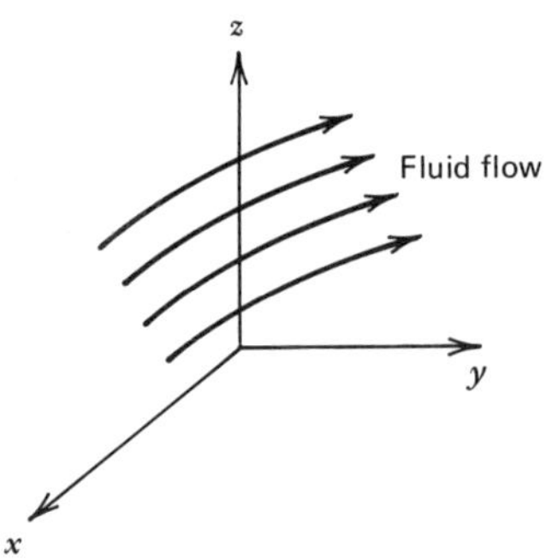

Figure 1. Fluid flow in a Cartesian coordinate system.

Each point in the flow can be described by the values of x, y and z at that point. Under the assumptions of Euler, then, the flow properties may be described by functional relations of the form

$$\mathbf{V} = \mathbf{V}(x,y,z,t), \qquad T = T(x,y,z,t), \qquad \rho = \rho(x,y,z,t)$$

and so on. By filling in numbers for x, y, z, and t, the value of each property at a particular point in space at a particular time is determined.

It should be pointed out that this is not the only possible approach to the problem which may be taken. An alternative would be to describe the properties of the fluid not as functions of a point in space, but as functions of a particular fluid particle. A particle of fluid might be identified, for example, by its position in space at a particular time, say, $t = 0$. At that time the particle may be located in space at a point $x = \xi, y = \eta, z = \zeta$. The properties of the flow may then be described by functional relations of the form

$$\mathbf{V} = \mathbf{V}(\xi,\eta,\zeta,t), \qquad T = T(\xi,\eta,\zeta,t), \qquad \rho = \rho(\xi,\eta,\zeta,t)$$

This is the approach taken by Lagrange [2].

The approach taken by Euler assumes that the fluid is a continuum. Strictly speaking, this assumption is not correct since, according to the molecular theory of matter, a fluid is made up of discrete particles called molecules. Thus, one cannot speak with accuracy about fluid properties at a point. As long as the average distance between molecules is small compared to the scale of the problem, however, this difficulty is more apparent than real. If, for example, we are dealing with the flow over a wing with a ten foot chord, and the average distance between molecules is a millionth of an inch, then, for all practical purposes, the fluid may be considered a continuum. If, on the other hand, the distance between molecules is on the order of a foot, then the situation is no longer so clear. In applying the continuum assumption, we must be careful that the average distance between molecules is small compared to the scale of the problem.

Let us now derive the equations governing the flow of a homogeneous, inviscid, compressible fluid. The first law to be applied is the law of the conservation of mass. This law states that the change in the amount of mass in a given volume over a period of time equals the difference between the amount of mass which has entered the volume during that time, and the amount which has left. Alternatively, one may say that the rate at which the mass in the volume is changing equals the difference between the rate at which mass is entering the volume and the rate at which mass is leaving. One may use the analogy of the bucket with a hole in it (Figure 2). The rate of change of the amount of fluid

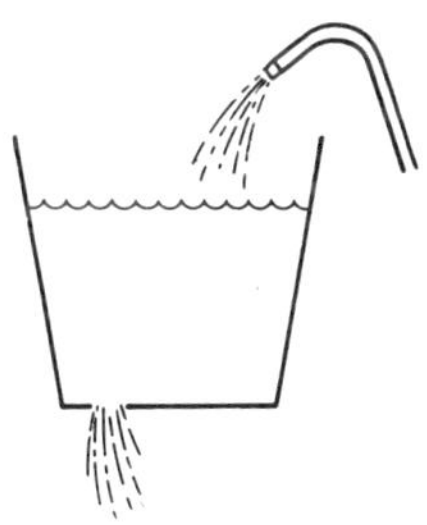

Figure 2. An example of the conservation of mass.

in the bucket is equal to the difference between the rate at which fluid is flowing into the bucket through the hose and the rate at which it is flowing out through the hole in the bottom.

Consider now a fixed volume $\mathscr{V}$ in a cartesian coordinate system through which the fluid is flowing (Figure 3). Let S be the surface of the volume and let $\mathbf{n}$

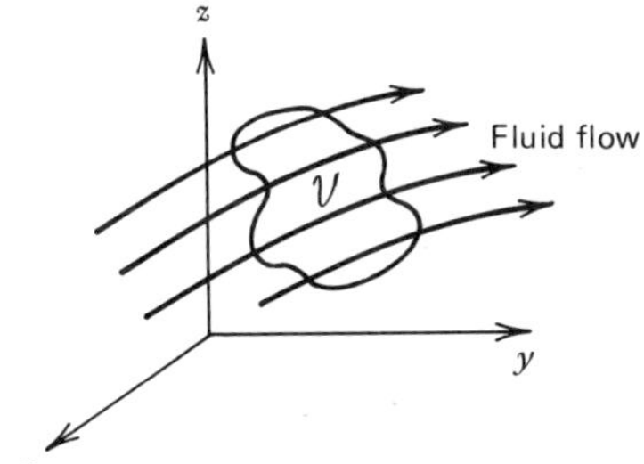

Figure 3. Control volume in three-dimensional flow.

be the unit outward normal at any point of the surface. The net inflow of mass into the volume is then given by,

$$-\int_S \rho \mathbf{V}\cdot\mathbf{n}\, dS$$

where the subscript S indicates that the integral is taken over the surface.

The mass of the fluid in the volume $\mathscr{V}$ at any time is given by

$$\int_{\mathscr{V}} \rho\, d\mathscr{V}$$

where the subscript $\mathscr{V}$ indicates that the integral is taken over the volume.

The rate of change of the mass within the volume is, thus,

$$\frac{\partial}{\partial t}\int_{\mathscr{V}} \rho\, d\mathscr{V}$$

Hence, according to the law of the conservation of mass,

$$\frac{\partial}{\partial t}\int_{\mathscr{V}} \rho\, d\mathscr{V} = -\int_S \rho \mathbf{V}\cdot\mathbf{n}\, dS$$

From vector calculus we know that

$$\int_S \rho \mathbf{V}\cdot\mathbf{n}\, dS = \int_{\mathscr{V}} \nabla(\rho \mathbf{V})\, d\mathscr{V}$$

Also, since $\mathscr{V}$ is not a function of t,

$$\frac{\partial}{\partial t}\int_{\mathscr{V}} \rho\, d\mathscr{V} = \int_{\mathscr{V}} \frac{\partial \rho}{\partial t}\, d\mathscr{V}$$

Hence,

$$\int_{\mathscr{V}} \frac{\partial \rho}{\partial t}\, d\mathscr{V} = -\int_{\mathscr{V}} \nabla(\rho \mathscr{V})\, d\mathscr{V}$$

Now since $\mathscr{V}$ is arbitrary, the equation must be valid for any $\mathscr{V}$, no matter how small. This requirement can only be satisfied if the integrands are equal at every point. Thus,

$$\frac{\partial \rho}{\partial t} + \nabla(\rho \mathbf{V}) = 0 \tag{5}$$

This is the continuity equation for a homogeneous, compressible fluid as it is most commonly written.

We now turn to the second fundamental law, that of the conservation of momentum. This law was first enunciated by Newton [3] and nowadays is most commonly written in the form,

$$\mathbf{F} = m\mathbf{a} \tag{6}$$

As stated in equation (6), this law applies to a closed system. A closed system is one which always consists of the same particles of mass or molecules. In fluid mechanics, however, we are usually interested in open systems. An open system is one into which mass is entering and/or from which mass is leaving. A pipe with water flowing through it would be an open system. We are, thus, faced with the task of restating the law in a form valid for open systems.

The easiest way of applying laws valid only for closed systems to open systems is to apply them to each particle of the open system as it moves along. As long as we stay with the same particle, that particle represents a closed system. If we consider a particle of fluid enclosed by a volume $d\mathscr{V}$, the mass of that particle will be

$$\rho\, d\mathscr{V}$$

When it comes to expressing the acceleration, however, we encounter a problem in the Eulerian system. As will be recalled, in the Eulerian system, the velocity was expressed as a function of x, y, z, and t.

$$\mathbf{V} = \mathbf{V}(x, y, z, t)$$

This represents a velocity at a particular point in space at a particular time. Thus, the expression

$$\frac{\partial \mathbf{V}}{\partial t}$$

does not represent the acceleration of the particle; rather, it represents the rate of change of the velocity at a particular point in space.

Consider, on the other hand, the Lagrangian system. In this system, the velocity is given by

$$\mathbf{V} = \mathbf{V}(\xi, \eta, \zeta, t)$$

This represents the velocity of a particular particle rather than the velocity at a particular point in space. Hence, the expression

$$\frac{\partial \mathbf{V}}{\partial \mathrm{t}}$$

in Lagrangian coordinates represents the acceleration of a particular particle. It might be considered that because of the ease with which the acceleration may be expressed in the Lagrangian system, this system is preferable to the Eulerian one. In most practical problems, however, other factors, such as the formulation of boundary conditions, enter in, which make the Eulerian system preferable.

Let us, therefore, attempt to express the acceleration of a particle in the Eulerian system. Given that

$$\mathbf{V} = \mathbf{V}(x, y, z, t)$$

it follows that

$$d\mathbf{V} = \frac{\partial \mathbf{V}}{\partial x}dx + \frac{\partial \mathbf{V}}{\partial y}dy + \frac{\partial \mathbf{V}}{\partial z}dz + \frac{\partial \mathbf{V}}{\partial t}dt \tag{7}$$

This represents the difference between the velocity at the point x, y, z at time t and the velocity at the point $x + dx, y + dy, z + dz$ at time $t + dt$. It should be pointed out that this statement is accurate only in the limit as dx, dy, dz, and dt go to zero. Otherwise, it is approximate. Let us now divide both sides of equation (7) by dt. Then,

$$\frac{d\mathbf{V}}{dt} = \frac{\partial \mathbf{V}}{\partial x}\frac{dx}{dt} + \frac{\partial \mathbf{V}}{\partial y}\frac{dy}{dt} + \frac{\partial \mathbf{V}}{\partial z}\frac{dz}{dt} + \frac{\partial \mathbf{V}}{\partial t}$$

Since the ratios of dx, dy, dz, and dt to each other are arbitrary, let us choose them in such a way that the moving particle of fluid located at x, y, z at time t will be located at $x + dx, y + dy, z + dz$ at time $t + dt$. dV will then represent the change in velocity of that particular particle in time dt. If we now take the limit as dx, dy, dz, and dt go to zero, we get

$$\frac{dx}{dt} = u \qquad \frac{dy}{dt} = v \qquad \frac{dz}{dt} = w$$

where

$$\mathbf{V} = \mathbf{i}u + \mathbf{j}v + \mathbf{k}w$$

and $d\mathbf{V}/dt$ will then represent the acceleration of a particular particle in the Eulerian system.

$$\frac{d\mathbf{V}}{dt} = u\frac{\partial \mathbf{V}}{\partial x} + v\frac{\partial \mathbf{V}}{\partial y} + w\frac{\partial \mathbf{V}}{\partial z} + \frac{\partial \mathbf{V}}{\partial t} = \frac{D\mathbf{V}}{Dt} = \mathbf{a} \tag{8}$$

When the particular values

$$\frac{dx}{dt} = u, \qquad \frac{dy}{dt} = v, \qquad \frac{dz}{dt} = w$$

are used, as in equation (8), then $d\mathbf{V}/dt$ is usually written

$$\frac{D\mathbf{V}}{Dt}$$

to distinguish it from other values of $d\mathbf{V}/dt$. $D\mathbf{V}/Dt$ is called the substantial derivative.

It may be noted that the value of the operator

$$\frac{D}{Dt} = u\frac{\partial}{\partial x} + v\frac{\partial}{\partial y} + w\frac{\partial}{\partial z} + \frac{\partial}{\partial t}$$

is not restricted to the velocity. It may be applied to any property of the fluid and will then give the rate of change of that property for a particular particle of fluid. Thus,

$$\frac{D\rho}{Dt}$$

represents the rate of change of the density of a particular particle, as it moves along,

$$\frac{DT}{Dt}$$

represents the rate of change of the temperature of a particular particle, and so on.

Having expressed the mass and acceleration of a particular particle in the Eulerian system, we may now apply the law of the conservation of momentum, equation (6), to that particle:

$$\mathbf{F} = m\mathbf{a} = \rho\, d\mathscr{V}\frac{D\mathbf{V}}{Dt} = \mathbf{f}\, d\mathscr{V} \tag{9}$$

In order to evaluate the force **F**, it will be easier for us to work in terms of integrals. Let us, therefore, return to the situation shown in Figure 3. It is important to note that the volume $\mathscr{V}$ is fixed in space. Applying equation (9) to each particle of fluid in the volume $\mathscr{V}$ at a given time we have,

$$\int_{\mathscr{V}} \mathbf{f}\, d\mathscr{V} = \int_{\mathscr{V}} \rho\frac{D\mathbf{V}}{Dt}\, d\mathscr{V}$$

In the absence of viscous, chemical, electrical, magnetic or body forces on the fluid in volume $\mathscr{V}$, the only force acting on the fluid is that of pressure. Since the pressure of adjacent fluid elements within the volume concels out, only the pressure acting on the surface of the volume $\mathscr{V}$ need be taken into account. If **n** is the unit outward normal, the force acting on the volume $\mathscr{V}$ will be given by

$$\int_{\mathscr{V}} f\, d\mathscr{V} = -\int_{S} p\mathbf{n}\, dS$$

From vector calculus,

$$\int_{S} p\mathbf{n}\, dS = \int_{\mathscr{V}} \nabla p\, d\mathscr{V}$$

Hence we have

$$\int_{\mathscr{V}} \rho\frac{D\mathbf{V}}{Dt}\, d\mathscr{V} = -\int_{\mathscr{V}} \nabla p\, d\mathscr{V}$$

and since this relation must be valid for any volume $\mathscr{V}$ no matter how small,

$$\frac{D\mathbf{V}}{Dt} = -\nabla p \tag{10}$$

This equation is sometimes referred to as the Euler equation.

If a body force $\rho\mathbf{g}$ is present, the total force acting on the volume $\mathscr{V}$ becomes

$$\int_{\mathscr{V}} \mathbf{f}\,d\mathscr{V} = \int_{\mathscr{V}} \nabla p\,d\mathscr{V} + \int_{\mathscr{V}} \rho\mathbf{g}\,d\mathscr{V}$$

and equation (10) becomes,

$$\rho\frac{D\mathbf{V}}{Dt} = -\nabla p + \rho\mathbf{g} \tag{10a}$$

Let us now turn our attention to the third fundamental law, that of the conservation of energy. As pointed out earlier, the first law of thermodynamics may be written for a closed system in the form

$$đQ = dE + đW \tag{11}$$

Once again we must rewrite this law for an open system in order for it to be useful in compressible flow. If the heat transfer $đQ$, the change in energy dE, and the work $đW$ take place in time dt, we may write

$$\frac{dQ}{dt} = \frac{dE}{dt} + \frac{đW}{dt} \tag{12}$$

Let us now take for our system the mass inside volume $\mathscr{V}$ as shown in Figure 3, at some particular time t. If the heat enters or leaves the system only through conduction across its surface S, then we may write, according to Fourier's Law,

$$\frac{đQ}{dt} = -\int_S k\nabla T\cdot\mathbf{n}\,dS \tag{13}$$

where k is the thermal conductivity and $\mathbf{n}$ the unit outward normal on the surface. From vector calculus we know that

$$\int_S k\nabla T\cdot\mathbf{n}\,dS = \int_{\mathscr{V}} \nabla\cdot k\nabla T\,d\mathscr{V} \tag{14}$$

Now let e be the energy per unit mass of the system. Then,

$$\frac{dE}{dt} = \int_{\mathscr{V}} \rho\frac{De}{Dt}\,d\mathscr{V} \tag{15}$$

In the absence of chemical, electrical, viscous, magnetic, or body forces, the work done by the system will be entirely against the surrounding pressure forces. Hence,

$$\frac{đW}{dt} = \int_S p\mathbf{V}\cdot\mathbf{n}\,dS \tag{16}$$

Again, from the divergence theorem,

$$\int_S p\mathbf{V}\cdot\mathbf{n}\,dS = \int_{\mathscr{V}} \nabla(p\mathbf{V})\,d\mathscr{V} \tag{17}$$

Combining equations (12) through (17)

$$-\int_{\mathscr{V}} \nabla \cdot k \nabla T \, d\mathscr{V} = \int_{\mathscr{V}} \rho \frac{De}{Dt} \, d\mathscr{V} + \int_{\mathscr{V}} \nabla (p\mathbf{V}) \, d\mathscr{V} \tag{18}$$

Since this equation must be valid for any volume $\mathscr{V}$ no matter how small,

$$-\nabla \cdot k \nabla T = \rho \frac{De}{Dt} + \nabla (p\mathbf{V}) \tag{19}$$

This is one form of the energy equation.

The energy equation is sometimes written in a different form, more convenient for use in some engineering calculations. The energy may be broken up into its component parts for a moving system. For example, if velocity and gravity are present,

$$e = \tfrac{1}{2} V^2 + e_t + gZ \tag{20}$$

where V represents the magnitude of the velocity, e_t the internal energy per unit mass, and gZ the potential energy per unit mass due to gravity. Z represents the distance in the $-\mathbf{g}$ direction. Also,

$$\nabla (p\mathbf{V}) = p(\nabla \cdot \mathbf{V}) + \mathbf{V} \cdot \nabla p$$

and from the momentum equation (10a)

$$\mathbf{V} \cdot \rho \mathbf{g} - \mathbf{V} \cdot \nabla p = \mathbf{V} \cdot \rho \frac{D\mathbf{V}}{Dt} = \rho \frac{D}{Dt} (\tfrac{1}{2} V^2) = -\rho g \frac{DZ}{Dt} - \mathbf{V} \cdot \nabla p$$

Also, we need not restrict heat addition to conduction across the boundary. Heat may also be added or subtracted by chemical reaction. Let us, therefore, indicate the heat added or subtracted per unit volume per unit time by the more general symbol $\dot{q}$. Thus we may write,

$$\dot{q} = \rho \frac{D}{Dt} [e_t + \tfrac{1}{2} V^2 + gZ] + p \nabla \mathbf{V} - \rho \frac{D}{Dt} [\tfrac{1}{2} V^2 + gZ] \tag{21}$$

It is of interest to observe here that the pressure work, $\nabla(pV)$, divides naturally into two parts: $\mathbf{V} \nabla p$, which is the work of increasing or decreasing the kinetic energy of the system, and $p \nabla \mathbf{V}$, which is the work of compressing or expanding the system. This may be illustrated by the use of a somewhat crude example. Consider a slug of gas flowing in a pipe of cross section A:

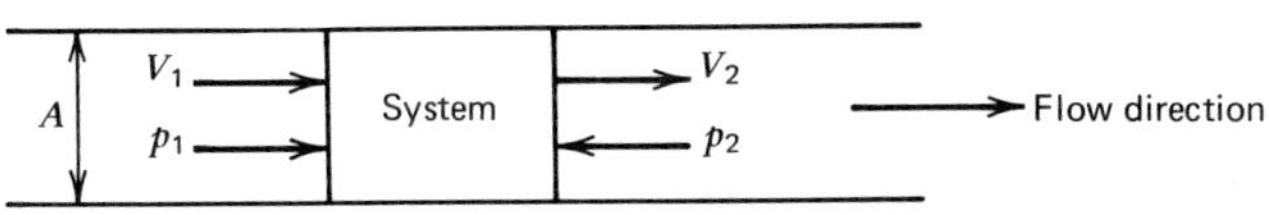

Figure 4. A slug of gas in one-dimensional flow.

In this case,

$$\int_S p\mathbf{V}\cdot\mathbf{n}\,dS = (p_2 V_2 - p_1 V_1)A \tag{22}$$

The right-hand side of equation (22) may be rewritten

$$(p_2 V_2 - p_1 V_1)A = A(p_2 - p_1)\left(\frac{V_1 + V_2}{2}\right) + A\left(\frac{p_1 + p_2}{2}\right)(V_1 - V_2) \tag{23}$$

$A(p_2 - p_1)$ represents the net force on the slug, and $(V_1 + V_2)/2$ the average velocity. Hence, the first term on the right-hand side of equation (23) represents the rate of work going to accelerate the slug. $(p_1 + p_2)/2$ represents the average pressure on the slug and $V_1 - V_2$ represents the rate of deformation. Hence, the second term on the right-hand side of equation (23) represents the work going to deform the slug.

Returning now to equation (21), the kinetic and potential energy cancel and we have,

$$\dot{q} = \rho\frac{De_t}{Dt} + p\nabla\mathbf{V} \tag{24}$$

From the equation of continuity,

$$\nabla\mathbf{V} = -\frac{1}{\rho}\frac{D\rho}{Dt}$$

and, adding and subtracting $\rho(D/Dt)(p/\rho)$,

$$p\nabla\mathbf{V} = \frac{p}{\rho}\frac{D\rho}{Dt} + \rho\frac{D}{Dt}\left(\frac{p}{\rho}\right) - \rho\frac{D}{Dt}\left(\frac{p}{\rho}\right)$$

$$= -\frac{Dp}{Dt} + \rho\frac{D}{Dt}\left(\frac{p}{\rho}\right) \tag{24a}$$

Substituting in equation (24)

$$\dot{q} = \rho\frac{De_t}{Dt} + \rho\frac{D}{Dt}\left(\frac{p}{\rho}\right) - \frac{Dp}{Dt} \tag{25}$$

The specific enthalpy h is defined by

$$h = e_t + \frac{p}{\rho} \tag{26}$$

Hence

$$\dot{q} = \rho\frac{D}{Dt}\left(e_t + \frac{p}{\rho}\right) - \frac{Dp}{Dt}$$

or

$$\rho\frac{Dh}{Dt} = \frac{Dp}{Dt} + \dot{q} \tag{27}$$

It is in this form that the energy equation is most frequently encountered.

The Navier-Stokes Equation

The equations of Euler, which we have derived above, are deficient in one important respect: they neglect viscosity. This does not mean that they are not useful. In many problems in gas dynamics, the effect of the viscosity is small. However, there are certain cases in which the effect of viscosity is important and hence cannot be neglected. We are thus led to seek an amendation of the Euler equations to include the effects of viscosity. Such an amendation was first made by M. Navier [4] for incompressible flow. In a paper presented to the Royal Academy of Sciences (Paris) on March 18, 1822, Navier pointed out the deficiencies of the equations which Euler had derived sixty-five years earlier, citing as an example the flow from a long-necked vessel which, observation indicated, was considerably slower than predicted by the equations of Euler (the distinction between viscous and inviscid flow was, of course, known to Euler, as well as to d'Alembert twenty years earlier). In order to correct for these discrepancies, Navier proposed a molecular model of matter which is not in accord with modern views on the subject. Nevertheless, using this model, he was able to derive the equations which now bear the name Navier-Stokes equations in the incompressible case, which experiment has proven to be valid and which, still today, may be described as the fundamental equations of fluid dynamics.

In 1845, G. G. Stokes [5] rederived the equations of Navier, using however a macroscopic rather than a molecular hypothesis regarding viscosity, and extending the equations to the case of compressible flow.

Before proceeding to the derivation of the Navier-Stokes equations, as the equations derived by Stokes are now called, let us review the modern theory regarding the origin of viscosity. Viscosity may be defined as that quality of a fluid which makes tangential stresses in the fluid possible. Consider first a two-dimensional flow with a velocity gradient, as shown in Figure 5.

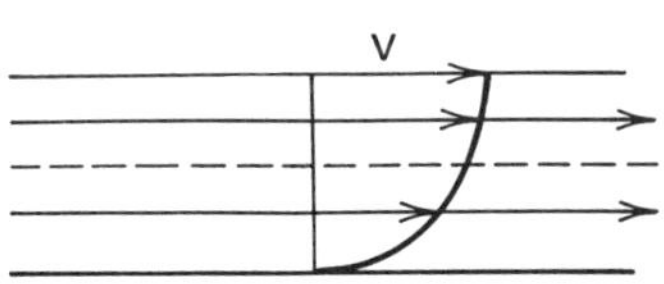

Figure 5. Velocity profile in two-dimensional flow.

The dashed line represents an imaginary dividing line between the upper and lower parts of the fluid. Due to the random motion of the molecules, molecules are always crossing the dashed line from above and below. However, those molecules crossing the line from below will have, on the average, less momentum in the direction of the flow than those crossing the line from above. There is, thus, a net transfer of momentum across the dashed line, the momentum below the dashed line being increased and the momentum above the line being decreased. It follows from Newton's law of motion

$$\mathbf{F} = \frac{d(m\mathbf{V})}{dt} \tag{28}$$

that the fluid above the dashed line is exerting a force on the fluid below, dragging it forward as it were, while the fluid below the dashed line is exerting a force on the fluid above it, tending to retard it.

While the random motion of the molecules explains the presence of tangential stresses in the fluid, it is important to realize that the same random motion may also cause normal stresses. In order to see this let us take the case of a one-dimensional flow with a velocity variation in the direction of the flow, as shown in Figure 6. Let us now examine the flow from the vantage point of an

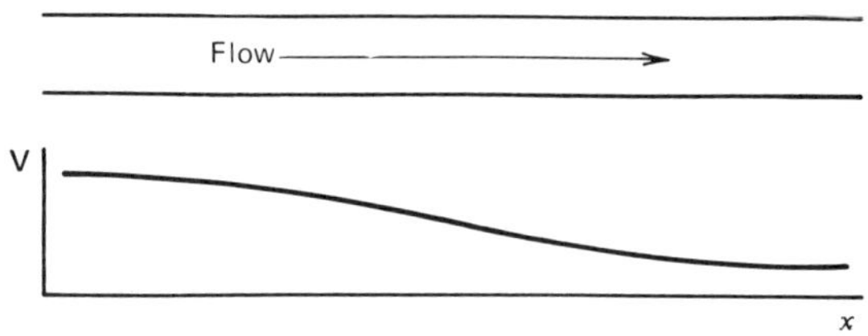

Figure 6. Velocity gradient in the flow direction.

observer traveling with the flow. The observer will see molecules continuously passing from the fluid in front of him to the fluid behind him, and vice versa, due to the random motion of the molecules. However, the molecules passing him from front to back will have less forward momentum than the molecules passing him from back to front. Thus, the fluid in front of the observer is gaining momentum while the fluid behind the observer is losing it. From equation (28), therefore, it is apparent that the fluid in back of the observer is exerting an accelerating force on the fluid in front of him, and the fluid in front of the observer is exerting a retarding force on the fluid in back of him.

It might also be pointed out that the effect of this phenomenon in the present case is to flatten the velocity profile, an effect which is of importance in the study of shock waves.*

Having examined the viscosity qualitatively, we may now inquire into the magnitude of the forces thus produced. It was first proposed by Newton [3] that for the case shown in Figure 5, the magnitude of the viscous force is proportional to the rate of strain. Consider the coordinate system shown in Figure 7.

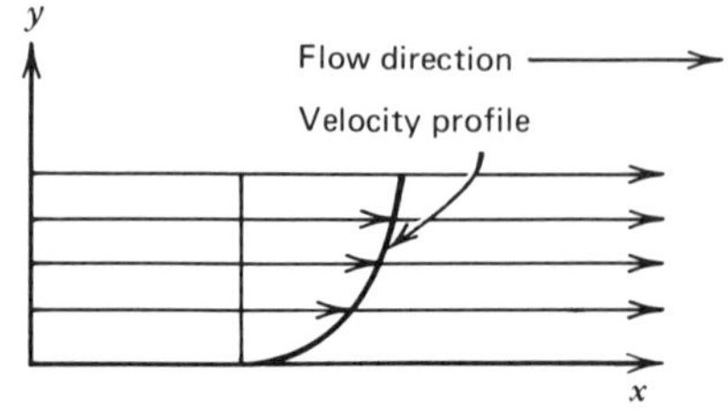

Figure 7. Velocity profile in x-y coordinate system.

*In liquids, intermolecular forces also play an important role in viscosity. In gases, this effect is less important.

Let

$$\mathbf{V} = \mathbf{i}u$$

in the present case, where $\mathbf{i}$ is the unit vector in the x direction. Then, according to Newton,

$$\tau \sim \frac{\partial u}{\partial y}$$

where τ is the shear stress per unit area. If we call the constant of proportionality μ, we get

$$\tau = \mu \frac{\partial u}{\partial y}$$

where μ is called the coefficient of viscosity, or viscosity for short.

Newton thus ennunciated the principle for the case of Figure 7: stress is proportional to rate of strain. A fluid which obeys this principle is called a Newtonian fluid. Most, but not all, fluids are Newtonian in their behavior.

What about the case of Figure 6, however? Can we say that the horizontal stress is proportional to $\partial u/\partial x$? And if so, is the constant of proportionality equal to μ? The issue here is by no means so clear. We will come back to this point below.

In order to contribute to the understanding of the problem, we will derive the Navier-Stokes equations in a number of different ways. Let us begin by a derivation of the Navier-Stokes equations for two-dimensional, incompressible flow, using scalar components.

Consider a rectangular element of fluid, as shown in Figure 8. The reader may

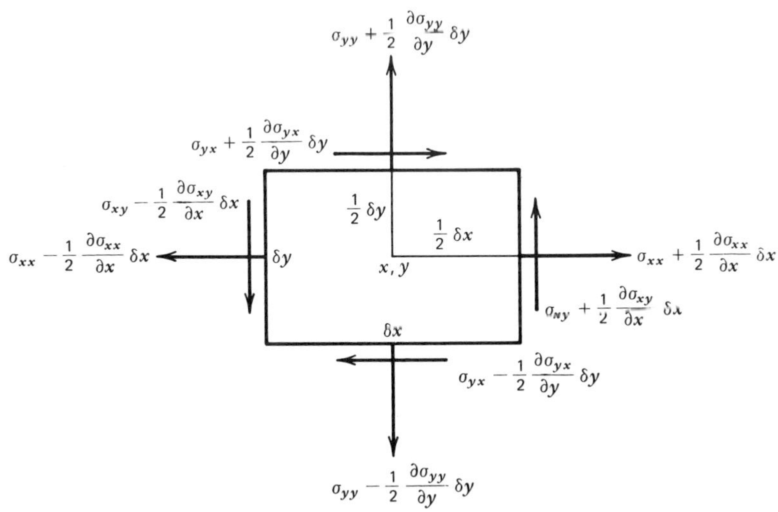

Figure 8. Stress components on a fluid element in two dimensions.

well ask if, by considering a rectangular element of fluid, the analysis will be valid only for rectangular elements of fluid. What, for example, if the fluid element is round? The answer is that any shape may be subdivided into rectangles, with the leftover nonrectangular portions shrinking in size as the rectangles grow smaller and their number grows larger. In the limit as the size of the rectangles shrinks to zero and their number increases to infinity, the remaining nonrectangular portions vanish. Since, in the course of our derivation, we shall be passing to the limit as our rectangle shrinks to zero, the analysis will be exact. At the same time it should be understood that so long as we are dealing with a finite rectangle, the analysis remains approximate.

On each side of our rectangle of fluid in two dimensional, viscous flow, forces may be acting with both normal and tangential components. For the sake of convenience, let us take these components as positive in the direction shown by the arrows in Figure 8. Let σ_{xx}, σ_{yy}, σ_{xy}, and σ_{yx} be the normal and tangential stress components at the point x, y. Then, at the points $x + \frac{1}{2}\delta x, y$; $y + \frac{1}{2}\delta y, x$; $x - \frac{1}{2}\delta x, y$; $y - \frac{1}{2}\delta y, x$, the normal and tangential stress components will be given approximately by the values shown in Figure 8. Since δx and δy are small, we may make the further assumption that these values represent the average values over the faces on which they act.

The net force on the rectangle in the x direction will then be

$$\frac{\partial \sigma_{xx}}{\partial x}\delta x \delta y + \frac{\partial \sigma_{yx}}{\partial y}\delta y \delta x$$

Similarly, the net force on the rectangle in the y direction will be

$$\frac{\partial \sigma_{yy}}{\partial y}\delta y \delta x + \frac{\partial \sigma_{xy}}{\partial x}\delta x \delta y$$

Now let ρ be the density of the fluid within the element, and let $\mathbf{V} = \mathbf{i}u + \mathbf{j}v$ be the average velocity, where $\mathbf{i}$ and $\mathbf{j}$ are the unit vectors in the x and y directions, and hence u and v are the x and y components of the velocity. From Newton's law, $\mathbf{F} = m\mathbf{a}$, therefore, we have

$$\begin{aligned} \rho \delta x \delta y \frac{Du}{Dt} &= \left(\frac{\partial \sigma_{xx}}{\partial x} + \frac{\partial \sigma_{yx}}{\partial y}\right)\delta x \delta y \\ \rho \delta x \delta y \frac{Dv}{Dt} &= \left(\frac{\partial \sigma_{xy}}{\partial x} + \frac{\partial \sigma_{yy}}{\partial y}\right)\delta x \delta y \end{aligned} \tag{29}$$

Because δx and δy are not zero, we may divide both sides by $\delta y\, \delta x$ and pass to the limit as $\delta x, \delta y \longrightarrow 0$. The equations will then be exact.

Consider now the torque on the element $\delta x \, \delta y$ about its center. Taking moments about the point x, y due to the shear stresses (the normal stresses do not contribute to the torque) and adding, we get

$$\text{Torque} = (\sigma_{xy} - \sigma_{yx})\delta x \delta y$$

taking counterclockwise as positive. Recall from elementary mechanics,

$$\text{Torque} = I\frac{d\omega}{dt}$$

where I is the moment of inertia and $d\omega/dt$ the angular acceleration. For our rectangle,

$$I = \rho\,\delta x\,\delta y \frac{(\delta x)^2 + (\delta y)^2}{12}$$

Hence,

$$(\sigma_{xy} - \sigma_{yx})\delta x\,\delta y = \rho\,\delta x\,\delta y \frac{(\delta x)^2 + (\delta y)^2}{12}\frac{d\omega}{dt}$$

Dividing both sides by $\delta x\,\delta y$ and passing to the limit as $\delta x, \delta y \longrightarrow 0$, we get

$$\sigma_{xy} = \sigma_{yx} \tag{30}$$

since $d\omega/dt$ is finite. This is sometimes expressed in words by saying the stress tensor is symmetric.

It should be pointed out that while the above procedure is correct mathematically, there may be a problem physically, since, if we take our rectangle small enough, the continuum assumption will no longer be valid. In practice, however, it is found that the limitation on the symmetry of the stress tensor is the same as the limitation on the continuum assumption—namely, that the scale of the problem be large compared to the average distance between the molecules.

Returning then to equations (29), dividing both sides by $\delta x\,\delta y$ and passing to the limit as $\delta x\,\delta y \to 0$, we get,

$$\begin{aligned} \rho\frac{Du}{Dt} &= \frac{\partial\sigma_{xx}}{\partial x} + \frac{\partial\sigma_{xy}}{\partial y} \\ \rho\frac{Dv}{Dt} &= \frac{\partial\sigma_{xy}}{\partial x} + \frac{\partial\sigma_{yy}}{\partial y} \end{aligned} \tag{31}$$

These equations are exact within the limits of the continuum assumption.

We must now turn to the evaluation of the stresses $\sigma_{xx}, \sigma_{xy}, \sigma_{yy}$. Here we take as our point of departure the hypothesis of Newton [3] stated in book II, section IX, of the Principia:

> The resistance arising from the want of lubricity in the parts of the fluid is, other things being equal, proportional to the velocity with which the parts of the fluid are separated from one another.

Newton here was talking about the entrainment of a fluid by a cylinder rotating about its own axis. Thus, his hypothesis applies, strictly speaking, only to an evaluation of σ_{xy}. However, the statement of Newton may be interpreted more broadly—by taking "the velocity with which the parts are separated" as a vector quantity, namely, the difference in the vector velocity between the two particles of fluid, divided by the distance between them. Then, there would be only one proportionality constant and the components of the stress would then equal the corresponding components of the rate of strain times the proportionality constant in every direction. We will come back to this point below.

Returning now to our rectangle of Figure 8, let us place the coordinate axes coinciding with the sides of the rectangle at time t, as shown in Figure 9.

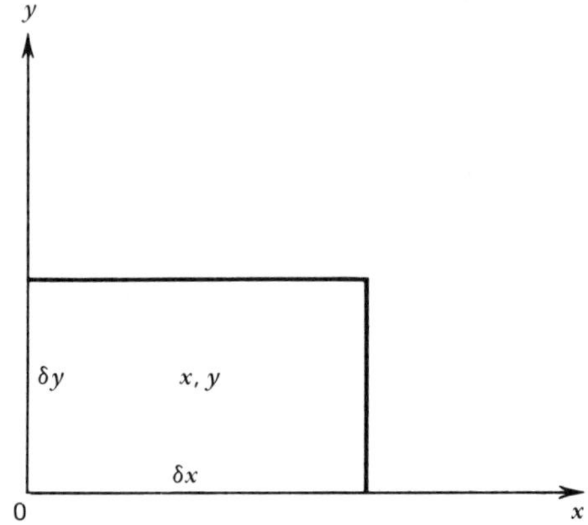

Figure 9. x-y coordinate system coinciding with sides of fluid element at time t.

Let the origin of the coordinates remain always attached to the same particle of fluid as the fluid moves along, but let the orientation of the x and y axes remain fixed. Consider now the displacement of the fluid particle located at x,y at time t, during some time Δt, with respect to our coordinate system. At time $t + \Delta t$, the particle will be located at

$$x + \xi, \qquad y + \eta$$

If x and y are small, then, approximately,

$$\xi = \frac{\partial \xi}{\partial x} x + \frac{\partial \xi}{\partial y} y$$

$$\eta = \frac{\partial \eta}{\partial x} x + \frac{\partial \eta}{\partial y} y$$

The displacement vector may be divided into two components: one of rotation and one of elongation (Figure 10). The rotation component is a and the elonga-

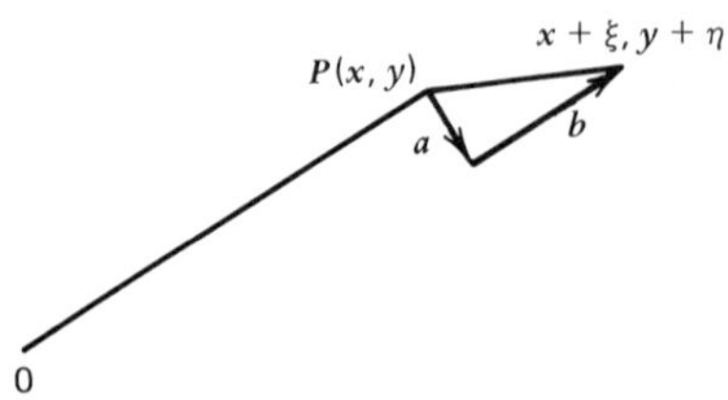

Figure 10. Rotation and elongation components of the displacement vector.

tion component is b. We cannot, however, divide these vectors by the length OP to obtain the components of strain. This is because if all the points in the fluid element have the same rotation vector a, the element is rotating as a solid body and hence the vector a will not represent a displacement of the fluid particles with respect to each other. In other words, it will not be a strain. Let us, therefore, take the rotation vector as the sum of two parts: the average rotation of all the particles of fluid in time Δt inside our rectangle of fluid, and the difference between that average and the actual rotation at the point under consideration. Only the second part may then contribute to the actual strain due to the rotation. The average rotation of the fluid in the rectangle in time Δt is given by

$$\frac{1}{2}\left(\frac{\partial \xi}{\partial y}-\frac{\partial \eta}{\partial x}\right)$$

Thus, we must subtract from the x component of our rotation vector,

$$\frac{1}{2}\left(\frac{\partial \xi}{\partial y}-\frac{\partial \eta}{\partial x}\right)y$$

and from the y component,

$$-\frac{1}{2}\left(\frac{\partial \xi}{\partial y}-\frac{\partial \eta}{\partial x}\right)x$$

Rewriting ξ and η accordingly, we have

$$\xi=\frac{1}{2}\left(\frac{\partial \xi}{\partial y}-\frac{\partial \eta}{\partial x}\right)y+\frac{\partial \xi}{\partial x}x+\frac{1}{2}\left(\frac{\partial \xi}{\partial y}+\frac{\partial \eta}{\partial x}\right)y$$

$$\eta=-\frac{1}{2}\left(\frac{\partial \xi}{\partial y}-\frac{\partial \eta}{\partial x}\right)x+\frac{\partial \eta}{\partial y}y+\frac{1}{2}\left(\frac{\partial \eta}{\partial x}+\frac{\partial \xi}{\partial y}\right)x$$

Only the last two terms in each expression can contribute to the strain. The same effect could have been achieved by writing the displacements with respect to a coordinate system rotating with average rate of rotation of the fluid.

Now let

$$\xi=\xi_1+\xi_2$$

$$\eta=\eta_1+\eta_2$$

where

$$\xi_1=\frac{1}{2}\left(\frac{\partial \xi}{\partial y}-\frac{\partial \eta}{\partial x}\right)y, \qquad \xi_2=\frac{\partial \xi}{\partial x}x+\frac{1}{2}\left(\frac{\partial \xi}{\partial y}+\frac{\partial \eta}{\partial x}\right)y$$

$$\eta_1=-\frac{1}{2}\left(\frac{\partial \xi}{\partial y}-\frac{\partial \eta}{\partial x}\right)x, \qquad \eta_2=\frac{\partial \eta}{\partial y}y+\frac{1}{2}\left(\frac{\partial \xi}{\partial y}+\frac{\partial \eta}{\partial x}\right)x$$

and let

$$\xi_2=\varepsilon_{xx}x+\varepsilon_{xy}y$$

$$\eta_2=\varepsilon_{yy}y+\varepsilon_{yx}x$$

The quantities $\varepsilon_{xx}, \varepsilon_{xy}, \varepsilon_{yy}, \varepsilon_{yx}$ represent the components of strain. The rates of strain, then, may be written

$$\dot{\varepsilon}_{xx} = \frac{\partial u}{\partial x} \qquad \dot{\varepsilon}_{xy} = \frac{1}{2}\left(\frac{\partial u}{\partial y} + \frac{\partial v}{\partial x}\right)$$

$$\dot{\varepsilon}_{yy} = \frac{\partial v}{\partial y} \qquad \dot{\varepsilon}_{yx} = \frac{1}{2}\left(\frac{\partial u}{\partial y} + \frac{\partial v}{\partial x}\right)$$

These equations may be obtained simply by dividing the equations for the strains by Δt and passing to the limit as $\Delta t \longrightarrow 0$. Now in the case of two-dimensional, incompressible flow it is observed by experiment that

$$\sigma'_{xx} \sim \dot{\varepsilon}_{xx} \qquad \sigma'_{xy} \sim \dot{\varepsilon}_{xy}$$

$$\sigma'_{yy} \sim \dot{\varepsilon}_{yy} \qquad \sigma'_{yx} \sim \dot{\varepsilon}_{yx}$$

where the primes indicate viscous stresses. Taking the proportionality constant as 2μ, we get,

$$\begin{aligned} \sigma'_{xx} &= 2\mu\frac{\partial u}{\partial x} \qquad \sigma'_{xy} = \mu\left(\frac{\partial u}{\partial y} + \frac{\partial v}{\partial x}\right) \\ \sigma'_{yy} &= 2\mu\frac{\partial v}{\partial y} \qquad \sigma'_{yx} = \mu\left(\frac{\partial u}{\partial y} + \frac{\partial v}{\partial x}\right) \end{aligned} \tag{32}$$

It will be observed that the previously demonstrated requirement that $\sigma_{yx} = \sigma_{xy}$ is satisfied by the above relations.

In the case of tangential stresses, only viscous forces are present. Hence,

$$\sigma_{xy} = \sigma'_{xy} \qquad \sigma_{yx} = \sigma'_{yx} \tag{33}$$

In the case of normal stresses, however, there are, in addition to the viscous stresses what might be described as pressure stresses. These result from the exchange of momentum due to the actual collisions of molecules, as opposed to viscous stresses, which result from the diffusion of molecules among each other. In the absence of viscous stresses, these pressure stresses are simply the conventional static pressure. Let us designate the pressure stresses by $-p'$. The normal stresses then become

$$\begin{aligned} \sigma_{xx} &= -p' + 2\mu\frac{\partial u}{\partial x} \\ \sigma_{yy} &= p' + 2\mu\frac{\partial v}{\partial y} \end{aligned} \tag{34}$$

Substituting for the stresses in equations (31) using equations (32), (33), and (34), and keeping in mind that for incompressible flow

$$\frac{\partial u}{\partial x} + \frac{\partial v}{\partial y} = 0$$

we get

$$\rho\frac{Du}{Dt} = -\frac{\partial p'}{\partial x} + \mu\left(\frac{\partial^2 \mu}{\partial x^2} + \frac{\partial^2 u}{\partial y^2}\right)$$

$$\rho\frac{Dv}{Dt} = -\frac{\partial p'}{\partial y} + \mu\left(\frac{\partial^2 v}{\partial x^2} + \frac{\partial^2 v}{\partial y^2}\right) \tag{35}$$

It was the feeling of Stokes that the pressure p ought to be defined as the average normal stress at a point. Thus, in two dimensions,

$$-p = \tfrac{1}{2}(\sigma_{xx} + \sigma_{yy}) \tag{36}$$

Adding equations (34),

$$-2p' = \sigma_{xx} + \sigma_{yy} - 2\mu\left(\frac{\partial u}{\partial x} + \frac{\partial v}{\partial y}\right) \tag{37}$$

Hence, for incompressible flow,

$$p' = p$$

and equations (34) become,

$$\rho\frac{Du}{Dt} = -\frac{\partial p}{\partial x} + \mu\left(\frac{\partial^2 u}{\partial x^2} + \frac{\partial^2 u}{\partial y^2}\right)$$

$$\rho\frac{Dv}{Dt} = -\frac{\partial p}{\partial y} + \mu\left(\frac{\partial^2 v}{\partial x^2} + \frac{\partial^2 v}{\partial y^2}\right) \tag{38}$$

Equations (38) are the Navier-Stokes equations for two-dimensional, incompressible flow.

The Navier-Stokes Equations for the Three-Dimensional, Compressible Case

Consider now the stresses on a volume element of fluid, as shown in Figure 11.

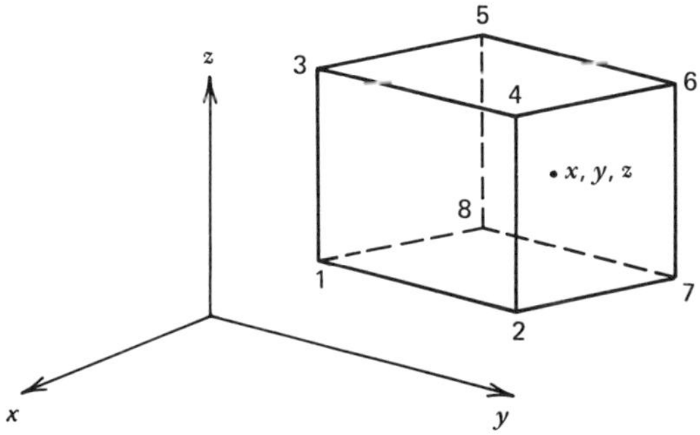

Figure 11. Fluid element in three-dimensional, rectangular coordinate system.

The normal stresses at the point x, y, z, in the x, y and z directions will be called σ_{xx}, σ_{yy}, and σ_{zz}, respectively, and the tangential stresses will be called $\sigma_{xy}, \sigma_{yz}, \sigma_{zx}, \sigma_{yx}, \sigma_{zy}$ and σ_{xz}. Taking x, y, z at the center of the volume element and assuming that the element is small, the average normal and tangential stresses on the face 2467 will be given, approximately, by

$$\sigma_{yy} + \frac{\partial \sigma_{yy}}{\partial x}\frac{\delta x}{2}$$

$$\sigma_{yx} + \frac{\partial \sigma_{yx}}{\partial x}\frac{\delta x}{2}$$

$$\sigma_{yz} + \frac{\partial \sigma_{yz}}{\partial x}\frac{\delta x}{2}$$

as shown in Figure 12. It will be observed that the first subscript, y in this case,

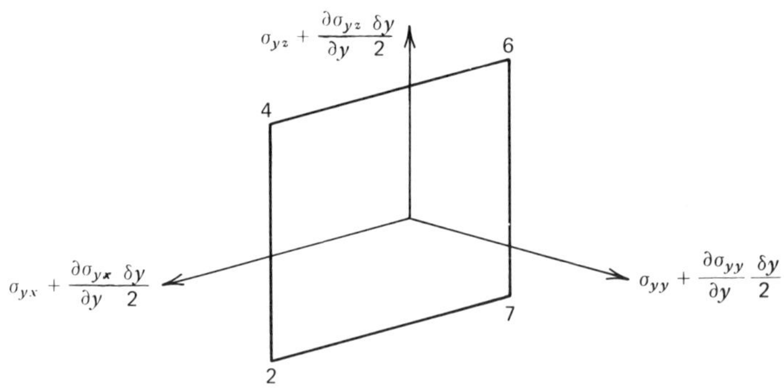

Figure 12. Stresses on face of fluid element perpendicular to y axis.

indicates the axis to which the face is perpendicular. The second subscript indicates the direction of the stress component.

For the face 3456, the normal stress component will be, approximately,

$$\sigma_{zz} + \frac{\partial \sigma_{zz}}{\partial z}\frac{\delta z}{2}$$

and the tangential stresses,

$$\sigma_{zx} + \frac{\partial \sigma_{zx}}{\partial z}\frac{\delta z}{2}, \qquad \sigma_{zy} + \frac{\partial \sigma_{zy}}{\partial z}\frac{\delta z}{2}$$

Similarly for the remaining faces.

The force on the volume is, then, in the x direction,

$$\left[\frac{\partial \sigma_{xx}}{\partial x} + \frac{\partial \sigma_{yx}}{\partial y} + \frac{\partial \sigma_{zx}}{\partial z}\right]\delta x\,\delta y\,\delta z \tag{39}$$

in the y direction,

$$\left[\frac{\partial \sigma_{xy}}{\partial x}+\frac{\partial \sigma_{yy}}{\partial y}+\frac{\partial \sigma_{zy}}{\partial z}\right]\delta x\,\delta y\,\delta z \tag{40}$$

and in the z direction,

$$\left[\frac{\partial \sigma_{xz}}{\partial x}+\frac{\partial \sigma_{yz}}{\partial y}+\frac{\partial \sigma_{zz}}{\partial z}\right]\delta x\,\delta y\,\delta z \tag{41}$$

By the same argument as in the two-dimensional case, that the angular acceleration must remain finite as the volume shrinks to zero, it follows that,

$$\sigma_{xy}=\sigma_{yx}, \qquad \sigma_{xz}=\sigma_{zx}, \qquad \sigma_{yz}=\sigma_{zy}$$

Substituting in Newton's equation $\mathbf{F}=m\mathbf{a}$ for the element of volume, we get

$$\begin{aligned}
\rho\frac{Du}{Dt}\delta x\,\delta y\,\delta z &= \left[\frac{\partial \sigma_{xx}}{\partial x}+\frac{\partial \sigma_{xy}}{\partial y}+\frac{\partial \sigma_{xz}}{\partial z}\right]\delta x\,\delta y\,\delta z\\
\rho\frac{Dv}{Dt}\delta x\,\delta y\,\delta z &= \left[\frac{\partial \sigma_{xy}}{\partial x}+\frac{\partial \sigma_{yy}}{\partial y}+\frac{\partial \sigma_{yz}}{\partial z}\right]\delta x\,\delta y\,\delta z\\
\rho\frac{Dw}{Dt}\delta x\,\delta y\,\delta z &= \left[\frac{\partial \sigma_{xz}}{\partial x}+\frac{\partial \sigma_{yz}}{\partial y}+\frac{\partial \sigma_{zz}}{\partial z}\right]\delta x\,\delta y\,\delta z
\end{aligned} \tag{42}$$

where

$$\mathbf{V}=\mathbf{i}u+\mathbf{j}v+\mathbf{k}w$$

represents the average velocity in the volume. Dividing both sides of equation (42) by $\delta x\,\delta y\,\delta z$ and passing to the limit as $\delta x\,\delta y\,\delta z \to 0$, the relations become exact. In order to find the stresses we again invoke Stokes' hypothesis that stress is proportional to rate of strain. Let us, therefore, find the rate of strain in a more general, three dimensional system.

Let P be a point of the volume element under consideration and choose now the origin O of the coordinate system the point 8 in Figure 11, as shown in Figure 13. It should be emphasized that the choice of the origin is purely

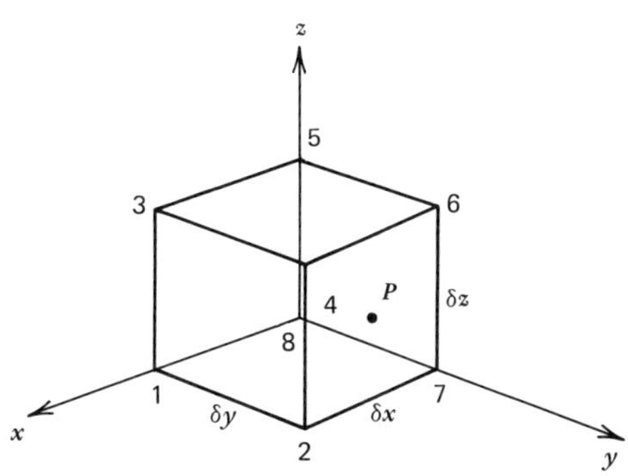

Figure 13. Coordinate system attached to fluid particle at some time t.

arbitrary. Any other point of the volume element could have been chosen just as well (except the point P). The system under consideration consists of the fluid occupying the volume element at some time t. The fluid is presumed to be moving in some arbitrary way. Let us further stipulate that the origin O shall always remain attached to the same fluid particle, and hence be moving with the fluid. The coordinate directions x, y, z, however, are presumed to remain the same. In other words, the axes are moving, but not rotating.

Let ξ, η, and ζ designate the distance moved by the point P in the x, y, and z directions relative to our coordinate system in time Δt. Then, if $x, y,$ and z are small, we may say, approximately,

$$\begin{aligned}\xi &= \frac{\partial \xi}{\partial x}x + \frac{\partial \xi}{\partial y}y + \frac{\partial \xi}{\partial z}z \\ \eta &= \frac{\partial \eta}{\partial x}x + \frac{\partial \eta}{\partial y}y + \frac{\partial \eta}{\partial z}z \\ \zeta &= \frac{\partial \zeta}{\partial x}x + \frac{\partial \zeta}{\partial y}y + \frac{\partial \zeta}{\partial z}z\end{aligned} \tag{43}$$

Once again, these displacements are not proportional to the actual strains, since the actual strains can depend only on the displacement of the particles of the fluid relative to each other. The average rate of rotation of the fluid about the point O is given by

$$\boldsymbol{\omega} = \mathbf{i}\frac{1}{2}\left(\frac{\partial \zeta}{\partial y} - \frac{\partial \eta}{\partial z}\right) + \mathbf{j}\frac{1}{2}\left(\frac{\partial \xi}{\partial z} - \frac{\partial \zeta}{\partial x}\right) + \mathbf{k}\frac{1}{2}\left(\frac{\partial \eta}{\partial x} - \frac{\partial \xi}{\partial y}\right)$$

That this represents the average rate of rotation may be seen by examining each component. For example, the z component of rotation represents the average rate of rotation about the z axis, as shown in Figure 14. Similarly for the compo-

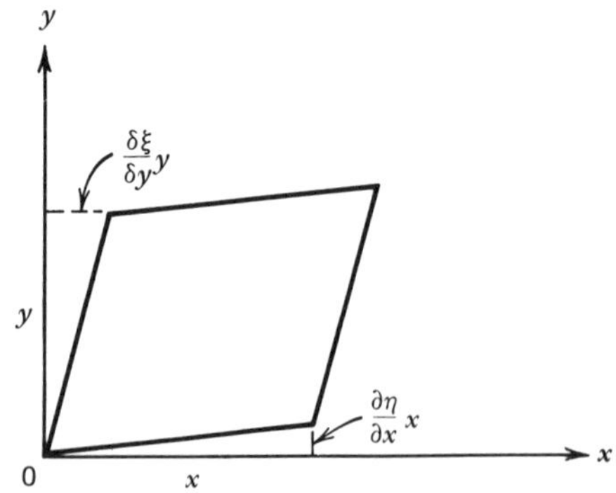

Figure 14. Displacement of particle in the x-y plane.

nents about the x and y axes. The displacement of the point P in time Δt due to the average rate of rotation of the body is given by

$$\boldsymbol{\omega} \times \mathbf{r} = \boldsymbol{\omega} \times (\mathbf{i}x + \mathbf{j}y + \mathbf{k}z)$$

where r is the vector OP. The x, y, and z components of $\boldsymbol{\omega} \times \mathbf{r}$ are, then,

$$
\begin{aligned}
\xi_r &= \frac{1}{2}\left(\frac{\partial \xi}{\partial y} - \frac{\partial \eta}{\partial x}\right)y + \frac{1}{2}\left(\frac{\partial \xi}{\partial z} - \frac{\partial \zeta}{\partial x}\right)z \\
\eta_r &= \frac{1}{2}\left(\frac{\partial \eta}{\partial x} - \frac{\partial \xi}{\partial y}\right)x + \frac{1}{2}\left(\frac{\partial \eta}{\partial z} - \frac{\partial \zeta}{\partial y}\right)z \qquad (44) \\
\zeta_r &= \frac{1}{2}\left(\frac{\partial \zeta}{\partial x} - \frac{\partial \xi}{\partial z}\right)x + \frac{1}{2}\left(\frac{\partial z}{\partial y} - \frac{\partial \eta}{\partial z}\right)z
\end{aligned}
$$

The portion of the displacement of the point P due to the average rate of rotation of the fluid about the point O cannot contribute to the strain since this represents a solid body rotation and not a displacement of the point P relative to adjacent point. In order to get the strain at the point P, therefore, we must subtract from the total displacement of the point P that portion which is due to the average rate of rotation of the fluid about O. If ξ_d, η_d, ζ_d represent the portion of the x, y and z components of the displacement which contribute to the strain, then, using equations (43) and (44), we have,

$$
\begin{aligned}
\xi_d &= \xi - \xi_r = \frac{\partial \xi}{\partial x}x + \frac{1}{2}\left(\frac{\partial \xi}{\partial y} + \frac{\partial \eta}{\partial x}\right)y + \frac{1}{2}\left(\frac{\partial \zeta}{\partial x} + \frac{\partial \xi}{\partial z}\right)z \\
\eta_d &= \eta - \eta_r = \frac{1}{2}\left(\frac{\partial \xi}{\partial y} + \frac{\partial \eta}{\partial x}\right)x + \frac{\partial \eta}{\partial y}y + \frac{1}{2}\left(\frac{\partial \eta}{\partial z} + \frac{\partial \zeta}{\partial y}\right)z \qquad (45) \\
\zeta_d &= \zeta - \zeta_r = \frac{1}{2}\left(\frac{\partial \zeta}{\partial x} + \frac{\partial \xi}{\partial y}\right)x + \frac{1}{2}\left(\frac{\partial \eta}{\partial z} + \frac{\partial \zeta}{\partial y}\right)y + \frac{\partial \zeta}{\partial z}z
\end{aligned}
$$

Let us rewrite these equations in the form

$$
\begin{aligned}
\xi_d &= \varepsilon_{xx}x + \varepsilon_{xy}y + \varepsilon_{xz}z \\
\eta_d &= \varepsilon_{yx}x + \varepsilon_{yy}y + \varepsilon_{yz}z \\
\zeta_d &= \varepsilon_{zx}x + \varepsilon_{zy}y + \varepsilon_{zz}z
\end{aligned}
$$

where

$$
\begin{aligned}
\varepsilon_{xx} &= \frac{\partial \xi}{\partial x} & \varepsilon_{xy} &= \frac{1}{2}\left(\frac{\partial \xi}{\partial y} + \frac{\partial \eta}{\partial x}\right) = \varepsilon_{yx} \\
\varepsilon_{yy} &= \frac{\partial \eta}{\partial y} & \varepsilon_{xz} &= \frac{1}{2}\left(\frac{\partial \zeta}{\partial x} + \frac{\partial \xi}{\partial z}\right) = \varepsilon_{zx} \\
\varepsilon_{zz} &= \frac{\partial \zeta}{\partial z} & \varepsilon_{zy} &= \frac{1}{2}\left(\frac{\partial \eta}{\partial z} + \frac{\partial \zeta}{\partial y}\right) = \varepsilon_{yz}
\end{aligned}
$$

We now proceed from Hooke's law that stress is proportional to strain. However, we are not justified in writing

$$\sigma_{xx} \sim \varepsilon_{xx} \qquad \sigma_{xy} \sim \varepsilon_{xy}$$

etc., since each component of stress may result in a variety of strains. The easiest way to see this is to stretch a rubber band. Not only does it get longer, but it gets narrower as well. Thus, both ε_{xx} and ε_{yy} depend on σ_{xx} in this simple case. If we have stresses on each face of a cube, and in different directions, the situation is clearly more complicated. So we must now try to establish some relation between the stresses

$$\sigma_{nm}, \qquad n,m = x,y,z$$

and the strains

$$\varepsilon_{nm}, \qquad n,m = x,y,z$$

Let

$$\mathbf{s}_d = \mathbf{i}\xi_d + \mathbf{j}\eta_d + \mathbf{k}\zeta_d$$

and consider the expression

$$\mathbf{s}_d \cdot \mathbf{r} = \varepsilon_{xx}x^2 + 2\varepsilon_{xy}xy + \varepsilon_{yy}y^2 + \dots = f(x,y,z) \tag{46}$$

Clearly, the value of the function $\mathbf{s}_d \cdot \mathbf{r}$ for any given point of the body must be independent of the coordinate system used. We observe, however, that the function $f(x,y,z) =$ constant is a surface of the second order (ellipsoid, hyperboloid, etc., depending on the values of the coefficients). By choosing our coordinate system to coincide with the principal axes of the surface, we may eliminate the cross terms. Let x_1, x_2, x_3, be the principal axes. Then,

$$\mathbf{s}_d \cdot \mathbf{r} = \varepsilon_1 x_1^2 + \varepsilon_2 x_2^2 + \varepsilon_3 x_3^2 = f(x_1, x_2, x_3) \tag{47}$$

The $\varepsilon_1, \varepsilon_2$, and ε_3 in the new system of coordinates correspond to $\varepsilon_{xx}, \varepsilon_{yy}$, and ε_{zz} in the old. It follows that

$$\xi'_d = \varepsilon_1 x_1, \qquad \eta'_d = \varepsilon_2 x_2, \qquad \zeta'_d = \varepsilon_3 x_3$$

where ξ'_d, η'_d, and ζ'_d are the displacements in the x_1, x_2, and x_3 directions. This means that if we orient our volume with the faces perpendicular to the principal axes, the deformation of the volume will involve only pure extension or contraction. There will be no shear.

Let us now try to establish the relation between the old coefficients ε_{nm}, and the new ones $\varepsilon_i, i = 1,2,3$. Let $\alpha_i, \beta_i, \gamma_i (i = 1,2,3)$ represent the direction cosines between the old and the new axes. Thus,

$$x_i = \alpha_i x + \beta_i y + \gamma_i z$$

$i = 1,2,3$. It follows that

$$\begin{aligned} f(x_1, x_2, x_3) &= \sum_i \varepsilon_i x_i^2 = \sum_i \varepsilon_i (\alpha_i x + \beta_i y + \gamma_i z)^2 \\ &= x^2 \sum_i \varepsilon_i \alpha_i^2 + 2xy \sum_i \varepsilon_i \alpha_i \beta_i + y^2 \sum_i \varepsilon_i \beta_i^2 + \dots \end{aligned} \tag{48}$$

Comparison with equation (46) shows that

$$\begin{aligned}
\varepsilon_{xx} &= \alpha_1^2\varepsilon_1 + \alpha_2^2\varepsilon_2 + \alpha_3^2\varepsilon_3 \\
\varepsilon_{yy} &= \beta_1^2\varepsilon_1 + \beta_2^2\varepsilon_2 + \beta_3^2\varepsilon_3 \\
\varepsilon_{zz} &= \gamma_1^2\varepsilon_1 + \gamma_2^2\varepsilon_2 + \gamma_3^2\varepsilon_3 \\
\varepsilon_{xy} &= \alpha_1\beta_1\varepsilon_1 + \alpha_2\beta_2\varepsilon_2 + \alpha_3\beta_3\varepsilon_3 \\
\varepsilon_{yz} &= \beta_1\gamma_1\varepsilon_1 + \beta_2\gamma_2\varepsilon_2 + \beta_3\gamma_3\varepsilon_3 \\
\varepsilon_{zx} &= \gamma_1\alpha_1\varepsilon_1 + \gamma_2\alpha_2\varepsilon_2 + \gamma_3\alpha_3\varepsilon_3
\end{aligned} \tag{49}$$

If our body is now oriented so that the sides are parallel to the principal axes, and if, as assumed, the body is homogeneous and isotropic, then the stresses on each of the six faces will be normal to those faces. However, each stress will still be a function of all three strains, and vice versa. Assuming, in accordance with Hooke's law, that the dependence is linear, we may write,

$$\sigma_1 = a\varepsilon_1 + b\varepsilon_2 + c\varepsilon_3 \tag{50}$$

Because of the isotropy of the body, it must then be equally true that

$$\sigma_2 = a\varepsilon_2 + b\varepsilon_1 + c\varepsilon_3 \tag{51}$$

and

$$\sigma_3 = a\varepsilon_3 + b\varepsilon_2 + c\varepsilon_1 \tag{52}$$

Similarly, isotropy requires that $b = c$. Equations (50)–(52) may be rewritten,

$$\sigma_i = 2\mu\varepsilon_i + \lambda(\varepsilon_1 + \varepsilon_2 + \varepsilon_3) \tag{53}$$

$i = 1, 2, 3$ where $2\mu = a - b$ and $\lambda = b$.

We must now establish the relation between the components of the stress tensor in the old and new coordinate systems. In order to do this, let us consider a triangle through the point P, perpendicular to the x direction (Figure 15),

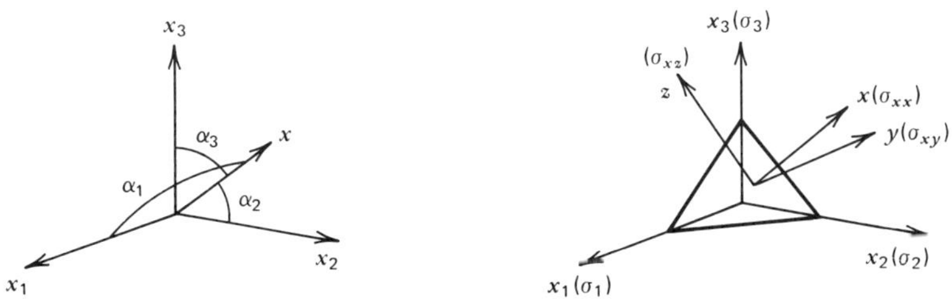

Figure 15. Relation of the principal axes to the original coordinate system.

with the directions of the principal axes as shown. The force on this triangle must be the same regardless of which coordinate system we measure it in.

Let A_n be the area of the triangle, and let α_1, α_2, and α_3 be the direction cosines of the x direction in the x_1, x_2, x_3, coordinate system. The force on the triangle in the x direction is, then,

$$A_n\sigma_{xx} = \sigma_1\alpha_1 A_1 + \sigma_2\alpha_2 A_2 + \sigma_3\alpha_3 A_3$$

where A_1, A_2, A_3 represent the areas of the projections of the triangle in the x_1x_2, x_3x_1, and x_1x_2 planes, respectively. But

$$\frac{A_1}{A_n} = \alpha_1, \qquad \frac{A_2}{A_n} = \alpha_2, \qquad \frac{A_3}{A_n} = \alpha_3$$

Thus,

$$\sigma_{xx} = \alpha_1^2\sigma_1 + \alpha_2^2\sigma_2 + \alpha_3^2\sigma_3 \tag{54}$$

Now let β_i and $\gamma_i, i = 1, 2, 3$, be the direction cosines of the y and z directions in the x_1, x_2, x_3 coordinate system. The force on the triangle in the y direction is, then,

$$A_n\sigma_{xy} = \beta_1\sigma_1A_1 + \beta_2\sigma_2A_2 + \beta_2\sigma_3A_3$$

or

$$\sigma_{xy} = \beta_1\alpha_1\sigma_1 + \beta_2\alpha_2\sigma_2 + \beta_3\alpha_3\sigma_3 \tag{55}$$

Similarly,

$$\sigma_{xz} = \gamma_1\alpha_1\sigma_1 + \gamma_2\alpha_2\sigma_2 + \gamma_3\alpha_3\sigma_3$$

Repeating the procedure for triangles perpendicular to the y and z directions, we get

$$\begin{aligned}
\sigma_{yy} &= \beta_1^2\sigma_1 + \beta_2^2\sigma_2 + \beta_3^2\sigma_3 \\
\sigma_{yx} &= \beta_1\alpha_1\sigma_1 + \beta_2\alpha_2\sigma_2 + \beta_3\alpha_3\sigma_3 \\
\sigma_{yz} &= \beta_1\gamma_1\sigma_1 + \beta_2\gamma_2\sigma_2 + \beta_3\gamma_3\sigma_3 \\
\sigma_{zz} &= \gamma_1^2\sigma_1 + \gamma_2^2\sigma_2 + \gamma_3^2\sigma_3 \\
\sigma_{zx} &= \gamma_1\alpha_1\sigma_1 + \gamma_2\alpha_2\sigma_2 + \gamma_3\alpha_3\sigma_3 \\
\sigma_{zy} &= \gamma_1\beta_1\sigma_1 + \gamma_2\beta_2\sigma_2 + \gamma_3\beta_3\sigma_3
\end{aligned} \tag{56}$$

We observe that $\sigma_{xy} = \sigma_{yx}$ and $\sigma_{xz} = \sigma_{zx}$. Introducing equations (53) into equations (54) and (55), we get

$$\begin{aligned}
\sigma_{xx} &= 2\mu\varepsilon_{xx} + \lambda(\alpha_1^2 + \alpha_2^2 + \alpha_3^2)(\varepsilon_1 + \varepsilon_2 + \varepsilon_3) \\
\sigma_{xy} &= 2\mu\varepsilon_{xy} + \lambda(\alpha_1\beta_1 + \alpha_2\beta_2 + \alpha_3\beta_3)(\varepsilon_1 + \varepsilon_2 + \varepsilon_3)
\end{aligned}$$

Now recall from analytic geometry,

$$\begin{aligned}
\alpha_1^2 + \alpha_2^2 + \alpha_3^2 &= 1 \\
\alpha_1\alpha_2 + \beta_1\beta_2 + \gamma_1\gamma_2 &= 0
\end{aligned} \tag{57}$$

Adding the first three equations (49) and using equations (57) we see that

$$\varepsilon_{xx} + \varepsilon_{yy} + \varepsilon_{zz} = \varepsilon_1 + \varepsilon_2 + \varepsilon_3 \tag{58}$$

Hence,

$$\begin{aligned}
\sigma_{xx} &= 2\mu\varepsilon_{xx} + \lambda(\varepsilon_{xx} + \varepsilon_{yy} + \varepsilon_{zz}) \\
\sigma_{xy} &= 2\mu\varepsilon_{xy}
\end{aligned}$$

Similarly,

$$\begin{aligned}
\sigma_{yy} &= 2\mu\varepsilon_{yy} + \lambda(\varepsilon_{xx} + \varepsilon_{yy} + \varepsilon_{zz}) \\
\sigma_{zz} &= 2\mu\varepsilon_{zz} + \lambda(\varepsilon_{xx} + \varepsilon_{yy} + \varepsilon_{zz}) \\
\sigma_{xz} &= 2\mu\varepsilon_{xz} \qquad \sigma_{yz} = 2\mu\varepsilon_{yz}
\end{aligned} \tag{59}$$

Equations (59) represent Hooke's law for a homogeneous, isotropic, deformable solid. These equations may be applied to a fluid by substituting rate of strain for strain. Thus, in place of

$$\varepsilon_{xx} = \frac{\partial \xi}{\partial x}, \qquad \varepsilon_{xy} = \frac{1}{2}\left(\frac{\partial \xi}{\partial y} + \frac{\partial \eta}{\partial x}\right)$$

we would write

$$\dot{\varepsilon}_{xx} = \frac{\partial u}{\partial x}, \qquad \dot{\varepsilon}_{xy} = \frac{1}{2}\left(\frac{\partial u}{\partial y} + \frac{\partial v}{\partial x}\right)$$

and so on. Equations (59) then become

$$\begin{aligned}
\sigma'_{xx} &= 2\mu\frac{\partial u}{\partial x} \qquad \sigma'_{xy} = \mu\left(\frac{\partial u}{\partial y} + \frac{\partial v}{\partial x}\right) \\
\sigma'_{yy} &= 2\mu\frac{\partial v}{\partial y} \qquad \sigma'_{yz} = \mu\left(\frac{\partial v}{\partial z} + \frac{\partial w}{\partial y}\right) \\
\sigma'_{zz} &= 2\mu\frac{\partial w}{\partial z} \qquad \sigma'_{zx} = \mu\left(\frac{\partial w}{\partial y} + \frac{\partial v}{\partial z}\right)
\end{aligned} \tag{60}$$

According to the assumptions of Stokes, then, equations (60) represent the viscous stresses on an element of fluid. In the case of shear stresses, the viscous stresses will be the only stresses. For normal stresses, however, we must add the isotropic pressure stress $-p'$. Thus, the total stresses will be given by

$$\begin{aligned}
\sigma_{xx} &= -p' + \sigma'_{xx} \qquad \sigma_{xy} = \sigma'_{xy} \\
\sigma_{yy} &= -p' + \sigma'_{yy} \qquad \sigma_{yz} = \sigma'_{yz} \\
\sigma_{zz} &= -p' + \sigma'_{zz} \qquad \sigma_{xz} = \sigma'_{xz}
\end{aligned} \tag{61}$$

Let us now return to equations (42). Dividing both sides by $\delta x\,\delta y\,\delta z$ and passing to the limit as $\delta x, \delta y, \delta z, \to 0$, we have

$$\begin{aligned}
\rho\frac{Du}{Dt} &= \frac{\partial \sigma_{xx}}{\partial x} + \frac{\partial \sigma_{xy}}{\partial y} + \frac{\partial \sigma_{xz}}{\partial z} \\
\rho\frac{Dv}{Dt} &= \frac{\partial \sigma_{xy}}{\partial x} + \frac{\partial \sigma_{yy}}{\partial y} + \frac{\partial \sigma_{yz}}{\partial z} \\
\rho\frac{Dw}{Dt} &= \frac{\partial \sigma_{xz}}{\partial x} + \frac{\partial \sigma_{yz}}{\partial y} + \frac{\partial \sigma_{zz}}{\partial z}
\end{aligned} \tag{62}$$

Substituting equations (60) and (61) into equations (62),

$$\rho\frac{Du}{Dt} = -\frac{\partial p'}{\partial x} + \frac{\partial}{\partial x}\left[2\mu\frac{\partial u}{\partial x} + \lambda\nabla\mathbf{V}\right] + \frac{\partial}{\partial y}\left[\mu\left(\frac{\partial u}{\partial y} + \frac{\partial v}{\partial x}\right)\right] + \frac{\partial}{\partial z}\left[\mu\left(\frac{\partial w}{\partial x} + \frac{\partial u}{\partial z}\right)\right]$$

$$\rho\frac{Dv}{Dt} = -\frac{\partial p'}{\partial y} + \frac{\partial}{\partial y}\left[2\mu\frac{\partial v}{\partial y} + \lambda\nabla\mathbf{V}\right] + \frac{\partial}{\partial z}\left[\mu\left(\frac{\partial v}{\partial z} + \frac{\partial w}{\partial y}\right)\right] + \frac{\partial}{\partial x}\left[\mu\left(\frac{\partial u}{\partial y} + \frac{\partial v}{\partial x}\right)\right]$$

$$\rho\frac{Dw}{Dt} = -\frac{\partial p'}{\partial z} + \frac{\partial}{\partial z}\left[2\mu\frac{\partial w}{\partial z} + \lambda\nabla\mathbf{V}\right] + \frac{\partial}{\partial x}\left[\mu\left(\frac{\partial w}{\partial x} + \frac{\partial u}{\partial z}\right)\right] + \frac{\partial}{\partial y}\left[\mu\left(\frac{\partial v}{\partial z} + \frac{\partial w}{\partial y}\right)\right] \tag{63}$$

While μ is called the coefficient of viscosity, or viscosity for short, λ is called the second coefficient of viscosity, or second viscosity. Stokes assumed in effect that

$$\lambda = -\tfrac{2}{3}\mu \tag{64}$$

Stokes based this assumption on kinetic theory considerations and, indeed, it may be shown that his assumption is correct for an ideal, monoatomic gas. In practice, however, it is found that this assumption does not restrict the applicability of the equations to monoatomic gases only. Equation (64) is usually referred to as the Stokes hypothesis.

We must also address ourselves to the problem of what we have called the pressure stresses. While in a fluid at rest the definition of the pressure is a simple matter, in a moving, viscous, compressible fluid the matter is no longer so simple. Stokes decided to define the pressure as the average normal force on a small element of fluid; that is,

$$p = \tfrac{1}{3}(\sigma_{xx} + \sigma_{yy} + \sigma_{zz})$$

Adding equations (61), we get

$$\sigma_{xx} + \sigma_{yy} + \sigma_{zz} = -3p' + \sigma'_{xx} + \sigma'_{yy} + \sigma'_{zz}$$

Substituting for σ'_{xx}, σ'_{yy}, and σ'_{zz} from equations (60) and using equation (64) for λ, we find that

$$\sigma'_{xx} + \sigma'_{yy} + \sigma'_{zz} = 0$$

Hence,

$$p' = \tfrac{1}{3}(\sigma_{xx} + \sigma_{yy} + \sigma_{zz}) = p \tag{65}$$

Substituting equations (64) and (65) into equations (63) we get

$$\rho\frac{Du}{Dt} = -\frac{\partial p}{\partial x} + \frac{\partial}{\partial x}\left[\mu\left(2\frac{\partial u}{\partial x} - \frac{2}{3}\nabla\mathbf{V}\right)\right] + \frac{\partial}{\partial y}\left[\mu\left(\frac{\partial u}{\partial y} + \frac{\partial v}{\partial x}\right)\right] + \frac{\partial}{\partial z}\left[\mu\left(\frac{\partial w}{\partial x} + \frac{\partial u}{\partial z}\right)\right]$$

$$\rho\frac{Dv}{Dt} = -\frac{\partial p}{\partial y} + \frac{\partial}{\partial y}\left[\mu\left(2\frac{\partial v}{\partial y} - \frac{2}{3}\nabla\mathbf{V}\right)\right] + \frac{\partial}{\partial z}\left[\mu\left(\frac{\partial v}{\partial z} + \frac{\partial w}{\partial y}\right)\right] + \frac{\partial}{\partial x}\left[\mu\left(\frac{\partial u}{\partial y} + \frac{\partial v}{\partial x}\right)\right]$$

$$\rho\frac{Dw}{Dt} = -\frac{\partial p}{\partial z} + \frac{\partial}{\partial z}\left[\mu\left(2\frac{\partial w}{\partial z} - \frac{2}{3}\nabla\mathbf{V}\right)\right] + \frac{\partial}{\partial x}\left[\mu\left(\frac{\partial w}{\partial x} + \frac{\partial u}{\partial z}\right)\right] + \frac{\partial}{\partial y}\left[\mu\left(\frac{\partial v}{\partial z} + \frac{\partial w}{\partial y}\right)\right]. \tag{66}$$

Equations (66) are commonly referred to as the Navier-Stokes equations.

The Navier-Stokes equations may also be derived by the use of vector calculus. This form of the derivation is helpful in providing a clearer understanding of the meaning of the equations. Consider a spherical element of fluid as shown in Figure 16. dS is an element of area around the point P on the surface of the

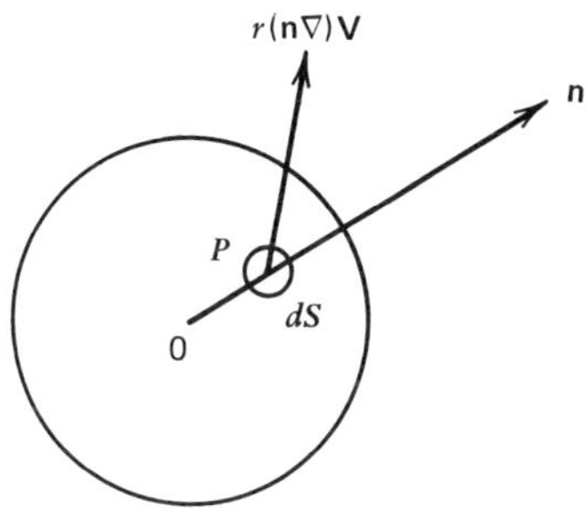

Figure 16. Definition of symbols for a spherical fluid element.

sphere, and $\mathbf{n}$ the unit outward normal on dS. r is the radius of the sphere and is assumed small. The velocity of the fluid at the point P relative to the fluid at point O is given by

$$r(\mathbf{n}\nabla)\mathbf{V}$$

The velocity component at P relative to O due to the average rotation of the fluid in the sphere about O is

$$\tfrac{1}{2}r(\nabla \times \mathbf{V}) \times \mathbf{n}$$

This component of the relative velocity cannot contribute to the strain since it represents a solid body rotation. Hence, we may say that the strain must depend on the relative velocity component,

$$r(\mathbf{n}\cdot\nabla)\mathbf{V} - \tfrac{1}{2}r(\nabla \times \mathbf{V}) \times \mathbf{n} \tag{67}$$

Dividing by r, we get the rate of strain,

$$(\mathbf{n}\cdot\nabla)\mathbf{V} - \tfrac{1}{2}(\nabla \times \mathbf{V}) \times \mathbf{n} \tag{68}$$

We now assume that the stress (force per unit area) on dS is proportional to, and in the direction of, the vector (68).

In order to understand the meaning of this assumption compared to those in our previous derivation, let us take $\mathbf{n}$ along the y axis and expand (68) in components. Then $\mathbf{n} = \mathbf{j}$ and

$$(\mathbf{n}\cdot\nabla)\mathbf{V} + \tfrac{1}{2}\mathbf{n} \times (\nabla \times \mathbf{V}) = \mathbf{i}\frac{\partial u}{\partial y} + \mathbf{j}\frac{\partial v}{\partial y} + \mathbf{k}\frac{\partial w}{\partial y} + \tfrac{1}{2}\left[\mathbf{k}\left(\frac{\partial v}{\partial z} - \frac{\partial w}{\partial y}\right) + \mathbf{i}\left(\frac{\partial v}{\partial x} - \frac{\partial u}{\partial y}\right)\right] \tag{69}$$

If the viscous stress is proportional to this we have

$$\sigma'_{yy} \sim \frac{\partial v}{\partial y}, \qquad \sigma'_{yx} \sim \tfrac{1}{2}\left(\frac{\partial v}{\partial x} + \frac{\partial u}{\partial y}\right), \qquad \sigma'_{yz} \sim \tfrac{1}{2}\left(\frac{\partial v}{\partial z} + \frac{\partial w}{\partial y}\right) \tag{70}$$

and if the proportionality constant is 2μ,

$$\sigma'_{yy} = 2\mu\frac{\partial v}{\partial y}, \qquad \sigma'_{yx} = \mu\left(\frac{\partial v}{\partial x} + \frac{\partial u}{\partial y}\right), \qquad \sigma'_{yz} = \mu\left(\frac{\partial v}{\partial z} + \frac{\partial w}{\partial y}\right) \tag{71}$$

Comparing equations (71) to equations (60), we see that they are identical except for the term $\lambda\nabla\mathbf{V}$ in σ'_{yy}. Similar results are obtained for $\sigma'_{xx}, \sigma'_{xz}$, and σ'_{zz}. The normal components differ by the term $\lambda\nabla\mathbf{V}$, while the shear components are identical.

Returning now to our derivation, we have

$$\boldsymbol{\sigma}' = 2\mu(\mathbf{n}\cdot\nabla)\mathbf{V} + \mu\mathbf{n} \times (\nabla \times \mathbf{V}) \tag{72}$$

This may also be written

$$\boldsymbol{\sigma}' = 2(\mathbf{n}\cdot\mu\nabla)\mathbf{V} + \mathbf{n} \times \mu(\nabla \times \mathbf{V})$$

The viscous force on the fluid element about O is, then,

$$\int_S 2(\mathbf{n}\cdot\mu\nabla)\mathbf{V}\,dS + \int_S \mathbf{n} \times \mu(\nabla \times \mathbf{V})\,dS \tag{73}$$

Reasoning as before, we must add a pressure strip p, in order to find the total force, $\bar{p}$ being assumed constant in magnitude over the surface S and being everywhere in the $-\mathbf{n}$ direction. The total force on the spherical element is, then,

$$-\int_S \bar{p}\mathbf{n}\,dS + \int_S 2(\mathbf{n}\cdot\mu\nabla)\mathbf{V}\,dS + \int_S \mathbf{n} \times \mu(\nabla \times \mathbf{V})\,dS \tag{74}$$

and, from Newton's law of motion, this force must equal

$$\int_{\mathscr{V}} \rho\frac{D\mathbf{V}}{Dt}\,d\mathscr{V}$$

where $\mathscr{V}$ is the volume of the element of fluid. Transforming the surface integral to volume integrals by means of the divergence theorem, we get

$$\int_{\mathscr{V}} \rho\frac{D\mathbf{V}}{Dt}\,d\mathscr{V} = -\int_{\mathscr{V}} \nabla\bar{p}\,d\mathscr{V} + \int_{\mathscr{V}} [\nabla(2\mu\nabla)]\mathbf{V}\,d\mathscr{V} + \int_{\mathscr{V}} \nabla \times \mu(\nabla \times \mathbf{V})\,d\mathscr{V} \tag{75}$$

Dividing both sides of (75) by $\mathscr{V}$ and taking the limit as $\mathscr{V} \to 0$, we get

$$\rho\frac{D\mathbf{V}}{Dt} = -\nabla\bar{p} + 2(\nabla\cdot\mu\nabla)\mathbf{V} + \nabla \times \mu(\nabla \times \mathbf{V}) \tag{76}$$

In order to compare this result with equation (66) let us write out the components of equation (76):

$$\begin{aligned}
\rho\frac{Du}{Dt} &= -\frac{\partial\bar{p}}{\partial x} + 2\frac{\partial}{\partial x}\left(\mu\frac{\partial u}{\partial x}\right) + \frac{\partial}{\partial y}\left[\mu\left(\frac{\partial v}{\partial x} + \frac{\partial u}{\partial y}\right)\right] + \frac{\partial}{\partial z}\left[\mu\left(\frac{\partial w}{\partial x} + \frac{\partial u}{\partial z}\right)\right] \\
\rho\frac{Dv}{Dt} &= -\frac{\partial\bar{p}}{\partial y} + 2\frac{\partial}{\partial y}\left(\mu\frac{\partial u}{\partial y}\right) + \frac{\partial}{\partial z}\left[\mu\left(\frac{\partial v}{\partial z} + \frac{\partial w}{\partial y}\right)\right] + \frac{\partial}{\partial x}\left[\mu\left(\frac{\partial u}{\partial y} + \frac{\partial v}{\partial x}\right)\right] \\
\rho\frac{Dw}{Dt} &= -\frac{\partial\bar{p}}{\partial z} + 2\frac{\partial}{\partial z}\left(\mu\frac{\partial w}{\partial z}\right) + \frac{\partial}{\partial x}\left[\mu\left(\frac{\partial w}{\partial x} + \frac{\partial u}{\partial z}\right)\right] + \frac{\partial}{\partial y}\left[\mu\left(\frac{\partial v}{\partial z} + \frac{\partial w}{\partial y}\right)\right]
\end{aligned} \tag{77}$$

We observe that equations (77) are identical to equations (66) except for the factor $-\frac{2}{3}\mu\nabla\mathbf{V}$ appearing in each of the equations (67). Recall, however, that p was defined as the average normal force around a point. In the present case,

$$\sigma_{xx} = -\bar{p} + 2\mu\frac{\partial u}{\partial x}$$
$$\sigma_{yy} = -\bar{p} + 2\mu\frac{\partial v}{\partial y} \tag{78}$$
$$\sigma_{zz} = -\bar{p} + 2\mu\frac{\partial w}{\partial z}$$

Adding,

$$\sigma_{xx} + \sigma_{yy} + \sigma_{zz} = -3\bar{p} + 2\mu\nabla\mathbf{V}$$

Hence,

$$-p = \tfrac{1}{3}(\sigma_{xx} + \sigma_{yy} + \sigma_{zz}) = -\bar{p} + \tfrac{2}{3}\mu\nabla\mathbf{V}$$

or

$$\bar{p} = p + \tfrac{2}{3}\mu\nabla\mathbf{V} \tag{79}$$

Substituting for $\bar{p}$ from equation (79) into equations (77) we find that these equations are identical to equations (66).

Equation (72) states that if the velocity is taken with respect to a coordinate system which is rotating with the average angular velocity of the fluid around some point, then the viscous stress on any surface near that point will be proportional to the derivative of the velocity in a direction normal to that surface. Unfortunately, this simple principle cannot be valid in general. The reason is that it neglects the effect of the stresses in the two perpendicular directions on each normal stress components. Consider a cube with stress σ_{yy} on the top plane (Figure 17). According to equation (71), σ'_{yy} depends only

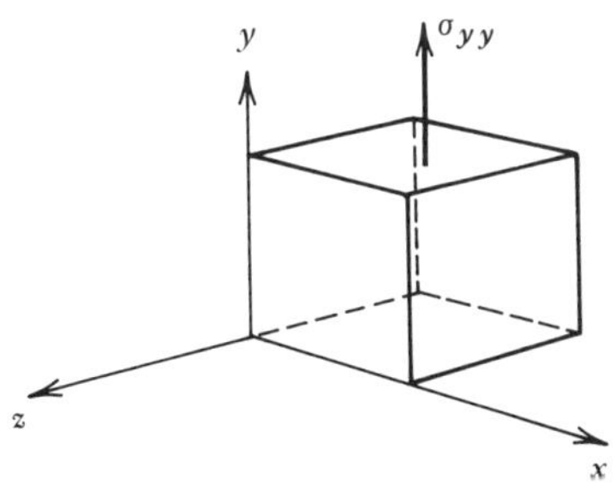

Figure 17. Cubic element of fluid with stress in the y direction.

on the displacement in the y direction. But we know this is not true. Returning to equations (60) we see that σ'_{yy} depends on the displacements in the other directions too. How is it, then, that the derivation based on such a false assumption can yield the right result? The answer lies in the form of the normal stress components. The normal stress components may be written [cf. equations (59)]

$$\sigma'_{xx} = a\dot{\varepsilon}_{xx} + b(\dot{\varepsilon}_{yy} + \dot{\varepsilon}_{zz})$$
$$\sigma'_{yy} = a\dot{\varepsilon}_{yy} + b(\dot{\varepsilon}_{xx} + \dot{\varepsilon}_{zz})$$
$$\sigma'_{zz} = a\dot{\varepsilon}_{zz} + b(\dot{\varepsilon}_{yy} + \dot{\varepsilon}_{xx})$$

This is the most natural way to write these equations on physical grounds. The strain on each face is affected by the strains on the other two faces. However, we may rewrite the equations in the form

$$\sigma'_{xx} = (a-b)\dot{\varepsilon}_{xx} + b(\dot{\varepsilon}_{xx} + \dot{\varepsilon}_{yy} + \dot{\varepsilon}_{zz})$$
$$\sigma'_{yy} = (a-b)\dot{\varepsilon}_{yy} + b(\dot{\varepsilon}_{xx} + \dot{\varepsilon}_{yy} + \dot{\varepsilon}_{zz})$$
$$\sigma'_{zz} = (a-b)\dot{\varepsilon}_{zz} + b(\dot{\varepsilon}_{xx} + \dot{\varepsilon}_{yy} + \dot{\varepsilon}_{zz})$$

This puts the equations in a form in which each normal stress varies with the normal strain plus another term which is the same for each of the three normal components. As a result, this term may be lumped with the "pressure stress," which is also the same in each direction. This is, in effect, what we did in writing p in equation (74). It should be kept in mind, however, that p is not the pressure as we have defined it, but a combination of pressure and viscous stresses. By the same token,

$$2\mu\frac{\partial v}{\partial y}$$

is not, in general, the normal component of the viscous stress, nor even the entire contribution of the normal component of the strain to that component of the stress, that entire contribution being,

$$2\mu\frac{\partial v}{\partial y} + \lambda\frac{\partial v}{\partial y}$$

The qualification "in general" has been inserted because in the incompressible case, $\nabla \mathbf{V} = 0$, and then we see from equation (60) that

$$\sigma'_{yy} = 2\mu\frac{\partial v}{\partial y}, \qquad \sigma'_{xx} = 2\mu\frac{\partial u}{\partial x}, \qquad \sigma'_{zz} = 2\mu\frac{\partial w}{\partial z}$$

In this case, the assumption of equation (72) becomes valid and $\bar{p}$ reduces to p.

Substituting equation (79) into equation (76) we get,

$$\rho\frac{D\mathbf{V}}{Dt} = -\nabla p - \tfrac{2}{3}\nabla[\mu(\nabla\mathbf{V})] + [\nabla\cdot(2\mu\nabla)]\mathbf{V} + \nabla \times [\mu\nabla \times \mathbf{V}] \qquad (80)$$

This is the Navier-Stokes equation in vector form.

For incompressible flow, where μ may generally be considered constant, equation (80) reduces to

$$\rho\frac{D\mathbf{V}}{Dt} = -\nabla p + \mu\nabla^2\mathbf{V} \qquad (81)$$

In his original derivation, Stokes included what he called the "external forces" or what we would now call body forces. The body force most usually encountered is gravity, but, in the case of electrically or magnetically charged fluids, the body forces may also be electric or magnetic in nature. If $\mathbf{f}$ is the body force per unit mass, we need only add $\rho\mathbf{f}$ to the right-hand side of equation (80).

2 Elementary One Dimensional Flow

Many practical problems in gas dynamics are one-dimensional, or quasi-one-dimensional, in nature. Such problems may be treated by considering only the x component of the velocity, or the average value of the x component, across the flow.

The Basic Equations

A quasi-one-dimensional flow problem is one in which the y and z components of the velocity are small compared to the x component. Flow through a pipe with a slowly varying cross section would be an example of such a flow (Figure 18). Let us see how the governing equations may be simplified in such a case.

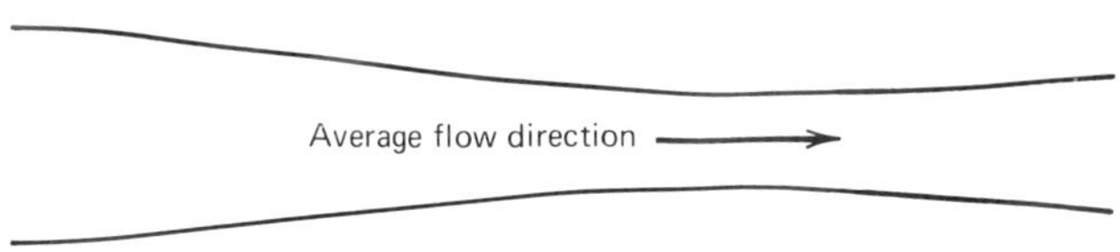

Figure 18. Quasi-one-dimensional flow.

Recall the continuity equation (5):

$$\frac{\partial \rho}{\partial t} + \nabla(\rho \mathbf{V}) = 0$$

Let us now apply this equation to a section of the pipe in Figure (18), enclosing a volume $\mathcal{V}$ (Figure 19). Taking the coordinate system as shown, the average

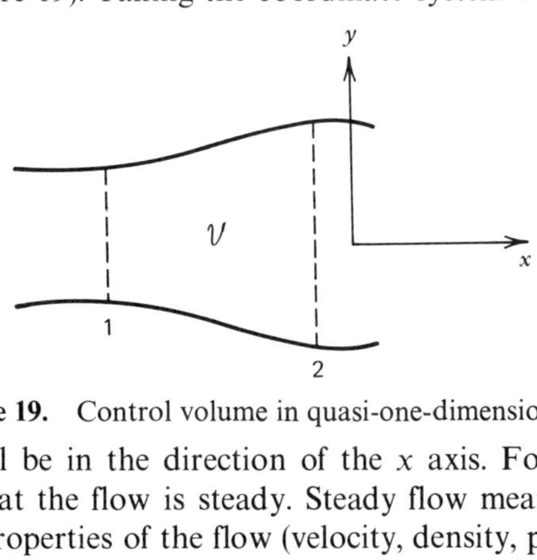

Figure 19. Control volume in quasi-one-dimensional flow.

flow direction will be in the direction of the x axis. For simplicity, we will assume further that the flow is steady. Steady flow means that at any given point x, y, z, the properties of the flow (velocity, density, pressure, etc.) are not changing with time. Since the rate of change of a property of the flow at a point fixed in space is given by the partial derivative with respect to time of the property evaluated at that point, if the flow is steady, all the partial derivatives with respect to time are zero. Equation (5) then becomes

$$\nabla(\rho \mathbf{V}) = 0 \tag{82}$$

Integrating equation (82) over the volume $\mathscr{V}$, and applying the divergence theorem, we have

$$\int_{\mathscr{V}} \nabla(\rho \mathbf{V})\, d\mathscr{V} = \int_{S} \rho \mathbf{V} \cdot \mathbf{n}\, dS = 0 \tag{83}$$

where S represents the surface area of the volume $\mathscr{V}$ and $\mathbf{n}$ the unit outward normal. If A_1 and A_2 are the cross sectional areas of the pipe at 1 and 2 as indicated in Figure 19, equation (83) becomes

$$\int_{A_1} \rho \mathbf{V} \cdot \mathbf{n}\, dS + \int_{A_2} \rho \mathbf{V} \cdot \mathbf{n}\, dS = 0 \tag{84}$$

since there is no flow across the boundary of the pipe. Now let $\bar{\rho}_1, \overline{\mathbf{V}_1 \cdot \mathbf{n}_1}$ be the average density and normal velocity component on A_1 and $\bar{\rho}_2, \overline{\mathbf{V}_2 \cdot \mathbf{n}_2}$ be the average density and normal velocity component on A_2. Then, by the mean value theorem,*

$$\bar{\rho}_1 \overline{\mathbf{V}_1 \cdot \mathbf{n}_1} \int_{A_1} dS + \bar{\rho}_2 \overline{\mathbf{V}_2 \cdot \mathbf{n}_2} \int_{A_2} dS = 0$$

Let

$$V_1 = -\overline{\mathbf{V}_1 \cdot \mathbf{n}_1}, \qquad V_2 = \overline{\mathbf{V}_2 \cdot \mathbf{n}}$$

---

*The mean value theorem says that given $G(x, y)$ is continuous over some area A, then there exists a point $x = \xi, y = \eta$, on A, such that

$$\int_A G(x, y)\, dA = G(\xi, \eta) \int_A dA$$

In the present case, $G(x, y) = \rho \mathbf{V} \cdot \mathbf{n}(x, y)$ and hence there will exist some point ξ, η, on A, such that

$$\overline{\rho \mathbf{V} \cdot \mathbf{n}} A = \rho \mathbf{V} \cdot \mathbf{n}(\xi, \eta) A = \int_A \rho \mathbf{V} \cdot \mathbf{n}\, dA$$

Moreover,

$$\rho \mathbf{V} \cdot \mathbf{n}(\xi, \eta) = \rho(\xi, \eta) \mathbf{V}(\xi, \eta) \cdot \mathbf{n}$$

It should be noted that if

$$\bar{\rho} = \rho(\xi', \eta') A = \int_A \rho(x, y)\, dA, \qquad \overline{\mathbf{V} \cdot \mathbf{n}} A = \mathbf{V}(\xi'', \eta'') \cdot \mathbf{n} A = \int_A \mathbf{V}(x, y) \cdot \mathbf{n}\, dA$$

then it is not necessarily true that

$$\xi' = \xi'' = \xi, \qquad \eta' = \eta'' = \eta$$

Hence, it does not follow that

$$\bar{\rho} \overline{\mathbf{V} \cdot \mathbf{n}} = \overline{\rho \mathbf{V} \cdot \mathbf{n}} \tag{a}$$

However, if neither ρ nor $\mathbf{V}$ vary much across A, then equation (a) may be taken as approximately true. This is the basis of the one-dimensional flow assumption.

Then

$$\rho_1 V_1 A_1 = \rho_2 V_2 A_2 \tag{85}$$

and since 1 and 2 are entirely arbitrary, we may say, in general,

$$\rho V A = \text{constant} \tag{86}$$

Equation (86) is sometimes referred to as the continuity equation for one-dimensional steady flow. It should be kept in mind, however, that ρ and V in this equation represent average quantities.

Since $\mathbf{V}\cdot\mathbf{n} = u$, the x component of the velocity in this particular case, we can also write equation (86) in the form

$$\rho \bar{u} A = \text{constant}$$

where $\bar{u}$ represents the average x component of the velocity on A.

Consider now the momentum equation in the inviscid case.

$$\rho \frac{D\mathbf{V}}{Dt} = -\nabla p$$

We wish to write this equation in a simplified form for the quasi-one-dimensional case of Figure 18. The continuity equation may be written in the form

$$\frac{D\rho}{Dt} + \rho(\nabla\cdot\mathbf{V}) = 0 \tag{87}$$

Hence,

$$\begin{aligned} \rho \frac{D\mathbf{V}}{Dt} &= \rho \frac{D\mathbf{V}}{Dt} + \left[\frac{D\rho}{Dt} + \rho(\nabla\cdot\mathbf{V})\right]\mathbf{V} \\ &= \frac{D(\rho\mathbf{V})}{Dt} + (\rho\mathbf{V})(\nabla\cdot\mathbf{V}) \\ &= \frac{\partial(\rho\mathbf{V})}{\partial t} + \rho\mathbf{V}(\nabla\cdot\mathbf{V}) + (\mathbf{V}\cdot\nabla)(\rho\mathbf{V}) \end{aligned} \tag{88}$$

Substituting (88) in (10) and integrating over the volume $\mathscr{V}$ in Figure 19, we get, in the case of steady flow,

$$\int_{\mathscr{V}} [\rho\mathbf{V}(\nabla\cdot\mathbf{V}) + (\mathbf{V}\cdot\nabla)(\rho\mathbf{V})]\, d\mathscr{V} = -\int_{\mathscr{V}} \nabla p\, d\mathscr{V} \tag{89}$$

Applying Gauss's theorem to both sides of equation (89) we get

$$-\int_S p\mathbf{n}\, dS = \int_S \rho\mathbf{V}(\mathbf{V}\cdot\mathbf{n})\, dS \tag{90}$$

Referring again to Figure 19, let A_1 and A_2 be the cross-sectional areas of the flow at 1 and 2, and let A_3 be the circumferential area of the duct between 1 and 2. Then (90) becomes

$$-\int_{A_1} p\mathbf{n}\, dS - \int_{A_2} p\mathbf{n}\, dS - \int_{A_3} p\mathbf{n}\, dS = \int_{A_1} \rho\mathbf{V}(\mathbf{V}\cdot\mathbf{n})\, dS + \int_{A_2} \rho\mathbf{V}(\mathbf{V}\cdot\mathbf{n})\, dS$$

Taking the x component,

$$\int_{A_1} p\,dS - \int_{A_2} p\,dS - \int_{A_3} p\mathbf{n}\cdot\mathbf{i}\,dS = -\int_{A_1} \rho u\cdot u\,dS + \int_{A_2} \rho u\cdot u\,dS$$

Now let p_1, p_2, and p_3 be the average pressures over A_1, A_2, and A_3, respectively, and let ρ_1, u_1 and ρ_2, u_2 be the average density and x-velocity component over A_1 and A_2. Then,

$$A_1 p_1 - p_2 A_2 + p_3(A_2 - A_1) = -\rho_1 u_1^2 A_1 + \rho_2 u_2^2 A_2 \tag{91}$$

Because of axial symmetry, $u_1 = V_1, u_2 = V_2$. Further,

$$\rho_1 u_1 A_1 = \rho_2 u_2 A_2 = \dot{m} \tag{92}$$

where $\dot{m}$ is the mass flow rate. $-p_3(A_2 - A_1)$ is the x component of the force exerted by the fluid in the volume $\mathscr{V}$ on the wall of the duct. Let us call this force F. Then,

$$(A_1 p_1 + \dot{m} V_1) - (A_2 p_2 + \dot{m} V_2) = F \tag{93}$$

The function $pA + \dot{m}V$ is sometimes called the impulse function.

Returning now to equation (10), let $\mathbf{s}$ be the unit vector in the direction of the streamline at any point. Equation (10) may be written, for steady flow,

$$\rho \frac{D\mathbf{V}}{Dt} = \rho[\tfrac{1}{2}\nabla V^2 - \mathbf{V} \times (\nabla \times \mathbf{V})] = -\nabla p \tag{94}$$

where V is the magnitude of the velocity. Taking the dot product of both sides with $\mathbf{s}$,

$$\tfrac{1}{2}\mathbf{s}\cdot\nabla V^2 - \mathbf{s}\cdot[\mathbf{V} \times (\nabla \times \mathbf{V})] = -\mathbf{s}\cdot\frac{\nabla p}{\rho} \tag{95}$$

Since $\mathbf{s}$ is perpendicular to $\mathbf{V} \times (\nabla \times \mathbf{V})$,

$$\mathbf{s}\cdot[\mathbf{V} \times (\nabla \times \mathbf{V})] = 0$$

Moreover, it will be recalled that a unit vector, dotted into the gradient of a scalar, gives the rate of change of the scalar in the direction of the vector. Hence, (95) becomes

$$\rho V \frac{dV}{ds} = -\frac{dp}{ds} \tag{96}$$

or, along a streamline,

$$\rho V\,dV = -dp \tag{97}$$

Recall now that $\frac{1}{2}(\nabla \times \mathbf{V})$ represents the average rate of rotation of the fluid in a small volume about a point. If $\nabla \times \mathbf{V} = 0$, the fluid is said to be irrotational. In such a case, (94) becomes

$$\tfrac{1}{2}\rho \nabla V^2 = -\nabla p \tag{98}$$

and hence,

$$\rho V\,dV = -\,dp \tag{99}$$

in any direction.

The energy equation. Recall equation (27),

$$\rho\frac{Dh}{Dt} = \frac{Dp}{Dt} + \dot{q}$$

Let us now apply this equation to the case of Figure 19.

$$\int_{\mathscr{V}} \rho\frac{Dh}{Dt}\,d\mathscr{V} = \int_{\mathscr{V}} \rho\mathbf{V}\cdot\nabla h\,d\mathscr{V} = \int_{\mathscr{V}} [\nabla(h\mathbf{V}\rho) - h\nabla(\rho\mathbf{V})]\,d\mathscr{V}$$

in the steady case. But $\nabla(\rho\mathbf{V}) = 0$. Hence, we may apply the divergence theorem to get

$$\int_{\mathscr{V}} \rho\frac{Dh}{Dt}\,d\mathscr{V} = \int_S \mathbf{n}\cdot h\mathbf{V}\,dS$$

Also, using the momentum and continuity equations,

$$\int_{\mathscr{V}} \frac{Dp}{Dt}\,d\mathscr{V} = \int_{\mathscr{V}} \mathbf{V}\cdot\nabla p\,d\mathscr{V} = -\int_{\mathscr{V}} \rho\frac{D\frac{1}{2}V^2}{Dt}\,d\mathscr{V} - \int_{\mathscr{V}} \rho g\frac{DZ}{Dt}\,d\mathscr{V}$$

Now,

$$-\int_{\mathscr{V}} \rho\frac{D\frac{1}{2}V^2}{Dt}\,d\mathscr{V} = -\int_{\mathscr{V}} \nabla(\tfrac{1}{2}\rho V^2\mathbf{V})\,d\mathscr{V} = -\int_S \mathbf{n}\cdot\mathbf{V}\rho\frac{V^2}{2}\,dS \tag{100}$$

Also,

$$\int_{\mathscr{V}} \rho g\frac{DZ}{Dt}\,d\mathscr{V} = \int_{\mathscr{V}} \rho g\mathbf{V}\cdot\nabla Z\,d\mathscr{V} = \int_{\mathscr{V}} \nabla(\rho g\mathbf{V}Z)\,d\mathscr{V} = \int_S \mathbf{n}\cdot\mathbf{V}\rho g Z\,dS \tag{101}$$

Substituting the results of equations (98)–(100) into equation (27), we get

$$\int_S \mathbf{n}\cdot h\mathbf{V}\rho\,dS = -\int_S \mathbf{n}\cdot\mathbf{V}\rho\frac{V^2}{2}\,dS + \int_{\mathscr{V}} \dot{q}\,d\mathscr{V} - \int_S \mathbf{n}\cdot\mathbf{V}\rho g Z\,dS$$

Now let the subscripts 1 and 2 indicate the average value over the areas A_1 and A_2. Then,

$$\dot{m}(h_2 - h_1) = \dot{m}\left(\frac{V_1^2}{2} - \frac{V_2^2}{2}\right) + \dot{Q} - \dot{m}g(Z_2 - Z_1) \tag{102}$$

where $\dot{Q}$ represents the net rate of heat flux into the volume $\mathscr{V}$.

Equations (86), (93), (97), (99), and (102) represent the most useful, and most frequently encountered forms of the one-dimensional flow equations.

In problems in gas dynamics, the effect of gravity is usually small compared to other terms, and hence the last term of equation (102) is often omitted. Moreover, in many practical problems, the amount of heat entering or leaving the flow is also small compared to the other terms and hence may be neglected. Where heat transfer and gravity may be neglected, equation (102) reduces to

$$h_1 + \frac{V_1^2}{2} = h_2 + \frac{V_2^2}{2} \tag{103}$$

Since 1 and 2 are entirely arbitrary, it may be said in general that

$$h + \frac{V^2}{2} = h_0 \tag{104}$$

where h and V represent the average enthalpy and velocity magnitude over any cross section of the flow and h_0 is a constant called the stagnation enthalpy or total enthalpy. A flow in which $\dot{Q}$ is zero is said to be adiabatic.

Recall, now, from elementary thermodynamics, the relation,

$$T\,dS = dh - \frac{dp}{\rho} \tag{105}$$

This equation gives the entropy change for a closed system consisting of a pure substance in the absence of nonpressure forces. A pure substance is one whose chemical composition is the same everywhere, and is not changing with time.

Now recall that along a streamline, in inviscid flow,

$$\rho V\,dV = -\,dp$$

If the flow is steady, this equation describes not only the variation of the pressure with the velocity from point to point, but also the variation of the pressure with the velocity of a particle of fluid as it flows along a streamline. Differentiating equation (104),

$$dh + V\,dV = 0 \tag{106}$$

Substituting (106) and (97) into (105),

$$T\,dS = dh - \frac{dp}{\rho} = dh + V\,dV \tag{107}$$

Thus we have the result that in inviscid, steady, adiabatic flow of a pure substance in the absence of nonpressure forces, the entropy is constant along a streamline. It might be noted that this result is not restricted to streamlines in a quasi-one-dimensional flow. Any streamline can be regarded locally as a quasi-one-dimensional flow if we confine our attention to a small tabular section of the

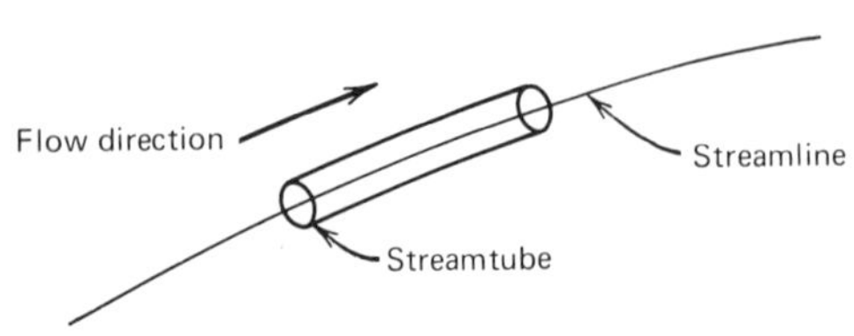

Figure 20. Streamtube.

flow around the streamline (Figure 20). Such a tabular section is called a streamtube. Neither equation (104) nor equation (97) is affected by the fact that the streamtube may be curving.

The Speed of Sound

Physical evidence indicates that sound waves consist of a series of small pressure pulses, which propagate through a medium with some finite velocity. Consider a gas at rest which is slightly disturbed in some way, say by banging a gong. The gong vibrates, causing small changes in the pressure, density, and velocity in the gas. Let ρ_0, p_0 and zero be the values of the density, pressure and velocity in the gas at rest, and let ρ', p', and $\mathbf{V}'$ be the changes in density, pressure and velocity caused by the disturbance. Substituting in the continuity equation we have

$$\frac{\partial}{\partial t}(\rho_0 + \rho') + \nabla((\rho_0 + \rho')\mathbf{V}') = 0$$

Omitting higher order terms this reduces to

$$\frac{\partial \rho'}{\partial t} + \rho_0 \nabla \mathbf{V}' = 0 \tag{108}$$

The momentum equation, neglecting viscosity, becomes

$$\frac{\partial \mathbf{V}'}{\partial t} + \mathbf{V}' \cdot \nabla \mathbf{V}' + \frac{1}{\rho_0 + \rho'} \nabla(p_0 + p') = 0$$

Again neglecting higher order terms, we have

$$\frac{\partial \mathbf{V}'}{\partial t} + \frac{1}{\rho_0} \nabla p' = 0 \tag{109}$$

Now let us assume that p varies in some known way with ρ. That is,

$$p = p(\rho) \tag{110}$$

Such a relationship may be obtained by eliminating h from the equation of state

$$p = p(\rho, h)$$

by introduction of the energy equation. For the moment we simply assume that an equation of the form (110) exists. Then,

$$\frac{dp}{d\rho} = \frac{dp'}{d\rho'} = (c_0 + c')^2 \tag{111}$$

say. Then,

$$\frac{\partial \rho'}{\partial t} = \frac{\partial p'}{\partial t}\frac{d\rho'}{dp'} = \frac{1}{(c_0 + c')^2}\frac{\partial p'}{\partial t} \tag{112}$$

and, from (108),

$$\frac{\partial p'}{\partial t} + \rho_0 (c_0 + c')^2 \nabla \mathbf{V}' = 0$$

or, omitting higher order terms,

$$\frac{\partial p'}{\partial t} + \rho_0 c_0^2 \nabla \mathbf{V}' = 0 \tag{113}$$

Differentiating (113) with respect to t,

$$\frac{\partial^2 p'}{\partial t^2} + \rho_0 c_0^2 \nabla \frac{\partial \mathbf{V}'}{\partial t} = 0 \tag{114}$$

Taking the divergence of (109),

$$\nabla \frac{\partial \mathbf{V}'}{\partial t} + \frac{1}{\rho_0} \nabla^2 p' = 0 \tag{115}$$

Combining (114) and (115),

$$\nabla^2 p' = \frac{1}{c_0^2} \frac{\partial^2 p'}{\partial t^2} \tag{116}$$

This will be recognized as the wave equation. In order to give physical meaning to this equation, let us suppose that the disturbance is spherical. That is, suppose that at some time $t_0, p = p' + p_0$ for $r \leq R$, and $p = p_0$ for $r > R$ (Figure 21)

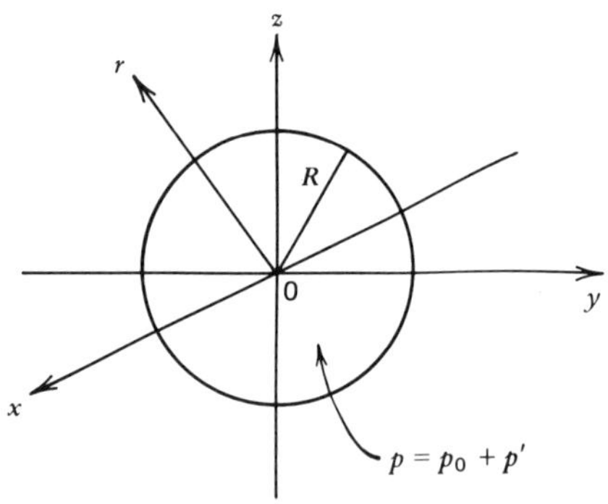

Figure 21. Spherical pressure wave.

where $r^2 = x^2 + y^2 + z^2$. The question is, how will the pressure vary with time in x, y, z space?

For spherical symmetry,

$$\nabla^2 p' = \frac{\partial^2 p'}{\partial r^2} + \frac{2}{r} \frac{\partial p'}{\partial r} = \frac{1}{c_0^2} \frac{\partial^2 p'}{\partial t^2} \tag{117}$$

A solution of equation (117) is of the form

$$p' = \frac{1}{r} [f(r - c_0 t)] \tag{118}$$

From the boundary conditions,

$$f(r - c_0 t_0) = 0$$

when $r > R$, and

$$f(r - c_0 t_0) = rp'$$

when $r \leq R$. Thus, it follows that

$$f(r - c_0 t) = 0$$

when $r - c_0 t > R - c_0 t_0$ and

$$f(r - c_0 t) = rp'$$

when $r - c_0 t \leq R - c_0 t_0$. In short, what we have is a pressure wave traveling spherically outward from the origin with constant velocity c_0. It remains to evaluate c_0.

We will show subsequently that an inviscid flow which is irrotational and adiabatic is also isentropic. These conditions are closely approximated in the present case. Hence, we may assume that the process is isentropic.

The Perfect Gas

Before we can proceed to evaluate the speed of sound, we will need an equation of state. In engineering applications, the equation of state most frequently encountered is

$$p = \rho RT \tag{119}$$

where R is the gas constant per unit mass. A gas which obeys equation (119) and for which

$$e_t = e_t(T) \tag{120}$$

that is, the internal energy is a function of temperature only, is called a perfect gas. A perfect gas may be distinguished from an ideal gas, which is simply a gas with zero viscosity. These definitions are the traditional ones; however, in the course of time they have sometimes become confused, and the term ideal gas is used on occasion to mean a perfect gas, and vice versa. In the present work, we shall stick to the original definitions.

Recall the definition of specific heat at constant volume:

$$c_v = \left(\frac{\partial e_t}{\partial T}\right)_{\mathscr{V}}$$

For a perfect gas, since $e_t = e_t(T)$ only,

$$c_v = \frac{de_t}{dT}$$

Similarly, the specific heat at constant pressure may be defined by

$$c_p = \left(\frac{\partial h}{\partial T}\right)_p$$

Since

$$h = e_t + \frac{p}{\rho}$$

it follows that for a perfect gas, $h = h(T)$ only, and

$$c_p = \frac{dh}{dT} = \frac{de_t}{dT} + R = c_v + R \tag{121}$$

The ratio of specific heats, c_p/c_v will be called γ. That is,

$$\frac{c_p}{c_v} = \gamma \tag{122}$$

In the present chapter we shall assume that for a perfect gas, c_p and c_v are constant.

Many real gases may be approximated by perfect gases with good accuracy provided that the temperature is well below the dissociation temperature and the pressure is low compared to the critical pressure. Recall now from thermodynamics that for a pure substance,

$$T\,dS = de_t + p\,d\left(\frac{1}{\rho}\right)$$

For a perfect gas, then,

$$\frac{dS}{c_v} = \frac{dp}{p} - \gamma\frac{d\rho}{\rho}$$

Hence, for an isentropic process,

$$\frac{p}{\rho^\gamma} = \text{constant} \tag{123}$$

Speed of Sound in a Perfect Gas

Recall now that the speed of propagation of a weak, spherical pressure wave was found to be

$$c_0 = \sqrt{\frac{dp}{d\rho}}$$

Furthermore, under the inviscid flow assumption, the process was found to be essentially isentropic. For a perfect gas, then,

$$c_0 = \sqrt{\frac{dp}{d\rho}} = \sqrt{\frac{\gamma p}{\rho}} = \sqrt{\gamma R T} \tag{124}$$

The validity of this equation may easily be tested by experiment. Its accuracy is by no means a priori evident. Assuming that sound waves propagate outward spherically from their source and that real gases may be approximated by perfect gases, there still remains the question of viscosity, which was neglected in the derivation. This is not a minor point. For in leaving the viscous terms out of the momentum equation, we have changed the order of the equation. Hence we may not say, a priori, that the limit of the solution of the full equation, including viscosity, as the viscosity goes to zero, is equal to the solution of the abbreviated equation, with the viscosity neglected altogether. Thus, to establish the validity of equation (124) we must either resort to more sophisticated mathematical means, or to experiment.

Because of the relative slowness of the speed of sound, it is easily tested by experiment. All it requires is two people standing within sight of each other, five hundred to a thousand feet apart, one with a source of sharp noise, the other with a stop watch. It is found that equation (124) predicts the velocity of sound in air to within 1% accuracy. For air at normal temperature and pressure, $\gamma = 1.4$, $R = 53.3$ ft-lb-°F, and $T = 530$°R on a 70°F day. Thus, according to equation (124), the velocity of sound will be

$$c_0 = 1128 \text{ ft/sec}$$

Sound Wave in One Dimension

Consider a constant area duct filled with gas and suppose that at time t_0 a small pressure discontinuity exists in the duct such that on one side of the discontinuity the pressure is p_0 and on the other side $p_0 + p'$ (Figure 22).

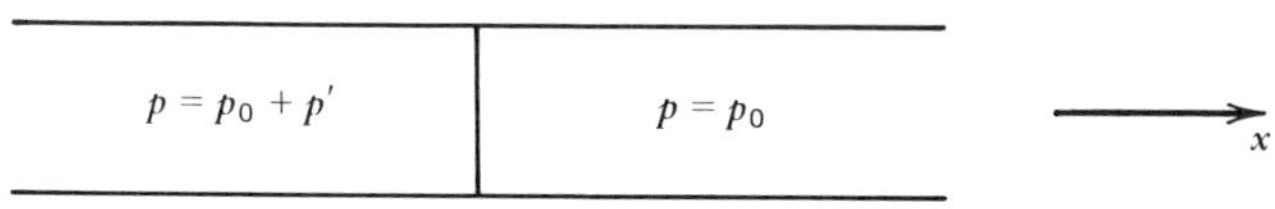

Figure 22. One-dimensional pressure wave.

Equation (116) becomes

$$\frac{\partial^2 p'}{\partial x^2} = \frac{1}{c_0^2}\frac{\partial^2 p'}{\partial t^2}$$

This equation has a solution,

$$p' = f(x - c_0 t)$$

thus indicating that the discontinuity moves to the right with velocity c_0. The equation is also satisfied by the solution,

$$p' = f(x + c_0 t)$$

Thus, the wave may move either to the right or to the left. On physical grounds one may reason that if a sound occurs at the middle of the pipe, there will be two waves, one moving to the left, the other to the right, both away from the source of the sound.

It will be observed that the speed of the wave is the same in the one-dimensional case as in the spherically symmetric case. However, we found that the wave attenuates as $1/r$ in the spherically symmetric case, while in the one-dimensional case, the wave appears unattenuated.

Effect of Viscosity on a One-Dimensional Sound Wave

In the one-dimensional case, the effect of viscosity on a small pressure wave is easily demonstrated analytically. We consider again a small pressure wave propagating into a medium at rest. In such a case, the velocity, pressure, and density fluctuations in the medium caused by the passage of the wave may be assumed to be small. Recall that

$$\frac{\partial \rho}{\partial t} + \nabla(\rho \mathbf{V}) = 0$$

Now let

$$\rho = \rho_0(1 + s) \tag{125}$$

where ρ_0 is constant and s is small. Then, neglecting higher order terms,

$$\frac{\partial s}{\partial t} + \nabla \mathbf{V} = 0 \tag{126}$$

Now define $\phi(x,t)$ by the equation

$$u = \frac{\partial \phi}{\partial x} \tag{127}$$

Then (126) becomes

$$\frac{\partial s}{\partial t} + \frac{\partial^2 \phi}{\partial x^2} = 0 \tag{128}$$

recalling that we are dealing now with the one-dimensional case. Also, for the one dimensional case, the Navier-Stokes equation reduces to

$$\rho \frac{\partial u}{\partial t} = \frac{\partial p}{\partial x} - \frac{2}{3}\frac{\partial}{\partial x}\left(\mu \frac{\partial u}{\partial x}\right) + 2\frac{\partial}{\partial x}\left(\mu \frac{\partial u}{\partial x}\right) \tag{129}$$

Since the fluctuation in μ, if any, will be small, we may take μ as constant. Then (129) becomes

$$\rho_0 \frac{\partial^2 \phi}{\partial x \partial t} = -\frac{\partial p}{\partial x} + \frac{4}{3}\mu \frac{\partial^3 \phi}{\partial x^3} \tag{130}$$

Also,

$$\frac{\partial p}{\partial x} = \frac{dp}{\partial \rho}\frac{\partial \rho}{\partial x} = \rho_0 c_0^2 \frac{\partial s}{\partial x} \tag{131}$$

Thus,

$$\frac{\partial^2 \phi}{\partial x \partial t} = -c_0^2 \frac{\partial s}{\partial x} + \frac{4}{3}\nu \frac{\partial^3 \phi}{\partial x^3} \tag{132}$$

where $\nu = \mu/\rho_0$.

Equation (132) may be integrated to give

$$\frac{\partial \phi}{\partial t} = -c_0^2 s + \frac{4}{3}\nu \frac{\partial^2 \phi}{\partial x^2} \tag{133}$$

where the arbitrary function of t has been absorbed into the ϕ. Using (128) to eliminate s, we get

$$c_0^2 \frac{\partial^2 \phi}{\partial x^2} - \frac{\partial^2 \phi}{\partial t^2} + \frac{4}{3}\nu \frac{\partial^3 \phi}{\partial x^2 \partial t} = 0 \tag{134}$$

This equation has a solution of the form

$$\phi = \psi(t) e^{ikx} \tag{135}$$

Substituting (135) into (134) we get the ordinary differential equation

$$\frac{d^2\psi}{dt^2} + \frac{4}{3}\nu k^2 \frac{d\psi}{dt} + c_0^2 k^2 \psi = 0 \tag{136}$$

which has the solution

$$\psi(t) = \exp\left[-\tfrac{2}{3}\nu k^2 t + i\omega t\right] \tag{137}$$

where

$$\omega^2 = c_0^2 k^2 \left[1 - \left(\frac{2\nu k}{3c_0}\right)^2\right] \tag{138}$$

In order to get some feel for the effect of viscosity, let us consider the case of air at normal temperature and pressure. For this case,

$$\tfrac{2}{3}\nu \cong 0.1\ \text{cm}^2/\text{sec}, \qquad c_0 \cong 3.4 \times 10^4\ \text{cm/sec}.$$

Since our solution is periodic in nature, let

$$k = \frac{2\pi}{\lambda}$$

In the audible range,

$$1.7\,\text{cm} < \lambda < 1.7 \times 10^3\,\text{cm}$$

Hence,

$$10^{-8} < \frac{2\nu k}{3c_0} < 10^{-5} \ll 1 \tag{139}$$

Hence,

$$\omega \cong \pm c_0 k \tag{140}$$

and

$$\phi = \exp[-\tfrac{2}{3}\nu_0 k^2 t][\cos k(x \pm c_0 t) + i \sin k(x \pm c_0 t)] \tag{141}$$

In a physical solution, only the real part of ϕ can have meaning. Also, in the argument of the cosine, we may choose either the plus or the minus sign. Thus, we may say

$$\phi = \exp[-\tfrac{2}{3}\nu_0 k^2 t] \cos k(x - c_0 t) \tag{142}$$

This represents a wave traveling to the right with a velocity c_0. So we see that in air, at any rate, the viscosity has very little effect on the velocity at which the sound wave propagates. This conclusion is valid for gases in general. On the other hand, we see that the effect of the viscosity is to decrease the amplitude of the wave. Likewise, the condition (139) indicates that the wavelength also has very little effect on the speed of sound.

The validity of the solution is not, of course, limited to cosine shaped waves. Since equation (134) is linear, waves of any shape may be built up by means of Fourier series. While the conclusions reached here have been based on a one dimensional analysis, they are equally valid in three dimensions.

Insentropic, Steady Flow

Consider the experimental set-up shown in Figure 23. A large reservoir is pumped up to a certain pressure p_0 while the temperature in the reservoir is

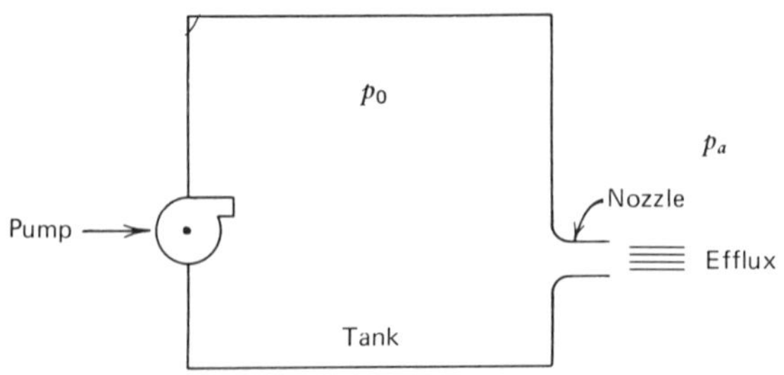

Figure 23. Pressure vessel with converging nozzle.

kept constant. A nozzle in the shape shown in attached to the side of the tank and the speed of the efflux is measured. It was found in early experiments that while the velocity of the efflux increased with the pressure p_0 initially, as p_0 was gradually raised from the outside air pressure p_a, after the pressure p_0 reached a certain value the exhaust velocity could no longer be increased further by increasing p_0. This raises two questions: 1. Why does not the exhaust velocity continue to rise as p_0 is increased? and 2. Is there some way to overcome this problem and obtain a higher exhaust velocity?

Let us analyze the problem from the standpoint of quasi one dimensional inviscid flow. Consider a small section of the duct shown in Figure 18 (Figure 24). Referring then to equation (91), we have

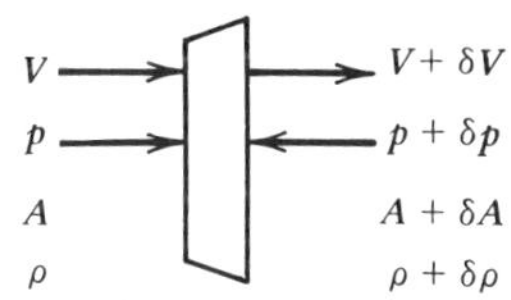

Figure 24. Control volume in quasi-one-dimensional flow.

$$\begin{array}{lll} p_1 = p & p_2 = p + \delta p & p_3 \cong p + \frac{1}{2}\delta p \\ u_1 = V & u_2 = V + \delta V & \rho_2 = \rho + \delta\rho \\ A_1 = A & A_2 = A + \delta A & \rho_1 = \rho \end{array}$$

Then

$$pA + (p + \tfrac{1}{2}\delta p)\delta A - (p + \delta p)(A + \delta A) = (\rho + \delta\rho)(V + \delta V)^2 (A + \delta A) - \rho V^2 A$$

Dividing both sides by δA and passing to the limit as $\delta A \to 0$, we get

$$\rho V\, dV = -\, dp \tag{143}$$

where it will be noted that ρ, V and p are now average values across the flow and discontinuities have been excluded.

Recall now equation (86)

$$\rho V A = \text{constant}$$

Differentiating,

$$\frac{d\rho}{\rho} + \frac{dV}{V} + \frac{dA}{A} = 0 \tag{144}$$

Combining (143) and (144)

$$\frac{dA}{dp} = \frac{A}{\rho}\left(\frac{1}{V^2} - \frac{d\rho}{dp}\right)$$

If we now assume that the flow is isentropic,

$$\frac{d\rho}{dp} = \frac{1}{c^2}$$

where c is the speed of sound. Then

$$\frac{dA}{dp} = \frac{A}{\rho V^2}\left(1 - \frac{V^2}{c^2}\right) \tag{145}$$

The ratio V/c is called the Mach number, after the Austrian physicist, Ernst Mach, and is usually designated by the letter M. Thus,

$$\frac{V}{c} = M \tag{146}$$

Equation (145) reveals a rather striking result. It says that whether or not an increase (or decrease) in area will cause an increase (or decrease) in pressure, depends on whether the average velocity across the flow at a given cross section is greater or less than the speed of sound.

It will be noted that the speed of sound here has been designated by the letter c and not c_0. In the derivation for the value of the speed of sound, only the disturbance made by the sound wave itself was considered. Since the disturbance was small, the speed of sound could be assumed constant. In the present case, however, the equations indicate that substantial variations in the state properties may take place as the cross-sectional area changes. This does not invalidate our previous derivation. It simply means that the speed of sound must be taken relative to the local conditions at a point. For example, for a perfect gas we found that

$$c = \sqrt{\gamma RT}$$

Thus, the speed of sound in this case is determined by the local temperature at any given point in the flow. A sound wave at that point will then travel with a speed c relative to an observer traveling with the flow at that point.

Recall now the equation

$$dp = -\rho V\,dV$$

Thus, the derivative dp/dV is always negative, since V here is the magnitude of a vector $\mathbf{V}$. Taking equations (143) and (145) together we see that when

$$M < 1, \qquad \frac{dA}{dV} < 0$$

when

$$M > 1, \qquad \frac{dA}{dV} > 0$$

Thus, for isentropic flow in a purely contracting nozzle, such as that shown in Figure 24, the velocity can never exceed the local velocity of the speed of sound. Experiments indicate that the limiting velocity spoken of above is, in fact, equal to this velocity. We have, therefore, answered the first of the two questions raised at the beginning of this discussion.

Equation (145) also gives a clue as to how we might answer the second question. Let us suppose that instead of a purely contracting nozzle, we make one which will contract until the local velocity is equal to the local speed of sound, and then expand. According to equations (145) and (143), such an arrangement should make it possible to attain velocities exceeding the speed of sound. Experiments indicate that this is indeed the case. A nozzle of this type is called a converging-diverging, or deLaval, nozzle (Figure 25).

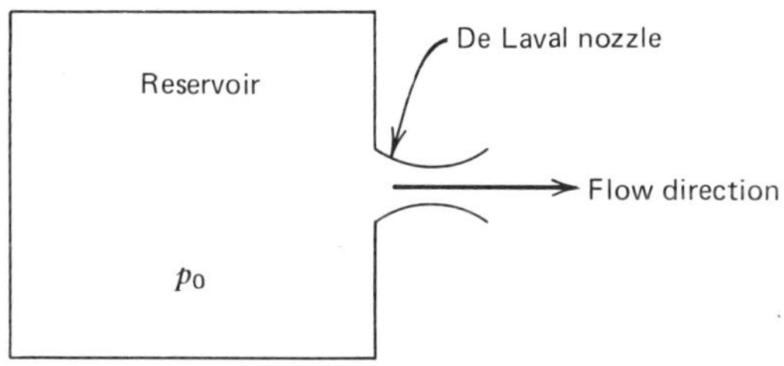

Figure 25. Pressure vessel with converging-diverging nozzle.

Let us now turn our attention to how the different properties of the flow vary along the nozzle. Assume that there is no heat transfer across the walls of the nozzle and that the axis of the nozzle is parallel to the ground. Then the energy equation (104) applies and we have

$$h + \frac{V^2}{2} = h_0 = \text{constant}$$

Carrying our assumption one step further, let us assume that we are dealing with a perfect gas. Then

$$c_p T + \frac{V^2}{2} = c_p T_0 \tag{147}$$

Dividing both sides of (147) by $c_p T$ we have

$$1 + \frac{V^2}{2c_p T}\frac{\gamma R}{\gamma R} = \frac{T_0}{T}$$

and, keeping in mind that

$$M = \frac{V}{c}, \qquad c^2 = \gamma R T \qquad \frac{\gamma R}{c_p} = \gamma - 1$$

we have

$$T_0 = T\left(1 + \frac{\gamma - 1}{2} M^2\right) \tag{148}$$

The quantity T_0 is called the stagnation, or total, temperature. Now recall from elementary thermodynamics, for a perfect gas,

$$T\,ds = c_p\,dT - \frac{1}{\rho}dp$$

Hence,

$$ds = c_p\frac{dT}{T} - R\frac{dp}{p}$$

since $p = \rho RT$. For an isentropic process, then,

$$\frac{p}{p_0} = \left(\frac{T}{T_0}\right)^{R/c_p} = \left(\frac{T}{T_0}\right)^{\gamma/(\gamma-1)}$$

It follows directly from equation (148), therefore, that

$$p_0 = p\left(1 + \frac{\gamma-1}{2}M^2\right)^{\gamma/(\gamma-1)} = \text{constant} \tag{149}$$

The constant p_0 is called the stagnation, or total, pressure. Similarly,

$$\rho_0 = \rho\left(1 + \frac{\gamma-1}{2}M^2\right)^{1/(\gamma-1)} = \text{constant} \tag{150}$$

where ρ_0 is called the stagnation, or total, density.

We have yet to inquire into what is perhaps the most important point, the area variation. Recall that

$$\dot{m} = \rho VA$$

Hence,

$$\frac{\dot{m}}{A} = \rho V = \frac{p}{RT}V = \frac{p_0 V}{\sqrt{\gamma RT}}\sqrt{\frac{V}{R}}\sqrt{\frac{T_0}{T}}\frac{1}{\sqrt{T_0}}\frac{p}{p_0}$$

and, making use of equation 148

$$\frac{\dot{m}}{A} = \sqrt{\frac{\gamma}{R}}\frac{p_0}{\sqrt{T_0}}M\left[1 + \frac{\gamma-1}{2}M^2\right]^{(1/2)-\gamma/(\gamma-1)} \tag{151}$$

Equations (148) through (151) refer the local conditions to the stagnation conditions. In the case of the area and the velocity this is not convenient, since at the stagnation condition the velocity is zero and the area, as may be seen from equation (151), is infinite. It is, therefore, more convenient to refer these quantities to the conditions where $M = 1$. In order to establish these relations, let us first establish the relation between the properties at any two points in the flow in terms of the Mach number. From equation (148).

$$\frac{T_1}{T_2} = \frac{1 + \frac{\gamma-1}{2}M_2^2}{1 + \frac{\gamma-1}{2}M_1^2} \tag{152}$$

From equation (149),

$$\frac{p_1}{p_2} = \left[\frac{1 + \frac{\gamma - 1}{2} M_2^2}{1 + \frac{\gamma - 1}{2} M_1^2}\right]^{\gamma/(\gamma - 1)} \tag{153}$$

From equation (150),

$$\frac{\rho_1}{\rho_2} = \left[\frac{1 + \frac{\gamma - 1}{2} M_2^2}{1 + \frac{\gamma - 1}{2} M_1^2}\right]^{1/(\gamma - 1)} \tag{154}$$

since it has been assumed that T_0, p_0, and ρ_0 are constant along the flow. Since $\dot{m}, \gamma$, and R are also assumed constant we have, from equation (151)

$$\frac{A_1}{A_2} = \frac{M_2}{M_1}\left[\frac{1 + \frac{\gamma - 1}{2} M_2^2}{1 + \frac{\gamma - 1}{2} M_1^2}\right]^{(1/2) - \gamma/(\gamma - 1)} \tag{155}$$

All the quantities in the flow may now be referred to the conditions where $M = 1$. Simply set $M_2 = 1$ and $M_1 = M$ in the above equations and designate the conditions at $M = 1$ by an asterisk. Then

$$\frac{T}{T^*} = \frac{\frac{\gamma + 1}{2}}{1 + \frac{\gamma - 1}{2} M^2} \tag{156}$$

$$\frac{p}{p^*} = \left[\frac{\frac{\gamma + 1}{2}}{1 + \frac{\gamma - 1}{2} M^2}\right]^{\gamma/(\gamma - 1)} \tag{157}$$

$$\frac{\rho}{\rho^*} = \left[\frac{\frac{\gamma + 1}{2}}{1 + \frac{\gamma - 1}{2} M^2}\right]^{1/(\gamma - 1)} \tag{158}$$

$$\frac{A}{A^*} = \frac{1}{M}\left[\frac{\frac{\gamma + 1}{2}}{1 + \frac{\gamma - 1}{2} M^2}\right]^{(\gamma + 1)/2(1 - \gamma)} \tag{159}$$

The conditions at $M = 1$ are called the critical conditions. It will be observed that for any given perfect gas, the ratio of the stagnation conditions to the critical conditions is a fixed number depending only on γ. Thus,

$$\frac{T_0}{T^*}=\frac{\gamma+1}{2}, \qquad \frac{p_0}{p^*}=\left[\frac{\gamma+1}{2}\right]^{\gamma/(\gamma-1)}, \qquad \frac{\rho_0}{\rho^*}=\left[\frac{\gamma+1}{2}\right]^{1/(\gamma-1)} \tag{160}$$

The velocity may be found directly from the energy equation (147). Solving for V we get

$$V=\sqrt{2c_p(T_0-T)}$$

or

$$V=\sqrt{2c_pT_0\left(1-\frac{T}{T_0}\right)}=\sqrt{2c_pT_0\left(1-\frac{1}{1+\frac{\gamma-1}{2}M^2}\right)} \tag{161}$$

The critical velocity is, then

$$V^*=\sqrt{2c_pT_0\left(1-\frac{2}{\gamma+1}\right)}=c^* \tag{162}$$

since $V=V^*$ when $M=1$. c^* is called the critical speed of sound. From (161)

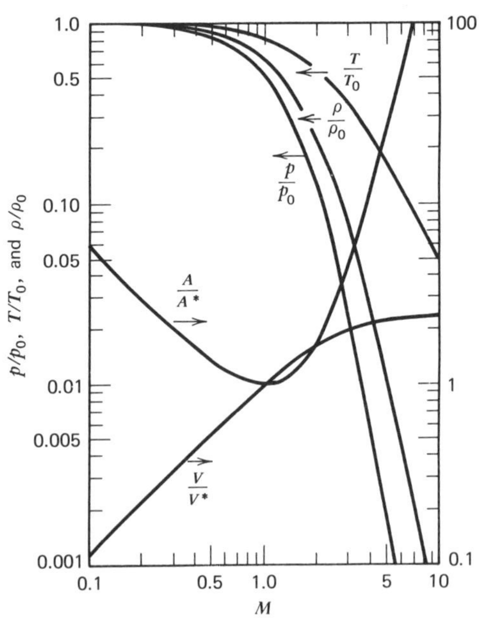

Figure 26. Variation of flow properties with Mach number in quasi-one-dimensional flow [234].

and (162)

$$\frac{V}{V^*} = \sqrt{\frac{\frac{\gamma+1}{2}M^2}{1+\frac{\gamma-1}{2}M^2}} \tag{163}$$

Equations (148), (149), (150), (159), and (163) are plotted in Figure 26. In order to gain a clearer understanding of the meaning of these equations, let us consider two experiments. First, consider a converging nozzle connected to a large chamber, the pressure in which may be lowered by a pump (Figure 27). Let p_b

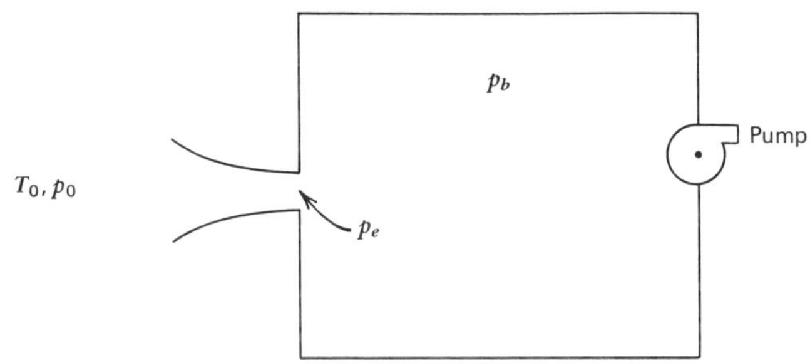

Figure 27. Low pressure vessel with converging nozzle.

be the pressure in the chamber, called the back pressure. Let p_e be the pressure at the exit of the nozzle into the chamber, called the exit pressure, and let p_0 be the outside or atmospheric pressure to which the nozzle opens. Beginning with the back pressure at p_0, let p_b gradually be lowered. Flow will then occur in the nozzle. Let us record the variation of the pressure in the nozzle as the back pressure is lowered (Figure 28).

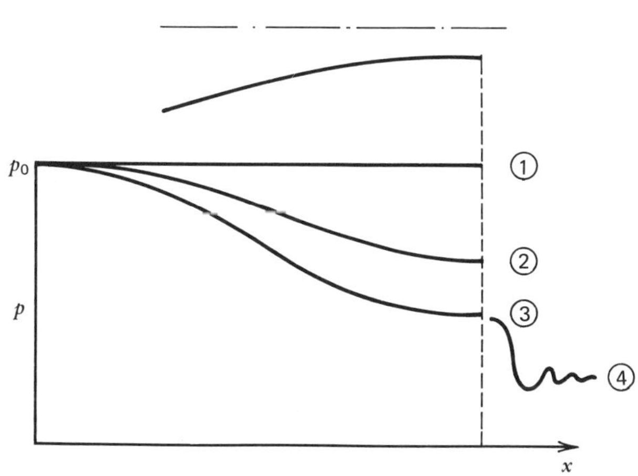

Figure 28. Variation of pressure in converging nozzle.

x represents the axial distance along the nozzle. Line 1 represents the condition for which $p_b = p_0$. The pressure in the nozzle is constant and there is no flow. Line 2 represents the condition where $p_0 > p_b > p^*$. The experimental result indicates that the pressure variation along the nozzle is closely predicted by equations (149) and (151) if it is assumed that the stagnation temperature and pressure in the nozzle are equal to the temperature and pressure in the outside atmosphere and the exit pressure equals the back pressure. Line 3 represents the condition $p_b = p^*$ and the experimental result agrees with the equations (149) and (151) as in line 2. Line 4 represents the condition $p_b < p^*$. Now it is observed that the pressure distribution in the nozzle is the same as for line 3. Thus, p_b no longer equals p_e and a pressure adjustment takes place in the chamber beyond the nozzle. The exit pressure p_e remains equal to p^*. This result is not surprising in view of the fact that equation (149) indicates that a further lowering of the pressure in the nozzle would mean a Mach number greater than one. As we established at the outset of this discussion, however, the Mach number cannot go from less than one, or subsonic, to greater than one, or supersonic, in a strictly converging nozzle.

A further physical explanation of this phenomenon is possible. In the section on the speed of sound we established that a small pressure wave travels with the speed of sound relative to the gas. Thus, if a differential between the exit pressure and the back pressure were to occur, a pressure wave would propagate up the nozzle from the exit with a speed relative to the nozzle equal to the speed of sound minus the velocity of the gas in the nozzle, thereby equalizing the pressure at the exit. If, however, the velocity of the gas in the nozzle is equal to the speed of sound, it is clear that the pressure wave can make no headway against the flow, thus making it possible for the exit pressure to exceed the back pressure. Of course pressure waves can and do propagate the other way, into the chamber, but here multidimensional effects enter in which will be discussed in a later chapter.

If the temperatures involved in the experiment are well below the dissociation temperature of the gas and the pressures involved are well below the critical pressure, it is found that not only the pressure, but the density, temperature and flow rate are well predicted by the equations which we have derived, in spite of

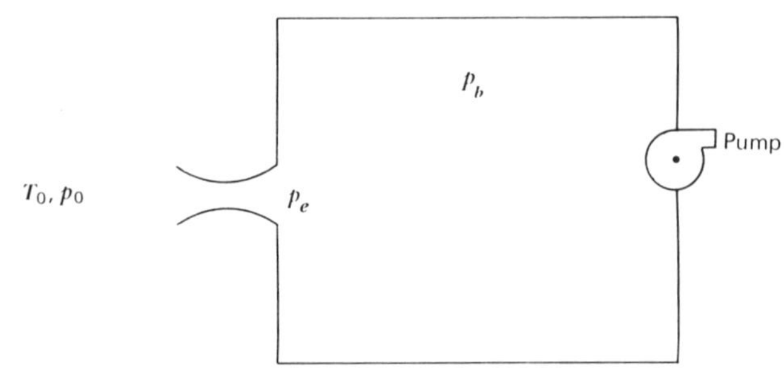

Figure 29. Low pressure vessel with converging-diverging nozzle.

the assumptions which we have made: quasi-one-dimensional flow, perfect gas, no heat transfer and constant entropy.

Now let us consider another experiment (Figure 29). This experiment is similar to the previous one, except that we have now replaced the converging nozzle by a converging-diverging one. Again we begin with $p_b = p_0$ and gradually lower the pressure in the tank, measuring the pressure distribution in the nozzle. The results are shown in Figure 30. Again, x represents the axial distance along

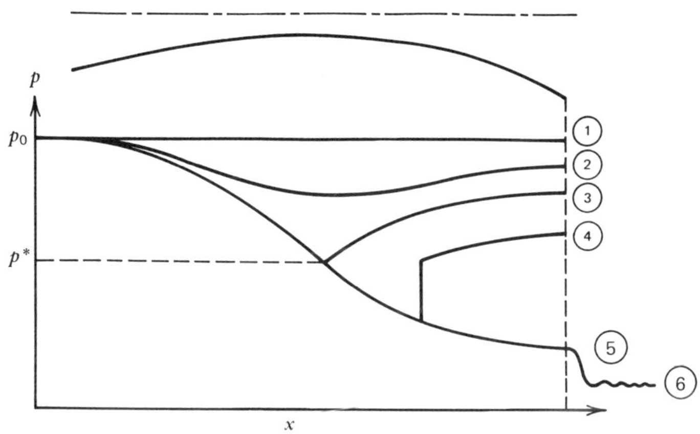

Figure 30. Pressure variation in converging-diverging nozzle.

the nozzle. Line 1 represents a back pressure equal to p_0. For this case the pressure is the same everywhere and no flow occurs. Line 2 represents the pressure distribution when the back pressure is slightly lower than p_0. In this case we observe that $p_e = p_b$ and the flow is subsonic throughout. The pressure reaches a minimum at the throat and then rises again toward the exit. This curve is well predicted by equations (149) and (151). Line 3 represents the pressure distribution when the back pressure is lowered just enough to cause the velocity to reach the sonic value at the throat. Once again we find that $p_e = p_b$ and the curve is well predicted by equations (149) and (151). Line 4 represents the pressure distribution when the back pressure is lowered somewhat below the value for line 3. Once again we find that the exit pressure equals the back pressure. However, a strange phenomenon now occurs. The flow goes supersonic past the throat, but then there is a discontinuity in the pressure. The pressure suddenly rises and the flow becomes subsonic. This phenomenon is not predicted by any of the equations which we have derived thus far. It would be reasonable to suppose that one of the assumptions which we have made in deriving our equations does not apply in the present instance. As we will see later, this is indeed the case.

When the back pressure is lowered to the exit pressure indicated by line 5, we observe that the discontinuity present in line 4 disappears. The Mach number at the throat is now one and subsequent to the throat the flow is entirely supersonic. Further, the exit pressure equals the back pressure and the entire curve is well predicted by equations (149) and (151). We see then that when the velocity in the throat reaches the sonic value, there are two possible solutions of equations (149) and (151) for the subsequent flow in the nozzle: the one indicated by line 3 and the one indicated by line 5.

If the back pressure is lowered further, below the value of p_e for line 5, no change is observed in the flow in the nozzle. Once again, because the flow at the exit is supersonic, pressure disturbances cannot travel upstream, and hence a further lowering of the back pressure does not affect the flow in the nozzle.

This physical argument should not be carried too far. It does not answer the question of what happens if the difference between the exit pressure and the back pressure becomes large, or what effect the presence of viscosity might have. Discussion of these factors will be taken up in subsequent chapters, specifically those on shock waves and boundary layers.

If the stagnation enthalpy is constant in a quasi-one-dimensional flow, the variation of the flow parameters may be presented in a convenient graphical form. Recall from elementary thermodynamics for a perfect gas,

$$ds = c_p \frac{dT}{T} - R\frac{dp}{p} \tag{164}$$

On a graph of temperature versus entropy the lines of constant pressure may be drawn using equation (164). These are indicated as p_1, p_2, p_3 on Figure 31.

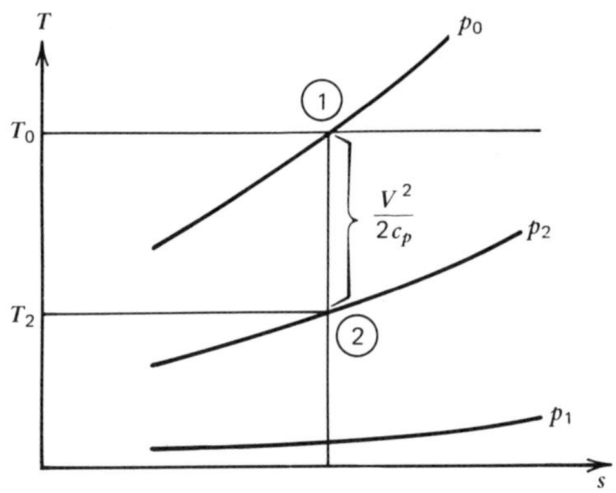

Figure 31. Temperature entropy diagram.

If the stagnation temperature is known, it too may be marked on the figure, as a line parallel to the x axis. If the stagnation pressure p_0 is known, the intersection of the line $T = T_0$ with the line $p = p_0$ represents the stagnation conditions, marked as point 1 on the figure. If, moreover, the flow is isentropic, then all possible conditions of the flow will lie on a vertical line drawn downward from the point 1. Then, if the temperature (or pressure) at any point in the flow is known, the pressure (or temperature) and the velocity at that point may be read off the graph. For example, if the temperature is T_2, the pressure will be given by the isobar passing through the point 2 on the graph, and the velocity may be found from the distance between the points 1 and 2.

Normal Shock Waves

The discontinuity observed in line 4 of Figure 30, which was found experimentally, leads us to ask whether such a phenomenon can be predicted and explained theoretically. Let us first examine this question in a very simplistic way. Suppose we have a one-dimensional, adiabatic flow. Can this flow support a discontinuity as shown in Figure 32? We may reason that if we can find values of

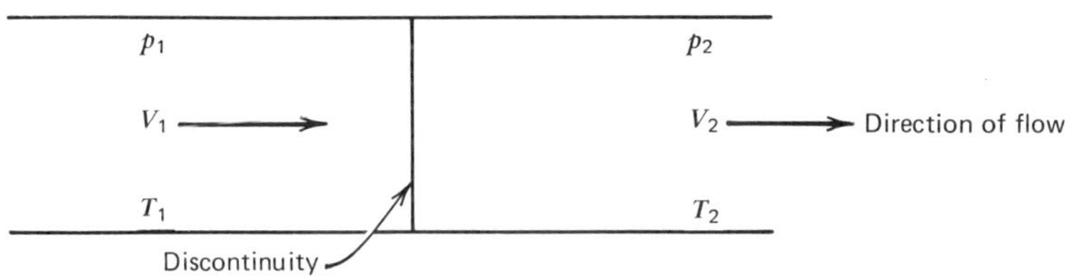

Figure 32. Discontinuity in one-dimensional flow.

the flow parameters $p_1, p_2, V_1, V_2, T_1, T_2$ which satisfy the governing equations, and if the governing equations are valid, then the answer must be yes.

The governing equations in this case are,

The energy equation:

$$c_p T_1 + \frac{V_1^2}{2} = c_p T_2 + \frac{V_2^2}{2} \tag{165}$$

The momentum equation:

$$p_1 + \rho_1 V_1^2 = p_2 + \rho_2 V_2^2 \tag{166}$$

The continuity equation:

$$\rho_1 V_1 = \rho_2 V_2 \tag{167}$$

The equation of state:

$$p_1 = \rho_1 R T_1, \qquad p_2 = \rho_2 R T_2 \tag{168}$$

assuming a perfect gas. Let us postulate a set of conditions p_1, T_1, ρ_1, V_1, and let us see if we can find a set of values p_2, T_2, ρ_2, V_2, which differ by a finite amount from p_1, T_1, ρ_1, V_1 and which satisfy equations (165)–(168). If we accept equation (165)–(168) as valid, the possible existence of such a discontinuity will then have been established, for these equations contain no restrictions regarding the distance between regions 1 and 2.

Because of the algebraic complexity of attempting to solve equations (165)–

(168) simultaneously, we resort to a graphical method. For a perfect gas,

$$s_2 - s_1 = c_p \ln \frac{T_2}{T_1} - R \ln \frac{p_2}{p_1} \tag{169}$$

Combining equations (165), (167), (168) and (169), we may eliminate ρ, V, and p to get s as a function of T:

$$s_2 - s_1 = c_p \ln \frac{T_2}{T_1} - R \ln \frac{T_2}{T_1} \sqrt{\frac{h_0 - c_p T_1}{h_0 - c_p T_2}} \tag{170}$$

A plot of this equation is called a Fanno line. Combining equations (166)–(169) we get

$$s_2 - s_1 = c_p \ln \frac{T_2}{T_1} - R \ln \left[b_1 + \sqrt{b_1^2 - a_1 T_2} \right] \tag{171}$$

where

$$b_1 = \frac{1}{2}\left[1 + \frac{\dot{m}^2 R T_1}{A^2 p_1^2}\right], \qquad a_1 = \frac{\dot{m}^2 R}{A^2 p_1^2}$$

A plot of equation (171) is called a Rayleigh line. A plot of equations (170) and (171) is shown in Figure 33. At the points where the two lines intersect,

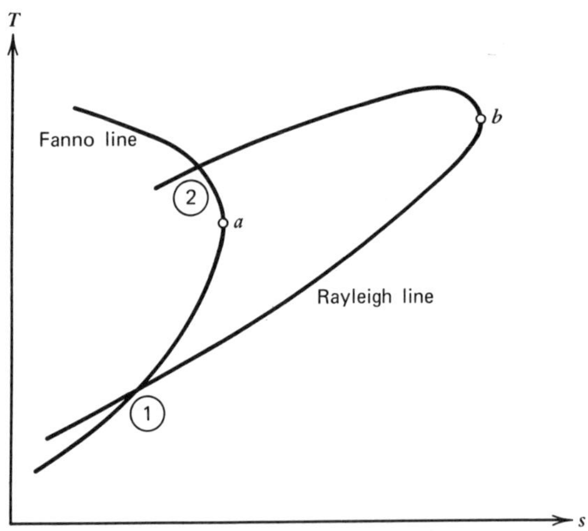

Figure 33. Fanno and Rayleigh lines.

all five equations (165)–(169) will be satisfied. The plot indicates that for a given set of conditions 1, there will indeed be another set of conditions 2, for which all the equations will be satisfied, thus lending credence to the supposition that discontinuities in a flow may indeed be possible.

It will be observed that no direction is built into the governing equations. That is, the equations do not specify whether the flow is going from 1 to 2 or from 2 to 1. An examination of Figure 33, however, indicates that the entropy at

2 is higher than the entropy at 1. Since the flow in Figure 32 is isolated from the surroundings, in the sense that no heat or work is crossing the boundaries of the system, it follows that a decrease in entropy would be a violation of the second law of thermodynamics. Thus, a discontinuity can only proceed from 1 to 2 and not from 2 to 1.

For different initial properties, different curves are obtained (Figure 34).

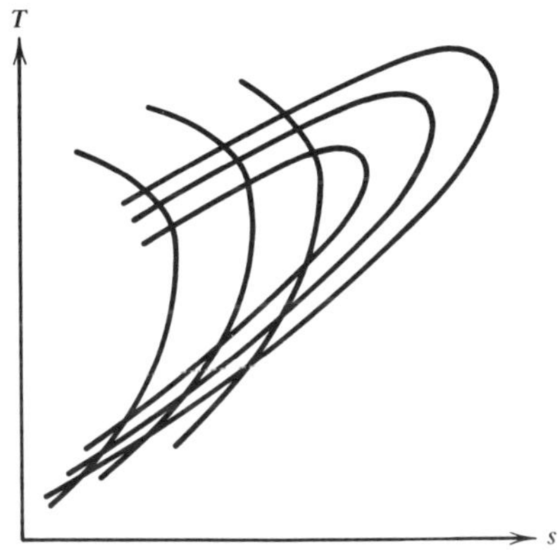

Figure 34. Fanno and Rayleigh lines for different initial values.

However, it will be observed that the two intersecting points of any two curves always share the same characteristics. The point at higher T always has a higher entropy than the point at lower T; and the two points always lie on opposite sides of the point of maximum entropy on each curve.

Let us see if we can find the significance of the point of maximum entropy on the Fanno and Rayleigh lines. On the Fanno line,

$$dh + d\left(\frac{V^2}{2}\right) = \rho\, dV + V d\rho = 0 \tag{172a}$$

Recall that

$$TdS = dh - \frac{dp}{\rho} \tag{172b}$$

At the point of maximum entropy, point a on Figure 33, $ds = 0$. Hence,

$$dh = \frac{dp}{\rho} = \frac{dp}{d\rho}\frac{d\rho}{\rho} = -VdV \tag{173}$$

from equations (172a) and (172b). Thus,

$$\frac{dp}{d\rho} = -V\rho\frac{dV}{d\rho} = V^2 \tag{174}$$

So the velocity of the gas at point a is equal to the square root of $dp/d\rho$. But since $ds = 0$ at this point, it follows that the velocity is equal to the speed of sound there. The point b on the Rayleigh line (Figure 33) has the same significance. On the Rayleigh line,

$$dp + \rho V dV = 0$$

Hence,

$$\frac{dp}{dV} = \frac{dp}{d\rho}\frac{d\rho}{dV} = -\rho V$$

or

$$\frac{dp}{d\rho} = -\rho V \frac{dV}{d\rho}$$

But from the equation of continuity,

$$\frac{dV}{d\rho} = -\frac{V}{\rho}$$

and hence

$$\frac{dp}{d\rho} = V^2$$

But $ds = 0$ at point b; thus, $V^2 = c^2$, where c is the speed of sound.

Now since h_0 is constant along the Fanno line, it follows from the energy equation,

$$h_0 = h + \frac{V^2}{2} \tag{175}$$

that as T increases, V decreases. Hence, the velocity will be lower at point 2 than at point 1. Now

$$M = \frac{V}{\sqrt{\gamma R T}}$$

for a perfect gas. It follows that when equation (175) is valid, an increase in V means an increase in M and vice versa. Thus the velocity at point 1 must be supersonic and the velocity at point 2 must be subsonic. It follows from our discussion that discontinuities of the type we have been considering can only exist in supersonic flow and will always reduce the velocity to subsonic flow.

It has been argued that the above derivation regarding the possible existence of discontinuities in fluid flow is not really valid because the viscosity has been neglected. We have already seen that when we took the viscosity into account in deriving the momentum equation, we obtained a set of differential equations of the second order, the Navier Stokes equations. Comparison with the momentum equation derived when viscosity is neglected shows that when the viscous terms are dropped, the order of the equation is reduced from second to first. In general, such a change in a differential equation will change the basic character of the solution. That fact that the viscosity is small, therefore, is not a guarantee that it may be neglected altogether. This is because the limit of the solution of the viscous equation, as the viscosity goes to zero, is not necessarily the same as the solution of the inviscid equation. Let us, therefore, see if we can establish the possible existence of discontinuities while retaining the viscous terms in the governing equations. We have already derived the momentum equation including viscosity; however, we have not yet included the effects of viscosity in the energy equation. Before we can proceed, therefore, we we will have to turn our attention to this matter.

Energy Equation Including Viscosity

Recall the first law of thermodynamics,

$$\frac{đQ}{dt} = \frac{dE}{dt} + \frac{đW}{dt}$$

The only difference between the viscous case and the inviscid case is that in the viscous case, the work term must include the work done by the viscous forces as well as the work done by the pressure forces. Consider once again a rectangular element of fluid, as shown in Figure 12 with the components of the force which the surrounding fluid exerts on each face as shown in Figure 13. Let $\mathbf{f}_x, \mathbf{f}_y$, and $\mathbf{f}_z$ be the viscous forces on the faces perpendicular to the x, y, and z axes, respectively. Then the work per unit time which the viscous forces are doing on the element of fluid is given by, approximately,

$$\begin{aligned}\frac{đW_v}{dt} &= \frac{\partial}{\partial x}[\mathbf{V}\cdot\mathbf{f}_x]\delta x\,\delta y\,\delta z + \frac{\partial}{\partial y}[\mathbf{V}\cdot\mathbf{f}_y]\delta x\,\delta y\,\delta z + \frac{\partial}{\partial z}[\mathbf{V}\cdot\mathbf{f}_z]\delta x\,\delta y\,\delta z \\ &= \left[\frac{\partial}{\partial x}(\mathbf{V}\cdot\mathbf{f}_x) + \frac{\partial}{\partial y}(\mathbf{V}\cdot\mathbf{f}_y) + \frac{\partial}{\partial z}(\mathbf{V}\cdot\mathbf{f}_z)\right]\delta\mathscr{V} \qquad (176)\end{aligned}$$

In the notation which we used to derive the Navier Stokes equations, this becomes

$$\begin{aligned}\frac{đW_v}{dt} = \Bigg[&\frac{\partial}{\partial x}(\sigma'_{xx}u + \sigma_{xy}v + \sigma_{xz}w) + \frac{\partial}{\partial y}(\sigma_{yx}u + \sigma'_{yy}v + \sigma_{yz}w) \\ &+ \frac{\partial}{\partial z}(\sigma_{zx}u + \sigma_{zy}v + \sigma'_{zz}w)\Bigg] \qquad (177)\end{aligned}$$

This may be expanded to give

$$\frac{đW_v}{dt} = \mu\Psi\delta\mathscr{V} + \mu\Phi\delta\mathscr{V} \qquad (178)$$

where

$$\begin{aligned}\mu\Psi = {}& u\left(\frac{\partial\sigma'_{xx}}{\partial x} + \frac{\partial\sigma_{yx}}{\partial y} + \frac{\partial\sigma_{zx}}{\partial z}\right) + v\left(\frac{\partial\sigma_{xy}}{\partial x} + \frac{\partial\sigma'_{yy}}{\partial y} + \frac{\partial\sigma_{zy}}{\partial z}\right) \\ &+ w\left(\frac{\partial\sigma_{xz}}{\partial x} + \frac{\partial\sigma_{yz}}{\partial y} + \frac{\partial\sigma'_{zz}}{\partial z}\right) \qquad (179)\end{aligned}$$

and

$$\mu\Phi = \left(\sigma'_{xx}\frac{\partial u}{\partial x} + \sigma_{xy}\frac{\partial v}{\partial x} + \sigma_{xz}\frac{\partial w}{\partial x}\right) + \left(\sigma_{yx}\frac{\partial u}{\partial y} + \sigma'_{yy}\frac{\partial v}{\partial y} + \sigma_{yz}\frac{\partial w}{\partial y}\right) + \left(\sigma_{zx}\frac{\partial u}{\partial z} + \sigma_{zy}\frac{\partial v}{\partial z} + \sigma'_{zz}\frac{\partial w}{\partial z}\right)^* \tag{180}$$

Multiplying equations (181) by u, v, and w, respectively, and adding, we get,

$$\rho\frac{D\frac{1}{2}V^2}{Dt} = -\mathbf{V}\cdot\nabla\mathbf{p} + \mu\Psi + \rho\mathbf{V}\cdot\mathbf{g}$$

or (182)

$$\rho\frac{D\frac{1}{2}V^2}{Dt} + \mathbf{V}\cdot\nabla\mathbf{p} - \mu\Psi - \rho\mathbf{V}\cdot\mathbf{g} = 0^\dagger$$

Equation (19) now becomes

$$-\nabla\cdot k\nabla T = \rho\frac{D\frac{1}{2}V^2}{Dt} + \rho\frac{De_t}{Dt} + \rho g\frac{DZ}{Dt} + \nabla(p\mathbf{V}) - \mu\Psi - \mu\Phi \tag{183}$$

and, using (182),

$$-\nabla\cdot k\nabla T = \rho\frac{De_t}{Dt} + p(\nabla\mathbf{V}) - \mu\Phi \tag{184}$$

Using (24a),

$$-\nabla\cdot k\nabla T = \rho\frac{De_t}{Dt} - \frac{Dp}{Dt} + \rho\frac{D}{Dt}\left(\frac{p}{\rho}\right) - \mu\Phi \tag{185}$$

Letting

$$-\nabla\cdot k\nabla T = \dot{q}$$

* In the presence of a body force

$$\mathbf{g} = \mathbf{i}g_x + \mathbf{j}g_y + \mathbf{k}g_z$$

equations (62) may be written

$$\rho\frac{Du}{Dt} = -\frac{\partial p}{\partial x} + \rho g_x + \frac{\partial\sigma'_{xx}}{\partial x} + \frac{\partial\sigma_{xy}}{\partial y} + \frac{\partial\sigma_{xz}}{\partial z}$$

$$\rho\frac{Dv}{Dt} = -\frac{\partial p}{\partial y} + \rho g_y + \frac{\partial\sigma_{xy}}{\partial x} + \frac{\partial\sigma'_{yy}}{\partial y} + \frac{\partial\sigma_{yz}}{\partial z} \tag{181}$$

$$\rho\frac{Dw}{Dt} = -\frac{\partial p}{\partial z} + \rho g_z + \frac{\partial\sigma_{xz}}{\partial x} + \frac{\partial\sigma_{yz}}{\partial y} + \frac{\partial\sigma'_{zz}}{\partial z}$$

† As in equation (21) we may write $\rho g DZ/Dt$ instead of $-\rho\mathbf{V}\cdot\mathbf{g}$. Then,

$$\rho\frac{D\frac{1}{2}V^2}{Dt} + \mathbf{V}\cdot\nabla p - \mu\Psi + \rho g\frac{DZ}{Dt} = 0$$

and

$$e_t + \frac{p}{\rho} = h$$

we have

$$\rho \frac{Dh}{Dt} = \frac{Dp}{Dt} + \dot{q} + \mu\Phi \tag{186}$$

It will be observed that equation (186) is identical to equation (27) except for the term $\mu\Phi$. Φ may be evaluated by substituting equations (60) into equation (180). Thus,

$$\Phi = 2\left[\left(\frac{\partial u}{\partial x}\right)^2 + \left(\frac{\partial v}{\partial y}\right)^2 + \left(\frac{\partial w}{\partial z}\right)^2\right] + \left(\frac{\partial v}{\partial x} + \frac{\partial u}{\partial y}\right)^2 + \left(\frac{\partial w}{\partial y} + \frac{\partial v}{\partial z}\right)^2 + \left(\frac{\partial u}{\partial z} + \frac{\partial w}{\partial x}\right)^2 - \frac{2}{3}\left(\frac{\partial u}{\partial x} + \frac{\partial v}{\partial y} + \frac{\partial w}{\partial z}\right) \tag{187}$$

Equation (186) with Φ defined by (187) constitutes the energy equation for viscous, compressible flow. It will be noted that the perfect gas assumption has not been made in deriving equation (186).

It will be observed from equation (182) that of the work done by the viscous forces, only Ψ contributes to the kinetic and potential energy of the system. It follows that Φ can contribute only to the internal energy of the system. Because, according to the second law of thermodynamics, while mechanical work may be turned entirely into heat, heat may never be turned entirely into mechanical work, the function Φ is sometimes called the viscous dissipation function. The name is somewhat misleading since the mechanical work $\mu\Phi$ is entirely "dissipated" by heat transfer only if the boundary conditions permit. If, for example, the system is insulated, as in flow through an insulated pipe, the work $\mu\Phi$ is not really dissipated but is retained by the system as internal energy.

Shock Wave Relations Including Viscosity and Thermal Conductivity

We now consider once again the situation shown in Figure 32. We wish to examine the question of whether a plane, stationary discontinuity can exist in a one-dimensional flow of a viscous, heat-conducting fluid.

For the one-dimensional case, the continuity equation may be written,

$$\frac{\partial \rho}{\partial t} + \frac{\partial}{\partial x}(\rho u) = 0 \tag{188}$$

The Navier Stokes equation for one-dimensional flow reduces to

$$\rho \frac{\partial u}{\partial t} + \rho u \frac{\partial u}{\partial x} = -\frac{\partial p}{\partial x} + \frac{\partial}{\partial x}\left(2\mu \frac{\partial u}{\partial x}\right) - \frac{2}{3}\frac{\partial}{\partial x}\left(\mu \frac{\partial u}{\partial x}\right) \tag{189}$$

Combining equations (188) and (189) we may write the momentum equation in the form,

$$\frac{\partial}{\partial t}(\rho u) + \frac{\partial}{\partial x}\left(\rho u^2 + p - \tfrac{4}{2}\mu\frac{\partial u}{\partial x}\right) = 0 \tag{190}$$

For one-dimensional flow, the energy equation (184) becomes,

$$-\frac{\partial}{\partial x}\left(k\frac{\partial T}{\partial x}\right) = \rho\frac{\partial e_t}{\partial t} + \rho u\frac{\partial e_t}{\partial x} + p\frac{\partial u}{\partial x} - \frac{4}{3}\left(\frac{\partial u}{\partial x}\right)^2 \tag{191}$$

Using equations (188) and (189), we may write the energy equation in the form

$$\frac{\partial}{\partial t}\left[\rho(\tfrac{1}{2}u^2 + e_t)\right] + \frac{\partial}{\partial x}\left[\rho u(\tfrac{1}{2}u^2 + h) - \tfrac{4}{3}\mu u\frac{\partial u}{\partial x} - k\frac{\partial T}{\partial x}\right] = 0 \tag{192}$$

For the case of steady flow, equations (188), (190) and (192) become,

$$\frac{d}{dx}(\rho u) = 0 \tag{193}$$

$$\frac{d}{dx}\left(\rho u^2 + p - \tfrac{4}{3}\mu\frac{du}{dx}\right) = 0 \tag{194}$$

$$\frac{d}{dx}\left[\rho u(\tfrac{1}{2}u^2 + h) - \tfrac{4}{3}\mu u\frac{du}{dx} - k\frac{dT}{dx}\right] = 0 \tag{195}$$

Integrating between any two points in the flow,

$$\rho_1 u_1 = \rho_2 u_2 \tag{196}$$

$$\rho_1 u_1^2 + p_1 - \tfrac{4}{3}\mu_1\left(\frac{du}{dx}\right)_1 = \rho_2 u_2^2 + p_2 - \tfrac{4}{3}\mu\left(\frac{du}{\partial x}\right)_2 \tag{197}$$

$$\rho_1 u_1(\tfrac{1}{2}u_1^2 + h_1) - \tfrac{4}{3}\mu u_1\left(\frac{du}{dx}\right)_1 - k\left(\frac{dT}{dx}\right)_1$$

$$= \rho_2 u_2(\tfrac{1}{2}u_2^2 + h_2) - \tfrac{4}{3}\mu u_2\left(\frac{du}{dx}\right)_2 - k\left(\frac{dT}{dx}\right)_2 \tag{198}$$

It will be observed that a solution of the system of equations (165)–(168) will also satisfy the system of equations (196)–(198) if, at the points 1 and 2, the derivatives of u and T with respect to x are zero. Because the system of equations (193)–(195) represents a system of differential equations, however, we are not justified in assuming, a priori, that such a solution is valid. Instead, we can specify the conditions at one point in the flow and then try to solve the system of differential equations (193)–(195) to find u, p, ρ, and T as functions of x. If we can find such a solution, it will then be apparent whether or not at some point in the flow other than the initial point the derivatives of u and T go to zero and u and T assume the values u_2 and T_2 of the inviscid solution.

For a perfect gas, equation (198) may be rewritten,

$$\frac{u^2}{2} + c_p T - \frac{k}{c_p \rho u}\left[\frac{4\mu c_p}{3k} u \frac{du}{dx} + c_p \frac{dT}{dx}\right] = c_1 \tag{199}$$

where c_1 is a constant which may be evaluated by assigning values to u, T, ρ, du/dx and dT/dx at some point in the flow. For air at normal (room) temperature and pressure, the Prandtl number, $c_p\mu/k$, is approximately equal to 3/4. Assuming c_p constant and $Pr = 3/4$, (199) becomes,

$$h_0 - \frac{k}{c_p \dot{m}} \frac{dh_0}{dx} = c_1 \tag{200}$$

where

$$h_0 = \frac{u^2}{2} + c_p T$$

the stagnation enthalpy, and $\dot{m} = \rho u =$ constant, which may be evaluated from the given conditions at some point in the flow.

Equation (200) may be integrated to give

$$h_0 = c_1 + c_2 \exp\left[\frac{c_p}{k} \dot{m} x\right] \tag{201}$$

Thus, if h_0 is to remain bounded as $x \to \infty, c_2$ must be zero. Hence we find that in the viscous case, as in the inviscid case,

$$h_0 = \text{constant} \tag{202}$$

Returning now to the momentum equation we may write

$$\dot{m}u + p - \tfrac{4}{3}\mu \frac{du}{dx} = c_3 \tag{203}$$

where c_3 is a constant which may be evaluated by specifying the conditions at some point in the flow. For a perfect gas we may write

$$p = \frac{\dot{m}RT}{u} \tag{204}$$

If we eliminate p and T from equation (203) using equations (202) and (204) we arrive at an ordinary differential equation for u which has a closed-form solution if the constants c_1, c_3 and $\dot{m}$ are known and if μ is assumed constant. If the conditions at some point in the flow are specified, c_1, c_3, and $\dot{m}$ are determined.

The solution to equation (203) for the case of constant μ and k has the form

$$\frac{1 - \dfrac{u}{u_1}}{\left(\dfrac{u}{u_1} - \alpha\right)^{\alpha}} = c_4 \exp\left[K(1 - \alpha)M_1 x\right] \tag{205}$$

where

$$\alpha = \frac{\gamma - 1}{\gamma + 1} + \frac{2}{\gamma + 1}\frac{1}{M_1^2} \tag{206}$$

$$K = \frac{3}{8}\sqrt{\frac{p_1 \rho_1}{\mu \gamma}}(\gamma + 1) \tag{207}$$

and c_4 is an arbitrary constant which depends on where x is taken as zero. u_1, p_1, ρ_1, and M_1 are the properties of the flow at minus infinity and are presumed given. The form of equation (205) indicates that α must be less than one and hence, from equation (206), M_1 must be greater than one. We may choose c_4 by taking $x = 0$ when $d^2u/dx^2 = 0$. A plot of x versus u then yields the result shown in Figure 35. The scale of x has been greatly expanded, as 90% of the

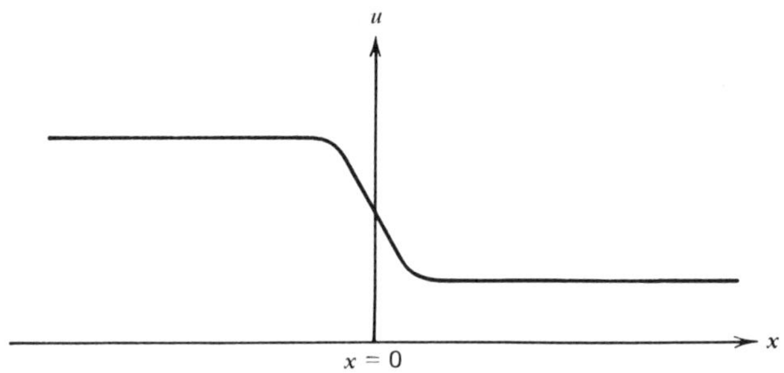

Figure 35. Velocity variation through a shock wave according to Becker. [6]

velocity change occurs over a distance on the order of 10^{-6} cm for air at normal temperature and pressure. This solution, which was first obtained by Becker [6], thus appears to confirm the possibility of rapid changes in flow properties in a uniforms stream with initially supersonic velocity. In addition, it appears to give the velocity variation in the "discontinuity." The solution approaches the inviscid solution of the system of equations (165)–(168) at $x = \pm\infty$. Thus, inclusion of the viscosity and thermal conductivity in the equations does not appear to affect the initial and final flow parameters.

Remarkable as the solution of Becker appears, it was immediately attacked on the grounds that the distance over which most of the variation of the flow parameters occurs in the solution is so short as to render the continuum assumption of doubtful validity: 10^{-6} cm is less than one mean free path for air at normal temperature and pressure. Becker himself recognized this shortcoming and suggested the use of the Boltzmann equation instead. As Thomas [7] pointed out, however, Becker neglected the variation of μ and k in the shock wave, a variation which may not be negligible. The method of Becker may be modified by taking this variation into account. The solution of equation (203) can then no longer be found in closed form but must be obtained numerically. The flow parameters at $x = \pm\infty$, however, are not affected. Only the thickness over which most of the variation in u occurs is increased. Thus, Morduchow and Libby [8], using the assumptions

$$\mu = \mu_1 \left(\frac{T}{T_1}\right)^n \qquad Pr = \tfrac{3}{4}, \qquad n = 0.768$$

found shock wave thicknesses several times as great as those of Becker.

We have seen that both viscid and inviscid analyses lead to the conclusion that discontinuities, or at least sharp changes in the parameters of a supersonic flow, may be possible. Further, both viscous and inviscid analyses predict the same change in the flow parameters for a given set of initial conditions. Experiments not only bear out the validity of the conclusion qualitatively, but in many cases the quantitative predictions of the above theories are found to be quite accurate. In so far as the initial and final values of the flow parameters across a shock are concerned, the main limitation on the theory appears to be the perfect gas assumption. As long as the fluid involved behaves reasonably like a perfect gas, the theoretical predictions tend to be quite good. Departures from the theory tend to be found in cases where higher temperatures are involved, causing departures from the perfect gas assumptions. Such higher temperatures may be caused by the shock waves themselves. The accuracy of the results is thus limited to the range of lower supersonic Mach numbers when the initial gas temperature is not well below ambient. In a subsequent chapter we will deal with cases of flow at higher temperatures above, say, 2000°R. For the present, however, we will confine ourselves to the region in which the perfect gas assumption is reasonably valid.

While both the viscid and inviscid theories correctly predict changes in the flow parameters across a shock, subject to the perfect gas assumption, they are less successful in predicting shock wave thickness. The most common definition of shock wave thickness is that of Prandtl. This definition is shown in Figure 36. A tangent is drawn to the velocity curve at the point where $d^2u/dx^2 = 0$.

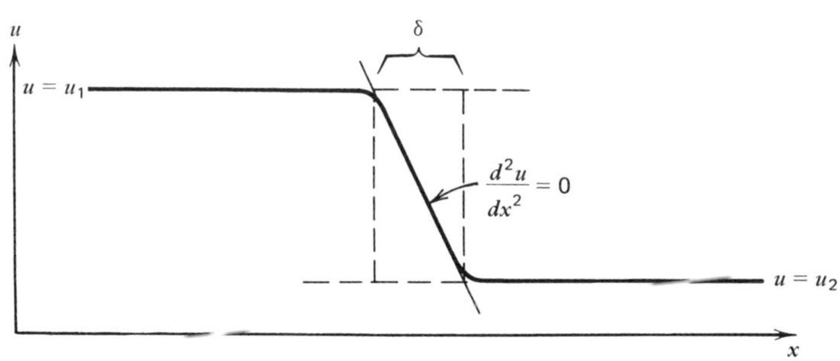

Figure 36. Definition of shock-wave thickness.

The points of intersection of this tangent with the lines $u = u_1$ and $u = u_2$ are found. The shock wave thickness δ is defined as the difference in the x coordinate of the two points of intersection. Figure 37 shows the results of the calculation of Becker ($n = 0$) and Morduchow and Libby ($n = 0.768$). The thickness,

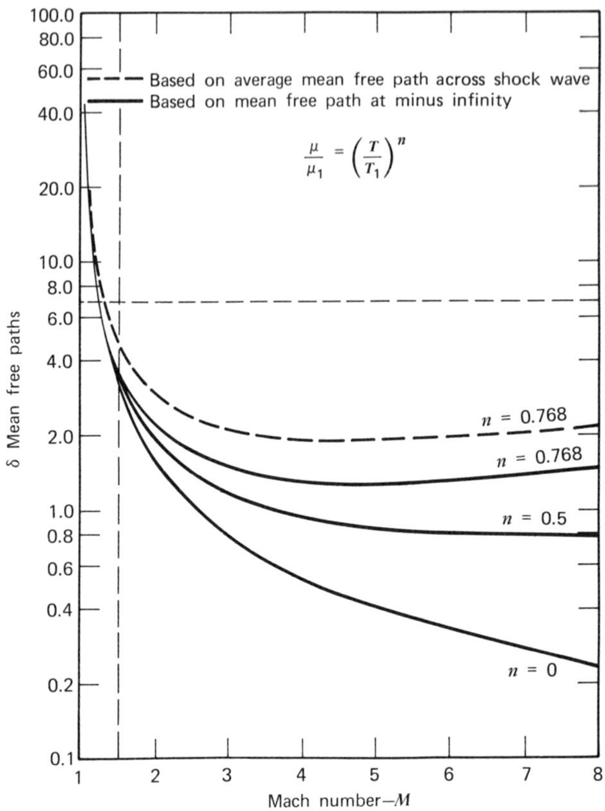

Figure 37. Shock-wave thickness for fixed and variable μ [8].

based on Prandtl's definition, is shown in average mean free paths across the wave. The fluid is presumed to be a perfect gas with $\gamma = 1.4$.

In general, agreement between the Navier-Stokes type of solution, such as that of Morduchow and Libby, and experiment is good up to a Mach number of about 2, after which it deteriorates. At higher Mach numbers, experimental results as to shock thickness tend to vary widely. There is some indication, however, that shock thicknesses reach a minimum at a certain Mach number and then increase. For example, Ballard and Venable [9] obtained the following results for Argon, using electron beam densitometry:

Mach number	Shock thickness (in mean free paths)
3.0	4.4
4.0	3.0
5.8	8.5

A more detailed discussion of this question may be found in Bradley [10].

There is a simple physical explanation regarding the origin of shock waves. Consider an isentropic pressure wave propagating to the right, as shown in Figure 38. Such a wave may be approximated to any desired accuracy by a

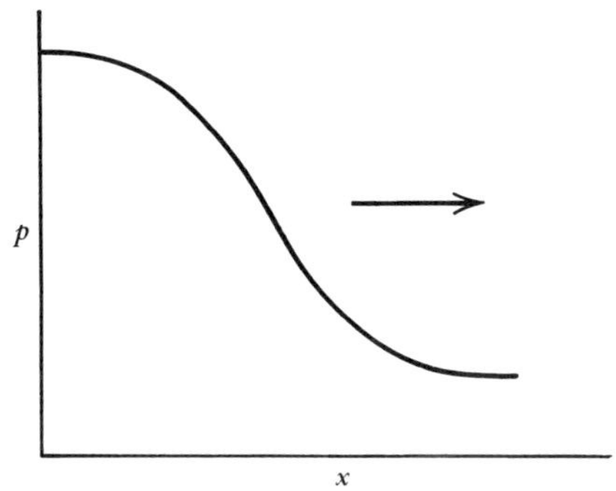

Figure 38. Isentropic pressure wave.

series of small pressure waves such as those shown in Figure 39. As we have

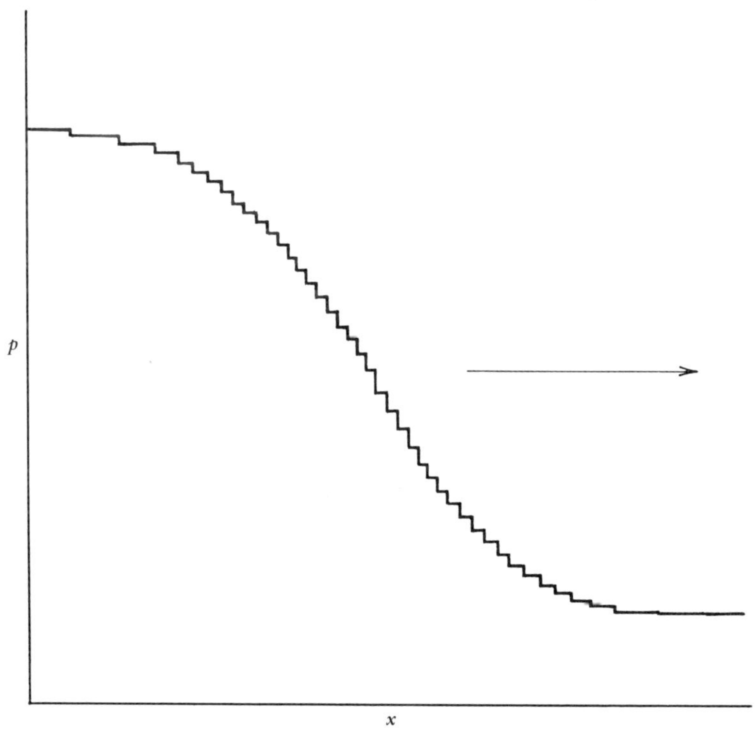

Figure 39. Isentropic pressure wave approximated by a series of small steps.

shown above in the section on sound waves, such waves propagate with the local speed of sound relative to the fluid. Now for an isentropic process,

$$p = \text{constant} \cdot T^{\gamma/(\gamma-1)}, \qquad c = \sqrt{\gamma g R T}$$

Thus, as the pressure rises, so does the temperature and so does the speed of sound. Hence, the back of the wave is traveling more rapidly than the front and therefore the wave must steepen as it goes along (Figure 40). When the

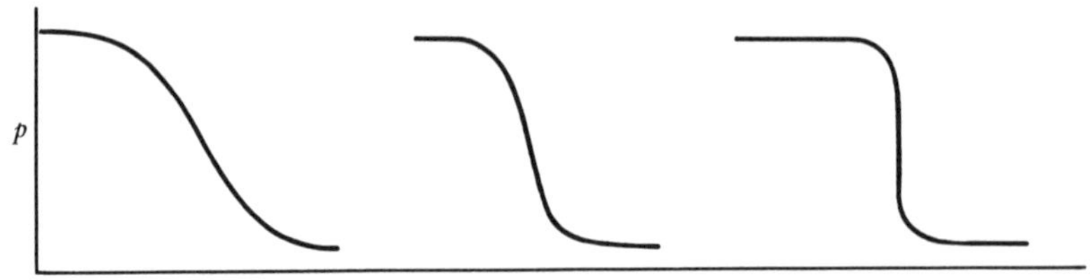

Figure 40. Steepening of pressure wave.

wave becomes sufficiently steep, however, the inviscid analysis is no longer adequate to describe the situation. The viscous terms then become sufficiently large to counteract the steepening of the wave and a balance is established whereby the wave attains some equilibrium thickness.

In the case of rarefaction waves, the situation is just the opposite (Figure 41).

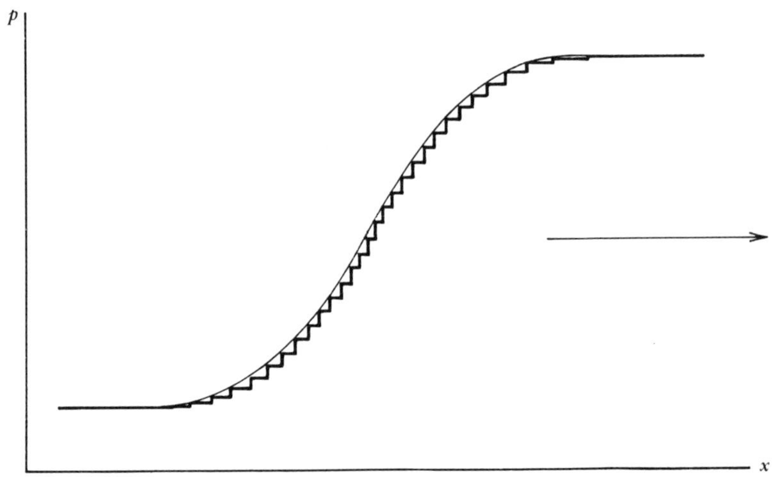

Figure 41. Rarefaction wave and approximation by a series of small steps.

Once again, the wave may be approximated by a series of small waves traveling with the local speed of sound. The small waves in the region of higher pressure are traveling faster than the small waves in the region of lower pressure. Thus, the wave as a whole tends to spread out as it goes along. The general rule is:

compression waves steepen, rarefaction waves flatten

Let us now return to the system of equations, (165)–(168). As in the case of quasi-one-dimensional flow, it is found that the variation of all the parameters may be expressed in terms of the Mach number, specifically the upstream

Mach number. Since $T_{01} = T_{02}$ across the wave, we have, from equation (148)

$$\frac{T_2}{T_1} = \frac{1 + \dfrac{\gamma - 1}{2} M_1^2}{1 + \dfrac{\gamma - 1}{2} M_2^2} \tag{208}$$

From the equations of state (168) and continuity (167), the definition of Mach number (146) and the value of the speed of sound in a perfect gas (124), we have

$$\frac{T_2}{T_1} = \frac{p_2 \rho_1}{p_1 \rho_2} = \frac{p_2 V_2}{p_1 V_1} = \frac{p_2 M_2}{p_1 M_1} \sqrt{\frac{T_2}{T_1}} \tag{209}$$

Combining equations (209) and (208) and solving for the pressure ratio, we get

$$\frac{p_2}{p_1} = \frac{M_1}{M_2} \sqrt{\frac{1 + \dfrac{\gamma - 1}{2} M_1^2}{1 + \dfrac{\gamma - 1}{2} M_2^2}} \tag{210}$$

Another expression for the pressure ratio may be obtained using the momentum equation (166),

$$p_1 + \rho_1 V_1^2 = p_2 + \rho_2 V_2^2$$

Using once again the equations of state (168), the definition of Mach number (146) and the speed of sound for a perfect gas (124), we get

$$\frac{p_2}{p_1} = \frac{1 + \gamma M_1^2}{1 + \gamma M_2^2} \tag{211}$$

This equation is sometimes referred to as the Rayleigh equation. Eliminating p_2/p_1 between equations (210) and (211) and solving for M_2, we get

$$M_2^2 = \frac{M_1^2 + \dfrac{2}{\gamma - 1}}{\dfrac{2\gamma}{\gamma - 1} M_1^2 - 1} \tag{212}$$

Thus we see that once the upstream Mach number is given, the downstream Mach number is uniquely determined.

Substituting equation (212) into equations (208) and (211) we have

$$\frac{T_2}{T_1} = \frac{\left(1 + \dfrac{\gamma - 1}{2} M_1^2\right)\left(\dfrac{2\gamma}{\gamma - 1} M_1^2 - 1\right)}{\dfrac{(\gamma + 1)^2}{2(\gamma - 1)} M_1^2} \tag{213}$$

and

$$\frac{p_2}{p_1} = \frac{2\gamma}{\gamma + 1} M_1^2 - \frac{\gamma - 1}{\gamma + 1} \tag{214}$$

The initial Mach number thus also fixes the temperature and pressure ratios across the shock and hence it follows from the equation of state that the density ratio is also fixed.

$$\frac{\rho_2}{\rho_1} = \frac{p_2}{p_1}\frac{T_1}{T_2} = \frac{\dfrac{\gamma+1}{2}M_1^2}{1+\dfrac{\gamma-1}{2}M_1^2} = \frac{V_1}{V_2} \tag{215}$$

By the continuity equation (167), equation (215) also fixes the velocity ratio. Eliminating M_1 between equations (214) and (215), we obtain a relation between the density and pressure ratios.

$$\frac{p_2}{p_1} = \frac{\dfrac{\gamma+1}{\gamma-1}\dfrac{\rho_2}{\rho_1} - 1}{\dfrac{\gamma+1}{\gamma-1} - \dfrac{\rho_2}{\rho_1}} \tag{216}$$

Equation (216) is known as the Rankine-Hugoniot equation.

Recall now the definition of the critical speed of sound, c^*. From equation (162)

$$c^* = \sqrt{2c_p T_0\left(1 - \frac{2}{\gamma+1}\right)}$$

Combining equation (162) with equations (165)–(168), we find

$$V_1 V_2 = c^{*2} \tag{217}$$

This equation is known as the Prandtl relation.

The stagnation pressure ratio across the wave may be found by combining the equation for the stagnation pressure (149) with the equation for the static pressure ratio across the shock (214). Thus,

$$\frac{p_{02}}{p_{01}} = \frac{p_{02}}{p_2}\cdot\frac{p_1}{p_{01}}\cdot\frac{p_2}{p_1} = \left[\frac{\dfrac{\gamma+1}{2}M_1^2}{1+\dfrac{\gamma-1}{2}M_1^2}\right]^{\gamma/(\gamma-1)} \left[\frac{1}{\dfrac{2\gamma}{\gamma+1}M_1^2 - \dfrac{\gamma-1}{\gamma+1}}\right]^{1/(\gamma-1)} \tag{218}$$

where equation (212) has been introduced to eliminate M_2. The entropy change across the shock may be found in terms of the initial Mach number by substituting equations (213) and (214) for the temperature and pressure ratios in equation (169).

$$s_2 - s_1 = c_p \ln\frac{T_2}{T_1} - R\ln\frac{p_2}{p_1}$$

Equation (169) may also be written

$$s_2 - s_1 = c_p \ln\frac{T_2/T_1}{(p_2/p_1)^{(\gamma-1)/\gamma}} \tag{219}$$

keeping in mind that $R/c_p = (\gamma - 1)/\gamma$. In terms of stagnation values, (217) becomes

$$s_2 - s_1 = c_p \ln \frac{\dfrac{T_{02}}{1+\dfrac{\gamma-1}{2}M_2^2} \cdot \dfrac{1+\dfrac{\gamma-1}{2}M_1^2}{T_{01}}}{\dfrac{p_{02}^{(\gamma-1)/\gamma}}{1+\dfrac{\gamma-1}{2}M_2^2} \cdot \dfrac{1+\dfrac{\gamma-1}{2}M_1^2}{p_{01}^{(\gamma-1)/\gamma}}} = c_p \ln \frac{\dfrac{T_{02}}{T_{01}}}{\left(\dfrac{p_{02}}{p_{01}}\right)^{(\gamma-1)/\gamma}} \tag{220}$$

Since $T_{01} = T_{02}$ across a shock, we see that there is a direct relation between entropy rise and stagnation pressure ratio.

$$s_2 - s_1 = c_p \ln \left(\frac{p_{01}}{p_{02}}\right)^{(\gamma-1)/\gamma} = R \ln \frac{p_{01}}{p_{02}} \tag{221}$$

A number of general conclusions may be drawn from equations (212)–(219). For example, as $M_1 \longrightarrow \infty$, certain of the property ratios increase without limit while others approach finite limits.
From equation (212), as $M_1 \to \infty, M_2 \to (\gamma - 1)/2\gamma$.
From equation (213), as $M_1 \to \infty, T_2/T_1 \to \infty$.
From equation (214), as $M_1 \to \infty, p_2/p_1 \to \infty$.
From equation (215), as $M_1 \to \infty$,

$$\frac{\rho_2}{\rho_1} = \frac{V_1}{V_2} \longrightarrow \frac{\gamma+1}{\gamma-1}$$

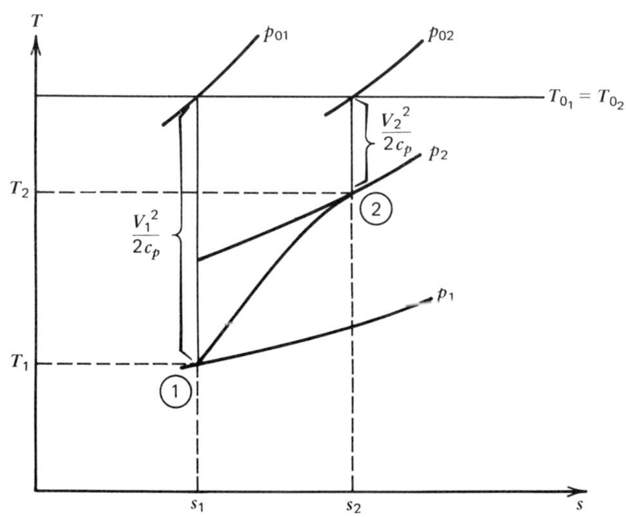

Figure 42. Nonisentropic deceleration shown in the *T-s* diagram.

From equation (218), as $M_1 \to \infty$, $p_{02}/p_{01} \to 0$.
From equation (221), as $M_1 \to \infty$, $s_2 - s_1 \to \infty$.

These conclusions can be regarded as qualitative only, since, as T_2 rises, the perfect gas assumption becomes increasingly less valid.

The relation between the parameters may also be illustrated on a $T-s$ diagram (Figure 42). In engineering practice, an increase in entropy or a decrease in stagnation pressure is generally said to imply a decrease in the available energy. Such a statement may be interpreted by considering an isentropic expansion of the flow from the point 2 to the pressure p_1. From the $T-s$ diagram and the energy equation,

$$c_p T + \frac{V^2}{2} = c_p T_0$$

it is seen that the kinetic energy of the stream after such an expansion will be less than $V_1^2/2$. In this sense, then, the available kinetic energy of the flow may be said to have decreased.

Flow in Supersonic Nozzle with Shocks

We now return to the problem illustrated in Figure 43, which we touched

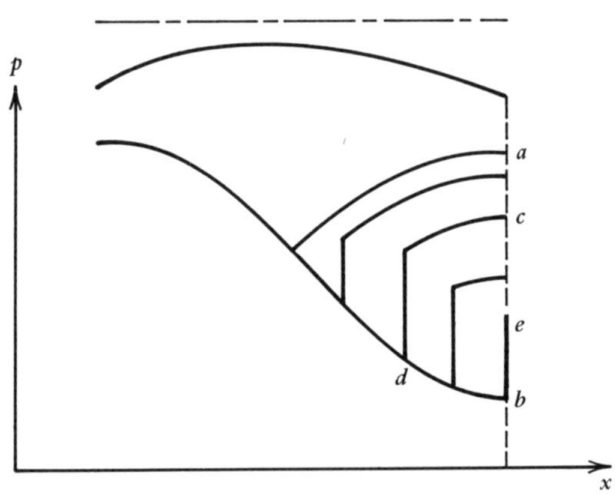

Figure 43. Shock waves in a converging-diverging nozzle.

upon in the section on isentropic flow. The curves ending at a and b represent the possible isentropic solutions when the flow in the throat is choked. In the case where the back pressure lies between a and b, as at the point c, we are now able to calculate the flow in the nozzle more exactly.

If a single discontinuity exists at the point d, the flow on either side of that point may still be considered isentropic. Since the flow downstream of the shock is subsonic, and since the nozzle at that point is diverging, it follows that the velocity at the exit will be subsonic. The pressure in the exit of the nozzle must then equal the back pressure, that is, the pressure outside the nozzle, since a subsonic flow cannot support any pressure discontinuities. The location of the shock is thus fixed.

In practice, the location of the shock for a given overall pressure ratio and area ratio may be determined by trial and error.

1 Postulate a location at which the shock is fixed.

2 Calculate the mach number at that point using equation (159)

$$\frac{A}{A^*} = \frac{1}{M}\left[\frac{\dfrac{\gamma+1}{2}}{1+\dfrac{\gamma-1}{2}M^2}\right]^{(\gamma+1)/2(1-\gamma)}$$

3 Find the Mach number after a shock with the initial Mach number given by equation (159), using equation (212)

$$M_2 = \frac{M_1^2 + \dfrac{2}{\gamma-1}}{\dfrac{2\gamma}{\gamma-1}M_1^2 - 1}$$

4 Find the stagnation pressure after the shock using equation (218)

$$\frac{p_{01}}{p_{02}} = \left[\frac{2\gamma}{\gamma+1}M_1^2 - \frac{\gamma-1}{\gamma+1}\right]^{1/(\gamma-1)}\left[\frac{1+\dfrac{\gamma-1}{2}M_1^2}{\dfrac{\gamma+2}{2}M_1^2}\right]^{\gamma/(\gamma-1)}$$

5 Find the Mach number at the exit of the nozzle using equation (155)

$$\frac{A_c}{A_d} = \frac{M_d}{M_c}\left[\frac{1+\dfrac{\gamma-1}{2}M_d^2}{1+\dfrac{\gamma-1}{2}M_c^2}\right]^{(1/2)-\gamma/(\gamma-1)}$$

6 Find the pressure at the nozzle exit using equation (149)

$$p_d = \frac{p_{02}}{\left[1+\dfrac{\gamma-1}{2}M_d^2\right]^{\gamma/(\gamma-1)}}$$

7 Compare this pressure to the given back pressure. If the two are not equal, repeat the entire procedure starting with a new location for the shock.

In practice, such calculations may be much simplified by the use of tables for the given equations. Such tables may be found in the back of the book for $\gamma = 1.4$.

As is shown on Figure 43, as the back pressure is lowered from a to b, the shock gradually moves downstream until it stands at the exit of the nozzle. At that point, however, the exit pressure is still above its isentropic, supersonic value. A region exists, therefore, in which the exit pressure cannot be made to coincide with the back pressure either by isentropic flow or by a normal shock. It is found experimentally that in such cases, the flow remains supersonic at the exit of the nozzle and that the pressure subsequently adjusts itself by oblique shocks outside the nozzle. Such shocks will be discussed in a subsequent chapter.

It must be pointed out that solutions of the above type will occur in practice only if the viscous effects are small. Viscous effects are generally most pronounced in a region near the wall where the velocity gradients perpendicular to the wall are large. In such regions it may be seen that by Newton's hypothesis

$$\tau \sim \frac{\partial u}{\partial y}$$

that the viscous forces may be large. Since in our analysis we have neglected both viscous forces and vertical velocity gradients, one cannot expect the solution to be valid in cases where viscous forces are important over a substantial portion of the cross section of the flow. The region in which viscous forces are important near the wall is generally referred to as the boundary layer, and the effect of such regions will be discussed in a subsequent chapter.

One-Dimensional Flows with the Addition of Machanical Work

As will be recalled, we originally derived the energy equation for a compressible fluid, equation (27),

$$\rho \frac{Dh}{Dt} = \frac{Dp}{Dt} + \dot{q}$$

from the first law of thermodynamics,

$$\frac{đQ}{dt} = \frac{dE}{dt} + \frac{đW}{dt}$$

under the assumption that the only work done by the fluid within a given volume at a given time was work against the surrounding pressure forces. Subsequently, we extended the analysis to include viscous forces (equation (186)). Such analyses are valid provided the region over which we are integrating contains only fluid. In the case where there are moving solid objects present in the flow, such as compressors or turbines, the application of equations (27) or (186) becomes extremely difficult since the boundaries of the fluid may be moving rapidly in various directions and have complicated shapes, as for example the surfaces of compressor and turbine blades. The problem may thus be greatly simplified by including the moving solid objects within the volume to be integrated and considering only their net effect.

A simple example will illustrate the point. Consider a flow with a propeller in it (Figure 44). Rather than attempt to use the moving propeller blades as

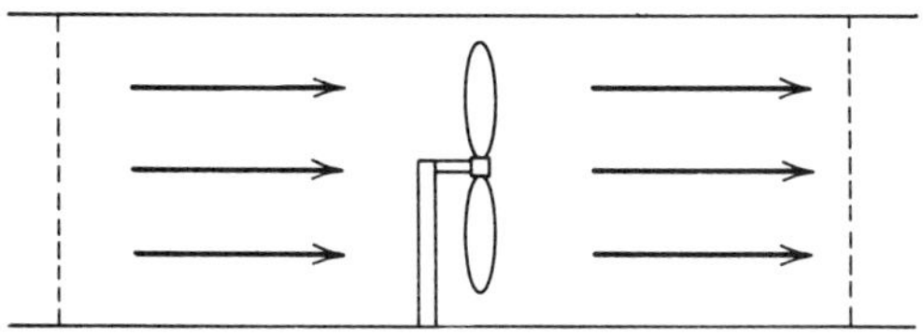

Figure 44. Addition of mechanical work to a flow.

boundaries of our control volume, we may take as our volume the region bounded by the dotted lines and the walls of the duct. The effect of the propeller on the flow may then be taken into account by simply adding another work term to the energy equation, the work done by the propeller on the flow, Moreover, the work done by the propeller will usually be simple to determine in practice if the power transmitted to the propeller is known, since nearly all this power will be transmitted to the flow, either in the form of increased kinetic energy, or heat. Thus, equation (27) may be rewritten

$$\rho\frac{Dh}{Dt}=\frac{Dp}{Dt}+\dot{q}+\dot{w} \tag{222}$$

where $\dot{w}$ represents the rate at which work is being done on the fluid, per unit volume, by moving solid objects within the control volume. Such work may be referred to as mechanical work.

In the case of quasi-one-dimensional flow, equation (102) may be written

$$h_2-h_1=\frac{V_1^2}{2}-\frac{V_2^2}{2}+\frac{\dot{Q}}{\dot{m}}+\frac{\dot{W}}{\dot{m}}-g(z_2-z_1) \tag{223}$$

where $\dot{W}$ represents the rate at which work is being done on the fluid between points 1 and 2.

Let us apply equation (223) to an example. Suppose we wish to construct a wind tunnel with a cross-sectional area of one square foot in the test section, with a Mach number of 2 at normal temperature (530°R) and pressure (1 atm). Air is to be taken from the atmosphere at rest and is to be given sufficient energy by means of a pump or compressor to meet the required conditions in the test section. How much power will be required from the pump or compressor?

The following conditions prevail in the test section: $\rho=0.00238$ slugs ft^3, $V=2260$ ft/sec, $A=1$ ft^2, $\dot{m}=\rho VA=5.4$ slugs/sec. Let the subscript 1 indicate the outside atmosphere and the subscript 2 indicate the condition in the test section. Then let

$$V_1=0,\qquad h_1=h_2,\qquad V_2=2260\,\text{ft/sec}$$

and from equation (223),

$$\dot{W} = \dot{m}\frac{V_2^2}{2}$$

Substituting numbers, we find

$$\dot{W} = 25{,}070\ hp$$

Since this is a very large amount, let us see if we cannot find a more efficient way of creating the same conditions in the test section. One way to do this would be to have a closed loop. In the absence of losses in available energy, once the air in the loop has been accelerated to Mach 2, it would simply continue to go round and round at Mach 2 and no further power would be required to keep the tunnel going. Unfortunately, in order to accelerate the flow to Mach 2, a pump or compressor must be located somewhere in the circuit. Furthermore, since pumps and compressors do not work very well with supersonic speeds at the inlet, it will be necessary to slow the flow down to subsonic speed somewhere in the circuit. Even if this can be done without shock waves, the pump or compressor could not be removed once Mach 2 had been attained due to the necessity to overcome friction at the walls to keep the flow going.

Consider the circuit shown in Figure 45. C indicates the compressor and

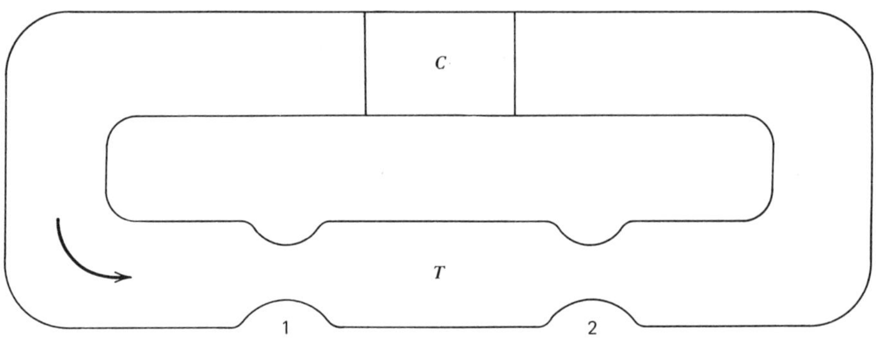

Figure 45. Closed circuit supersonic tunnel with Kantrowitz-Donaldson diffuser.

T the test section. The arrow indicates the direction of the flow. If the cross-sectional areas at 1 and 2 are the same, the possibility exists in theory for a steady shock-free flow around the circuit with supersonic flow in the test section and subsonic flow at the compressor. In fact, however, such a flow could not be established if the air in the circuit is initially at rest. As the compressor is started, it tends to lower the pressure on its upstream side and to raise the pressure on its downstream side. Thus, as the flow accelerates in nozzle 1, the pressure at the exit of that nozzle must pass through the region between a and b in Figure 43, causing the formation of a shock in the diverging section. Recall now equation (151),

$$\frac{\dot{m}}{A} = \sqrt{\frac{\gamma}{R}}\frac{p_0}{\sqrt{T_0}}M\left[1 + \frac{\gamma - 1}{2}M^2\right]^{-(\gamma+1)/2(\gamma-1)}$$

If the flow through the first nozzle is choked we have,

$$\frac{\dot{m}}{A_1}\frac{\sqrt{T_{01}}}{p_{01}} = \sqrt{\frac{\gamma}{R}}\left[\frac{\gamma+1}{2}\right]^{-(\gamma+1)/2(\gamma-1)} \tag{224}$$

Since equation (224) is only a restatement of the law of the conservation of mass for a perfect gas, it follows that for a steady flow,

$$\frac{\dot{m}}{A^*}\frac{\sqrt{T_0}}{p_0} = \text{constant} \tag{225}$$

where A^* is the cross-sectional area at any point where the Mach number is one. A^* thus represents the minimum area through which the flow can pass for a given $\dot{m}$, T_0 and p_0. If we assume that the flow in Figure 45 is adiabatic and steady we have

$$A^* p_0 = \text{constant} \tag{226}$$

since T_0 and $\dot{m}$ then do not change from point to point in the flow. Now if there is a shock in the diverging section of the first nozzle, it follows from equation (218) that the stagnation pressure of the flow is decreased. It then follows from equation (226) that the minimum area through which the flow can pass downstream of the shock is greater than A_1. If $A_1 = A_2$, therefore, no steady-state solution exists with a shock in the diverging section of the first nozzle. If we are to attain supersonic flow in the test section, the downstream throat area A_2 must be sufficiently large to accommodate the flow for the maximum decrease in p_0 to be encountered, namely, when the shock stands in the exit of the first nozzle. A further increase in the pressure ratio across the compressor will then cause the shock to be drawn through the test section and "swallowed" by the second throat. The diffuser would operate with maximum efficiency if the shock could be positioned at the second throat. Unfortunately, this turns out in practice to be impossible due to inherent instability. Once the shock is swallowed, however, a slight increase in the back pressure will move it forward until it is fixed just short of the second throat where it will be stable. In this condition, the diffuser may be operated with reasonable efficiency. A further gain in efficiency may be obtained if the second throat is of a variable area. The area may then be decreased after the tunnel has started, causing the shock to stand at a lower Mach number, or even to disappear altogether, though the latter is difficult to realize in practice. Supersonic tunnels with variable area downstream throats are, in fact, used in a number of experimental installations.

Quasi-One-Dimensional Flow with Heat Addition, Mechanical Work and Area Change

Let dQ represent an increment of heat added to the flow and dW represent a small amount of mechanical work done on the flow. Since Q and T occur together in the energy equation (223), we may treat them as a single term. Let

$$dG = dQ + dW \tag{227}$$

Neglecting body forces, we get, from equation (223),

$$dG = dh + d\frac{V^2}{2} \tag{228}$$

For a perfect gas, this may be rewritten

$$\frac{dG}{c_p T} = \frac{dT}{T} + \frac{\gamma - 1}{2} M^2 \frac{dV^2}{V^2} \tag{229}$$

The presence of mechanical work and heat transfer does not affect the momentum equation. Hence,

$$dp = -\rho V\, dV$$

For a perfect gas, this is equivalent to

$$\frac{dp}{p} + \frac{\gamma M^2}{2} \frac{dV^2}{V^2} = 0 \tag{230}$$

The continuity equation may be written

$$\frac{d\rho}{\rho} + \frac{dV}{V} + \frac{dA}{A} = 0 \tag{231}$$

The equation of state for a perfect gas

$$\frac{dp}{p} = \frac{d\rho}{\rho} + \frac{dT}{T} \tag{232}$$

The Mach number for a perfect gas

$$\frac{dM^2}{M^2} = \frac{dV^2}{V^2} - \frac{dT}{T} \tag{233}$$

The stagnation temperature change may be written

$$\frac{dT_0}{T_0} = \frac{dT}{T} + \frac{\dfrac{\gamma - 1}{2} M^2}{1 + \dfrac{\gamma - 1}{2} M^2} \frac{dM^2}{M^2} \tag{234}$$

From equation (228) we see that

$$dG = c_p\, dT_0 \tag{235}$$

Using equations (227) through (235), we may write the equations governing the quasi-one-dimensional flow of a perfect gas with heat transfer and mechanical work in the form

$$\frac{dM^2}{M^2} = -\frac{2\left(1+\frac{\gamma-1}{2}M^2\right)}{1-M^2}\frac{dA}{A} + \frac{1+\gamma M^2}{1-M^2}\frac{dG}{c_pT} \tag{236}$$

$$\frac{dV}{V} = \frac{-1}{1-M^2}\frac{dA}{A} + \frac{1}{1-M^2}\frac{dG}{c_pT} \tag{237}$$

$$\frac{dT}{T} = \frac{(\gamma-1)M^2}{1-M^2}\frac{dA}{A} + \frac{1-\gamma M^2}{1-M^2}\frac{dG}{c_pT} \tag{238}$$

$$\frac{dp}{p} = \frac{\gamma M^2}{1-M^2}\frac{dA}{A} - \frac{\gamma M^2}{1-M^2}\frac{dG}{c_pT} \tag{239}$$

We may refer to G as the energy transfer function since it represents the energy transferred across the boundary of the flow. If there is no energy transfer, equations (236) through (239) reduce to the equations for quasi-one-dimensional adiabatic flow. Equations (236) through (239) are best evaluated numerically. That is, given the conditions at some point in the flow, the value of the coefficients on the right-hand side of each equation is determined at that point. For given increments in A and G, then, the corresponding changes in $M, V, T,$ and p may be determined.

It is of particular interest that the effects of energy transfer and area change are linearly additive. Thus, each effect may be considered separately. The effect of area change alone has been considered previously. Let us consider the effect of energy transfer alone, then. This may be done by setting $dA = 0$ in each of the equations (236) through (239). Then,

$$\frac{dM^2}{M^2} = \frac{1+\gamma M^2}{1-M^2}\frac{dG}{c_pT} \tag{240}$$

$$\frac{dV}{V} = \frac{1}{1-M^2}\frac{dG}{c_pT} \tag{241}$$

$$\frac{dT}{T} = \frac{1-\gamma M^2}{1-M^2}\frac{dG}{c_pT} \tag{242}$$

$$\frac{dp}{p} = -\frac{\gamma M^2}{1-M^2}\frac{dG}{c_pT} \tag{243}$$

Let us examine each of these equations in turn. Equation (240) indicates that the addition of energy to the flow tends to drive the Mach number towards one. Thus, when M is less than one and dG is positive, dM^2 is also positive. When M is greater than one and dG is positive, dM^2 is negative. Clearly, the subtraction of energy from the flow, as with heat loss across the boundary or the presence of a turbine, will have the opposite effect; that is, drive the Mach number away from one.

A similar effect is noted in equation (241). The addition of energy will increase the velocity if the flow is subsonic and decrease the velocity if the flow is supersonic. In this connection it may be noted that in the present case the velocity and Mach number rise and fall together. When V goes up, M goes up, and when V goes down, M goes down. The same correlation was observed in isentropic flow with variable area.

Equation (242) reveals a rather more complicated effect. When M is less than $1/\sqrt{\gamma}$, the numerator and denominator are both positive. Thus, the addition of energy raises the temperature of the stream. When M lies between $1/\sqrt{\gamma}$ and 1, however, the numerator is negative while the denominator remains positive. In this region, then, the addition of energy causes a drop in the temperature. Since dG may consist entirely of heat addition, this means that the flow may be cooled by heating it! Of course it should be kept in mind that we are dealing here with an average temperature across the flow. When M is greater than 1, numerator and denominator are both negative and hence energy addition once again tends to increase the temperature.

Equation (243) reveals that when M is less than one, the addition of energy to the flow causes the pressure to drop while when M is greater than one, energy addition causes the pressure to rise. Comparison with equation (240) shows that pressure and Mach number vary inversely in the present case, a relation also observed in quasi-one-dimensional, isentropic flow.

3 Two- and Three-Dimensional Subsonic Flow

Irrotationality and Velocity Potential

The complete Navier–Stokes equation in two and three dimensions is one of great complexity and complete solutions can, in general, be obtained only by numerical means. Thus, it is desirable to introduce simplifying assumptions whenever the introduction of such assumptions will not reduce the accuracy of the solution to an unacceptable degree. One assumption which it is frequently possible to make, to great advantage, is that of irrotationality.

We have already seen in Chapter 1 that rate of rotation of a fluid at a point is given by

$$\tfrac{1}{2}\nabla \times \mathbf{V}$$

As was pointed out previously, this "rate of rotation" is really a somewhat arbitrary definition. The expression $\tfrac{1}{2}\nabla \times \mathbf{V}$ is, in fact, the average rate of rotation of three mutually perpendicular lines of fluid through the point of the flow where the expression in evaluated, in the limit as the length of the lines shrinks to zero. The expression is invariant under transformation of the coordinates, meaning that it does not matter which three mutually perpendicular lines we choose. A visual concept of the idea of rotation may be obtained by fastening two matches together at their midpoint, perpendicular to each other, and floating them on the surface of a stream. The rate at which the matches rotate will be approximately equal to $\tfrac{1}{2}\nabla \times \mathbf{V}$ at their point of intersection. The idea may be extended to three dimensions by imagining three mutually perpendicular matches tied together at their midpoint and having the same specific gravity as the fluid in which they are suspended. Their rate and direction of rotation would then approximately equal $\tfrac{1}{2}\nabla \times \mathbf{V}$.

The *circulation* in a fluid is defined in the following way. Consider any closed curve C in the fluid. The circulation, Γ, is defined by

$$\Gamma = \oint_C \mathbf{V}\cdot d\mathbf{s} \tag{244}$$

If the circulation is zero around every closed curve in the fluid, the flow is irrotational. This follows directly from Stokes' Theorem:

$$\oint_C \mathbf{V}\cdot d\mathbf{s} = \int_A \mathbf{n}\cdot\nabla \times \mathbf{V}\, dA \tag{245}$$

where A is any area within the fluid bounded by the closed curve C. Similarly, if the flow is irrotational, the circulation around every closed curve is zero.

Consider now a flow in which the circulation Γ is zero around every closed curve. It follows that the value of the line integral

$$\int_a^b \mathbf{V}\cdot d\mathbf{s}$$

between any two points a and b in the fluid must be independent of the path. This may readily be seen from Figure 46. Let C be the solid closed curve. Let

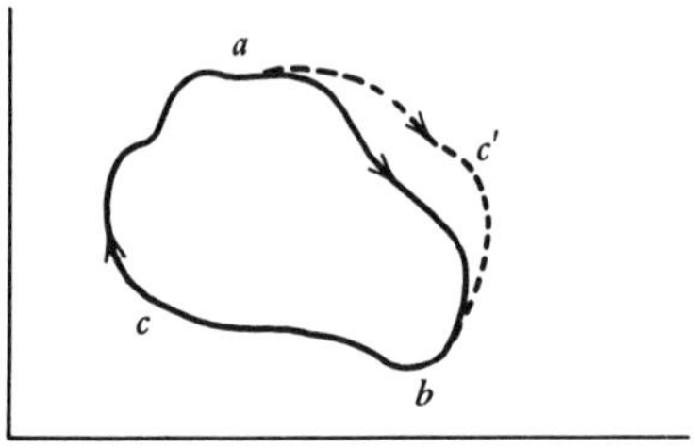

Figure 46. Closed paths in fluid flow.

C' be another closed curve through a and b which coincides with C going clockwise from b to a but differs from C going clockwise from a to b, as shown by the dotted line. Then

$$\oint_C \mathbf{V}\cdot d\mathbf{s} = \oint_{C'} \mathbf{V}\cdot d\mathbf{s} = 0 \tag{246}$$

But

$$\oint_C \mathbf{V}\cdot d\mathbf{s} = \int_{C,\,a}^{b} \mathbf{V}\cdot d\mathbf{s} + \int_{C,\,b}^{a} \mathbf{V}\cdot d\mathbf{s} = 0$$

and

$$\oint_C \mathbf{V}\cdot d\mathbf{s} = \int_{C',\,a}^{b} \mathbf{V}\cdot d\mathbf{s} + \int_{C,\,b}^{a} \mathbf{V}\cdot d\mathbf{s} = 0$$

where the C and C' through the integral sign indicate the path along which the integrals are to be taken. It follows that

$$\int_{C,\,b}^{a} \mathbf{V}\cdot d\mathbf{s} = \int_{C',\,b}^{a} \mathbf{V}\cdot d\mathbf{s} \tag{247}$$

Since a and b as well as C and C' are entirely arbitrary, it follows that the value of the line integral

$$\int_a^b \mathbf{V}\cdot d\mathbf{s}$$

between any two points in the flow is independent of the path and depends only on the end points a and b. In this connection, the following theorem is relevant:

Theorem. Let $\mathbf{V}$ be continuous on an open region R and let a be a fixed point in R. For each point b in R and for every pair of polygonal curves C and C' in R joining a to b, assume that we have

$$\oint_a^b \mathbf{V}\cdot d\mathbf{s} = \oint_a^b \mathbf{V}\cdot d\mathbf{s}$$

Then there exists a function ϕ such that $\mathbf{V} = \nabla\phi$ at each point of R.

For the proof of this theorem, the reader is referred to Apostal [11]. The function ϕ is called the *velocity potential* and is dependent for its existence on the irrotationality of the flow. It follows from the theorem that $\mathbf{V}\cdot d\mathbf{s} = d\phi$ and hence that

$$\int_a^b \mathbf{V}\cdot d\mathbf{s} = \phi(b) - \phi(a) \tag{248}$$

This may be seen as follows: Let

$$d\mathbf{s} = \mathbf{i}\,dx + \mathbf{j}\,dy + \mathbf{k}\,dz$$

$$\mathbf{V} = \nabla\phi = \mathbf{i}\frac{\partial\phi}{\partial x} + \mathbf{j}\frac{\partial\phi}{\partial y} + \mathbf{k}\frac{\partial\phi}{\partial z}$$

Then

$$\mathbf{V}\cdot d\mathbf{s} = \frac{\partial\phi}{\partial x}dx + \frac{\partial\phi}{\partial y}dy + \frac{\partial\phi}{\partial z}dz$$

It also follows that the derivative of ϕ in any direction is the component of the $\mathbf{V}$ in that direction. Let

$$\mathbf{V} = \mathbf{i}u + \mathbf{j}v + \mathbf{k}w$$

where u, v and w are the x, y and z components of the velocity. Hence,

$$u = \frac{\partial\phi}{\partial x}$$

But since the x direction is entirely arbitrary, the derivative of ϕ in any direction must equal the component of velocity in that direction. In particular, it follows that

$$v = \frac{\partial\phi}{\partial y}, \qquad w = \frac{\partial\phi}{\partial z}$$

The existence of a velocity potential makes possible a considerable simplification in the solution of problems in compressible flow in certain cases. We might, therefore, inquire under what circumstances a flow may be considered irrotational in practice. The Navier–Stokes equation may be written in the form

$$\rho\frac{D\mathbf{V}}{Dt} = -\rho\mathbf{F} - \nabla p - \mu\nabla\times\zeta + \tfrac{4}{3}\mu\nabla(\nabla\mathbf{V}) \tag{249}$$

where the vorticity $\zeta = \nabla \times \mathbf{V}$, $\mathbf{F}$ is a body force and μ is assumed constant. If we are dealing with a conservative force field we can write

$$\mathbf{F} = \nabla \Omega$$

where Ω is the force potential. Taking the curl of both sides of equation (249) under the assumption of a conservative force field and keeping in mind that the curl of a gradient is zero, we get, for constant ρ,

$$\frac{D\zeta}{Dt} = (\zeta \nabla)\mathbf{V} + \nu \nabla \times (\nabla^2 \mathbf{V}) \tag{250}$$

or, since

$$\nabla \times (\nabla^2 \mathbf{V}) = -\nabla \times (\nabla . \times \zeta) = \nabla^2 \zeta$$

$$\frac{D\zeta}{Dt} = (\zeta \nabla)\mathbf{V} + \nu \nabla^2 \zeta \tag{251}$$

If a motion starts from rest, ζ is initially zero and hence, initially

$$\frac{D\zeta}{Dt} = \nu \nabla^2 \zeta \tag{252}$$

Thus, it follows that if $\nu = 0$, an incompressible motion which is initially irrotational will remain irrotational. A similar result may be demonstrated for compressible flow. Since compressible fluids (gases) generally are of small viscosity, the assumption of irrotationality is often justified. Exceptions are very slow motions, where the inertia term $(\zeta \nabla)\nabla$ and the viscous term $\nu(\nabla^2 \zeta)$ are of approximately equal order of magnitude, and flow near walls or solid boundaries, where velocity gradients tend to be large, thus increasing the size of the viscous term, even for small viscosity, to a magnitude comparable to the inertia term. The region of rotational flow near a solid boundary (boundary layer) is often very thin and may, in many cases, be neglected. The subject of boundary layers will be taken up in more detail in subsequent chapters.

The question of rotationality is also intimately related to the variation of the enthalpy and entropy in a compressible flow. Recall from Chapter 2 the energy equation in the form

$$\int_S \mathbf{n} \cdot h\mathbf{V} \rho \, dS = -\int_S \mathbf{n} \cdot \mathbf{V} \rho \frac{V^2}{2} dS + \int_{\mathscr{V}} \dot{q} \, d\mathscr{V} - \int_S \mathbf{n} \cdot \mathbf{V} \rho g Z \, dS \tag{253}$$

In Chapter 2 we integrated this equation over the volume $\mathscr{V}$ in the quasi-one-dimensional flow shown in Figure 19. Since the volume $\mathscr{V}$ of surface S over which the integrals in equation (253) are taken is arbitrary, let us, instead, integrate the equation (253) over the stream tube shown in Figure 20. Designating the entrance of the streamtube by the subscript 1 and the exit by subscript 2, we observe that we once again obtain equation (102)

$$\dot{m}(h_2 - h_1) = \dot{m}\left(\frac{V_1^2}{2} - \frac{V_2^2}{2}\right) + \dot{Q} - \dot{m}q(Z_2 - Z_1) \tag{102}$$

Dividing by $\dot{m}$ and taking the limit as the cross-sectional area of the stream tube goes to zero, we get

$$h_2 - h_1 = \frac{V_1^2}{2} - \frac{V_2^2}{2} + q - g(Z_2 - Z_1) \tag{254}$$

along a streamline in steady flow. If now the heat transfer and changes in elevation are neglected, we get

$$h_1 + \frac{V_1^2}{2} = h_2 + \frac{V_2^2}{2} \tag{255}$$

or

$$h_0 = h + \frac{V^2}{2} = \text{constant} \tag{256}$$

along a streamline.

It may, in addition, be postulated that h_0 is the same along every streamline. Such a situation would be approximated in many physical instances. Take for example the flow over an object in uniform stream; h_0 would be the same along every streamline at $-\infty$ and, therefore, the same everywhere if q and Z are negligible. And indeed, in the absence of sources of heat (or cold) the heat transfer rates in compressible flows tend to be small compared to the other terms in equation (102). Exceptions occur at very high speeds or in special situations involving large temperature differences. Similarly, changes in elevation in most engineering applications of compressible flow tend to be small. Thus, the assumption $h_0 = \text{constant}$ is justified in many practical applications.

We have already observed in Chapter 2 that in the case where $h_0 = \text{constant}$ throughout the flow field, the entropy will be constant along a streamline in inviscid flow. We must now prove this more rigorously.

Recall

$$T\,ds = dh - \frac{dp}{\rho}$$

and

$$h_0 = h + \frac{V^2}{2}$$

hence

$$\frac{Dh_0}{Dt} = \frac{Dh}{Dt} + \frac{D}{Dt}\left(\tfrac{1}{2}V^2\right) = \frac{Dh}{Dt} + \mathbf{V}\cdot\frac{D\mathbf{V}}{Dt}$$

and

$$T\frac{Ds}{Dt} = \frac{Dh}{Dt} - \frac{1}{\rho}\frac{Dp}{Dt} = \frac{Dh}{Dt} - \frac{1}{\rho}\left[\frac{\partial p}{\partial t} + (\mathbf{V}\cdot\nabla)p\right]$$

Subtracting

$$\frac{Dh_0}{Dt} - T\frac{Ds}{Dt} = \frac{1}{\rho}\frac{\partial p}{\partial t} + \mathbf{V}\cdot\left\{\frac{D\mathbf{V}}{Dt} + \frac{1}{\rho}\nabla p\right\} = \frac{1}{\rho}\frac{\partial p}{\partial t}$$

since

$$\frac{D\mathbf{V}}{Dt} + \frac{1}{\rho}\nabla p = 0$$

If the flow is steady and h_0 is constant, it follows that

$$\frac{Ds}{Dt} = 0$$

That is, the entropy of a particle does not change as it moves along.

These results may now be related directly to the rotation of the fluid. Consider two streamlines a distance dn apart as shown in Figure 47. Let R be the radius

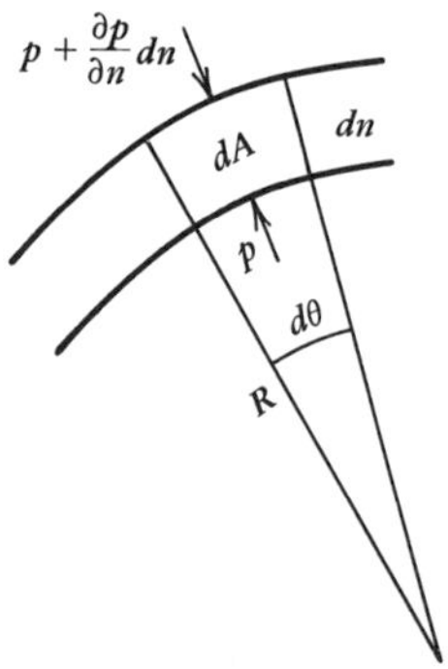

Figure 47. Fluid element in streamline coordinates.

curvature of the lower streamline. Let us now find the circulation $d\Gamma$ about the element of area dA:

$$d\Gamma = VRd\theta - \left(V + \frac{\partial V}{\partial n} dn\right)(R + dn)d\theta = -\left(R\frac{\partial V}{\partial n} + V\right)d\theta\, dn \qquad (257)$$

where V is the magnitude of the velocity. Now recall from Stokes' Theorem

$$\Gamma = \oint \mathbf{V}\cdot d\mathbf{s} = \int_A \mathbf{n}\cdot\nabla - \mathbf{V}\, dA = \int_A \zeta dA$$

where s is the circumference of the area A. Taking the differential of both sides,

$$d\Gamma = \zeta\, dA$$

or,

$$\frac{d\Gamma}{dA} = \zeta$$

In the present case, $dA = R\, d\theta\, dn$, and hence,

$$\frac{d\Gamma}{dA} = \zeta = \frac{d\Gamma}{Rd\theta\, dn}$$

Substituting for $d\Gamma$ from equation (257) we get

$$\zeta = \frac{-\left(R\frac{\partial V}{\partial n} + V\right)d\theta\, dn}{R\, d\theta\, dn} = -\frac{\partial V}{\partial n} - \frac{V}{R} \tag{258}$$

Consider now the force on the element dA. Only pressure force is present in the present instance. Thus we have

$$\text{Force} = \left(p + \frac{\partial p}{\partial n}dn\right)(R + dn)d\theta - pRd\theta - 2\left(pdn\frac{d\theta}{2}\right)$$

being the force in the radial direction on the element dA, due to the pressure gradient $\partial p/\partial n$. Adding terms we have

$$\text{Force} = \frac{\partial p}{\partial n}R\, dn\, d\theta + \frac{\partial p}{\partial n}dn^2\, d\theta$$

Because the flow is steady, the pressure force in the n direction must equal the mass of the element times the centrifugal acceleration. From elementary physics, we know that this quantity is given by

$$\frac{V^2}{R}\rho \cdot R\, dn\, d\theta$$

Hence,

$$\frac{\partial p}{\partial n}R\, dn\, d\theta + \frac{\partial p}{\partial n}dn^2\, d\theta = \frac{V^2}{R}\rho R\, dn\, d\theta$$

Dividing both sides by $dn\, d\theta$ and passing to the limit as $dn \to 0$, we get

$$\frac{\partial p}{\partial n} = \frac{\rho V^2}{R} \qquad \text{or} \qquad \frac{V}{R} = \frac{1}{\rho V}\frac{\partial p}{\partial n}$$

Substituting for V/R in equation (258) we get

$$\zeta = -\frac{\partial V}{\partial n} - \frac{1}{\rho V}\frac{\partial p}{\partial n} \tag{259}$$

Recall

$$h_0 = h + \frac{V^2}{2}$$

Differentiating in the n direction,

$$\frac{\partial h_0}{\partial n} = \frac{\partial h}{\partial n} + V\frac{\partial V}{\partial n} \tag{260}$$

Recall further that

$$T\, ds = dh - \frac{1}{\rho}dp$$

Again differentiating in the n direction,

$$T\frac{\partial s}{\partial n}=\frac{\partial h}{\partial n}-\frac{1}{\rho}\frac{\partial p}{\partial n} \tag{261}$$

or

$$\frac{\partial p}{\partial n}=-\rho T\frac{\partial s}{\partial n}+\rho\frac{\partial h}{\partial n} \tag{262}$$

Substituting for $\partial h/\partial n$ from equation (260)

$$\frac{\partial p}{\partial n}=-\rho T\frac{\partial s}{\partial n}+\rho\frac{\partial h_0}{\partial n}-\rho V\frac{\partial V}{\partial n} \tag{263}$$

Substituting equation (263) for $\partial p/\partial n$ in equation (259) for ζ we get

$$\zeta=-\frac{\partial V}{\partial n}-\frac{1}{\rho V}\left\{-\rho T\frac{\partial s}{\partial n}+\rho\frac{\partial h_0}{\partial n}-\rho V\frac{\partial V}{\partial n}\right\}$$

or

$$\zeta=-\frac{1}{V}\left\{T\frac{\partial s}{\partial n}-\frac{\partial h_0}{\partial n}\right\} \tag{264}$$

This is known as Crocco's Theorem. It states that if the entropy and stagnation enthalpy of a compressible, inviscid fluid are constant throughout the flow field, the vorticity will be zero. Of course the vorticity could also be zero if $T\,\partial s_0/\partial n=\partial h_0/\partial n$ everywhere even if they are not both zero, but such a situation is extremely unlikely from a physical point of view.

We have already pointed out, in Chapter 2, equations (94) through (99), that for an inviscid compressible flow, the equation

$$\rho V\,dV=-dp \tag{97}$$

is valid along every streamline and, in the case of irrotational flow, is valid in every direction.

Derivation of the Velocity Potential Equation

We have stated previously that the existence of a velocity potential is of great help in solving certain problems in compressible flow. Recall that in general, to solve a problem in compressible flow requires the simultaneous solution of four different equations—energy, momentum, continuity and state. To do this is, in general, very difficult. If, therefore, the system of four simultaneous equations could be reduced to a single equation with a single unknown, the solution might be greatly simplified. One way to reduce the system of four simultaneous equations to a single equation is by use of the velocity potential.

Recall the continuity equation for steady flow,

$$\nabla\cdot(\rho\mathbf{V})=0$$

But

$$\mathbf{V}=\nabla\phi$$

Hence

$$\nabla\cdot(\rho\nabla\phi)=\nabla\rho\cdot\nabla\phi+\rho\nabla^2\phi=0 \tag{265}$$

Now recall

$$dp=-\rho d\left(\frac{V^2}{2}\right)$$

for irrotational flow. In addition, s and h_0 are constant by Crocco's Theorem. Hence

$$dp=c^2d\rho=-\rho d\left(\frac{V^2}{2}\right)$$

in every direction. Since ∇ represents the rate of change in the direction of the maximum rate of change, it follows that

$$c^2\nabla\rho=-\rho\nabla\left(\frac{V^2}{2}\right)$$

Substituting for $\nabla\rho$ in equation (265),

$$-\frac{\rho}{c^2}\nabla\left(\frac{V^2}{2}\right)\cdot\nabla\phi+\rho\nabla^2\phi=0$$

or

$$-\frac{1}{2c^2}\nabla\phi\cdot\nabla(\nabla\phi\cdot\nabla\phi)+\nabla^2\phi=0 \tag{266}$$

We may observe here that in the case of incompressible flow, where the speed of sound is infinite, equation (266) reduces to the Laplace equation $\nabla^2\phi=0$. Returning to the case of compressible flow, recall that in the present instance,

$$h_0=h+\frac{V^2}{2}=\text{constant}$$

For a perfect gas with constant c_p we may write

$$c_pT_0=c_pT+\frac{V^2}{2}$$

or

$$\frac{c_p}{\gamma R}\gamma RT_0=\frac{c_p}{\gamma R}\gamma RT+\frac{V^2}{2}$$

That is,

$$c_0^2=c^2+\frac{\gamma R}{c_p}\frac{V^2}{2}=c^2+\frac{\gamma-1}{2}V^2$$

Hence

$$c^2=c_0^2-\frac{\gamma-1}{2}\nabla\phi\cdot\nabla\phi$$

Substituting in equation (266) we have

$$2\left(c_0^2 - \frac{\gamma - 1}{2}\nabla\phi\cdot\nabla\phi\right)\nabla^2\phi = \nabla\phi\cdot\nabla(\nabla\phi\cdot\nabla\phi) \tag{267}$$

Since c_0 is presumed known, the only unknown in this equation is ϕ. Notice that in deriving equation (267) we have used all four governing equations: energy, momentum, continuity and state. We have thus succeeded in reducing a system of four simultaneous equations with four unknowns $(\mathbf{V}, \rho, p, h)$ to a single differential equation with only one unknown, ϕ. Equation (266) is often written out in components, in which case it takes the form

$$\begin{aligned}\phi_{xx}(c^2 - \phi_x^2) + \phi_{yy}(c^2 - \phi_y^2) + \phi_{zz}(c^2 - \phi_z^2)\\ = 2\phi_x\phi_y\phi_{xy} + 2\phi_x\phi_z\phi_{xz} + 2\phi_y\phi_z\phi_{yz}\end{aligned} \tag{268}$$

where the subscripts denote partial differentiation. The equation may also be written in terms of the velocity components in which case it becomes,

$$u_x(c^2 - u^2) + v_y(c^2 - v^2) + w_z(c^2 - w^2) = 2uvu_y + 2uww_x + 2vwv_z \tag{269}$$

The equation takes a particularly simple form in the case of two-dimensional flow,

$$\phi_{xx}(c^2 - \phi_x^2) + \phi_{yy}(c^2 - \phi_y^2) = 2\phi_x\phi_y\phi_{xy} \tag{270}$$

or

$$u_x(c^2 - u^2) + v_y(c^2 - v^2) = 2uvu_y \tag{271}$$

Equation (266) may also be rewritten in the particularly simple form

$$2c^2\nabla^2\phi = \nabla\phi.\nabla(\nabla\phi.\nabla\phi) \tag{272}$$

Some comments on the nature of equation (272)

We have already seen in Chapter 2, how, in the case of isentropic, quasi-one-dimensional flow, the behavior of the flow depends markedly on whether the flow is subsonic or supersonic. In subsonic flow a diverging duct decreased the velocity of the flow, in supersonic flow a diverging duct increased the velocity; in subsonic flow a diverging duct increased the temperature, in supersonic flow a diverging duct decreased the temperature, in subsonic flow no discontinuities were possible, in supersonic flow shock waves were observed, and so on. At the point $M = 1$, many of the derivatives of the flow such as dM/dA, dV/dA, and dT/dA went to infinity. So the point $M = 1$ represented a kind of dividing line, with the nature of the solution of the governing equation for $M < 1$, differing markedly from the solution of $M > 1$. In the case of two- and three-dimensional potential flow, the same phenomenon is observed. The nature of the solution of the potential equation for $M < 1$ differs markedly from the solution for $M > 1$. That this must be so follows directly from the theory of partial differential equations.

For the sake of simplicity, let us consider the two-dimensional case of equation

(272). We may write this equation in the form

$$A\phi_{xx} + B\phi_{xy} + C\phi_{yy} = 0 \tag{273}$$

This is a homogeneous, quasilinear partial differential equation of the second order. It is quasilinear (and not linear) because the first-order derivatives ϕ_x and ϕ_y appear in the coefficients A, B and C. It is well known from the theory of partial differential equations that the nature of the solution of an equation of this type will depend on whether the equation is elliptic, parabolic, or hyperbolic.

In order to find out if equation (273) is elliptic, hyperbolic, or parabolic, we proceed as follows. We solve the following three equations simultaneously for ϕ_{xy}:

$$\begin{aligned} A\phi_{xx} + B\phi_{xy} + C\phi_{yy} &= 0 \\ dx\phi_{xx} + dy\phi_{xy} \qquad\quad &= d\phi_x \\ dx\phi_{xy} + dy\phi_{yy} &= d\phi_y \end{aligned}$$

The second and third of the equations are simply the equations for the differentials of the functions ϕ_x and ϕ_y from elementary calculus. Solving the system of three equations for ϕ_{xy} we get

$$\phi_{xy} = \frac{\begin{vmatrix} A & 0 & C \\ dx & d\phi_x & 0 \\ 0 & d\phi_y & dy \end{vmatrix}}{\begin{vmatrix} A & B & C \\ dx & dy & 0 \\ 0 & dx & dy \end{vmatrix}} = \frac{A\,dy\,d\phi_x + C\,dx\,d\phi_y}{A\,dy^2 - B\,dx\,dy + C\,dx^2} \tag{274}$$

Whether the equation (273) is elliptic, hyperbolic or parabolic depends on whether real values of dy/dx exist for which the denominator of equation (274) goes to zero. This may be determined by solving the quadratic equation

$$A\left(\frac{dy}{dx}\right)^2 - B\frac{dy}{dx} + C = 0 \tag{275}$$

The solution is

$$\frac{dy}{dx} = \frac{B \pm \sqrt{B^2 - 4AC}}{2A}$$

If $B^2 - 4AC > 0$, a real solution to equation (275) exists and the equation (273) is said to be hyperbolic. If $B^2 - 4AC = 0$, a real solution to equation (275) also exists and the equation (273) is said to be parabolic. If $B^2 - 4AC < 0$, no real solution to equation (275) exists and the equation (273) is said to be elliptic.

Let us see in which category the potential equation (271) falls. For this equation

$$A = c^2 - u^2, \qquad B = 2uv, \qquad C = c^2 - v^2$$

Hence

$$\begin{aligned} B^2 - 4AC &= 4u^2v^2 - 4(c^2 - u^2)(c^2 - v^2) \\ &= 4[u^2v^2 - (c^4 - c^2v^2 - u^2c^2 + u^2v^2)] \\ &= 4c^2[(u^2 + v^2) - c^2] \end{aligned}$$

We may rewrite this

$$B^2 - 4AC = 4c^4[M^2 - 1]$$

Hence equation (271) is elliptic when $M < 1$ and hyperbolic when $M > 1$. Thus we see that the value $M = 1$ once again represents a sharp dividing line in the nature of the solution in two-dimensional flow just as in one-dimensional flow. A similar result can be demonstrated for three-dimensional flow.

A detailed analysis of the nature of partial differential equations is beyond the scope of this work. For such an analysis the reader is referred to texts on the subject such as Courant–Hilbert [12], Snedden [13] or Petrovskii [14]. Here we simply wish to make use of the results. One of the chief differences in the nature of the solution of elliptic and hyperbolic equations lies in the effect of changing the boundary conditions. Elliptic equations are characterized by the smoothness of their solutions. No discontinuities are present. Thus, a change in the boundary conditions must result in a change in the solution throughout the flow field. Let us consider as an example the subsonic flow over a cylinder. A change in the shape of the cylinder will result in a change in the magnitude and direction of the velocity at every point of the flow field. It must be remembered, however, that we are dealing here with steady flow. Thus we are speaking here of the steady flow over two cylinders of different shape, not of the transient changes which may result from the sudden change in the shape of the cylinder.

Hyperbolic equations will be dealt with in the next chapter. It may be remarked here, however, that in the hyperbolic case, a change in the boundary conditions may result in a change in only a portion of the flow field—the so-called region of influence. In a hyperbolic flow field, discontinuities are also possible.

Since the potential equation is parabolic only when M is identically equal to one, this case is of little practical interest. The case of mixed multidimensional flow, however, when M is both less than and greater than one in the same flow field, presents considerable difficulties and will be treated in a separate chapter, the chapter on transonic flow.

In the present chapter, we shall confine ourselves to a discussion of the solution of the potential equation in the elliptic case; that is, when the flow is everywhere subsonic. Unfortunately, there is no simple general method of solving the elliptic potential equation. It is, therefore, necessary to resort to a variety of approximate and/or numerical techniques, depending on the particular problem to be solved.

Linearization of the Potential Equation

In the case of a uniform flow with small perturbations, such as the flow over an airfoil, the potential equation may be linearized and then, by a suitable change of variables, transformed into the incompressible potential equation, that is, the Laplace equation. Thus, in problems where the incompressible solution is known, the compressible solution may be obtained by simply applying the inverse transformation. In those cases where the incompressible solution is not known, we may still make use of the known techniques for solving the Laplace equation.

We proceed as follows. Consider a thin object lying in a uniform stream as shown in Figure 48. The velocity at infinity is V_∞ and Mach number at infinity

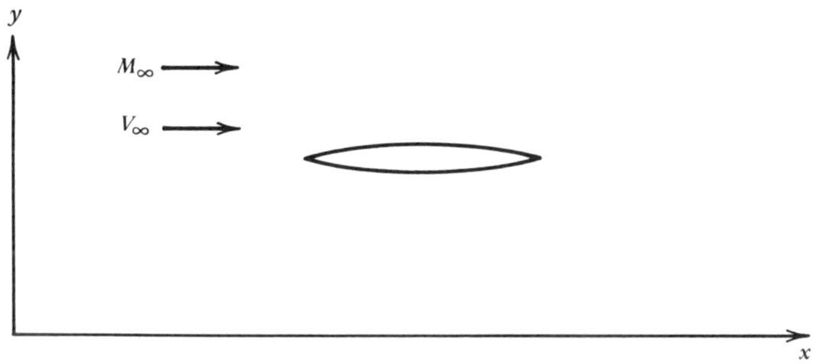

Figure 48. Flow over a thin body.

is M_∞. We now define the perturbation velocity components u and v so that the x component of the velocity at any point in given by $V_\infty + u$ and the y component of the velocity is given by v. Both u and v are assumed small compared to V_∞.

Recall the speed of sound is given by

$$c_0^2 = c^2 + \frac{\gamma - 1}{2}(\phi_x^2 + \phi_y^2) \tag{276}$$

Hence

$$c^2 + \frac{\gamma - 1}{2}[(V_\infty + u)^2 + v^2] = c_\infty^2 + \frac{\gamma - 1}{2}V_\infty^2 \tag{277}$$

where the subscript ∞ denotes the conditions at infinity. Solving equation (277) for c_∞^2/c^2 and expanding the result in a Taylor series, we get

$$\frac{c_\infty^2}{c^2} = 1 + \frac{\gamma - 1}{2}M_\infty^2\left[\frac{2u}{V_\infty} + \frac{u^2 + v^2}{V_\infty^2}\right] + \dots \tag{278}$$

Now recall the potential equation (270)

$$\left(1 - \frac{\phi_x^2}{c^2}\right)\phi_{xx} - 2\frac{\phi_x\phi_y}{c^2}\phi_{xy} + \left(1 - \frac{\phi_y^2}{c^2}\right)\phi_{yy} = 0 \tag{270}$$

Substituting equation (278) for c_∞^2/c^2 into the coefficients of equation (270) we get

$$1 - \frac{\phi_x}{c^2} = 1 - M_\infty^2 - \frac{u}{V_\infty}[2M_\infty^2 + (\gamma - 1)M_\infty^4] + \dots$$

$$2\frac{\phi_x\phi_y}{c^2} = 2M_\infty^2\frac{v}{V_\infty} + 2M_\infty^2\frac{uv}{V_\infty^2} + \dots$$

$$1 - \frac{\phi_y^2}{c^2} = 1 - \frac{v^2}{V_\infty^2}M_\infty^2\left[1 + (\gamma - 1)M_\infty^2\frac{u}{V_\infty}\right] + \dots$$

Thus, if $M_\infty < 1$ while $u/V_\infty, v/V_\infty \ll 1$, it follows that

$$1 - \frac{\phi_y^2}{c^2} \cong 1, \qquad 1 - \frac{\phi_x^2}{c^2} \cong 1 - M_\infty^2, \qquad 2\frac{\phi_x\phi_y}{c^2} \cong 0$$

and the potential equation (270) becomes

$$\beta^2\phi_{xx} + \phi_{yy} = 0 \tag{279}$$

where $\beta^2 = 1 - M_\infty^2$. If we make the change of variables.

$$x' = x, \qquad y' = \beta y \tag{280}$$

equation (279) reduces to

$$\phi_{x'x'} + \phi_{y'y'} = 0 \tag{281}$$

Equation (281) is the Laplace equation, which is the governing equation for incompressible flow. The transformation (280) thus raises the possibility of using known incompressible flow solutions to find corresponding solutions in compressible flow. Such a process would be desirable, in view of the many known solutions of incompressible flow problems, as well as the well known methods for solving Laplace's equation for various boundary conditions. Before we can apply the transformation (280) to solving compressible flow problems, however, we must also transform the boundary conditions. Consider, then, the problem illustrated in Figure 49. A slender body whose shape is

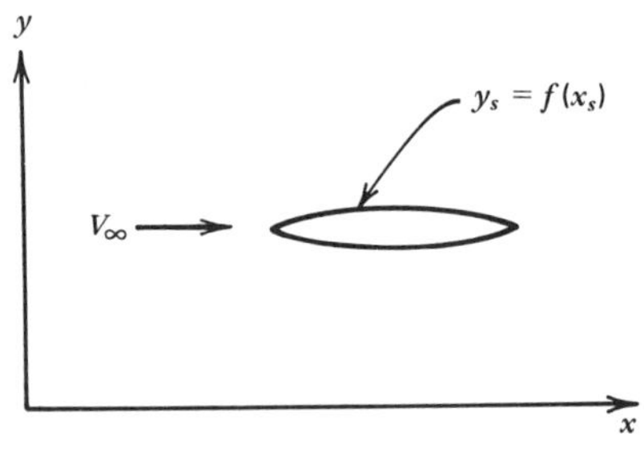

Figure 49. Slender body of given shape in a uniform stream.

given lies in a uniform compressible stream whose velocity is V_∞ at an infinite distance from the body. Let the surface of the body be designated by $y_s = f(x_s)$. Then it follows that

$$\frac{dy_s}{dx_s} = \frac{v}{V_\infty + u} = \frac{\phi_y}{V_\infty + u} \cong \frac{\phi_y}{V_\infty} \tag{282}$$

Let us now apply the transformation (280) to the body in Figure 49. Then

$$y'_s = \beta y_s, \qquad x'_s = x_s, \qquad \frac{dy'_s}{dx'_s} = \beta \frac{dy_s}{dx_s}$$

The transformed plane, hereafter called the incompressible plane, is shown in Figure 50. If we now consider the incompressible flow over the transformed

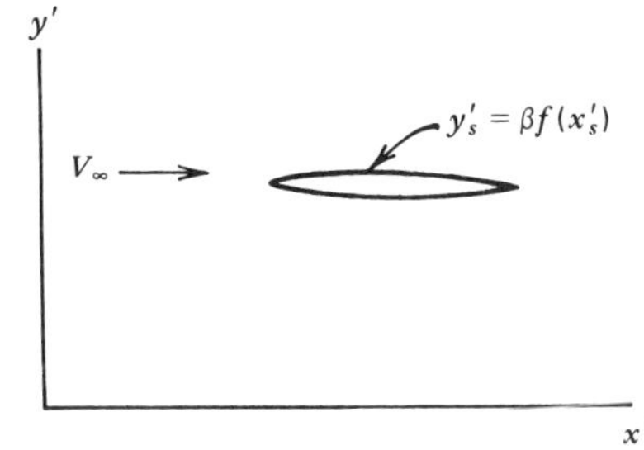

Figure 50. Transformed profile of slender body in incompressible plane.

profile, we require

$$\frac{dy'_s}{dx'_s} = \frac{\phi_{y'}}{V_\infty} \tag{283}$$

Performing the inverse transformation,

$$\frac{dy_s}{dx_s} = \frac{1}{\beta}\frac{dy'_s}{dx'_s} = \frac{1}{\beta}\frac{\phi'_y}{V_\infty} = \frac{1}{\beta^2}\frac{\phi_y}{V_\infty} \neq \frac{v}{V_\infty} \tag{284}$$

The inverse transformation of (280) thus does not satisfy the boundary conditions in the compressible plane.

To get around this problem, let us try the following transformation. Let

$$\phi = V_\infty x + \bar{\phi}, \qquad \phi' = V_\infty x + \bar{\phi}', \qquad \bar{\phi}' = \beta^2 \bar{\phi}, \qquad y' = \beta y, \qquad x' = x \tag{285}$$

The quantity $\bar{\phi}$ is known as the perturbation potential. It follows from the transformation (285) that

$$u = \bar{\phi}_{x}, \qquad v = \bar{\phi}_{y}, \qquad \beta^2 \bar{\phi}_{xx} + \bar{\phi}_{yy} = 0, \qquad \bar{\phi}'_{x'x'} + \bar{\phi}'_{y'y'} = 0 \tag{286}$$

The boundary conditions in the compressible plane (Figure 49) become,

$$x, y = \infty: \quad \bar{\phi}_x = 0, \quad \bar{\phi}_y = 0; \qquad x = x_s, \quad y = y_s: \quad \bar{\phi}_y \cong V_\infty \frac{dy_s}{dx_s} \tag{287}$$

Applying the transformation (286) to the boundary conditions (287) we get
At the body:

$$\frac{dy_s}{dx_s} = \frac{1}{\beta}\frac{dy'_s}{dx'_s} \cong \frac{1}{\beta}\frac{\bar{\phi}'_{y'}}{V_\infty} = \frac{\bar{\phi}_y}{V_\infty} = \frac{v}{V_\infty} \tag{288}$$

At infinity:

$$\bar{\phi}'_{x'} = \bar{\phi}'_{y'} = 0 \tag{289}$$

Thus we find that under the transformation (285), the boundary conditions are satisfied in both the compressible and incompressible planes.

To sum up, then, in order to find the compressible flow over the profile shown in Figure 49, we solve for the incompressible flow over the similar, but somewhat thinner profile of Figure 50 and then perform the inverse of the transformation (285).

The use of the linearized equation (281) is not limited to the calculation of flow over bodies. It is equally valid for flow over walls, provided the walls do not substantially differ in shape from a flat plane. For example, let us consider

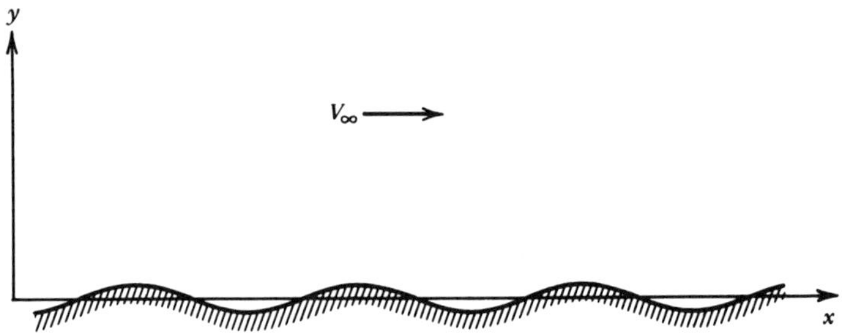

Figure 51. Flow over a wavy wall.

the flow over a wave-shaped wall (Figure 51). The equation of the wall is given by

$$y_s = h\cos\frac{2\pi x}{l}$$

where h represents the amplitude of the wave and l the wavelength. It follows that in order for the linearized equation to be applicable, h will have to be small compared to l.

In the y', x' plane, the wall will have the shape

$$y'_s = \beta h\cos\frac{2\pi x'_s}{l} \tag{290}$$

We must now solve Laplace's equation

$$\phi_{x'x'} + \phi_{y'y'} = 0 \tag{281}$$

with the boundary condition (290). We begin by defining the perturbation potential $\bar{\phi}$ by the equation

$$\phi = V_\infty x + \bar{\phi} \tag{291}$$

Then

$$\bar{\phi}_{x'x'} + \bar{\phi}_{y'y'} = 0 \tag{292}$$

The boundary conditions in the compressible plane at the wall may be established with the help of Figure 52. At the wall,

$$\frac{dy_s}{dx_s} = \frac{v}{V_\infty + u} \cong \frac{v}{V_\infty} \tag{293}$$

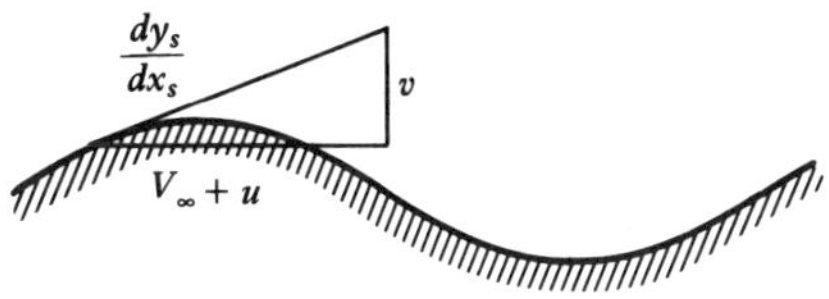

Figure 52. Boundary condition at wavy wall.

since u is assumed small compared to V_∞. Thus,

$$\frac{dy_s}{dx_s} = \frac{1}{V_\infty}(\phi_y)_{y=y_s} \tag{294}$$

In the transformed plane, thus,

$$\frac{1}{\beta}\frac{dy'_s}{dx'_s} = \frac{\beta}{V_\infty}(\phi_{y'})_{y'=y'_s} \tag{295}$$

The boundary condition at the wall of the transformed plane is, then,

$$\bar{\phi}_{y'} = \phi_{y'} = \frac{V_\infty}{\beta^2}\frac{dy'_s}{dx'_s} = -\frac{V_\infty}{\beta^2}\beta h \sin\frac{2\pi x'_s}{l}\cdot\frac{2\pi}{l} \tag{296}$$

At $y = \infty$,

$$\bar{\phi}_y = \bar{\phi}_{y'} = 0, \qquad \bar{\phi}_x = \bar{\phi}_{x'} = 0 \tag{297}$$

We must, therefore, solve equation (292) with the boundary conditions (296) and (297). Applying the method of separation of variables, let

$$\bar{\phi} = \phi_1(x)\phi_2(y) \tag{298}$$

Substituting into Laplace's equation, we get

$$\frac{1}{\phi_1}\frac{d^2\phi_1}{dx'^2} = -\frac{1}{\phi_2}\frac{d^2\phi_2}{dy'^2} = -k^2 \tag{299}$$

where k is an arbitrary constant. Integrating,

$$\phi_1 = A_1 \cos kx' + B_1 \sin kx'$$
$$\phi_2 = A_2 e^{-ky'} + B_2 e^{ky'}$$

where A_1, A_2, B_1 and B_2 are arbitrary constants. Since $\phi \to 0$ as $y \to \infty$, $B_2 = 0$. Thus we have

$$\bar{\phi} = A_2 e^{-ky'}[A_1 \cos kx' + B_1 \sin kx'] \tag{300}$$

and

$$\bar{\phi}_{y'} = -kA_2 e^{-ky'}[A_1 \cos kx' + B_1 \sin kx'] \tag{301}$$

The boundary condition at infinity is automatically satisfied by this equation. At the wall,

$$\bar{\phi}_{y'} = -\frac{V_\infty}{\beta}\frac{2\pi h}{l}\sin\frac{2\pi x'_s}{l} \tag{302}$$

Comparison with equation (301) shows that $A_1 = 0$. The factor $e^{-ky'}$ in equation (301) prevents the equation from satisfying the boundary conditions (302) exactly. However, since the amplitude of the wave of the wall has been assumed small, we may assume

$$e^{-ky'_s} \cong 1 \tag{303}$$

and hence

$$-kA_2B_1 = -\frac{V_\infty}{\beta}\frac{2\pi h}{l}, \qquad k = \frac{2\pi}{l} \tag{304}$$

Thus,

$$\bar{\phi} = -\frac{V_\infty h}{\beta}e^{-(2\pi/l)y'}\sin\frac{2\pi}{l}x' \tag{305}$$

$$\phi = V_\infty\left[x' - \frac{h}{\beta}e^{-(2\pi/l)y'}\sin\frac{2\pi}{l}x'\right]$$

or

$$= V_\infty\left[x - \frac{h}{\beta}e^{-(2\pi\beta/l)y}\sin\frac{2\pi}{l}x\right] \tag{306}$$

giving us the solution for the compressible case.

Once the velocity potential has been found, the magnitude and direction of the velocity at every point are, of course, determined, as are the enthalpy and pressure at every point: $V = \sqrt{\phi_x^2 + \phi_y^2}$, $\mathbf{V} = \mathbf{i}\phi_x + \mathbf{j}\phi_y$.

The enthalpy may be found from the equation

$$h_0 = h + \frac{V^2}{2} \tag{307}$$

where h_0 is presumed known. In the case of a perfect gas with constant c_p this equation also fixes the temperature since $h = c_p T$ in that case. The pressure may be found from the equation

$$\rho V\,dV + dp = 0 \tag{308}$$

with the help of an equation of state. For a perfect gas,

$$\frac{p}{p_0} = \left(\frac{\rho}{\rho_0}\right)^\gamma \tag{309}$$

in isentropic flow. Substituting equation (309) into equation (308) and integrating we get

$$\frac{V^2}{2} + \frac{\gamma}{\gamma - 1}\frac{p_0^{1/\gamma}}{\rho_0}p^{(\gamma-1)/\gamma} = \frac{\gamma}{\gamma - 1}\frac{p_0}{\rho_0} \tag{310}$$

which fixes the pressure if p_0 and ρ_0 are known.

It is of interest to observe that equations (307) and (310) can also be expressed in the form

$$T_0 = T\left(1 + \frac{\gamma - 1}{2} M^2\right) \tag{311}$$

and

$$p_0 = p\left(1 + \frac{\gamma - 1}{2} M^2\right)^{\gamma/(\gamma - 1)} \tag{312}$$

familiar from quasi-one-dimensional flow. It is now seen that these equations are also valid in two- (and three-) dimensional flow if h_0 and s are constant.

The solution of the flow over a wave-shaped wall may be extended to the case of a wall with arbitrary surface fluctuations by means of Fourier series, as long as the fluctuations are small [15]. The solution then takes the form

$$\bar{\Phi}(x,\eta) = -\frac{V_\infty}{\beta\pi}\int_0^\infty \int_{-\infty}^{\infty} \lambda^{-1} e^{-\lambda\eta} F'(\xi) \cos \lambda(\xi - x)\, d\xi\, d\lambda \tag{313}$$

where the shape of the wall is given by $y_w = F(x)$ and $\eta = \beta y$. The solution may also be extended to the case of axially symmetric flow [15]. In polar coordinates, equation (270) then becomes

$$(c^2 - \phi_x^2)\phi_{xx} - 2\phi_x\phi_r\phi_{xr} + (c^2 - \phi_r^2)\phi_{rr} + \frac{c^2}{r}\phi_r = 0 \tag{314}$$

Figure 53. Polar coordinate system.

where the coordinate system is shown in figure 53. Then

$$V_\infty + u = \phi_x, \qquad v = \phi_r \tag{315}$$

$$\phi = V_\infty x + \bar{\Phi} \tag{316}$$

Proceeding as before, we find the linearized equation for the perturbation potential

$$\beta^2 \bar{\Phi}_{xx} + \bar{\Phi}_{rr} + \frac{1}{r}\bar{\Phi}_r = 0 \tag{317}$$

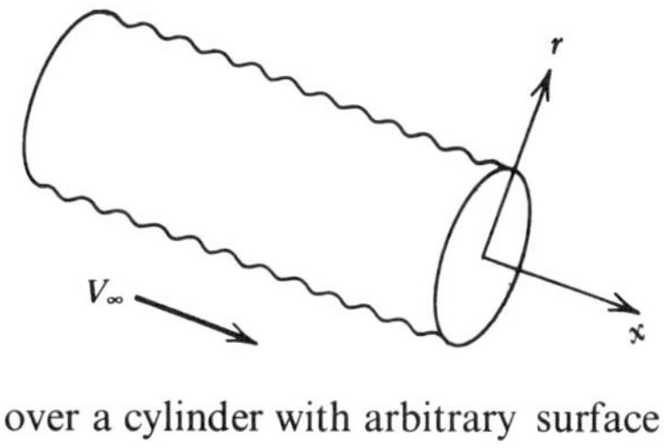

Figure 54. Flow over a cylinder with arbitrary surface perturbation.

We consider now the axial flow over a cylinder whose surface is perturbed by a small amount in an axially symmetric manner (Figure 54). The shape of the surface is given by

$$r_s(x) = \sum_{n=1}^{\infty} \left[a_n \cos\left(2\pi n \frac{x}{l}\right) + b_n \sin\left(2\pi n \frac{x}{l}\right) \right] \tag{318}$$

The boundary conditions now become

$$u(x, \infty) = 0, \qquad v(x, r_m) \cong V_\infty \frac{dr_s}{dx_s} \tag{319}$$

where r_m is the mean radius of the cylinder. From the theory of partial differential equations [16] it is known that an equation of the form of equation (317) has a solution of the form

$$\bar{\phi}(x,r) = \sum_{n=1}^{\infty} \left[A_n \cos\left(2\pi n \frac{x}{l}\right) + B_n \sin\left(2\pi n \frac{x}{l}\right) \right] \times \left[C_n I_0\left(\beta 2\pi n \frac{r}{l}\right) + D_n K_0\left(\beta 2\pi n \frac{r}{l}\right) \right] \tag{320}$$

where I_0 and K_0 are the modified Bessel functions of order zero and A_n, B_n, C_n, and D_n are arbitrary constants. Using the boundary conditions (319) to evaluate the constants, we find

$$\bar{\phi} = \frac{V_\infty}{\beta} \sum_{n=1}^{\infty} \left[a_n \sin\left(2\pi n \frac{x}{l}\right) - b_n \cos\left(2\pi n \frac{x}{l}\right) \right] \frac{K_0\left(\beta 2n\pi \frac{r_m}{l}\right)}{K_1\left(\beta 2n\pi \frac{r_m}{l}\right)} \tag{321}$$

where K_0 and K_1 are the Bessel functions of order zero and one.

Evaluation of the Pressure Coefficient

The equation for the pressure in terms of the velocity in the case of irrotational flow with constant entropy and stagnation enthalpy has already been given (equation (310)). If the pressure is expressed in terms of the pressure coefficient, a particularly simple relationship exists in the case of flow of a perfect gas with small perturbations. Recall

$$2c_p(T_\infty - T) = V^2 - V_\infty^2 = [(V_\infty + u)^2 + v^2] - V_\infty^2$$

Hence,

$$\frac{T}{T_\infty} = 1 - \frac{V_\infty^2}{T_\infty}\left[\frac{2uV_\infty + u^2 + v^2}{2c_p V_\infty^2}\right] = 1 - \frac{\gamma - 1}{2} M_\infty^2 \left[\frac{2u}{V_\infty} + \frac{u^2 + v^2}{V_\infty^2}\right] \tag{322}$$

Also

$$c_p = \frac{p - p_\infty}{\frac{1}{2}\rho_\infty V_\infty^2} = \frac{\frac{p}{p_\infty} - 1}{\frac{1}{2}\gamma M_\infty^2} = \frac{\left(\frac{T}{T_\infty}\right)^{\gamma/(\gamma-1)} - 1}{\frac{1}{2}\gamma M_\infty^2} \tag{323}$$

Substituting equation (322) into equation (323) and omitting higher order terms, we have

$$c_p \cong -2\frac{u}{V_\infty} \tag{324}$$

Hence, the pressure itself is given by

$$p = \tfrac{1}{2}\rho_\infty V_\infty^2 c_p + p_\infty = -\rho_\infty V_\infty u + p_\infty \tag{325}$$

The Prandtl–Glauert Rule

Recall from equation (286)

$$u = \Phi_x \tag{286}$$

Consider now the incompressible flow over a slender profile with small perturbations and let the pressure coefficient for this case be designated by c_p'. Now for incompressible flow,

$$p - p_\infty = \tfrac{1}{2}\rho(V_\infty^2 - V^2) \tag{326}$$

where $V^2 = (V_\infty + u')^2 + v'^2$, the primes designating the perturbation velocities in the incompressible flow. Then

$$c_p' = \frac{p - p_\infty}{\frac{1}{2}\rho V_\infty^2} = \frac{\frac{1}{2}\rho(V_\infty^2 - V^2)}{\frac{1}{2}\rho V_\infty^2} = \frac{-2u'V_\infty - u'^2 - v'^2}{V_\infty^2} = -\frac{2u'}{V_\infty} \tag{327}$$

Thus, the same approximation may be used for the pressure coefficient in incompressible as in compressible flow with small perturbations.

Referring back to the transformation (285) and equations (286) with the boundary conditions (287) and (288) we have

$$u' = \Phi'_{x'} \tag{328}$$

and hence

$$c_p' = -2\frac{\Phi'_{x'}}{V_\infty} \tag{329}$$

Comparing then the c_p' for the incompressible flow of Figure 50 with the c_p for the compressible flow of Figure 49 we have

$$\frac{c_p}{c_p'} = \frac{\Phi_x}{\Phi'_{x'}} = \frac{1}{\beta^2} \tag{330}$$

Thus we may find the pressure coefficient on a slender profile in compressible flow directly from the pressure coefficient on a somewhat thinner profile in incompressible flow. Equation (330) is sometimes referred to as Gothert's rule.

It would be more convenient still if we could find a relation between the compressible and incompressible pressure coefficients on identical profiles. consider again the two profiles shown in Figure 49 and 50, and suppose now that both profiles are in incompressible flow. Let the pressure coefficient for the profile in Figure 50 be designated by c'_p as before, and let the pressure coefficient for the incompressible flow over the profile in Figure 49 be designated by c''_p. From the theory of incompressible flow we know that

$$\frac{c'_p}{c''_p} = \beta \tag{331}$$

where β is the ratio of the slopes of the two profiles, i.e.

$$\frac{dy'_s}{dx'_s} = \frac{1}{\beta}\frac{dy''_s}{dx''_s} \tag{332}$$

From equations (330) and (331) then, we have

$$\frac{c_p}{c''_p} = \frac{c_p}{c'_p} \cdot \frac{c'_p}{c''_p} = \frac{1}{\beta^2} \cdot \beta = \frac{1}{\beta} \tag{333}$$

This equation states that the pressure coefficients of the compressible and incompressible flow over the same profile are related by the equation

$$c_p = \frac{c''_p}{\sqrt{1 - M_\infty^2}} \tag{334}$$

This is called the Prandtl–Glauert rule.

Notice that equation (334) states that the effect of the compressibility is to increase the pressure change at a given point. The effect of neglecting the compressibility, therefore, and proceeding on the assumption of incompressible flow, is to underestimate the pressure coefficient at any given point. Moreover, this defect is seen to be proportional to the square root of $1 - M_\infty^2$. Again it must be emphasized that this is an approximate correction formula based on the assumption of small perturbations or, more precisely, the assumptions $M < 1, u/V_\infty, v/V_\infty \ll 1$. Nevertheless, the formula works well in many practical applications such as airplane wings and compressor blades.

Rayleigh–Janzen Method

So far we have restricted ourselves to the case of small perturbations. We wish now to consider a method which is not restricted to small perturbations. Consider the flow over a body which is not necessarily thin (Figure 55). Once again the flow is assumed to be uniform and parallel at infinity with velocity V_∞. We now propose a solution of the potential equation (270) of the form

$$\phi = V_\infty(\phi_0 + M_\infty^2 \phi_1 + M_\infty^4 \phi_2 + \ldots) \tag{335}$$

In short we are proposing a power series solution in powers of M_∞^2.

For convenience we may write the potential equation (270) in the form

$$\phi_{xx} + \phi_{yy} = \frac{V_\infty^2}{c^2}\left[\frac{u^2}{V_\infty^2}\phi_{xx} + 2\frac{uv}{V_\infty^2}\phi_{xy} + \frac{v^2}{V_\infty^2}\phi_{yy}\right] \tag{336}$$

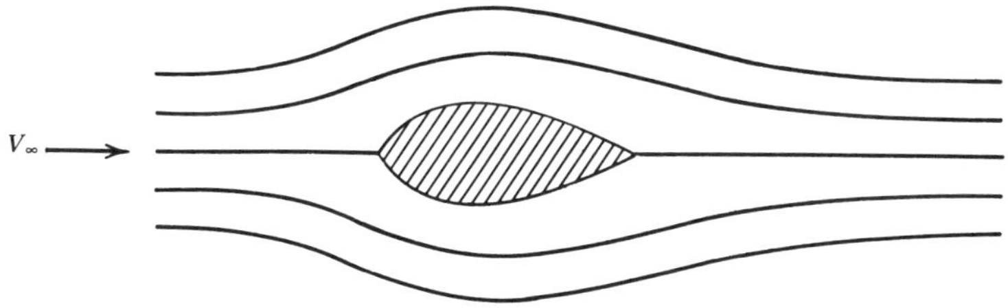

Figure 55. Nonslender body in a uniform stream.

where u and v now represent the actual, not perturbation, velocity components. Recall

$$c^2 + \frac{\gamma - 1}{2} V^2 = c_\infty^2 + \frac{\gamma - 1}{2} V_\infty^2 \tag{337}$$

This may be rewritten in the form

$$\frac{c^2}{c_\infty^2} = 1 - \frac{\gamma - 1}{2} M_\infty^2 \left[\frac{u^2 + v^2}{V_\infty^2} - 1 \right] \tag{338}$$

and hence

$$\frac{V_\infty^2}{c^2} = \frac{V_\infty^2}{c_\infty^2} \cdot \frac{c_\infty^2}{c^2} = \frac{M_\infty^2}{1 - \dfrac{\gamma - 1}{2} M_\infty^2 \left[\dfrac{u^2 + v^2}{V_\infty^2} - 1 \right]} \tag{339}$$

From equation (335),

$$\frac{u}{V_\infty} = \frac{1}{V_\infty} \phi_x = \phi_{0x} + M_\infty^2 \phi_{1x} + M_\infty^4 \phi_{2x} + \dots \tag{340}$$

$$\frac{v}{V_\infty} = \frac{1}{V_\infty} \phi_y = \phi_{0y} + M_\infty^2 \phi_{1y} + M_\infty^4 \phi_{2y} + \dots \tag{341}$$

Substituting equations (339), (340), and (341) into (336) and equating powers of M_∞^2, we get

$$\phi_{0xx} + \phi_{0yy} = 0 \tag{342}$$

$$\phi_{1xx} + \phi_{1yy} = \phi_{0x}^2 \phi_{0xx} + 2\phi_{0y}\phi_{0x}\phi_{0xy} + \phi_{0y}^2 \phi_{0yy} \tag{343}$$

and so on; or, in general

$$\phi_{mxx} + \phi_{myy} = f(\phi_0, \phi_1, \dots, \phi_{m-1}) \tag{344}$$

Since in incompressible flow, $c_\infty = \infty$ and hence $M_\infty = 0$, we are taking perturbations of the incompressible flow due to the compressibility. It follows, therefore, that ϕ_0 must satisfy the Laplace equation, since it represents the solution where $M_\infty = 0$; that is, the solution in the incompressible case. Once the Laplace equation has been solved for the given set of boundary conditions, the succeeding equations (344) may be solved sequentially. The smaller the Mach number, the more rapidly the series (335) will converge. The higher order terms, $\phi_1, \phi_2, \ldots$ represent corrections for the compressibility.

The boundary conditions are, at infinity,

$$\phi_x = V_\infty, \qquad \phi_y = 0$$

Hence

$$\begin{aligned} \phi_{0x} &= 1, \qquad \phi_{mx} = 0, \qquad m \geq 1 \\ \phi_{0y} &= 0, \qquad \phi_{my} = 0, \qquad m \geq 1 \end{aligned}$$

At the body surface, the component of the velocity normal to the surface must be zero. Hence, if n is the direction normal to the surface,

$$\phi_n = 0$$

or

$$\phi_{0n} = 0, \qquad \phi_{mn} = 0, \qquad m \geq 1$$

In many cases, the series converges quite rapidly. For example, for the flow over a cylinder at $M_\infty = 0.4$, the first four terms of the series (335) give the solution to within 1% accuracy for the inviscid case.

Rayleigh–Ritz Method

The Rayleigh–Ritz method falls under the more general heading of variational techniques. A brief review of the fundamentals of variational calculus is, therefore, in order.

Consider the integral

$$J(y) = \int_{x_0}^{x_1} F(x, y, y')dx, \qquad y' = \frac{dy}{dx} \tag{345}$$

$x_0, x_1, y(x_0), y(x_1)$ are given. It is desired to find a function $y(x)$ which will satisfy the boundary conditions and for which $J(y)$ will be a minimum. Let the desired function be $y = f(x)$ and let $\eta(x)$ be an arbitrary function of x which, however, satisfies the conditions $\eta(x_0) = \eta(x_1) = 0$. Then in general we may say that

$$y(x) = f(x) + \varepsilon\eta(x)$$

where ε is a parameter. Thus if ε is sufficiently small [and $\eta(x)$ is bounded in $x_0 < x < x_1$] $|f(x) - y(x)|$ will be arbitrarily small in the interval $x_0 < x < x_1$. Hence,

$$J(y) = J(f + \varepsilon\eta) \equiv \Phi(\varepsilon)$$

has a minimum at $\varepsilon = 0$, and $\Phi'(0) = 0$. Substituting into (345),

$$\Phi(\varepsilon) = \int_{x_0}^{x_1} F(x_1 f + \varepsilon\eta, \quad f' + \varepsilon\eta')dx$$

and

$$\Phi'(\varepsilon) = \int_{x_0}^{x_1} (F_f \eta + F_{f'} \eta')\, dx = \Phi'(0) = 0$$

Integrating by parts,

$$\Phi'(\varepsilon) = \int_{x_0}^{x_1} F_f \eta \, dx + F_f \eta \Big|_{x_0}^{x_1} - \int_{x_0}^{x_1} \eta \frac{dF_{f'}}{dx} dx$$

But

$$F_f \eta \Big|_{x_0}^{x_1} = 0$$

Hence

$$\Phi'(0) = \int_{x_0}^{x_1} \left(F_f - \frac{dF_{f'}}{dx} \right) \eta \, dx \tag{346}$$

Equation (346) will be satisfied if

$$F_f - \frac{dF_{f'}}{dx} = 0 \tag{347}$$

We now introduce the notation

$$\varepsilon \eta = \delta y$$

where δy is called the variation of $y(x)$. Also, let

$$\delta J = \varepsilon \Phi'(0) = \varepsilon \int_{x_0}^{x_1} (\eta F_y + \eta' F_{y'}) dx$$

$$= \varepsilon \int_{x_0}^{x_1} \eta \left(F_y - \frac{d}{dx} F_{y'} \right) dx$$

δJ is called the first variation of the integral J. It follows that if the first variation of J is zero, J will be an extremum.

Extension to Three Dimensions

We seek the function $\phi(x, y, z)$, which will satisfy certain boundary conditions and which will make the integral

$$J = \int_{\mathscr{V}} F(x, y, z, \phi, \phi_x, \phi_y, \phi_z)\, dx\, dy\, dz$$

an extremum (maximum, minimum, or saddle point). Let $\phi = f(x, y, z)$ be the function we seek. Let $\eta(x, y, z)$ be an arbitrary function which vanishes on the boundary of the volume $\mathscr{V}$, and let

$$\phi = f + \varepsilon \eta$$

where ε is a parameter. Then

$$J(\phi) = J(f + \varepsilon \eta) = \Phi(\varepsilon)$$

has an extremum at $\varepsilon = 0$, and hence $\Phi'(0) = 0$.

$$\Phi(\varepsilon) = \int_{\mathscr{V}} F(x, y, z, f + \varepsilon\eta, f_x + \varepsilon\eta_x, f_y + \varepsilon\eta_y, f_z + \varepsilon\eta_z)\, d\mathscr{V}$$

and

$$\Phi'(\varepsilon) = \int_{\mathscr{V}} (F_\phi \eta + F_{\phi_x}\eta_x + F_{\phi_y}\eta_y + F_{\phi_z}\eta_z)\, d\mathscr{V} \tag{348}$$

Now according to Gauss's integral theorem,

$$\int_{\mathscr{V}} (F_\phi \eta + F_{\phi_x}\eta_x + F_{\phi_y}\eta_y + F_{\phi_z}\eta_z)\, d\mathscr{V} = \int_{\mathscr{V}} \eta(F_{\phi_x} x + F_{\phi_y} y + F_{\phi_z} z)\, dx\, dy\, dz$$

since η vanishes on the boundary of $\mathscr{V}$. Now let

$$\delta\phi = \varepsilon\eta \tag{350}$$

$\delta\phi$ is called the variation of ϕ.
Let

$$\delta J = \varepsilon \int_{\mathscr{V}} (F_\phi \eta + F_{\phi_x}\eta_x + F_{\phi_y}\eta_y + F_{\phi_z}\eta_z)\, d\mathscr{V} \tag{351}$$

Since ε is not a function of x, y, or z, this may also be written

$$\delta J = \int_{\mathscr{V}} [F_\phi \varepsilon\eta + F_{\phi x}(\varepsilon\eta)_x + F_{\phi y}(\varepsilon\eta)_y + F_{\phi z}(\varepsilon\eta)_z]\, d\mathscr{V} \tag{352}$$

or, using equation (350),

$$\delta J = \int_{\mathscr{V}} [F_\phi \delta\phi + F_{\phi_x}(\delta\phi)_x + F_{\phi_y}(\delta\phi)_y + F_{\phi_z}(\delta\phi)_z]\, d\mathscr{V} \tag{353}$$

δJ is called the first variation of J. Thus, if the first variation of J vanishes the integral J will be an extremum.

Application to Compressible Flow

It was Bateman [17] who first observed that for the steady, irrotational, inviscid flow of a perfect gas inside a closed volume,

$$J = \int_{\mathscr{V}} p(\phi)\, d\mathscr{V} \tag{354}$$

is an extremum. In the present case

$$p = C_1 (V_{max}^2 - V^2)^{\gamma/(\gamma - 1)} \tag{355}$$

where C_1 is a constant and $V_{\max}$ is the velocity when $T = 0$. Also, since for isentropic flow,

$$\rho \sim p^{1/\gamma} \tag{356}$$

we may say

$$\rho = C_2 (V_{\max}^2 - V^2)^{1/(\gamma - 1)} \tag{357}$$

Now since

$$V^2 = \sum_i \phi_{x_i}^2, \qquad i = 1,2,3 \tag{358}$$

where $x_1 = x, x_2 = y, x_3 = z$, the integral

$$J = \int_{\mathscr{V}} p(\phi) d\mathscr{V} \tag{359}$$

will be an extremum if

$$\delta J = \delta \int_{\mathscr{V}} \left(V_{\max}^2 - \sum_i \phi_{x_i}^2 \right)^{\gamma/(\gamma-1)} d\mathscr{V} = 0 \tag{360}$$

Taking the first variation of J and making use of equation (356), (357), and (358), and keeping in mind that $\varepsilon\eta = \delta\phi$, we get

$$\delta J = C_3 \int_{\mathscr{V}} \rho \sum_i \phi_{x_i} (\delta\phi)_{x_i} d\mathscr{V} \tag{361}$$

where C_3 is a constant.

Equation (359) may be rewritten

$$\delta J = \int_{\mathscr{V}} \sum_i [(\rho\,\delta\phi\,\phi_{x_i})_{x_i} - \delta\phi(\rho\phi_{x_i})_{x_i}] d\mathscr{V} \tag{362}$$

From Gauss's Theorem

$$\int_{\mathscr{V}} \sum_i (\rho\delta\phi\phi_{x_i})_{x_i} = -\int_S \rho\delta\phi\phi_n dS \tag{363}$$

where ϕ_n is the derivative of ϕ normal to the surface S. If S is a solid boundary, $\phi_n = 0$ and (362) becomes

$$\delta J = \int_{\mathscr{V}} \delta\phi \sum_i (\rho\phi_{x_i})_{x_i} d\mathscr{V}$$

But if ϕ is the velocity potential,

$$\sum_i (\rho\phi_{x_i})_{x_i} = 0$$

is the continuity equation for steady flow; hence, $\delta J = 0$ and

$$\int_{\mathscr{V}} p(\phi) d\mathscr{V}$$

is an extremum when ϕ is the velocity potential. q.e.d. The problem of finding ϕ for the steady, inviscid, irrotational flow inside a closed container can, therefore, be approached as a problem in variational calculus.

Method of Solution of the Variational Problem

The method which was first used by Lord Rayleigh [18] and later formalized by Ritz [19] is based on the following considerations. Given a function

$$J = \int_{x_0}^{x_1} F(x, y, y', \ldots, y^k)\, dx \tag{364}$$

where F is a function of $x, y(x)$, and the derivatives of $y(x)$ up to the kth order, with the boundary conditions on $y(x)$ at $x = x_0$ and $x = x_1$ specified, we seek the function $y(x)$, which will satisfy the b.c.'s and for which J will be an extremum, that is, a maximum, minimum, or saddle point. Let us express the function $y(x)$ in the form

$$y(x) = \sum_{i=1}^{n} a_i \phi_i(x) \tag{365}$$

where the $\phi_i(x)$ are continuously differentiable and the expression $\sum_{i=1}^{n} a_i \phi_i(x)$ satisfies the given boundary conditions at $x = x_0$ and $x = x_1$. The functions $\phi_i(x)$ are otherwise arbitrary. Having chosen suitable functions $\phi_i(x)$ we substitute in equation (364) and integrate. The resulting integral, $J(a_1, a_2, \dots a_n)$ will then have an extremum when

$$\frac{\partial J}{\partial a_1} = \frac{\partial J}{\partial a_2} = \dots = \frac{\partial J}{\partial a_n} = 0 \tag{366}$$

The equations (366) may be solved simultaneously for the requisite coefficients. If the function J has an extremum, $\bar{J}$ say, and if the series $\sum_{i=1}^{n} a_i \phi_i(x)$ converges as $n \to \infty$, it may be shown that for every $\varepsilon > 0$ there exists a δ such that whenever $n > \delta$,

$$|J_n - \bar{J}| < \varepsilon$$

where

$$J_n = \int_{x_0}^{x_1} F(x, y_n, y_n', \dots, y_n^k)dx, \qquad y_n = \sum_{i=1}^{n} a_i \phi_i(x)$$

and where the a_i satisfy equations (366). That is,

$$\lim_{n \to \infty} J_n = \bar{J} \tag{367}$$

Consider now a variational problem

$$I = \int_{x_0}^{x_1} F(x, \phi, \phi')\, dx \tag{368}$$

We seek the function ϕ for which I will be an extremum. Let

$$\phi = \sum_i A_i \phi_i(x), \qquad \delta\phi = \sum_i \delta A_i \phi_i(x) \tag{369}$$

where δA_i represents some arbitrary perturbation on A_i. Then J will be an extremum when

$$\delta I = \sum_i \frac{\partial I}{\partial A_i} \delta A_i = 0 \tag{370}$$

for arbitrary δA_i; this is, when $\partial I / \partial A_i = 0$ for all i.

Now

$$\delta I = \int_{x_0}^{x_1} \sum_i [F_\phi \phi_i + F_{\phi'} \phi_i'] \, dx \, \delta A_i$$

$$= \int_{x_0}^{x_1} \left[F_\phi \sum_i \phi_i \delta A_i + F_{\phi'} \frac{\partial}{\partial x} \sum_i \phi_i \delta A_i \right] dx$$

$$= \int_{x_0}^{x_1} \left[F_\phi \delta\phi + F_{\phi'} \frac{\partial}{\partial x} \delta\phi \right] dx \tag{371}$$

Thus, with ϕ and $\delta\phi$ as defined by equation (359), δI is the first variation of I, and if δI vanishes I will be an extremum. When $\phi = \phi(x, y, z)$ this result extends naturally to equation (353).

A full discussion of the Rayleigh–Ritz method is beyond the scope of this work. For such questions as whether the extremum will be a maximum, minimum, or saddle point, whether or not the series (365) will converge for a given set of function $\phi_i(x)$, and if so how fast, and so on, the reader is referred to books on variational calculus.

The utility of the Rayleigh–Ritz method lies in the fact that while the function $y(x)$ which will make J an extremum may be complicated, it is often possible to approximate it by a series of simple functions, analogous to Fourier or Taylor series.

The Rayleigh–Ritz method is not restricted to single integrals. It works equally well on multiple integrals, in particular, surface and volume integrals. Thus we may have

$$J = \int_{\mathscr{V}} F(x, y, z, \phi(x, y, z), \phi_x, \phi_y, \phi_z, \ldots) \, dx \, dy \, dz$$

We seek the function ϕ which will make J an extremum. We then postulate

$$\phi = \sum_{i=1}^{n} a_i \phi_i(x, y, z)$$

and proceed as before.

Wang [20] has extended the work of Bateman to the infinite domain. Thus, while Bateman essentially considered the flow of a gas inside a body, Wang considered the flow of a gas outside the body, as, for instance, the flow over a body in a uniform stream.

Bateman's integral

$$\int_{\mathscr{V}} p(\phi) \, d\mathscr{V}$$

becomes infinite as $\mathscr{V} \to \infty$. Thus, the variational problem becomes meaningless. Wang surmounted this difficulty by dividing the velocity potential into two parts:

$$\phi = \phi_1 + \phi_2$$

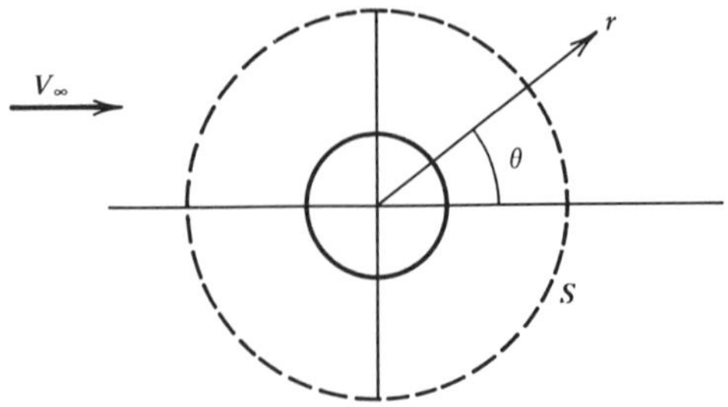

Figure 55(a) Coordinate system and nomenclature for flow over a circular cylinder.

where ϕ_1 is the incompressible velocity potential for the given problem and ϕ_2 the correction due to compressibility. ϕ_1 is presumed known, and, therefore, fixed. The variational problem

$$\int_{\mathscr{V}} p(\phi)\,d\mathscr{V} = \max \tag{372}$$

is, therefore, equivalent to

$$\int_{\mathscr{V}} [p(\phi) - p(\phi_1)]\,d\mathscr{V} = \max \tag{373}$$

The limit of the integral (373) as $\mathscr{V} \to \infty$ is, however, finite. Thus, the problem becomes amenable to solution.

Example: Flow over a Circular Cylinder

We consider first the flow between the surface of the cylinder and a cylindrical surface S around the cylinder and at a great distance from it. On this surface, the boundary condition becomes

$$\frac{\partial \phi}{\partial n} = -\frac{\partial \phi}{\partial r} = -V_\infty \cos\theta + 0\left(\frac{1}{r^2}\right) \tag{374}$$

We now postulate a solution of the form

$$\phi = \phi_1 + \frac{1}{r}\sum_{n=1}^{N} (A_{1n}\cos n\theta + B_{1n}\sin n\theta) + 0\left(\frac{1}{r^2}\right) \tag{375}$$

and set

$$\delta\phi = \frac{1}{r}\sum_{n=1}^{N} (\delta A_{1n}\cos n\theta + \delta B_{1n}\sin n\theta) + 0\left(\frac{1}{r^2}\right) \tag{376}$$

The first variation of the integral

$$\int_{\mathscr{V}} [p(\phi) - p(\phi_1)]\,d\mathscr{V} \tag{377}$$

will then be given by

$$\delta J = -\int_S \rho\,\delta\phi \frac{\partial \phi}{\partial n}\,dS - \int_{\mathscr{V}} \delta\phi \nabla\cdot(\rho\nabla\phi)\,d\mathscr{V} \tag{378}$$

where ϕ and $\delta\phi$ are given by equations (375) and (376), respectively. This is the same as the first variation of the integral

$$\int_{\mathscr{V}} p(\phi)\, d\mathscr{V}$$

because ϕ_1 is presumed known and hence

$$\int_{\mathscr{V}} p(\phi_1)\, d\mathscr{V}$$

is constant. For isentropic flow in the present case we also have

$$\rho = \left(\frac{\gamma - 1}{2\gamma}\right)^{1/(\gamma-1)} \left[V_{\max}^2 - \phi_r^2 - \frac{1}{r}\phi_\theta^2\right]^{1/(\gamma-1)} \tag{379}$$

For large r it is convenient to rewrite this as

$$\rho = \left(\frac{\gamma - 1}{2\gamma}\right)^{1/(\gamma-1)} \left[V_{\max}^2 - V_\infty^2 + O\left(\frac{1}{r^2}\right)\right] \tag{380}$$

Substituting (375), (376), and (380) into equation (378) and integrating, we get

$$\delta J = \int_{\theta=0}^{2\pi} \rho\,\delta\phi \frac{\partial \phi}{\partial n} r\, d\theta = c_1 (V_{\max}^2 - V_\infty^2) V_\infty \pi \delta A_{11} \tag{381}$$

in the limit as $r \to \infty$, where c_1 is a constant depending on γ. Thus the first variation of equation (377) is not necessarily zero. It follows that the velocity potential in the present case will not necessarily satisfy the criterion which assumes that

$$\int_{\mathscr{V}} [p(\phi) - p(\phi_1)]\, d\mathscr{V}$$

will be a maximum. This is due to the difference in boundary conditions. The velocity potential will, however, satisfy the criterion that

$$J = \int_{\mathscr{V}} [p(\phi) - p(\phi_1)]\, d\mathscr{V} + K \tag{382}$$

be a maximum where

$$K = c_1 (V_{\max}^2 - V_\infty^2) V_\infty \pi A_{11}$$

The value of K obviously depends on the form of the assumed solution. That is, had we assumed a different form for ϕ and $\delta\phi$ from equations (375) and (376), we would have gotten a different expression for K.

The velocity potential ϕ_1 for incompressible flow about a cylinder is given by

$$\phi_1 = V_\infty \left[r + \frac{b^2}{r}\right] \cos\theta - \frac{\kappa}{2\pi}\theta$$

where b is the radius of the cylinder and κ the circulation. We now postulate $\phi_2 = \phi - \phi_1$ in the form

$$\phi_2 = \sum_{m=1}^{M} \sum_{n=1}^{n} [A_{mn} R_m(r) \cos n\theta + B_{mn} R_m(r) \sin n\theta] \tag{383}$$

where

$$R_m(r) = \frac{b^{m+1}}{mr^m} - \frac{b^{m+3}}{(m+2)r^{m+2}} \tag{384}$$

The assumed form of the solution is seen to satisfy the boundary conditions

$$\begin{aligned} r = \infty: \quad & \frac{\partial \phi}{\partial r} = V_\infty \cos\theta, \qquad \frac{1}{r}\frac{\partial \phi}{\partial \theta} = -V_\infty \sin\theta \\ r = b: \quad & \frac{\partial \phi}{\partial n} = 0 \end{aligned} \tag{385}$$

For the case $\kappa = 0$, B_{nm} must be zero since the flow will be symmetric. Taking the radius of the cylinder as 1, which we may do without loss of generality, we then have

$$\begin{aligned} \phi = {} & V_\infty\left(r + \frac{1}{r}\right)\cos\theta + A_{11}\left(\frac{1}{r} - \frac{1}{3r^3}\right)\cos\theta \\ & + A_{13}\left(\frac{1}{r} + \frac{1}{3r^3}\right)\cos 3\theta + A_{31}\left(\frac{1}{3r^3} - \frac{1}{5r^5}\right)\cos\theta + \ldots \end{aligned} \tag{386}$$

Substituting in equation (382), integrating, and differentiating with respect to each coefficient A_{mn} in turn in accordance with the Rayleigh–Ritz method, we arrive at a system of simultaneous equations for the A_{mn}. Wang has carried out the numerical solution for the first six coefficients. Compared to the experimental results of Taylor [21], the results are found to give an accuracy of 5% or better. It might be noted, however, that the Rayleigh–Janzen method carried out to the third approximation, gives slightly better results in the present case. The computations involved are quite lengthy for both methods. The Rayleigh–Ritz method may be extended to more complicated shapes through use of conformal transformation [22].

Relaxation Method

The relaxation method is a numerical, trial-and-error method and may be used to solve for either the stream function or the velocity potential with more or less arbitrary boundary conditions. The method may be applied to supersonic and transonic flow as well as subsonic flow. In the present chapter, however, we limit ourselves to subsonic flow.

Relaxaton method using the stream function

Let us take as an example the steady, adiabatic, irrotational, two-dimensional, subsonic flow over a body in a uniform stream. Recall the continuity equation and the irrotationality condition

$$\nabla \cdot (\rho \mathbf{V}) = 0 \tag{387}$$

$$\nabla \times \mathbf{V} = 0 \tag{388}$$

Define now the stream function ψ by the equations

$$\frac{\rho u}{\rho_0} = \psi_y, \qquad \frac{\rho v}{\rho_0} = -\psi_x \tag{389}$$

where the subscript 0 denotes the stagnation conditions. It will be observed that the continuity equation is automatically satisfied by this definition of ψ in the two-dimensional, steady case. In the two-dimensional case, the irrotationality condition becomes

$$v_x - u_y = 0 \tag{390}$$

Substituting equation (389) for ψ into equation (390), we get

$$\psi_{xx} + \psi_{yy} - \psi_x\left(\ln\frac{\rho}{\rho_0}\right)_x - \psi_y\left(\ln\frac{\rho}{\rho_0}\right)_y = 0 \tag{391}$$

We now propose to solve this equation by the method of finite differences. To do so we set up a grid of points in the flow field as shown in Figure 56. We now pick

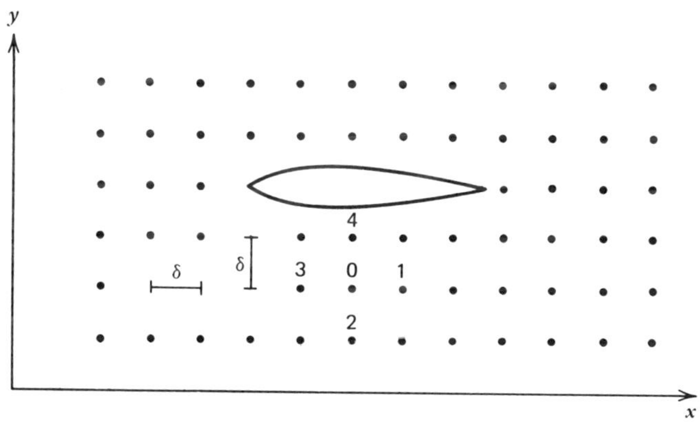

Figure 56. Grid for flow calculation by relaxation method.

any five adjacent points, 0, 1, 2, 3, 4 say, as shown in the figure. The points must, however, be located with respect to each other in the manner shown. Let the distance between adjacent points be δ. Then we may approximate the derivatives of ψ at the point 0 as follows:

$$\psi_x \cong \frac{\psi_1 - \psi_0}{\delta} \cong \frac{\psi_1 - \psi_3}{2\delta}; \qquad \psi_y \cong \frac{\psi_4 - \psi_0}{\delta} \cong \frac{\psi_4 - \psi_2}{2\delta}$$

$$\psi_{xx} \cong \frac{1}{\delta}\left[\frac{\psi_1 - \psi_0}{\delta} - \frac{\psi_0 - \psi_3}{\delta}\right] = \frac{\psi_1 - 2\psi_0 + \psi_3}{\delta^2} \tag{392}$$

$$\psi_{yy} \cong \frac{1}{\delta}\left[\frac{\psi_4 - \psi_0}{\delta} - \frac{\psi_0 - \psi_2}{\delta}\right] = \frac{\psi_4 - 2\psi_0 + \psi_2}{\delta^2}$$

The derivatives of $\ln(\rho/\rho_0)$ may be approximated by

$$\left[\ln\frac{\rho}{\rho_0}\right]_x = \frac{\ln\left(\dfrac{\rho}{\rho_0}\right)_1 - \ln\left(\dfrac{\rho}{\rho_0}\right)_3}{2\delta}$$

$$\left[\ln\frac{\rho}{\rho_0}\right]_y = \frac{\ln\left(\dfrac{\rho}{\rho_0}\right)_4 - \ln\left(\dfrac{\rho}{\rho_0}\right)_2}{2\delta} \tag{393}$$

Before we can apply the relaxation method, we must express the density in terms of ψ. Since the flow is adiabatic and isentropic, we have

$$\frac{\rho}{\rho_0} = \left[1 - \frac{\gamma-1}{2}\frac{V^2}{c_0^2}\right]^{1/(\gamma-1)} = \left[1 - \frac{\gamma-1}{2}\frac{\left(\dfrac{\rho_0}{\rho}\psi_y\right)^2 + \left(\dfrac{\rho_0}{\rho}\psi_x\right)^2}{c_0^2}\right]^{1/(\gamma-1)} \tag{394}$$

where c_0 is the stagnation speed of sound. Both ρ_0 and c_0 are presumed known. The density may thus be found as a function of the velocity. For the purposes of the present method, it is best to solve equation (394) in tabular or graphical form, plotting or listing ρ/ρ_0 versus V^2/c_0^2. Substituting equations (392) and (393) into equation (391) we get the finite difference form of the differential equation

$$\psi_1 + \psi_2 + \psi_3 + \psi_4 - 4\psi_0 - \tfrac{1}{4}(\psi_1 - \psi_3)\left[\ln\left(\frac{\rho}{\rho_0}\right)_1 - \ln\left(\frac{\rho}{\rho_0}\right)_3\right]$$

$$- \tfrac{1}{4}(\psi_4 - \psi_2)\left[\ln\left(\frac{\rho}{\rho_0}\right)_4 - \ln\left(\frac{\rho}{\rho_0}\right)_2\right] = 0 \tag{395}$$

We now begin by guessing the value of ψ at every point of the grid of Figure 56. This is not as hard as it might seem. If the incompressible solution is known, this provides a good starting point. Even if the incompressible solution is not known, the shape of the streamlines may be approximated given the boundary conditions. At the surface of the body, the velocity must be tangent to the surface. Far from the body, the velocity must approach the free stream value. Once the value of ψ at each point has been guessed, the corresponding values of ψ_x, ψ_y, and ρ are found and the results substituted into equation (395). If the right-hand side of the equation (395) is not zero for each set of points, the values of ψ_i are adjusted until equation (395) is satisfied at every point. This process is referred to as "relaxation," in the sense of taking the tension out of a net which has been pulled askew. Although this method was originally developed in the thirties and forties [23–26], the advent of the modern high-speed computer has made it far more practical than it was originally. Obviously, the finer the grid is taken, the smaller δ, the more accurate the solution will become but the more work will be involved (or more computer time).

Relaxation method using the velocity potential

The equation for the velocity potential may be written in the form

$$\phi_{xx} + \phi_{yy} = \frac{1}{c^2}\phi_x^2\phi_{xx} + \frac{1}{c^2}\phi_y^2\phi_{yy} - \frac{1}{c^2}\phi_x\phi_y 2\phi_{xy} \tag{396}$$

where

$$c^2 = c_0^2 - \frac{\gamma - 1}{2}V^2 = c_0^2 - \frac{\gamma - 1}{2}(\phi_x^2 + \phi_y^2) \tag{397}$$

Referring once again to Figure 56, we have,

$$\phi_x \cong \frac{\phi_3 - \phi_0}{\delta} \cong \frac{\phi_3 - \phi_1}{2\delta}; \qquad \phi_y \cong \frac{\phi_4 - \phi_0}{\delta} \cong \frac{\phi_4 - \phi_2}{2\delta}$$

$$\phi_{xx} \cong \frac{(\phi_3 - \phi_0) - (\phi_0 - \phi_1)}{\delta^2} = \frac{\phi_3 - 2\phi_0 + \phi_1}{\delta^2}$$

$$\phi_{yy} \cong \frac{(\phi_4 - \phi_0) - (\phi_0 - \phi_2)}{\delta^2} = \frac{\phi_4 - 2\phi_0 + \phi_2}{\delta^2} \tag{398}$$

In order to express the cross derivatives in finite difference form we must take

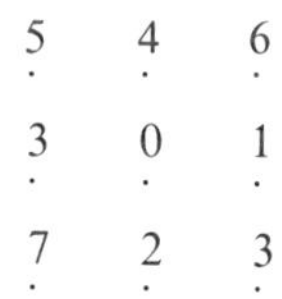

Figure 57. Grid points for computing second-order derivatives.

in two or more additional points, as show in Figure 57. Then

$$\phi_{xy} = \phi_{yx} \cong \frac{\dfrac{\phi_5 - \phi_6}{2\delta} - \dfrac{\phi_7 - \phi_8}{2\delta}}{2\delta} \cong \frac{\dfrac{\phi_6 - \phi_8}{2\delta} - \dfrac{\phi_5 - \phi_7}{2\delta}}{2\delta} \tag{399}$$

Substitution of equations (398) and (399) into equations (396) and (397) once again yields the equations in finite difference form:

$$\phi_1 + \phi_3 + \phi_4 + \phi_5 - 4\phi_0 - \frac{1}{4c^2\delta^4}[(\phi_3 - \phi_1)^2(\phi_3 - 2\phi_0 - \phi_1)$$
$$+ (\phi_4 - \phi_2)^2(\phi_4 - 2\phi_0 + \phi_2) - \tfrac{1}{2}(\phi_3 - \phi_1)(\phi_4 - \phi_2)(\phi_5 - \phi_6 - \phi_7 + \phi_8)] \tag{400}$$

The procedure is the same as when using the stream function. We guess at the value of ϕ at each point of the grid, substitute in equation (400) and adjust the values of each point until the equation everywhere is satisfied.

The boundary conditions are the same as before. The speed of sound c is found by putting equation (397) in finite difference form:

$$c^2 = c_0^2 - \frac{\gamma - 1}{2}\left[\left(\frac{\phi_3 - \phi_1}{2\delta}\right)^2 + \left(\frac{\phi_4 - \phi_2}{2\delta}\right)^2\right] \tag{401}$$

Once again the incompressible solution may be used as a starting point if it is known.

Comparison of results with other methods and experiment

Emmons [27] applied the relaxation method to computing the flow over an airfoil (NACA 0012) and compared the results to those obtained by other calculation techniques. Amick [28] gave experimental results for the same airfoil. Figure 58 shows the comparison between the pressure coefficient

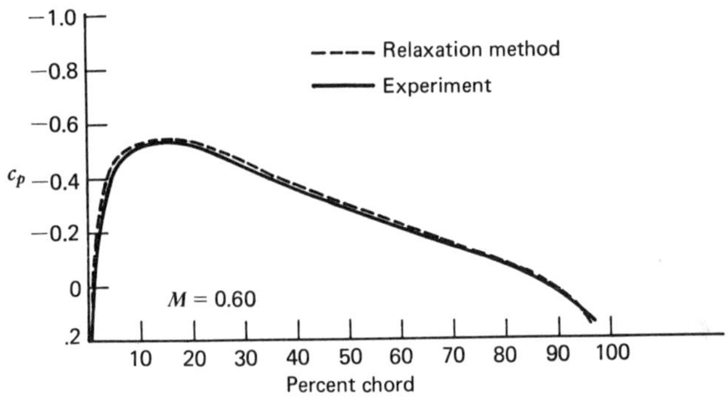

Figure 58. Pressure coefficient from relaxation method and experiment over NACA 0012 airfoil [28].

obtained by relaxation theory and experiment for the airfoil in a wind tunnel at a free stream Mach number of 0.60. Agreement is seen to be within about 5%. For this airfoil, 0.60 is below the critical Mach number. Thus, the flow is subsonic throughout. Figure 59 shows the results of experiment, relaxation, and the Karman–Tsien approximation (see below) for the same airfoil at $M_\infty = 0.7$.

The von Karman–Tsien method is seen to agree more closely with experiment than the relaxation method. A modified relaxation method corrected for tunnel constriction gives better results.

As the free stream Mach number rises the results of the various methods diverge. This is shown in Figure 60 where c_p as found by various methods is plotted against free stream Mach number for the same airfoil at 20% chord. In the present case, it is seen that while the Karman–Tsien and linearized theories give results which are too low, the relaxation method gives results which are too high.

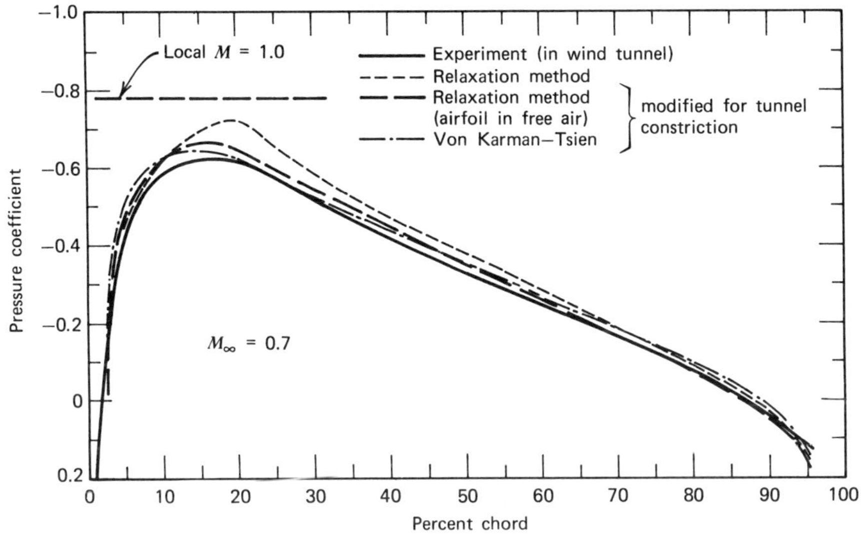

Figure 59. Pressure coefficient computed by various theories over NACA 0012 airfoil [28].

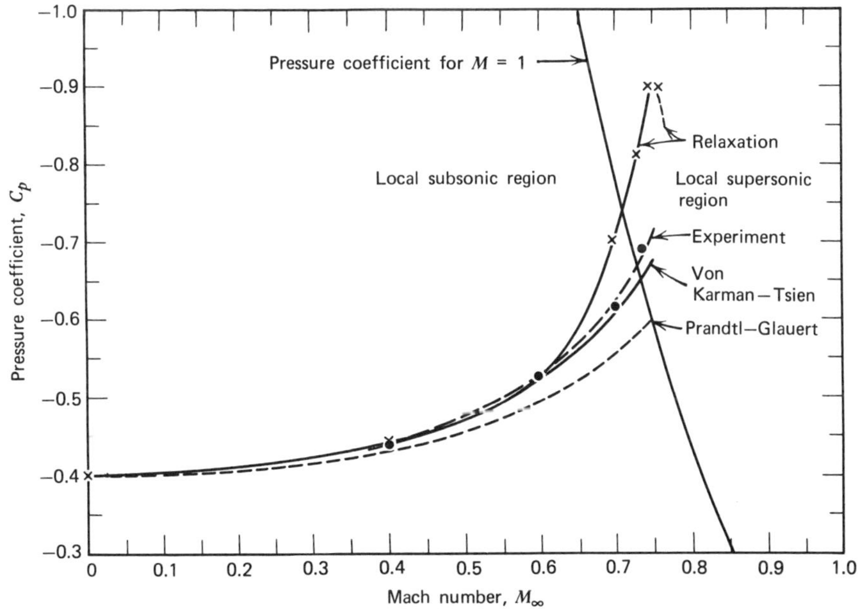

Figure 60. Pressure coefficient at 20% chord versus free stream Mach number computed by various methods [27].

The Stream Function Equation

Even in the case of irrotational flow, where a velocity potential exists, the equation for the velocity potential is highly nonlinear and, therefore, difficult to solve. Most of the methods of solving two- and three-dimensional problems which we have discussed so far attempt to find solutions for the velocity potential. In general, any attempt to find higher order approximations to the velocity potential quickly becomes very tedious. In the previous section, we introduced the concept of the stream function. Just as a knowledge of the velocity potential represents a complete solution to a given problem in irrotational isentropic flow, so does a knowledge of the stream function. It is natural to inquire, therefore, if anything is to be gained in general by seeking solutions to the stream function rather than the velocity potential.

We have already derived two equations, (391) and (394), which combined yield a single equation for the stream function. This equation also appears to be highly nonlinear. Further insight into this question may be gained by deriving an equation for ψ which is similar in form to the velocity potential equation (270). Substituting ψ into the irrotationality condition for two-dimensional flow we may write

$$\sigma(\psi_{xx} + \psi_{yy}) = \rho_x \psi_x + \rho_y \psi_y \tag{402}$$

Now

$$d\rho = \frac{d\rho}{dp}dp = -\frac{1}{c^2}\rho d(\tfrac{1}{2}V^2) = -\frac{1}{c^2}\rho d\left(\frac{1}{2}\left(\frac{\rho_0}{\rho}\right)^2(\psi_x^2 + \psi_y^2)\right)$$

Hence,

$$\rho_{x_i} = -\frac{\rho}{2c^2}\left[\left(\frac{\rho_0}{\rho}\right)^2(\psi_x^2 + \psi_y^2)\right]_{x_i}, \qquad i = 1, 2 \quad x_1 = x \quad x_2 = y$$

Solving for ρ_{x_i}, we get

$$\rho_{x_i} = -\frac{\dfrac{1}{\rho}\left(\dfrac{\rho_0}{c}\right)^2(\psi_x^2 + \psi_y^2)_{x_i}}{1 - \dfrac{1}{c^2}\left(\dfrac{\rho_0}{c}\right)^2(\psi_x^2 + \psi_y^2)} \tag{403}$$

Substituting equation (403) into equation (402), we get

$$\left[1 - \left(\frac{\rho_0}{\rho}\right)^2 \frac{\psi_y}{c^2}\right]\psi_{xx} + \left[1 - \left(\frac{\rho_0}{\rho}\right)^2 \frac{\psi_x^2}{c^2}\right]\psi_{yy} + 2\left(\frac{\rho_0}{\rho}\right)^2 \frac{\psi_x \psi_y}{c^2}\psi_{xy} = 0 \tag{404}$$

where

$$c^2 = c_0^2 - \frac{\gamma - 1}{2}\left(\frac{\rho_0}{\rho}\right)(\psi_x^2 + \psi_y^2) \tag{405}$$

and

$$\frac{\rho_0}{\rho} = \left[1 + \frac{\gamma - 1}{2}\left(\frac{\rho_0}{\rho}\right)^2 \frac{\psi_x^2 + \psi_y^2}{c^2}\right]^{1/(\gamma - 1)} \tag{406}$$

Equation (404) is analogous to the equation for the velocity potential. Thus we see that nothing has been gained by writing the equation in terms of the stream function. If anything, the stream function equation is even more nonlinear than the velocity potential equation. This does not mean, however, that the stream function concept is useless in compressible flow. As will be seen in the next section, a simpler equation in ψ may be obtained by changing the independent variables from x and y to V and θ.

Hodograph Method

The hodograph method involves solving a problem using the velocity components u and v (or V and θ) as independent variables rather than the physical coordinates x and y. Thus we must first express the dependent variables, ϕ or ψ as functions of u and v or V and θ.

Recall

$$\mathbf{V} = \mathbf{i}u + \mathbf{j}v$$

and

$$u = V\cos\theta, \qquad v = V\sin\theta$$

where V is the magnitude of the velocity and θ is measured counterclockwise from the positive x axis. Again we assume that ϕ exists and define ψ as before. Since ϕ and ψ are functions of x and y we have

$$\begin{aligned} d\phi &= \phi_x\,dx + \phi_y\,dy = V\cos\theta\,dx + V\sin\theta\,dy \\ d\psi &= \psi_x\,dx + \psi_y\,dy = -\frac{\rho}{\rho_0}V\sin\theta\,dx + \frac{\rho}{\rho_0}V\cos\theta\,dy \end{aligned} \tag{407}$$

We now assume that ϕ and ψ are continuous functions of V and θ as well as x and y. We can then say that

$$\begin{aligned} d\phi &= \phi_V\,dV + \phi_0\,d\theta \\ d\psi &= \psi_V\,dV + \psi_0\,d\theta \end{aligned} \tag{408}$$

Solving for dx and dy from equation (407) we find

$$\begin{aligned} dx &= \frac{\cos\theta}{V}d\phi - \frac{\rho_0\sin\theta}{\rho V}d\psi \\ dy &= \frac{\sin\theta}{V}d\phi + \frac{\rho_0\cos\theta}{\rho V}d\psi \end{aligned} \tag{409}$$

Substituting for $d\phi$ and $d\psi$ from (408),

$$\begin{aligned} dx &= \left(\frac{\cos\theta}{V}\phi_V - \frac{\rho_0\sin\theta}{\rho V}\psi_V\right)dV + \left(\frac{\cos\theta}{V}\phi_\theta - \frac{\rho_0\sin\theta}{\rho V}\psi_\theta\right)d\theta \\ dy &= \left(\frac{\sin\theta}{V}\phi_V - \frac{\rho_0\sin\theta}{\rho V}\psi_V\right)dV + \left(\frac{\sin\theta}{V}\phi_\theta + \frac{\rho_0\cos\theta}{\rho V}\psi_\theta\right)d\theta \end{aligned} \tag{410}$$

Now we assume further the x and y may be expressed as continuous functions of

V and θ. Then

$$dx = x_V\,dV + x_\theta\,d\theta$$
$$dy = y_V\,dV + y_\theta\,d\theta \qquad (411)$$

In making this statement it is well to keep in mind that while V and θ are single-valued functions of x and y, x and y are not necessarily single-valued functions of V and θ. That is, given the solution to an actual flow problem, and given a point x, y in the flow field, the value of V and θ at that point will be uniquely determined. On the other hand, given a value for V and θ, that value may exist at more than one point x, y in the flow field. Figure 61 will illustrate this point.

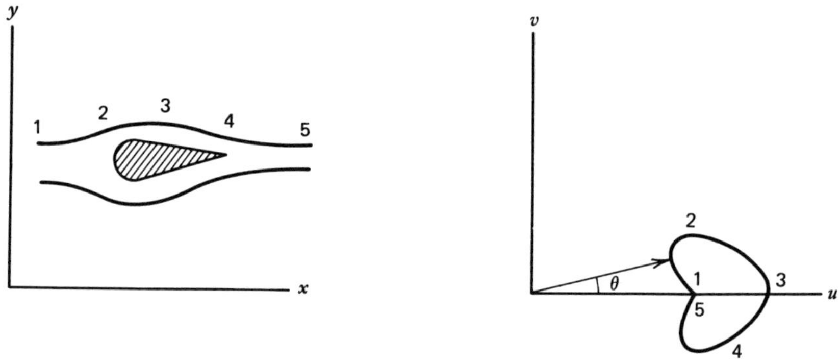

Figure 61. Physical and hodograph planes for flow over an airfoil.

Consider the streamline 1, 2, 3, 4, 5 in the flow over an object in the x, y plane, also called the physical plane. We now map this streamline onto the u, v plane, also called the hodograph plane. We notice that the points 1 and 5, representing $x = \infty$ in the physical plane, map into the same point, namely $V = V_\infty$, $\theta = 0$, in the hodograph plane. Thus, given that $V = V_\infty$ and $\theta = 0$, we cannot tell whether we are at $x = +\infty$ or at $x = -\infty$. In more complicated flow fields, there may be several points where the functions $x = x(\theta, V)\, y = y(\theta, V)$ are not single-valued. Care must, therefore, be exercised in applying the hodograph method to flow fields where x and y are not unique functions of V and θ.

Comparing equations (410) and (411), we see that

$$x_V = \frac{\phi_V}{V}\cos\theta - \frac{\rho_0\psi_V}{\rho V}\sin\theta$$
$$x_\theta = \frac{\phi_\theta}{V}\cos\theta - \frac{\rho_0\psi_\theta}{\rho V}\sin\theta \qquad (412)$$

and

$$y_V = \frac{\phi_V}{V}\sin\theta + \frac{\rho_0\psi_V}{\rho V}\cos\theta$$
$$y_\theta = \frac{\phi_\theta}{V}\sin\theta + \frac{\rho_0\psi_\theta}{\rho V}\cos\theta \qquad (413)$$

We now assume further that

$$x_{V\theta} = x_{\theta V}, \qquad y_{\theta V} = y_{V\theta} \tag{414}$$

This will be true at a point V, θ, if $x_V, x_\theta, x_{V\theta}$ and $y_\theta, y_V, y_{\theta V}$ exist and if $x_{V\theta}$ and $y_{V\theta}$ are continuous at the point in question. Substituting equations (412, 413) into equation (414) we get

$$-\frac{\phi_V}{V}\sin\theta - \frac{\rho_0\psi_V}{\rho V}\cos\theta = -\frac{\phi_\theta}{V^2}\cos\theta + \frac{\rho_0\psi_\theta}{\rho V^2}\sin\theta - \frac{\psi_\theta}{V}\left(\frac{\rho_0}{\rho}\right)_V \sin\theta$$
$$\frac{\phi_V}{V}\cos\theta - \frac{\rho_0\psi_V}{\rho V}\sin\theta = -\frac{\phi_\theta}{V^2}\sin\theta - \frac{\rho_0\psi_\theta}{\rho V^2}\cos\theta + \frac{\psi_\theta}{V}\left(\frac{\rho_0}{\rho}\right)_V \cos\theta \tag{415}$$

Multiplying the first of these equations by $\cos\theta$, the second by $\sin\theta$, and adding, we get

$$\phi_\theta = \frac{\rho_0 V \psi_V}{\rho} \tag{416}$$

Multiplying the first of equations (415) by $\sin\theta$ and the second by $\cos\theta$ and subtracting, we get

$$\phi_V = V\left(\frac{\rho_0}{\rho V}\right)_V \psi_\theta \tag{417}$$

This may be rewritten

$$\phi_V = -\frac{\rho_0(1-M^2)\psi_\theta}{\rho V} \tag{418}$$

We now assume

$$\phi_{V\theta} = \phi_{\theta V} \tag{419}$$

Differentiating equation (418) with respect to θ and equation (416) with respect to V and equating, we get

$$V(1+M^2)\psi_V + V^2\psi_{VV} + (1-M^2)\psi_{\theta\theta} = 0 \tag{420}$$

This is the hodograph equation for the stream function. It is observed to be linear in the dependent variable.

The Tangent Gas Approximation

Equation (420) may be rewritten

$$\left(\frac{\rho_0 V\psi_V}{\rho}\right)_V + \frac{\rho_0(1-M^2)\psi_{\theta\theta}}{\rho V} = 0 \tag{421}$$

or

$$\frac{\rho}{\rho_0}\frac{V}{1-M^2}\left(\frac{\rho_0}{\rho}V\psi_V\right)_V + \psi_{\theta\theta} = 0 \tag{422}$$

We now assume

$$\frac{\rho}{\rho_0} = \sqrt{1 - M^2} \tag{423}$$

This is a reasonable assumption if M is not too large since

$$\frac{\rho}{\rho_0} = \left[1 + \frac{\gamma - 1}{2} M^2\right]^{-1/(\gamma-1)} = 1 - \tfrac{1}{2}M^2 + \ldots \cong (1 - M^2)^{1/2} \tag{424}$$

Equation (423) also implies that

$$p = A - \frac{B}{\rho} \tag{425}$$

This may be seen as follows: Equation (423) may be rewritten

$$\left[1 - \left(\frac{\rho}{\rho_0}\right)^2\right]\frac{dp}{d\rho} = V^2 \tag{426}$$

Also,

$$V\frac{dV}{d\rho} + \frac{1}{\rho}\frac{dp}{d\rho} = 0 \tag{427}$$

Differentiating equations (426) with respect to ρ and inserting equation (427) for $dV/d\rho$ we get

$$\left[1 - \left(\frac{\rho}{\rho_0}\right)^2\right]\left[\frac{d^2p}{d\rho^2} + \frac{2}{\rho}\frac{dp}{d\rho}\right] = 0 \tag{428}$$

which is satisfied by

$$p = A - \frac{B}{\rho}$$

where A and B are constants.

An interpretation of this equation is the so-called "tangent gas," which may be seen from Figure 61 (a). In other words, the constants A and B may be so

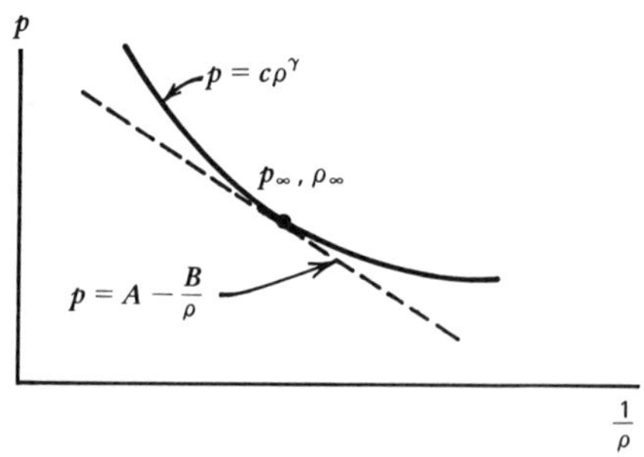

Figure 61(a). The tangent gas approximation.

chosen that the line $p = A - B/\rho$ is tangent to the actual p versus ρ curve at the free stream condition. The equation $p = A - B/\rho$ then describes the behavior of a fictitious "tangent gas" whose p, ρ variation is similar to that of the actual gas as long as p and ρ do not depart too much from their free stream values.

Substituting equation (423) into equation (422) we have

$$\frac{V}{\sqrt{1-M^2}}\left[\frac{V}{\sqrt{1-M^2}}\psi_V\right]_V + \psi_{\theta\theta} = 0 \tag{429}$$

We now introduce the transformation

$$\frac{dV_i}{V_i} = \sqrt{1-M^2}\,\frac{dV}{V} \tag{430}$$

Substituting into equation (429) we get

$$V_i(V_i\psi_{V_i})_{V_i} + \psi_{i_{\theta\theta}} = 0 \tag{431}$$

Thus it is seen that V_i represents that solution of equation (429) for incompressible flow. Thus, equation (430) represents another transformation from incompressible to compressible flow. Integrating equation (430) and making use of the tangent gas approximation, we get

$$V = \frac{4c_0^2 V_i}{4c_0^2 - V_i^2} \tag{432}$$

Evaluation of the pressure coefficient.

Recall

$$c_p = \frac{p - p_\infty}{\frac{1}{2}\rho_\infty V_\infty^2} \tag{423}$$

Introducing the tangent gas approximation,

$$c_p = \frac{c_\infty^2\left(1 - \dfrac{\rho_\infty}{\rho}\right)}{\frac{1}{2}V_\infty^2} = \frac{2}{M_\infty^2}\left[1 - \sqrt{1 - M_\infty^2 + M_\infty^2\frac{V^2}{V_\infty^2}}\right] \tag{433}$$

Now for incompressible flow,

$$c_{p_i} = 1 - \frac{V_i^2}{V_{i\infty}^2} \tag{434}$$

Combining equations (432), (433), and (434), we get

$$c_p = \frac{c_{pi}}{\sqrt{1-M_\infty^2} + \dfrac{M_\infty^2}{1+\sqrt{1-M_\infty^2}}\dfrac{c_{pi}}{2}} \tag{435}$$

This is called the Karman–Tsien pressure correction formula.

It is of interest to compare this formula with the Prandtle–Glauert Rule which states

$$c_p = \frac{c_{pi}}{\sqrt{1 - M_\infty^2}} \tag{334}$$

The Karman–Tsien pressure correction formula is thus seen to be a higher order approximation for c_p.

Profile Shape Correction

In deriving the Prandtle–Glauert Rule, we specified that c_p and c_{pi} be taken over the same profile. For the Karman–Tsien formula, the matter is not so simple. Let us suppose we have a solution ψ_i to equation (431) for the flow over a body of given shape. Applying the transformation (432), we obtain the corresponding compressible flow. How are the boundary conditions in the physical plane for this compressible flow related to those for the incompressible flow?

For the incompressible flow we have, from equation (409)

$$dx_i = \frac{\cos\theta}{V_i} d\phi_i - \frac{\sin\theta}{V_i} d\psi_i, \qquad dy_i = \frac{\sin\theta}{V_i} d\phi_i + \frac{\cos\theta}{V_i} d\psi_i \tag{436}$$

since $\rho_0/\rho_i = 1$. For the compressible flow, we may introduce the tangent gas approximation for ρ_0/ρ into equations (409). Then

$$\begin{aligned} dx &= \frac{4c_0^2 - V_i^2}{4c_0^2 V_i} \cos\theta\, d\phi - \frac{4c_0^2 + V_i^2}{4c_0^2 V_i} \sin\theta\, d\psi \\ dy &= \frac{4c_0^2 - V_i^2}{4c_0^2 V_i} \sin\theta\, d\phi + \frac{4c_0^2 + V_i^2}{4c_0^2 V_i} \cos\theta\, d\psi \end{aligned} \tag{437}$$

We now set $d\psi = d\psi_i$ and $d\phi = d\phi_i$ and determine the corresponding difference between dx and dx_i and between dy and dy_i. To do this it is easier to work in complex variables. Let

$$dz = dx + i\,dy, \qquad dz_i = dx_i + i\,dy_i \tag{438}$$

Then

$$dz = \frac{e^{i\theta}}{V_i} dF - \frac{e^{-i\theta}}{V_i} \frac{V_i^2}{4c_0^2} d\bar{F} \tag{439}$$

where $F = \phi + i\psi$, the complex potential, and $\bar{F}$ is the complex conjugate of F. In the incompressible case we have

$$dz_i = \frac{e^{i\theta}}{V_i} dF \tag{440}$$

since $c_0 = \infty$ in incompressible flow.

Thus

$$dz - dz_i = -\frac{V_i e^{i\theta}}{4c_0^2} d\bar{F} \tag{441}$$

Since $V_i e^{i\theta} = d\bar{F}/d\bar{z}_i$, this may be rewritten

$$dz - dz_i = -\frac{1}{4c_0^2}\left(\frac{d\bar{F}}{d\bar{z}_i}\right)^2 d\bar{z}_i \tag{442}$$

or

$$z - z_i = -\frac{1}{4c_0^2}\int\left(\frac{d\bar{F}}{d\bar{z}_i}\right)^2 d\bar{z}_i \tag{443}$$

Thus, when F is known for the incompressible case, we can calculate the distortion caused by the Karman–Tsien transformation in the physical plane. In practice, the distortion turns out to be small. Moreover, the error involved in neglecting the distortion tends to be cancelled out by the error caused by the tangent gas approximation. Thus we may neglect the distortion in practice. Used in this way, the Karman–Tsien pressure correction formula yields remarkably good results, equal or superior to the laborious relaxation method in many cases (see Figures 59 and 60).

Direct solution of the hodograph equation (420) is seldom practical due to the difficulty of formulating the boundary conditions. In the physical plane, formulation of the boundary conditions is generally simple. The surface of a body is usually a streamline. Thus, given the shape of the body, $y_s = y_s(x_s)$, $\psi(y_s, x_x) =$ constant. In the hodograph plane, while the direction of the velocity on the surface of a body is known, the magnitude is generally not known a priori. Thus we do not have a relation $V_s = V_s(\theta_s)$ to begin with and, therefore, cannot formulate a boundary condition $\psi(V_s, \theta_s) =$ constant. There are a small number of exceptions to this, as for example, the boundary of a free jet. But in general, the difficulty in formulating the boundary conditions precludes direct solution of the hodograph equation.

Expansion Solution

For incompressible 2-dimensional flow, the velocity potential may be expressed in the form

$$\phi = V_\infty r\cos\theta + \frac{m}{2\pi}\ln r + \frac{\Gamma}{2\pi}\theta + \sum_{n=1}^{\infty} a_n(\theta) r^{-n} \tag{444}$$

where r and θ are the polar coordinates, m the source strength at the origin, if any, Γ the circulation about the origin (if any). Equation (444) satisfies the Laplace equation and the boundary conditions for the case of the flow of a uniform stream over a body. The form of the functions $a_n(\theta)$ will depend on the boundary conditions but will, in any case, be periodic. Ludford [29] has shown that for the corresponding compressible case, the velocity potential may be expressed in the form

$$\phi = V_\infty r\cos\theta + \frac{m}{2\pi\beta}\ln r + \frac{\Gamma}{2\pi}\tan^{-1}[\beta\tan\theta] + \sum_{n=0}^{\infty}\sum_{p=0}^{\infty} a_{np}(\theta)\left[\frac{\ln r}{r}\right]^n \frac{1}{r^p} \tag{445}$$

where $a_{np}(\theta)$ is periodic and $\beta^2 = 1 - M_\infty^2$, provided m and Γ are not both zero. If m and Γ are both zero, the solution becomes

$$\phi = V_\infty r \cos\theta + \sum_{n=1}^{\infty} a_n(\theta) r^{-n} \tag{446}$$

with a_n periodic. The expansions (444) and (446) are complete, in the sense that they cover all cases of the disturbance of a uniform stream. Furthermore, the series are uniformly and absolutely convergent so long as $M < 1$. The solution is thus restricted to subsonic flow.

Whalen and Mulholland [30] have determined the functions $a_n(\theta)$ in equation (446) explicitly. Since the solution is written in polar coordinates, the equation to be satisfied is

$$(c^2 - \phi_r^2)\phi_{rr} + \left(c^2 - \frac{\phi_\theta^2}{r^2}\right)\frac{\phi_{\theta\theta}}{r^2} - 2\phi_r\phi_\theta\frac{\phi_{r\theta}}{r^2} + \frac{\phi_r}{r}\left(c^2 + \frac{\phi_\theta^2}{r^2}\right) = 0 \tag{447}$$

the boundary conditions are

$$\lim_{r\to\infty} \phi_r = -V_\infty \sin\theta$$

$$\lim_{r\to\infty} \frac{1}{r}\phi_\theta = V_\infty \cos\theta$$

and on the body surface,

$$\phi_n = 0$$

In addition, the solution must approach the inviscid solution as $M_\infty \to 0$. Substituting the proposed form of the solution (446) into the differential equation (447) and equating powers of r, we find

$$\begin{aligned} a_p'' + p^2 a_p = M_0\{&a_n a_j a_k[(\eta + 1)n + 1]nkj + a_n a_j' a_k'[\eta n + 2j + 1]n \\ &+ a_n a_j'' a_k[\eta nk] + a_n' a_j'' a_k'[\eta + 1] \end{aligned} \tag{448}$$

where $\eta = (\gamma - 1)/2$ and $M_0 = V_\infty/c_0$. The subscripts $p, n, j, k = 1, 2, 3, \ldots$ subject to the constraint

$$p = n + j + k + 2 \tag{449}$$

The multiple subscripts indicate summation in the tensor notation. That is, the right-hand side of the equation (448) represents the sum of all terms whose subscripts n, j, and k satisfy the relation (449). The functions $a_n(\theta), n = 1, 2, 3, \ldots$ may thus be obtained by solving the system of ordinary differential equations (448). The technique used is to first find the complementary solution and then the complete solution of equation (448). As will be recalled, the complementary solution is the solution of the homogeneous part of the differential equation. The solution may be rewritten in the form

$$\phi(r, \theta) = V_\infty \sum_{n=1}^{\infty} a_n(\theta) r^{-n} \tag{450}$$

Comparison with equation (446), or solution of equation (448) for a_{-1}, then shows that

$$a_{-1}(\theta) = \cos\theta, \qquad a_0(\theta) = 0 \tag{451}$$

The system of equations (448) may then be solved sequentially, only the highest order term, $a_i(\theta)$, being unknown in each equation. The homogeneous part of the equation will consist of those terms involving $a_i(\theta)$ where i is the maximum value of the indexes p, n, j, k present. (Note, however, that n, j, and k are never larger than p.) An inspection of equation (448) subject to condition (449) shows that starting with $p = 2$, some of the terms of the equation will not contain $a_i(\theta)$ [e.g., $a_2(\theta)$ when $p = 2$]. These are the inhomogeneous terms. The remaining terms constitute the corresponding homogeneous equation.

Making use of equation (449), it is found that the corresponding homogeneous equations may be expressed in the form

$$y_n''(b^2 - \sin^2\theta) - 2(n+1)\sin\theta\cos\theta\, y_n' + y_n[n^2 b^2 + n - (n^2 + 2n)\cos^2\theta] = 0 \tag{452}$$

where

$$b^2 = \frac{1}{M_0^2} - \eta$$

and $y_n(\theta)$ is the complementary part of the function $a_n(\theta)$. Equation (452) may be rewritten in the form

$$y_n'' - \left[\frac{k_1}{Z} + \frac{k_2}{Z-1} + \frac{k_3}{Z-b^2}\right] y_n' + \frac{k_4 k_5 Z + k_6}{Z(Z-1)(Z-b^2)} y_n = 0 \tag{453}$$

where $Z = \sin^2\theta$, $y = y(Z)$ and

$$k_1 = \frac{1}{2} \qquad k_4 = \frac{n}{2}$$

$$k_2 = \frac{1}{2} \qquad k_5 = \frac{n}{2} + 1$$

$$k_3 = n + 1 \qquad k_6 = \frac{n(nb^2 - n - 1)}{4}$$

Equation (453) is known as Heun's differential equation and is usually written in the form

$$y'' + \left(\frac{\gamma}{x} + \frac{\varepsilon}{x-1} + \frac{\delta}{x-a}\right) y' + \frac{(\alpha\beta x + b)y}{x(x-1)(x-a)} = 0 \tag{454}$$

where

$$\alpha + \beta + 1 = \gamma + \delta + \varepsilon \tag{455}$$

If γ is not an integer and $a \neq 0$, two independent solutions of Heun's equation are

$$y = F(a, b; \alpha, \beta, \gamma, \delta; x) = \sum_{n=1}^{\infty} c_n x^n \tag{456}$$

and

$$y = x^{1-\gamma} F(a, b - (1-\gamma)(\delta + a\varepsilon); 1 + \alpha - \gamma, 1 + \beta - \gamma, 2 - \gamma, \delta; x) \quad (457)$$

where c_n is given by $c_0 = 1, c_1 = -b/\gamma a$ and

$$(n+2)(n+1+\gamma)ac_{n+2} = -(n+\alpha)(n+\beta)c_n + \{(n+1)^2(a+1) + (n+1)[(\alpha+\beta-\delta)a + \gamma + \delta - 1] - b\}c_{n+1} \quad (458)$$

The complete solution for the $a_n(\theta)$ in equation (448) may now be found by letting

$$a_n(\theta) = y_n(\theta)v(\theta) \quad (459)$$

The inhomogeneous equation for $a_n(\theta)$ (448) may be written in the form

$$a_n'' + P(\theta)a_n' + Q(\theta)a_n = \frac{R(\theta)}{(b^2 - \sin^2\theta)} \quad (460)$$

Substituting equation (459) into equation (460),

$$y_n v'' + (2y_n' + p(\theta)y_n)v'' = \frac{R(\theta)}{b^2 - \sin^2\theta} \quad (461)$$

This is a linear, first-order equation for V with the solution

$$v(\theta) = \int \left[\frac{e^{\int P d\theta}}{y_n^2} \int \frac{y_n R(\theta) e^{\int P d\theta} d\theta}{b^2 - \sin^2\theta} \right] d\theta \quad (462)$$

where

$$P(\theta) = \frac{2(n+1)\sin\theta\cos\theta}{\sin^2\theta - b^2} \quad (463)$$

$$R(\theta) = \sum_{l=-1}^{n-1} \sum_{j=-1}^{n-1} \sum_{k=-1}^{n-1} a_l a_j a_k [(\eta+1)l + 1]ljk + a_l a_j' a_k' [\eta l + 2j + 1]l + a_l a_j'' a_k \eta jk + a_l a_j'' a_k' (\eta + 1) \quad (464)$$

and

$$y_n(\theta) = F\left[b^2, \frac{n(nb^2 - n + 1)}{4}; \frac{n}{2}, \frac{n+2}{2}, \frac{1}{2}, n+1; \sin^2\theta \right] \quad (465)$$

Noting that $b^2 = 1/M_\infty^2$, the solution for a_n becomes

$$\alpha_n(\theta) = -A_n M_\infty^{2n+2} y_n(\theta) \int \frac{d\theta}{y_n^2 (M_\infty^2 \sin^2\theta - 1)^{n+1}} - B_n y_n - y_n M_\infty^2 \int \frac{[\int y_n R(\theta)(M_\infty^2 \sin^2\theta - 1)\, d\theta]}{y_n^2 (M_\infty^2 \sin^2\theta - 1)^{n+1}} d\theta \quad (466)$$

where A_n and B_n are constants of integration to be determined by the boundary conditions. The complete solution for the velocity potential is then given by

$$\phi(r,\theta) = U_\infty r\cos\theta - U_\infty \sum_{n=0}^{\infty} r^{-n} A_n y_n(\theta) \int \frac{M_\infty^{2n+2}\, d\theta}{y_n^2[M_\infty^2 \sin^2\theta - 1]^{n+1}} - U_\infty \sum_{n=0}^{\infty} B_n r^{-n} y_n(\theta) - U_\infty \sum_{n=0}^{\infty} r^{-n} M_\infty^2 \int \frac{[\int y_n R(\theta)(M_\infty^2 \sin^2\theta - 1)^n\, d\theta]}{y_n^2 (M_\infty^2 \sin^2\theta - 1)^{n+1}}\, d\theta \qquad (467)$$

Equation (467) thus represents the complete solution to the velocity potential equation (447), for the case of two-dimensional, compressible, subsonic, flow, uniform at infinity with zero source strength and zero circulation.

As a test of the solution (467), Whalen and Mulholland [30] have applied it to the case of flow over a circular cylinder. Retaining only the first three terms of the expansion, it is found that

$$\phi = U_\infty \cos\theta \left[r + \frac{1}{r(1 - M_\infty^2 \sin^2\theta)} \right] \qquad (468)$$

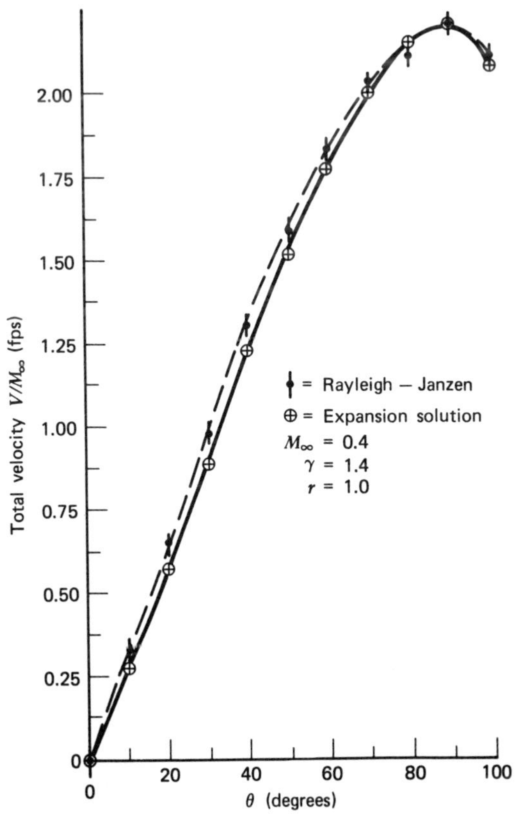

Figure 62. Velocity distribution over a circular cylinder by two methods compared [30].

The factor $(1 - M_\infty^2 \sin^2 \theta)$ thus represents a first-order correction for the compressibility. It is of interest to compare this result with the solution to the same problem obtained by the Rayleigh–Janzen method, retaining only the first three terms:

$$\phi(r, \theta) = U_\infty \cos\theta \left[\left(r + \frac{1}{r} \right) + M_\infty^2 \left(\frac{13}{12r} - \frac{1}{2r^3} + \frac{1}{12r^5} \right) \right]$$

$$+ U_\infty \cos 3\theta \cdot M_\infty^2 \left[\frac{1}{12r^3} - \frac{1}{4r} \right] \tag{469}$$

A comparison of the results of the two methods is shown in Figure 62.

4 Two- and Three-Dimensional Steady Supersonic Flow

Method of Characteristics

We now consider the case of two-dimensional, steady, potential flow with the Mach number greater than 1. Recall the potential equation for two-dimensional, steady flow:

$$\left(1-\frac{\phi_x^2}{c^2}\right)\phi_{xx}-2\frac{\phi_x\phi_y}{c^2}\phi_{xy}+\left(1-\frac{\phi_y^2}{c^2}\right)\phi_{yy}=0 \tag{470}$$

In Chapter 3 we established that in the case of supersonic flow, this equation is hyperbolic. That is, if the equation (470) is written is the form

$$A\phi_{xx}+B\phi_{xy}+C\phi_{yy}=0 \tag{471}$$

then

$$B^2-4AC>0 \tag{472}$$

whenever $M>1$. We wish now to solve equation (471) for ϕ_x and ϕ_y, given some boundary condition, say, given the values ϕ_x and ϕ_y along some curve in the x, y plane, say $y=f(x)$.

Since ϕ_x and ϕ_y are presumed to be continuous values of x and y, we may write

$$d\phi_x=\phi_{xx}\,dx+\phi_{xy}\,dy \tag{473}$$

$$d\phi_y=\phi_{yx}\,dx+\phi_{yy}\,dy \tag{474}$$

Equations (471), (473), and (474) may be considered a system of three simultaneous equations for the three unknowns ϕ_{xy}, ϕ_{xy}, and ϕ_{yy}. Using the method of elementary algebra, we may solve the three equations simultaneously for any one of the unknowns in terms of the coefficients. Solving for ϕ_{xy}, we get

$$\phi_{xy}=\frac{\begin{vmatrix} A & 0 & C \\ dx & d\phi_x & 0 \\ 0 & d\phi_y & dy \end{vmatrix}}{\begin{vmatrix} A & B & C \\ dx & dy & 0 \\ 0 & dx & dy \end{vmatrix}}=\frac{A\,dy\,d\phi_x+C\,dx\,d\phi_y}{A\,dy^2-B\,dx\,dy+C\,dx^2} \tag{475}$$

Since ϕ_{xy} is presumed to be bounded throughout the flow field, if the denominator of equation (475) is zero, then the numerator must be zero also. Notice that in equations (473) and (474), the choice of dy/dx is entirely arbitrary. We may, therefore, choose dy/dx in such a way that the denominator of equation (475) is zero. Thus, let

$$A\,dy^2 - B\,dx\,dy + C\,dx^2 = 0 \tag{476}$$

Then

$$A\left(\frac{dy}{dx}\right)_c^2 - B\left(\frac{dy}{dx}\right)_c + C = 0 \tag{477}$$

where

$$\left(\frac{dy}{dx}\right)_c = \frac{B \pm \sqrt{B^2 - 4AC}}{2A} \tag{478}$$

Thus we see that the slope dy/dx for which the denominator of equation (475) is zero is real only when equation (471) is hyperbolic; that is, when $B^2 - 4AC > 0$.

If $(dy/dx)_c$ is real, it follows that

$$\left(\frac{d\phi_y}{d\phi_x}\right)_c = -\frac{A}{C}\left(\frac{dy}{dx}\right)_c \tag{479}$$

since ϕ_{xy} is everywhere finite. This result enables us to find a solution for $\phi_x(x, y)$, $\phi_y(x, y)$ for a given set of boundary conditions. Consider the curve S in the x, y plane and suppose that ϕ, ϕ_x and ϕ_y are given along S (Figure 63). The coefficients A, B, and C are functions of x, y, ϕ_x, and ϕ_y in the most general case. Thus, if ϕ, ϕ_x, and ϕ_y are given along S, A, B, and C are also known along S. At each point of S, therefore, equation (478) gives two slopes, $(dy/dx)_\pm$. If lines having these slopes are drawn through the points 1 and 2 in Figure 63,

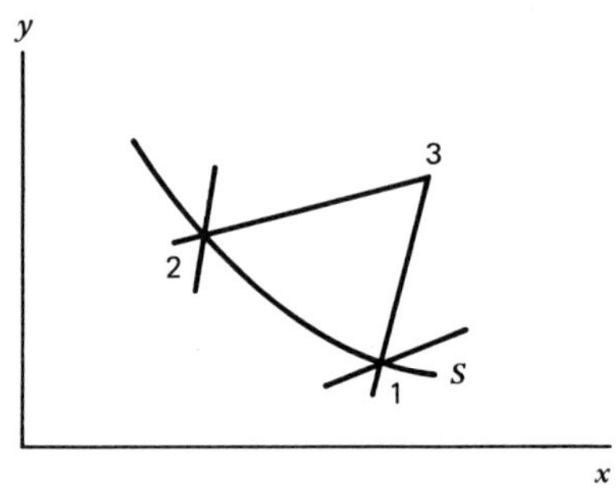

Figure 63. Characteristics in the physical plane.

they will intersect at some point 3. The location of the point three may be found by the method of finite differences. That is,

$$\frac{y_3 - y_1}{x_3 - x_1} = \left(\frac{dy}{dx}\right)_{1+}, \qquad \frac{y_3 - y_2}{x_3 - x_2} = \left(\frac{dy}{dx}\right)_{2-} \tag{480}$$

the subscripts + and − indicating the sign to be used in equation (478). Equation (480) may be solved simultaneously for x_3 and y_3.

Equation (479) may also be written in finite-difference form

$$\frac{\phi_{y_3} - \phi_{y_1}}{\phi_{x_3} - \phi_{x_1}} = -\frac{A}{C}\left(\frac{dy}{dx}\right)_{1+}, \qquad \frac{\phi_{y_3} - \phi_{y_2}}{\phi_{x_3} - \phi_{x_2}} = -\frac{A}{C}\left(\frac{dy}{dx}\right)_{2-} \tag{481}$$

Equation (481) may be solved simultaneously for ϕ_{y_3} and ϕ_{x_3}, thus giving the x and y velocity components at the point 3. The process may be repeated for successive points until the solution is found throughout an entire region. Suppose, for example, that the solution is known along a line $a - b$ as shown in Figure 64.

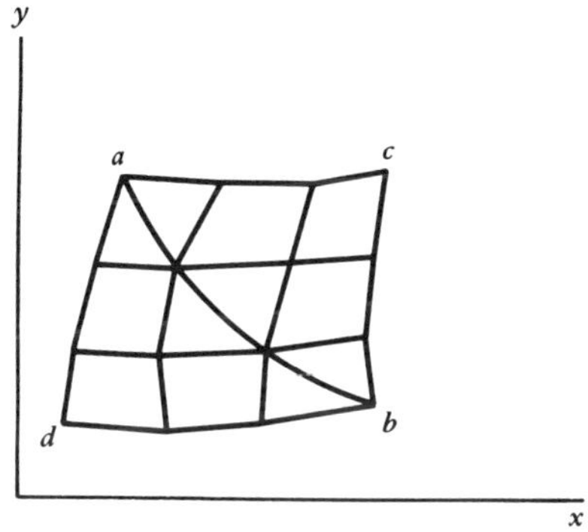

Figure 64. Domain of dependence.

Applying the above method to find the solution at successive points indicates that for the given boundary condition the solution may be found only inside the region *abcd*. Inside this region, however, the solution is uniquely determined. This indicates an important property of hyperbolic equations. A change in the value of ϕ outside of the region *abcd* will have no effect on the solution inside the region *abcd*. This is in contrast to the property of elliptic equations, for which change in the boundary condition at a single point will cause a change in the solution everywhere.

The lines whose slopes are given at each point by the value of $(dy/dx)_\pm$ at that point are called the physical characteristics, and hence the method of solution outlined above is called the *method of characteristics.*

Consider now the effect of the variation of the boundary condition at a single point p. Drawing the characteristics passing through point p, Figure 65, we see

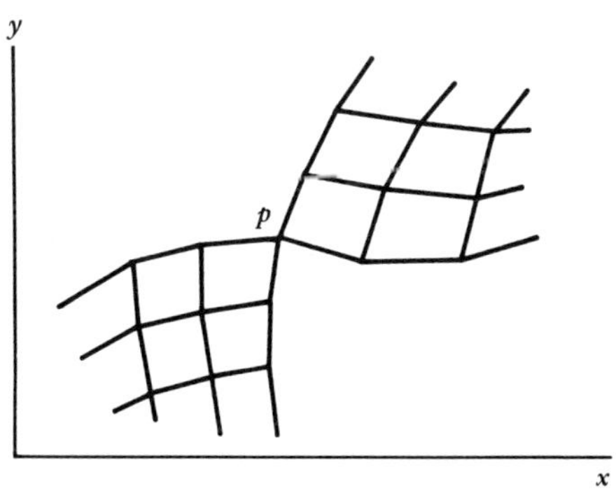

Figure 65. Region of influence.

that a change in the condition at p can affect the solution only in a region bounded by the characteristic passing through p. Earlier, we have seen that the potential equation for subsonic flow is elliptic, while the potential equation for supersonic flow is hyperbolic. Given the markedly different effect of the boundary conditions in these two cases, we may expect the solutions of subsonic flow problems to be fundamentally different from those of supersonic flow problems. For example, one may expect that a body placed in a subsonic stream would alter the flow everywhere in the stream, while a body placed in a supersonic stream would alter the flow only in a part of the stream.

Simple Wave Theory and Prandtl–Meyer Flow

Recall in Chapter 3 we linearized the equation for the velocity potential. The linearized equation for the velocity potential was found to be

$$\phi_{yy} = (M_\infty^2 - 1)\phi_{xx} \tag{482}$$

The assumptions involved in deriving this equation were

$$\frac{u}{V_\infty} M_\infty^2 \ll 1, \qquad \frac{v}{V_\infty} M_\infty^2 \ll 1, \qquad \frac{M_\infty^2}{1 - M_\infty^2}\frac{u}{V_\infty} \ll 1 \tag{483}$$

where u and v are the x and y components of the perturbation velocity. For subsonic flow with small perturbations these conditions are met if M_∞ is not too close to 1. The conditions may also be met for supersonic flow if the perturbations are sufficiently small.

For the case of $M_\infty > 1$, equation (482) is the wave equation with general solution

$$\phi = \phi_1(x + \beta y) + \phi_2(x - \beta y) \tag{484}$$

where ϕ is the perturbation potential, $\beta = \sqrt{M_\infty^2 - 1}$ and ϕ_1 and ϕ_2 are functions of $x + \beta y$ and $x - \beta y$, respectively.

Let us now consider the case in which $\phi_1 = 0$. Then

$$\phi = \phi_2(x - \beta y), \qquad u = \phi_x = \phi_2', \qquad v = \phi_y = -\beta\phi_2' \tag{485}$$

where the prime indicates differentiation with respect to the argument of the function. The slope of the streamlines is given by

$$\left[\frac{dy}{dx}\right]_{str.} = \frac{v}{V_\infty + u} = -\frac{\beta\phi_2'}{V_\infty + \phi_2'} \cong -\frac{\beta}{V_\infty}\phi_2'(x - \beta y)$$

Thus we see that the streamlines have constant slopes along the lines

$$x = \beta y + \text{constant} \tag{486}$$

These lines have slopes

$$\frac{dy}{dx} = \frac{1}{\beta} = \frac{1}{\sqrt{M_\infty^2 - 1}} \tag{487}$$

and are called Mach lines. The angle, called α, which the Mach lines make with the horizontal, is called the Mach angle. Thus,

$$\tan \alpha = \frac{1}{\sqrt{M_\infty^2 - 1}} \tag{488}$$

Let us now consider an example of this type of flow. Suppose we have supersonic flow over a wall with a small ramp on it (Figure 66). Every particle of

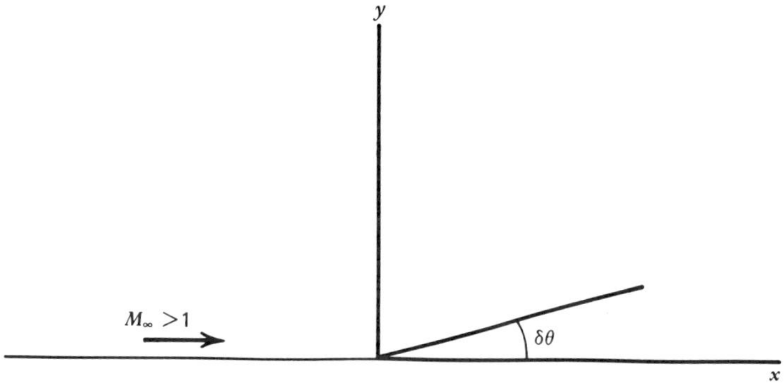

Figure 66. Flow over a ramp-nomenclature.

fluid flowing along the wall is deflected slightly on hitting the ramp, resulting in a slight change in its velocity and hence in its pressure if the flow is isentropic. A slight variation of the pressure at a point of a flow results in the propagation of a pressure wave outward from the point, moving with the speed of sound relative to the flow, as discussed in Chapter 2. We may trace the path of such a wave by adding the velocity of the flow to the velocity of the wave.

In Figure 67 we have drawn the wave in successive locations as it is swept

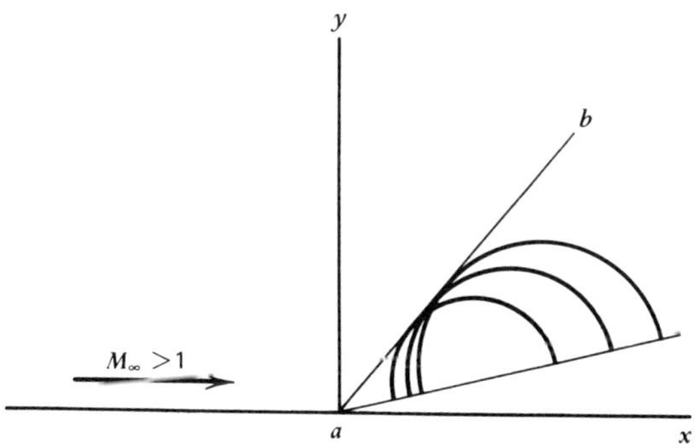

Figure 67. Flow over a ramp-Mach wave.

downstream. Since the flow is presumed steady, one may also regard Figure 67 as showing some of the waves which are continuously being generated by the flow as it hits the ramps. Of course in the actual flow, the waves are not really distinct, but one may think of the actual flow as the limit when the waves are spaced ever closer and closer together. In this case it will be observed that the effects of successive waves cancel each other everywhere except along the line *ab*, which forms the envelope of the waves. In the limit as $\delta\theta \to 0$, the line *ab* makes an angle α with the flow direction, as may be observed in Figure 68. Thus, $\cot\alpha = \sqrt{M_\infty^2 - 1}$.

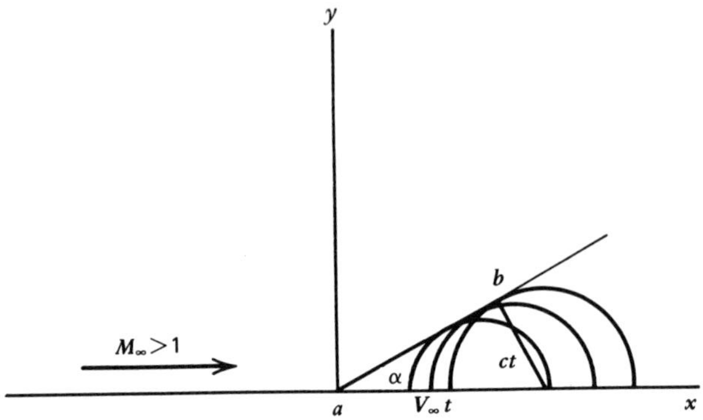

Figure 68. Flow over a ramp-determination of Mach angle.

If the angle $\delta\theta$ is small, we may apply the theory of small perturbations to this case. From the above discussion, we infer that the presence of the ramp cannot affect the flow ahead of the line *ab*. Thus we may postulate a solution of the form

$$\phi = \phi_2(x - \beta y) \tag{489}$$

This says the streamlines of the flow shall have constant slope along lines making an angle α with the x axis. Thus, the flow pattern would be as shown in Figure 69.

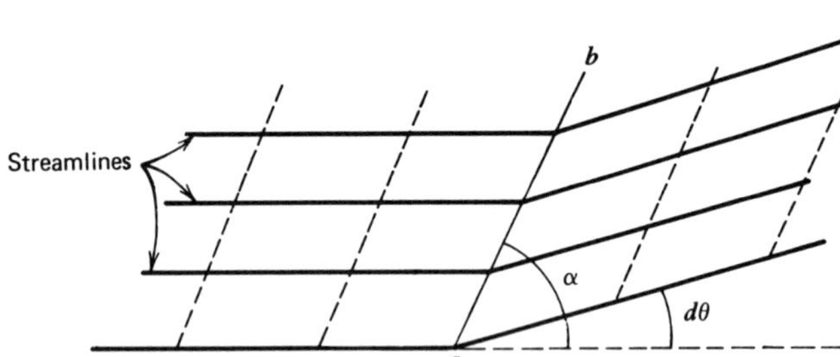

Figure 69. Flow over a ramp-streamlines and Mach wave.

This flow pattern satisfies the equation for the perturbation potential and the boundary condition that the flow be parallel to the wall at the wall. In actual fact, the slope of the line *ab* in Figure 69 would not in fact be α, first due to the deflection of the flow by the ramp, and second due to the change in the speed of sound caused by the flow deflection. As long as $d\theta$ is small, however, the difference between the angle of the line *ab* and α is small, and hence the solution (489) is correct within the assumptions of the small perturbation theory.

Let us now see if we can determine the actual change in velocity which takes place in the flow as it crosses the Mach line *ab*. To do this, we refer to Figure 70.

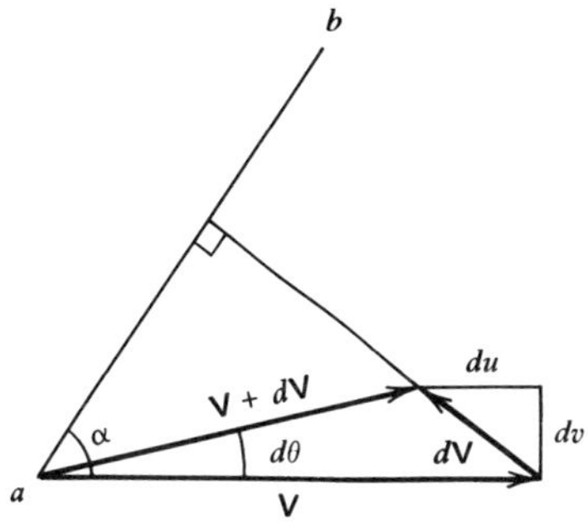

Figure 70. Flow over a ramp-change in velocity.

V is the velocity vector of the flow ahead of the Mach line *ab* and $\mathbf{V} + d\mathbf{V}$ is the velocity vector of the flow after it passes the Mach line *ab*. This may be verified as follows. Since the solution is of the form

$$\phi = \phi_2(x - \beta y)$$

and since

$$u = \phi_x = \phi_2'(x - \beta y)$$

and

$$v = \phi_y = -\beta\phi_2'(x - \beta y)$$

where the prime indicates differentiation with respect to the argument of the function. Thus the Mach lines must also be lines along which the velocity is constant. Since the flow is adiabatic and reversible this means that all the state properties, i.e. pressure, temperature, density, etc., are constant along the Mach lines. It follows from the momentum equation that the component of the velocity tangent to the line *ab* cannot change as the flow crosses the line. This may be seen by writing the momentum equation tangent to the wave. Since there is no net force in the direction parallel to the line *ab*,

$$(\rho V_n)V_t = (\rho + d\rho)(V_n + dV_n)(V_t + dV_t) \tag{490}$$

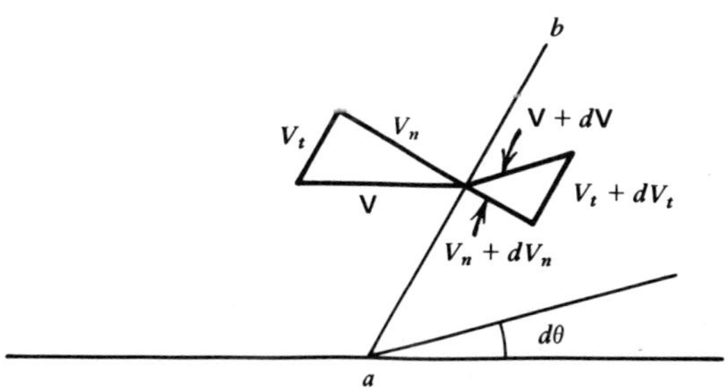

Figure 71. Flow over a ramp-change in velocity components.

where V_n and V_t represent the velocity components normal and tangent to the wave respectively as shown in Figure 71. But continuity requires that

$$\rho V_n = (\rho + d\rho)(V_n + dV_n) \tag{491}$$

Hence,

$$dV_t = 0 \tag{492}$$

Thus, the vector $\mathbf{V} + d\mathbf{V}$ must be as shown in Figure 70. The line *ab* is a wave front, as shown in Figure 71. For that reason, it is sometimes called a wave (for short) or Mach wave.

Returning now to Figure 70, we observed that

$$dv \cong V\,d\theta, \qquad du \cong dV \tag{493}$$

where V is the magnitude of the velocity. Furthermore, we observe that

$$-\frac{du}{dv} = \tan\alpha = \frac{1}{\sqrt{M_\infty^2 - 1}} \tag{494}$$

If $d\theta$ is sufficiently small we may treat it as a differential. Hence, using equation (493),

$$\frac{1}{V}\frac{dV}{d\theta} = -\frac{1}{\sqrt{M_\infty^2 - 1}} \tag{495}$$

Up to now, our theory has been approximate. Our solution, ϕ_2, represented the solution of the linearized (i.e. approximate) equation for the perturbation potential. Equations (493) and (495) are also approximate, becoming exact only in the limit of $d\theta \to 0$.

Let us now consider the supersonic flow over a curved wall (Figure 72).

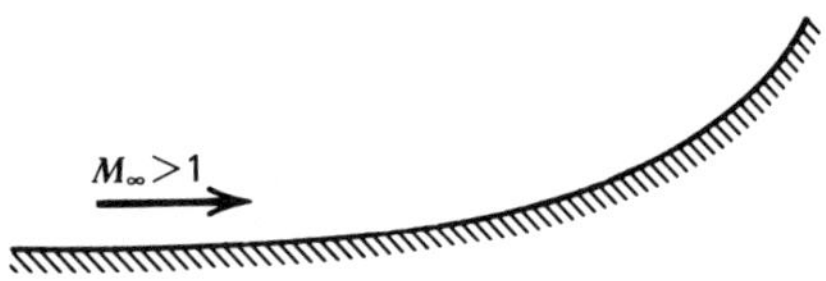

Figure 72. Flow over a curved wall.

This wall may be approximated arbitrarily closely by substituting for it a succession of ramps, as in Figure 73. Using our approximate method, we may

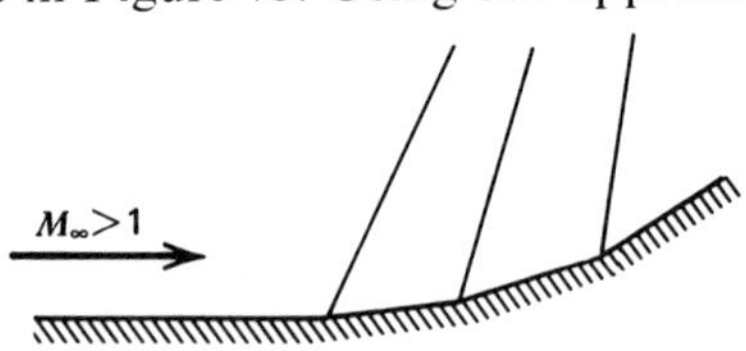

Figure 73. Flow over a curved wall approximated by a series of ramps.

find the change in velocity for each ramp in turn, since the flow after passing through each Mach wave is uniform and parallel. The formula (495) may be made valid for the entire wall be rewriting it

$$\frac{1}{V}\frac{dV}{d\theta} = -\frac{1}{\sqrt{M_i^2 - 1}} \tag{496}$$

where M_i is the Mach number ahead of each wave. Moreover, the wall may be approximated with arbitrary closeness by decreasing the turning angle of each ramp while increasing the number of ramps. The decrease in the turning angle at the same time makes the approximate solution more accurate for each ramp. It is logical, therefore, to consider whether, by passing to the limit as each turning angle goes to zero while the number of ramps becomes infinite, the solution does not become exact. In this case we would have

$$\frac{1}{V}\frac{dV}{d\theta} = -\frac{1}{\sqrt{M^2 - 1}} \tag{497}$$

where M is the local Mach number and θ the local angle of inclination of the wall. In such a case we would be justified in integrating equation (497). For an isentropic flow we have

$$\frac{dV}{V} = \frac{dM^2}{2M^2\left(1 + \frac{\gamma - 1}{2}M^2\right)} \tag{498}$$

This follows directly from the fact that

$$dh + V\,dV, \qquad dp + \rho V\,dV = 0$$

for a reversible adiabatic flow. Substituting in equation (497) we have

$$d\theta = -\frac{\sqrt{M^2 - 1}\,dM^2}{2M\left(1 + \frac{\gamma - 1}{2}M^2\right)} \tag{499}$$

which may be integrated to give

$$\theta = -\sqrt{\frac{\gamma + 1}{\gamma - 1}}\arctan\sqrt{\frac{\gamma - 1}{\gamma + 1}(M^2 - 1)} + \arctan\sqrt{M^2 - 1} + C \tag{500}$$

where C is an arbitrary constant of integration.

Of course it is not as yet clear whether or not this is a valid solution. While it is true that the error for each ramp decreases as the ramp angle decreases, if, at the same time we increase the number of ramps, may not the cumulative error remain large? In Figure 73 it also is seen that the Mach waves tend to intersect if the wall is curving upward. What is the consequence of this? In order to answer these questions, we must investigate the problem from another angle.

Recall the equation for the slopes of the physical characteristics:

$$\frac{dy}{dx} = \frac{-\frac{uv}{c^2} \pm \sqrt{\frac{u^2 + v^2}{c^2} - 1}}{1 - \frac{u^2}{c^2}} \tag{501}$$

where u and v represent the actual (not perturbation) velocity components. Since the orientation of the x and y axes may be chosen arbitrarily, let us choose the x axis in the flow direction at some point along our curved wall. At that point then, v will be zero, and

$$\frac{dy}{dx} = \frac{\pm\sqrt{\dfrac{V^2}{c^2} - 1}}{1 - \dfrac{V^2}{c^2}} = \frac{\pm 1}{\sqrt{M^2 - 1}} = \mp \tan\alpha \tag{502}$$

Thus we see that the physical characteristics with the + sign coincide with the Mach waves.

Recall, now, the equation for the hodograph characteristics

$$\frac{dv}{du} = \frac{\dfrac{uv}{c^2} \pm \sqrt{\dfrac{u^2 + v^2}{c^2} - 1}}{1 - \dfrac{v^2}{c^2}} \tag{503}$$

Now along our curved wall

$$\frac{du}{dv} = \frac{1}{V}\frac{dV}{d\theta} \tag{504}$$

as shown previously. Again, taking the x axis in the flow direction at a given point, we have, at that point,

$$\frac{1}{V}\frac{dV}{d\theta} = \mp\frac{1}{\sqrt{M^2 - 1}} = \mp \tan\alpha \tag{505}$$

Since the point is arbitrary, equation (505) must be true at every point along the wall. It will be observed, however, that when the minus sign is chosen, equation (505) is identical to equation (497). This does not yet prove that equation (500) is an exact solution. Its significance, however, may be seen in the following way:

The full equation for the velocity potential may be written in the form

$$\left(1 - \frac{u^2}{c^2}\right)\frac{\partial u}{\partial x} - \frac{uv}{c^2}\left(\frac{\partial v}{\partial x} + \frac{\partial u}{\partial y}\right) + \left(1 - \frac{v^2}{c^2}\right)\frac{\partial v}{\partial y} = 0 \tag{506}$$

since the irrotationality condition requires that

$$\frac{\partial v}{\partial x} - \frac{\partial u}{\partial y} = 0 \tag{507}$$

In the case of flow over a monotonically curved wall, there is a one to one relationship between the magnitude of the velocity and the flow direction. That is, V is a function of θ only or

$$V = V(\theta) \tag{508}$$

Keeping in mind that $u = V\cos\theta$ and $v = V\sin\theta$ and substituting equation (508) into equations (506) and (507), we get

$$\left[(1-M^2)\cos\theta\frac{dV}{d\theta} - V\sin\theta\right]\frac{\partial\theta}{\partial x} + \left[(1-M^2)\sin\theta\frac{dV}{d\theta} + V\cos\theta\right]\frac{\partial\theta}{\partial y} = 0$$

$$\left[\sin\theta\frac{dV}{d\theta} + V\cos\theta\right]\frac{\partial\theta}{\partial x} + \left[-\cos\theta\frac{dV}{d\theta} + V\sin\theta\right]\frac{\partial\theta}{\partial y} = 0 \qquad (509)$$

These two equations may be regarded as simultaneous algebraic equations for $\frac{\partial\theta}{\partial x}$ and $\frac{\partial\theta}{\partial y}$ in the form

$$a_1\theta_x + b_1\theta_y = 0$$
$$a_2\theta_x + b_2\theta_y = 0$$

Hence

$$\theta_x = \frac{\begin{vmatrix} 0 & b_1 \\ 0 & b_2 \end{vmatrix}}{\begin{vmatrix} a_1 & b_1 \\ a_2 & b_2 \end{vmatrix}}, \qquad \theta_y = \frac{\begin{vmatrix} a_1 & 0 \\ a_2 & 0 \end{vmatrix}}{\begin{vmatrix} a_1 & b_1 \\ a_2 & b_2 \end{vmatrix}}$$

Thus, if θ_x and θ_y are not zero,

$$\begin{vmatrix} a_1 & b_1 \\ a_2 & b_2 \end{vmatrix} = \begin{vmatrix} (1-M^2)\cos\theta\frac{dV}{d\theta} - V\sin\theta & (1-M^2)\sin\theta\frac{dV}{d\theta} + V\cos\theta \\ \sin\theta\frac{dV}{d\theta} + V\cos\theta & -\cos\theta\frac{dV}{d\theta} + V\sin\theta \end{vmatrix}$$

$$= (1-M^2)\left(\frac{dV}{d\theta}\right)^2 + V^2 = 0$$

or

$$\frac{dV}{d\theta} = \pm\frac{V}{\sqrt{M^2-1}} \qquad (510)$$

Thus, in the case where $V = V(\theta)$, the velocity potential equation reduces to the equation for the hodograph characteristics. Since either the plus or minus sign may be used, this means that equation (500) is an exact solution of the isentropic flow over a curved wall.

Notice that the solution is equally valid whether the wall curves up or down. Only the sign of du and dv are reversed, as seen in Figure 74. Thus the entire solution for the flow over a curved wall maps into a single hodograph characteristic. Such a flow is called a Prandtl–Meyer flow, and equation (500) is sometimes called the Prandtl–Meyer equation. Because all the Mach waves present correspond to only one of the two possible characteristic directions at each point, such a flow is sometimes called a simple wave flow. Mathemati-

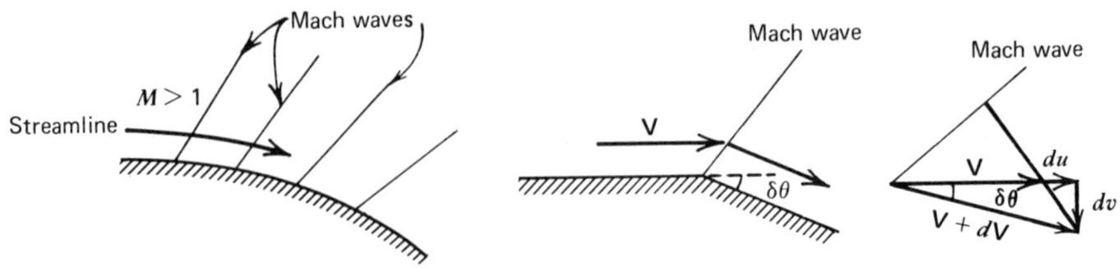

Figure 74. Mach waves for flow over a curved wall, showing change in velocity.

cally speaking, the flow is also referred to as singly degenerate. Again, the flow may be referred to as flow with waves of one family, the waves of one family being those whose slopes all correspond to $+1/\sqrt{M^2-1}$ or all correspond to $-1/\sqrt{M^2-1}$. In the above example, whether wall is curving up or down, only waves with slope $+1/\sqrt{M^2-1}$ are present, the slope being taken with respect to the local flow direction.

It should be noted that the solution does not include the distance over which the variation of θ occurs. Thus, $V(\theta)$ remains the same in cases (a), (b) and (c) as shown in Figure 75. In each case, the initial and final flow directions are the

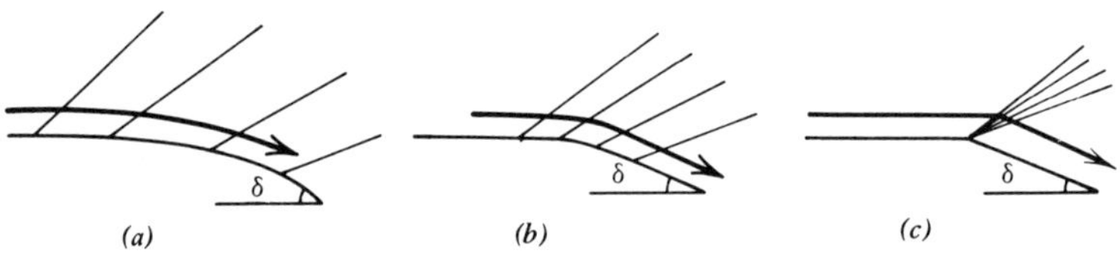

Figure 75. Mach waves for different radii of curvature.

same. Notice that in Figure 75(c), the limiting case of the change in flow direction taking place in zero length is shown. In this limit, the Mach waves then originate at a single point, but the solution remains valid. Such a wave pattern is called a Prandtl–Meyer fan or expansion fan.

Flow with Waves of Both Families

Consider first the flow under a curving wall as shown in Figure 76(a). If such

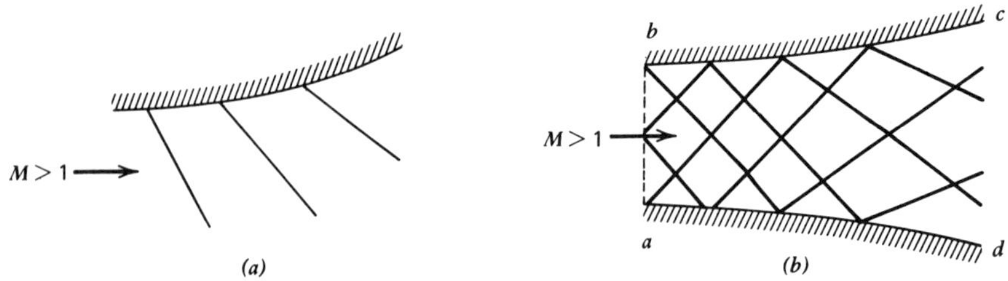

Figure 76. Flow with waves of one and two families.

a wall is placed over the curving wall of Figure 75 then it is seen that the Mach waves generated by both walls must intersect [Figure 76(b)]. Such a flow is called a flow with waves of both families. Let us now consider the following problem. Given the magnitude and direction of the velocity along the line *ab* in Figure 76(b), and the shape of the walls *bc* and *ad*, it is desired to find the magnitude and direction of the velocity in the region between the two walls. The entropy and stagnation enthalpy are presumed constant and the stagnation enthalpy is presumed known. The problem, then, may be solved by the method of characteristics. In practice, problems in the method of characteristics may be solved numerically, graphically, or by a combination of both methods.

We have already outlined the numerical method above. Let us now consider how we may solve the problem graphically.

Define the dimensionless velocity M^* as

$$M^* = \frac{V}{c^*} \tag{511}$$

Then M^* is related to M by the equation

$$M^2 = \frac{\dfrac{2}{\gamma+1}M^{*2}}{1 - \dfrac{\gamma-1}{\gamma+1}M^{*2}} \tag{512}$$

Substituting equation (512) into the Prandtl–Meyer equation, we get

$$\theta(M^*) = \pm\left\{\sqrt{\frac{\gamma+1}{\gamma-1}}\arctan\sqrt{\frac{M^{*2}-1}{\dfrac{\gamma+1}{\gamma-1} - M^{*2}}} - \arctan\sqrt{\frac{M^{*2}-1}{1-\dfrac{\gamma-1}{\gamma+1}M^{*2}}}\right\} + \text{constant} \tag{513}$$

As we have seen above, this equation is also the integral of the equation (505) for the slope of the hodograph characteristics. Let us plot equation (513) in the hodograph plane for different values of the constant of integration (Figure 77). In this plot,

$$u^* = \frac{u}{c^*} \qquad v^* = \frac{v}{c^*} \qquad \mathbf{V} = \mathbf{i}u + \mathbf{j}v \tag{514}$$

The curves marked I correspond to the minus sign in front of the bracket in equation (513), the curves marked II correspond to the plus sign. The curves all end on the circle $M^* = 1$ since the method is not valid for subsonic flow. A vector from the origin to a particular point in the hodograph plane represents

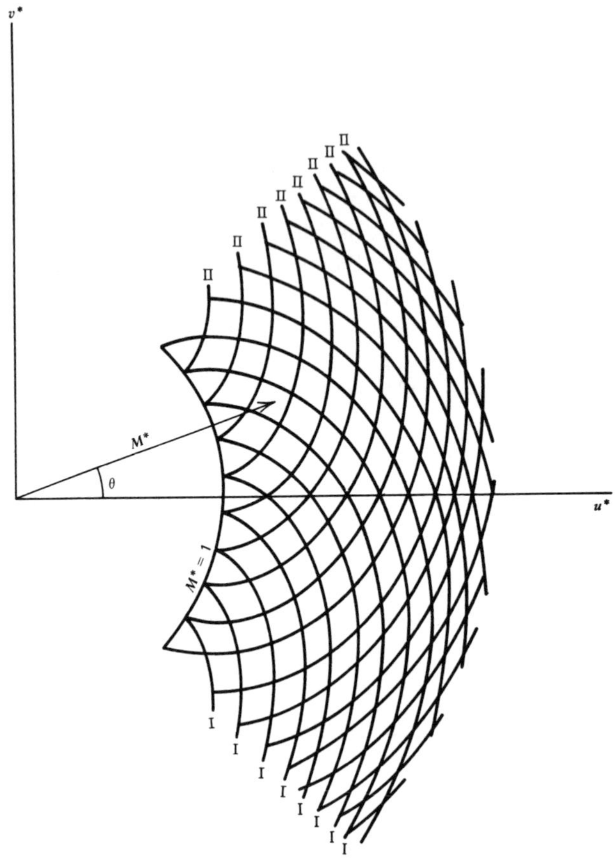

Figure 77. Hodograph characteristics.

the magnitude and direction of the velocity at the corresponding point in the physical plane, divided by c^*.

Let us see how we can use Figure 77 to solve the problem of Figure 76(b). In order to simplify the problem, let us suppose the flow is symmetric. Then we need solve only half the flow (Figure 78). In Figure 78 we have sketched in a rough indication of the physical characteristics. As yet we do not know the precise locations of the numbered points, nor the slopes of the characteristics at those points. As will be recalled, whether the plus or minus sign is used in equation (513) depends on which of the two slopes of the physical characteristics is used in the equation (479) for the hodograph characteristics. In Figure 78 we have marked the physical characteristics I and II. The use of the slope of a I characteristic in equation (479) for the slope of the hodograph characteristic results in a minus sign in equation (513) for $\theta(M^*)$ and hence in a I hodograph characteristic in Figure 77. This corresponds to the use of the minus sign in

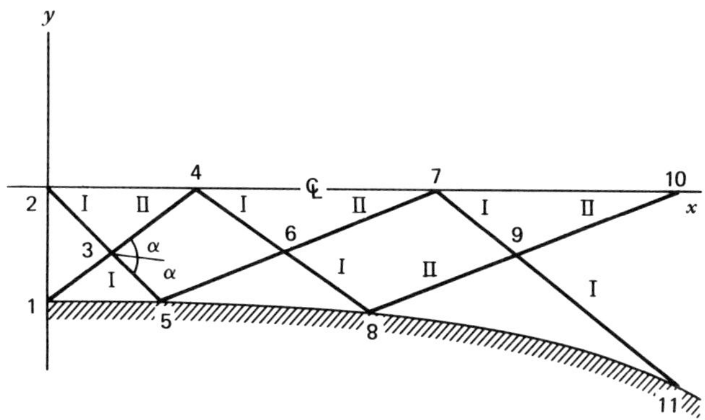

Figure 78. Physical characteristics in diverging nozzle.

front of the slope of the physical characteristic. Thus

$$\left(\frac{dy}{dx}\right)_{\mathrm{I}} = -\tan\alpha = -\frac{1}{\sqrt{M^2-1}} \tag{515}$$

$(dy/dx)_I$ being the slope of the physical characteristics marked I, M the local Mach number and α the local Mach angle, the angle which the physical characteristics make with the local streamline. Similarly, if the plus sign is used in the physical characteristics, the plus sign results in front of the bracket in equation (513) for $\theta(M^*)$ and the equation plots into the curve marked II in the hodograph plane. It will be observed that a physical characteristic of family I is perpendicular to a hodograph characteristic of family II. From equations (504, 505)

$$\frac{du}{dv} = \frac{1}{V}\left(\frac{dV}{d\theta}\right)_{\mathrm{I,II}} = \mp\tan\alpha$$

From equation (502)

$$\left(\frac{dy}{dx}\right)_{\mathrm{I,II}} = \mp\tan\alpha$$

Thus

$$\left(\frac{dv}{du}\right)_{\mathrm{II}}\left(\frac{dy}{dx}\right)_{\mathrm{I}} = \left(\frac{dv}{du}\right)_{\mathrm{I}}\left(\frac{dy}{dx}\right)_{\mathrm{II}} = -1 \tag{516}$$

and hence the two slopes are perpendicular.

Let us now assume that the flow is uniform and parallel along the line 1–2 in the physical plane, Figure 78. The flow along this line then maps into a single point in the hodograph plane, say 1, Figure 79. To get from point 2 to point 3 in the physical plane we must go along a line of family I. Therefore, to get from point 2 to point 3 in the hodograph plane we must also go along a line of family I. To get from point 1 to point 3 in the physical plane, we must go along a line of family II. Therefore, to get from point 1 to point 3 in the hodograph plane, we must also go along a line of family II. However, the distance moved along the line in the hodograph plane may be zero. Since points 1 and 2 coincide in the hodograph plane, the only way these conditions may be satisfied is if the point 3 coincides with the point 1 in the hodograph plane. Thus, the magnitude and direction of the velocity at point 3 is the same as at the points 1 and 2.

To get from the point 3 to the point 5 in the physical plane, we must go along a line of family I. Thus, to get from 3 to 5 in the hodograph plane, we must go to along a line of family I. Moreover, the direction of the velocity at point 5 is known—it must be parallel to the wall there. Thus we can locate the point 5 on the hodograph plane (Figure 79).

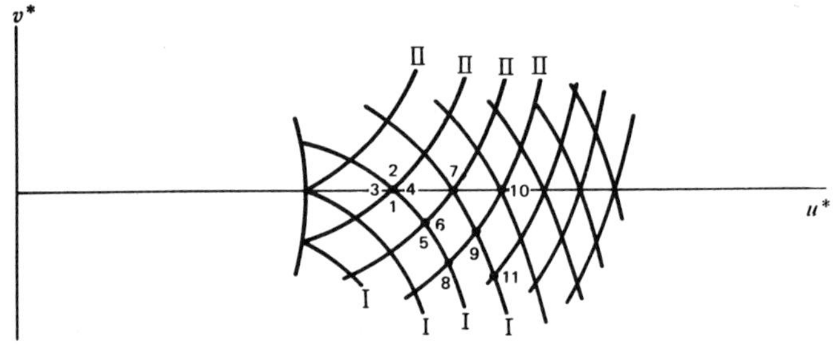

Figure 79. Hodograph characteristics in diverging nozzle.

To get from 3 to 4 we must gò along a line of family II in both planes. But the velocity at 4 is axial, due to the symmetry of the flow between the walls. Hence, the point 4 must coincide with the point 1, 2, 3 in the hodograph plane.

To get from the point 4 to the point 6 we must go along a line of family I in both planes. To get from 5 to 6 we must go along a line of family II in both planes. Both these conditions can be satisfied only if the points 5 and 6 coincide in the hodograph plane. Thus, the magnitude and direction of the velocity at 6 will be the same as at 5.

In a similar way, the magnitude and direction of the velocity may be found at points 7, 8, 9, 10, and 11, as shown in Figure 79.

It will be noted that in using the hodograph characteristics in the hodograph plane we have not had to resort to finite difference methods as we did in the physical plane. This is because the hodograph equation in plane, two-dimensional, irrotational flow is directly integrable. It will be seen subsequently that for axially symmetric, three-dimensional, or rotational flow, this will not be the case.

Since we have resorted to finite difference methods in the physical plane, (i.e., the slope of the physical characteristics has been taken as constant between points) the accuracy of the solution will depend on the fineness of the net; that is, on the closeness of the points of intersection of the characteristic lines used to each other. To illustrate the method we have used a very course net.

A finer net would look as shown in Figure 80. Naturally, the amount of work

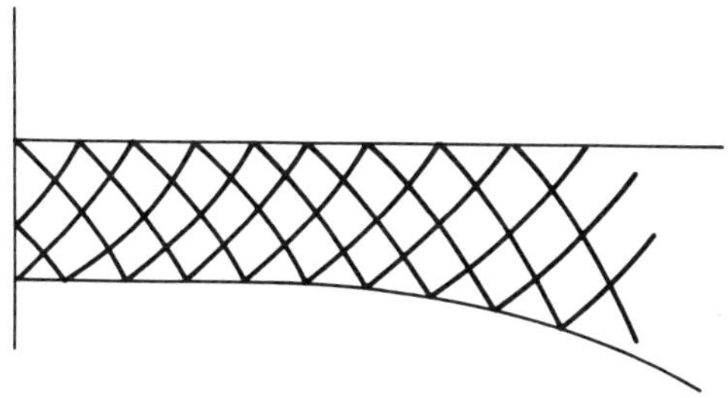

Figure 80. Finer mesh physical characteristics in diverging nozzle.

involved in calculating the points of such a net is considerably greater than for a coarse net. The work may be somewhat facilitated by the use of computers.

The above method is sometimes referred to as the lattice point method. An alternative approach would be the so-called field method (Figure 81). In this

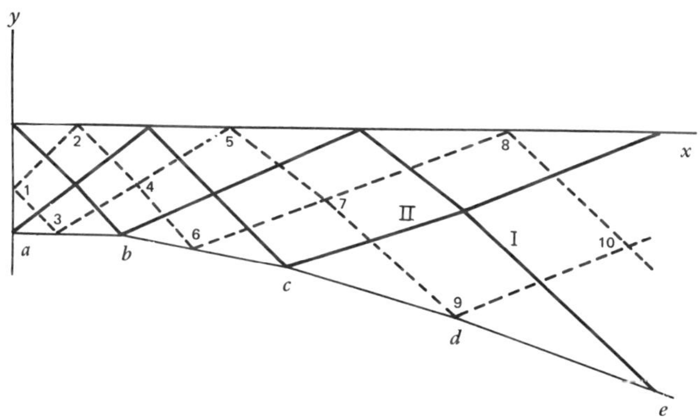

Figure 81. Field method versus lattice point method in diverging nozzle.

approach, the flow is assumed to have constant properties in each of the fields 1, 2, 3, 4, etc., bounded by the physical characteristics indicated by the solid lines. This method is, in fact, equivalent to using the lattice point method with the network of characteristics indicated by the dashed lines. In using the field method, the curved wall is replaced by a series of straight lines from a to b, b to c, c to d, and so on. The inclination of the flow in region 3 is then taken parallel to the line ab, in region 6 parallel to the line bc, and so on. In the regions 2, 5, and 8, the flow is taken to be in the x direction.

In going from one region, or "field," to the next, one goes *across* a characteristic line. This is in contrast to the lattice point method in which one went *along* a characteristic line from one point to the next. An examination of the dashed lines in Figure 81 indicates however, that going across a characteristic line of family II is equivalent to going along a characteristic line of family I, and vice versa. Thus, the conditions in the various fields may easily by found on the hodograph plane (Figure 82).

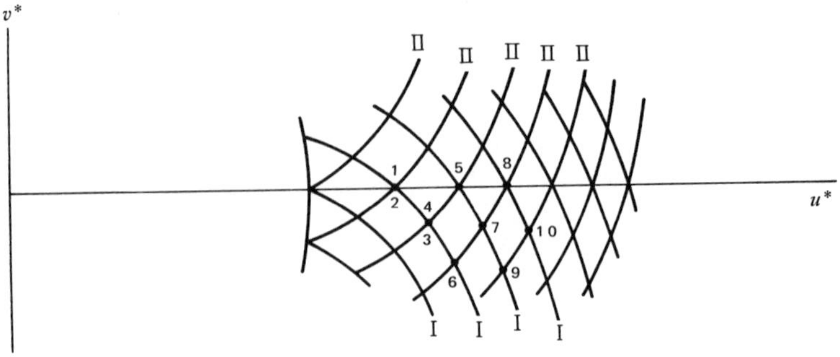

Figure 82. Hodograph characteristics in diverging nozzle—field method.

For the present example, the field method is not so good as the lattice point method since the slopes of the lines *ab*, *bc*, *cd*, etc. are not known, a priori. In certain cases, however, the field method is superior. For example, in the case of a free jet (Figure 83). Consider a slightly underexpanded supersonic

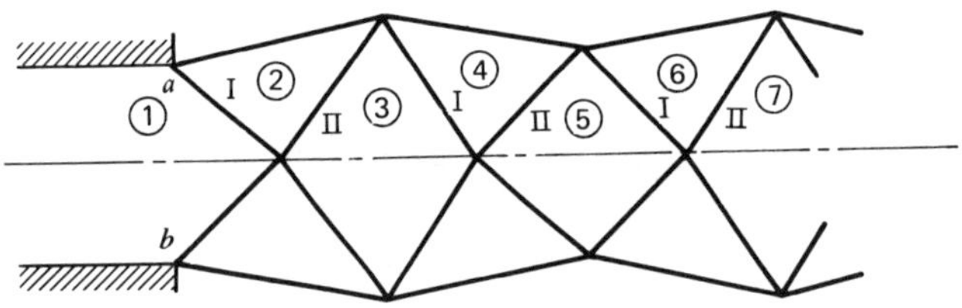

Figure 83. Compression and expansion waves in a free jet.

nozzle. The pressure in region 1 is slightly higher than the external pressure p_a. In region 2, the pressure must be equal to p_a since no pressure discontinuities are possible across a free boundary. Thus, an expansion wave separates the regions 1 and 2. To be strictly accurate, an expansion, or Prandtl–Meyer fan must occur at points *a* and *b*. If the pressure difference across the fan is small however, we may approximate it by a single expansion wave inclined at the Mach angle to the flow in region 1, or by a wave at the average wave angle between regions 1 and 2.

The conditions in region 1 are given and the pressure in region 2 is known. Thus, the equations

$$\theta = \theta(M), \qquad p_0 = p\left(1 + \frac{\gamma - 1}{2} M^2\right)^{\gamma/\gamma - 1)} \tag{517}$$

where $\theta = \theta(M)$ is the Prandtl–Meyer equation, fix the Mach number and flow direction in region 2. The second of equations (517) may be solved for M and the corresponding θ may then be found from Figure 84, the hodograph characteristics. In going from region 1 to region 2 we go across a wave of family I and, therefore, along a wave of family II. In this connection it is helpful to

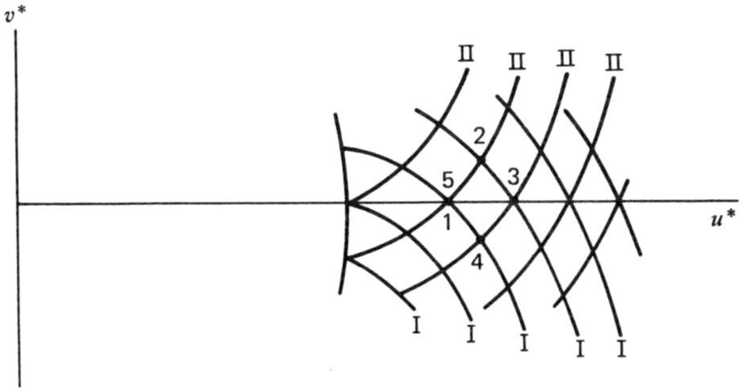

Figure 84. Hodograph characteristics for free jet.

remember a simple rule: expansion waves turn the flow *away* from the wave; compression waves turn the flow *toward* the wave. This is illustrated in Figure 85.

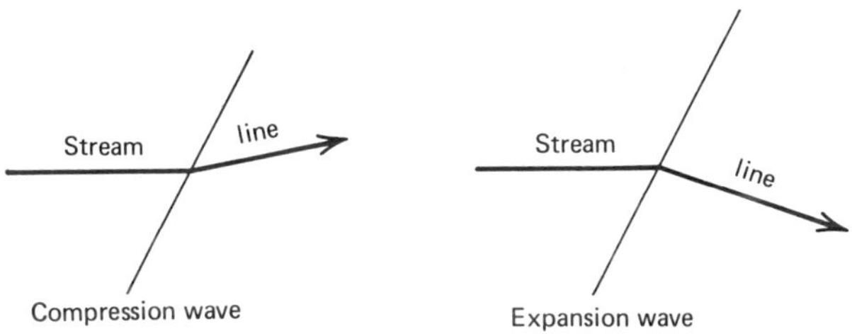

Figure 85. Compression and expansion waves, showing paths of streamlines.

Moreover, $p_2 < p_1, M_2 > M_1$. Thus, the point 2 is easily located in the hodograph plane. In region 3 the flow must be parallel to the x axis due to the symmetry of the flow. Since, in going from 2 to 3 we go across a characteristic of family II, we must go along a characteristic of family I in the hodograph plane. The point 3 in this plane is thus located. Since $M_3 > M_2$, it follows that $p_3 < p_a$. Thus, in going from region 3 to 4, the flow must go through a compression wave. To get from 3 to 4 we go across a wave of family I and hence along a wave of family II. Moreover, $M_4 = M_2$ since $p_4 = p_2 = p_a$, and the flow is presumed isentropic; i.e., $p_0 =$ constant. The point 4 is thus located on the hodograph plane. In region 5 the flow must be axial, due to the symmetry of the flow. Thus, in going from 4 to 5, the flow is turned toward the wave separating the two regions, and hence this wave is a compression wave. It is also a wave of family II, and hence to go from 4 to 5 in the hodograph plane we must go along a wave of family I. This locates the point 5 in the hodograph plane. The point 5 is seen to coincide with the point 1, and so the cycle is repetitive. That is, the conditions in region 7 will be the same as the conditions in region 2, the next region will be equivalent to region 3, and so on.

This solution points up another useful rule: Waves reflect in like manner off solid boundaries and in unlike manner off free boundaries. This means that off a straight solid wall, a compression wave will reflect as a compression wave and an expansion wave will reflect as an expansion wave (Figure 86). Off a

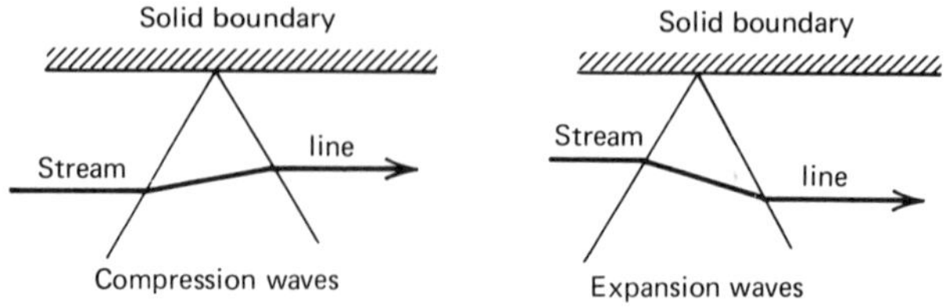

Figure 86. Reflection of waves off a solid boundary.

free boundary, a compression wave reflects as an expansion wave while an expansion wave will reflect as a compression wave. The reason is that along a solid boundary, the flow direction is the boundary condition, while off a free boundary the pressure is the boundary condition. If a solid wall is bent at the point where a wave touches it by the same amount as the streamline, no reflection will result. This technique is sometimes used to design supersonic nozzles with uniform flow at the exit (Figure 87).

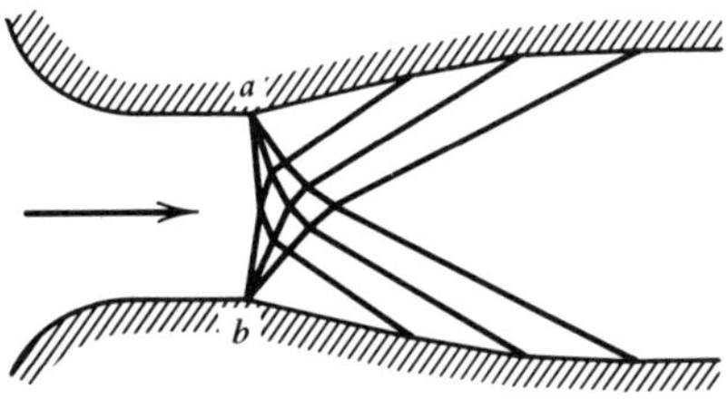

Figure 87. Supersonic nozzle with uniform flow at exit.

If a sharp corner at a, b is not desired one of the interior streamlines may be used as the wall.

Axially Symmetric Characteristics

The method of characteristics may be applied to axially symmetric flow as well as plane flow. Consider the coordinate system shown in Figure 88. If the

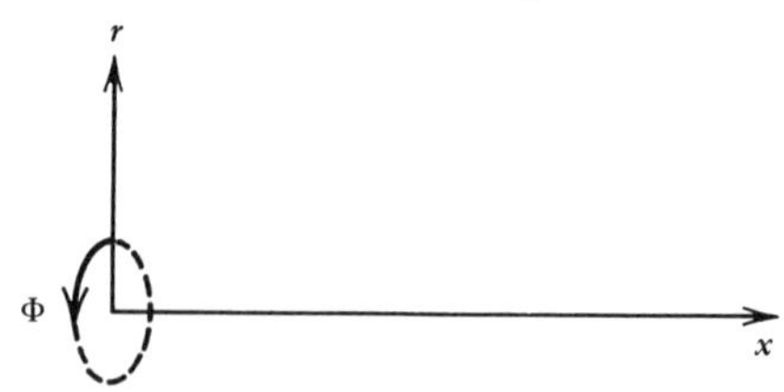

Figure 88. Polar coordinate system for axially symmetric flow.

flow is axially symmetric, the flow properties are independent of Φ. If u is the velocity component in the x direction, and v is the velocity component in the r direction, then,

$$u = \phi_x, \qquad v = \phi_r, \qquad \mathbf{V} = \nabla\phi$$

where ϕ is the velocity potential. The equation for the velocity potential then becomes

$$(c^2 - \phi_x^2)\phi_{xx} - 2\phi_x\phi_r\phi_{xr} + (c^2 - \phi_r^2)\phi_{rr} + \frac{c^2}{r}\phi_r = 0 \tag{518}$$

Applying the method of characteristics, we find, for the slope of the physical characteristics

$$\left(\frac{dr}{dx}\right)_{\mathrm{I,II}} = \frac{\phi_x\phi_r \mp c\sqrt{\phi_x^2 + \phi_r^2 - c^2}}{\phi_x^2 - c^2} = \tan(\theta \mp \alpha) \tag{519}$$

where θ is the flow direction and α the Mach angle. It will be observed that this is the same result which we obtained for plane flow.

The equation for the hodograph characteristics becomes

$$(c^2 - \phi_x^2)\left(\frac{dr}{dx}\right)_{\mathrm{I,II}} d\phi_x + (c^2 - \phi_r^2)d\phi_r + \frac{c^2\phi_r}{r}dr = 0 \tag{520}$$

If we now convert to the polar coordinate system in the r, x plane,

$$\phi_x = u = V\cos\theta, \qquad \phi_r = v = V\sin\theta \tag{521}$$

where V is the magnitude of the velocity, the equation for the hodograph characteristics becomes,

$$\sin\theta\frac{dr}{r} + \left[\cos\theta \mp \frac{\sqrt{V^2 - c^2}}{c}\sin\theta\right]\left[d\theta \pm \frac{\sqrt{V^2 - c^2}}{c}\frac{dV}{V}\right] = 0 \tag{522}$$

In terms of the Mach angle $\alpha = \text{arc}\sin 1/\mathrm{M}$, this becomes

$$d\theta \pm \cot\alpha\frac{dV}{V} \mp \frac{\sin\theta\sin\alpha}{\sin(\alpha \mp \theta)}\frac{dr}{r} = 0 \tag{523}$$

The first two terms of the equation are identical with the plane case (equation (505)). However, we see that there is now a third term. This makes impossible the integration of the equation once for all cases which enabled us to draw Figure 77 for the hodograph characteristics in the plane case. In the axially symmetric case the equation must be integrated separately and numerically for each individual case, making such calculations more tedious in the axially symmetric case than in the plane case.

Three-Dimensional Characteristics

In three-dimensional, cartesian coordinates, the equation for the velocity potential becomes

$$\begin{aligned}(u_1^2 - c^2)\phi_{x_1x_1} &+ (u_2^2 - c^2)\phi_{x_2x_2} + (u_3^2 - c^2)\phi_{x_3x_3} \\ &+ 2u_1u_2\phi_{x_1x_2} + 2u_2u_3\phi_{x_2x_3} + 2u_1u_3\phi_{x_1x_3} = 0\end{aligned} \tag{524}$$

where u_1, u_2, and u_3 are the velocity components in the x_1, x_2, and x_3 directions. This may be written in the form

$$\sum_{i,j=1}^{3} a_{ij}\phi_{x_i x_j} = 0$$

where

$$\begin{aligned} a_{11} &= (u_1^2 - c^2) & a_{12} &= u_1 u_2 & a_{21} &= u_2 u_1 \\ a_{22} &= (u_2^2 - c^2) & a_{13} &= u_1 u_3 & a_{31} &= u_3 u_1 \\ a_{33} &= (u_3^2 - c^2) & a_{23} &= u_2 u_3 & a_{32} &= u_3 u_2 \end{aligned}$$

Choose now any surface

$$\bar{x}_1 = f(\bar{x}_2, \bar{x}_3) \tag{525}$$

Then

$$\frac{\partial \phi_{x_i}}{\partial \bar{x}_2} = \sum_j \phi_{x_i x_j} \frac{\partial x_i}{\partial \bar{x}_2}, \qquad \frac{\partial \phi_{x_i}}{\partial \bar{x}_3} = \sum_j \phi_{x_i x_j} \frac{\partial x_i}{\partial \bar{x}_3} \tag{526}$$

$i, j = 1, 2, 3$. If we now consider the six $\phi_{x_i x_j}$ to be the unknowns, we have seven equations. The six equations (526) and the potential equation (524). Thus, only five of the equations (526) can be independent. Using five of the equations (526) and the potential equation we can solve for any of the six unknowns $\phi_{x_i x_j}$. The solutions will have the form of the quotient of two determinants

$$\phi_{x_i x_j} = \frac{Y_{ij}}{Z} \tag{527}$$

As in the two-dimensional case, we assume that the $\phi_{x_i x_j}$ are everywhere bounded. Thus, when $Z = 0$, $Y_{ij} = 0$ also.

$$\begin{aligned} Z = (u_1^2 - c^2)L_1^2 + (u_2^2 - c^2)L_2^2 + (u_3^2 - c^2)L_3^2 + 2u_1 u_2 L_1 L_2 \\ + 2u_2 u_3 L_2 L_3 + 2u_3 u_1 L_3 L_1 \end{aligned} \tag{528}$$

keeping in mind that $\phi_{x_i} = u_i$, $i = 1, 2, 3$ and

$$L_i = \frac{\partial x_j}{\partial \bar{x}_2} \cdot \frac{\partial x_k}{\partial \bar{x}_3} - \frac{\partial x_k}{\partial \bar{x}_2} \frac{\partial x_j}{\partial \bar{x}_3} \tag{529}$$

$i, j, k = 1, 2, 3$. The L_i are proportional to the direction cosines of the normal to the surface $\bar{x}_1 = f(\bar{x}_2, \bar{x}_3)$. Equation (528) is of the form

$$Z = \sum_{i,j} a_{ij} L_i L_j \tag{530}$$

$i, j = 1, 2, 3$. Thus, the equation $Z = 0$ will be satisfied by an infinite number of direction numbers L_i, $i = 1, 2, 3$, each set of L_i corresponding to a different surface $\bar{x}_1 = f(\bar{x}_2, \bar{x}_3)$. If these L_i are real, the equation (524) is said to be hyperbolic. If the L_i are imaginary, equation (524) is said to be elliptic. If equation (530) is a perfect square, equation (524) is said to be parabolic. The L_i will be real provided

$$u_1^2 + u_2^2 + u_3^2 > c^2 \tag{531}$$

Thus, the potential equation will be hyperbolic if $M > 1$, elliptic if $M < 1$, and parabolic if $M = 1$, just as in the two-dimensional case. If equation (524) is hyperbolic, all the lines defined by the direction numbers L_i, which satisfy the relation $Z = 0$ at a given point lie on a cone whose vertex coincides with that point. Thus an infinite number of surfaces $x_1 = f(x_2, x_3)$ satisfy the relation $Z = 0$ at a given point. However, the envelope of the tangent planes to these surfaces at the given point will likewise be a cone since the locus of the normals to the surfaces at the point is a cone. Figure 89 illustrates this point.

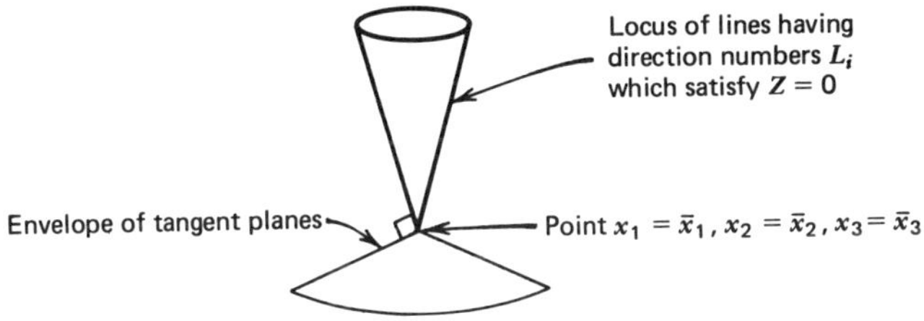

Figure 89. Characteristic surfaces in three-dimensional, supersonic flow.

The tangent planes have the equation

$$L_1(x_1 - \bar{x}_1) + L_2(x_2 - \bar{x}_2) + L_3(x_3 - \bar{x}_2) = 0$$

and hence the envelope of the tangent planes is given by

$$A_{ij}(x_i - \bar{x}_i)(x_j - \bar{x}_j) = 0 \tag{532}$$

$i, j = 1, 2, 3$, and A_{ij} is the cofactor of a_{ij} in the determinant

$$\begin{vmatrix} a_{11} & a_{12} & a_{13} \\ a_{21} & a_{22} & a_{23} \\ a_{31} & a_{32} & a_{33} \end{vmatrix}$$

The envelope of the tangent planes given by equation (532) is called the characteristic cone. It is the three-dimensional equivalent of the characteristic lines in the two-dimensional case. Substituting for a_{ij} in equation (532), the equation for the characteristic cone becomes

$$\begin{aligned} &(u_2^2 + u_3^2 - c^2)(x - x_1)^2 + (u_3^2 + u_1^2 - c^2)(x_2 - \bar{x}_2)^2 + (u_1^2 + u_2^2 - c^2)(x_3 - \bar{x}_3)^2 \\ &- 2u_2u_3(x_2 - \bar{x}_2)(x_3 - \bar{x}_3) - 2u_3u_1(x_1 - \bar{x}_1)(x_3 - \bar{x}_3) \\ &- 2u_2u_1(x_1 - \bar{x}_1)(x_2 - \bar{x}_2) = 0 \end{aligned} \tag{533}$$

If we take the point $\bar{x}_1, \bar{x}_2, \bar{x}_3$ as the origin of our coordinate system and our axes so that u_2 and u_3 are zero, we get

$$-c^2x_1^2 + (u_1^2 - c^2)x_2^2 + (u_1 - c^2)x_3^2 = 0$$

$$x_2^2 + x_3^2 = \frac{x_1^2}{(M^2 - 1)} \tag{534}$$

This is the equation of a cone, vertex at the origin, axis along the x_1 axis, i.e., in the flow direction (Figure 90).

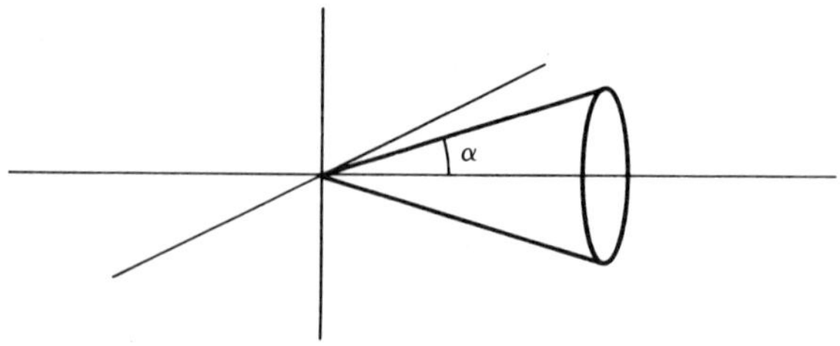

Figure 90. Mach cone in three-dimensional, supersonic flow.

The semivertex angle α is given by

$$\tan\alpha = \frac{x_1}{\sqrt{x_2^2 + x_3^2}} = \frac{1}{\sqrt{M^2 - 1}} \qquad (535)$$

Thus, α is seen to be the Mach angle and the cone given by equation (532) is called the Mach cone as well as the characteristic cone. Equation (535) states that a disturbance at a point in a three-dimensional, steady, supersonic potential flow can affect the flow only inside the Mach cone downstream of the point. This result is analogous to that obtained in the two-dimensional, plane case, in which a disturbance at a point was found to affect the flow only in the region downstream of the point between the two characteristic lines passing through the point.

While, strictly speaking, it is possible to calculate the entire flow field in a region, given suitable boundary conditions, by setting the numerator of equation (527) equal to zero and solving for the $\phi_{x_i\bar{x}_j}$, $i, j = 1, 2, 3$, in practice the procedure is too cumbersome to be of much practical value. Details of the numerical method are given by Thornhill [31] and Ferri [32]. In practice, simplifying assumptions are usually made (see, for example, Ferri [32, 33]) such as the linearization of the potential equation in the case of small perturbations.

Oblique Shock Waves

In Chapter 2 we demonstrated theoretically the possibility of the existence of discontinuities in the flow properties of a one-dimensional flow. While in such a flow, the flow properties can vary in only one direction, we pictured the flow in the x, y plane consisting of a number of identical streamlines, one on top of the other. In such a flow, the discontinuity appears as a straight line perpendicular to the flow direction.

We now consider the question of whether it is possible to have a flow, which is initially uniform and parallel, with a straight line discontinuity set at an angle other than 90° to the flow direction (Figure 91).

That such a case is possible can be demonstrated by a simple process of reasoning. Consider a flow with a normal shock wave. In Figure 92 such a flow is shown with respect to an observer at rest. Consider, however, how this flow

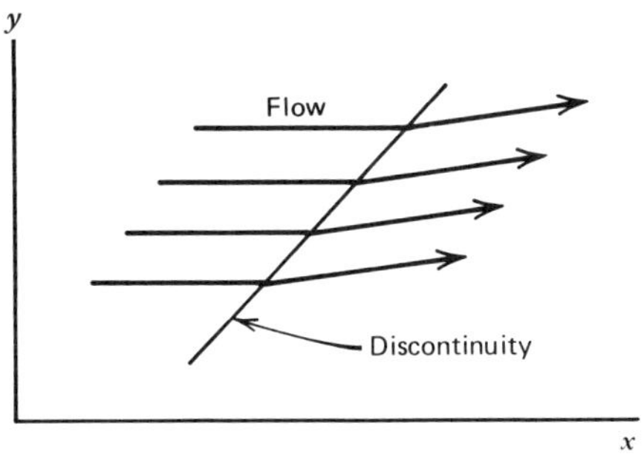

Figure 91. Oblique shock with streamline.

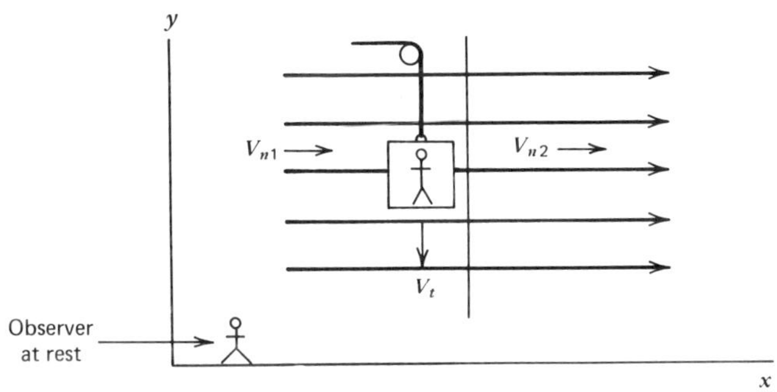

Figure 92. Shock wave with stationary and moving coordinate systems.

appears to an observer traveling downward along the shock wave in a elevator with a velocity V_t. Let V_{n1} and V_{n2} be the velocity before and after the shock wave as seen by the observer at rest. To the observer in the elevator, the velocity before the shock will appear as V_1 and the velocity after the shock as V_2 (Figure 93).

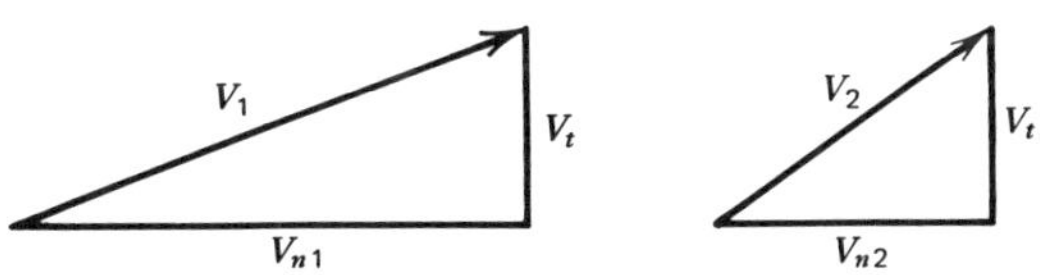

Figure 93. Change in flow direction through an oblique shock.

Thus, the flow with respect to the moving observer appears as shown in Figure 94. By tilting his head to one side a little bit, the moving observer may see the flow as in Figure 95. The word moving is relative, however. The observer in the elevator is "moving" with respect to the observer "at rest." One may just as well say that the observer "at rest" is moving with respect to the observer in the elevator. Thus, if the flow in Figure 92 is physically possible, so is the flow in Figure 95.

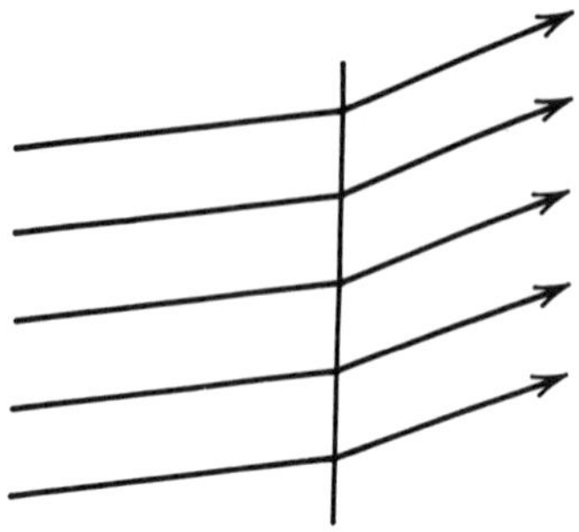

Figure 94. Oblique shock as seen by man in elevator.

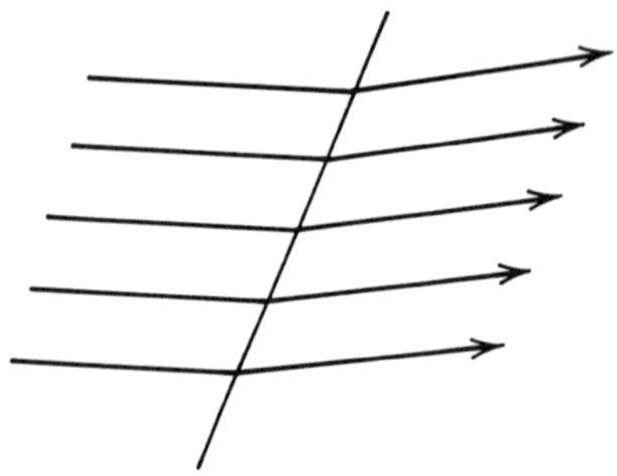

Figure 95. Oblique shock as seen by man in elevator leaning to one side.

The case may be stated in more mathematical terms by recalling that our fundamental equations of motion are valid in any "inertial" or "Newtonian" coordinate system. This means one that is not accelerating. If the elevator is descending with a steady speed, it represents the origin of a perfectly valid inertial coordinate system. Let us, therefore, rewrite the equations for the normal shock in terms of the new coordinate system. Consider the control volume as shown in Figure 96. The continuity equation becomes

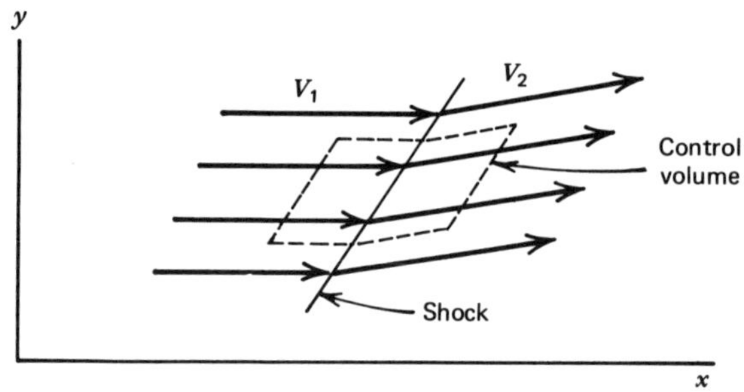

Figure 96. Control volume around an oblique shock wave.

$$\oint \rho \mathbf{V} \cdot \mathbf{n} dS = 0 \qquad \text{or} \qquad \rho_1 V_{n1} = \rho_2 V_{n2} \tag{536}$$

The momentum equation becomes

$$\oint \rho \mathbf{V}(\mathbf{V} \cdot \mathbf{n}) dS = \oint p \mathbf{n}\, dS \tag{537}$$

We may write $\mathbf{V}$ in terms of V_n and V_t. Thus $\mathbf{V} = \mathbf{i}_1 V_n + \mathbf{i}_2 V_t$ where $\mathbf{i}_1$ and $\mathbf{i}_2$ represent the unit vectors in the direction normal and tangential to the shock. Equating $\mathbf{i}_1$ and $\mathbf{i}_2$ components we get

$$(\rho_1 V_{n1})V_t = (\rho_2 V_{n2})V_t, \qquad p_1 - p_2 = \rho_2 V_{n2}^2 - \rho_1 V_{n1}^2 \tag{538}$$

where the subscripts 1 and 2 designate the upstream and downstream side of the shock, respectively. Notice that the state properties, p, ρ, T, etc., are unaffected by the choice of the coordinate system.

The energy equation. If we neglect thermal conductivity and heat transfer, we may write equation (253)

$$\oint \mathbf{n}\cdot h\mathbf{V}\rho\, dS = -\oint \mathbf{n}\cdot \mathbf{V}\rho \frac{V^2}{2} dS \tag{539}$$

or

$$c_p(T_1 - T_2) = \frac{(V_2^2 - V_1^2)}{2} \tag{540}$$

It will be observed here that stagnation enthalpy is preserved across the shock in the new coordinate system; however, its value is not the same as in the case of the previous, or "normal shock" coordinate system.

Recall now the normal shock equations (211) to (215). In order to rewrite these equations in our new coordinate system it is only necessary to substitute M_{n1} for the M_1 and M_{n2} for the M_2 in these equations. Thus we now have

$$\frac{p_2}{p_1} = \frac{1 + \gamma M_{n1}^2}{1 + \gamma M_{n2}^2} \tag{541}$$

$$M_{n2}^2 = \frac{M_{n1}^2 + \dfrac{2}{\gamma - 1}}{\dfrac{2\gamma}{\gamma - 1} M_{n1}^2 - 1} \tag{542}$$

$$\frac{T_2}{T_1} = \frac{\left(1 + \dfrac{\gamma - 1}{2} M_{n1}^2\right)\left(\dfrac{2\gamma}{\gamma - 1} M_{n1}^2 - 1\right)}{\dfrac{(\gamma + 1)}{2(\gamma - 1)} M_{n1}^2} \tag{543}$$

$$\frac{p_2}{p_1} = \frac{2\gamma}{\gamma + 1} M_{n1}^2 - \frac{\gamma - 1}{\gamma + 1} \tag{544}$$

$$\frac{\rho_2}{\rho_1} = \frac{V_{n1}}{V_{n2}} = \frac{\dfrac{\gamma + 1}{2} M_{n1}^2}{1 + \dfrac{\gamma - 1}{2} M_{n1}^2} \tag{545}$$

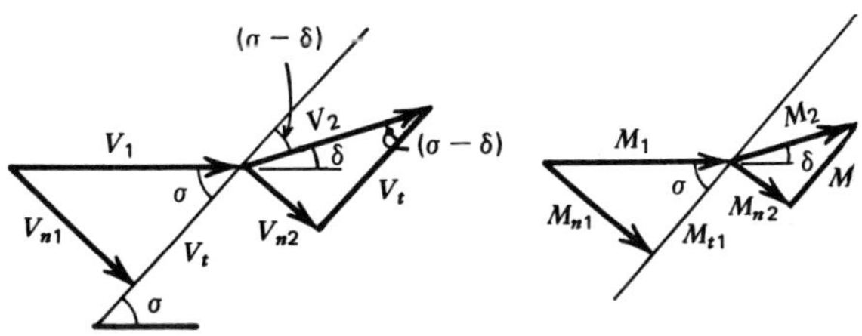

Figure 97. Variation of velocity components and flow direction through an oblique shock with symbols defined.

In practice it is more convenient to rewrite these equations in terms of the *new* M_1, M_2, V_1 and V_2 (Figure 97).

Notice that $M_{t1} > M_{t2}$ since the speed of sound rises across the wave. Referring to Figure 97 we see that

$$\begin{aligned} V_{n1} &= V_1 \sin\sigma & M_{n1} &= M_1 \sin\sigma \\ V_{n2} &= V_2 \sin(\sigma-\delta) & M_{n2} &= M_2 \sin(\sigma-\delta) \\ V_t &= V_1 \cos\sigma & V_{t_2} &= V_{t_1} = V_2\cos(\sigma-\delta) \end{aligned} \tag{546}$$

Hence

$$\frac{p_2}{p_1} = \frac{1+\gamma M_1^2 \sin^2\sigma}{1+\gamma M_2^2\sin^2(\sigma-\delta)} \tag{547}$$

$$M_2^2\sin^2(\sigma-\delta) = \frac{M_1^2\sin^2\sigma + \dfrac{2}{\gamma-1}}{\dfrac{2\gamma}{\gamma-1}M_1^2\sin^2\sigma - 1} \tag{548}$$

$$\frac{p_2}{p_1} = \frac{2\gamma}{\gamma+1}M_1^2\sin^2\sigma - \frac{\gamma-1}{\gamma+1} \tag{549}$$

$$\frac{\rho^2}{\rho_1} = \frac{V_1\sin\sigma}{V_2\sin(\sigma-\delta)} = \frac{\dfrac{\gamma+1}{2}M_1^2\sin^2\sigma}{1+\dfrac{\gamma-1}{2}M_1^2\sin^2\sigma} \tag{550}$$

Also, since $V_{t1} = V_{t2}$, we have

$$\frac{V_1}{V_2} = \frac{\cos(\sigma-\delta)}{\cos\sigma} \tag{551}$$

and, from the continuity equation,

$$\frac{\rho_2}{\rho_1} = \frac{V_{n1}}{V_{n2}} = \frac{V_1}{V_2}\frac{\sin\sigma}{\sin(\sigma-\delta)} = \frac{\cos(\sigma-\delta)\cdot\sin\sigma}{\cos\sigma\cdot\sin(\sigma-\delta)}$$

$$\frac{\rho_2}{\rho_1} = \frac{\tan\sigma}{\tan(\sigma-\delta)} \tag{552}$$

In addition, the Rankine–Hugoniot relation must remain valid

$$\frac{p_2}{p_1} = \frac{\dfrac{\gamma+1}{\gamma-1}\dfrac{\rho_2}{\rho_1} - 1}{\dfrac{\gamma+1}{\gamma-1} - \dfrac{\rho_2}{\rho_1}} \tag{553}$$

We now have four independent equations, (548), (549), (552), (553), and six unknowns: M_1, M_2, σ, δ, p_2/p_1, and ρ_2/ρ_1. Thus, we can plot any two of these variables against each other, using a third as a parameter. For example, if we plot M_1 versus σ using δ as a parameter, we get (Figure 98). This figure reveals

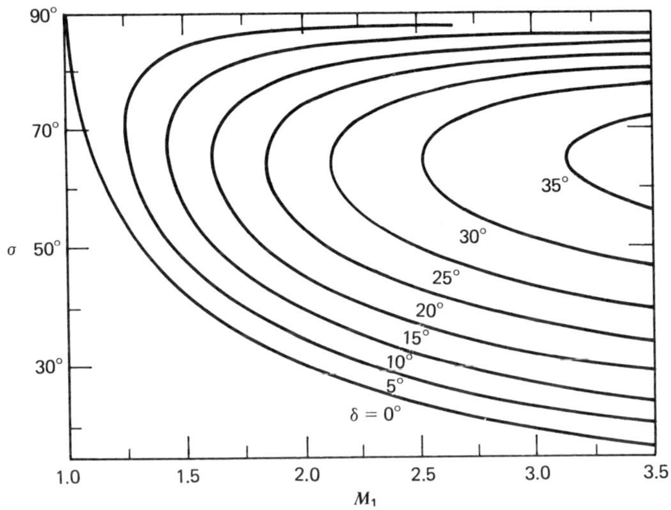

Figure 98. Shock wave angle as a function upstream Mach number and flow deflection angle.

a number of interesting features. First, it is seen that for each M_1 above a certain minimum, there are two solutions for σ for each given δ. Below the minimum M_1, however, there is no solution. This result has been investigated experimentally by placing a wedge in a supersonic flow (Figure 99). It is found

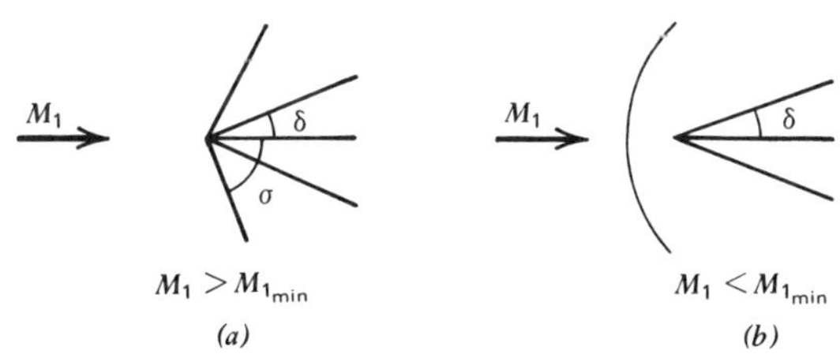

Figure 99. Attached and detached shocks.

that as long as M_1 is above its minimum value for a given δ, a shock is attached the wedge at the angle predicted by the lower part of the curve of Figure 99. When M_1 is below its minimum value, however, it is found that a curved shock stands in front of the wedge instead [Figure 99(b)]. Such a shock is called a detached shock. One may inquire why the shock wave detaches when the Mach number drops below a certain value, or the wedge angle increases beyond a certain value. The most obvious answer to this question would be that the normal component of the Mach number drops below one. An examination of Figure 98 reveals that this is not the case. For every minimum M_1, M_{n1} is still well above one. There does not appear to be any simple physical explanation of shock detachment. One can only say that no simultaneous solution of the above equations exists when M_1 is below $M_{1\min}$ and hence no attached shock can exist under this circumstance without violating one or more of the fundamental laws of nature.

Prandtl Relation

For a normal shock, we found the Prandtl relation

$$V_{n1} V_{n2} = c_n^{*2} \tag{554}$$

The c_n^* was based on the velocity normal to the shock, relative to the shock. Relative to an oblique shock, a new c^* must be defined. Thus

$$c_n^{*2} \frac{\gamma+1}{2(\gamma-1)} = \frac{1}{\gamma-1} c_1^2 + \frac{V_{n1}^2}{2} = \frac{1}{\gamma-1} c_2^2 + \frac{V_{n2}^2}{2} \tag{555}$$

On the other hand, if c^* is the critical speed of sound relative to the oblique shock,

$$c^{*2} \frac{\gamma+1}{2(\gamma-1)} = \frac{1}{\gamma-1} c_1^2 + \frac{V_{n1}^2 + V_t^2}{2} = \frac{1}{\gamma-1} c_2^2 + \frac{V_{n2}^2 + V_t^2}{2} \tag{556}$$

$$= c_n^{*2} \frac{\gamma+1}{2(\gamma-1)} + \frac{V_t^2}{2}$$

Thus,

$$c_n^{*2} = c^{*2} - V_t^2 \frac{\gamma+1}{\gamma-1} \tag{557}$$

and hence,

$$V_{n1} V_{n2} = c^{*2} - V_t^2 \frac{\gamma+1}{\gamma-1} \tag{558}$$

Shock Polar

A useful relationship exists between the velocities before and after an oblique shock in terms of the x and y components of the velocity. Consider the coordinate system shown in Figure 100. Let u and v be the x and y components of the

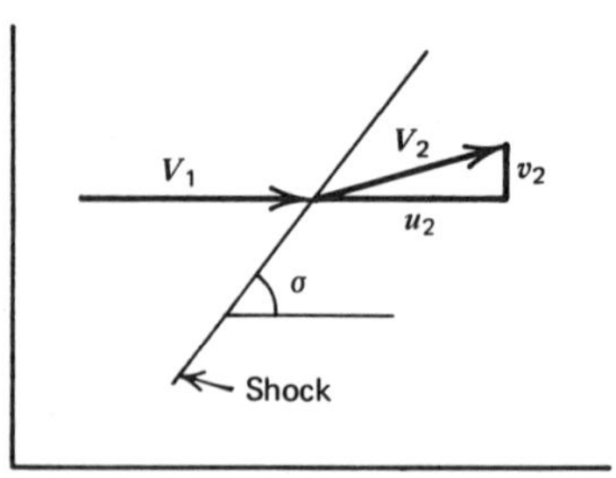

Figure 100. x and y components of velocity after oblique shock.

velocity. Then $V_{n1} = u_1 \sin\sigma$ and $V_t = u_1 \cos\sigma$, $V_1 = u_1$. Let us now draw the velocity components in the hodograph (u, v) plane (Figure 101):

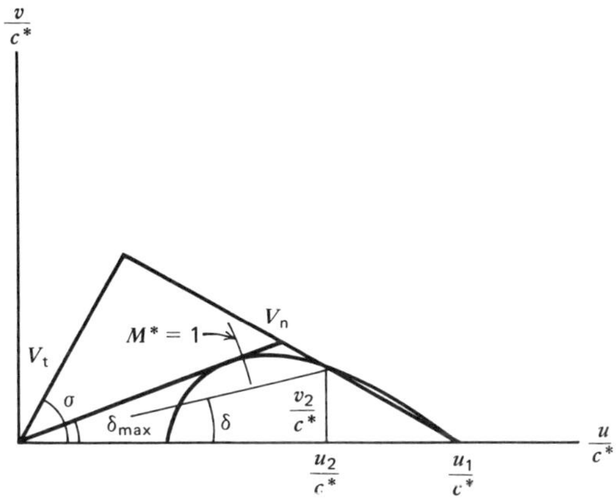

Figure 101. Oblique shock velocity components in hodograph plane.

An examination of the figure shows that

$$V_{n2} = V_{n1} - \sqrt{v_2^2 - (u_2 - u_1)^2} = u_1 \sin\sigma - \sqrt{v_2^2 + (u_2 - u_1)^2} \tag{559}$$

Substituting in equation (558)

$$V_{n2} V_{n1} = (u_1 \sin\sigma)(u_1 \sin\sigma - \sqrt{v_2 + (u_2 - u_1)^2}) = c^{*2} - \frac{\gamma - 1}{\gamma + 1} u_1^2 \cos^2\sigma \tag{560}$$

Again referring to Figure 101, we see that

$$\sin\sigma = \cos\sigma' = \frac{u_2 - u_1}{\sqrt{v_2^2 + (u_2 - u_1)^2}} \tag{561}$$

and

$$\cos\sigma = \sin\sigma' = \frac{v_2}{\sqrt{v_2^2 + (u_2 - u_1)^2}} \tag{562}$$

Substituting equations (561) and (562) into equation (560) and rearranging terms, we get,

$$(u_1 - u_2)^2 \left(u_2 - \frac{c^{*2}}{u_1} \right) = v_2^2 \left(\frac{c^{*2}}{u_1} + \frac{2}{\gamma + 1} u_1 - u_2 \right) \tag{563}$$

A plot of this equation in the hodograph plane for various values of u_1/c^* is shown in Figure 102:

The curves shown in Figure 102 are known as shock polars. It will be noted that when $M_1 = \infty$, the shock polar becomes a circle. The shock polars inside this circle, then, cover all cases from $M_1 = 1$ to $M_1 = \infty$. Notice that the distance from the origin in Figure 102 is M^*. Recall that when $M = \infty$,

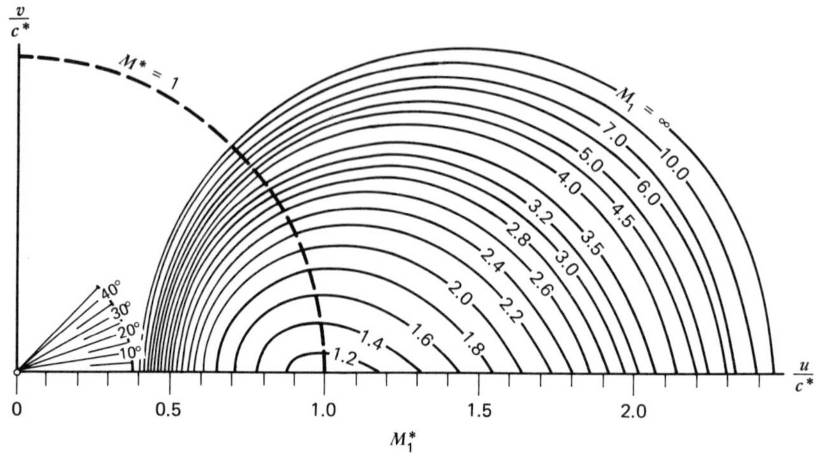

Figure 102. Shock polars.

$M^* = \sqrt{(\gamma+1)/(\gamma-1)}$. Thus, the circle $M_1 = \infty$ intercepts the u/c^* axis at $u/c^* = \sqrt{(\gamma+1)(\gamma-1)}$, this being the value of u_1/c^* when $M_1 = \infty$.

For every shock polar, there are two solutions when $\delta = 0$. The "weak" solution, corresponding to no change in the flow at all, $u_2/c^* = u_1/c^*$, and the "strong" solution, corresponding to a normal shock. u_2/c^* for the latter is given by the second point where the shock polar intercepts the u/c^* axis, the first being u_1/c^*. For $M_1^* = \sqrt{(\gamma+1)/(\gamma-1)}$, $M_2^* = \sqrt{(\gamma-1)/(\gamma+1)}$ (from the normal shock relations). These are the two intercepts of the circular shock polar for $M_1 = \infty$ on the u/c^* axis:

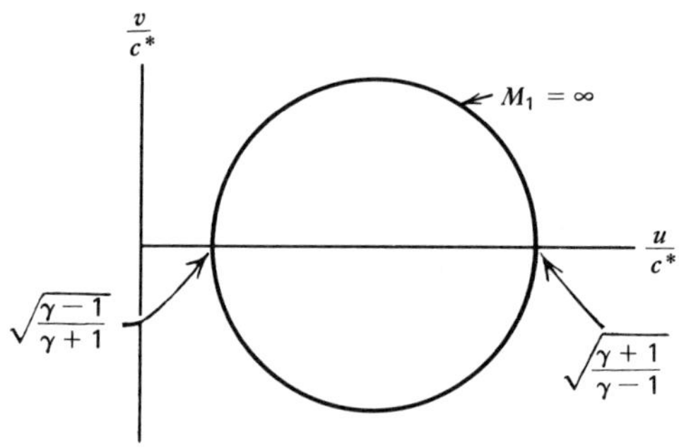

Figure 103. Shock polar for $M_1 = \infty$.

The equation of the circle may be obtained by direct substitution into equation (563). It is

$$\left(\frac{v_2}{c^{*2}}\right)^2 + \left(\frac{u_2}{c^*} - 1.429\right)^2 = 1.041 \tag{564}$$

for air, $\gamma = 1.4$.

The shock polar is a useful tool for oblique shock calculations. Referring to Figure 104, we see that given any two parameters, we may read off the rest.

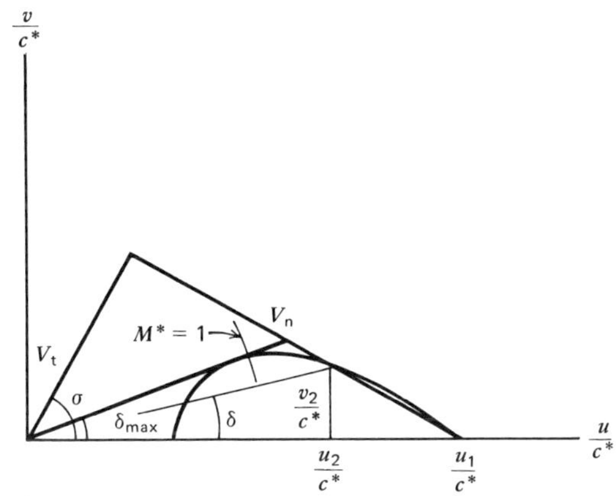

Figure 104. Velocity components, flow angle, and wave agle determined from shock polar.

For example, if we are given u_1/c^* and δ we may read off V_2/c^*, v_2/c^*, u_2/c^*, V_{n1}, V_{n2}, V_t and σ. The shock polar reveals a number of other interesting features. For a given δ it may be seen that there are two solutions so long as $\delta < \delta_{max}$. These correspond to the two solutions previously revealed in Figure 98. Generally, one solution is supersonic, the other is subsonic, downstream of the shock. However, it is observed that there is a small region between $M_2 = 1$ and δ_{max} where both solutions are subsonic. Which solution occurs in practice has been found experimentally to depend on the downstream pressure. In the flow over a wedge in open air, the supersonic, or "weak" solution is found to prevail, since the pressure far downstream of the wedge is essentially atmospheric. If the flow is in a duct, however, such as a jet engine diffuser, where the back pressure may be controlled, the "strong" or subsonic solution may prevail if the back pressure is sufficiently high.

In practice, oblique shock waves are generated by sudden changes in the flow direction, as caused by a wedge or ramp (Figure 105). One may inquire, there-

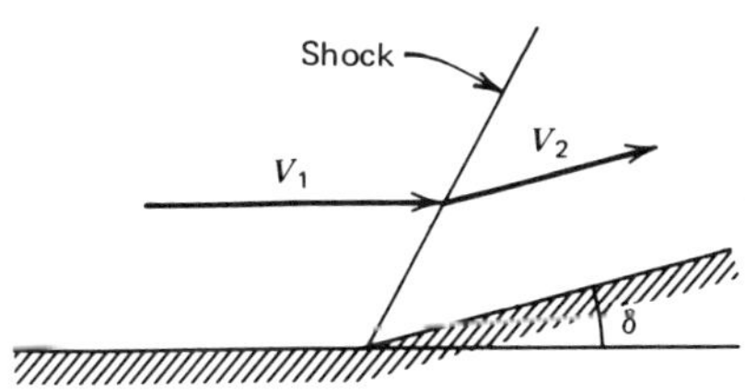

Figure 105. Oblique shock generated by a ramp.

fore, what the relation is between oblique shock waves and the Mach waves discussed earlier in this chapter. Recall that for a small change in flow direction, the angle α of the Mach wave to the flow was given by

$$\sin \alpha = \frac{1}{M} \tag{565}$$

Referring to Figure 106, the angle of the shock wave σ is given by $\sin\sigma = V_{n1}/V_1$.

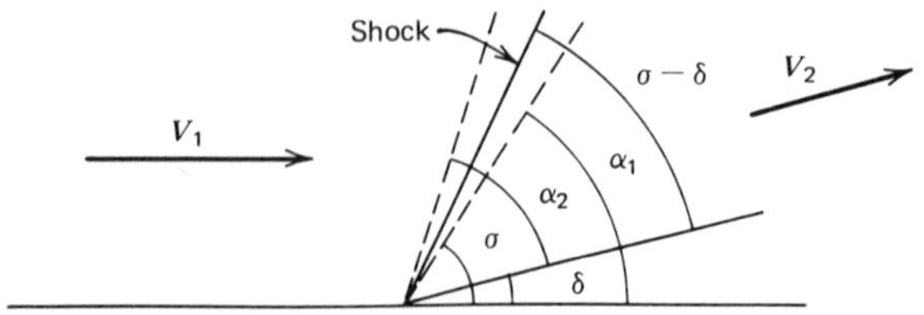

Figure 106. Shock wave and Mach waves for flow over a ramp.

The angle of the Mach wave for the flow ahead of the shock is given by $\sin\alpha_1 = c_1/V_1$ and the angle of the Mach wave for the flow downstream of the shock is given by $\sin\alpha_2 = c_2/V_2$. Also, $\sin(\sigma-\delta) = V_{n2}/V_2$. Now since $V_{n1} > c_1$ while $V_{n2} < c_2$, follows that the shock wave lies between the upstream and the downstream Mach waves. As $\delta \to 0$, we observed from the shock polar that $V_{n1} \to V_{n2}$ for the "weak" solution; hence, in the limit as $\delta \to 0$, $V_{n1} = V_{n2} = c_1 = c_2$, and hence the shock wave angle approaches the Mach angle. The "strong" solution for the case $\delta = 0$ corresponds to a normal shock. Since a shock wave becomes a Mach wave in the limit as $\delta \to 0$ for the "weak" solution, the shock polar must be tangent to the hodograph characteristics at the point $\delta = 0$ (Figure 107). Also, the tangent to the shock polar at this point will be

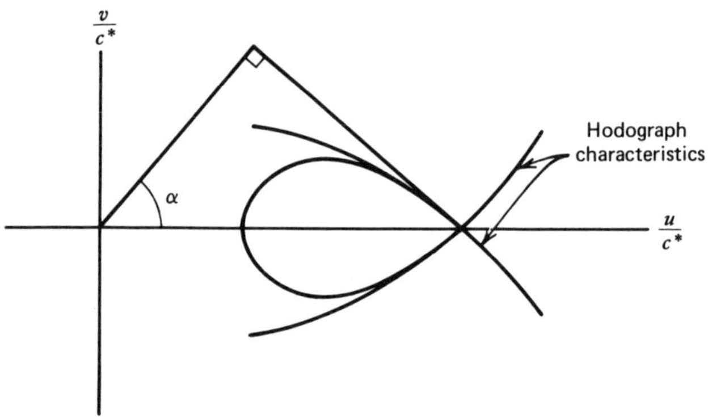

Figure 107. Shock polar and hodograph characteristics on the hodograph plane.

perpendicular to the Mach wave.

We may now return to a question which was left in abeyance earlier—the case of a supersonic nozzle in which the exit pressure is only slightly lower than the back pressure. Figure 108 shows the pressure distribution in a deLavel nozzle with supersonic flow. p_1 is the exit pressure, p_a the back pressure, and p_g the exit pressure for the case of a normal shock at the exit. We consider the case $p_1 < p_a < p_g$. In such a case, the flow pattern shown in Figure 109 is found to prevail. Oblique shocks form at the exit of the nozzle. The angle of the shocks is such that the pressure in regions 2, p_2, is equal to p_a. This is necessary since no pressure discontinuities are possible across a free boundary. This example

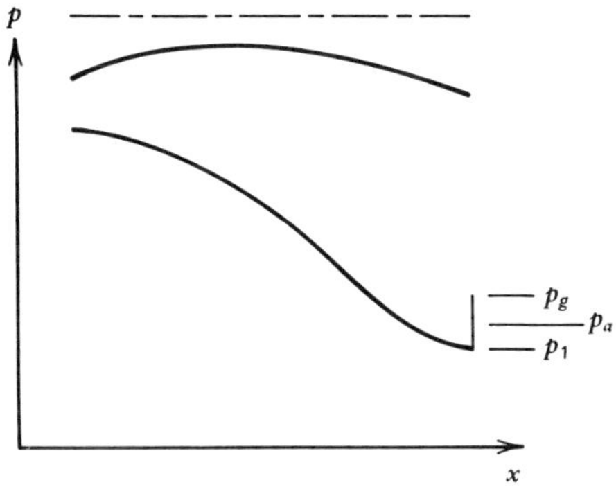

Figure 108. Pressure distribution in a deLaval nozzle with shock at exit.

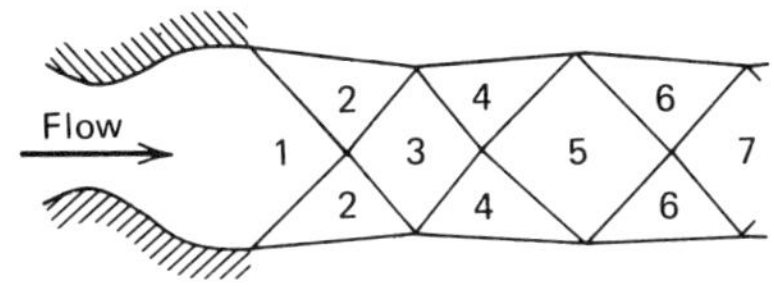

Figure 109. Flow pattern for overexpanded nozzle.

illustrates another phenomenon associated with oblique shock waves—reflection off a free boundary and intersection of waves. Since the flow in regions 2 converges, the flow must pass through another shock to get from region 2 to region 3, in order to make the streamlines parallel in region 3. Put another way, shock waves are bent by intersection but not broken. As a result, the pressure in region 3 will exceed the back pressure. Since in region 4 the pressure must equal p_a (free boundary), the shock waves must reflect as expansion waves from a free boundary. The flow in regions 4 now diverges and hence must pass through another expansion wave to make the flow parallel in region 5. This in turn causes the expansion waves to reflect as compression waves off the free boundary, making the pressure in region 6 atmospheric.

It should be noted that there are no isentropic compression waves. Any compression wave which causes a finite change in the flow direction is an oblique shock. On the other hand, if δ is not too large, oblique shock waves may be *approximated* by isentropic compression waves. This is most readily seen by examining Figure 107. It is seen that for a short distance on either side of $\delta = 0$, the shock polar and the hodograph characteristics almost coincide. In this range, therefore, the approximation of oblique shocks by compression waves is reasonably good. As δ becomes larger, however, the shock polar and the hodograph characteristics diverge and the error caused by assuming isentropic flow increases. Expansion waves, on the other hand, are isentropic so long as the flow is assumed to be inviscid and adiabatic. In practice, however, shock waves will not reflect as expansion waves of a single discontinuity but as expansion fans, as discussed in the section on isentropic flow, thereby complicating the flow pattern (Figure 110). This does not affect the flow properties in regions 4 and 5 since they depend only on the total turning angle. The properties in region

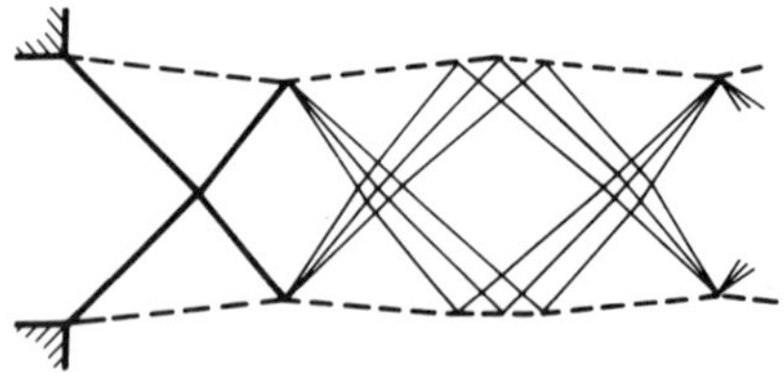

Figure 110. Flow pattern for over expanded nozzle showing detail of expansion waves.

6 will be affected, however, since the entropy change across a series of small compression (shock) waves will not be the same as the entropy change across a single shock wave causing the same change in direction. On the other hand, converging compression waves generally meld into a single shock. Of course the wave pattern shown in Figure 110 is only an approximation for purposes of illustration. In actuality, the flow properties vary continuously between regions 3 and 4 and between regions 4 and 5 over finite regions bounded by the upstream and downstream Mach waves. Also, the flow properties will vary continuously between regions 5 and 6, at least up to point where the compression region melds into a single shock. In practice, the complexities caused by the actual flow pattern are circumvented by treating all waves, expansion and compression, as single discontinuities, as in Figure 109. Reasonably accurate results may be obtained in this way.

Qualitatively the same rules apply to oblique shocks as to isentropic compression waves; viz., oblique shocks turn the flow *toward* the wave, reflect in unlike manner off free boundaries, i.e., as expansion waves, and reflect in like manner off solid boundaries, i.e., as compression waves. It must again be emphasized that there are no isentropic compression waves of finite strength. They are merely a convenient fiction. All compression waves of finite strength, in "real life," are shock waves.

Method of computation. The flow properties in Figure 109 may readily be computed graphically. To get from region (1) to region (2) we must satisfy the condition $p_2/p_1 = pa/p_e$. Recall now the equation

$$\frac{p_2}{p_1} = \frac{2\gamma}{\gamma+1} M_1^2 \sin^2 \sigma_1 - \frac{\gamma-1}{\gamma+1} \tag{566}$$

Since p_2, p_1, and M_1 are presumed known, we may solve for σ. Having σ_1, we may read δ_2 and M_2^* off the shock polar (Figure 111). Using M_2^* as the initial M^* and keeping in mind that in going from 2 to 3 the flow must be turned back through the same angle as δ_1, we may read off M_3^* and σ_{2-3} off the corresponding shock polar, keeping in mind that σ is always the angle the shock makes with the flow direction just before the shock.

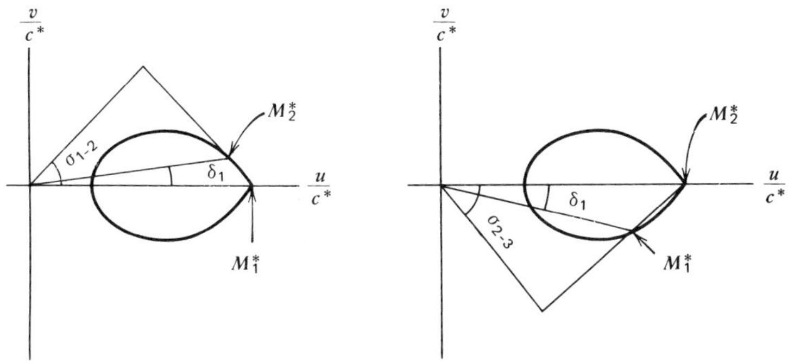

Figure 111. Computation of flow properties using shock polar.

In going from 3 to 4, the flow is isentropic. Thus, $p_{03} = p_{04}$ and hence

$$p_{03} = p_3\left(1 + \frac{\gamma - 1}{2} M_3^2\right)^{\gamma/(\gamma-1)} = p_4\left(1 + \frac{\gamma - 1}{2} M_4^2\right)^{\gamma/(\gamma-1)}, \quad p_4 = p_a \quad (567)$$

Since p_3, M_3, and $p_4 = p_a$ are known, we may solve for M_4. The flow direction in region 4 may then be read off the hodograph characteristics, Figure 112.

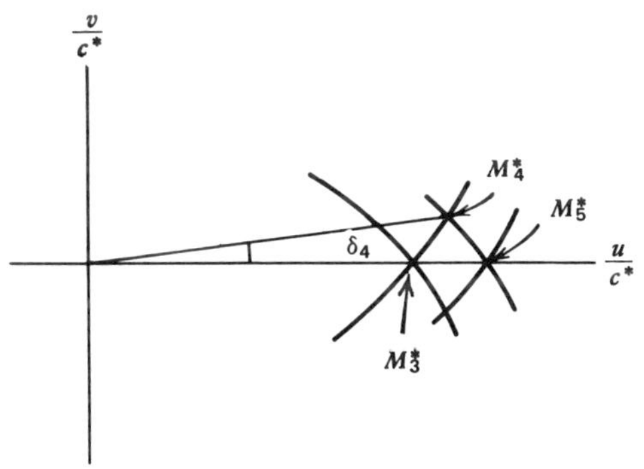

Figure 112. Computation of flow properties using hodograph characteristics.

In going from region 4 to region 5 to flow must be turned back through an angle δ_4; therefore M_5^* may also be read off Figure 112.

By the same methods, the computations may be continued through as many regions as desired. Notice that in the present case, the flow pattern is not repetitive as in the isentropic case. The conditions in region 5 are not the same as the conditions in region 1.

Reflection off a solid boundary. An oblique shock striking a solid boundary must reflect in like manner in order to keep the flow direction at the wall parallel to the wall (Figure 113). It may be that the first shock slows the flow down to such an extent that the flow cannot be turned back parallel to the wall by a reflected oblique shock. In such a case a configuration known as a lambda shock occurs (Figure 114). In any actual flow, a region exists near the wall where

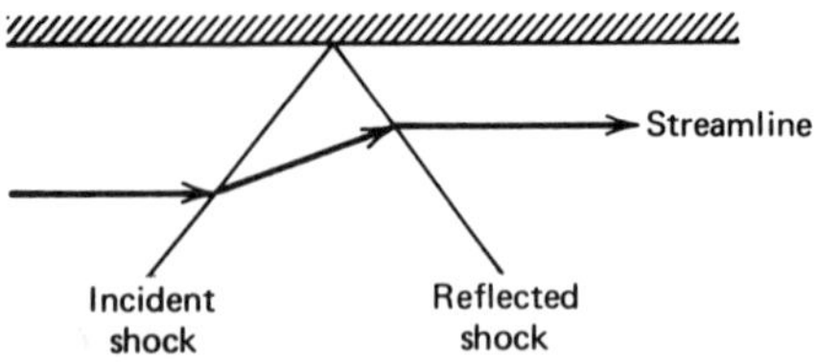

Figure 113. Shock reflecting off a solid boundary.

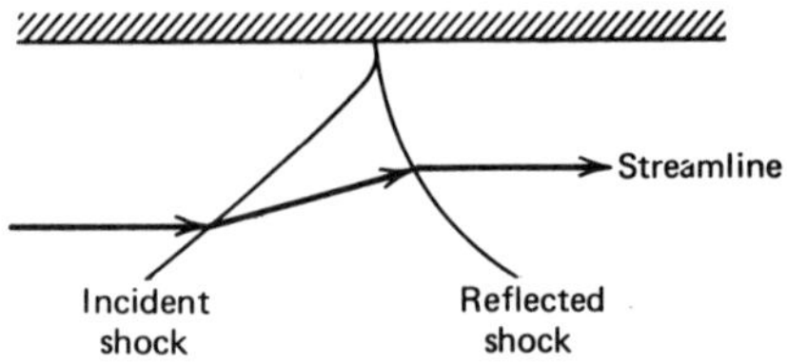

Figure 114. Lambda shock reflection.

viscous effects cannot be neglected. If this region is thin, it is known as a boundary layer. At the wall, in any real flow, the velocity is zero. As one approaches the wall, therefore, the velocity decreases. In a small region near the wall, the Mach number will be less than one. Thus, no shock wave can ever actually extend to the wall. As long as the subsonic layer is thin, however, the flow patterns predicted by the inviscid theory will essentially prevail. Thick boundary layers, on the other hand, may lead to substantial modifications of the flow patterns. These questions will be discussed further in a subsequent chapter.

Axially Symmetric Flow Over a Cone

In a previous section, we demonstrated the existence of plane oblique shock waves by a simple physical argument. The governing equations for plane oblique shock waves may also be derived directly from the fundamental equations of momentum, continuity, and energy. Experiments verify the accuracy of the results. On the other hand, experiments also indicate the existence of conical shock waves. Such waves are generated, for example, by a pointed cone set in a supersonic stream at zero angle of attack (Figure 115).

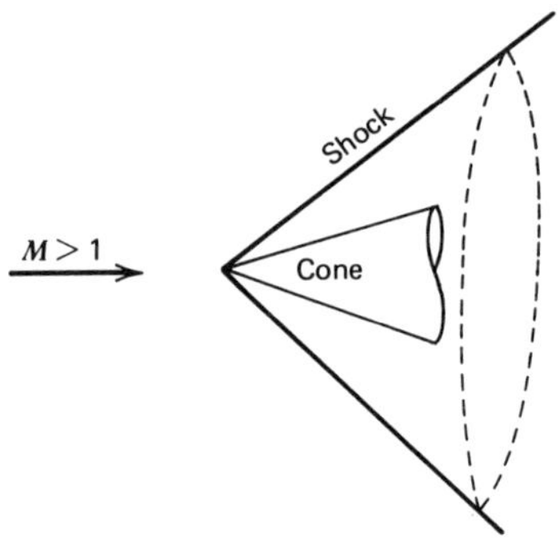

Figure 115. Cone in supersonic flight.

In this case, however, there appears to be no simple physical argument proving the feasibility of such waves, either *ab initio* or starting with the existence of plane waves. Nor does there appear to be as yet any mathematical solution of the governing equations in polar or spherical coordinates giving the conditions before and after shock, supposing such a shock to exist. Rather, authors have begun with the assumption that conical shock behave locally like plane shocks, and gone on from there. Where any argument is attempted at all to justify this assumption, it is usually said that as shock waves are very thin compared to the radius of curvature of the cone, conical shock waves must behave locally like plane shock waves. Whatever may be said for the virtues of this argument, experiments indicate that conical shock waves do indeed behave locally like plane waves. Thus, if one takes a section of a conical shock wave, it is found that for a given initial Mach number M_1 and a given wave angle σ_1 the deflection angle δ is the same as in the plane case (Figure 116). These results were first

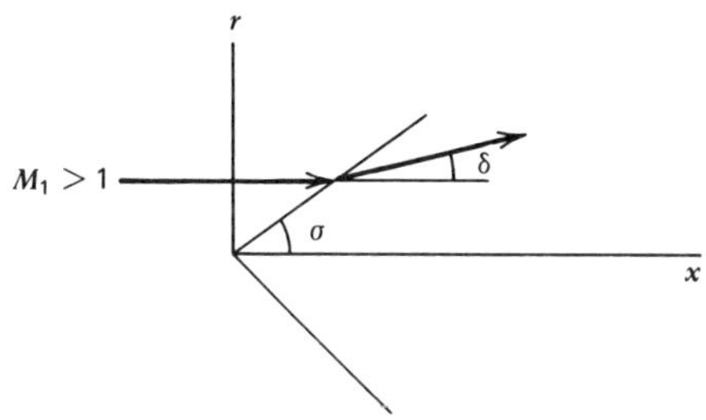

Figure 116. Wave angle and deflection angle for conical shock.

presented by Busemann [34] in 1929 and Taylor and Macoll [35] in 1933.

On the other hand, in the case of a cone as shown in Figure 115, the deflection angle δ is not equal to cone angle θ. Thus, the flow pattern downstream of the wave cannot be uniform and parallel as in the plane case. The problem, then, is to find the flow pattern between the shock wave and the surface of the cone. Since the shock wave is straight and has everywhere the same angle of inclination to the flow, the flow downstream of the wave will be irrotational, under the assumption of inviscid flow. The most obvious approach, therefore, would be to use the method of characteristics. We may observe, however, that axially symmetric flow over a cone is similar, in certain respects, to Prandtl–Meyer flow. The flow is essentially two-dimensional, x or r being the only independent variables. On physical grounds we may reason that Mach waves of only one family are present i.e., disturbances cannot propagate upstream. Thus it may be that the flow is singly degenerate, i.e., that the entire physical plane maps into a single curve in the hodograph plane, as was the case in Prandtl–Meyer flow. This implies that u and v are not independent of each other.

We are thus led to seek a solution for which $v = v(u)$. Recall the potential equation for axisymmetric flow

$$\left(1-\frac{u^2}{c^2}\right)u_x-\frac{uv}{c^2}(v_x+u_r)+\left(1-\frac{v^2}{c^2}\right)v_r+\frac{v}{r}=0 \tag{569}$$

where u and v are the x and r velocity components in the polar coordinate system of Figure 116. The irrotationality condition is

$$v_x-u_r=0 \tag{570}$$

We now introduce the assumption that v and u are functions only of $\sigma \equiv x/r$. We justify this assumption *a posteriori*; that is, we will show that this assumption leads to a solution which satisfies the equations (569), (570) and the boundary conditions. Equation (570) the becomes

$$v_\sigma + \sigma u_\sigma = 0 \tag{571}$$

Equation (569) becomes

$$\left(1 - \frac{u^2}{c^2}\right)u_\sigma - 2\frac{uv}{c^2}V_\sigma - \left(1 - \frac{v^2}{c^2}\right)\sigma v_\sigma + v = 0 \tag{572}$$

If $v = v(u)$, equation (571) becomes

$$\sigma = -v_u \tag{573}$$

Differentiating with respect to σ,

$$1 = -v_{uu}u_\sigma \qquad \text{or} \qquad u_\sigma = -\frac{1}{v_{uu}} \tag{574}$$

From equations (571), (573) and (574),

$$v_\sigma = -\sigma u_\sigma = \frac{v_u}{v_{uu}} \tag{575}$$

Hence, equation (572) becomes

$$\left(1 - \frac{u^2}{c^2}\right) + \left(1 - \frac{v^2}{c^2}\right)v_u^2 - v v_{uu} - 2\frac{uv}{c^2}v_u = 0 \tag{576}$$

or

$$v v_{uu} = 1 + v_u^2 - \frac{(u + v v_u)^2}{c^2} \tag{577}$$

No closed-form solution exists for this nonlinear second-order ordinary differential equation. The equation may, however, be integrated numerically. The integration may be performed in two different ways depending on the boundary conditions which are specified—upward from the cone or downward from the shock. If the cone angle, surface Mach number and speed of sound are specified, u, v, c and σ are known on the surface of the cone, v_u is given by $\sigma = -v_u$. The integration may then be carried forward until a value of σ is reached for which the oblique shock wave relations are satisfied. Alternatively, the free stream Mach number, stagnation temperature and shock wave angle may be specified, and the integration carried downward from the shock until a value of σ is reached for which $v/u = \sigma$. At this point one has attained the surface of the cone. Notice that one must actually solve the equation to find the relation between the cone angle and the shock angle. In practice, it is thus best to work from tables or graphs of solutions. The best such tables are those by Sims (NASA SP 3004).

Busemann [34] has given a graphical method for solving equation (572). While Busemann presented his work at a meeting in Prague in 1929, the details of his derivation were never published. Only a summary appears in ZAMM [34]. Nevertheless, the details are easily reconstructed. The coordinate system

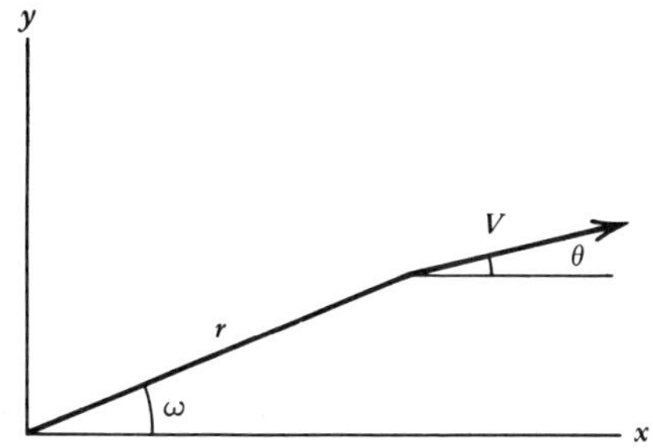

Figure 117. Coordinate system for method of Busemann.

used is shown in Figure 117. We take as our starting point equation (572)

$$\left(1-\frac{u^2}{c^2}\right)u_\sigma - 2\frac{uv}{c^2}v_\sigma - \left(1-\frac{v^2}{c^2}\right)\sigma v_\sigma + v = 0 \tag{578}$$

and observe that

$$v = V\sin\theta \qquad \sigma = \tan\omega \qquad \frac{d}{d\sigma} = \cos^2\omega\frac{d}{d\omega} \tag{579}$$

$$u = V\cos\theta \qquad d\omega = \cos^2\omega\, d\sigma \qquad c^2 = \frac{\gamma-1}{2}(V_m^2 - V^2) \tag{580}$$

Hence,

$$u_\sigma = \frac{du}{d\sigma} = \cos^2\omega\frac{d}{d\omega}(V\cos\theta) = \cos^2\omega\left(\frac{dV}{d\omega}\cos\theta - V\sin\theta\frac{d\theta}{d\omega}\right) \tag{581}$$

$$v_\sigma = \frac{dv}{d\sigma} = \cos^2\omega\frac{d}{d\omega}(V\sin\theta) = \cos^2\omega\left(\frac{dV}{d\omega}\sin\theta + V\cos\theta\frac{d\theta}{d\omega}\right) \tag{582}$$

We introduce the irrotationality condition

$$\Gamma = \oint \mathbf{V}\cdot d\mathbf{s} = 0 \tag{583}$$

taking V_r as the velocity in the r direction and V_ω as the velocity in the ω direction, as shown in Figure 118.

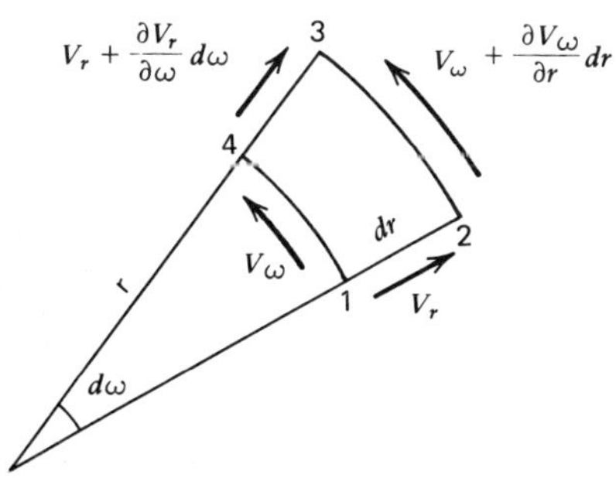

Figure 118. Circulation around a fluid element in the r-ω plane.

Taking the circulation around 1234 and simplifying, we have

$$V_\omega = \frac{dV_r}{d\omega} \tag{584}$$

Now,

$$V_r = V\cos(\omega - \theta),\ V_\omega = -V\sin(\omega - \theta) \qquad (585),(586)$$

Hence

$$\frac{dV_r}{d\omega} = -V\left(1 - \frac{d\theta}{d\omega}\right)\sin(\omega - \theta) + \frac{dV}{d\omega}\cos(\omega - \theta) \tag{587}$$

$$\frac{dV_\omega}{d\omega} = -V\left(1 - \frac{d\theta}{d\omega}\right)\cos(\omega - \theta) - \frac{dV}{d\omega}\sin(\omega - \theta) \tag{588}$$

Substituting in (584) for V_ω from (586) and for $dV_r/d\omega$ from (587), we get,

$$\frac{d\theta}{d\omega} = -\frac{dV/d\omega}{V\tan(\omega - \theta)} \tag{589}$$

Substituting equations (579), (580), (581), (582) and (589) into equation (578), simplifying, we get

$$\frac{dV}{d\omega} = \frac{V\sin(\omega - \theta)\dfrac{\sin\theta}{\sin\omega}}{1 - \dfrac{2V^2\sin^2(\omega - \theta)}{(\gamma - 1)(V_m^2 - V^2)}} \tag{590}$$

Geometrical interpretation. Equations (589) and (590) admit of a number of interesting geometrical interpretations. Consider first the hodograph streamline of the flow between the shock and the cone (Figure 119). All the points of the flow must map into this streamline since V and θ are functions of ω only.

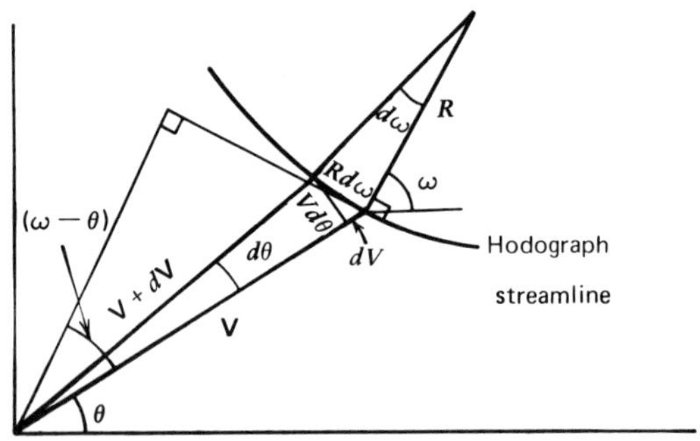

Figure 119. Finding hodograph streamline by graphical method.

From equation (589)

$$\frac{1}{V}\frac{dV}{d\theta} = -\tan(\omega - \theta) \tag{591}$$

Reference to Figure 119 shows that this means that the radius of curvature of the hodograph streamline makes an angle ω with the horizontal. Also,

$$R\,d\omega = \frac{dV}{\sin(\omega - \theta)} \qquad \text{or} \qquad \frac{dV}{d\omega} = R \sin(\omega - \theta) \tag{592}$$

Substituting for $dV/d\omega$ from equation (590), we have

$$R = \frac{V \dfrac{\sin\theta}{\sin\omega}}{1 - \dfrac{2V^2 \sin^2(\omega - \theta)}{(\gamma - 1)(V_m^2 - V^2)}} \tag{593}$$

Recalling that $\quad \dfrac{\gamma - 1}{2}(V_{max}^2 - V^2) = c^2 = \dfrac{\gamma + 1}{2} c^{*2} - \dfrac{\gamma - 1}{2} V^2$

this equation may also be written in the form

$$R = \frac{V \dfrac{\sin\theta}{\sin\omega}}{1 - \dfrac{\sin^2(\omega - \theta)}{[(\gamma+1)/2](c^{*2}/V^2) - [(\gamma-1)/2]}} \tag{594}$$

which is the form in which it was originally given by Busemann. Knowing the magnitude and direction of R, we may construct the hodograph streamline geometrically. We begin at the point a, Figure 120, just behind the shock.

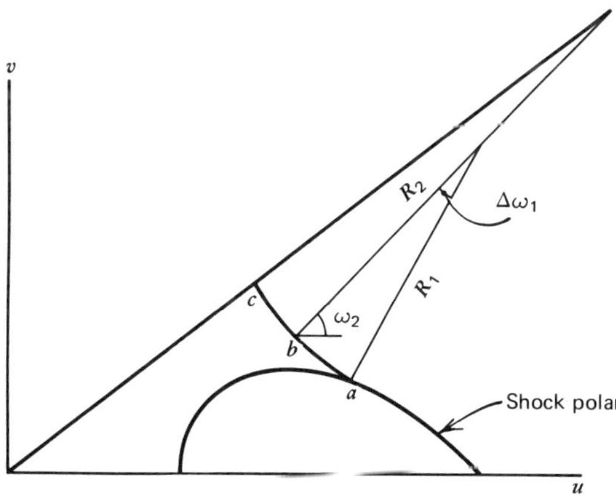

Figure 120. Finding hodograph streamline by graphical method.

Using equation (594) we evaluate R there. Then, with a compass, we swing R over until we reach point b, giving us the approximate hodograph streamline between a and b. The choice of the angle $\Delta\omega$, is arbitrary. The smaller the angle is chosen, the greater the accuracy. At the point b we have a new θ, a new V, and a new ω, ω_2; hence we have a new R, R_2. We repeat the process until R coincides with a line through the origin, at which point we have reached the surface of the cone. Surprising accuracy can be obtained with just two increments $\Delta\omega$, as shown in the figure. Busemann, however, used as many as eight increments, giving very high accuracy.

Since shock polars are usually plotted u/c^*, v/c^* in practice it is easier to work with $R^* = R/c^*$. From equation (593)

$$R^* = \frac{M^* \dfrac{\sin\theta}{\sin\omega}}{1 - M^2 \sin^2(\omega - \theta)} \tag{595}$$

since

$$(V_m^2/V^2) - 1 = \frac{2}{M^2(\gamma - 1)}$$

If, for a given shock polar, the hodograph streamline is constructed for every possible shock angle, a figure is obtained such as that shown in Figure 121.

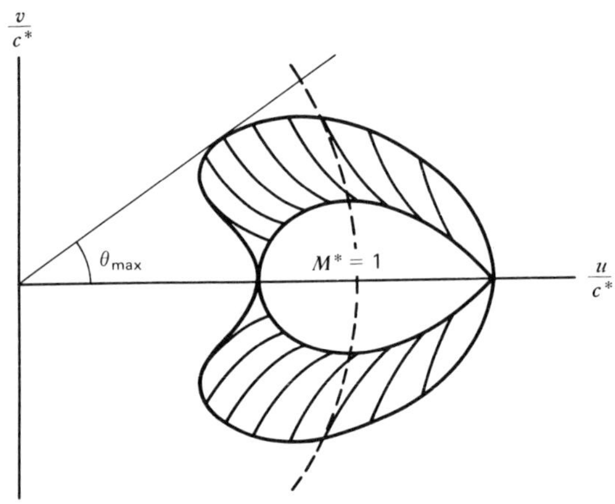

Figure 121. Busemann apple curve [235].

Such a figure is called a Busemann apple curve. The figure reveals that for every initial Mach number, there is a maximum cone angle θ_{max} beyond which no solution exists. This is analogous to the case of the wedge. When the cone semiangle exceeds θ_{max}, a curved shock, shaped rather like a saucer, or bowl, depending on the Mach number, stands in front of the cone. No attached shock is possible in this case. For a given Mach number, the permissible cone angle turns out to be larger than the permissible wedge angle (Figure 122).

It will be observed that only the "weak" solutions are meaningful under the assumption of conical flow. This is because if the flow downstream of the shock is subsonic, as would be the case for the "strong" solution, the flow would be affected by the downstream conditions. Thus, the assumption that the flow properties are independent of the distance from the nose of the cone would no longer be valid. This does not preclude the existence of a strong solution with nonconical flow downstream of the shock. As yet, however, no such flow appears to have been observed experimentally. On the other hand, subsonic flow may exist between the shock and the cone with a "weak" solution. This is because

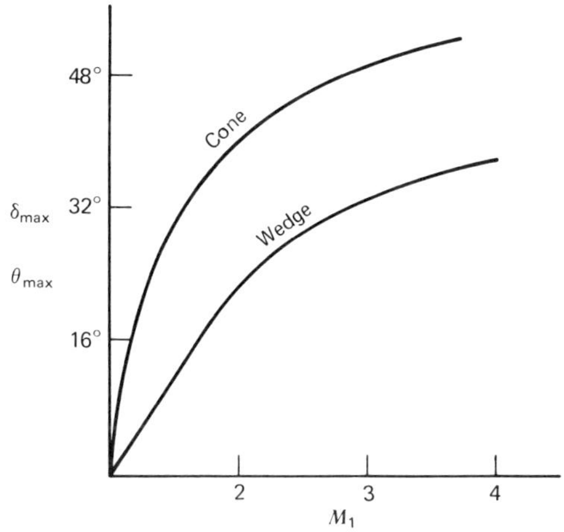

Figure 122. Flow deflection angles for wedge and cone [234].

the flow slows down after passing through the shock, as may be observed in Figure 121. Such flows have been observed experimentally. As might be expected, however, the flow is no longer conical in the subsonic region.

Curved Shocks

The shock associated with the supersonic flow over an object of finite length must, sooner or later, be curved. This may be seen most easily by considering the case of the double wedge. The flow in region 2 is given by the ordinary oblique shock solution. At point *a*, however, and expansion fan originates, which is approximated in the Figure 123 by two expansion waves. These intersect at points *b* and *c* with the shock waves. Since there can be no discontinuities in pressure or flow direction behind the shock, the expansion wave must reflect at *b*, and the shock wave must bend at *b*, in such a way that the flow in regions 4 and 5 will have the same direction and pressure. Under these circumstances, it will not be possible to equalize the velocity in regions 4 and 5. Thus, a slip line develops between these two regions. It follows that the flow will no longer be irrotational behind the bent shock. Likewise, it may be seen by

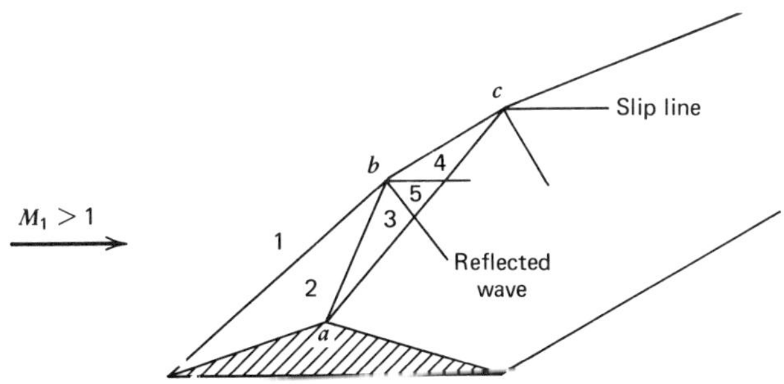

Figure 123. Flow pattern over a triangular airfoil.

Crocco's Theorem that the entropy behind the shock is no longer uniform. The variation of the flow parameters through the shock wave, for a given M_1, depends only on the *local* shock wave angle. This is because both μ and k, the viscosity and the thermal conductivity, are assumed to be zero behind the shock. Thus, each streamline passes through the shock independent of its fellows. The change in the flow properties along each streamline as it passes through a curved shock is thus the same as it would be if the streamline passed through a straight shock having the same angle of inclination as the tangent to the curved shock at the point where the streamline crosses it. The variation of the entropy along the back of the shock front is equally apparent from this consideration. These considerations apply at the point c as well as at every other point where an expansion wave intersects a shock wave. In the case of an expansion fan intersecting a shock wave, the variation is, in reality, continuous. Thus, the shock wave becomes a smooth curve.

It will be observed from Figure 123 that an exact calculation of the flow behind the shock is very complex. Even if the expansion fan is approximated by only two waves, the reflection and slip line from the first wave intersect the second wave, and numerous different regions are formed. In the case of several expansion waves, the calculation becomes even lengthier. In practice, it is found that the reflected waves are usually weak and may be ignored without undue sacrifice in accuracy.

Figure 123 also shows a shock at the trailing edge, needed to turn the flow parallel at that point. All the waves caused by the profile ultimately interact in such a way that the shock waves degenerate into Mach waves at a great distance from the profile, while the remainder of the flow becomes uniform and parallel. It is of interest to observe that were this not the case, supersonic wave drag would be very large and hence supersonic flight impractical. This may be seen in the following way:

Consider a control volume around the profile, Figure 124, the boundaries of the volume being at a large distance from the profile. The pressure at these

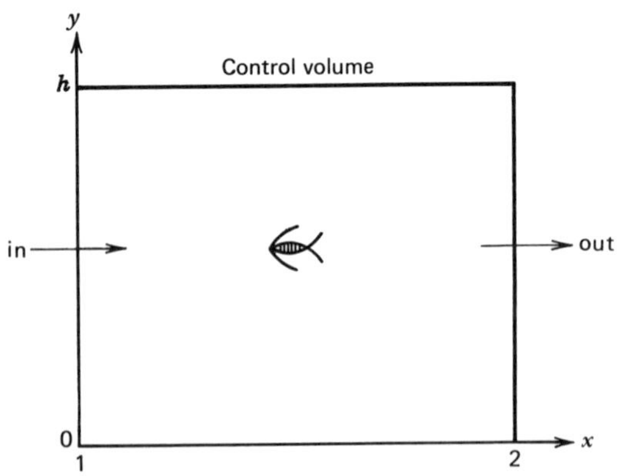

Figure 124. Control volume for flow over an airfoil.

boundaries will have essentially its atmospheric value. On the other hand, the stagnation pressure behind the profile will be less than ahead of the profile due to the presence of the shock. Thus, the momentum flow rate will also be less. Recall the drag D is given by

$$D = \int_0^h \rho_1 V_1^2\, dy - \int_0^h \rho_2 V_2^2\, dy \tag{596}$$

if we take zero at the ground level and h at the outer edge of the atmosphere. Also recall

$$p_0 = p\left(1 + \frac{\gamma - 1}{2} M^2\right)^{\gamma/(\gamma - 1)} \tag{597}$$

Hence

$$\rho V^2 = \frac{2\gamma}{\gamma - 1} p \left[\frac{p_0}{p} - 1\right]^{(\gamma - 1)/\gamma} \tag{598}$$

and

$$D \cong \frac{2\gamma}{\gamma - 1} \int_0^h p_a \left\{ \left[\frac{p_{01}}{p_a} - 1\right]^{(\gamma - 1)/\gamma} - \left[\frac{p_{02}}{p_a} - 1\right]^{(\gamma - 1)/\gamma} \right\} dy \tag{599}$$

Thus, the drag depends on the integral of the change in the stagnation pressure across the leading and trailing edge shocks. The drag is kept within reasonable bounds only because the integrand decreases rapidly as we move away from the profile.

It is often said that the drag depends on the entropy rise across the shocks. The relation

$$s_2 - s_1 = -R \ln \frac{p_{02}}{p_{01}} \tag{600}$$

for adiabatic flow shows why this is so. Infinite entropy rise means infinite drag.

The foregoing remarks are not limited to double-wedge profiles, but apply to all profiles. For a smooth profile, the flow may be approximated by a series of expansion waves (Figure 125). It will be noted that the reflected waves, where

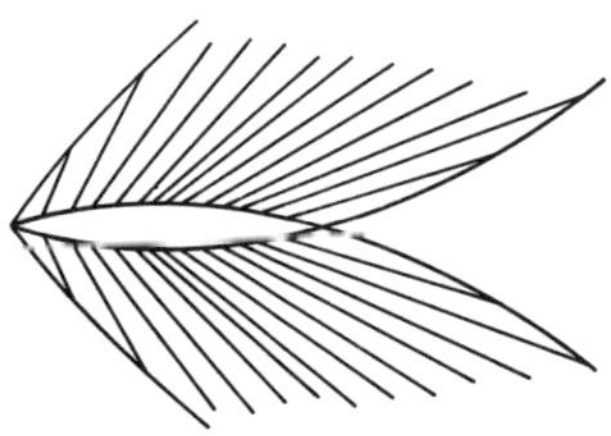

Figure 125. Mach waves and shock waves over an airfoil, neglecting reflections.

the expansion waves intersect with the shocks, have been omitted. This may usually be done without undue sacrifice in accuracy and saves an enormous amount of computation. The flow between the shocks is thus simply a Prandtl–Meyer flow. Also to be noted is the fact that the trailing edge shocks are initially steepened by intersection with the expansion waves. They too, however, ultimately degenerate into Mach waves.

Consider the flow in Figure 126. The rear portion of the body in Figure 126

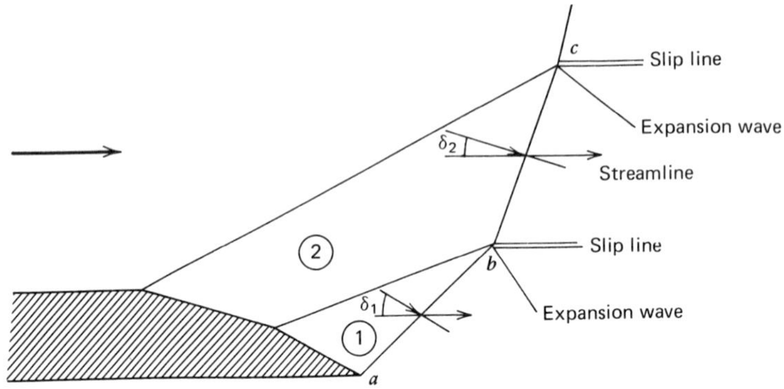

Figure 126. Development of trailing edge shock.

is approximated by a series of straight lines. Notice that the Mach number in region 1 is greater than the Mach number in region 2, and the turning angle δ_1 is greater than the turning angle δ_2. Thus, the shock wave *ab* must be stronger than the shock wave *bc*. Thus, even though the shock wave *abc* appears to be curving upward, it is actually getting weaker. Since pressure and flow direction cannot have discontinuities downstream of the shock, the expansion waves continue through the shock but are sharply deflected. Since velocity magnitude cannot also be made continuous, slip lines result.

An upward curving shock wave may, however, get stronger instead of weaker. This happens when shock waves intersect another shock, as in Figure 127.

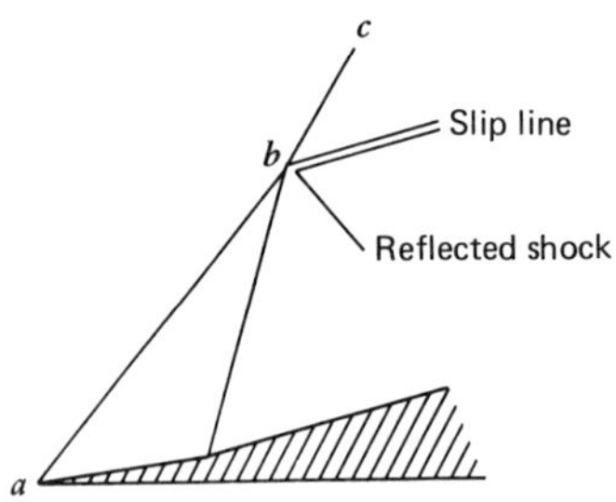

Figure 127. Detail of shock intersection.

To illustrate the process, a double ramp is considered. The shock wave generated by the second ramp intersects the shock wave generated by the first ramp. Again, continuity of pressure and flow direction must be maintained behind the shock. Thus, the first shock curves upward while the second shock is reflected. Continuity in velocity cannot be maintained and hence a slip line results. Notice that the shock wave *bc* is now stronger than the shock wave *ab*. In the case of continuous upward curvature of a surface, the flow may be approximated by a series of shock–shock interactions (Figure 128). If the body involved is of finite

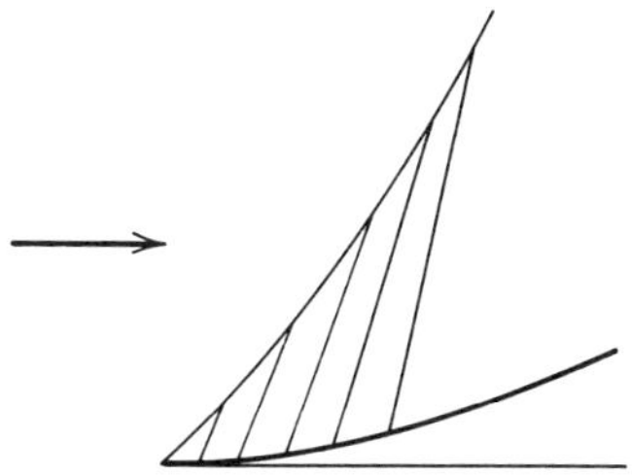

Figure 128. Coalescing of pressure waves over an upward curving surface.

thickness, the surface of the body must eventually curve downward, creating expansion waves which, interacting with the shock, cause the shock to curve back the other way and ultimately decay into a Mach wave as before.

Shock Equations in Curvilinear Coordinates

It is sometimes advantageous to write the equations governing the flow across a curved shock in an orthogonal coordinate system consisting of the streamlines and their normals. Recall in Cartesian coordinates,

$$u\frac{\partial u}{\partial x} + v\frac{\partial u}{\partial y} = -\frac{1}{\rho}\frac{\partial p}{\partial x} \tag{601}$$

$$u\frac{\partial v}{\partial x} + v\frac{\partial v}{\partial y} = -\frac{1}{\rho}\frac{\partial p}{\partial y} \tag{602}$$

$$\frac{\partial u}{\partial x}\left(1 - \frac{u^2}{c^2}\right) + \frac{\partial v}{\partial y}\left(1 - \frac{v^2}{c^2}\right) - \frac{uv}{c^2}\left(\frac{\partial u}{\partial y} + \frac{\partial v}{\partial x}\right) = 0 \tag{603}$$

Now if at any point along a streamline, we take x in the flow direction and y perpendicular to it, then $V = u$, $v = 0$ there. Let s be the distance along the streamline, n be the distance normal to it, and θ be the flow direction. Then equations (601), (602) and (603) become

$$V\frac{\partial V}{\partial s} = -\frac{1}{\rho}\frac{\partial p}{\partial s} \qquad \rho V^2\frac{\partial \theta}{\partial s} = -\frac{\partial p}{\partial n} \tag{604}$$

$$(1 - M^2)\frac{\partial V}{\partial s} + V\frac{\partial \theta}{\partial n} = 0 \tag{605}$$

where we note that dv becomes $V\,d\theta$. In addition, recall the oblique shock wave equations

$$\rho_1 V_1 \sin\sigma = \rho_2 V_2 \sin(\sigma - \delta) \tag{606}$$

$$p_1 + \rho_1 V_1^2 \sin^2\sigma = p_2 + \rho_2 V_2^2 \sin(\sigma - \delta) \tag{607}$$

$$V_1 \cos\sigma = V_2 \cos(\sigma - \delta) \tag{608}$$

$$V_1 V_2 \sin\sigma \sin(\sigma - \delta) = c^{*2} - \frac{\gamma - 1}{\gamma + 1} V_1^2 \cos^2\sigma \tag{609}$$

From equations (606) through (609) we have

$$p_2 = p_1 + \rho_1\left[V_1^2\left(1 - \frac{2}{\gamma+1}\cos^2\sigma\right) - c^{*2}\right] \tag{610}$$

and

$$p_2 = p_1 + \rho_1 V_1^2[\sin^2\sigma - \sin\sigma\cos\sigma\tan(\sigma-\delta)] \tag{611}$$

Let l be the distance along the shock. Then

$$\frac{\partial p_2}{\partial l} = \frac{\partial p_2}{\partial n}\sin(\sigma-\delta) + \frac{\partial p_2}{\partial s}\cos(\sigma-\delta) \tag{612}$$

Differentiating equations (610) and (611) with respect to l and setting them equal to equation (612), we have two equations in two unknowns.

Now recall equation (604)

$$\rho V^2\frac{\partial\theta_2}{\partial s} = -\frac{\partial p_2}{\partial n} \tag{613}$$

Hence, given the conditions ahead of the shock and the shock wave shape, we now have $\partial p_2/\partial n$ and $\partial\theta_2/\partial s$ at every point of the shock.

For uniform conditions upstream of a curved shock, every point downstream of the shock lies along a single shock polar. Given $\partial\theta/\partial s$, we may draw the direction of the hodograph streamline at every point of the shock polar. Carrying out the calculations and plotting the result we get Figure 129. This figure is

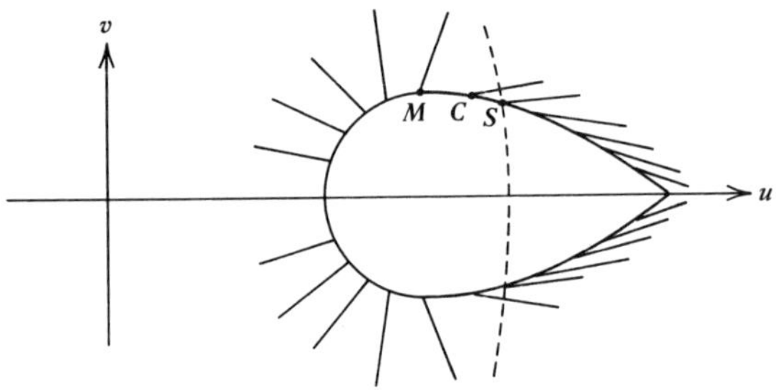

Figure 129. Hedgehog [238].

generally referred to as a hedgehog. The point C is called the Crocco point. It is point at which $\partial\theta_2/\partial s = 0$. Notice that the Crocco point lies between the sonic point and the point of maximum deflection angle, M.

Detached Shocks

The problem of determining the precise shape and location of detached shock waves and the flow behind them is extremely complex. This is because part of the flow must be subsonic behind the shock, making the problem a mixed subsonic, supersonic, transonic flow problem. Thus, a study of this problem more properly belongs in the chapter on transonic flow. We will confine ourselves here to outlining a few methods of finding the shock stand off distance ε (Figure 130). No exact solution of this problem as yet exists. The most rigorous

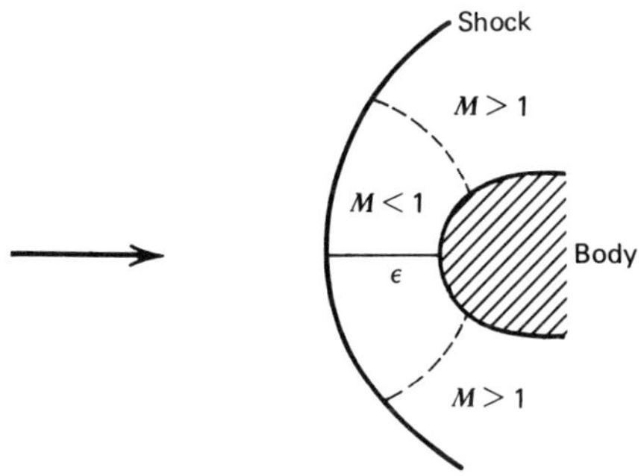

Figure 130. Blunt body with detached shock.

methods are of such complexity as to be too lengthy for routine engineering application. Attention has, therefore, focused on semi-empirical methods making various simplifying assumptions.

Moeckel [36] has given a method based on the assumption that the detached shock is of hyperbolic shape and that the sonic line is straight (Figure 131).

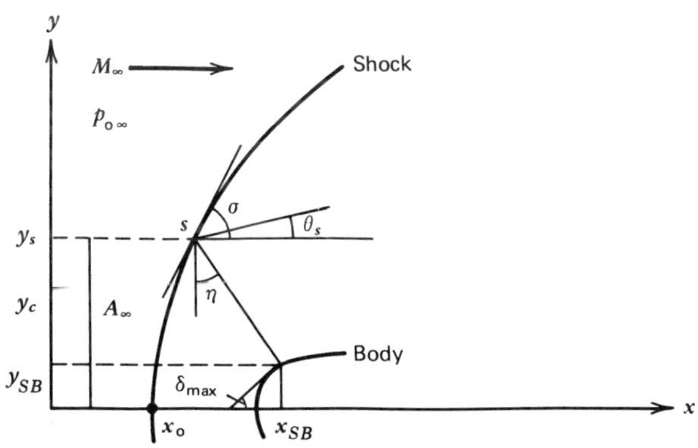

Figure 131. Detail and nomenclature of detached shock for the method of Moeckel [36].

Thus, the angle σ at the point s is known. The shape of the wave is assumed to be

$$\beta y = \sqrt{M_\infty^2 - 1}\, y = \sqrt{x^2 - x_0^2} \tag{614}$$

The local shock angle is then

$$\tan \sigma = \frac{x}{\beta\sqrt{x^2 - x_0^2}} = \frac{\sqrt{x_0^2 + \beta^2 y^2}}{\beta^2 y} \tag{615}$$

The angle η is found by assuming that the average flow direction across the sonic line is normal to the sonic line. The point SB where the sonic line intersects the body is assumed to be the point where the tangent to the body makes an angle δ_{max} with the x axis, δ_{max} being the maximum possible flow deflection angle for an attached shock at the given free stream Mach number. Thus,

$$\eta = \tfrac{1}{2}(\theta_s + \delta_{max}) \tag{616}$$

We assume further that the velocity across the sonic line, V_s, is constant and equal to its value downstream of the sonic point s. Thus, we have

$$\frac{A_\infty}{A_s} = \frac{\rho_s V_s}{\rho_\infty V_\infty} = \frac{\rho^* V^*}{\rho_\infty V_\infty} \cdot \frac{\rho_s V_s}{\rho^* V^*} = \left(\frac{A}{A^*}\right)_{is} \frac{p_{0c}}{p_{0\infty}} \tag{617}$$

where p_{0c} is the average stagnation pressure along the sonic line and $(A/A^*)_{is}$ is the isentropic area ratio for the free stream Mach number M_∞. p_{0c} is taken on the centroid streamline downstream of the shock. For plane flow, the centroid streamline intersects the shock at $y = y_s/2$, for axisymmetric flow at $y = 2y_s/3$. These assumptions serve to fix the location of the shock. Using analytic geometry, it is found that

$$\frac{y_{SB}}{y_s} = \left[1 - \frac{p_{0\infty}}{p_c}\left(\frac{A^*}{A}\right)_{is} \cos\eta\right]^n \tag{618}$$

where $n = 1$ for plane flow and $n = \frac{1}{2}$ for axisymmetric flow. This formula is found to be reasonably dependable for free stream Mach numbers between one and two and body shapes having rounded, though not necessarily round, noses.

In the case of cones, Ward and Pugh (37) have demonstrated experimentally that blunting of the cone does not affect the location or shape of the bow shock, at least to long as real gas effects do not enter in. They show further that a linear correlation exists between cone angle and shock stand off distance. The nomenclature is given in Figure 132. Figure 133 gives the variation of Δ/D with θ for

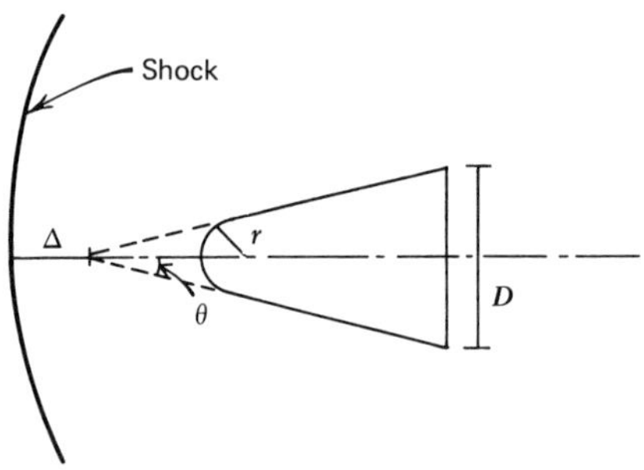

Figure 132. Nomenclature for detached shock used in the method of Ward and Pugh [37].

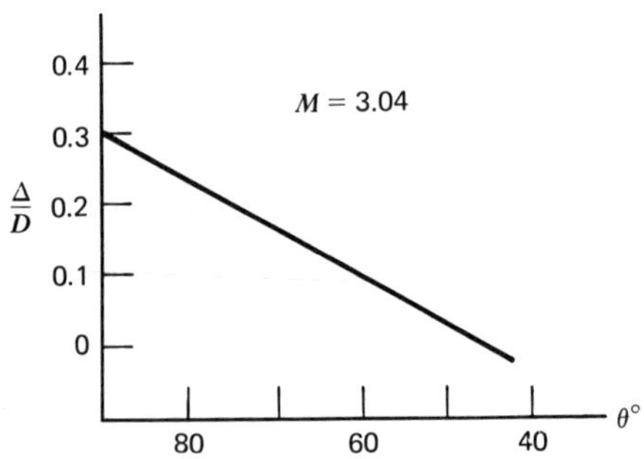

Figure 133. Shock stand off distance as computed by Ward and Pugh [37] for $M = 3.04$.

$M = 3.04$. This variation is found to be independent of the nose radius r, at least so long as r is not of the same order of magnitude as D. The correlation is valid all the way from the maximum cone angle for separation to $\theta = 90°$, where $\Delta = 0.305D$, the stand-off distance for a disk. While Ward and Pugh made tests only at $M = 3.04$, data from other authors, notably Kaatari [38] and Owen [39], indicate that the conclusions are valid at other Mach numbers as well. Using known stand-off distances for a disk along with the known cone angles for detachment, θ_d, Ward and Pugh have constructed Figure 134

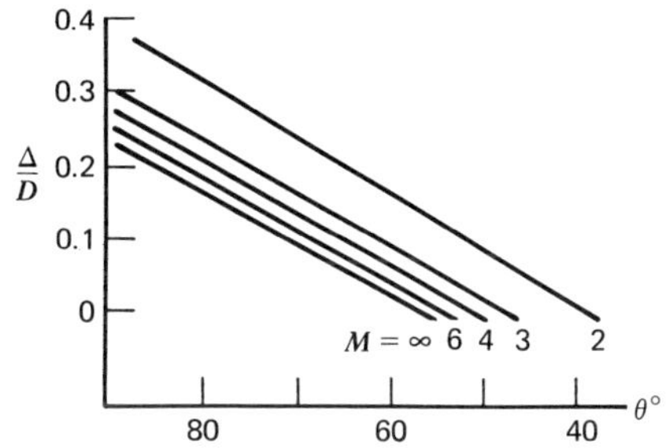

Figure 134. Shock stand off distance as computed by Ward and Pugh [37] for various Mach numbers.

by simply connecting the known Δ/D for $\theta = \theta_d$ and $\theta = 90°$ by straight lines. While experimental verification is not available for the entire range of the Figure 134, existing data support the accuracy of the projections.

Analytically, the line of Figure 133 is given by

$$\frac{\Delta}{D} = 0.00924(\theta - 43°) \tag{619}$$

The lines of Figure 134 have the general equation

$$\frac{\Delta}{D} = 0.00924[\theta - f(M_\infty)] \tag{620}$$

Thus, by a judicious choice of $f(M_\infty)$, it is possible to calculate the stand-off distance for all cones, sharp or rounded, for all Mach numbers for which a detached shock exists and where real gas effects are not prominent, using equation (620).

Guy [40] has studied the stand-off distance of the shock in front of a sphere (Figure 135). He assumes the shock wave to be spherical in the vicinity of the forward stagnation point. Using as a basis the theory of Witham [41], Guy proposes the formula

$$\frac{\delta}{r} = \left[\frac{(\gamma - 1)M_1^2 + 2}{4(M_1^2 - 1)} + 1\right]^{1/2K(Z)} - 1 \tag{621}$$

where

$$\frac{1}{K(Z)} = \frac{1}{2}\left[\frac{1 + 2(1 - \mu)^2}{\gamma + 1}\right]\left[2\mu + 1 + \frac{2\gamma}{(\gamma - 1) + (\gamma + 1)Z}\right] \tag{622}$$

and

$$\mu = \frac{(\gamma+1)+(\gamma-1)Z}{2\gamma Z}, \qquad Z = \frac{2\gamma M_1^2 - (\gamma-1)}{\gamma+1} \tag{623}$$

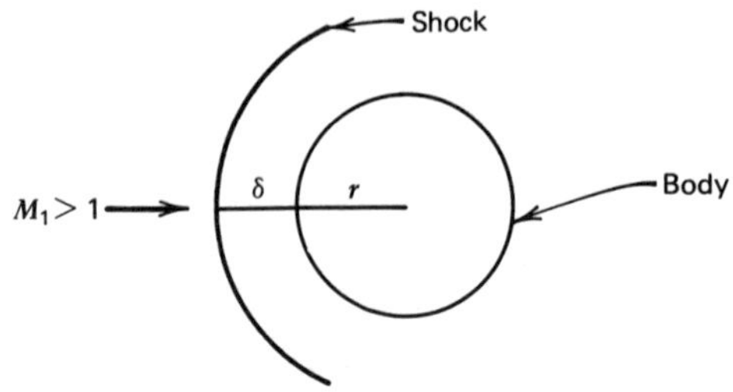

Figure 135. Sphere with detached shock.

Comparison with the experiments of Schwartz and Eckerman [42] for $\gamma = \frac{5}{3}$ and Heberle, Wood and Gooderum [43] and Van Dyke and Milton [44] for $\gamma = \frac{7}{5}$ show good agreement for a range of Mach numbers from 1.5 to 10 (Figure 136).

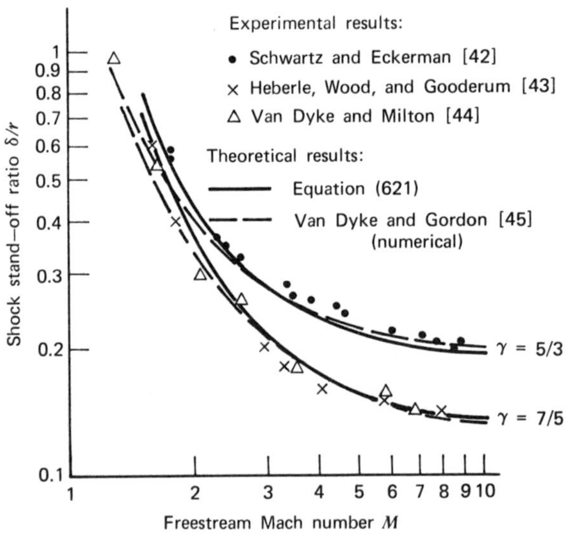

Figure 136. Shock stand off distance from various investigators compared [40].

It is of interest to observe that the shock stand-off distance decreases markedly as γ decreases, and as M increases, and levels off as M approaches 10.

Radvogin [46] has studied the shock stand-off distance in front of various axisymmetric blunt bodies as well as ogives and symmetric plane bodies. Consider the case of Figure 137. We attach to a blunt body a tangent cone

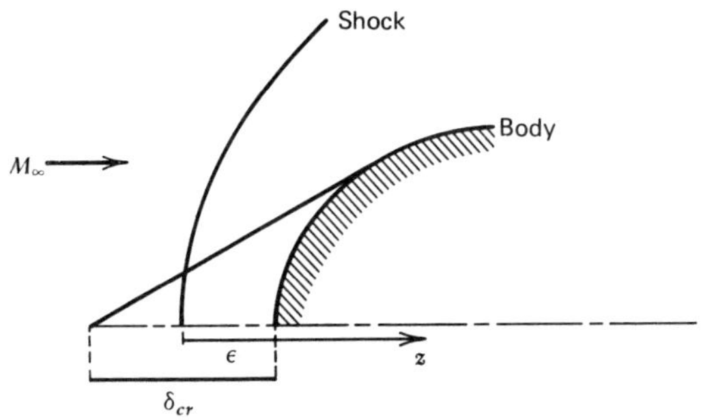

Figure 137. Axially symmetric body with detached shock showing notation of Radvogin [46].

whose half angle equals the maximum flow deflection angle for the given free stream Mach number M_∞. The distance from the apex of this cone to the nose of the body we call δ_{cr}. The distance from the nose of the body to the shock we call ε. Radvogin finds that

$$\delta_{cr} = k\varepsilon \tag{624}$$

where k is a constant depending only on the shape of the body. The accuracy of equation (624) may be determined by comparison with known data. For example, for a sphere:

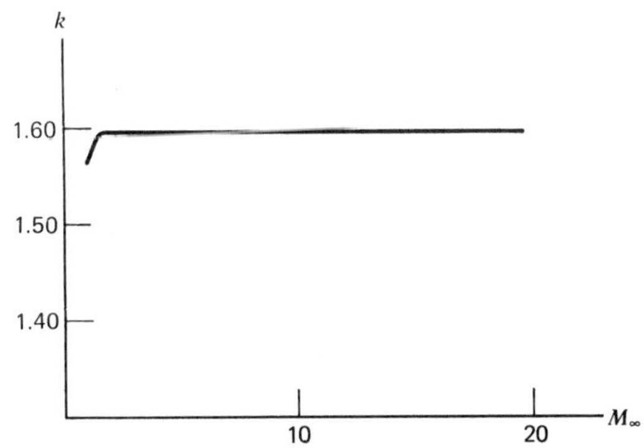

Figure 138. Variation of the parameter k with Mach number in equation (624).

At $M = 1.2$, $k = 1.57$, at $M = 1.3$, $k = 1.58$, at $M = 1.5$, $k = 1.59$; and from $M = 2$ to $M = 20$, $k = 1.60$. All data here assumes an ideal gas with $\gamma = 1.4$.

The equation (624) has also been tested for bodies of revolution of the form

$$r^2 = 2z - qz^2 \tag{625}$$

Table 1 shows the values of k for various bodies.

Table 1

body \ M	2	4	6	10	20
A	1.65	1.63	1.63	1.62	1.62
B	1.63	1.63	1.62	—	1.60
C	1.65	1.64	1.63	—	1.62
D	1.68	1.63	1.63	—	1.63
E	—	1.68	1.79	—	—
F	1.43	1.43	1.43	—	1.43

where body A is a paraboloid, $q = 0$
B is an ellipsoid, $q = 0.25$
C is an hyperboloid with asymptotic slope of 10°
D is an hyperboloid with asymptotic slope of 25°
E is an hyperboloid with asymptotic slope of 35°
F is a body of the form $r = z^{1/3}$.

The available data shows a remarkable consistency for spheres, paraboloids, ellipsoids and hyperboloids, with k nearly invariant with changes in body shape and Mach number. The variation for body F, however, indicates that caution is required in applying the results to bodies of other shapes. Even here, there is no variation with Mach number. The value of $k = 1.79$ for body E at $M = 6$ appears to be an anomaly requiring further investigation.

Radvogin has also established a correlation for ogival bodies. We deal with an ogive of nose half-angle β formed by attaching a cone to a spherical nosed blunt body at a Mach number M_∞ below the critical Mach number; i.e., with a detached shock. β_{cr} is the maximum cone half angle for which an attached shock is possible at the given Mach number. As before, ε is the distance between the nose of the body and the shock. δ is the distance between the nose of the actual body and the nose of a hypothetical body having nose half angle β_{cr} as shown in figure 139. The hypothetical nose cone is tangent to the actual body where the two meet. δ_{cr} is the distance from the nose of the original blunt body to the nose of the hypothetical cone. Now let $\tilde{\varepsilon} = \varepsilon/\delta_{cr}$. Radvogin finds

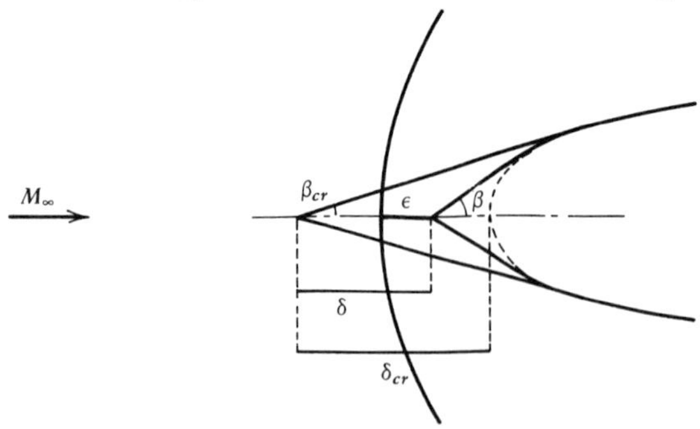

Figure 139. Notation of Radvogin [46] for ogival bodies.

that over the range $0 < \xi < 1$, $\tilde{\varepsilon}$ may be approximated by a relation of the form

$$\tilde{\varepsilon} = a\xi^2 + b\xi^3, \qquad \xi = \delta/\delta_{cr} \tag{626}$$

where $a = (3/k) - 1$ and $b = 1 - (2/k)$; $k = 1.6$. Comparison with known data for $\gamma = 1.4$ shows equation (626) to be a good approximation over the entire range of ξ. Radvogin has also determined the correlation of δ_{cr}/ε for plane flows. The results are shown in Table 2. Although these values are not as

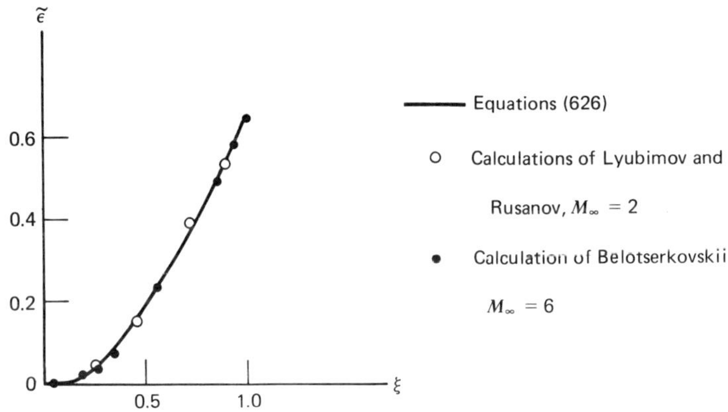

Figure 140. Results of Radvogin compared with those of other investigators [46].

consistent as those for the axisymmetric case, they nevertheless provide a handy

Table 2

Body \ M_∞	2	4	6	10	20
Circular cylinder	1.17	1.10	1.08	1.07	1.06
Parabolic cylinder		1.09	1.06	—	1.05

rule of thumb. Taking k as 1.1 in all cases listed results in a maximum error of less than six percent.

A general limiting criterion for the applicability of the above correlations in both the plane and axisymmetric cases is that the sonic point not be displaced by the attachment of the tangential critical cone.

Rotational Characteristics

The method of characteristics may also be applied to rotational, adiabatic, inviscid flow. Such conditions would be met approximately, for example, in the flow behind a curved shock.

Recall that the condition for the existence of a velocity potential was that the flow be irrotational. In the present case, therefore, we cannot take as our point of departure the equation for the velocity potential. Instead, we must revert to the original equations governing the flow. We confine ourselves here to the plane, two-dimensional, steady case.

The continuity and momentum equations for this case:

$$\rho u_x + u\rho_x + \rho v_y + v\rho_y = 0 \tag{627}$$

$$uu_x + vu_y + \frac{1}{\rho} p_x = 0 \tag{628}$$

$$uv_x + vv_y + \frac{1}{\rho} p_y = 0 \tag{629}$$

Recall now that for adiabatic inviscid flow, the entropy is constant along a streamline. It is thus more convenient to work in terms of entropy rather than pressure. Since entropy is constant along a streamline, we have

$$\mathbf{V}\cdot\nabla s = 0 \tag{630}$$

or

$$us_x + vs_y \tag{631}$$

since

$$p = p(\rho, s), \qquad dp = (p_s)_\rho\, ds + (p_\rho)_s\, d\rho \tag{632}$$

where the subscript outside the parenthesis indicates the variable held constant during differentiation. Thus, $(p_s)_\rho$ is the rate of change of p with respect to s while ρ is held constant.

Recall for a perfect gas,

$$\frac{ds}{c_v} = \frac{dp}{p} - \gamma\frac{d\rho}{\rho} \tag{633}$$

and hence,

$$(p_s)_\rho = \frac{p}{c_v}$$

Also,

$$(p_\rho)_s = c^2 = \frac{\gamma p}{\rho} \tag{634}$$

and

$$p = \frac{\rho c^2}{\gamma}$$

Thus,

$$dp = \frac{p}{c_v} ds + c^2 d\rho = \frac{\rho c^2}{\gamma c_v} ds + c^2 d\rho \tag{635}$$

Hence,

$$p_x = \frac{p}{c_v} s_x + c^2 \rho_x = \frac{\rho c^2}{\gamma c_v} s_x + c^2 \rho_x \tag{636}$$

and

$$p_y = \frac{p}{c_v} s_y + c^2 \rho_y = \frac{\rho c^2}{\gamma c_v} s_y + c^2 \rho_y \tag{637}$$

Thus, our governing equations become

$$(\rho u)_x + (\rho v)_y = 0 \tag{638}$$

$$uu_x + vu_y + \frac{1}{\rho}\left(\frac{\rho c^2}{\gamma c_v} s_x + c^2 \rho_x\right) = 0 \tag{639}$$

$$uv_x + vv_y + \frac{1}{\rho}\left(\frac{\rho c^2}{\gamma c_v} s_y + c^2 \rho_y\right) = 0 \tag{640}$$

$$us_x + vs_y = 0 \tag{641}$$

In the irrotational case, we wrote down the velocity potential equation and treated ϕ_{xx}, ϕ_{xy}, and ϕ_{yy} as the unknowns. In the same way we may now write down equations (638) through (641) and treat u_x, u_y, v_x, v_y, ρ_x, ρ_y, s_x, s_y, as the unknowns. In the irrotational case, we wrote down the expansion for $d\phi_x$ and $d\phi_y$ to make the problem determinate, that is, to provide us with a number of equations equal to the number of unknowns. In the same way we may now write down the expressions for the differentials for u, v, ρ and s:

$$du = u_x\,dx + u_y\,dy \tag{642}$$

$$dv = v_x\,dx + v_y\,dy \tag{643}$$

$$d\rho = \rho_x\,dx + \rho_y\,dy \tag{644}$$

$$ds = s_x\,dx + s_y\,dy \tag{645}$$

Thus we have eight equations, (638) through (645), and eight unknowns; u_x, u_y, v_x, v_y, ρ_x, ρ_y, s_x, and s_y. Solving the eight equations simultaneously for $\partial u/\partial x$, we obtain the solution in the form of the quotient of two determinants:

$$\frac{\partial u}{\partial x} = \frac{\begin{vmatrix} 0 & 0 & 0 & \rho & u & v & 0 & 0 \\ 0 & v & 0 & 0 & c^2/\rho & 0 & c^2/\gamma & 0 \\ 0 & 0 & u & v & 0 & c^2/\rho & 0 & c^2/\gamma \\ 0 & 0 & 0 & 0 & 0 & 0 & u & v \\ du & dy & 0 & 0 & 0 & 0 & 0 & 0 \\ dv & 0 & dx & dy & 0 & 0 & 0 & 0 \\ d\rho & 0 & 0 & 0 & dx & dy & 0 & 0 \\ d(s/c_v) & 0 & 0 & 0 & 0 & 0 & dx & dy \end{vmatrix}}{\begin{vmatrix} \rho & 0 & 0 & \rho & u & v & 0 & 0 \\ u & v & 0 & 0 & c^2/\rho & 0 & c^2/\gamma & 0 \\ 0 & 0 & u & v & 0 & c^2/\rho & 0 & c^2/\gamma \\ 0 & 0 & 0 & 0 & 0 & 0 & u & v \\ dx & dy & 0 & 0 & 0 & 0 & 0 & 0 \\ 0 & 0 & dx & dy & 0 & 0 & 0 & 0 \\ 0 & 0 & 0 & 0 & dx & dy & 0 & 0 \\ 0 & 0 & 0 & 0 & 0 & 0 & dx & dy \end{vmatrix}} \tag{646}$$

Since $\partial u/\partial x$ is supposed to be bounded (finite) everywhere, if the denominator is zero, the numerator must also be zero. For the denominator we get

$$[(v^2 - c^2)\,dx^2 - 2uv\,dx\,dy + (u^2 - c^2)\,dy^2][u\,dy - v\,dx] \tag{647}$$

The denominator will thus be equal to zero if either

$$u\,dy - v\,dx = 0 \tag{648}$$

or if

$$(v^2 - c^2)dx^2 - 2uv\,dx\,dy + (u^2 - c^2)dy^2 = 0 \tag{649}$$

Equation (648) is simply the equation of the streamlines. Thus, in rotational flow, the streamlines are also characteristic lines. The significance of the second equation may best be seen by using polar coordinates. Setting $u = V\cos\theta$ and $v = V\sin\theta$, we get

$$\frac{dy}{dx} = \tan(\theta \pm \alpha) \tag{650}$$

where α is the Mach angle. These characteristic directions are thus the same as in irrotational flow.

Using the result $dy/dx = v/u$, and setting the numerator of equation (646) (determinant) equal to zero, we get

$$ds = 0, \qquad u\,du + v\,dv + c^2\frac{d\rho}{\rho} = 0 \tag{651}$$

These results must follow of necessity since the entropy is constant along the streamlines, and the second of equations (651) is simply the Euler equation $V\,dV + dp/\rho = 0$. Using equation (650) for dy/dx, and setting the numerator equal to zero, we get, in polar coordinates,

$$\frac{d\theta}{\sin\alpha\cos\alpha} + \frac{d\rho}{\rho} + \frac{1}{\gamma}d\left(\frac{s}{c_v}\right) = 0 \tag{652}$$

Recall, for constant stagnation enthalpy,

$$V\,dV + c_p\,dT = 0 \tag{653}$$

Since $c = \sqrt{\gamma RT}$, $\gamma = c_p/c_v$, $R = c_p - c_v$ for a perfect gas, we have

$$V\,dV + \frac{2}{\gamma - 1}c\,dc = 0 \tag{654}$$

Also for a perfect gas,

$$T\,ds = c_v\,dT + pd\left(\frac{1}{\rho}\right) \tag{655}$$

Hence,

$$\frac{d\rho}{\rho} = -\frac{1}{\gamma - 1}\frac{ds}{c_v} - \frac{V\,dV}{c^2} \tag{656}$$

Substituting equation (656) into equation (652) we get

$$\frac{1}{V}\frac{dV}{d\theta} = \pm \tan\alpha - \frac{\sin^2\alpha}{\gamma(\gamma-1)}\cdot\frac{d(s/c_v)}{d\theta} \tag{657}$$

It will be observed that this result differs from that obtained in the plane case only by the last term on the right-hand side. This, however, makes it impossible to draw a generalized graph for the hodograph characteristics and makes it necessary to solve the equation separately for each step in the characteristics net.

Method of Solution. Assume the flow conditions are given along some non-characteristic curve *ab* in the flow (Figure 141). The two characteristic lines 1–3

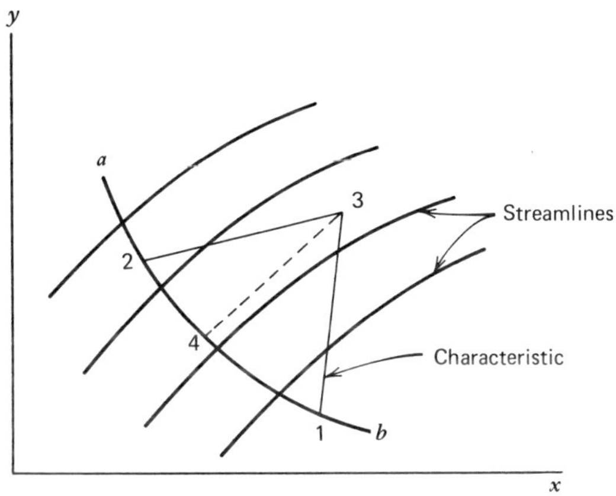

Figure 141. Physical characteristics in rotational flow.

and 2–3 are laid off using equation (650). That is, they make an angle $\pm\alpha$ with flow direction at points 1 and 2. Equation (657) may then be written in finite difference form:

$$V_3 - V_1 = -V_1 \tan\alpha_1(\theta_3 - \theta_1) - V_1\frac{\sin^2\alpha_1}{\gamma(\gamma-1)}\cdot\frac{s_3 - s_1}{c_v} \tag{658}$$

$$V_3 - V_2 = +V_2 \tan\alpha_2(\theta_3 - \theta_2) - V_2\frac{\sin^2\alpha_2}{\gamma(\gamma-1)}\cdot\frac{s_3 - s_2}{c_v} \tag{659}$$

Notice that the minus sign in front of the first term on the right-hand side of equation (658) corresponds to the physical characteristic making an angle $+\alpha$ with the flow direction, while the plus sign in front of the first term on the right-hand side of equation (659) corresponds to the physical characteristic making an angle $-\alpha$ with the flow direction. At the point 4, a line through that point having the flow direction of $\mathbf{V}_4$ passes through the point 3. We assume that the entropy at 3 equals the entropy at 4. (We are thus neglecting the curvature of the streamline between 3 and 4.) The two equations (658) and (659) may now be solved simultaneously for V_3 and θ_3, keeping in mind that $\sin\alpha = 1/M$ and that because the flow is adiabatic, V and M are uniquely related through the relation

$$\frac{dM}{M} = \left(1 + \frac{\gamma - 1}{2} M^2\right) \frac{dV}{V} \tag{660}$$

Notice that the noncharacteristic line *ab* may not coincide with a streamline since streamlines are also characteristic lines in the present case. It is easy to see that if initial data is given only along a streamline, the problem cannot be solved. Suppose initial data is given only along the streamline *cd* (Figure 142).

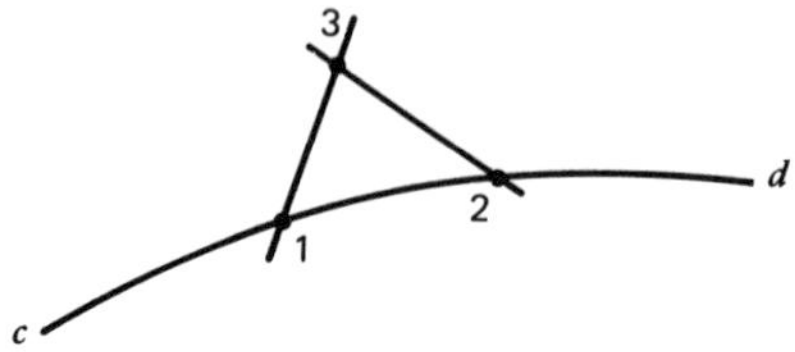

Figure 142. Initial data given along a streamline.

While the point 3 may be located in the physical plane, the entropy cannot be estimated there, since no streamline passes from *cd* to 3 and no other initial data is given. Thus, we have two equations, (658) and (659) and three unknowns, V_3, θ_3, and s_3, and hence the problem is indeterminate. This restriction does not apply in irrotational flow since the entropy is the same everywhere.

Axially Symmetric Flow with Rotation

The method of characteristic may also be applied to axially symmetric flow with rotation. In this case the equations for the physical and hodograph characteristics are found to be

$$\frac{dr}{dx} = \tan(\theta \mp \alpha) \tag{661}$$

$$d\theta \pm \frac{\cot\alpha}{V} dV \mp \frac{\sin\theta \sin\alpha}{\sin(\theta \pm \alpha)} \frac{dr}{r} \pm \frac{\sin\alpha \cos\alpha}{R\gamma} ds = 0 \tag{662}$$

where r is the radial coordinate, x the axial coordinate, R the gas constant and s the entropy. The upper sign is for the physical characteristics making an angle $-\alpha$ with the flow direction and the lower sign is used for the physical characteristics making an angle $+\alpha$ with the flow direction. The equations may be solved by the same finite difference method as in the plane case.

Some Linearized Solutions for Supersonic Flow

While the method of characteristics, together with the shock wave theory, gives us the means to find solutions of supersonic flow problems which are "exact," in the sense that the fluid is assumed inviscid, adiabatic and a perfect gas, such calculations are often long and tedious or require complicated computer programs. In the case of small perturbations, good approximations may often be obtained to the solution by making certain assumptions. We will discuss here three examples of such an approach.

Ackeret First-Order Theory

Given a thin body such as an airfoil in supersonic flow (Figure 143). The "exact"

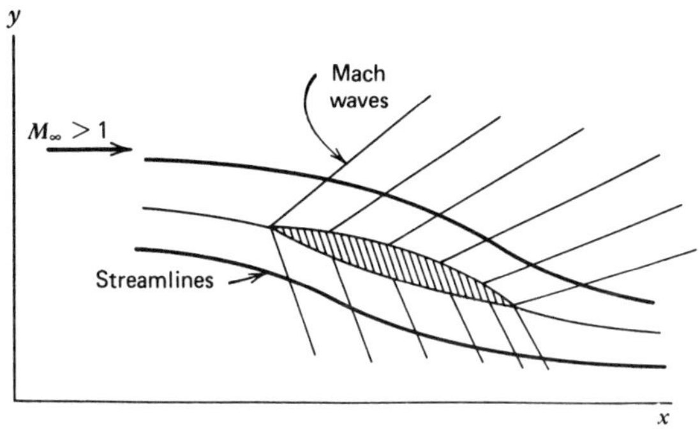

Figure 143. Flow pattern over an airfoil according to first-order (Ackeret) theory.

solution to such a problem would involve curved shocks attached to the nose and tail of the airfoil and the method of characteristics in between. The solution may be simplified by assuming Prandtl–Meyer flow between the leading and trailing edge shocks. A further simplification may be obtained by ignoring the shock wave altogether and assuming potential flow throughout. This would be justified if the shocks are sufficiently weak, either because the profile is very thin, or because the free stream Mach number is low (supersonic) or both. In this connection it may be recalled that oblique shock waves approach Mach waves as the pressure difference across the wave approaches zero.

Recall the Prandtl–Meyer equation,

$$d\theta = \sqrt{M^2 - 1}\,\frac{dV}{V} \tag{663}$$

Since the flow is presumed irrotational, $dp + \rho V d\,V = 0$ in every direction. This may be rewritten

$$dp = -\rho V\,dV = -\rho V^2 \frac{dV}{V} = -\frac{p}{RT}\frac{\gamma}{\gamma} V^2 \frac{dV}{V} = -p\gamma M^2 \frac{dV}{V} \tag{664}$$

Hence from equation (663)

$$dp = -\frac{\gamma p M^2\,d\theta}{\sqrt{M^2 - 1}} \tag{665}$$

If the flow is at all times perturbed only slightly from the free stream Mach number M_∞, equation (665) may be approximated by

$$\Delta p = -\frac{\gamma p M_\infty^2\,\Delta\theta}{\sqrt{M_\infty^2 - 1}} \tag{666}$$

Notice that over the top of the airfoil in Figure 143 a downward deflection of the flow results in a decrease in pressure, while on the lower surface of the airfoil a downward deflection of the flow results in an increase in the pressure. Thus, on the top surface we have

$$p - p_\infty = \frac{\gamma p_\infty M_\infty \theta}{\sqrt{M_\infty^2 - 1}} \tag{667}$$

while on the bottom surface,

$$p - p_\infty = \frac{-\gamma p_\infty M_\infty^2 \theta}{\sqrt{M_\infty^2 - 1}} \tag{668}$$

where $\theta = 0$ in the free stream and is measured positive counter clockwise as is customary. In terms of c_p this may be written

$$c_p = \frac{p - p_\infty}{\frac{1}{2}\rho_\infty V_\infty^2} = \frac{p - p_\infty}{\frac{1}{2}\gamma p_\infty M_\infty^2} = \pm \frac{2\theta}{\sqrt{M_\infty^2 - 1}} \tag{669}$$

The same result may also be obtained directly from the linearized potential equation. In the previous chapter it was shown that for small perturbations,

$$c_p = -\frac{2u}{V_\infty} \tag{670}$$

where u is the perturbation velocity component in the x direction. This result is equally valid whether the flow is subsonic or supersonic. Now recall further that the general solution for the perturbation velocity potential in the supersonic case was shown to be

$$\phi = f(x - \beta y) + g(x + \beta y) \tag{671}$$

where $\beta = \sqrt{M_\infty^2 - 1}$. Thus

$$\phi_x = u = f'(x - \beta y) + g'(x + \beta y); \qquad \phi_y = v = -\beta f'(x - \beta y) + \beta g'(x + \beta y) \tag{672}$$

where u and v are the perturbation velocity components in the x and y directions and the primes indicate differentiation with respect to the argument of the function.

Referring again to Figure 143, notice that over the top of the airfoil, the flow has constant properties along Mach lines of positive slope, whereas on the bottom of the airfoil, the flow has constant properties on Mach lines of negative slope. This follows from our previous discussion of Mach waves, Prandtl–Meyer flow and the small perturbation solution. Thus, on the top of the airfoil in Figure 143, the perturbation potential has the solution

$$\phi = f(x - \beta y), \qquad u = f', \qquad v = -\beta f' \tag{673}$$

while on the bottom surface, the solution is given by

$$\phi = g(x - \beta y), \qquad u = g', \qquad v = \beta y' \tag{674}$$

Now recall at the surface of a thin body with small flow perturbation,

$$\frac{v}{V_\infty} \cong \frac{dy}{dx} \tag{675}$$

where V_∞ is the free stream velocity and dy/dx the slope of the surface at any given point. From equations (673, 674) $v = \pm \beta u$ and hence $dy/dx = \pm u\beta/V_\infty$ or $u/V_\infty = (\pm 1/\beta)(dy/dx)$.
Hence

$$c_p = -\frac{2u}{V_\infty} = \pm\frac{2}{\beta}\frac{dy}{dx} = \pm\frac{2dy/dx}{\sqrt{M_\infty^2 - 1}} \tag{676}$$

But $dy/dx = \tan\theta \cong \theta$ when θ is small. Thus equation (676) becomes

$$c_p = \pm\frac{2\theta}{\sqrt{M_\infty^2 - 1}} \tag{677}$$

where the + sign refers to the top surface and the minus sign refers to the bottom surface. This is the same result obtained previously from the Prandtl–Meyer equation. If the shape of the airfoil is known, dy/dx is known at each point of the surface, and hence the pressure may be calculated at each point of the surface using the formula (676). By integrating over the surface of the airfoil, the lift and may be calculated. This is the basis of the linearized theory of Ackeret [233].

Supersonic Flow over a Wave-Shaped Wall

Another classic solution given by Ackeret is the supersonic flow past a wave shaped wall (Figure 144). We have already considered this problem for the

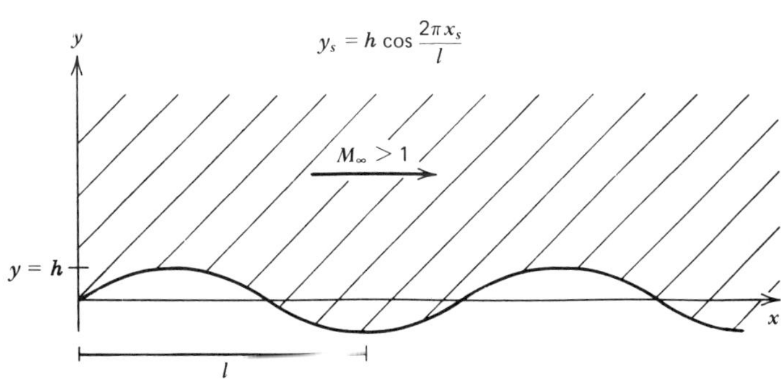

Figure 144. Supersonic flow over a wavy wall according to first-order theory.

subsonic case. As before, we have a wall whose shape is given by

$$y_s = h\cos\frac{2\pi x_s}{l}$$

For flow over the wall, the flow properties must be constant along Mach waves of positive slope. Thus,

$$\phi = f(x - \beta y), \qquad u = \phi_x = f'(x - \beta y), \qquad v = \phi_y = -\beta f'(x - \beta y) \quad (678)$$

At the wall, the flow must be parallel to the wall. Thus we have,

$$\frac{dy_s}{dx_s} = \frac{v}{V_\infty} = -\frac{\beta u}{V_\infty} = -h\frac{2\pi}{l}\sin\frac{2\pi x_s}{l} = \frac{-\beta f'(x - \beta y)_{y=y_s}}{V_\infty} \quad (679)$$

Thus,

$$f'(x - \beta y)_{y=y_s} = \frac{2\pi h V_\infty}{\beta l}\sin\frac{2\pi x_s}{l} \cong f'(x) \quad (680)$$

since y_s is assumed small at the surface of the wall. The governing equation $\beta\phi_{xx} = \phi_{yy}$ and the boundary condition (680) will thus be satisfied by

$$\phi_x = \frac{2\pi h V_\infty}{\beta l}\sin\frac{2\pi(x - \beta y)}{l} \quad (681)$$

and hence

$$\phi = -\frac{hV_\infty}{\beta}\cos\frac{2\pi}{l}(x - \beta y) \quad (682)$$

The Karman–Moore Method for Axially Symmetric Flow

In the case of axially symmetric supersonic flow with small perturbations, Karman and Moore [47] developed the method of source distribution whereby the flow over an axially symmetric, slender body of arbitrary shape may be determined. Again, the shock waves are neglected and the flow is presumed isentropic. Recall for incompressible flow,

$$\nabla^2\phi = 0 \quad (683)$$

A solution of this equation is $\phi = q/4\pi R$ where $R = \sqrt{x^2 + y^2 + z^2}$. This is a source flow with q the volume flow rate from a source at $R = 0$.

For compressible flow with small perturbations, the equation for the perturbation velocity potential is

$$-\beta^2\phi_{xx} + \phi_{yy} + \phi_{zz} = 0 \quad (684)$$

where $-\beta^2 = 1 - M_\infty^2$. This equation may be reduced to the Laplace equation by the change of variable $y' = i\beta y, z' = i\beta z$ for $M_\infty > 1$, where $i = \sqrt{-1}$. Then

$$\phi_{xx} + \phi_{y'y'} + \phi_{z'z'} = 0 \quad (685)$$

A solution to this equation is thus

$$\phi = -\frac{q}{2\pi\sqrt{x^2 + y'^2 + z'^2}} = -\frac{q}{2\pi\sqrt{x^2 - \beta^2(y^2 + z^2)}} \quad (686)$$

A flow of this type is referred to as a "supersonic source flow." The shape of the perturbation equipotential lines and perturbation streamlines may be found by rewriting equation (636) in the form

$$\left(\frac{x}{\beta a}\right)^2 - \left(\frac{y}{a}\right)^2 - \left(\frac{z}{a}\right)^2 = 1, \qquad a = \frac{q}{4\pi\phi} \tag{687}$$

which, for $\phi =$ constant, are hyperboloids of revolution about the $\pm x$ axis (Figure 145). The angle of the asymptotes is the Mach angle $\alpha(\tan\alpha = 1/\beta)$.

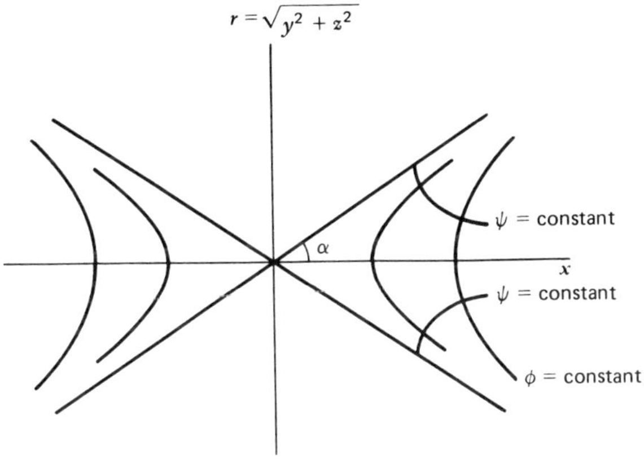

Figure 145. Supersonic source flow.

Since the free stream $M_\infty > 1$ is presumed to be from left to right, only the right side of Figure 145 has physical meaning.

A source need not, of course, be located at the origin. It could be located at some point $x = \xi$ on the x axis (Figure 146). In this case, the perturbation poten-

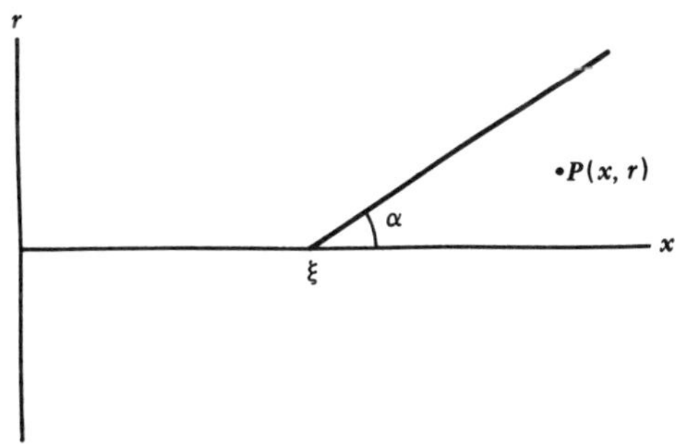

Figure 146. Region of influence of a source located at $x = \xi$.

tial at the point $P(x,r)$ due to the source at $x = \xi$ becomes

$$\phi = \frac{q}{4\pi\sqrt{(x - \xi)^2 - \beta^2 r^2}}, \qquad r = y^2 + z^2 \tag{688}$$

Because equation (685) is linear, if (688) is a solution, then

$$\phi = \sum_i \phi_i = \sum_i \frac{q_i}{4\pi\sqrt{(x - \xi_i) - \beta^2 r^2}} \tag{689}$$

is also a solution, implying a whole series of sources distributed along the x axis. It follows that an infinite number of sources of infinitesimal strength $dq(\xi) = f(\xi)d\xi$ distributed continuously along the x axis leads to the solution

$$\phi(x, r) = \int_0^{x-\beta r} \frac{f(\xi)\,d\xi}{4\pi\sqrt{(x-\xi)^2 - \beta^2 r^2}} \tag{690}$$

(For a rigorous proof of this statement, the reader is referred to books on linear partial differential equations). For $\xi > x - \beta r$, the solution is no longer real. Physically, this may be interpreted by noting that the flow at a point $P(x, r)$ cannot be influenced by sources lying downstream of $\xi = x - \beta r$. (The above argument may also be presented in reverse form; i.e., one may state that equation (690) is a known solution of equation (685), as may be verified by direct substitution, and then show that this solution may be interpreted as a continuous source distribution along the x axis.)

Application to Particular Cases. The form of the function of $f(\xi)$ for a particular application of the solution (690) is determined by the boundary conditions. Recall we are dealing with axially symmetric flow over a slender body. At the surface of the body,

$$\frac{dr}{dx} \cong \frac{v}{V_\infty} \tag{691}$$

and

$$v = \phi_r = -\int_0^{x-\beta r} \frac{f'(\xi)}{4\pi}\left(\frac{x-\xi}{r}\right)\frac{d\xi}{\sqrt{(x-\xi)^2 - \beta^2 r^2}} \tag{692}$$

For small r,

$$v \cong \int_0^x \frac{f'(\xi)\,d\xi}{4\pi r} = \frac{f(x)}{4\pi r} \tag{693}$$

Note that $f(0)$ must equal zero. Otherwise, there would be a stagnation point at the nose and hence a detached shock. The source distribution may now be found directly from the boundary condition:

$$\left(\frac{dr}{dx}\right)_s = \frac{f(x)}{4\pi r V_\infty} \tag{694}$$

Note that u, v, ϕ, and $c_p = -2u/V_\infty$ are thus determined.

Thompson [48] has applied the method to finding the pressure distribution over a V-2 missile at $M = 1.27$ and finds that the results agree well with experiment.

5 Unsteady One-Dimensional Flow

Isentropic Flow of Simple Waves

In Chapter 2 we determined the speed of a small pressure wave by deriving the wave equation. We also considered the steady, quasi-one-dimensional flow of a gas in a duct of varying cross section, using the one-dimensional flow equations in their simplest form. We may now apply the latter technique to the former problem.

Consider a steady one-dimensional flow of constant cross section with a small discontinuity in flow properties at some point of the flow (Figure 147).

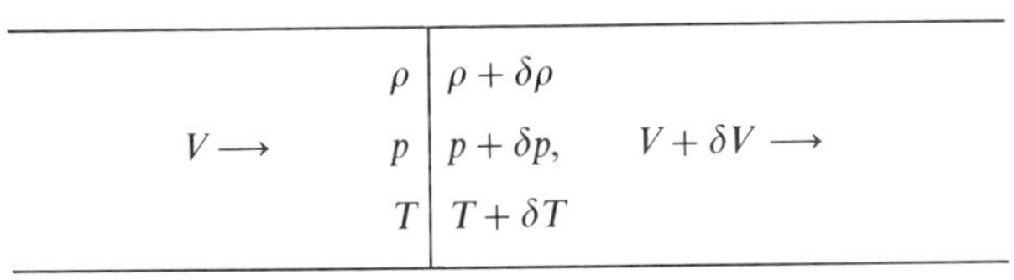

Figure 147. Small discontinuity in one-dimensional flow.

The discontinuity is presumed to be at rest relative to the observer. The continuity equation is then

$$\rho VA = (\rho + \delta\rho)(V + \delta V)A = \rho VA + \rho\delta VA + V\delta\rho A + \delta\rho\delta VA \qquad (695)$$

or

$$\rho\delta V + V\delta\rho + \delta\rho\delta V = 0 \qquad (696)$$

The momentum equation

$$Ap - A(p + \delta p) = \rho VA(V + \delta V - V) \qquad (697)$$

or

$$\delta p = \rho V\,\delta V \qquad (698)$$

Notice that these relations are equally valid if the two regions are separated by a region of finite width rather than a discontinuity. If we divide equation (696) by $\delta\rho$ and pass to the limit as $\rho V \rightarrow 0$, we get

$$\frac{dV}{d\rho} = -\frac{V}{\rho} \qquad (699)$$

Notice that the continuity equation requires $\delta\rho$ to go to zero as $\delta V \rightarrow 0$. From equations (699) and (698)

$$\frac{dV}{d\rho} = \frac{dV}{dp}\frac{dp}{d\rho} = -\frac{1}{\rho V}\frac{dp}{d\rho} = -\frac{V}{\rho} \tag{700}$$

If the flow is presumed isentropic, $dp/d\rho = c^2$ and we have

$$V^2 = c^2 \tag{701}$$

This is, of course, the same result we obtained previously by deriving the wave equation. It is usually put into words by saying that a small pressure wave propagates with the speed of sound relative to the flow. This is rather loose language. As long as δp is finite, equation (701) is only approximately true. It is exact only if $\delta p = 0$ in which case there is no wave. A more precise statement is that equation (701) becomes progressively more exact as δp becomes smaller.

In addition to the velocity of the wave, we may also determine the variation of the pressure, velocity and speed of sound across the wave relative to each other. From equations (700) and (124) we have

$$\frac{dp}{dV} = -\rho c = \frac{\gamma p}{c} \qquad \text{or} \qquad \frac{dp}{p} = -\frac{\gamma}{c}dV \tag{702}$$

for a perfect gas. Recall for an isentropic perfect gas,

$$\frac{dp}{p} = \frac{2\gamma}{\gamma - 1}\frac{dc}{c} \tag{703}$$

Thus

$$\frac{dc}{dV} = -\frac{\gamma - 1}{2} \tag{704}$$

Consider now a situation in which we have a continuous variation of the pressure rather than a small pressure discontinuity. The variation of the pressure with x at some time t is shown in Figure 148. The flow is not steady. All the flow

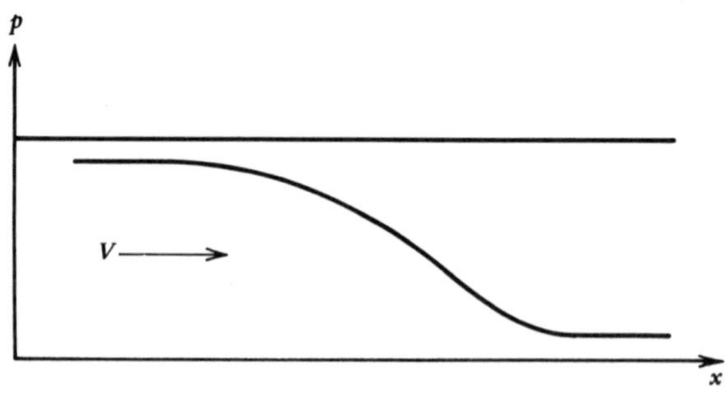

Figure 148. Pressure wave in one-dimensional flow.

properties (except entropy) vary with x and t. The problem is, given the conditions at time t_1, say, $p(x,t_1), T(x,t_1), V(x,t_1), c(x,t_1), \rho(x,t_1)$, how will these function vary with time? i.e., find

$$p(x,t),\ T(x,t),\ V(x,t),\ \rho(x,t),\ c(x,t)$$

In order to solve the problem, let us approximate the continuous pressure distribution by a series of small increments (Figure 149). Each increment will

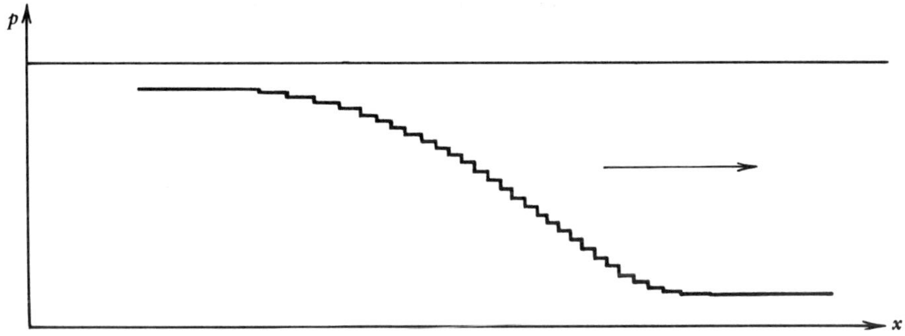

Figure 149. Pressure wave approximated by a series of small steps.

then be a small pressure wave. Although this is an unsteady flow, we may treat this flow as "locally steady." That is, we may calculate the change in the flow properties across each wave by fixing a coordinate system in the wave and applying the results of the previous section. The waves ahead of a given wave affect that wave only by determining the flow conditions upstream of the wave. The waves behind, or downstream, of a given wave cannot affect the wave at all unless they catch up to it. Thus, across each of the wavelets,

$$dc = +\frac{\gamma - 1}{2} dV \tag{705}$$

Now the continuous wave may be approximated evermore closely by decreasing the size of each increment while increasing the number of increments. This also increases the accuracy of equation (705) across each wavelet. In the limit as the increments approach zero while the number of waves becomes infinite, equation (705) becomes exact. In order to find the change between any two points in the flow, therefore, one may integrate equation (705) and obtain

$$c_2 - c_1 = \frac{\gamma - 1}{2}(V_2 - V_1) \tag{706}$$

or

$$\frac{c_2}{c_1} = 1 + \frac{\gamma - 1}{2}\left(\frac{V_2}{c_1} - \frac{V_1}{c_1}\right) \tag{707}$$

Moreover, since $c^2 \sim T$ and $T \sim p^{(\gamma - 1)/\gamma}$, we have

$$\frac{T_2}{T_1} = \left[1 + \frac{\gamma - 1}{2}\left(\frac{V_2}{c_1} - \frac{V_1}{c_1}\right)\right]^2 \quad \text{and} \quad \frac{p_2}{p_1} = \left[1 + \frac{\gamma - 1}{2}\left(\frac{V_2}{c_1} - \frac{V_1}{c_1}\right)\right]^{2\gamma/(\gamma - 1)} \tag{708}$$

These equations apply to a wave such as that shown in Figure 149; the pressure is increasing from right to left and the wave is presumed to be moving from left to right with respect to the flow. Such a wave is referred to as a right-running compression wave. The above technique could equally well be applied if the wave were moving from right to left with respect to the flow or if the pressure were decreasing from right to left. Only the signs change.

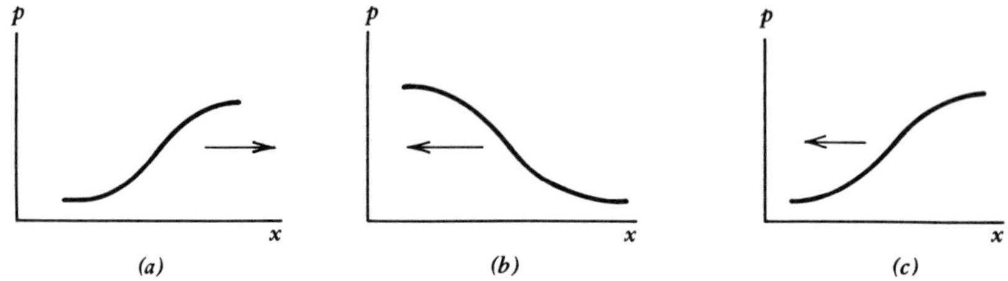

Figure 150. Right- and left-running compression and expansion waves.

Thus, for the right-running rarefaction wave [Figure 150(a)]

$$c_2 - c_1 = \frac{\gamma - 1}{2}(V_2 - V_1) \tag{709}$$

while for the left-running compression and expansion waves,

$$c_2 - c_1 = -\frac{\gamma - 1}{2}(V_2 - V_1) \tag{710}$$

The plus sign in front of the right-hand side of the equation (709) is correct for right-running waves, while the minus sign in front of the right-hand side of equation (710) is correct for left running waves.

At each point, the wave is propagating with the local speed of sound relative to the local flow velocity. Thus, for right-running waves, the velocity is $V + c$, while for left-running waves it is $V - c$. V is taken as positive in the plus x direction.

We may note here the resemblance between the approach taken in the case of Prandtl–Meyer flow and the approach taken here. In both cases we began with a small perturbation in the flow and then integrated the result over a large number of small perturbations to obtain the change in the flow variables for finite changes in the flow. In the present case, as in the former, such an approach is intuitive and does not constitute a proof. The procedure may seem reasonable, but we cannot really be sure that the results are valid unless we can find a more rigorous way to derive them.

Recall for unsteady one-dimensional flow,

$$\rho(u_t + uu_x) = -p_x \tag{711}$$

We now define ϕ by $u = \phi_x$. Although the function here is often spoken of as a velocity potential, there is no question here of irrotationality. All that is necessary is for the $\int u\,dx$ to be defined. On physical grounds, this is a reasonable assumption. Thus we have

$$\phi_{xt} + \phi_x\phi_{xx} = -\frac{1}{\rho}p_x = -\frac{c^2}{\rho}\rho_x \tag{712}$$

Integrating over x with t held constant we have

$$\phi_t + \tfrac{1}{2}\phi_x^2 + \int \frac{c^2}{\rho}d\rho = \phi_t + \tfrac{1}{2}\phi_x^2 + \int \frac{2c^2}{\gamma - 1}\frac{dc}{c} = \phi_t + \tfrac{1}{2}\phi_x^2 + \frac{c^2}{\gamma - 1} = f(t) \tag{713}$$

for isentropic flow. Differentiating equation (713) w.r.t. x and t in turn, we get

$$\phi_{xt} + \phi_x\phi_{xx} + \frac{c^2}{\rho}\rho_x = 0 \qquad \text{or} \qquad \rho_x = -\frac{\rho}{c^2}(\phi_{xt} + \phi_x\phi_{xx}) \tag{714}$$

$$\phi_{tt} + \phi_x\phi_{xt} + \frac{c^2}{\rho}\rho_t = f'(t) \qquad \text{or} \qquad \rho_t = -\frac{\rho}{c^2}(\phi_{tt} + \phi_x\phi_{xt}) + f'(t) \tag{715}$$

Substituting for u, ρ_x and ρ_t in the continuity equation

$$\rho_t + u\rho_x + \rho u_x = 0 \tag{716}$$

We get

$$(c^2 - \phi_x^2)\phi_{xx} - 2\phi_x\phi_{xt} - \phi_{tt} = -f'(t) \tag{717}$$

The $f(t)$ may be absorbed in the ϕ without loss of generality and hence it is equally valid to write

$$(c^2 - \phi_x^2)\phi_{xx} - 2\phi_x\phi_{xt} - \phi_{tt} = 0 \tag{718}$$

where from equation (713)

$$c^2 = (1 - \gamma)(\phi_t + \tfrac{1}{2}\phi_x^2) \tag{719}$$

Thus, we have, as in two-dimensional, potential flow, reduced the governing equations to a single equation in one unknown. The differential equation (718) is of the form

$$A\phi_{xx} + B\phi_{xt} + C\phi_{tt} = 0 \tag{720}$$

where $A = (c^2 - u^2)$, $B = -2u$ and $C = -1$. This equation will be hyperbolic if $B^2 - 4AC > 0$. Substituting for A, B, and C we get

$$B^2 - 4AC = 4u^2 + 4(c^2 - u^2) = 4c^2 > 0 \tag{721}$$

Thus equation (718) is always hyperbolic and hence may be solved by the method of characteristics.

Recall now the slope of the physical characteristics is given by

$$\frac{dt}{dx} = \frac{B \pm \sqrt{B^2 - 4AC}}{2A} \tag{722}$$

Substituting for A, B, and C, we get

$$\frac{dt}{dx} = \frac{1}{u \pm c} \tag{723}$$

Thus we see that the physical characteristics represent the "paths" of the pressure waves in the x, t plane.

The slope of the hodograph characteristics is given by

$$\frac{d\phi_t}{d\phi_x} = -\frac{B \pm \sqrt{B^2 - 4AC}}{2C} \tag{724}$$

Substituting for A, B, and C, we get

$$\frac{d\phi_t}{d\phi_x} = -(u \pm c) \tag{725}$$

The hodograph characteristics in unsteady one-dimensional flow do not lend themselves as readily to physical interpretation as do the hodograph characteristics in steady two-dimensional flow. This is because no simple physical interpretation of ϕ_t exists. We may rewrite the equation for the hodograph characteristics in physically more meaningful terms by substituting for ϕ_t from the equation for the speed of sound. Recall,

$$c^2 = (1 - \gamma)(\phi_t + \tfrac{1}{2}\phi_x^2) \tag{726}$$

Thus

$$\phi_t = \frac{c^2}{\gamma - 1} - \frac{1}{2}\phi_x^2 \tag{727}$$

and

$$d\phi_t = \frac{2c\,dc}{\gamma - 1} - u\,du \tag{728}$$

Substituting for $d\phi_t$ in equation (725), we get

$$\frac{2}{\gamma - 1}dc = \pm\, du \tag{729}$$

which may be integrated directly to give

$$\frac{2}{\gamma - 1}c = \pm\, u + \text{constant} \tag{730}$$

These lines are sometimes referred to as the state characteristics. As in the case of the hodograph characteristics in two-dimensional, steady flow, the state characteristics in one-dimensional unsteady flow may be plotted once and for all on a general graph valid for every problem (Figure 151). Numerical method of solution. The numerical method of solution is the same as in the two-dimensional steady case. We suppose the flow properties (velocity, temperature, pressure, etc.) are given along some noncharacteristic line in the x, t (physical) plane (Figure 152).

Notice that this line may be a line of constant t, but does not need to be. That is, the flow properties may be given as a function of x at some time t,

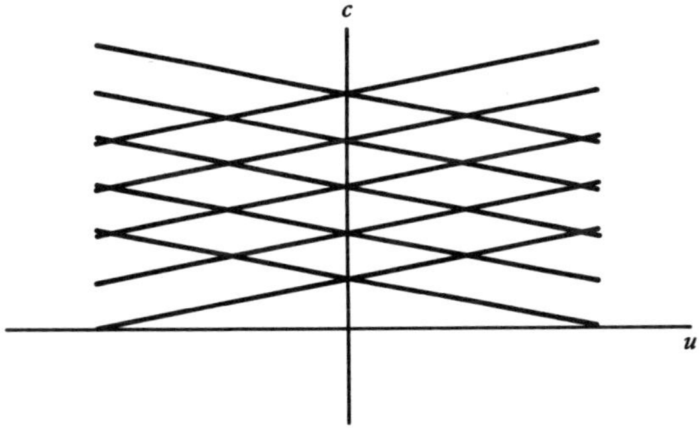

Figure 151. State characteristics for unsteady one-dimensional flow.

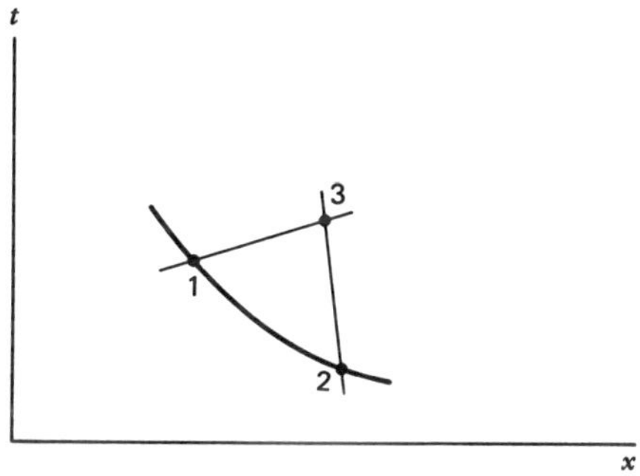

Figure 152. Physical characteristics in the $x-t$ plane.

or they may be given at different times t at each point x. The slopes of the physical characteristics 1–3 and 2–3 are given by $u + c$ and $u - c$, respectively. This locates the point 3. From 1 to 3 we have

$$\frac{2}{\gamma - 1} dc = - du \tag{731}$$

and in going from 2 to 3 we have

$$\frac{2}{\gamma - 1} dc = + du \tag{732}$$

The properties at the point 3 are thus established. The properties at the point 3 may also be found graphically. Referring once again to the state plane (Figure 153), we locate the points 1 and 2. Since du and dc have the same sign in going

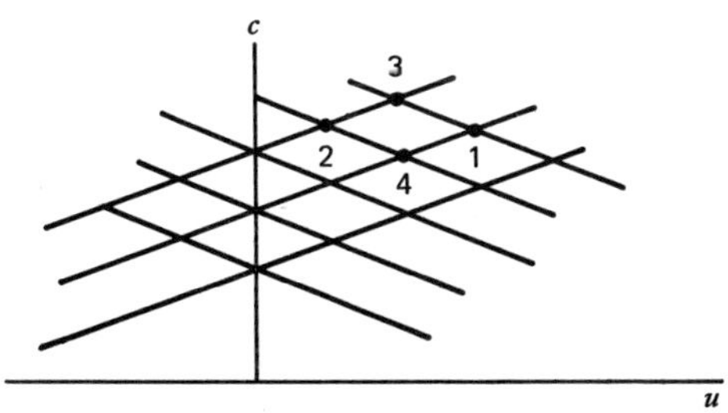

Figure 153. Determination of flow properties using state plane.

from 1 to 3, the point 3 must be at the intersection of the two state characteristics passing through the points 1 and 2, as shown in the figure.

Working either graphically or numerically, the flow properties may be found throughout a finite region bounded by the line along which the boundary conditions are given, and the two intersecting physical characteristics running through either end of the boundary, just as in the two-dimensional steady case (Figure 154).

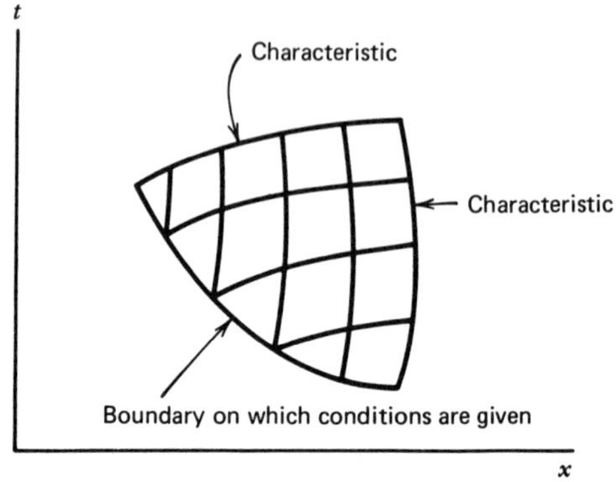

Figure 154. Region of influence in x–t plane.

Notice that in going from 2 to 3 in Figure 153, on the state plane, the changes in c and u are the same as those predicted by the small perturbation theory across a right-running compression wave. Similarly, in going from 1 to 3, the changes in c and u are the same as those predicted for a left-running compression wave by the small perturbation theory. In going from 2 to 4, the changes in c and u are the same as those predicted for a left-running expansion wave and in going from 1 to 4 the changes are the same as those predicted for a right-running expansion wave. Thus, all the changes in the flow may be thought of as being caused by a series of infinitesimal right n- and left-running compression and expansion waves.

Consider now the special case in which ϕ_x and ϕ_t are not independent but functions of each other. We have already considered similar cases in two- and three-dimensional steady flow. In the two-dimensional case where u and v were functionally related we had Prandtl–Meyer flow and, in the three-dimensional axially symmetric case, functional dependence of the velocity components led to the Taylor–Macoll solution.

Recall the equation

$$(c^2 - \phi_x^2)\phi_{xx} - 2\phi_x\phi_{xt} - \phi_{tt} = 0 \tag{733}$$

If $\phi_x = f(\phi_t)$, then

$$\frac{\partial\phi_x}{\partial x} = \frac{d\phi_x}{d\phi_t}\frac{\partial\phi_t}{\partial x}; \qquad \frac{\partial\phi_t}{\partial t} = \frac{d\phi_t}{d\phi_x}\frac{\partial\phi_x}{\partial t} = \frac{d\phi_t}{d\phi_x}\frac{\partial\phi_t}{\partial x} \tag{734}$$

Substituting (734) in (733).

$$(c^2 - \phi_x^2)\frac{\partial \phi_t}{\partial x}\frac{d\phi_x}{d\phi_t} - 2\phi_x\frac{\partial \phi_t}{\partial x} - \frac{d\phi_t}{d\phi_x}\frac{\partial \phi_t}{\partial x} = 0 \tag{735}$$

or

$$(c^2 - \phi_x^2)\left(\frac{d\phi_x}{d\phi_t}\right)^2 - 2\phi_x\left(\frac{d\phi_x}{d\phi_t}\right) - 1 = 0 \tag{736}$$

Thus

$$\frac{d\phi_x}{d\phi_t} = \frac{-1}{\phi_x - c}, \qquad \frac{-1}{\phi_x + c} \tag{737}$$

Assume now that the flow is isentropic. That is, the entropy does not vary with x or t. Then, from (713)

$$\frac{dc}{d\phi_x} = -\frac{\gamma - 1}{2c}\left(u + \frac{d\phi_t}{d\phi_x}\right) \tag{738}$$

Substituting for $d\phi_t/d\phi_x$ from equation (737) and recalling that $\phi_x = u$,

$$\frac{dc}{du} = \pm\frac{\gamma - 1}{2} \tag{739}$$

If ϕ_x and ϕ_t are uniquely related, we must chose either the plus or the minus sign in equation (737). Thus, only the plus or minus sign will apply in equation (739). This means that if ϕ_x and ϕ_t are uniquely related, all the points of the flow will fall along a single state characteristic. Equation (739) may be integrated to give

$$c = \pm\frac{\gamma - 1}{2}u + \text{constant} \tag{740}$$

This is the same result we obtained in the previous section when we applied the small perturbation theory to right- or left-running expansion or compression waves of finite strength. Equation (740) indicates that these are indeed valid solutions since they satisfy equation (733). The solution

$$c = +\frac{\gamma - 1}{2}u + \text{constant} \tag{741}$$

thus indicates a flow in which only right-running waves are present; that is, waves propagating with velocity $u + c$. The waves may be either compression or expansion waves since u and c increase together in right-running compression waves and decrease together in right-running expansion waves. Similarly, the solution

$$c = -\frac{\gamma - 1}{2}u + \text{constant} \tag{742}$$

indicates a flow in which only left-running waves are present, that is, waves propagating with velocity $u - c$. The waves may be compression or expansion waves since for a left-running compression wave u decreases when c increases, while for a left-running expansion wave, c decreases while u increases.

An unsteady one-dimensional flow in which only right- or left-running waves are present may thus be compared to a Prandtl–Meyer flow, in which only right- or left-sloping Mach waves are present. In addition to being called a simple wave flow, such a solution is also sometimes referred to as "singly degenerate."

An example of such a flow would be that caused by a piston in a cylinder accelerating to the left, as shown in Figure 155. While the piston may accelerate

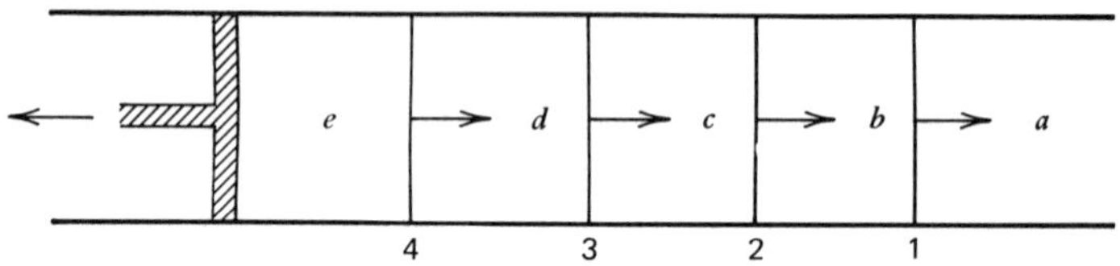

Figure 155. Expansion waves generated by accelerating piston.

smoothly, the flow may be approximated by a series of expansion waves, 1, 2, 3, 4, etc. traveling from left to right. This may be seen by examining the physical plane (Figure 156). The x axis gives the position in the cylinder and the t

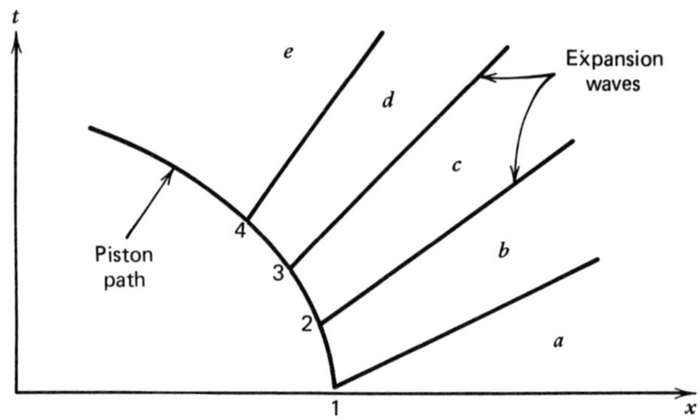

Figure 156. Piston path and Mach waves shown on the x–t plane.

axis the time. The path of the piston may be approximated by a series of straight lines. The inverse of the slope of these lines gives the velocity of the piston. Thus, if the piston is initially at rest and is suddenly given a small finite velocity, a velocity discontinuity is set up. At the face of the piston, the velocity of the air or gas must equal the velocity of the piston. In the rest of the cylinder, the gas is still at rest. If there is no viscosity or heat transfer, the entropy will remain everywhere the same. Thus the equations which we have derived indicate that a rarefaction wave will now propagate to the right with a velocity of the local speed of sound relative to the local flow velocity. Because the impulsive start of the piston is an approximation, we may take the wave velocity as being at the average speed of sound relative to the average flow velocity between the regions ahead of and behind the wave. In our straight-line approximation, the piston now travels at constant speed until it reaches point 2, when it again undergoes a sudden increment in velocity. This creates a new velocity discontinuity and hence a new expansion wave, just as at point 1. Similarly at points 3, 4, and so on. We may increase the accuracy of our approximation by replacing the piston wave by ever more and shorter straight lines yielding ever more and weaker expansion waves until, in the limit, the number of straight lines and waves becomes infinite at which point we have a continuous, exact solution. Thus, the approximate solution may be made as accurate as desired.

In dealing with problems of this type, it is sometimes more convenient to speak of the flow properties in regions a, b, c, d, e, etc., that is, in the regions between the waves, rather than speaking of the flow properties at points 1, 2, 3, 4, ... This distinction is analogous to that made between field method and lattice point method in the theory of characteristics for two-dimensional, steady flow. Given the velocity in each region (i.e., given the piston curve), the speed of sound in each region may be read directly off the state plane (Figure 157).

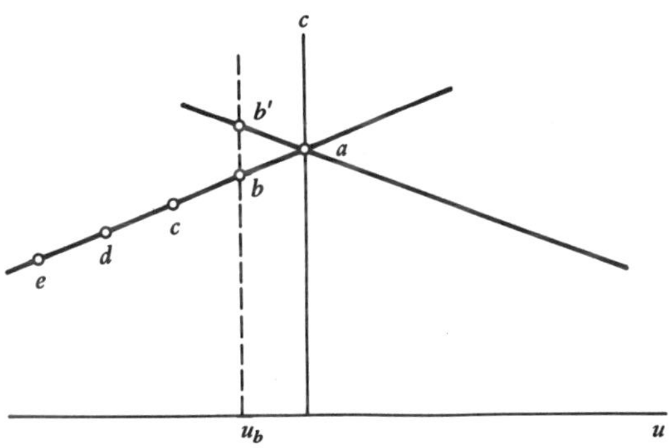

Figure 157. Determination of flow properties using the state plane.

Given the initial condition at $a\,(u=0)$, the point b must lie somewhere along the two lines intersecting at a. Given the velocity in region b, u_b, the point b must be to the left of the ordinate if the piston is moving to the left. This leaves only two possibilities: b and b'. But since the wave is an expansion wave, the speed of sound is lower in region b than in region a. This fixes the point b in Figure 157. That the wave must be an expansion wave may be seen from the following considerations. From equation (708)

$$\frac{p_b}{p_a}=\left[1\pm\frac{\gamma-1}{2}\left(\frac{u_b}{c_a}-\frac{u_a}{c_a}\right)\right]^{2\gamma/(\gamma-1)} \tag{743}$$

where the $\pm$ sign has been added to cover all four possibilities: left-running expansion wave, left-running compression wave, right-running expansion wave, right-running compression wave. u is taken as positive to the right. As was pointed out in connection with equations (709) and (710), the plus sign is correct for right-running waves, while the minus sign is correct for left-running waves. In the present case we have a right-running wave and hence

$$\frac{p_b}{p_a} = \left[1 + \frac{\gamma - 1}{2}\left(\frac{u_b}{c_a} - \frac{u_a}{c_a}\right)\right]^{2\gamma/(\gamma - 1)} \tag{744}$$

$u_a = 0$ and u_b is negative. Hence $p_b < p_a$ and the wave is an expansion wave. The temperature is given by the equation (708)

$$\frac{T_b}{T_a} = 1 + \frac{\gamma - 1}{2}\left(\frac{u_b}{c_a} - \frac{u_a}{c_a}\right) \tag{745}$$

Thus, all the properties in region b are known. In a similar way, the properties in regions $c, d, e, \ldots$ may be found, given the piston speed in each region.

Another way of looking at this problem is to regard the flow as continuous *ab initio*. In this case, the waves represent lines of constant properties. This may be seen in the following way. Observe the lines showing the paths of the expansion waves in the physical plane (Figure 156). Using the "field method," the regions between the waves represent regions of constant properties. Now increase the number of regions by dividing the piston curve into more and smaller velocity increments. This will bring successive expansion waves closer together. The regions between the waves remain regions of constant properties. In the limit as the waves come arbitrarily close together, the regions of constant properties become lines of constant properties whose slope equals that of the local wave path. Thus in a continuous simple wave flow, each wave path represents a line of constant properties. The slope of each wave path is, of course, given by

$$\left(\frac{dx}{dt}\right)_{\text{wave}} = c - u \tag{746}$$

in the present instance. That is, each wave travels to the right with a velocity equal to the local speed of sound relative to the local flow velocity. The local speed of sound may read directly off the state plane (Figure 157) given the local flow velocity, or it may be computed from.

$$c - c_a = \pm\frac{\gamma - 1}{2}(u - u_a) \tag{747}$$

The conditions in region a, where the gas is at rest, are presumed given. In the present case, $u_a = 0, u < 0$, and $c < c_a$ since we are dealing with expansion waves. Thus the plus sign must be used. In general, as was pointed out previously, the plus sign is used for right-running waves, the minus sign is used for left-running waves. The other conditions throughout the flow field may be determined from the equations

$$\frac{p}{p_a} = \left[1 \pm \frac{\gamma - 1}{2}\left(\frac{u - u_a}{c_a}\right)\right]^{2\gamma/(\gamma - 1)} \tag{748}$$

$$\frac{T}{T_a} = \left[1 \pm \frac{\gamma - 1}{2}\left(\frac{u - u_a}{c_a}\right)\right] \tag{749}$$

where the plus sign is used in the present case because we have a right-running wave.

Flow With Waves of Both Families

We have already seen, in a purely formal way, how we may solve problems in unsteady one-dimensional flow in general by method of characteristics. Let us now see how this method would work in a specific case. The case we will consider is that which would prevail if we close off the right side of the cylinder in the example considered in the previous section. As may be seen from Figure 158,

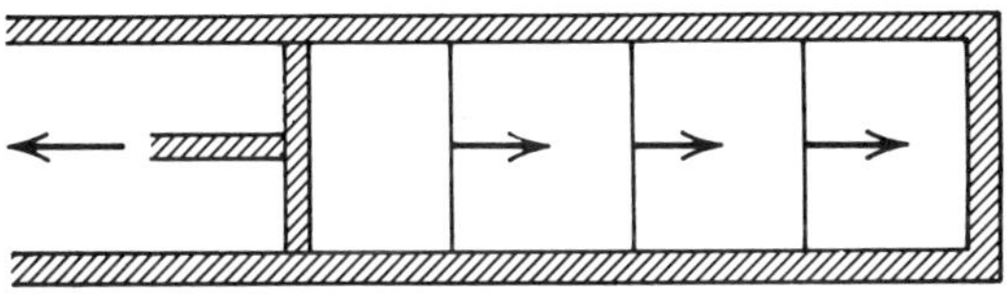

Figure 158. Piston accelerating in a closed cylinder.

everything will be the same as before until the first wave hits the closed end of the cylinder. It will now be found particularly advantageous to examine this problem using the field method. Notice that in the field adjoining the closed end of the cylinder, the velocity must always be zero. Behind the first wave, however, the gas has a finite velocity to the left. Thus, when the first wave hits the end of the cylinder, a velocity discontinuity is set up. The boundary condition at the wall may continue to be satisfied, however, if the first wave reflects off the wall as a like wave of equal strength going in the opposite direction; that is, as a left-running rarefaction wave. The velocity increment caused by the impinging wave is thus cancelled out by the reflected wave. Notice that if the impinging wave had been a compression wave, the boundary condition at the closed end would be satisfied if the wave reflected off the wall as a compression wave. In general, then, one may say that in unsteady, one-dimensional flow, waves reflect off a solid boundary in like manner; that is, compression waves reflect as compression waves, expansion waves reflect as expansion waves. It may be recalled that a similar rule prevailed in steady, two-dimensional flow.

After the wave has reflected off the closed end, it will encounter succeeding waves coming towards it. Experiments show that when two waves intersect, each wave continues on unchanged in character and direction, but changed in strength.

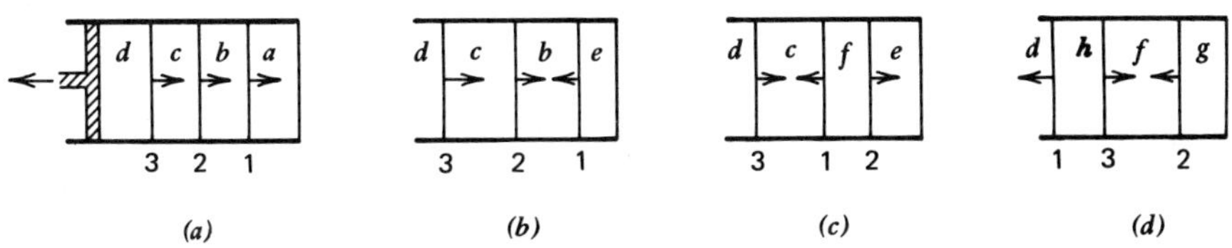

Figure 159. Reflection of expansion waves from a closed end.

The sequence of events is illustrated in Figure 159. Wave 2, after crossing wave 1, proceeds to the closed end and reflects in like manner. Meanwhile, wave 1 crosses wave 3, and so on. The paths of the waves may be shown on the physical plane (Figure 160). Notice that the position of the waves at any given

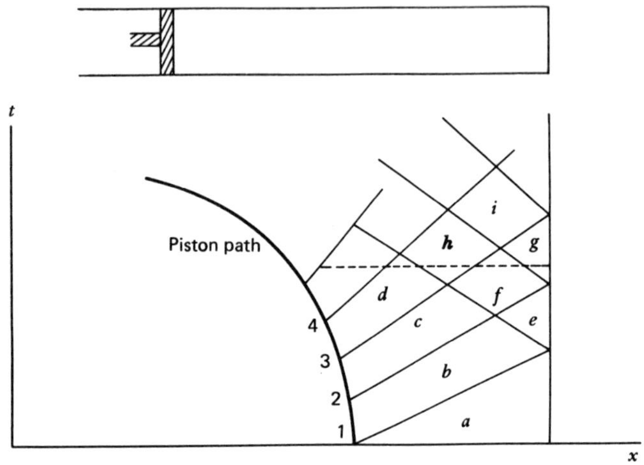

Figure 160. Reflection and intersection of Mach waves shown in the $x-t$ plane.

time may be established by drawing a horizontal line across the physical plane. For example, the dotted line in Figure 160 shows the location of the waves in Figure 159(d).

As yet we do not know the precise shape of the net in Figure 160 or the flow properties in regions $a, b, c, d \ldots$ These properties may, however, be determined graphically by use of the state plane. Referring to Figure 161, we

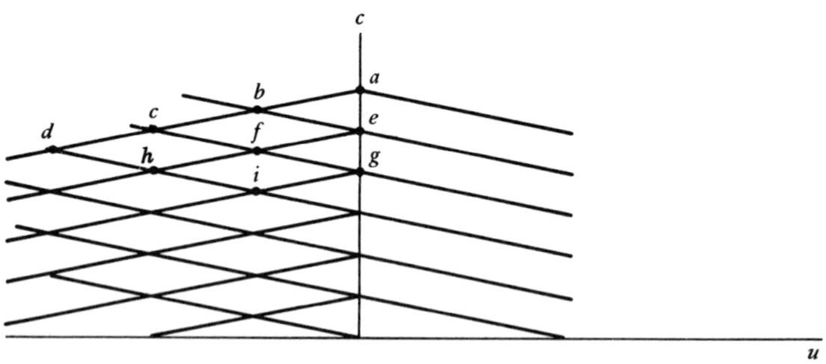

Figure 161. Determination of flow properties using the state plane.

locate the point *a* from the given initial data. The points *b*, *c*, and *d* are determined as in the case of simple wave flow. To get from *b* to *c*, we go across a left-running rarefaction wave. Furthermore, the velocity in region *e* must be zero. Thus we must follow the line

$$c - c_b = -\frac{\gamma - 1}{2}(u - u_b) \tag{750}$$

to the $u = 0$ axis. This fixes the point e. Numerically, we have

$$c_e - c_b = -\frac{\gamma - 1}{2}(0 - u_b) = \frac{\gamma - 1}{2}u_b \tag{751}$$

keeping in mind that u_b is negative and, hence, that $c_e < c_b$. To get from e to f we must go across a right-running rarefaction wave, while to get from c to f we must go across a left-running rarefaction wave. Thus we have

$$c_f - c_e = +\frac{\gamma - 2}{2}(u_f - u_e) \tag{752}$$

$$c_f - c_c = -\frac{\gamma - 1}{2}(u_f - u_c) \tag{753}$$

The intersection of the two state characteristics on the graph thus fixes the point f. In a similar way, the conditions in regions g, h, and i may be determined. Once the velocity in each region is known, the pressure and temperature may be found from equations (748) and (749). Notice that the flow is always locally a simple wave flow; that is, the flow only crosses one wave at a time. Thus, to find the pressure in region f, for example, we may write

$$\frac{p_f}{p_e} = \left[1 + \frac{\gamma - 1}{2}\left(\frac{u_f}{c_e} - \frac{u_e}{c_e}\right)\right]^{2\gamma/(\gamma - 1)} \tag{754}$$

since to get from e to f we go across a right-running wave. Alternatively, we may write

$$\frac{p_f}{p_c} = \left[1 - \frac{\gamma - 1}{2}\left(\frac{u_f}{c_c} - \frac{u_c}{c_c}\right)\right]^{2\gamma/(\gamma - 1)} \tag{755}$$

since to get from c to f we go across a left-running wave. Other than remembering the sign convention, this may also be verified by the following consideration: the pressure in f is less than in c; the velocity in f is "greater" (i.e., less negative) than in region c; hence, the minus sign must be used in equation (755). Similarly, the pressure in f is less than in e. The velocity in f is "less" (i.e., more negative) than in e. Hence, the plus sign must be used in equation (754). In general it will be found helpful to reason thus physically rather than merely remembering sign conventions in avoiding mistakes.

Reflection of waves from the open end of a tube. Consider, as another example, a small pressure wave approaching the open end of a tube. On reaching the open end, the wave must reflect in unlike manner so long as the flow behind the wave is subsonic, since there can be no pressure discontinuity at the exit in such a case. Thus, if the wave is initially a rarefaction wave it reflects as a compression wave and vice versa. This leads to the rule that pressure waves reflect in unlike manner off a free boundary. It will be recalled that a similar rule prevailed in the case of two-dimensional, steady flow.

Suppose two rarefaction waves are approaching the open end of the tube as shown in Figure 162. The conditions in regions 1, 2, and 3 are presumed given

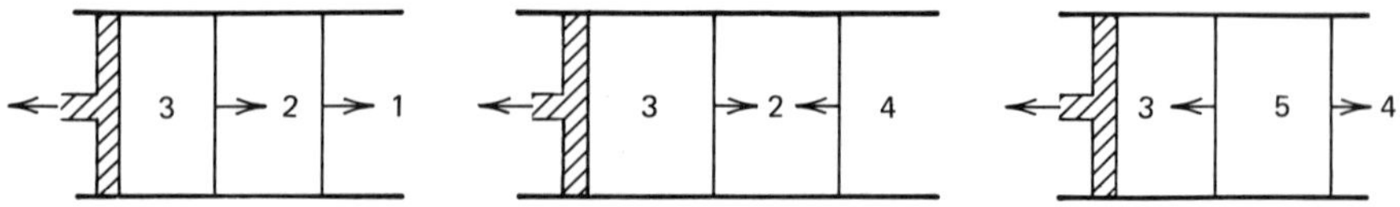

Figure 162. Reflection of waves from an open end.

and we desire to know the conditions in regions 4 and 5. Referring to the state plane, Figure 163, we may locate the points 1, 2, and 3. On reaching the open end,

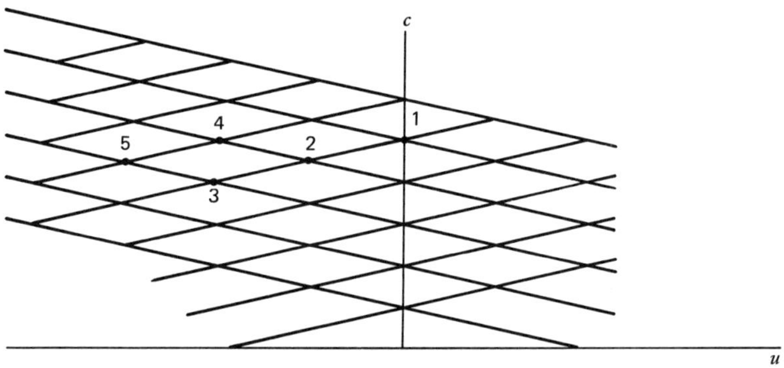

Figure 163. Determination of flow properties using the state plane.

the first wave reflects as a left-running compression wave. Thus we have

$$\frac{p_4}{p_2} = \left[1 - \frac{\gamma - 1}{2}\left(\frac{u_4}{c_2} - \frac{u_2}{c_2}\right)\right]^{2\gamma/(\gamma - 1)} \tag{756}$$

Since $p_4 = p_1^{\dagger}$ and u_2 is given, u_4 may be calculated. Notice that the velocity in region 2 is to the left. Since a compression wave entrains the gas, the velocity in region 4 must be even more negative than in region 2. On the other hand, the temperature, and hence the speed of sound, must be higher in region 4 than in region 2.

$$\frac{c_4}{c_2} = 1 - \frac{\gamma - 1}{2}\left(\frac{u_4 - u_2}{c_2}\right) \tag{757}$$

† Care must be taken in applying this condition. Schlieren photographs show that in addition to the formation of a reflected wave, the incident wave proceeds out the exit of the tube and into the space beyond, where it quickly weakens. Thus, the pressure at the exit is not, in fact, constant and atmospheric, but fluctuates somewhat. The exact variation of the exit pressure becomes a difficult problem in two-dimensional or axially symmetric unsteady flow. In practice it is usually found that assuming the exit pressure constant and atmospheric yields good results.

c_4 may be found from equation (757), thus fixing the point 4 in the state plane. Notice that $c_4 = c_1$. The reflected compression wave now crosses the second rarefaction wave. Each wave continues on after the intersection unchanged in character but changed in strength. To get from region 3 to region 5 we must go across a left-running compression wave. Thus the velocity increases to the left and the speed of sound rises. To get from region 4 to region 5 we must go across a right-running rarefaction wave. Thus, the velocity also increases to the left but the speed of sound drops. Thus, the point 5 is located by proceeding along the appropriate characteristics through the points 3 and 4, in the appropriate direction, to the point where the two characteristics intersect. Algebraically, the point 5 in the state plane is given by the simultaneous solution of the equations

$$\frac{c_5}{c_3} = 1 - \frac{\gamma - 1}{2} \frac{u_5 - u_3}{c_3}$$

$$\frac{c_5}{c_4} = 1 + \frac{\gamma - 1}{2} \frac{u_5 - u_4}{c_4} \tag{758}$$

The speed of each wave may be taken as

$$\frac{u_i + u_j}{2} \pm \frac{c_i + c_j}{2} \tag{759}$$

where the minus sign is taken for left-running waves and the plus sign for right-running waves. The formula (759) is of necessity approximate since there are, in reality, no rarefaction discontinuities and no isentropic compression waves. The analysis is thus valid only for waves of small strength; but, as we have pointed out, large pressure drops may be approximated by a series of small rarefaction waves. Large pressure rises, on the other hand, must be considered as shock waves. The analysis of moving shock waves will be dealt with in the next section.

As another example, let us consider a piston in a cylinder describing a simple harmonic motion. The path of the piston may be plotted as in Figure 164.

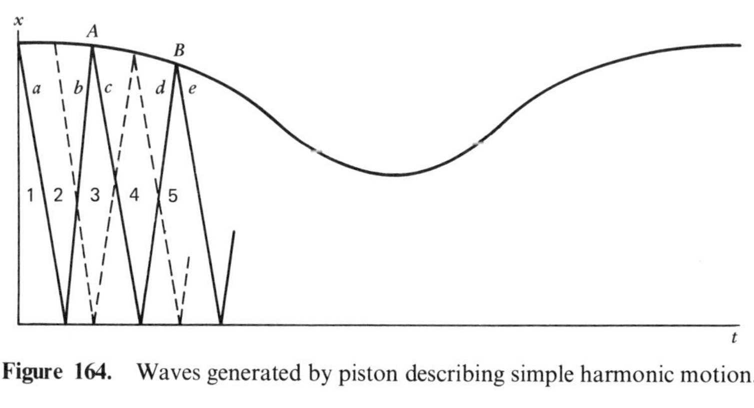

Figure 164. Waves generated by piston describing simple harmonic motion.

The piston curve in this case will be a sine curve. In addition, let us suppose the cylinder is closed at one end and that the piston moves back and forth between the dotted lines as shown in Figure 165. The dotted lines BDC and TDC

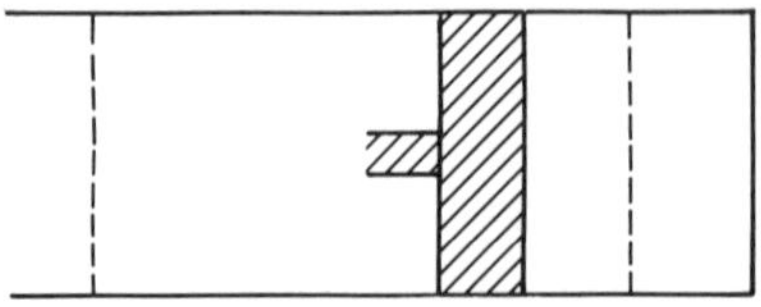

Figure 165. Reciprocating piston.

correspond to top dead center and bottom dead center in a reciprocating engine. It should be noted, however, that pistons in reciprocating engines do not describe simple harmonic motion. It is usual, in the analysis of reciprocating engines, to approximate the compression stroke by a quasisteady isentropic compression of an ideal gas. That is, it is presumed that throughout the compression process

$$p\mathcal{V}^{\gamma} = \text{constant}$$

In the present case we may retain the isentropic ideal gas assumption, but treat the problem as an unsteady one-dimensional flow. The spacing of the waves is critical here. If the waves are arbitrarily spaced, there will soon be so many waves intersecting with one another that the solution of the problem will become exceedingly tedious. The calculations may be greatly simplified, however, by considering only a single wave going back and forth, as shown in Figure 164. This will be a practical alternative if the wave bounces back and forth a sufficiently large number of times per quarter revolution. In reciprocating machinery this is usually the case. The approximation may be approached in a number of ways. Good results are obtained by the following method. Draw the first wave on the assumption that the initial piston velocity is zero. This will make the velocity in regions 1, 2 and 3 zero. Now take tangent to the piston curve at the point where the reflected wave hits the piston (point A in Figure 164). Taking the velocity in region 2 as one half the velocity at A, redraw the initial and reflected wave accordingly. This will shift the point of intersection A slightly. Taking the new velocity at A from the tangent to the piston curve at that point, draw the next two waves, c and d. Now take the velocity in region 4 as the average between A and B on the piston curve, redraw waves c and d accordingly. Repeat the procedure for succeeding waves. Alternatively, one may begin the calculation at a point halfway between TDC and A, taking the velocity behind the first wave as the velocity of the piston at that point, and the velocity behind succeeding wave as the velocity of the piston at the time of the origin of the wave. This pattern is indicated by the dotted lines in Figure 164. Both of these methods yield better results than starting at TDC and not iterating, causing the velocity in region 2 to remain zero and the wave pattern to be wrongly displaced. Notice that all the waves here are compression waves. When the pressure on the cylinder head is computed as a function of time, it is found to vary but little from that predicted by the quasi-steady-state calculation.

Moving Shocks

As was pointed out previously, there is no such thing as a finite, isentropic, compression wave. All finite compression waves are shock waves. On the other hand, if a compression wave is weak, the entropy rise across it will be small and may often be neglected. This point was discussed in a previous chapter (Chapter 4). Where the flow undergoes a continuous rise in pressure, the fluid may be thought of as passing through a series of small compression waves while still remaining isentropic. As an example, we may consider a piston accelerating to the right (Figure 166). Again dividing the piston curve into a series of straight

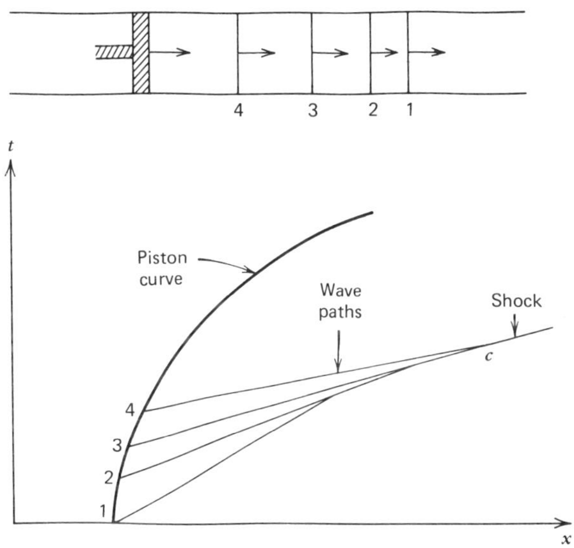

Figure 166. Compression waves generated by an accelerating piston.

lines we may approximate the flow by a series of compression waves launched by the successive changes in slope of the piston curve. Recall, however, in our discussion of the formation of normal shock waves (Chapter 2) we established the principle that rarefaction waves flatten while compression waves steepen. Thus, in our previous example (Figure 155), the distance between the waves grows larger as the waves progress. This may be seen by examining the physical plane (Figure 156), where it is seen that the paths of the waves diverge. In the case of compression waves, however, the opposite is true. Each waves travels faster than the one before it and hence, soon catches up to it. This is apparent from Figure 166, where the wave paths are seen to intersect. When one compression wave catches up to another compression wave, the two coalesce into a single wave whose strength will equal the sum of the strengths of two waves, so long as the entropy variation is not taken into account. If the piston accelerates smoothly up to point 4 in Figure 166, the subsequent development of the pressure profile will be as shown in Figure 167. Of course the entire profile is

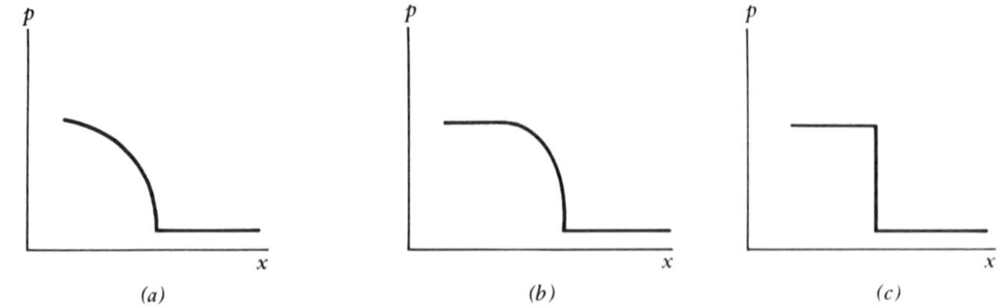

Figure 167. Steepening of a compression wave.

being swept downstream during this time. When the last wave catches up to the first, the entire pressure rise will have coalesced into a shock. The time and place of this coalescence is given by point c in Figure 166. The question now arises as to when the entropy variation must be taken into account. Recall the momentum and energy equations (189) and (191) for one-dimensional flow:

$$\rho\frac{\partial u}{\partial t}+\rho u\frac{\partial u}{\partial x}=-\frac{\partial p}{\partial x}+\frac{\partial}{\partial x}\left(2\mu\frac{\partial u}{\partial x}\right)-\frac{2}{3}\frac{\partial}{\partial x}\left(\mu\frac{\partial u}{\partial x}\right) \tag{189}$$

$$-\frac{\partial}{\partial x}\left(k\frac{\partial T}{\partial x}\right)=\rho\frac{\partial e_t}{\partial t}+\rho u\frac{\partial e_t}{\partial x}+p\frac{\partial u}{\partial x}-\frac{4}{3}\mu\left(\frac{\partial u}{\partial x}\right)^2 \tag{191}$$

Since the change in entropy is caused by the presence of viscosity and heat transfer, this change will not become significant until the terms $(\mu u_x)_x$, $(kT_x)_x$ and μu_x^2 become of significant size compared to the other terms in the equations. Since μ and k are small for gases, this will not occur until u_{xx} and T_{xx} are large, that is, until the changes in velocity and temperature take place over a very short distance. In a shock wave, the viscous and inertia terms are of the same order of magnitude since they balance each other out, so to speak. Thus, the isentropic flow assumption continues to be valid almost to the point where the shock wave has actually formed. Once the shock has formed, the flow may be considered steady with respect to the shock and, as was shown by Becker [6], the presence of viscosity and thermal conductivity does not affect the initial and final pressure and temperature across the shock for a given initial Mach number. The precise shape and development of the pressure wave in Figure 167 will depend on the manner in which the piston accelerates. There is no precise definition of where a small pressure wave becomes a shock wave. So long as the assumption of isentropic flow is maintained, however, the pressure distribution as a function of time may be calculated for any given piston curve, using the physical and state characteristics. The validity of the isentropic flow assumption may then be checked *a posteriori* by observing the variation of the local velocity and temperature gradients.

Where the compression waves are not weak, the shock wave relations must be resorted to. We have already derived the shock wave relations for steady, one-dimensional flow. In these relations, the shock wave was presumed to be at rest and the flow moving through it. Or, to put it another way, the velocities were taken with respect to an observer traveling with the shock wave. As long as the shock wave is moving with a constant speed, therefore, these equations are equally valid whether one is dealing with a shock wave at rest in a tube, say, with the flow going through it, or with a shock wave traveling down a tube into a medium at rest or having some other velocity. It is simply a matter of changing the coordinate system by adding or subtracting some fixed number to all velocities. In doing so one may transform a steady flow into an unsteady flow or vice versa. Thus, it is not necessary to rederive the shock wave equations to deal with moving shock waves. It is only necessary to remember how the variables in the shock wave equations were defined. Let $M_x = V_x/c_1$, where V_x is the velocity of the flow coming toward the shock wave, relative to the wave, that is, as seen by an observer attached to the wave, and c_1 is the speed of sound "ahead" of the wave, i.e., in the medium before it has passed through the wave. Let $M_y = V_y/c_2$, where V_y is the velocity leaving the wave, relative to the wave, that is, the velocity of the flow after it has passed through the wave as seen by an observer attached to the wave, and c_2 is the speed of sound in the medium after it has passed through the shock. Note that c_1 and c_2 are independent of the coordinate system.

Recall now the shock wave relations, with the variables as defined above,

$$M_y^2 = \frac{M_x^2 + \dfrac{2}{\gamma - 1}}{\dfrac{2\gamma}{\gamma - 1} M_x^2 - 1} \tag{212}$$

$$\frac{p_y}{p_x} = \frac{2\gamma}{\gamma + 1} M_x^2 - \frac{\gamma - 1}{\gamma + 1} \tag{214}$$

$$\frac{T_y}{T_x} = \frac{\left(1 + \dfrac{\gamma - 1}{2} M_x^2\right)\left(\dfrac{2\gamma}{\gamma - 1} M_x^2 - 1\right)}{\dfrac{(\gamma + 1)^2}{2(\gamma - 1)} M_x^2} \tag{213}$$

With the variables as defined above, these equations are equally valid in any inertial coordinate system. In dealing with moving shock problems, it is sometimes handy to have an expression for the velocity difference across the shock. Notice that this quantity, too, is independent of the coordinate system. Thus

$$u_2 - u_1 = M_y c_2 - M_x c_1 \tag{760}$$

or

$$\frac{u_2 - u_1}{c_1} = \frac{M_y}{M_x} M_x \frac{c_2}{c_1} - M_x = M_x\left(\frac{M_y}{M_x}\frac{c_2}{c_1} - 1\right) \tag{761}$$

Substituting for M_y/M_x and c_2/c_1 from equations (212) and (213), we get

$$\frac{u_2 - u_1}{c_1} = \pm\frac{2}{\gamma+1}\left(\frac{1}{M_x} - M_x\right) \tag{762}$$

where the plus or minus sign has been added to allow for sign convention. M_x is always taken as positive. Thus, if u is taken as positive to the right and

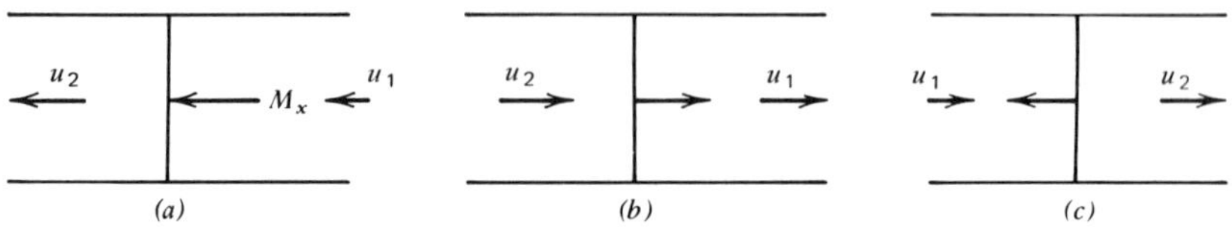

Figure 168. Flow velocities in stationary and moving coordinate systems.

the wave is moving to the right, as in Figure 168(b), u_2 will be greater than u_1 and the minus sign is in order. If the wave is moving to the left, as in Figure 168(c), u_2 will be less than u_1 and the plus sign is in order. Figure 168(a) shows the flow relative to the wave with u taken positive to the left. In this case also u_2 is less than u_1 and the plus sign is in order.

Calculation may be facilitated by plotting equation (762) (Figure 169).

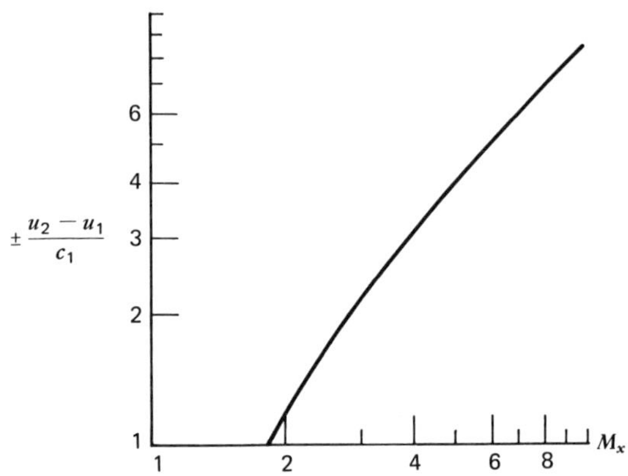

Figure 169. Velocity change versus upstream Mach number [234].

Alternatively, c_2/c_1 is sometimes plotted against $(u_2 - u_1)/c_1$, by eliminating M_x between equations (762) and (213). The corresponding value of M_x is shown at each point on the curve of Figure 170. Again, the $\pm$ sign is a matter of sign convention. The curve of Figure 170 is also called a shock polar.

Reflection and Interaction of Shocks

Consider a shock wave traveling down a tube as in Figure 171. The boundary conditions and method of calculation are analogous to the corresponding cases in isentropic flow. If the end of the tube is closed, the boundary condition

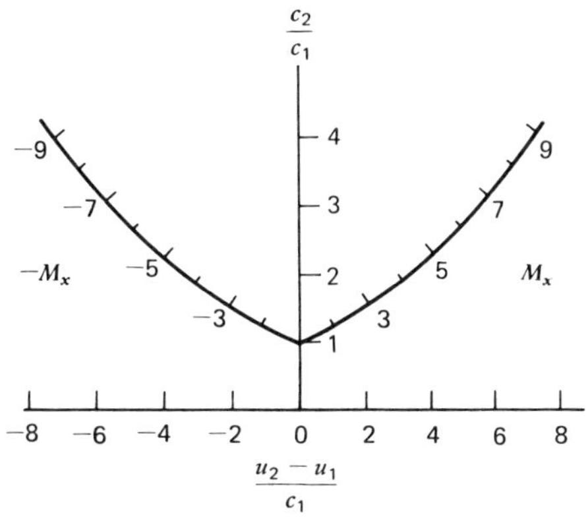

Figure 170. Shock polar for normal shock wave [234].

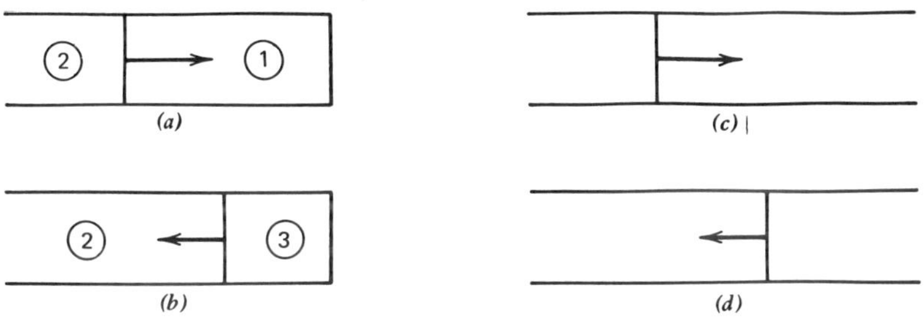

Figure 171. Shock reflection from closed and open ends.

$u = 0$ must always be satisfied there. Hence the incident shock reflects as a shock wave proceeding in the opposite direction. Given the conditions in region 1 and the M_x of the incident shock, the conditions in region 2 may be read off Figures (169) and (170). Using the u_2 and c_2 found in this way, and knowing $u_3 = 0$, gives us $(u_3 - u_2)/c_2$ which enables us to read the M_x of the reflected wave, and c_3/c_2, off the same two figures. The same procedure may be used for the case in which a shock wave strikes a moving piston. If the piston is moving to the right as in Figure 172(a), the incident wave will reflect as a shock, as in Figure 172(b), with the velocity in region 3 being equal to that of the piston.

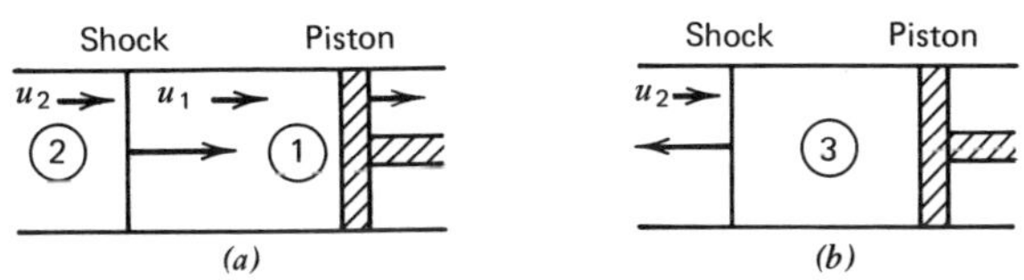

Figure 172. Shock reflection from moving piston.

The procedure for the case of a reflected shock may be applied to the case of a piston undergoing simple harmonic motion discussed in the previous section under the assumption of isentropic flow. Treating the compression waves as shocks will result in greater accuracy at the price of more work. As long as the compression waves are weak, however, the gain in accuracy will be small.

Consider now the case of a shock traveling down a tube with an open end. So long as the flow at the exit of the tube is subsonic, there can be no pressure discontinuity at that point. The first question to be addressed, therefore, is under what conditions subsonic flow will prevail at the exit. Schlieren photographs show that a shock, on reaching the open end of a tube, will continue out into the space beyond. If the exit of the tube is suspended in mid-air, the shock will quickly dissipate. If, on the other hand, the tube ends in a wall, the shock will dissipate more slowly, as shown in Figure 173. From the shock

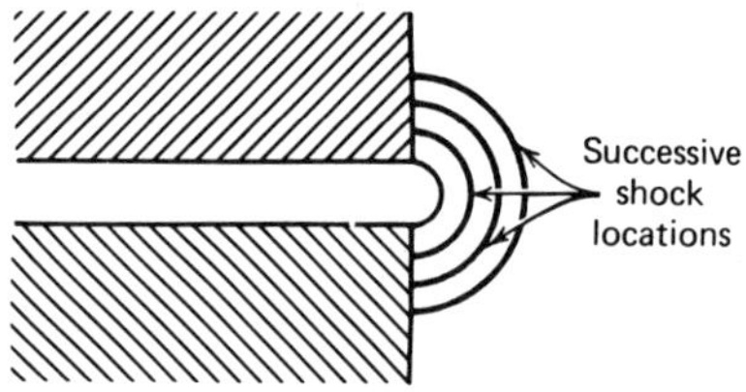

Figure 173. Shock emerging from duct.

polar, (Figure 170), we see that for a shock advancing into stagnant air, the velocity behind the shock will be supersonic when the Mach number M_x exceeds approximately 2.4. A rarefaction wave cannot proceed upstream against a supersonic flow. Thus, the flow in the tube will continue undisturbed for a length of time determined by the upstream boundary conditions. For weak shock waves, on the other hand, the pressure at the exit may be approximated by the atmospheric, outside or back pressure prevailing before the arrival of the shock. Subsonic flow at the exit will prevail. In this case no pressure discontinuity is permitted at the exit and hence a rarefaction wave will form at the exit and proceed upstream when the shock reaches the end of the tube. It is not quite correct to say that the shock reflects as a rarefaction wave, since, as has already been pointed out, it actually continues out the exit and into the space beyond. The actual variation of the exit pressure is a complicated problem in multidimensional, unsteady flow; but so long as the shock is weak, taking the exit pressure as the back pressure represents a good approximation.

Consider, then, a weak shock wave traveling down an open ended tube as in Figure 174. Given M_x, the pressure and temperature ratios across the shock

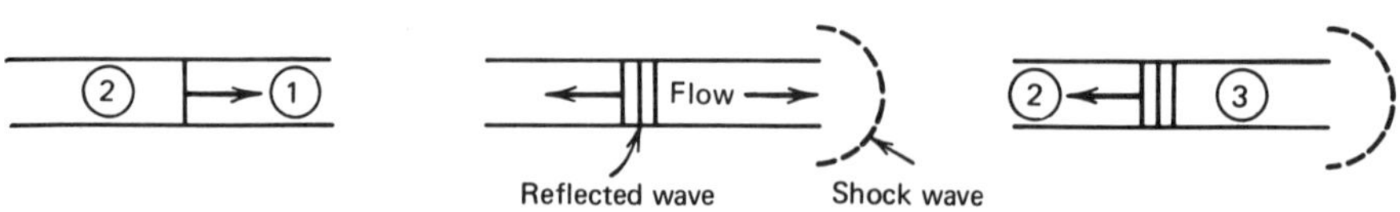

Figure 174. Details of shock reflection from an open end.

may be found from

$$\frac{p_2}{p_1} = \frac{2\gamma}{\gamma+1}M_x^2 - \frac{\gamma-1}{\gamma+1}, \qquad \frac{T_2}{T_1} = \frac{\left(1+\frac{\gamma-1}{2}M_x^2\right)\left(\frac{2\gamma}{\gamma-1}M_x^2-1\right)}{\frac{(\gamma+1)^2}{2(\gamma-1)}M_x^2} = \left(\frac{c_2}{c_1}\right)^2$$

Across the rarefaction wave,

$$\sqrt{\frac{T_3}{T_2}} = \frac{c_3}{c_2} = 1 - \frac{\gamma-1}{2}\left(\frac{u_3-u_2}{c_2}\right) = \left(\frac{p_3}{p_2}\right)^{(\gamma-1)/2\gamma} \cong \left(\frac{p_1}{p_2}\right)^{(\gamma-1)/2\gamma} \tag{763}$$

Thus, given M_x and the conditions in region 1, the conditions in regions 2 and 3 are determined. Actual calculations may, of course, be facilitated by the use of tables and graphs.

Intersection of shock waves

Another example of a problem in moving shock waves which may easily be solved is that of the intersection of two shock waves. Consider the situation in Figure 175. Two shock waves, *a* and *b*, approach each other, collide and pass on.

Figure 175. Intersection of shock waves.

Given the conditions in region 2 and the Mach numbers M_{xa} and M_{xb} of the two shocks, find the conditions in regions 4 and 5. Conservation of mass and momentum require that $u_4 = u_5$ and $p_4 = p_5$. On the other hand, if the two shock waves are of different strengths, the entropy rise across them will differ. Thus, the flow which has passed through shock wave *a* will have a different entropy and temperature from the flow which has passed through shock wave *b*. The interface separating the two regions is shown as a double line in Figure 175. Across wave *b* we have

$$\frac{u_4-u_1}{c_1} = -\frac{2}{\gamma+1}\left(M'_{xb} - \frac{1}{M'_{xb}}\right), \qquad \frac{p_4}{p_1} = \frac{2\gamma}{\gamma+1}M'^2_{xb} - \frac{\gamma-1}{\gamma+1} \tag{764}$$

where M'_{xb} is the Mach number of wave *b* after intersection and *u* is taken as positive to the right. The shock wave Mach numbers are taken as positive. Across wave *a* we have,

$$\frac{u_5-u_3}{c_3} = \frac{2}{\gamma+1}\left(M'_{xa} - \frac{1}{M'_{xa}}\right), \qquad \frac{p_5}{p_3} = \frac{2\gamma}{\gamma+1}M'^2_{xa} - \frac{\gamma-1}{\gamma+1} \tag{765}$$

But $u_5 = u_4, p_5 = p_4$. Thus we have four equations, (764) and (765), and four unknowns: $u_5 = u_4, p_5 = p_4, M'_{xa}$ and M'_{xb}, and the problem is rendered determinate.

The solution may also be obtained graphically by replotting the shock polars in dimensional coordinates (Figure 176). Eliminating M_x between equations

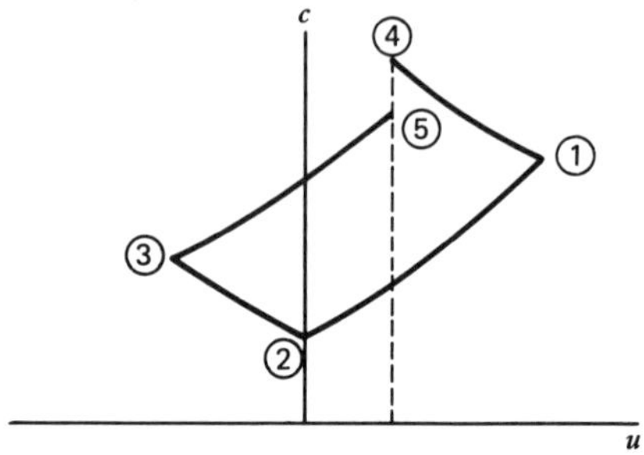

Figure 176. Determination of flow conditions in state plane for intersecting shocks.

(764) and (765), we may plot the pressure versus the velocity ratio. Applying

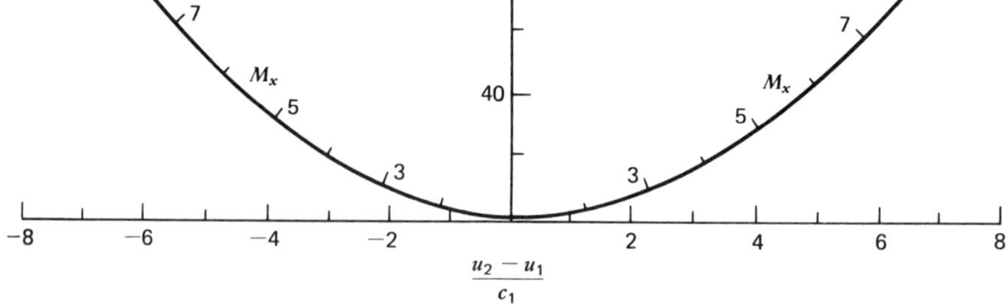

Figure 177. Pressure ratio versus velocity change across normal shocks [234].

Figure 177 in dimensional form to the present case we get Figure 178. The intersection of shock waves and rarefaction waves may be treated in a similar manner using the appropriate equations and graphs.

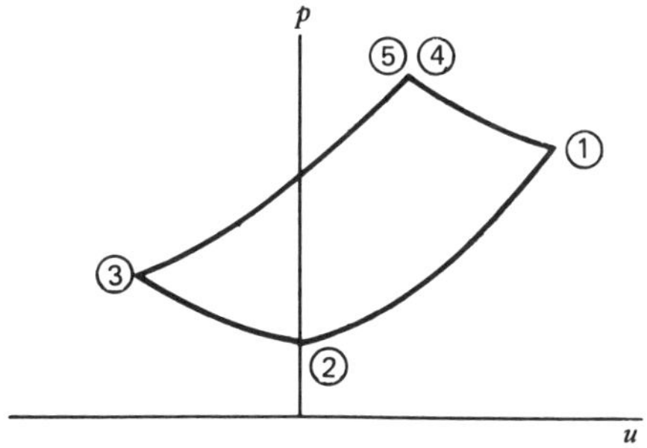

Figure 178. Determination of flow properties for intersecting shocks in the u–p plane.

Spherical and Cylindrical Shock Waves

Such waves may be considered an example of unsteady one-dimensional flow in the sense that only one spatial coordinate, r, is involved. Consider then shock waves of the form shown in Figure 179. In the cylindrical case, the shock

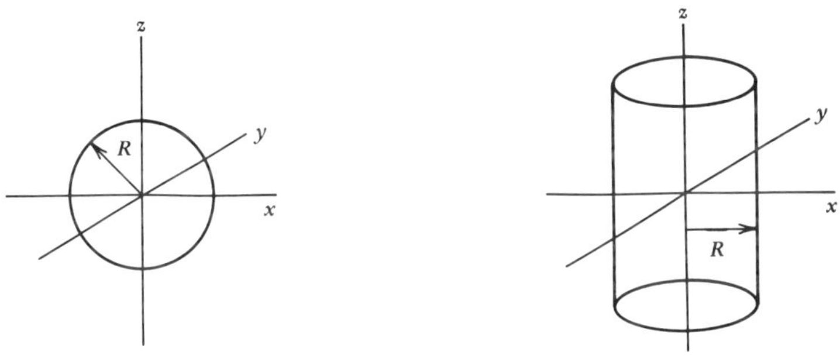

Figure 179. Spherical and cylindrical shocks.

is presumed to extend to $z = \pm\infty$. The shocks are presumed to be of equal strength all around and to behave locally like plane shocks. The entropy of individual fluid particles behind the shocks may be presumed to be constant. The governing equations for this region may be written for both the cylindrical and spherical case as

$$u_t + uu_r = -\frac{1}{\rho}p_r \tag{766}$$

$$\rho_t + (u\rho)_r + \delta\frac{\rho u}{r} = 0 \tag{767}$$

$$\sigma_t + u\sigma_r = 0 \qquad \text{where} \qquad \sigma = \frac{p}{\rho^\gamma} \tag{768}$$

where $\delta = 1$ for the cylindrical case and $\delta = 2$ for the spherical case, and r is the radial coordinate in polar and spherical coordinates. At the shock, $r = R$. We now postulate a solution for the region behind the shock of the form

$$\frac{\rho}{\rho_0} = \psi(\eta), \qquad \frac{p}{p_0} = \frac{f_1(\eta)}{R^{1+\delta}} \qquad u = \frac{\phi_1(\eta)}{R^{(1+\delta)/2}} \qquad \eta = \frac{r}{R} \tag{769}$$

Substituting the postulated form of the solution (769) into the governing equations (766), (767), (768), we get

$$\left(\frac{1+\delta}{2}\phi_1 + \eta\phi_1'\right)R^{-(3+\delta)/2}\frac{dR}{dt} - \phi_1\phi_1' R^{-(2+\delta)} - \frac{p_0}{\rho_0}\frac{f'}{\psi}R^{-(2+\delta)} = 0 \tag{770}$$

$$\eta\psi' R^{(1+\delta)/2}\frac{dR}{dt} - \phi_1\psi' - \psi\left(\phi_1' + \frac{\phi_1\delta}{\eta}\right) = 0 \tag{771}$$

$$\left[\gamma\eta f_1\frac{\psi'}{\psi} - \eta f_1' - (1+\delta)f_1\right]R^{(1+\delta)/2}\frac{dR}{dt} + \phi_1 f_1' - \frac{\gamma f_1\phi_1\psi'}{\psi} = 0 \tag{772}$$

where the primes indicate differentiation with respect to η. For the assumed form of the solution to be valid, therefore, it is necessary that

$$R^{(1+\delta)/2}\frac{dR}{dt} = \text{constant} = K \tag{773}$$

Let

$$f(\eta) = f_1(\eta)\frac{c_0^2}{K^2} = f_1(\eta)\frac{\gamma p_0}{\rho_0 K^2}, \qquad \phi(\eta) = \frac{1}{K}\phi_1(\eta) \tag{774}$$

Substituting equations (773) and (774) into equations (770), (771), and (772), and solving for f', ϕ', and ψ' we get

$$f' = \frac{2(1+\delta)(\eta-\phi)\eta\psi + (1-\delta)\gamma f\eta\psi\phi + 2\gamma f\psi\delta\phi^2}{2\eta[f - \psi(\eta-\phi)^2]} \tag{775}$$

$$\psi' = \frac{\psi[(\eta-\phi)f' + (1+\delta)f]}{\gamma f(\eta-\phi)} \tag{776}$$

$$\phi' = (\eta-\phi)\frac{\psi'}{\psi} - \frac{\phi\delta}{\eta} \tag{777}$$

For M_x large, the shock wave relations become

$$\frac{p_1}{p_0} \cong \frac{2\gamma}{\gamma+1}M_x^2, \qquad \frac{\rho_1}{\rho_0} \cong \frac{\gamma+1}{\gamma-1}, \qquad \frac{u_1}{U} \cong \frac{2}{\gamma+1} \tag{778}$$

where the subscripts 0 and 1 refer to the conditions immediately ahead of and behind the shock, and U is the shock wave velocity $= dR/dt$. In this case, the boundary conditions become

$$f(1) = \frac{2\gamma}{\gamma+1}, \qquad \psi(1) = \frac{\gamma+1}{\gamma-1}, \qquad \phi(1) = \frac{2}{\gamma+1} \tag{779}$$

Using the boundary conditions, equations (775), (776), and (777) may be integrated numerically from $\eta = 1$ to $\eta = 0$. The results are plotted in Figure 180,

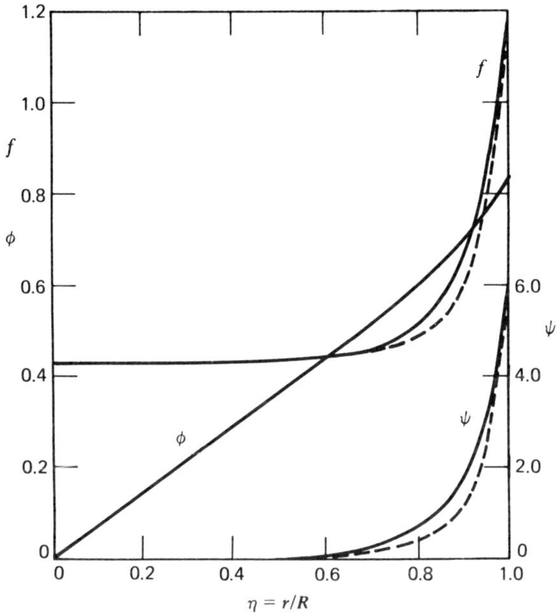

Figure 180. Results of Taylor [49] for spherical and cylindrical shocks [49].

where the solid lines represent the cylindrical solution ($\delta = 1$) and the dotted lines represent the spherical solution ($\delta = 2$). It should be kept in mind that when the shock is very strong, the ideal gas assumptions are liable to break down, whereas when the shock is weak, the assumed boundary conditions are no longer valid. Also, the assumed form of the solution implies that $p_1/p_0 = \infty$ where $R = 0$, implying a boundary condition near $t = 0$ which will most likely differ from reality. Nevertheless, Taylor [49] reports good agreement for the case of strong spherical explosions.

The total energy inside the shock wave for the spherical case and the total energy per unit length inside the shock wave for the cylindrical case is given by

$$E = 2\delta\pi \int_0^R \rho\left(c_v T + \frac{u^2}{2}\right) r^\delta\, dr = 2\pi\delta \int_0^R \left(\frac{p}{\gamma - 1} + \frac{1}{2}\rho u^2\right) r^\delta\, dr \tag{780}$$

Substituting for p, ρ, and u in terms of f, ψ, and ϕ we get

$$E = 2\pi\delta K^2 \rho_0 \int_0^1 \left(\frac{f}{\gamma(\gamma - 1)} + \frac{1}{2}\psi\phi^2\right)\eta^\delta\, d\eta \tag{781}$$

showing that the energy inside the wave is constant. Evaluating the integral we get $E = 3.85\,\rho_0 K^2$ for the cylindrical case, and $E = 5.36\,\rho_0 K^2$ for the spherical case, taking $\gamma = 1.4$. Recall now that

$$K = R^{(1+\delta)/2}\frac{dR}{dt}$$

so that

$$R^{(3+\delta)/2} = \frac{3+\delta}{2}Kt, \qquad R = \left[\frac{3+\delta}{2}Kt\right]^{2/(3+\delta)}, \qquad K = \frac{2}{3+\delta}\frac{R^{(3+\delta)/2}}{t}$$

Hence

$$E = n\rho_0 K^2 = n\rho_0\left(\frac{2}{3+\delta}\right)^2\frac{R^{3+\delta}}{t^2}$$

or

$$R = \left[\left(\frac{3+\delta}{2}\right)^2\frac{Et^2}{n\rho_0}\right]^{1/(3+\delta)}, u = \frac{dR}{dt} = \frac{2}{3+\delta}\left[\left(\frac{3+\delta}{2}\right)^2\frac{E}{n\rho_0}\right]^{1/(3+\delta)} t^{-(1+\delta)/(3+\delta)} \tag{782}$$

where $n = 3.85$ for the cylindrical case and $n = 5.36$ for the spherical case. Equation (782) thus gives us the rate of propagation of spherical and cylindrical shocks under the present assumptions.

It will be observed that the pressure behind the shock is now not constant as in the planar case, but decays with time. Recall,

$$\frac{p}{p_0} = \frac{f_1(\eta)}{R^{1+\delta}} = \frac{K^2\rho_0}{\gamma p_0}\frac{f(\eta)}{R^{1+\delta}} = \left(\frac{2}{3+\delta}\right)^2\frac{R^{3+\delta}}{t^2}\frac{\rho_0}{\gamma p_0}\frac{f(\eta)}{R^{1+\delta}}$$

Hence,

$$p = \left(\frac{2}{3+\delta}\right)^{(2+2\delta)/(3+\delta)}\frac{\rho_0}{\gamma}f(\eta)\left[\frac{E}{n\rho_0}\right]^{2/(3+\delta)} t^{-(2+2\delta)/(3+\delta)}$$

In the cylindrical case we find that the pressure is given by

$$p = 0.182(\rho_0 E)^{1/2} t^{-1} f(\eta) \tag{783}$$

showing that the pressure varies inversely with the time. $f(1) = 1.1667$, and hence immediately behind the shock,

$$p_1 = .212(\rho_0 E)^{1/2} t^{-1} \tag{784}$$

In the spherical case,

$$p = \frac{1}{4}\frac{\rho_0}{\gamma}f(\eta)\left[\frac{4E}{5.36\rho_0}\right]^{2/5} t^{-6/5} \tag{785}$$

For $\gamma = 1.4$, this works out to

$$p = 0.159\,\rho_0^{0.6}E^{0.4}t^{-1.2}f(\eta) \tag{786}$$

Immediately behind the shock,

$$f(\eta) = f(1) \cong \frac{2\gamma}{\gamma+1} = \frac{7}{6} \tag{787}$$

Hence

$$p_1 = 0.1855\,\rho_0^{0.6} E^{0.4} t^{-1.2} \tag{788}$$

Thus, the pressure behind the spherical shock decays somewhat more quickly than the pressure behind the cylindrical shock. As will be recalled, the above results are based on the assumption of a strong shock. Even if the shock is strong initially, however, it soon decays to the point where the strong shock assumption is no longer valid. A method for finding the pressure behind an advancing cylindrical or spherical shock which is not restricted by the strong shock assumption has been given by Brinkley and Kirkwood [50]. By making certain assumptions regarding the energy variation behind the shock wave, Brinkley and Kirkwood were able to reduce the governing system of partial differential equations to a system of ordinary differential equations which may be solved numerically. This numerical integration was performed by Sachdev [51]. In the form used by Sachdev, the equations to be integrated are

$$\frac{dp'}{dR} = \frac{1}{L}\left\{\frac{-(2R)^\delta \pi p'^3 G v}{p_0 D}\,\frac{1-y}{z-1} - \frac{p'\delta}{R}\left[2y + (1-y)G\right]\right\} \tag{789}$$

$$\frac{dD}{dR} = -(2R)^\delta \pi \rho_0 h(p') \tag{790}$$

where

$$h(p') = \frac{\gamma p_0}{\rho_0}\,\frac{z^2 - 1 - z\left(z + \dfrac{\gamma+1}{\gamma-1}\right)(1-z)^{-(\gamma-1)/\gamma}}{\gamma - 1 + (\gamma+1)z} \tag{791}$$

and

$$L = 2 + 2\left(1 - \frac{p'}{u}\frac{\partial U}{\partial p'}\right) - G \tag{792}$$

$$G = 1 - \left(\frac{\rho_0 U}{\rho_y c_y}\right)^2 \tag{793}$$

$p' = p_y - p_0$
$y = \rho_0/\rho_y$
$z = p_y/p_0$
U = velocity of the shock wave
R = distance of the shock wave from the origin

The subscript 0 indicates the condition ahead of the shock and the subscript y indicates the condition just behind the shock. $\delta = 1$ for the cylindrical case and $\delta = 2$ for the spherical case. v is a parameter whose value may be chosen to fit the solution to exact numerical or experimental results. Since U, ρ_y and c_y are functions of the pressure ratio across the shock and the given initial conditions, the above equations may be integrated to find the pressure behind the shock as a function of R.

In the limit of infinite shock strength, $h(p') \to 0$ and the solution of equations

(789) and (790) reduces to the solution of Taylor [49] given above. As $z \to 1$, equations (789) and (790) reduce to

$$\frac{dp'}{dR} + \frac{p'\delta}{2R} = -\frac{\gamma+1}{12\gamma^2}(2R)^{\delta}\frac{\pi p'^4}{Dp_0^2} \tag{794}$$

$$\frac{dD}{dR} = -\frac{\gamma+1}{12\gamma^2}(2R)^{\delta}\frac{\pi p'^3}{p_0^2} \tag{795}$$

These equations have a closed form solution which is

$$Rp' = p_1\left(\ln\frac{R}{R_1}\right)^{-1/2} \tag{796}$$

for the spherical case and

$$\sqrt{R}\,p' = p_1[2(\sqrt{R}-\sqrt{R_1})^{-1/2}] \tag{797}$$

for the cylindrical case, where p_1 and R_1 are constants of integration. This form of the weak shock solution agrees well with those predicted by Landau [52], Witham [53], and Sedov [54]. An alternative approach to the solution of the cylindrical and spherical shock wave problem is direct numerical integration of the original partial differential equations. This has been done by Plooster [55] for the cylindrical case and Lutzky and Lehto [56] for the spherical case.

A comparison of the results of the numerical solution of Plooster with the results of the numerical solution of Sachdev is shown in Figure 181. The "exact"

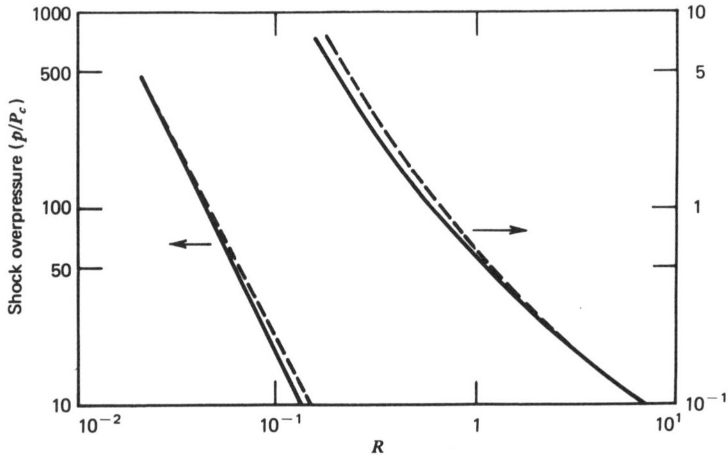

Figure 181. Results of Sachdev———[51] and Plooster--------[55] for cylindrical shocks. (P_c equals pressure at origin.)

numerical solutions of Plooster give the pressure distribution throughout the region behind the shock. Figure 182 shows how this pressure distribution varies

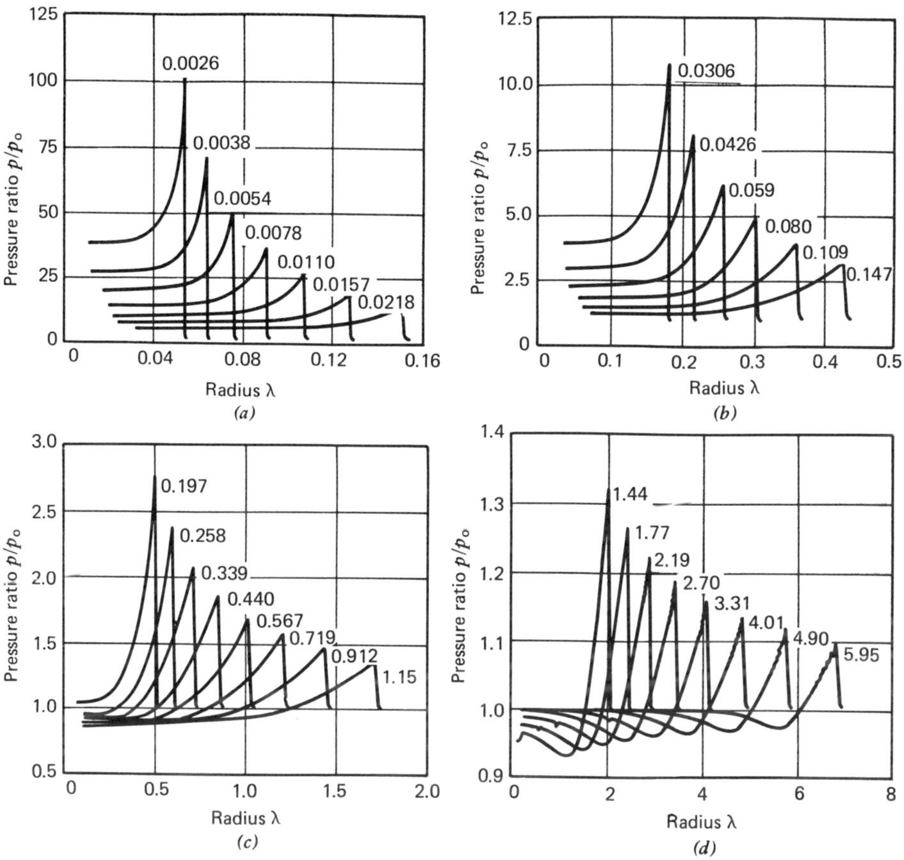

Figure 182. Results of Plooster [55] for cylindrical shock.

with time for the cylindrical case. The shock waves shown here were assumed to have been generated by an instantaneous energy release along a line source. p in Figure 182 is the pressure, p_0 the pressure ahead of the wave, $\lambda = r/(E_0/b\gamma p_0)^{1/2}$, where r is the radial coordinate, E_0 the energy released per unit length of the source, b is a constant depending upon the gas ($b = 3.84$ for air), and γ the ratio of specific heats. Of particular interest is the fact that the pressure falls below its original value p_0 and that both positive and negative pressure gradients are present behind the shock. Calculations of the velocity behind the shock show that the regions of negative pressure increment are also regions of negative (i.e., inward) velocity (Figure 183). The curves in Figures 182 and 183 are plotted for different nondimensional times τ as shown by the number near the peak of each curve. τ is defined as $c_0 t/R_0$ where $R_0 = E_0/b\gamma p_0$, c_0 is the speed of sound in the undisturbed gas ahead of the shock, and t is the time from the onset of the shock.

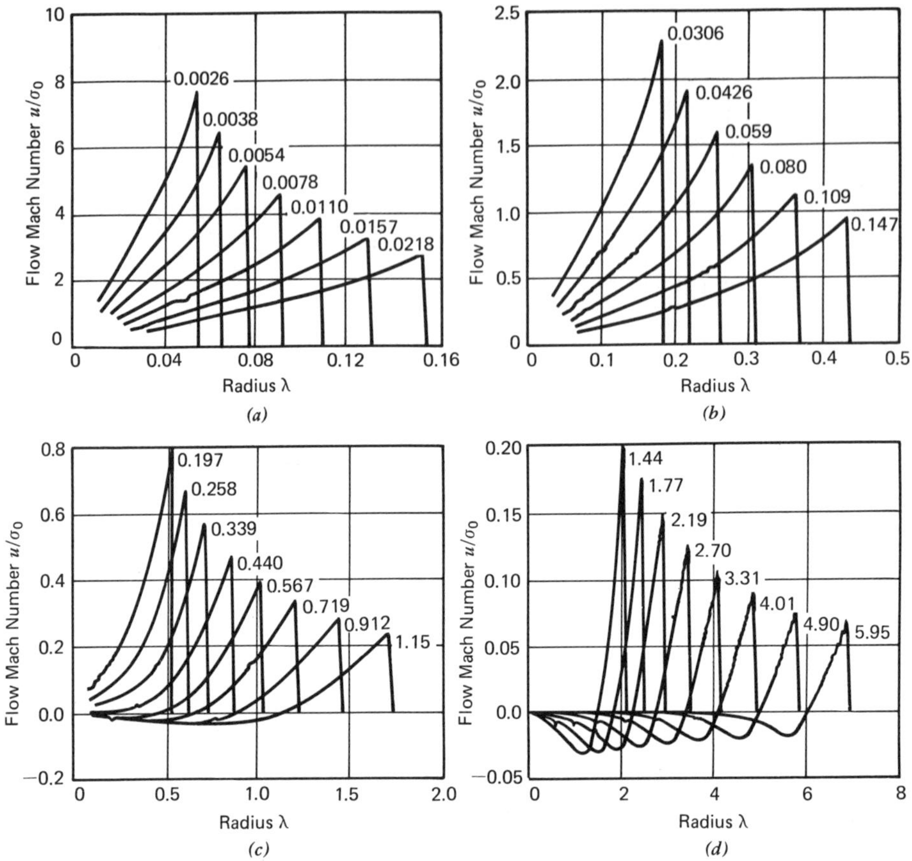

Figure 183. Results of Plooster [55] for cylindrical shock.

Riemann Problem

One of the best known of the unsteady one-dimensional flow problems is the Riemann, or shock tube problem. A tube is separated into two sections by a diaphragm (Figure 184). The pressure in section 3 in presumed to be much

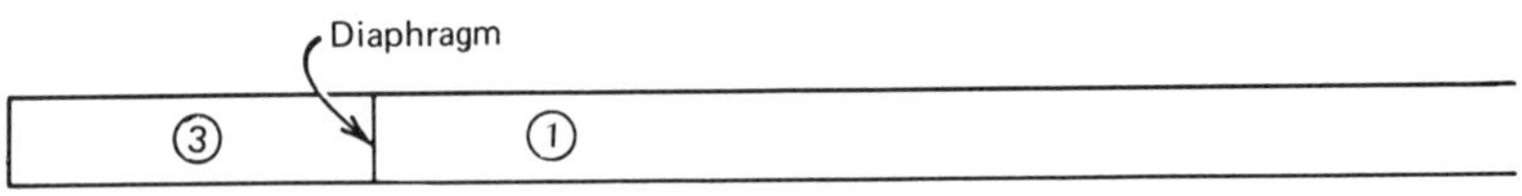

Figure 184. Shock tube.

higher than in section 1. The diaphragm is now caused to rupture by striking it with a nail or other means, thus, in effect, disappearing instantaneously. It is desired to predict the subsequent motion of the gas in the tube.

After the diaphragm bursts, a shock wave will proceed to the right while a rarefaction wave will proceed to the left (Figure 185). Remembering the rule

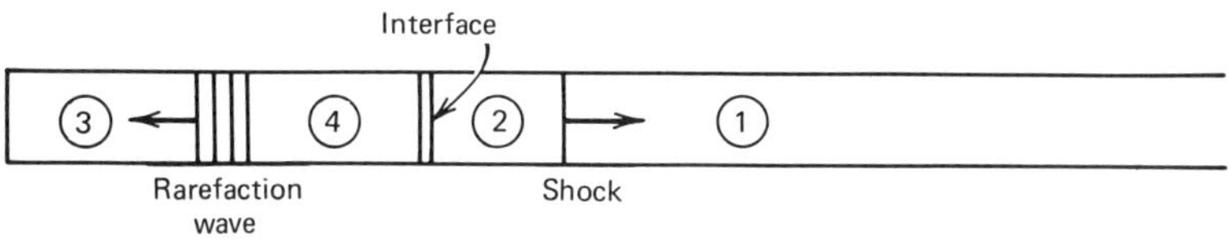

Figure 185. Wave pattern in shock tube.

compression waves steepen, rarefaction waves flatten, we note that the rarefaction wave is of a finite and increasing width while the shock wave remains essentially a discontinuity. The gas to the right of the interface is that which was originally on the right side of the diaphragm, while the gas to the left of the interface is that which was originally to the left of the diaphragm. Note that the gas in region 2 has passed through the shock wave while the gas in region 4 has passed through the rarefaction wave. Conservation of mass and momentum require that the pressure and velocity be the same in region 2 as in region 4. If, however, the state properties are initially the same in regions 1 and 3, it is clear that they cannot all be the same in regions 2 and 4. This is because the gas in region 2 has passed through a shock and has, therefore, a higher entropy than the gas in region 4, which has passed through an essentially isentropic expansion wave. Thus, the temperature in region 2 will be higher than in region 4.

The relations across the shock wave may be written

$$\frac{u_2 - u_1}{c_1} = \pm\frac{2}{\gamma+1}\left(M_x - \frac{1}{M_x}\right) \tag{798}$$

$$\frac{\rho_1}{\rho_2} = 1 - \frac{2}{\gamma+1}\left(1 - \frac{1}{M_x^2}\right) \tag{799}$$

$$\frac{p_2}{p_1} = 1 + \frac{2\gamma}{\gamma+1}\left(M_x^2 - 1\right) \tag{800}$$

Eliminating M_x between equations (798) and (800), and keeping in mind that $u_1 = 0$ in the present case, we have

$$\frac{u_2}{c_1} = \frac{\dfrac{p_2}{p_1} - 1}{\gamma\sqrt{1 + \dfrac{\gamma+1}{2\gamma}\left(\dfrac{p_2}{p_1} - 1\right)}} \tag{801}$$

Now across the left-running rarefaction wave with $u_3 = 0$,

$$\frac{u_4}{c_3} = \frac{2}{\gamma-1}\left(1 - \frac{c_4}{c_3}\right) = \frac{2}{\gamma-1}\left[1 - \left(\frac{p_4}{p_3}\right)^{(\gamma-1)/2\gamma}\right] \tag{802}$$

But $u_4 = u_2$, $p_4 = p_2$; assume further that $c_3 = c_1$; then

$$\frac{u_2}{c_1} = \frac{2}{\gamma-1}\left[1 - \left(\frac{p_2}{p_1}\right)^{(\gamma-1)/2\gamma}\left(\frac{p_1}{p_3}\right)^{(\gamma-1)/2\gamma}\right] \tag{803}$$

Equating equations (802) and (803) and solving for p_1/p_3, we get

$$\frac{p_1}{p_3} = \frac{p_1}{p_2}\left[1 - \frac{\gamma - 1}{2\gamma}\frac{\dfrac{p_2}{p_1} - 1}{\sqrt{1 + \dfrac{\gamma + 1}{2\gamma}\left(\dfrac{p_2}{p_1} - 1\right)}}\right]^{2\gamma/(\gamma - 1)} \tag{804}$$

giving us the required pressure ratio across the diaphragm for any desired pressure ratio across the shock. u_2 and M_x may then be found from equations (798) and (803).

The progress of the waves and interface may be shown graphically as in Figure 186. Notice that the precise location of the shock, interface, and expansion

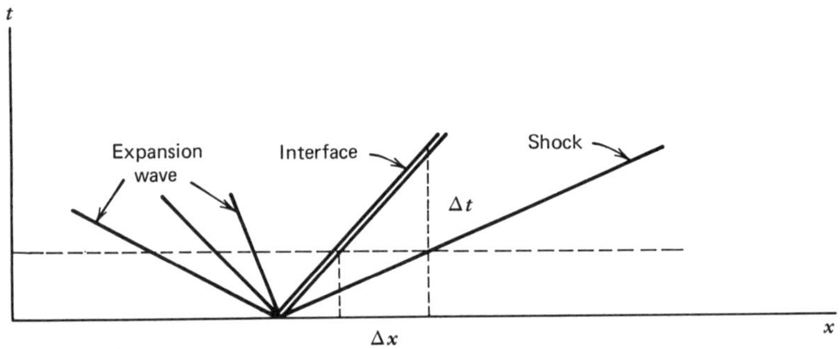

Figure 186. Wave pattern in shock tube shown on the x–t plane.

wave may be determined at any time t simply by drawing a horizontal line across the figure as illustrated by the dotted line.

Shock tubes are used to produce small quantities of high velocity, high temperature gas for experimental purposes. Usually, only the gas between the shock and the interface is used for test purposes since the state properties change abruptly at the interface. The gas between the shock and the interface is known as the test slug, whose width, Δx, is illustrated in Figure 186. The time which the test slug takes to pass a given point is known as the test time, Δt, and is also shown on the figure. Notice that the further downstream we go, the longer the test slug and test time becomes. For this reason, shock tubes are usually made as long as possible. There is a limit to how far it is profitable to extend a shock tube, however, since viscous and heat-transfer effects ultimately make themselves felt. If the shock tube is sufficiently long, the rarefaction waves, after having reflected from the closed end, will ultimately catch up to the shock, causing it to weaken. In this connection it may be noted that if a rarefaction wave follows a shock wave and the region between them has uniform properties, the rarefaction wave will always catch up to the shock wave, since the rarefaction wave travels with the speed of sound relative to the flow ahead of it, while a shock wave travels with less than the speed of sound relative to the flow behind it. In a shock tube, the situation is complicated by the fact that the region between the shock and the rarefaction wave is divided by a contact surface into two regions of different temperature. Nevertheless, the reflected rarefaction wave will catch up to the shock in the end.

In a shock tube, the high pressure section, section 3 in Figure 184, is called the driver, and the low pressure section, section 1, is called the driven section. So far we have tacitly assumed that the gas in the driver and driven section is the same. This is not necessarily, or even usually, the case. Nor are the driver and the driven section necessarily at the same initial temperature. Consider now the case where $c_1 \neq c_3, \gamma_1 \neq \gamma_3$. Across the shock we have

$$u_2 = c_1\left[\frac{p_2}{p_1} - 1\right]\sqrt{\frac{2}{\gamma_1\left[(\gamma_1+1)\dfrac{p_2}{p_1} + (\gamma_1 - 1)\right]}} \tag{805}$$

assuming $\gamma_1 = \gamma_2$. Across the rarefaction wave we have

$$u_4 = \frac{2c_3}{\gamma_3 - 1}\left[1 - \left(\frac{p_4}{p_3}\right)^{(\gamma_3 - 1)/2\gamma_3}\right] \tag{806}$$

Also, $u_2 = u_4$, $p_4 = p_2$. Combining (805) and (806)

$$\frac{p_3}{p_1} = \frac{p_2}{p_1}\left[1 - \frac{c_1}{c_3}(\gamma_3 - 1)\frac{\dfrac{p_2}{p_1} - 1}{\sqrt{2\gamma_1\left[(\gamma_1+1)\dfrac{p_2}{p_1} + (\gamma_1 - 1)\right]}}\right]^{-2\gamma_3/(\gamma_3 - 1)} \tag{807}$$

Recall now

$$\frac{p_2}{p_1} = \frac{2\gamma_1}{\gamma_1 + 1}M_x^2 - \frac{\gamma_1 - 1}{\gamma_1 + 1} \tag{808}$$

where M_x is the shock Mach number. Substituting in (807),

$$\frac{p_3}{p_1} = \frac{1}{\gamma_1 + 1}(2\gamma_1 M_x^2 + 1 - \gamma_1)\left[1 - \frac{c_1}{c_3}\left(\frac{\gamma_3 - 1}{\gamma_1 + 1}\right)\left(M_x - \frac{1}{M_x}\right)\right]^{-2\gamma_3/(\gamma_3 - 1)} \tag{809}$$

Notice that as $p_3/p_1 \to \infty$, p_2/p_1 does not increase indefinitely but approaches a finite limit. Thus, M_x also approaches a finite limit. Thus, in increasing driver pressure, one soon reaches a point of diminishing returns. Letting $p_3/p_1 \to \infty$, the limiting Mach number is

$$M_{x_{\max}} = \frac{1}{2}\left(\frac{\gamma_1 + 1}{\gamma_3 - 1}\right)\frac{c_3}{c_1} + \sqrt{\left[\frac{1}{2}\left(\frac{\gamma_1 + 1}{\gamma_3 - 1}\right)\frac{c_3}{c_1}\right]^2 + 1} \tag{810}$$

For gases and vapors, γ lies roughly between 1.1 and 1.8. The speed of sound, on the other hand, varies with molecular weight. In order to attain the maximum Mach number, therefore, a driver gas with a low molecular weight is desirable. Hydrogen has the lowest molecular weight of any gas, but hydrogen drivers are seldom used due to the danger posed by inflammability. Helium drivers, on the other hand, are quite common.

In addition to the use of inert gases pumped to high pressures, other types of drivers are also used. Among these are combustion drivers, in which a mixture of fuel and oxydizer is placed in the driver and ignited, causing a high pressure to be attained prior to the rupturing of the diaphragm. Alternatively, a powerful electric arc may be discharged through the driver gas to raise the pressure. Shock tubes powered in this way are said to be arc driven. Sometimes a magnetic field is used to accelerate the ionized gas created by the arc. In this case the shock tube is said to be magnetically driven.

Shock Tube with Area Change

Consider the case of a shock tube with area change at the end of the driver (Figure 187). If the velocity in 4 is supersonic, the right end of the left-running

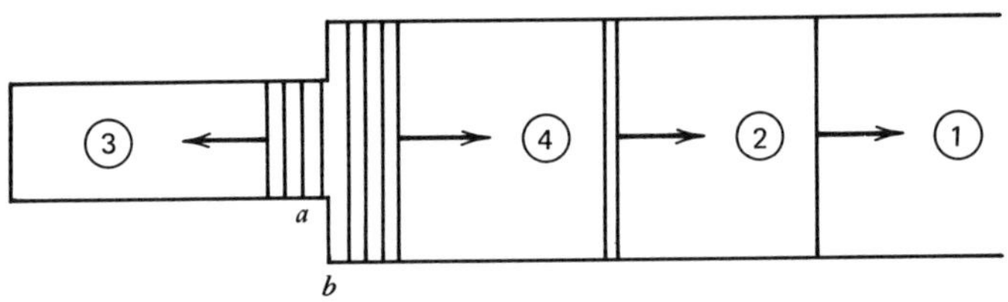

Figure 187. Wave pattern in shock tube with area change at driver.

rarefaction wave will proceed to the right, while the left end of the rarefaction wave will proceed to the left. Let a designate the end of the driver section and b the beginning of the driven section with the area change taking place between a and b. This area change may take place over zero length or over a small, finite length. In analyzing this problem we treat the flow between 3 and a as a left-running rarefaction wave in the usual manner. The flow between a and b, on the other hand, is treated as a quasi-steady, quasi-one-dimensional flow with area change. This may be justified as flows: shortly after the diaphragm bursts, the entire rarefaction wave will be concentrated in a region of small width δx. The pressure and velocity variation through the wave will be fully established and not vary subsequently (until the wave reaches the closed end of the driver). If the right end of the wave is moving to the right, somewhere inside the region δx the velocity will be sonic and the wave speed zero relative to the tube. Thus, the sonic point is not moving with time. In the limit as $t \to 0, \delta x \to 0$, and hence the sonic point must coincide with the location of the diaphragm. This, of course, is the ideal case where the diaphragm vanishes instantaneously at time $t = 0$.

We may now approximate the continuous rarefaction wave by a series of finite waves as before. The wave at the sonic point is stationary while the next wave to the right of the sonic point is moving to the right. If the change in area takes place abruptly or over a short distance, then the region of area change will quickly be free of waves, and hence a region of steady flow (Figure 188). (A more detailed analysis of unsteady flow with area change will be found in the next section.) Of course, if the area change is abrupt, the flow is not locally one-

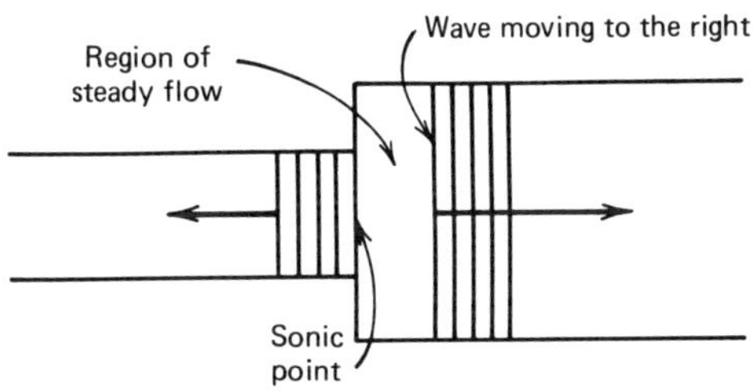

Figure 188. Detail of flow pattern at area change.

dimensional, but the approximation is found to give good results in practice. From b to 4 the flow is treated as a left running rarefaction wave in the usual manner. The flow in Figure 187 may best be visualized on the state (u, c) plane (Figure 189). Across the shock we follow the u, c shock polar, as previously

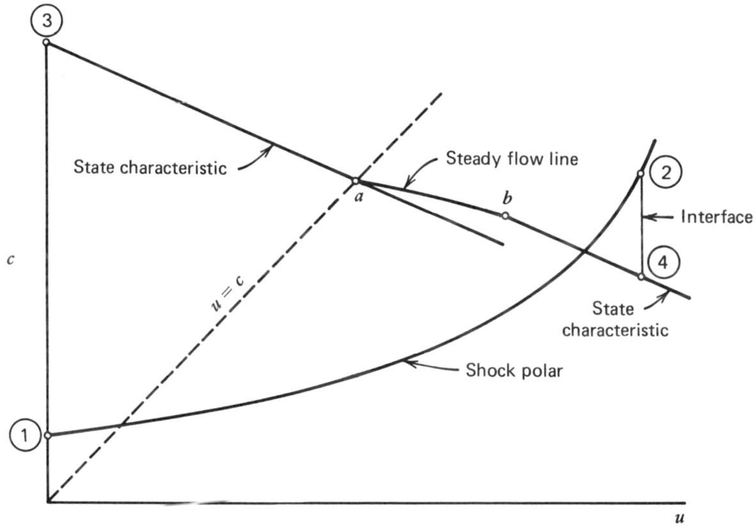

Figure 189. Flow pattern for shock tube with area change shown in the u–c plane.

shown in Figure 170. Across the interface there is no change in velocity but the temperature drops. From 4 to b we are crossing a left-running rarefaction wave and hence follow the appropriate state characteristic. From b to a we postulate steady adiabatic flow and hence we have

$$\frac{u^2}{2} + \frac{c^2}{\gamma - 1} = \text{constant} \tag{811}$$

and this equation is plotted as the steady flow line in Figure 189. From a to 3 we once again cross a left-running rarefaction wave and hence follow the appropriate state characteristic.

Recall that in going from 3 to a, the rarefaction wave cannot accelerate the flow beyond sonic speed. If the rarefaction wave continues after the expansion $a - b$, therefore, the velocity at a will be sonic and the right end of the rarefaction wave 3–a will be stationary. The flow then accelerates to supersonic speed across $a-b$. The left end of the rarefaction wave $b-4$ is thus moving to the right with velocity $u-c$.

Consider now the case of a constriction somewhere down the tube (Figure 190). This device is sometimes used in shock tubes to increase the pressure and

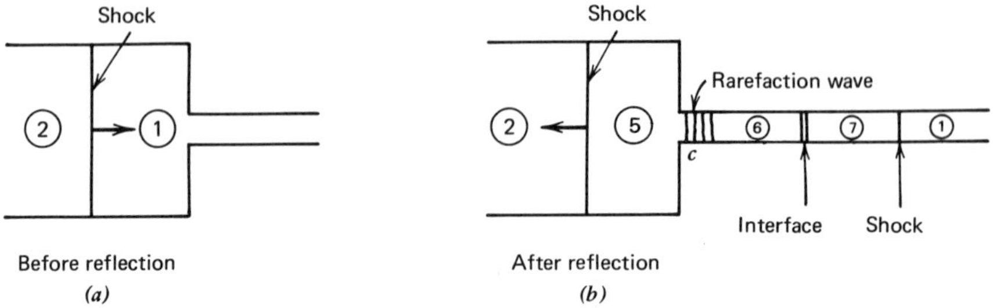

Figure 190. Flow pattern for shock tube with downstream area change.

test time. In such cases the tube is usually called a shock tunnel rather than a shock tube. Notice the large area ratio at the constriction. If the cross section of the continuing tube is sufficiently small, the pressure in region 5 will approach that caused by shock reflection off a closed end. We wish now to determine the conditions in region 5, 6, and 7. As the incident shock strikes the constriction, one part of it continues down the narrow tube while the rest is reflected back up the wider tube. Neither the continuing portion nor the reflected portion have the same strength as the original and the reflected portion does not have a hole in it like a donut but covers the entire cross section. Following the continuing portion of the shock there is an interface which separates the gas which has passed through the continuing shock from that which has passed through the reflected shock. As before, the effect of the change in area may be approximated by the quasi-steady-state analysis. Notice that the area change under consideration acts as a converging nozzle and hence cannot accelerate the flow beyond a Mach number of one. If the shock is of sufficient strength to

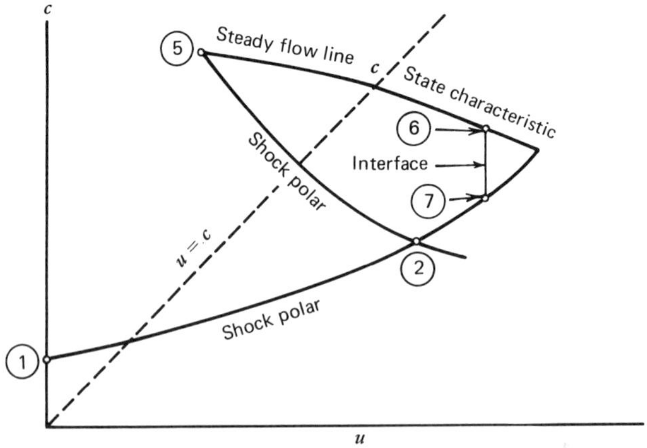

Figure 191. Flow pattern for shock tube with downstream area change shown in the $u-c$ plane.

generate supersonic flow in the narrow tube, a left-running rarefaction will be generated following the constriction which will accelerate the flow to its final value in region 6. The left end of the rarefaction wave will be stationary at the constriction where $u = c$. The right end will move to the right with velocity $u - c$. Once again, the process may be shown on a $u - c$ diagram (Figure 191).

Unsteady, Quasi-One-Dimensional Flow with Area and Entropy Change

Quasi-one-dimensional flow with area change will be understood here in the same sense as in Chapter 2; that is, flow in a duct with slowly varying cross section, all flow properties to be taken only as functions of distance along the duct and time. As before, the flow properties will be, in effect, average values across each cross section. Let x be the distance along the duct and $a(x)$ the cross section. Take as the control volume a section of the duct (Figure 192).

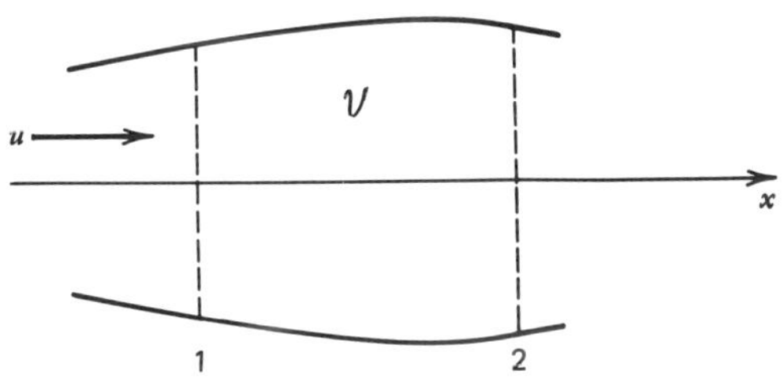

Figure 192. Control volume for unsteady quasi-one-dimensional flow.

Recall the equation of continuity

$$\frac{\partial \rho}{\partial t} + \nabla(\rho \mathbf{V}) = 0 \tag{5}$$

Integrating over the control volume,

$$\int_{\mathscr{V}} \frac{\partial \rho}{\partial t} d\mathscr{V} = \int_{x_1}^{x_2} \frac{\partial \rho}{\partial t} a(x)\, dx = -\int_{\mathscr{V}} \nabla(\rho \mathbf{V})\, d\mathscr{V} = -\int_S \rho \mathbf{V}\ \mathbf{n}\, dS$$
$$= \rho_1 u_1 a_1 - \rho_2 u_2 a_2 \tag{812}$$

Dividing both sides by $x_2 - x_1$ and taking the limit as $x_2 - x_1 \to 0$, we get

$$a\frac{\partial \rho}{\partial t} = \frac{\partial}{\partial x}(\rho u a) \tag{813}$$

Neglecting the term $v\,\partial u/\partial y$ under the assumption of quasi-one-dimensional flow, the x-momentum equation becomes

$$\frac{\partial u}{\partial t} + u\frac{\partial u}{\partial x} = -\frac{1}{\rho}\frac{\partial p}{\partial x} \tag{814}$$

The change in entropy of a particle is given by

$$T\, ds = dh - v\, dp \tag{815}$$

Introducing the perfect gas relations $p = \rho RT$, $dh = c_p\,dT$, $R = c_p - c_v$, $\gamma = c_p/c_v$, $c^2 = \gamma p/\rho$, this becomes

$$(\gamma - 1)T\,ds = \frac{1}{\rho}dp - \frac{c^2}{\rho}d\rho \tag{816}$$

The time rate of change of the entropy of a particle is then given by

$$(\gamma - 1)T\frac{Ds}{Dt} = \frac{1}{\rho}\frac{Dp}{Dt} - \frac{c^2}{\rho}\frac{D\rho}{Dt} \tag{817}$$

The speed of a small pressure discontinuity is still c, with respect to the flow, even though the cross-sectional area of the flow may be changing. This may be seen by observing that the cross-sectional area of the discontinuity is the same on either side of it and hence the derivation of equations (695) to (710) still applies. Thus, with respect to a coordinate system at rest, small disturbances still propagate with velocities $u \pm c$. The waves in the x,t plane whose slopes are everywhere $u \pm c$ are still called Mach lines or characteristics in the present case, since it may be shown that they are characteristic lines of the governing equations even when the entropy and area vary. This may be seen as follows: Recall the energy equation (24) combined with the continuity equation (5) yields

$$\dot{q} = \rho\frac{De_t}{Dt} - \frac{p}{\rho}\frac{D\rho}{Dt} \tag{818}$$

For a perfect gas, $e_t = c_v T = p/\rho(\gamma - 1)$. So

$$\frac{De_t}{Dt} = \frac{1}{(\gamma - 1)\rho}\left[\frac{Dp}{Dt} - \frac{p}{\rho}\frac{D\rho}{Dt}\right] \tag{819}$$

Substituting in (818) we get

$$(\gamma - 1)\dot{q} = \frac{Dp}{Dt} - c^2\frac{D\rho}{Dt} \tag{820}$$

Using this equation and the momentum (711) and continuity equations, we may set up a system of six equations in six unknowns:

$$\begin{aligned}
du &= u_x\,dx + u_t\,dt \\
dp &= \qquad\qquad p_x\,dx + p_t\,dt \\
d\rho &= \qquad\qquad\qquad\qquad \rho_x\,dx + \rho_t\,dt \\
-a_x\frac{\rho u}{a} &= u_x\rho \qquad\qquad\qquad\qquad \rho_x u + \rho_t \\
0 &= u_x u + u_t + p_x\frac{1}{\rho} \\
(\gamma - 1)\dot{q} &= \qquad\qquad p_x u + p_t - \rho_x u c^2 - \rho_t c^2
\end{aligned} \tag{821}$$

Solving for u_x we get

$$
u_x = \frac{\begin{vmatrix} du & dt & 0 & 0 & 0 & 0 \\ dp & 0 & dx & dt & 0 & 0 \\ d\rho & 0 & 0 & 0 & dx & dt \\ -\dfrac{\rho u}{a} a_x & 0 & 0 & 0 & u & 1 \\ 0 & 1 & \dfrac{1}{\rho} & 0 & 0 & 0 \\ (\gamma - 1)q & 0 & u & 1 & -uc^2 & -c^2 \end{vmatrix}}{\begin{vmatrix} dx & dt & 0 & 0 & 0 & 0 \\ 0 & 0 & dx & dt & 0 & 0 \\ 0 & 0 & 0 & 0 & dx & dt \\ \rho & 0 & 0 & 0 & u & 1 \\ u & 1 & \dfrac{1}{\rho} & 0 & 0 & 0 \\ 0 & 0 & u & 1 & -uc^2 & -c^2 \end{vmatrix}} \tag{822}
$$

The denominator works out to

$$
(dx - u\,dt)\left[(dx - u\,dt)^2 + c^2\,dt^2\right] \tag{823}
$$

This quantity will be zero when

$$
\frac{dx}{dt} = u \tag{824}
$$

$$
\frac{dx}{dt} = u \pm c \tag{825}
$$

Thus we see that in the present case, both the Mach lines and the streamlines are characteristics.

Consider now a plot of the Mach lines in the x, t plane (Figure 193). The

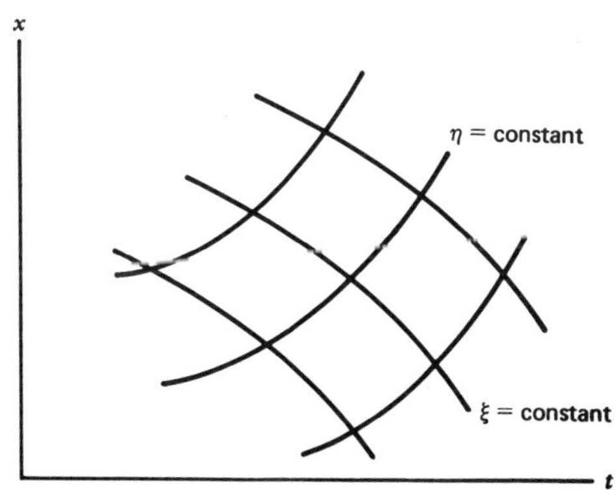

Figure 193. Physical characteristics for unsteady, quasi-one-dimensional flow.

lines ξ = constant have slope $u - c$ and the lines η = constant have slope $u + c$. In general, ξ and η will be functions of x and t. Thus we have

$$d\xi = \xi_x\,dx + \xi_t\,dt, \qquad d\eta = \eta_x\,dx + \eta_t\,dt \tag{826}$$

Along the line ξ and η equal constant, therefore,

$$\left(\frac{\partial x}{\partial t}\right)_\xi = -\frac{\xi_t}{\xi_x} = u - c, \qquad \left(\frac{\partial x}{\partial t}\right)_\eta = -\frac{\eta_t}{\eta_x} = u + c. \tag{827}$$

The change in any function $f(x,t)$ along the lines ξ = constant is given by

$$df_\xi = f_x\,dx_\xi + f_t\,dt_\xi \qquad \text{where} \qquad \frac{dx_\xi}{dt_\xi} = \left(\frac{\partial x}{\partial t}\right)_\xi \tag{828}$$

or

$$\frac{df_\xi}{dt_\xi} = \left(\frac{\partial f}{\partial t}\right)_\xi = f_t + f_x\left(\frac{\partial x}{\partial t}\right)_\xi = f_t + (u - c)f_x = \frac{Df}{Dt} - cf_x \tag{829}$$

Similarly,

$$\frac{df_\eta}{dt_\eta} = \left(\frac{\partial f}{\partial t}\right)_\eta = \frac{Df}{Dt} + cf_x \tag{830}$$

Consider now the case of constant entropy along a streamline. Then

$$\frac{Ds}{Dt} = \frac{1}{\rho}\frac{Dp}{Dt} - \frac{c^2}{\rho}\frac{D\rho}{Dt} = 0 \tag{831}$$

Expanding the continuity equation (813)

$$-a\frac{\partial \rho}{\partial t} = ua\frac{\partial \rho}{\partial x} + u\rho\frac{\partial a}{\partial x} + \rho a\frac{\partial u}{\partial x} \tag{832}$$

or

$$-\frac{a}{\rho}\frac{D\rho}{Dt} = u\frac{\partial a}{\partial x} + a\frac{\partial u}{\partial x} \tag{833}$$

Substituting for $D\rho/Dt$ from equation (831)

$$-\frac{1}{\rho}\frac{Dp}{Dt} = c^2\left[\frac{u}{a}\frac{\partial a}{\partial x} + \frac{\partial u}{\partial x}\right]$$

or

$$-\frac{1}{\rho}\frac{Dp}{Dt} - c^2\frac{\partial u}{\partial x} = c^2\frac{u}{a}\frac{\partial a}{\partial x} \tag{834}$$

Multiplying the momentum (711) by $\pm c$ and adding it to equation (834) we get

$$\pm c\left(\frac{Du}{Dt} \pm c\frac{\partial u}{\partial x}\right) + \frac{1}{\rho}\left(\frac{Dp}{Dt} \pm c\frac{\partial p}{\partial x}\right) = -c^2\frac{u}{a}\frac{\partial a}{\partial x} \tag{835}$$

But from equations (829) and (830) this is equivalent to

$$-\frac{du_\xi}{dt_\xi}+\frac{1}{\rho c}\frac{dp_\xi}{dt_\xi}=-c\frac{u}{a}\frac{\partial a}{\partial x} \tag{836}$$

and

$$+\frac{du_\eta}{dt_\eta}+\frac{1}{\rho c}\frac{dp_\eta}{dt_\eta}=-c\frac{u}{a}\frac{\partial a}{\partial x} \tag{837}$$

Cross multiplying by cdt, we have, along lines of constant ξ or η,

$$\mp c\,du+\frac{1}{\rho}dp=-c^2\frac{u}{a}\frac{da}{dx}dt \tag{838}$$

where the minus sign is used along a line of constant ξ and the plus sign is used along a line of constant η. da/dx may be written as an ordinary derivative since a is a function of x only. Eliminating $(1/\rho)dp$ using equation (815)

$$\mp c\,du+\frac{1}{\rho}dp=\mp c\,du+dh-T\,ds=c^2\frac{u}{a}\frac{da}{dx}dt \tag{839}$$

or

$$\mp c\,du+dh=-c^2\frac{u}{a}\frac{da}{dx}dt+T\,ds \tag{840}$$

Under the perfect gas assumption

$$dh=\frac{2c}{\gamma-1}dc, \qquad T=\frac{c^2}{c_p(\gamma-1)} \tag{841}$$

so

$$\mp du+\frac{2}{\gamma-1}dc=-c\frac{u}{a}\frac{da}{dx}dt+\frac{c}{\gamma-1}d\left(\frac{s}{c_p}\right) \tag{842}$$

It will be recalled that in the case of constant area and entropy,

$$\mp du+\frac{2}{\gamma-1}dc=0 \tag{843}$$

which could be integrated directly giving the state characteristics. The present case is similar to the case of two-dimensional steady flow with entropy variation considered earlier. The integration must be performed in a stepwise manner.

Assume that the flow conditions are known along the line ub in the x, t plane (Figure 194). It is desired to find the flow properties in the rest of the flow field. The location of the point 3 is found by drawing a line having slope $(u+c)_1$ through the point 1 and a line having slope $(u-c)_2$ through the point 2. The point of intersection of the two lines will be the point 3. The entropy at the point three may be found by locating the point 4 from which a line having slope u_4 intersects the point 3. Since the entropy of individual particles of fluid has been presumed constant, the entropy at 3 will have the same value as the entropy

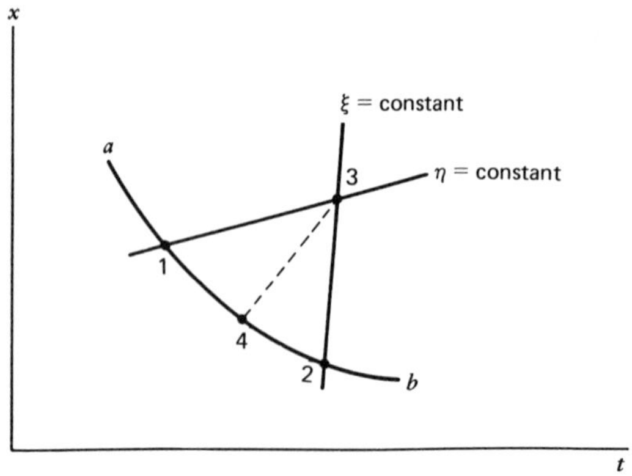

Figure 194. Characteristics calculation for nonisentropic flow.

at 4. Equation (842) may now be written in finite difference form:

$$(u_3 - u_1) + \frac{2}{\gamma - 1}(c_3 - c_1) = -c_1 \frac{u_1}{a_1}\left(\frac{da}{dx}\right)_1 (t_3 - t_1) + \frac{c_1/c_p}{\gamma - 1}(s_3 - s_1) \quad (844)$$

$$-(u_3 - u_2) + \frac{2}{\gamma - 1}(c_3 - c_2) = -c_2 \frac{u_2}{a_2}\left(\frac{da}{dx}\right)_2 (t_3 - t_2) + \frac{c_2/c_p}{\gamma - 1}(s_3 - s_2) \quad (845)$$

The equations may be solved simultaneously for u_3 and c_3. In the same way, the conditions at other lattice points may be found, until the flow field has been determined throughout a given region (Figure 195). The equations retain their

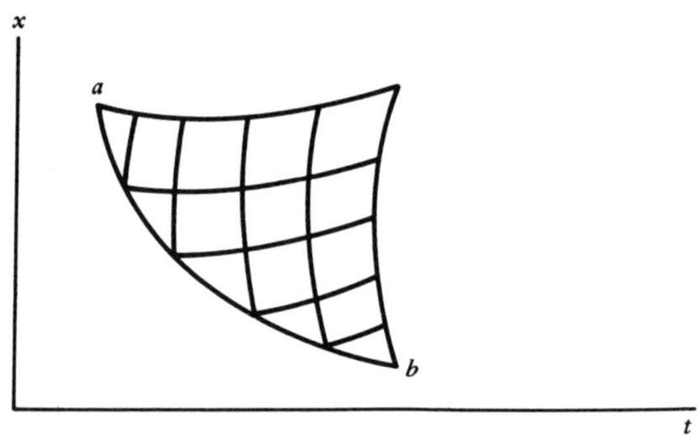

Figure 195. Solution in the x, t plane.

hyperbolic character. Thus, the flow field is determined only in a finite region, given the boundary conditions along a finite curve.

6 Transonic Flow

The problem of transonic flow is one of the most intriguing and difficult in the field of compressible flow. The difficulty arises from the fact that the governing equation for the velocity potential (270) changes from elliptic to hyperbolic as the Mach number goes from subsonic to supersonic. As will be recalled, the solutions to these two types of equations are entirely different. In a flow field where both subsonic and supersonic flow are present, a universally valid analytic solution is, therefore, hardly to be hoped for.

Because of the challenging nature of this problem, it has attracted the attention of mathematicians for many decades. Until recently, however, practical solutions had been found in only a very small number of cases. Only in the last 10 years have solutions been developed for the transonic flow over bodies of reasonably arbitrary shape, particularly over bodies of interest to engineers such as wings and cascades. Nevertheless, tribute must be paid to the early pioneers such as Tricomi [57] and Guderley [58], whose work laid the foundation for the solution of this problem.

Transonic Similarity Law

Recall the continuity and irrotationality conditions for steady, two-dimensional, irrotational flow:

$$(\rho u)_x + (\rho v)_y = 0 \tag{846}$$

$$u_y - v_x = 0 \tag{847}$$

We now consider the flow over a thin body; that is, one whose thickness is small compared to its length. In gas dynamics, the convention has arisen to call such a body "thin" if it is two-dimensional, and "slender" if it is three-dimensional. A "thin" body typically would be represented by the cross section of a wing, say, while a slender body might be a missile or rocket (Figure 196). We assume the flow to be isentropic. The body lies in a stream of free stream velocity u_∞ and the y component of the velocity, v, is assumed to be everywhere small compared to u_∞, which in turn is assumed to be close to c^*, the critical speed of sound.

Writing $\rho = \rho_\infty + \rho'$, where $\rho' \ll \rho_\infty$, equation (846) becomes

$$[(\rho_\infty + \rho')u]_x + [(\rho_\infty + \rho')v]_y = 0 \tag{848}$$

For the flow pattern under consideration, $(\rho' v)_y \ll (\rho' u)_x$ and hence, approximately,

$$(\rho u)_x + \rho_\infty v_y = 0 \tag{849}$$

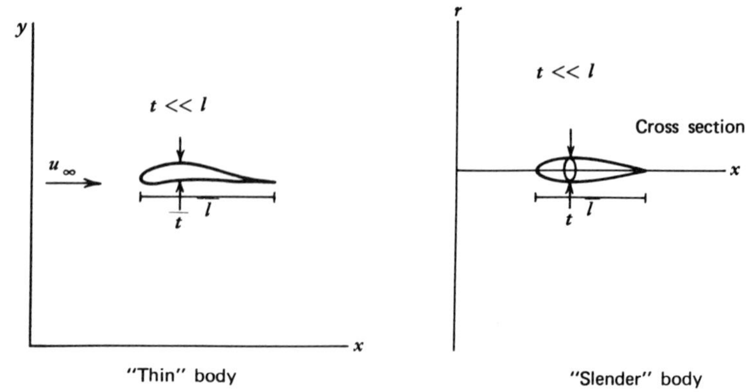

Figure 196 Thin body and slender body.

The quantity ρu, which is sometimes called the stream density, has a maximum where $u = c^*$ and decreases on either side. Near the critical point, we may express ρu as a power series,

$$\frac{\rho u}{\rho_\infty u_\infty} = 1 + \beta^2 \frac{u - u_\infty}{u_\infty} - \frac{1}{2}\beta^2 \frac{u_\infty}{c^* - u_\infty}\left(\frac{u - u_\infty}{u_\infty}\right)^2 + \ldots \tag{850}$$

where $\beta^2 = 1 - M_\infty^2$. We now introduce the following change of variables:

$$x' = x \qquad u' = \frac{u - u_\infty}{c^* - u_\infty}$$

$$y' = \beta y \qquad v' = \frac{v}{\beta(c^* - u_\infty)}$$

Equation (850) then becomes

$$\frac{\dfrac{\rho u}{\rho_\infty u_\infty} - 1}{\beta^2\left(\dfrac{c^*}{u_\infty} - 1\right)} = u' - \frac{u'^2}{2} + \ldots \tag{851}$$

Substituting in the continuity equation (849) and the irrotationality condition (847) and dropping higher order terms, we get

$$u'_{x'} + v'_{y'} = u' u'_{x'} \tag{852}$$

$$u'_{y'} - v'_{x'} = 0 \tag{853}$$

The solution to the system (852, 853) is rendered determinate by the boundary conditions,

$$\sqrt{x'^2 + y'^2} = \infty: \quad u' = v' = 0 \qquad y' = 0: \quad v' = v'_0(x') = \frac{\dfrac{v_0(x)}{u_\infty}}{\left(\dfrac{c^*}{u_\infty} - 1\right)} \tag{854}$$

The system of equations (852, 853) with the boundary conditions (854) thus constitute a similarity law for transonic flow; that is, all physical flows of the type under consideration whose transformed boundary conditions (854) are the same, will have the same solution in the transformed coordinate system. A single dimensionless solution of the system (852–854) leads to an infinite number of dimensional solutions. This form of the transonic similarity law was first derived by Guderley [59] and independently by Oswatitsch [60].

Consider now profiles which are affinely related in the y direction; that is, the y coordinate for each value of x varies by a constant factor. At a given value of x, the ratio of the values of $v_0(x)/u_\infty$ will be proportional to the thickness ratio τ. If we define a reduced thickness ratio, τ',

$$\tau' = \frac{\tau}{\beta\left(\dfrac{c^*}{u_\infty} - 1\right)} \tag{857}$$

then all profiles having the same τ' will have the same $v_0'(x')$ and hence the same $u_0(x')$. Figure 197 shows the comparison of this law with experiment. It may

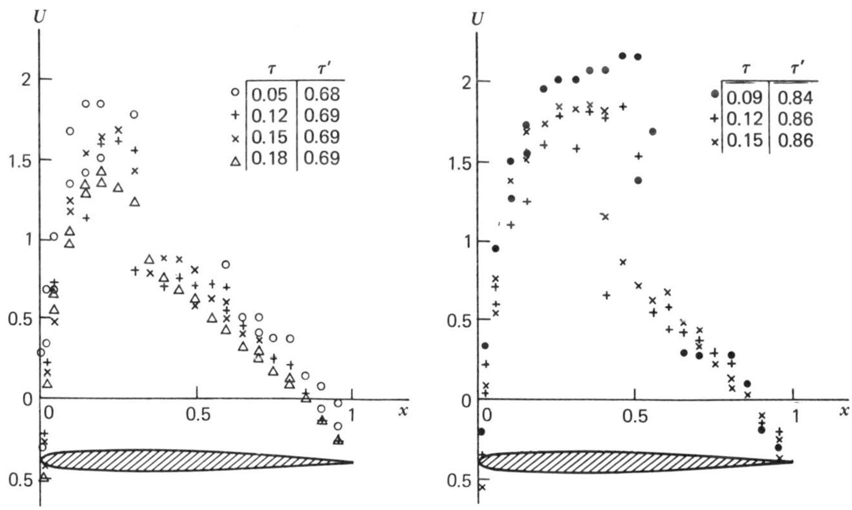

Figure 197 Correlation of similarity parameter by Oswatitsch [67].

be noted that the similarity law must still be valid even when shocks are present, provided the shocks are located at the same value of x for profiles affinely related in y. The discrepancies observed in Figure 197 may be accounted for by the fact that the shock waves move progressively upstream as the profile thickens. For normal shocks, ρu is constant across the shock, and hence

$$u_1' - \frac{u_1'^2}{2} = u_2' - \frac{u_2'^2}{2} \tag{858}$$

Hence

$$u_1' - 1 = 1 - u_2' \tag{859}$$

or

$$u_1 + u_2 = 2c^* \tag{860}$$

This is the same as taking a tangent to the Prandtl relation at $u_1 = u_2 = c^*$ (Figure 198).

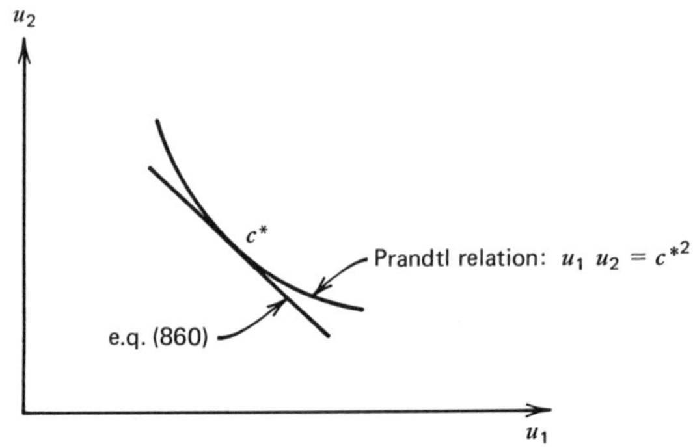

Figure 198 Approximation of Oswatitsch [60].

The transonic similarity law has been derived in a somewhat different way by von Karman [61]. Beginning with the full equation for the velocity potential

$$\left(1 - \frac{u^2}{c^2}\right)\phi_{xx} - 2\frac{uv}{c^2}\phi_{xy} + \left(1 - \frac{v^2}{c^2}\right)\phi_{yy} = 0 \tag{861}$$

and the energy equation

$$c^2 + \frac{\gamma - 1}{2}(u^2 + v^2) = \frac{\gamma + 1}{2}c^{*2} \tag{862}$$

we define a perturbation velocity potential ϕ':

$$\phi'_x = u - c^*, \qquad \phi'_y = v \tag{863}$$

where it is presumed that $\phi'_x \ll \phi_x$. Combining equations (861–863) and omitting higher order terms we get

$$\phi'_{yy} = \frac{\gamma + 1}{c^*}\phi'_x\phi'_{xx} + \frac{2}{c^*}\phi'_y\phi'_{xy} \tag{864}$$

The boundary conditions are

$$\text{at } \infty: \qquad \phi'_x = c^*(M^*_\infty - 1) \tag{865}$$

$$\text{at the body:} \qquad y = 0 \qquad \phi'_y = c^*\left(\frac{dy}{dx}\right)_b \tag{866}$$

where $(dy/dx)_b$ is the slope of the body surface. Notice that due to the thin body assumption and the neglect of higher order terms, we may take the surface of the body at $y = 0$.

Under our present assumptions, from equation (862),

$$\frac{c_\infty}{c^*} \cong 1 - \frac{\gamma - 1}{2}(M_\infty^* - 1) \tag{867}$$

and hence the boundary condition (865) becomes

$$\phi_x' = \frac{2}{\gamma + 1} c^*(M_\infty - 1) \tag{868}$$

Now let us define dimensionless space coordinates ξ and η. Let

$$x = l\xi, \qquad y = l(\tau\Gamma)^{-n}\eta \tag{869}$$

where l is the length of our body, τ the thickness ratio, and $\Gamma = (\gamma + 1)/2$. Further let us define a dimensionless stream function $F(\xi, \eta)$. Let

$$\phi' = lc^* \frac{1 - M_\infty}{\Gamma} F(\xi, \eta) \tag{870}$$

Substituting in equation (864), we get

$$F_{\eta\eta} = 2(\tau\Gamma)^{-2n}(1 - M_\infty)F_\xi F_{\xi\xi} + 2(1 - M_\infty)F_\eta F_{\xi\eta} \tag{871}$$

The boundary conditions (866) and (868) become

at ∞:

$$F_\xi = -1, \qquad F_\eta = 0 \tag{872}$$

at $\eta = \pm 0$:

$$F_\eta = (\tau\Gamma)^{1-n}(1 - M_\infty)^{-1} h(\xi) \tag{873}$$

where we have written $\eta = \pm 0$ to indicate that the slope on the bottom of the body may not be the same as the slope on the top. If now the coefficients in the equation and the boundary condition can be made functions of a single parameter, that parameter would be a similarity parameter. Let us first consider the coefficients $2(\tau\Gamma)^{-2n}(1 - M_\infty)$ in equation (871) and $(\tau\Gamma)^{1-n}(1 - M_\infty)^{-1}$ in the boundary condition (873). Let

$$(\tau\Gamma)^{-2n}(1 - M_\infty) = K \tag{874}$$

If we pick $n = \frac{1}{3}$, we see that

$$(\tau\Gamma)^{1-n}(1 - M_\infty)^{-1} = \frac{1}{K} \tag{875}$$

Substituting $n = \frac{1}{3}$ in equation (871)

$$\tau^{2/3} F_{\eta\eta} = 2\Gamma^{-2/3}(1 - M_\infty)F_\xi F_{\xi\xi} + 2\tau^{2/3}(1 - M_\infty)F_\eta F_{\eta\xi} \tag{876}$$

Since both τ and $1 - M_\infty$ are assumed small, the coefficient of the last term of equation (876) will be small compared to the coefficients of the other terms. If $F_\eta F_{\eta\xi}$ is of the same (or smaller) order of magnitude as $F_{\eta\eta}$ and $F_\xi F_{\xi\xi}$, the last term on the right-hand side may be neglected. Returning for a moment to dimensional variables, equation (876) may be written

$$(\gamma + 1)(c^* - u)u_x - 2vu_y + c^*v_y = 0 \tag{878}$$

In general in transonic flow it is observed that changes in the y direction occur more slowly than changes in the x direction. This may be explained mathematically by noting that as $M \to 1$ in a supersonic flow, the Mach angle approaches 90°. Hence, in the flow under consideration (Figure 196), the Mach lines become almost vertical. Since Mach lines are lines of constant flow properties, this would indicate that changes in the y direction proceed very slowly when $M = 1$. Noting that in the present case

$$\frac{u}{c^*} \cong 1, \qquad \frac{v}{c^*} \ll 1, \qquad \frac{c^*}{c^*} = 1, \qquad u_y, v_y \ll u_x$$

it seems reasonable to neglect the second term in equation (878) and hence the last term in equation (876). Then (876) becomes

$$F_{\eta\eta} - 2KF_{\xi}F_{\xi\xi} = 0 \tag{879}$$

where $K = (\tau\Gamma)^{-2/3}(1 - M_\infty)$. The boundary conditions become

at ∞:

$$F_\xi = -1, \qquad F_\eta = 0 \tag{880}$$

at $\eta = \neq 0$:

$$F_\eta = K^{-1}h(\xi)$$

Thus, transonic flows over bodies having the same K and the same $h(\xi)$ will have the same solution in the ξ, η plane. In this sense, then, K is a transonic similarity parameter.

Solutions to equation (879) may be found formally by means of the Legendre transformation in terms of Bessel functions. The difficulty lies, however, in finding solutions with boundary conditions of practical interest. This approach has not proved fruitful and the similarity law is used mainly to draw certain general conclusions about transonic flow.

Recall in our treatment of subsonic flow, we demonstrated that for small perturbations,

$$c_p = -2\frac{u'}{V_\infty} \tag{324}$$

where we have put a prime on the u to indicate that it is a perturbation velocity. A review of the derivation shows that the formula (324) is equally valid for isentropic transonic flow with small perturbations. In the present case we have

$$c_p = -2\left[\frac{1}{M_\infty}\left(1 + \frac{1 - M_\infty}{\Gamma}F_\xi\right) - 1\right] \tag{881}$$

If we assume $M_\infty \cong 1$, this reduces to

$$c_p = -\frac{2}{\Gamma}(1 - M_\infty)F_\xi \tag{882}$$

Recall $K = (1 - M_\infty)/(\tau\Gamma)^{2/3}$. Thus

$$c_p = -\frac{2}{\Gamma}\frac{1 - M_\infty}{(\tau\Gamma)^{2/3}}F_\xi = -2\frac{\tau^{2/3}}{\Gamma^{1/3}}KF_\xi \tag{883}$$

It follows that for a given K, the pressure coefficient will vary as $\tau^{2/3}$ and as $\Gamma^{-1/3}$. Similar criteria may be established for the lift and the drag coefficients. For the lift coefficient we have

$$c_l = \frac{1}{l}\oint c_p dx = \frac{\tau^{2/3}}{\Gamma^{1/3}}L(K) \tag{884}$$

where $L(K)$ indicates a function of K only, while for the drag coefficient we have

$$c_d = \frac{1}{l}\oint c_p \frac{dy}{dx}dx = \oint c_p \tau h(\xi)\, d\xi = \frac{\tau^{5/3}}{\Gamma^{1/3}}D(K) \tag{885}$$

where $D(K)$ indicates a function of K only and the integrals are taken around the surface of the body. Thus, for affinely related profiles having the same value of K, the lift and drag coefficients vary as $\tau^{2/3}/\Gamma^{1/3}$ and $\tau^{5/3}/\Gamma^{1/3}$, respectively.

Tricomi's Equation

Recall the hodograph equation (421):

$$\left(\frac{\rho_0 V\psi_V}{\rho}\right)_V + \frac{\rho_0(1 - M^2)\psi_{\theta\theta}}{\rho V} = 0 \tag{421}$$

Now define a new variable σ by the formula

$$\sigma = \int_{c^*}^{V} \frac{\rho}{\rho_0}\frac{dV}{V} \tag{886}$$

Equation (421) then becomes

$$\psi_{\sigma\sigma} + \left(\frac{\rho_0}{\rho}\right)^2(1 - M^2)\psi_{\theta\theta} = 0 \tag{889}$$

The coefficient $(\rho_0/\rho)^2(1 - M^2)$ is a function of the velocity only in isentropic flow and so is σ. It follows that the coefficient may also be expressed as a function of σ, at least in a table or graph, if not analytically. Designating this function by $k(\sigma)$, we have

$$\psi_{\sigma\sigma} + k(\sigma)\psi_{\theta\theta} = 0 \tag{890}$$

The function $k(\sigma)$ has been plotted by Tricomi and Ferrari [62], Figure 199. From the figure we observe that near $\sigma = 0\,(V = c^*)$, $k(\sigma) = \sigma$ is a good approximation. Hence equation (890) becomes

$$\sigma\psi_{\theta\theta} + \psi_{\sigma\sigma} = 0 \tag{891}$$

This is known as Tricomi's equation. For many years it was believed that the

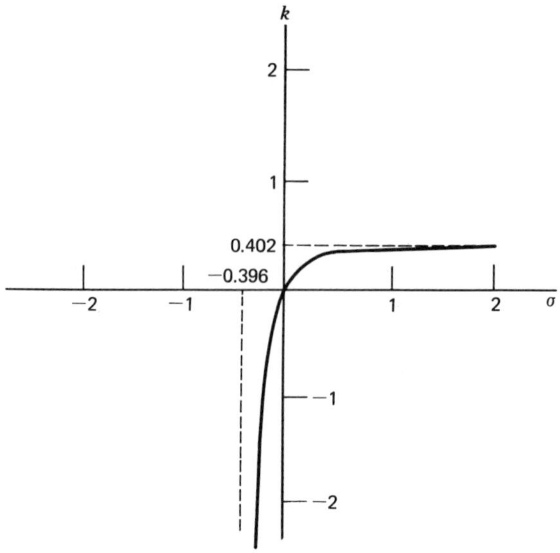

Figure 199 $k(\sigma)$, equation (890) from [26].

solution to the transonic flow problem lay in finding solutions to the Tricomi equation and a great deal of effort was expended in this direction. Unfortunately, this effort led only to solutions of great mathematical complexity but of little practical value. While a number of solutions of Tricomi's equation are known, the difficulty lies in finding solutions which satisfy boundary conditions of practical interest in the physical plane. Since the Tricomi equation is a hodograph equation, boundary conditions can be specified only in the hodograph plane. Usually, it is not known what boundary conditions these correspond to in the physical plane until after the solution has been found. Thus, it is not generally possible to use the Tricomi equation to solve particular physical problems. Only in certain special cases have physically meaningful solutions been found. For example, if at a boundary $\theta = \theta_0 =$ constant, we may specify $\psi = 0$ along that boundary and assume a solution of the form

$$\psi = g(\sigma, m) \sin\left(m\pi \frac{\theta}{\theta_0} \right) \tag{892}$$

where m is an appropriate constant. Substitution into Tricomi's equation yields the ordinary differential equation

$$g'' - \sigma \frac{m^2 \pi^2}{\theta_0^2} g = 0 \tag{893}$$

The solution to this equation is known to be

$$g(\sigma, m) = \sqrt{|\sigma|}\left[c_1 J_{1/3}\left(\frac{2}{3} \frac{m\pi}{\theta_0} |\sigma|^{3/2} \right) + c_2 J_{-1/3}\left(\frac{2}{3} \frac{m\pi}{\theta_0} |\sigma|^{3/2} \right) \right] \tag{894}$$

where $J_{1/3}$ and $J_{-1/3}$ are modified Bessel functions. Guderley and Yoshihara [63] have used this approach to find the solution for transonic flow over a wedge.

Transonic Flow Over Airfoils. General Considerations

It is well known that the velocity over the top of an airfoil placed in a subsonic free stream will exceed the free stream velocity. Thus, if the free stream velocity is sufficiently high, although subsonic, the velocity may exceed the speed of sound over at least a portion of the airfoil. The free stream Mach number at which the local velocity over the airfoil first reaches the speed of sound is called the critical Mach number. The critical Mach number thus depends on the shape of the airfoil.

If the free stream Mach number exceeds the critical Mach number but remains subsonic, a region of supersonic flow will form over the airfoil and the flow will become a "mixed flow"; that is, one containing both supersonic and subsonic regions. Alternatively, we may say that a region of supersonic flow will be imbedded in a subsonic flow. This is most simply illustrated by a simple airfoil at zero angle of attack (Figure 200). Inside the sonic line, where in flow

Figure 200 Supersonic region over airfoil in subsonic flow.

is supersonic, we have drawn in the right- and left-running characteristics. Notice that at the sonic line, the characteristics must meet in a cusp, since the potential flow equation there is parabolic and hence the right- and left-running characteristics have the same slope.

We may examine the flow inside the sonic line from the standpoint of the field method, as explained in our discussion of the method of characteristics. Notice that the waves with positive slope must be expansion waves since, in the absence of reflected waves, the flow over the surface of the airfoil is a Prandtl–Meyer flow in the supersonic region. Consider, however, what happens when one of these expansion waves reaches the sonic line (Figure 201). Since the flow is still presumed to be isentropic, the pressure along the sonic line must be constant and hence the pressures in region 1 and 3 must be equal. But since wave a is

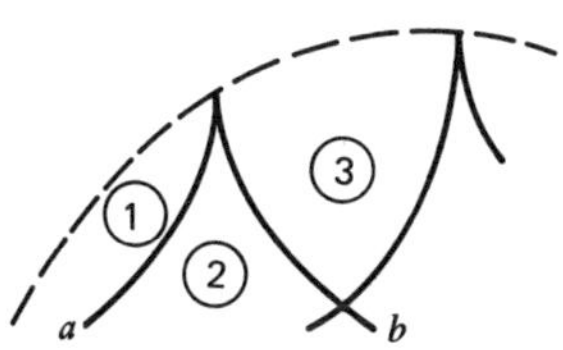

Figure 201 Detail of characteristics inside the supersonic region.

an expansion wave, the pressure in region 2 is less than the pressure in region 1. It follows that wave *b* must be a compression wave. We may say expansion wave *a* reflects as a compression wave *b* at the sonic line. The supersonic region may thus be approximated by a series of expansion and compression waves of finite strength.

As we move toward the rear of the supersonic region, the velocity must decrease as we move towards the sonic line. Thus, the right running compression waves progressively steepen. If the flow decelerates with sufficient rapidity the compression waves may intersect in the flow field, resulting in the formation of a shock (Figure 202). The formation of such shock waves was first observed

Figure 202 Shock formation in transonic flow.

in Schlieren photographs of the flow over wings of fighter aircraft taken during World War II. Such shock waves were also found to be unstable; that is, they had a tendency to fluctuate erratically, causing control difficulties when the aircraft exceeded the critical Mach number. The explanation of such shock waves as the intersection of characteristic waves was first proposed by von Karman [64]. Proceeding on this assumption, Nikolskii and Taganov (65) derived the criterion

$$\frac{dM^*}{d\theta} \leq M^* \tan \alpha \tag{895}$$

at the surface of the body in the supersonic region as the requirement for shock free flow, where $M^* = V/c^*$, α is the Mach angle and θ the flow direction. This condition was not fulfilled in the conventional airfoils of the time. In 1962, Pearcey [66] was able to demonstrate experimentally that surface shock waves could be delayed or even eliminated by suitable modifications of airfoil design. These modifications involved moving the adverse pressure gradient region on top of the airfoil as far to the rear as possible. This laid the foundation for the development of the so-called "supercritical airfoils," which are found on modern aircraft.

The theoretical solution of the transonic flow over airfoils had to await the advent of the modern, high-speed computer. There are basically two methods of approach to this problem: the integral method and the hodograph method. We shall deal with the integral method first.

Integral Method for the Transonic Flow Over Symmetric Airfoils

This method is due to Oswatitsch [67]. Recall equations (852) and (853)

$$u'_{x'} + v'_{y'} = u' u'_{x'} \tag{852}$$

$$u'_{y'} - v'_{x'} = 0 \tag{853}$$

If u' is everywhere small, the term $u'u'_{x'}$ in equation (852) may be neglected. Equations (852, 853) then reduce to

$$u'_{x'} + v'_{y'} = 0 \tag{896}$$

$$u'_{y'} - v'_{x'} = 0 \tag{897}$$

The system (896, 897) will be recognized as the Cauchy–Riemann equations. These possess the solution

$$\begin{aligned} u' &= \frac{q}{2\pi}\left[\frac{x-\xi}{(x-\xi)^2+(y-\eta)^2}\right] \\ v' &= \frac{q}{2\pi}\left[\frac{y-\eta}{(x-\xi)^2+(y-\eta)^2}\right] \end{aligned} \tag{898}$$

This will be recognized as analogous to an incompressible source flow, the equations (896, 897) having the form of the incompressible continuity and irrotationality conditions. The "source" is of strength z and located at ξ, η. Since the equations (896, 897) are linear, the sum of several solutions is also a solution. Hence, a series of "sources" distributed along the axis will also yield a solution. If the limit is taken as the number of sources increases and the strength of each source decreases, we have, in the limit,

$$u'_p = \frac{1}{\pi}\int_{-\infty}^{\infty} V_0(\xi)\frac{x-\xi}{(x-\xi)^2+y^2}\,d\xi \tag{899}$$

$$v'_p = \frac{1}{\pi}\int_{-\infty}^{\infty} V_0(\xi)\frac{y}{(x-\xi)^2+y^2}\,d\xi \tag{900}$$

where the subscript p indicates that these solutions are called Poisson integrals, corresponding to a continuous source distribution along the x axis of strength $2V_0(\xi)\,d\xi$.

By choosing the function $V_0(\xi)$ in such a way that the flow direction is tangent to a given profile at $y = 0$, the solution is made to represent the flow over that profile. Since we are dealing here with small perturbation, the slope at the wall

$$\left(\frac{dy}{dx}\right)_s = \frac{v}{u} \cong \frac{v}{u_\infty} \tag{901}$$

Thus, given u_∞, it is only necessary to specify $v(x)$ at the wall to render the problem determine. If the profile is symmetric,

$$v'_p(x,0) = V_0(x) \tag{902}$$

Hence, given the shape of the profile and the value of u_∞, the problem is solved.

If the term $u'u'_{x'}$ is not negligible in equation (852) then the system (852, 853) must be solved. Equation (852), however, is in effect the Poisson equation, as may be seen by substituting $u' = \phi'_x$, $v' = \phi'_y$ in the equation. The solution to the Poisson equation has the form

$$u' = u'_p + \frac{1}{2\pi}\int_{-\infty}^{\infty}\int_{-\infty}^{\infty} u'\frac{\partial u'}{\partial \xi}\left[\frac{x-\xi}{(x-\xi)^2+(y-\eta)^2}\right]d\xi\, d\eta \tag{903}$$

$$v' = v'_p + \frac{1}{2\pi}\int_{-\infty}^{\infty}\int_{-\infty}^{\infty} u'\frac{\partial u'}{\partial \xi}\left[\frac{y-\eta}{(x-\xi)^2+(y-\eta)^2}\right]d\xi\, d\eta \tag{904}$$

Again, the solution must be made to satisfy the boundary conditions. At $y = 0$ we have

$$u'(x,0) = u_p(x,0) + \frac{1}{2\pi}\int_{-\infty}^{\infty}\int_{-\infty}^{\infty} u'\frac{\partial u'}{\partial \xi'}\left[\frac{x-\xi}{(x-\xi)^2+\eta^2}\right]d\xi\, d\eta \tag{905}$$

$$v'(x,0) = v_p(x,0) - \frac{1}{2\pi}\int_{-\infty}^{\infty}\int_{-\infty}^{\infty} u'\frac{\partial u'}{\partial \xi}\left[\frac{\eta}{(x-\xi)^2+\eta^2}\right]d\xi\, d\eta \tag{906}$$

If the profile is symmetric about the x axis, the integral term in equation (906) is seen to vanish and hence the boundary conditions are satisfied by use of equation (902). For nonsymmetric profiles or symmetric profiles at angle of attack, the boundary conditions can no longer be so easily satisfied. Since u' is unknown, the equation (905) is an integral equation for the unknown function $u'(x', y')$.

In certain cases, the function $V_0(x)$ takes on a particularly simple form. For example, in the case of an airfoil of unit formed of two intersecting circular arcs,

$$v'_p(x,0) = V_0(x) = -4\tau' x \qquad \tau' = \frac{\tau}{\beta\left(\dfrac{c^*}{u_\infty} - 1\right)} \tag{907}$$

where τ the thickness ratio. The Poisson integral for u' at $y = 0$ may then be evaluated:

$$u'_p(x,0) = \frac{4}{\pi}\tau'\left[1 - x\ln\frac{1+2x}{1-2x}\right] \tag{908}$$

While the Poisson integrals are taken from $-\infty$ to $+\infty$, it should be noted that $V_0(x)$ has nonzero values only over the length of the airfoil. It will be observed that equation (908) does not give a correct result near the leading and trailing edges of the airfoil, where the assumption of small perturbations is violated since in reality these must be stagnation points. As in the Prandtl–Glauert rule, however, the solution remains valid except in a small region near the front and rear stagnation points.

The velocity distribution (908) will be observed to be symmetric and hence does not admit the possibility of a discontinuity (shock). Recall, however, equation (858)

$$u'_1 - \frac{u_1'^2}{2} = u'_2 - \frac{u_2'^2}{2} \tag{858}$$

across a shock. From equation (903) it follows that

$$u' - \frac{u'^2}{2} = u'_p - \frac{1}{2\pi}\int_{-\infty}^{\infty}\int_{-\infty}^{\infty}\frac{u'^2}{2}\left\{\frac{(\xi - x)^2 - (\eta - y)^2}{[(\xi - x)^2 + (\eta - y)^2]^2}\right\}d\xi\, d\eta \quad (909)$$

The integral on the right-hand side of equation (909) is thus continuous even if a shock is present. To be sure, the implicit assumption of a normal shock has been made, but when a shock appears over an airfoil in transonic flow this is usually a good assumption. Along the x axis we have then

$$u'(x,0) - \frac{[u'(x,0)]^2}{2} = u'_p(x,0) - \frac{1}{2\pi}\int_{-\infty}^{\infty}\int_{-\infty}^{\infty}\frac{u'^2}{2}\left\{\frac{(\xi - x)^2 - \eta^2}{[(\xi - x)^2 + \eta^2]^2}\right\}d\xi\, d\eta \quad (910)$$

The most obvious way to solve equation (909) would be to use the linearized solution u'_p in the integral and iterate. However, this procedure will be seen to be only a modification of the Rayleigh–Janzen method which does not give correct results when supersonic flow is present. On the other hand, the value of the integral in equations (909) and (910) is not very sensitive to variations in the function $u'(x, y)$. Thus, even a reasonable approximation should give good results. (A similar situation may be recalled in the case of the momentum integral method in boundary layer theory.) A good approximation to the function u' in the integral is found to be

$$u' = \frac{u_0}{\left(1 + \frac{y}{b}\right)^2} \quad (911)$$

The parameter $b(x)$ may be determined by applying the irrotationality condition. If $h(x)$ is the half thickness distribution of a plane, symmetric profile,

$$h = \int_{-\infty}^{x}\frac{V_0(x)}{u_\infty}\,dx \quad (912)$$

where $v_0(x)$ is the vertical velocity component at the surface of the profile. Now let H be defined by

$$H = \int_{-\infty}^{x} V_0(x)\,dx = \frac{h}{\beta\left(\frac{c^*}{u_\infty} - 1\right)} \quad (913)$$

The radius of curvature r of the surface of the profile is given by

$$\frac{1}{r} = -\frac{d^2h}{dx^2} = -\frac{d}{dx}\left(\frac{V_0}{u_\infty}\right) \quad (914)$$

Now define R by

$$\frac{1}{R} = -\frac{d^2H}{dx^2} = -\frac{dV_0}{dx} = \frac{1}{\beta r\left(\frac{c^*}{u_\infty} - 1\right)} \quad (915)$$

The irrotationality condition requires that

$$\left(\frac{\partial u'}{\partial y}\right)_0 = \frac{dV_0}{dx} \tag{916}$$

Thus,

$$\left(\frac{\partial u'}{\partial y}\right)_0 = -2\frac{u_0}{b} = \frac{dV_0}{dx} = \frac{d^2H}{dx^2} = -\frac{1}{R} \tag{917}$$

or

$$b = 2u_0 R = \frac{-2u_0}{dV_0/dx} = \frac{-2u_0}{d^2h/dx^2} \tag{918}$$

In fact, it is found that good results may be obtained by taking b constant and equal to

$$b = u_p'(x_m, 0)R(x_m) \tag{919}$$

where x_m is the point of maximum value of $u_p'(x_m, 0)$.

Inserting the approximation (911) in the integral in equation (910), we find

$$\frac{1}{2\pi}\int_{-\infty}^{\infty}\int_{-\infty}^{\infty}\frac{u'^2}{2}\left\{\frac{(\xi - x)^2 - \eta^2}{[(\xi - x)^2 + \eta^2]^2}\right\}d\xi\, d\eta = \int_{-\infty}^{\infty}\frac{u_0^2}{2b}E\left(\frac{\xi - x}{b}\right)d\xi \tag{920}$$

where

$$E(\pm\zeta) = \frac{4}{\pi(1+\zeta^2)^5}\left\{\pm\frac{\pi}{2}\zeta(5 - 10\zeta^2 + \zeta^4) - (1 - 10\zeta^2 + 5\zeta^4)\ln|\zeta| - \frac{1}{12}(1+\zeta^2)(25 - 71\zeta^2 - \zeta^4 - \zeta^6)\right\} \tag{921}$$

where ζ is simply the argument of the function. Equation (910) then becomes

$$u_0 - \frac{u_0^2}{2} = u_p'(x, 0) - \int_{-\infty}^{\infty}\frac{u_0^2}{2b}E\left(\frac{\xi - x}{b}\right)d\xi \tag{922}$$

Even with E and u_p' known, this equation is still difficult to solve for u_0. Because the integral is relatively insensitive to the exact shape of u_0, an approximation may be introduced there of the form

$$u_0 = u_m g\left(\frac{x - x_m}{l}\right) \tag{923}$$

where l is the length over which the approximation is taken, u_m the maximum velocity over the airfoil, and $x = x_m$ the point where the maximum velocity occurs. We shall assume a parabola for the function g when no shock is present and half a parabola in case a shock is present. Substituting (923) in the integral of equation (922) we get

$$\int_{-\infty}^{\infty}\frac{u_0^2}{2b}E\left(\frac{\xi - x}{b}\right)d\xi = \int_{-\infty}^{\infty}\frac{u_m^2}{2b}\left[g\left(\frac{\xi - x_m}{l}\right)\right]^2 E\left(\frac{\xi - x}{b}\right)d\xi = \frac{u_m^2}{2}f_1\left(\frac{x - x_m}{l}, \frac{b}{l}\right) \tag{924}$$

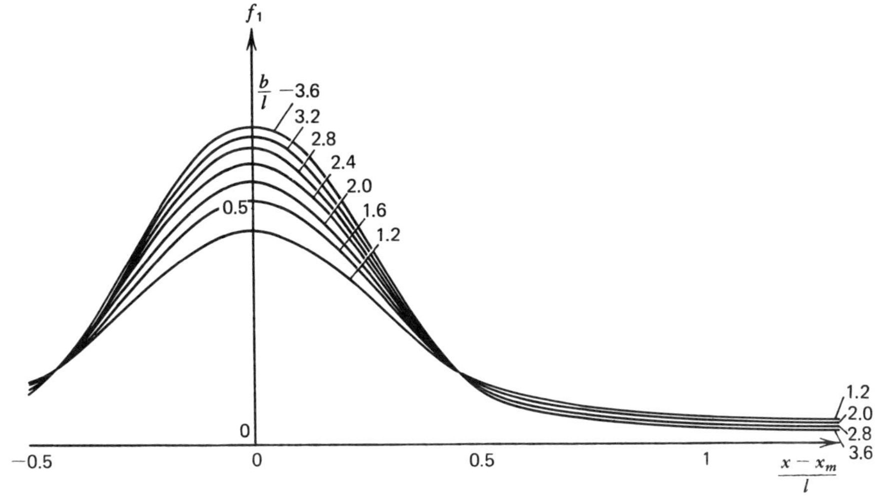

Figure 203 Function f_1, equation (924) from [67].

The function f_1 is plotted in Figure 203, where the value of b has been taken from equation (919). If the approximation (923) is extended to include the term $u_0^2/2$ in equation (922) we get

$$\frac{u_0}{2} - \int_{-\infty}^{\infty} \frac{u_0^2}{2b} E\left(\frac{\xi - x}{b}\right) d\xi = \frac{u_m^2}{2} f_2\left(\frac{x - x_m}{l}, \frac{b}{l}\right) = u_0 - u_p' \tag{925}$$

The function f_2 is plotted in Figure 204:

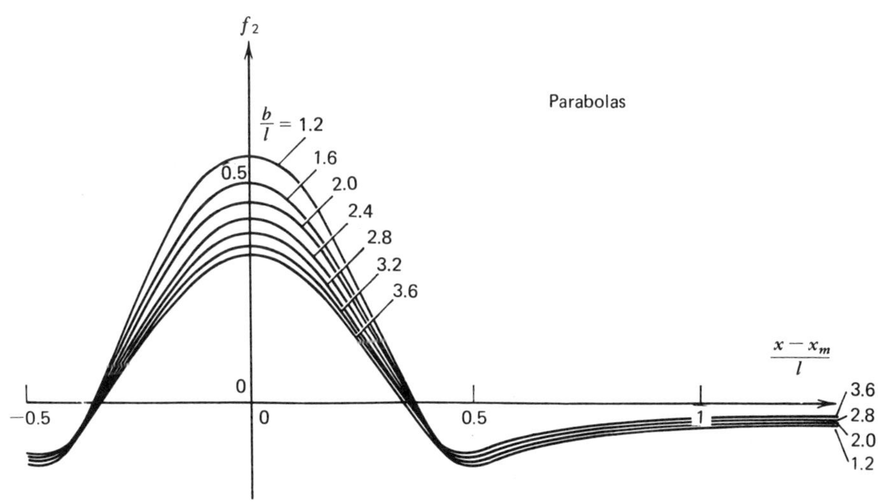

Figure 204 Function f_2, equation (925) [67].

We now have two possible methods of finding the velocity distribution over the airfoil. From (924) and (922) we have

$$u_0 - \frac{u_0^2}{2} = u'_p(x,0) - \frac{u_m^2}{2} f_1\left(\frac{x - x_m}{l}, \frac{b}{l}\right) \tag{926}$$

and from equation (925),

$$u_0 = u'_p(x,0) + \frac{u_m^2}{2} f_2\left(\frac{x - x_m}{l}, \frac{b}{l}\right) \tag{927}$$

In the vicinity of the sonic point, however, we observe that small fluctuations in the stream density $u_0 - u_0^2/2$ cause large fluctuations in the velocity. We conclude, therefore, that equation (927) is more suitable for finding the velocity distribution than equation (926).

Example: Application of the Method to a Circular Arc Airfoil

We will now illustrate the application of the method by applying it to a simple example—an airfoil consisting of two intersecting circular arcs (Figure 205).

Figure 205 Circular arc airfoil.

Case 1. Shock-Free-Flow. For the following, all lengths will be normalized with respect to the chord of the airfoil. The values of $u'_p(x,0)$ and $v'_p(x,0)$ have already been given in equations (907) and (908). A plot of $u'_p(x,0)$ is given in Figure 206. We observe that $u'_p(x,0) = 0$ at $x_1 = 0.085$ and $x_2 = 0.915$. We now

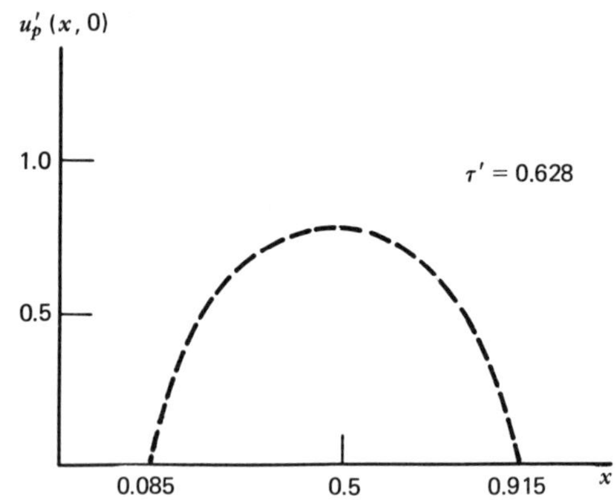

Figure 206 Poisson integral for circular arc air foil [67].

approximate the actual velocity distribution by a parabola intersecting the x axis at x_1 and x_2. The maximum velocity of the parabola will occur at $x_m = 0.5$

and $l = x_2 - x_1 = 0.83$. The velocity distribution becomes, then

$$u_0 = u'_p(x, 0) + \frac{u_m^2}{2} f_2\left(\frac{x - 0.5}{0.83}\right) \tag{927}$$

where it may be noted that

$$\frac{x - 0.5}{0.83} = \frac{x - x_m}{l} \tag{928}$$

The value of u_m is fixed by equation (927); for,

$$u_m = u'_{pm} + \frac{u_m^2}{2} f_2(0) \tag{929}$$

Equation (929) has two solutions for u_m of which the smaller is the correct one. The solutions for u_m are real only if $u_{pm} < 1/[2f_2(0)]$. Comparison with equation (908) shows that this places a limit on the magnitude of τ' for which the solution is valid. Since shock-free flow is assumed, the condition $u_{pm} = 1/[2f_2(0)]$ may also be regarded as the upper limit for which shock-free flow is possible. The value of b may be found from equation (919). For the circular arc profile, $R = 1/(4\tau')$ and hence $b = 2/\pi$. $l = x_2 - x_1 = 0.83$. The values of f_2 may then be read off Figure 204 for the given b/l. With $u'_p(x, 0)$ known from equation (908), the solution for u_0 is determined. The result has been plotted in Figure 207 for $\tau' = 0.707$.

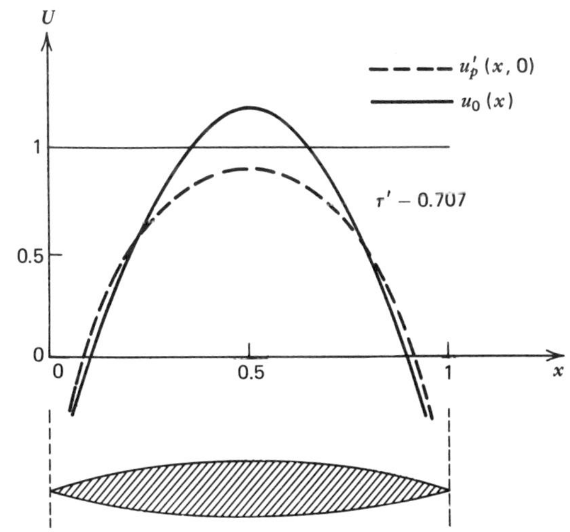

Figure 207 Velocity distribution over circular arc airfoil [67].

Case 2. Flow with Shock. For the case of flow with a shock, the velocity distribution in equation (922) may be approximated by two half parabolas, one before the shock and one after the shock, as shown in Figure 208. We now let

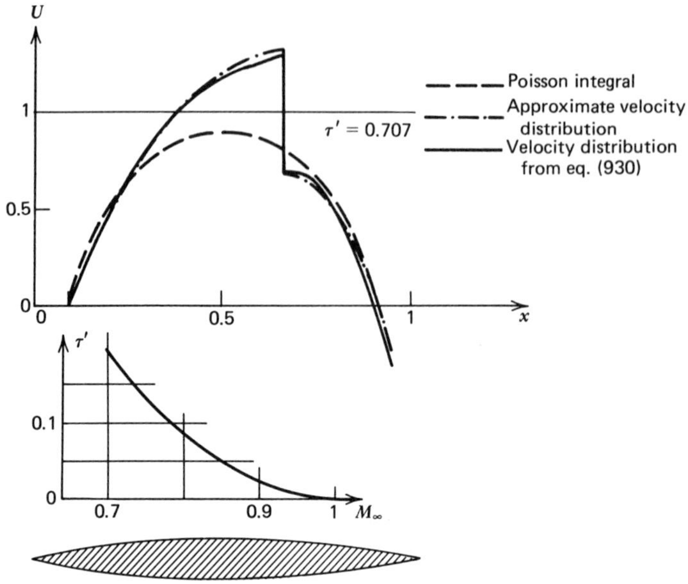

Figure 208 Velocity distribution over circular arc airfoil with shock [67].

$l = x_m$, where $u = u_m$, the velocity just before the shock, at $x = x_m$. The first half parabola intersects the x axis at $x = 0.085$, where $u'_p(x, 0) = 0$, and ends at $x = x_m$. The second half parabola begins at $x = x_m$ and ends at $x = 0.915$ where $u'_p(x, 0)$ is again zero. The second half parabola begins at its maximum where the velocity equals the velocity just downstream of the shock. From equation (858), it may be seen that the value of u_0 at this point is equal to $2 - u_m$. The assumption that u_0 may be approximated by these two half parabolas gives us a new value for f_2 in equation (925), which is plotted in Figure 209. The velocity distribution is then given by

$$u_0 = u'_p(x, 0) + \frac{u_m^2}{2} f_2\left(\frac{x}{l} - 1, \frac{2}{\pi l}\right) \tag{930}$$

where the f_2 is now read from Figure 209. This leaves us with two unknowns: u_m and l. These are, however, determined by the assumed form of the solution (930). At $x = x_m$ we have

$$u_m = u'_{pm} + \frac{u_m^2}{2} f_2\left(0, \frac{2}{\pi l}\right) \tag{931}$$

Now let α be any number between zero and 1. At $x = \alpha l$,

$$u_0 = u_m[1 - (1 - \alpha)^2] \tag{932}$$

simply by geometry. Thus we have

$$u_m[1 - (1 - \alpha)^2] = u'_p(\alpha l, 0) + \frac{u_m}{2} f_2\left(\alpha - 1, \frac{2}{\pi l}\right) \tag{933}$$

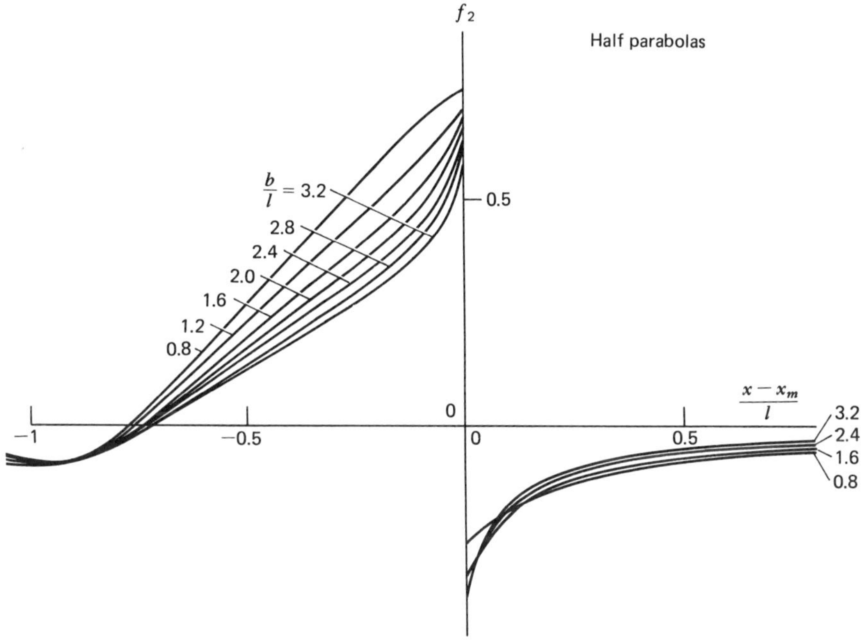

Figure 209 Function f_2, equation (925), for flow with shock [67].

Since $u_p'(x, 0)$ is presumed known, equation (931) and (933) may be solved simultaneously for u_m and l thus determining the shock location and the velocity before the shock. Equation (930) together with Figure 209 thus give the solution for the velocity distribution which is shown by the solid line in Figure 208.

It will be observed that the method gives two solutions, in the example just discussed, for the same boundary conditions—one with shock and one without. Indeed the method will give two solutions for all free stream Mach numbers above the critical value up to the point when $u'_{pm} = 1/[2f_2(0)]$ (as explained above), beyond which only the shock solution is possible. In practice it is observed that, for the type of profile under consideration, a shock will appear whenever the free stream Mach number exceeds the critical Mach number by more than a small amount. Both experimental and theoretical considerations indicate, however, that such flows may be unstable. Thus, both of the above solutions may be perfectly valid, given the assumption of steady, inviscid flow. In order to determine what actually happens in practice, both viscous and unsteady effects would have to be included.

Comparison with experimental results indicate that the solution of Figure 208 is only fair. The accuracy of the solution tends to improve with slightly higher thickness ratio. The chief difficulty, however, lies in the choice of the assumed form of the solution and its use, not only in the integral of equation (925), but in the term $u_0^2/2$ outside the integral as well. Improved results may be obtained dividing the profile into a series of increments and using equation (926) away from the sonic point and equation (927) in the region around the sonic point. Alternatively, better approximations to the actual velocity profile may be used than simple parabolas. This latter technique must rely to some extent on experimental data and therefore limits the value of the method. It also increases the complexity of the calculation. Oswatitsch [67] has applied these modifications to the flow over a series of NACA profiles with reasonably good results.

An iterative procedure which represents a slight improvement of the original method of Oswatitsch was worked out by Spreiter and Alksne [68]. The method has been extended to the case of lifting airfoils in supercritical, shock free flow by Nixon [69, 70, 71], Nixon and Patel [72], Nixon and Hancock [73] and Hansen [74]. The same authors have also proposed alternate adaptations of the integral method to flow over symmetric airfoils with shocks. Norstrud [75] has extended the method to three-dimensional flow over wings. Frohn [76] has proposed a modified integral method for flow with shocks. The details of these methods are too lengthy and complex to be given here. All require the use of computers.

In general it may be said that the integral method gives good results for transonic flow without shocks, but is less satisfactory when shocks are present. Proponents of the integral method believe that with further development, this problem will be overcome.

Hodograph Method

The other principal method of dealing with the problem of transonic flow over airfoils is the hodograph method. The modern application of this method to airfoils has been developed principally by Boerstel et al. in Holland [77], and Garabedian et al. in the United States [78]. We will follow here the method of Garabedian and Korn [79].

In the case of the integral method, the shape of the airfoil was presumed given and the method solved for the flow around the airfoil. This is called a direct method. In the hodograph method, the velocity distribution is specified at certain boundaries and the shape of the corresponding airfoil is calculated. This is called an inverse method.

As an aid in understanding the present method, a review of certain aspects of complex function theory will be helpful.

Suppose x and y are two complex variables $x_1 + iy_1$ and $x_2 + iy_2$ where $x_i, y_i, i = 1, 2$, are real. We now define z and z^* by

$$z = x + iy \qquad z^* = x - iy \tag{934}$$

Notice that when x_2 and y_2 are zero, $\bar{z} = z^*$, where the bar indicates the complex conjugate. Now let $\xi = F(z)$, $\eta = F(z^*)$, where $F(z)$ is an analytic function which is real when z is real. Then it follows from the Schwartz reflection principle that

$$\overline{F(\bar{z})} = F(z) \tag{935}$$

Hence, when x and y are real, $\xi = \bar{\eta}$ and vice versa. A pair of complex functions ξ and η of x and y for which x and y are real when $\xi = \bar{\eta}$ are called conjugate coordinates.

Suppose now that a function $G(\xi, \eta)$, where $\xi = \xi(x, y)$ and $\eta = \eta(x, y)$ are conjugate coordinates, is analytic in ξ and η. Then $G(\xi, \eta)$ may be written

$$G(\xi, \eta) = R(\xi, \eta) + iI(\xi, \eta) \tag{936}$$

where R and I are real on the real x, y plane. Thus if $G(\xi, \eta)$ is real when x and y are real, I must vanish identically there. But I can be written

$$I = \frac{1}{2i}[G(\xi, \eta) - \overline{G(\bar{\eta}, \bar{\xi})}] \tag{937}$$

from which it follows that

$$G(\xi, \eta) = \overline{G(\bar{\eta}, \bar{\xi})} \tag{938}$$

In particular, it follows that $x(\bar{\eta}, \bar{\xi}) = \overline{x(\xi, \eta)}$ and $y(\bar{\eta}, \bar{\xi}) = \overline{y(\xi, \eta)}$. It may also be shown that

$$G_\xi(\xi, \eta) = \overline{G_\eta(\bar{\eta}, \bar{\xi})} \tag{939}$$

A function which is real on the real x, y plane will be called a "real" function.

Recall now that the equation for the slope of the physical characteristics may be written

$$\left(\frac{dy}{dx}\right)_\pm = \frac{-\dfrac{uv}{c^2} \pm \sqrt{\dfrac{u^2 + v^2}{c^2} - 1}}{1 - \dfrac{u^2}{c^2}} \tag{501}$$

Because the radical became imaginary when $M < 1$, we rejected the use of this equation for subsonic flow. Suppose, however, that we now admit complex values. If x and y are allowed to be complex, i.e., $x = x_1 + ix_2$, $y = y_1 + iy_2$ then equation (501) will continue to have meaning in the subsonic range. Recall, moreover, equation (502)

$$\left(\frac{du}{dv}\right)_\pm = \mp \tan\alpha \qquad \left(\frac{dy}{dx}\right)_\pm = \pm \tan\alpha \tag{502}$$

Thus we have

$$\left(\frac{du}{dv}\right)_+ = \left(\frac{dy}{dx}\right)_- \quad \text{and} \quad \left(\frac{du}{dv}\right)_- = \left(\frac{dy}{dx}\right)_+ \tag{940}$$

If dy/dx is allowed to assume complex values, then du/dv, and hence u and v must also be allowed to assume complex values. Because u and v are assumed to be continuously differentiable function of x and y in the real domain, the function $u(x, y)$, $v(x, y)$ defined by equation (940) in the complex domain can be considered analytic functions. This means in particular that u and v will be functions of only the two independent complex variables x and y rather than of the four variables x_1, x_2, y_1, y_2 independently, and all the ordinary rules of differentiation with respect to x and y will apply.

Integrating equation (501) we get two functions of complex x versus complex y – one for the plus sign and one for the minus sign. These functions may be written as

$$\xi(x, y) = \text{constant} \qquad \eta(x, y) = \text{constant} \tag{941}$$

where ξ and η represent the characteristic coordinates in the complex domain, keeping in mind that ξ and η are complex and that ξ and η constant means their real parts are constant and their imaginary parts are constant. Since any point in the real or complex domain may be specified as the intersection of two characteristics or, more precisely, by specifying ξ and η, it follows that

$$x = x(\xi, \eta) \qquad y = y(\xi, \eta) \tag{942}$$

Thus

$$dy = \frac{\partial y}{\partial \xi} d\xi + \frac{\partial y}{\partial \eta} d\eta \qquad dx = \frac{\partial x}{\partial \xi} d\xi + \frac{\partial x}{\partial \eta} d\eta \tag{943}$$

Let us now specify the function $\eta =$ constant as the one on which $dy/dx = (dy/dx)_+$ in equation (501) and the surface $\xi =$ constant as the one on which $dy/dx = (dy/dx)_-$ in equation (501). Moreover, we will designate $(dy/dx)_+$ by λ_+ and $(dy/dx)_-$ by λ_-. Then on $\eta =$ constant,

$$y_\xi + \lambda_+ x_\xi = 0 \tag{944}$$

while for $\xi =$ constant,

$$y_\eta + \lambda_- x_\eta = 0 \tag{945}$$

where the subscripts denote partial differentiation. Since u and v are functions of x and y it follows that u and v may also be considered as functions of ξ and η. Mindful of equation (940), it follows that

$$u_\xi - \lambda_- v_\xi = 0 \tag{946}$$

$$u_\eta - \lambda_+ v_\eta = 0 \tag{947}$$

An examination of equation (501) shows that when u and v are real, λ_+ and λ_- are complex conjugates. ξ and η are conjugate coordinates, because when u and v are real,

$$\xi(u, v) = \bar{\eta}(u, v) \tag{948}$$

In order to find a solution of equations (944), and (945), we specify the boundary conditions on two characteristics in the real domain, $\xi = \xi_0$ and $\eta = \eta_0$ in the form

$$x(\xi, \eta_0) = f(\xi) + g(\eta_0) \tag{949}$$

$$x(\xi_0, \eta) = g(\eta) + f(\xi_0) \tag{950}$$

Since we are seeking the solution to a real physical problem, we would like to find real solutions to the system (944), (945) with the boundary conditions (949), (950). There are two ways to do this. One is to find a complex solution. Because equations (944) and (945) are linear, it can be shown that when $\xi = \bar{\eta}$, i.e. when u and v are real, the real part of the complex solution will be a solution which satisfies the equations and the boundary conditions in the real domain. Alternatively, we may try to pick the boundary conditions in such a way that the solution will be real in the real domain. Suppose $x(\xi,\eta)$, $y(\xi,\eta)$ are "real" functions. Recall $\xi_0 = \bar{\eta}_0$ and $\eta_0 = \bar{\xi}_0$ since ξ_0 and η_0 lie in the real domain. Also, recall from equation (938) that $x(\bar{\xi},\bar{\eta}) = \overline{x(\eta,\xi)}$ if x is a "real" function. It follows that $x(\xi,\eta_0) = \overline{x(\bar{\eta}_0,\bar{\xi})}$. Hence, if $x(\xi,\eta_0) = f(\xi) + g(\eta_0)$, then $\overline{x(\bar{\eta}_0,\bar{\xi})} = \overline{f(\bar{\eta}_0)} + \overline{g(\bar{\xi})}$. Now $x(\xi,\eta)$ will be real when $\xi = \bar{\eta}$, i.e., when u and v are real, if the imaginary part of x is zero there. But the imaginary part of x may be written

$$Ix(\xi,\eta_0) = \frac{1}{2i}\left[x(\xi,\eta_0) - \overline{x(\bar{\eta}_0,\bar{\xi})}\right] = \frac{1}{2i}\left[f(\xi) + g(\eta_0) - \overline{f(\bar{\eta}_0)} - \overline{g(\bar{\xi})}\right] \quad (951)$$

This will be zero if $f(\xi) = \overline{g(\bar{\xi})}$. Hence if $f(\xi) = \overline{g(\bar{\xi})}$, x will be a "real" function. It follows that u must then also be a "real" function. Such a condition may appear restrictive, but in fact the boundary conditions may always be expressed in this way if a "real" solution exists. For: suppose the boundary conditions are written in the form $x(\xi,\eta_0) = f(\xi) + g(\eta_0)$ where $f(\xi) \neq \overline{g(\bar{\xi})}$. Then the boundary conditions may be rewritten

$$x(\xi,\eta_0) = h(\xi) + j(\eta_0) \qquad x(\xi_0,\eta) = h(\xi_0) + j(\eta) \quad (952)$$

where

$$h(\xi) = \frac{f(\xi) + \overline{g(\bar{\xi})}}{2} \qquad j(\eta) = \frac{\overline{f(\bar{\eta})} + g(\eta)}{2} \quad (953)$$

The functions h and j satisfy the condition that

$$h(\xi) = \overline{j(\bar{\xi})}$$

thus assuring that the solution will be real in the real domain—that is, x and y will be real when u and v are. Moreover, the real part of the boundary condition (949) may be written

$$R[x(\xi,\eta_0)] = \tfrac{1}{2}\left[f(\xi) + \overline{g(\eta_0)} + f(\bar{\eta}_0) + \overline{g(\bar{\xi})}\right] \quad (954)$$

The boundary condition (951) may be written

$$x(\xi,\eta_0) = \tfrac{1}{2}\left[f(\xi) + \overline{g(\bar{\xi})} + \overline{f(\bar{\eta}_0)} + g(\eta_0)\right] \quad (955)$$

which is seen to be identical to (954). Thus, the real part of the solution of equations (944, 945) with boundary conditions (949, 950) satisfies the same boundary conditions as the "real" solution of the same equations with the boundary conditions (952, 953). Moreover, the real part of the first solution is also a solution of the equations (944, 945) in the real domain. Thus, the two solutions must be identical.

In order to illustrate the extension of the method of characteristics to subsonic flow, let us take a simple example: incompressible flow. For incompressible flow, the equations of continuity and irrotationality become

$$\begin{aligned} u_x + v_y &= 0 \\ v_x - u_y &= 0 \end{aligned} \tag{956}$$

Applying the method of characteristics to these two first-order equations we write

$$\begin{aligned} du &= u_x\,dx + u_y\,dy \\ dv &= \qquad\qquad\qquad v_x\,dx + v_y\,dy \\ 0 &= u_x \qquad\qquad\qquad\qquad + v_y \\ 0 &= \qquad - u_y + v_x \end{aligned} \tag{957}$$

This gives us a system of four equations and four unknown $-\, u_x, u_y, v_x, v_y$. Solving for v_x, we get

$$v_x = \frac{\begin{vmatrix} dx & dy & du & 0 \\ 0 & 0 & dv & dy \\ 1 & 0 & 0 & 1 \\ 0 & -1 & 0 & 0 \end{vmatrix}}{\begin{vmatrix} dx & dy & 0 & 0 \\ 0 & 0 & dx & dy \\ 1 & 0 & 0 & 1 \\ 0 & -1 & 1 & 0 \end{vmatrix}} \tag{958}$$

As in the supersonic case we obtain the characteristic directions by setting the denominator equal to zero, from which it follows that

$$\lambda_{\pm} = \pm i = \pm\sqrt{-1} \tag{959}$$

Thus equations (944, 945) and (946, 947) become

$$\begin{aligned} u_\xi - iv_\xi &= 0 \qquad & u_\eta - iv_\eta &= 0 \\ y_\xi - ix_\xi &= 0 \qquad & y_\eta + ix_\eta &= 0 \end{aligned} \tag{960}$$

Clearly, $w = u - iv$ and $w^* = u + iv$ are characteristic coordinates here. The reader will recognize w as the complex velocity from incompressible flow theory, keeping in mind the fact that in the present case, u and v may be complex. At the same time it will be noted that w and w^* are conjugate coordinates since $w = \bar{w}^*$ when u and v are real.

Consider now the flow over a circular cylinder. Recall from the theory of incompressible flow that the complex potential for this case was given by

$$F(z) = \phi + i\psi = z + \frac{1}{z} + 2ik \ln z \qquad k = \frac{\Gamma}{2\pi} \tag{961}$$

where $z = x + iy$, Γ represents the circulation around the body, and the velocity has been nondimensionalized with respect to u_∞. The complex velocity $w = u - iv$ is then

$$w = \frac{dF}{dz} = 1 - \frac{1}{z^2} + \frac{2ik}{z} \tag{962}$$

the inverse of which is

$$z = [\sqrt{1 - w - k^2} + ik]^{-1} \tag{963}$$

We may now define a new function, $\eta(w)$ such that

$$\eta = \sqrt{1 - w - k^2} + ik \tag{964}$$

Then for the circular cylinder,

$$z = \frac{1}{\eta} \tag{965}$$

Returning now to our method of characteristics, notice that if $w = \text{constant}$ is a characteristic, then $f(w) = \text{constant}$ will also be a characteristic, where $f(w)$ is any analytic function of w. Thus, $\eta = \text{constant}$ will also be a characteristic. We will now show that if η is a characteristic coordinate, then $\xi = \bar{\eta}$ is a conjugate characteristic coordinate. The proof will not be restricted to the present case, but will be made more general.

Suppose $\xi(w, w^*)$ is a complex analytic characteristic coordinate for a subsonic flow. We will prove that $\xi(\bar{w}^*, \bar{w})$, the complex conjugate of η when u and v are real, is then a conjugate characteristic coordinate. Let

$$\begin{aligned} \lambda_+(w, w^*) &= \lambda_1(w, w^*) + i\lambda_2(w, w^*) \\ \lambda_-(w, w^*) &= \lambda_1(w, w^*) - i\lambda_2(w, w^*) \end{aligned} \tag{966}$$

where λ_1 and λ_2 are real functions. It follows that

$$\lambda_+(w, w^*) = \overline{\lambda_-(\bar{w}^*, \bar{w})} \tag{967}$$

Since u and v may be written as functions of w and w^* it follows from equation (946) that

$$u_\xi(\bar{w}^*, \bar{w}) = \lambda_-(\bar{w}^*, \bar{w}) v_\xi(\bar{w}^*, \bar{w}) \tag{968}$$

from which it follows that

$$\overline{u_\xi(w, w^*)} = \overline{\lambda_+(w, w^*)}\, \overline{v(w, w^*)} \tag{969}$$

which is equivalent to

$$u_\xi = \lambda_+(w, w^*) v_\xi \tag{970}$$

which is equation (947), demonstrating that $\bar{\xi}$ (or an analytic function of $\bar{\xi}$) is the other characteristic coordinate.

Returning to our example of a circular cylinder, then, we see that if η as defined in equation (964) is one characteristic coordinate, then $\xi = \bar{\eta}$ represents the other characteristic coordinate.

We now treat the problem as a characteristic initial value problem; that is, we suppose that initial conditions are given along two characteristics: $\xi = \xi_0$ and $\eta = \eta_0$. In particular, we suppose that the initial values are given in the following form:

$$\begin{aligned} x(\xi, \eta_0) &= f(\xi) + g(\eta_0) \\ x(\xi_0, \eta) &= f(\xi_0) + g(\eta) \end{aligned} \tag{971}$$

where $f(\xi) = \overline{g(\bar{\xi})}$ thus assuring that x is a "real" function; that is, x is real when u and v are real. The problem is thus to satisfy the equations

$$y_\xi - ix_\xi = 0, \qquad y_\eta + ix_\eta = 0 \tag{972}$$

subject to the boundary conditions (971). Unlike characteristic solutions in the supersonic range which must be obtained at discreet points by stepwise integration, the present problem may be solved in closed form by inspection. The solution is seen to be

$$\begin{aligned} x(\xi, \eta) &= \overline{g(\bar{\xi})} + g(\eta) \\ y(\xi, \eta) &= i\overline{g(\bar{\xi})} - ig(\eta) \end{aligned} \tag{973}$$

When u and v are real, $\xi = \bar{\eta}$, and $x(\xi, \eta) = x(\bar{\eta}, \eta)$ is real. But $x(\bar{\eta}, \eta) = \overline{g(\eta)} + g(\eta)$. Hence $\overline{g(\eta)}$ must be the complex conjugate of $g(\eta)$ and hence

$$x(\bar{\eta}, \eta) = 2\,\mathrm{Re}[g(\eta)] \tag{974}$$

Similarly,

$$y(\bar{\eta}, \eta) = 2\,\mathrm{Im}[g(\eta)] \tag{975}$$

Now recall for the circular cylinder , $z = 1/\eta = x + iy$ where x and y are real. From equations (974) and (975)

$$x + iy = 2\,\mathrm{Re}\, g(\eta) + 2i\,\mathrm{Im}\, g(\eta) = 2g(\eta) = z = \frac{1}{\eta} \tag{976}$$

Thus if we take $g(\eta) = 1/2\eta$, the method of characteristics will give us the right answer.

One should not conclude that this example is invalid because the answer was known before hand. The example was given only to illustrate the method. In practice, one may postulate different forms of $g(\eta)$ and see what sort of flow they yield in the physical plane. For the case of compressible flow, a good starting point will be found to be taking $g(\eta)$ for the corresponding incompressible flow and seeing how the flow is modified by compressibility.

As a further example we may consider the flow over an elliptic cylinder without circulation. In this case, the complex potential is given by

$$F(z) = \hat{z} + \frac{1}{\hat{z}} \qquad \text{where} \qquad z = \hat{z} + \frac{E}{\hat{z}} \tag{977}$$

where the ellipse varies from circle to straight line as E varies from zero to 1. We now define η as before but with $k = 0$ since there is no circulation and the solution must reduce to that about a circle when $E \to 0$. Thus, let $\eta = \sqrt{1 - w}$ and hence

$$z = \frac{1}{\eta}(E\eta^2 + \beta^2)^{1/2} + \frac{E\eta}{(E\eta^2 + \beta^2)^{1/2}}, \qquad \beta^2 = 1 - E \tag{978}$$

The characteristics method will thus give the right answer for

$$g(\eta) = \frac{(E\eta^2 + \beta^2)^{1/2}}{2\eta} + \frac{E\eta}{2(E\eta^2 + \beta^2)^{1/2}} \tag{979}$$

Application of the method to the compressible case. In practice it will be found more expedient to work in polar coordinates. Let $q^2 = u^2 + v^2$ and $\theta = \arctan v/u$. Recall now the Prandtl–Meyer equation

$$\theta = \pm f(q) = \pm \left\{ \frac{1}{2}\sqrt{\frac{\gamma + 1}{\gamma - 1}} \arcsin\left[(\gamma - 1)\frac{q^2}{c^{*2}} - \gamma \right] + \frac{1}{2} \arcsin\left[(\gamma + 1)\frac{c^{*2}}{q^2} - \gamma \right] \right\} + C \tag{980}$$

As was shown in Chapter 4, this is simply the equation for the hodograph characteristics in polar coordinates. The hodograph characteristics in supersonic flow are thus given by

$$f(q) + \theta = \text{constant}, \qquad f(q_0) - \theta = \text{constant} \tag{981}$$

For $q < c^*$ in the real domain, $f(q)$ becomes complex. Following our method of extending the method of characteristics to the subsonic domain, we now allow q, θ, u, and v to become complex. In both the subsonic, complex domain, and the supersonic, real domain, the characteristics will then be given by equation (980). Since any analytic function of a characteristic coordinate will also be a characteristic coordinate, we may seek a function of $f(q) + \theta$, which reduces to the complex velocity w in the incompressible case, and a function of $f(q) - \theta$ which reduces to the conjugate of the complex velocity, w^*. Using these functions in the characteristic coordinates in a compressible solution for the flow about an elliptic cylinder, for example, we may find a solution which reduces to the incompressible solution found in the previous section as $c \to \infty$. Thus we seek functions $s = F(f(q) + \theta)$, $t = G(f(q) - \theta)$, such that

$$\lim_{c \to \infty} F(f(q) + \theta) = w = qe^{-i\theta} \tag{982}$$

$$\lim_{c \to \infty} G(f(q) - \theta) = w^* = qe^{i\theta} \tag{983}$$

This may be accomplished by setting

$$s = Ce^{-if(q)}e^{-i\theta}, \qquad t = Ce^{-if(q)}e^{i\theta} \tag{984}$$

Now define $h(q)$ by the equation

$$f(q) = i \ln \frac{h(q)}{C} \tag{985}$$

Then

$$h(q) = Cq\{[\sqrt{\phi_1^2 - 1} - \phi_1]^{1/\mu}[\sqrt{\phi_2^2 - q^4} + \phi_2]\}^{-1/2} \tag{986}$$

where

$$\phi_1 = \gamma - \frac{\gamma - 1}{c^{*2}} q^2, \qquad \phi_2 = (\gamma + 1)c^{*2} - q^2\gamma, \qquad \mu = \sqrt{\frac{\gamma - 1}{\gamma + 1}}$$

It may be seen that if C^2 reduces to the quantity in the bracket when $c^* \to \infty$, then $\lim_{c \to \infty} h(q) = q$. Since the terms containing q go out when $c^* \to \infty$, C may be chosen as the quantity in the bracket for any fixed value of q. Let us use $q = 1$. Then

$$C = \{[\sqrt{\phi_a^2 - 1} - \phi_a]^{1/\mu}[\sqrt{\phi_b^2 - 1} + \phi_b]\}^{1/2} \tag{987}$$

with

$$\phi_a = \gamma - \frac{\gamma - 1}{c^{*2}}, \qquad \phi_b = (\gamma + 1)c^{*2} - \gamma$$

s and t become

$$s = h(q)e^{-i\theta} \qquad t = h(q)e^{i\theta} \tag{988}$$

In the real domain, $q < c^*$ and hence $h(q)$ is real. Thus s and t are conjugate coordinates in subsonic flow. By analogy with the previous section we now define our characteristic coordinates, ξ and η, by the formulas

$$\eta = (1 - s + B^2)^{1/2} - B \tag{989}$$

$$\xi = (1 - t + \bar{B}^2)^{1/2} - \bar{B} \tag{990}$$

Notice that equation (989) reduces to equation (964) in the incompressible case and that ξ and η are conjugate coordinates in subsonic flow. Continuing our analogy with the incompressible case, we now assume a solution of the form

$$x = \frac{X_1(\xi, \eta)}{\eta} + X_3(\xi, \eta) + X_2(\xi, \eta) \ln \eta \tag{991}$$

$$y = \frac{Y_1(\xi, \eta)}{\eta} + Y_3(\xi, \eta) + Y_2(\xi, \eta) \ln \eta \tag{992}$$

where the X_i, Y_i are regular at $\xi = 0$, $\eta = 0$. The values of X_1, Y_1, X_3 and Y_3 for the incompressible flow about an elliptic cylinder may be seen by comparison with equation (978). The log terms are necessary to allow for circulation and hence lift. Recall now that x and y must satisfy the equation

$$y_\xi + \lambda_+ x_\xi = 0 \qquad y_\eta + \lambda_- x_\eta = 0 \tag{993}$$

subject to suitable boundary conditions. As we have already indicated, the real part of a complex solution will also satisfy equations (993) in the real domain; i.e., when u and v are real. Thus, if equations (991) and (992) represent complex solutions to the equations (993), then in the real domain,

$$x = \mathrm{Re}\left[\frac{X_1}{\eta} + X_2 \ln \eta + X_3\right] \tag{994}$$

$$y = \mathrm{Re}\left[\frac{Y_1}{\eta} + Y_2 \ln \eta + Y_3\right] \tag{995}$$

are also solutions. Substituting the assumed form of the solution (991), (992) into the equations (993) we get

$$Y_{3\xi} + \lambda_+ X_{3\xi} = -\frac{1}{\eta}(Y_{1\xi} + \lambda_+ X_{1\xi}) - (Y_{2\xi} + \lambda_+ X_{2\xi}) \ln \eta \tag{996}$$

$$\begin{aligned} Y_{3\eta} + \lambda_- X_{3\eta} = {} & \frac{1}{\eta^2}(Y_1 + \lambda_- X_1) - \frac{1}{\eta}(Y_{1\eta} + \lambda_- X_{1\eta} + Y_2 + \lambda_- X_2) \\ & - (Y_{2\eta} + \lambda_- X_{2\eta}) \ln \eta \end{aligned} \tag{997}$$

In order to make the X_i, Y_i regular at $\eta = 0$, we stipulate that

$$Y_{1\xi}(\xi, 0) + \lambda_+ X_{1\xi}(\xi, 0) = 0 \tag{998}$$

$$Y_1(\xi, 0) + \lambda_- X_1(\xi, 0) = 0 \qquad Y_2(\xi, 0) + \lambda_- X_2(\xi, 0) = (\lambda_-)_\eta X_1(\xi, 0)$$

$$Y_{2\xi}(\xi, \eta) + \lambda_+ X_{2\xi}(\xi, \eta) = 0 \qquad Y_{2\eta}(\xi, \eta) + \lambda_- X_{2\eta}(\xi, \eta) = 0$$

In addition we stipulate that X_2 and Y_2 shall be real when u and v (and hence q and θ) are real. Otherwise x and y will not be single-valued in the real physical plane, as may be seen by examining the terms $X_2 \ln \eta$ and $Y_2 \ln \eta$ in equations (994), (995). Since we may choose $X_1(0,0)$ at random, let us require $X_1(0,0) = \alpha$, where α is a constant which may be chosen at random. Now recall from equations (998) for $\eta = 0$,

$$Y_{1\xi} + \lambda_+ X_{1\xi} = 0, \qquad Y_{1\xi} + \lambda_- X_{1\xi} + X_1 \lambda_{-\xi} = 0 \tag{999}$$

Thus

$$(-\lambda_+ + \lambda_-)X_{1\xi} + X_1 \lambda_{-\xi} = 0 \tag{1000}$$

For $\eta = 0$, this is an ordinary differential equation for $X_1(\xi, 0)$ subject to the boundary conditions $X_1(0, 0) = \alpha$. The solution to equation (1000) may be written

$$X_1(\xi, 0) = \alpha \exp \int_0^\xi \frac{\lambda_{-\xi}(\xi, 0)}{\lambda_+(\xi, 0) - \lambda_-(\xi, 0)} d\xi \tag{1001}$$

where ξ is the variable of integration as well as the upper limit. Once $X_1(\xi, 0)$ has been found, $Y_1(\xi, 0)$ may be found from

$$Y_1(\xi, 0) = -\lambda_- X_1(\xi, 0) \tag{1002}$$

Any two functions $X_1(\xi, \eta)$, $Y_1(\xi, \eta)$ satisfying equations (1001) and (1002) may be used in the solution. In particular, meaningful solutions can be found by taking X_1 and Y_1 as functions of ξ only.

X_2 and Y_2 may be found by solving a characteristic initial value problem as discussed above. From (998) we see that X_2 and Y_2 must by themselves satisfy the characteristic equations

$$Y_{2\xi} + \lambda_+ X_{2\xi} = 0, \qquad Y_{2\eta} + \lambda_- X_{2\eta} = 0$$

subject to the boundary conditions that

$$Y_2 + \lambda_- X_2 = \lambda_{-\eta} X_1 \tag{1003}$$

along $\eta = 0$. Moreover, since X_2 and Y_2 are required to be real when q and θ are real, it follows that the boundary condition on $\xi = 0$ must be the reflection of the boundary condition on $\eta = 0$. That is

$$X_2(0, \eta) = \overline{X_2(\xi, 0)}, \qquad Y_2(0, \eta) = \overline{Y_2(\xi, 0)} \tag{1004}$$

Once the functions X_1, X_2, Y_1, Y_2 are determined, X_3, Y_3 may be found by solving another characteristic initial value problem. The equations to be satisfied are

$$Y_{3\xi} + \lambda_+ X_{3\xi} = f_1(\xi, \eta) \qquad \text{where} \qquad f_1(\xi, \eta) = -\frac{Y_{1\xi} + \lambda_+ X_{1\xi}}{\eta} \tag{a}$$

(1005)

$$Y_{3\eta} + \lambda_- X_{3\eta} = f_2(\xi, \eta) \quad \text{where} \quad f_2(\xi, \eta) = \frac{Y_1 + \lambda_- X_1}{\eta^2} - \frac{Y_{1\eta} + \lambda_- X_{1\eta} + \lambda_- X}{\eta} \tag{b}$$

To avoid computational difficulties, the boundary conditions are specified for $\xi = \xi_0 \neq 0$, $\eta = \eta_0 \neq 0$. The boundary conditions are specified, as before, in the form

$$x(\xi, \eta_0) = \overline{g(\xi)} + g(\eta_0), \qquad x(\xi_0, \eta) = g(\eta) + \overline{g(\xi_0)} \tag{1006}$$

For the case of flow without circulation, $(X_2, Y_2 = 0)$ meaningful flows can be formed by taking

$$g(\eta) = g_1(\eta) + \varepsilon_0 \ln(\alpha_0 - \eta) - \varepsilon_1 \ln(\eta - \alpha_1) \tag{1007}$$

where

$$g_1(\eta) = \frac{1}{\beta\eta}(E\eta^2 + \beta^2)^{1/2} + \frac{E\eta}{\beta(E\eta^2 + \beta^2)^{1/2}} - \frac{1}{\eta} \tag{1008}$$

For this initial condition it will be observed that the solution reduces to that for incompressible flow about an elliptic cylinder when ε_0 and ε_1 are taken as zero. With the ε_i not zero, the initial condition (1007) yields a series of symmetric airfoils. For flow about nonsymmetric cambered airfoils with lift, a more general form of the initial condition is used.

$$g(\eta) = E_M E\left[\frac{n - E_c}{Z + \sqrt{1 - E}} + \frac{\eta - E_c}{Z}\right] + E_T\eta + E_{Tsq}\eta^2 + E_{Tcu}\eta^3 \tag{1009}$$

$$+ \sum_{j=1}^{4} (T_{j1} + iT_{j2}) \ln(\eta - T_{j3} - iT_{j4}) + \sum_{j=1}^{4} (N_{j1} + iN_{j2}) \ln(\eta - N_{j3}$$

$$- iN_{j4}) + \sum_{j=1}^{4} (S_{j1} + S_{j2}) \ln(\eta - S_{j3} - iS_{j4})$$

$$Z = \sqrt{E(\eta - E_c)^2 + 1 - E}$$

By suitable variation of the parameters, airfoils of many different shapes may be generated.

Once the initial conditions are given, equations (1005) may be integrated numerically, just as in the supersonic case, except that it must be remembered that we are now dealing with complex variables. Suppose u, v, X_3 and Y_3 are known at points $P_1(\xi_1, \eta_1)$ and $P_2(\xi_2, \eta_2)$. The conditions at the point $P_3(\xi_2, \eta_1)$ can then be found from

$$\begin{aligned} u(P_3) - u(P_1) &= \tilde{\lambda}_- [v(P_3) - v(P_1)] \\ u(P_3) - u(P_2) &= \tilde{\lambda}_+ [v(P_3) - v(P_2)] \end{aligned} \tag{1010}$$

from equations (946, 947) where $\tilde{\lambda}_\pm$ represents the average $\lambda_\pm$ between points P_3 and P_1, and P_3 and P_2, respectively, and

$$Y_3(P_3) - Y_3(P_1) + \tilde{\lambda}_+ [X_3(P_3) - X_3(P_1)] = (\xi_2 - \xi_1)\tilde{f}_1 \tag{1011}$$

$$Y_3(P_3) - Y_3(P_2) + \tilde{\lambda}_- [X_3(P_3) - X_3(P_2)] = (\eta_1 - \eta_2)\tilde{f}_2 \tag{1012}$$

from equations (1005), where $\tilde{f}_1$ and $\tilde{f}_2$ represent the average values of f_1 between P_1 and P_3, and f_2 between P_2 and P_3, respectively. Recall, however, that if ξ and η are known, so are u and v, since η and ξ are functions of s and t, respectively, and s and t are functions of u and v.

Because computational difficulties are encountered when the initial values for X_3 and Y_3 are prescribed on $\xi = 0, \eta = 0$, the initial values for these functions should be prescribed on two non-zero characteristics, ξ_0 and η_0. By prescribing $\xi = \xi_0$, we are fixing the real and imaginary parts of ξ. However, $q = q_r + iq_i$, $\theta = \theta_r + i\theta_i$, where the subscripts r and i indicate the real and imaginary parts. Thus we have, in effect, four variables and two equations In order to prescribe a specific path on $\xi = \xi_0$ for numerical integration we need one more constraint. Such a constraint may be formulated by writing the real and imaginary parts of η on $\xi = \xi_0$ as functions of a single real parameter, τ. Thus, let $\eta_r = \eta_r(\tau)$ and $\eta_i = \eta_i(\tau)$, $\eta = \eta_r + i\eta_i$. When τ is specified, a given point on $\xi = \xi_0$ is fixed. A path on $\xi = \xi_0$ is thus traced by varying τ continuously. The choice of the function $\eta = \eta(\tau)$ is arbitrary. The path must start at $\eta = \eta_0$, however. The function

$X_3(\xi, \eta)$ is prescribed on the path and the function Y_3 is found by integrating equation (1005b).

In the same way, $\eta = \eta_0 \neq 0$ is chosen. This fixes the real and imaginary parts of η_0, giving two equations for four variables, $q_r, q_i, \theta_r, \theta_i$. A path on $\eta = \eta_0$ is determined by choosing two functions, $\xi_r = \xi_r(\sigma)$, $\xi_i = \xi_i(\sigma)$ on $\eta = \eta_0$, and varying σ. The functions $\xi_r(\sigma)$ and $\xi_i(\sigma)$ are arbitrary but the path must begin at $\xi = \xi_0$. $X_3(\xi, \eta)$ is prescribed along the path and equation (1005a) is used to find $Y_3(\xi, \eta)$. The numerical integration is then carried forward by taking stepwise increments of σ and τ as shown in Figure 210. The points may be found in the order indicated or in any other suitable order.

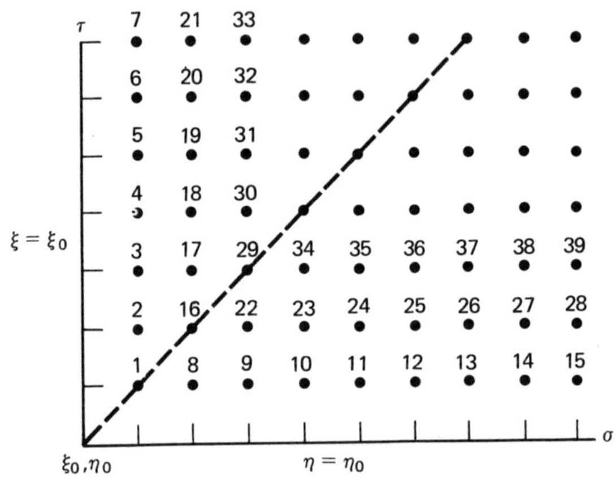

Figure 210 Order of numerical integration in the τ, σ plane [78].

Now it will be recalled that when $\xi = \bar{\eta}$, u and v are real. Since $\xi(\sigma)$ and $\eta(\tau)$ are arbitrary, we may choose these functions in such a way that when $\sigma = \tau$, $\xi(\sigma) = \bar{\eta}(\tau)$. Moreover, when u and v are real, x and y will be real. Thus, the solution along the diagonal 1, 16, 29, ... will give us points in the real physical plane corresponding to points in the real hodograph plane. Since we are interested only in real solutions, i.e., the solution along the diagonal, we need carry out the numerical integration only above (or below) the diagonal. Since $\xi(\sigma)$ and $\eta(\tau)$ are arbitrary, we may choose a succession of these functions and carry out the integration until the entire "area" $\xi = \xi_1$ and $\eta = \eta_0$ has been covered. This will give us the entire subsonic portion of the solution (Figure 211).

The solution may, however, be continued into the supersonic region. It is only on the sonic line itself that the method breaks down since $\lambda_+ = \lambda_-$. Everywhere else $\lambda_+ \neq \lambda_-$ and hence the method of characteristics may be applied. Recall

$$s = h(q)e^{-i\theta} = |1 - 2B\eta - \eta^2|e^{-i\theta} = Ce^{-if(q)}e^{-i\theta}$$
$$t = h(q)e^{-i\theta} = |1 - 2\bar{B}\xi - \xi^2|e^{i\theta} = Ce^{-if(q)}e^{+i\theta} \tag{1013}$$

When $q = c^*$,

$$|s| = |1 - 2B\eta - \eta^2| = |C|, \qquad |t| = |1 - 2\bar{B}\xi - \xi^2| = |C| \tag{1014}$$

But in the real supersonic domain, $f(q)$ and θ are real and hence

$$|s| = |t| = |C| \tag{1015}$$

Thus, those values of ξ and η which satisfy equation (1014) describe the entire supersonic domain. Consider now the points ξ on $\eta = \eta_0$ which satisfy equation (1014). These points shall be called the upper sonic locus. Similarly, the points η on $\xi = \xi_0$ which satisfy equation (1014) shall be called the lower sonic locus. By choosing $\eta(\tau)$ and $\xi(\sigma)$ properly, we may continue the numerical integration right across the sonic line and into the supersonic region.

Consider a path $\eta(\tau)$ on $\xi = \xi_0$ starting at $\eta = \eta_0$ and ending at $\eta = \eta_1$, where η_1 is on the lower sonic locus. Similarly, consider a path $\xi = \xi(\sigma)$ starting at $\xi = \xi_0$ and ending at $\xi = \xi_1$, where ξ_1 is on the upper sonic locus. Proceed now with the numerical integration. The farthest point from ξ_0, η_0 for which we can find a solution will be ξ_1, η_1. Since ξ_1 and η_1 satisfy equation (1014), the flow there may be either sonic or supersonic. By choosing $\xi(\sigma)$, $\eta(\tau)$ properly, we may assure that the point ξ_1, η_1 belongs to the real, supersonic domain. We now wish to extend the solution further into the supersonic domain. To do so, we extend the function $\xi(\sigma)$ in such a way that all the points to the right of ξ_1 on the σ axis (Figure 211) are on the upper sonic locus. Similarly, we extend the function $\eta(\tau)$ in such a way that all the points above η_1 on the τ axis (Figure 211) are on the lower sonic locus.

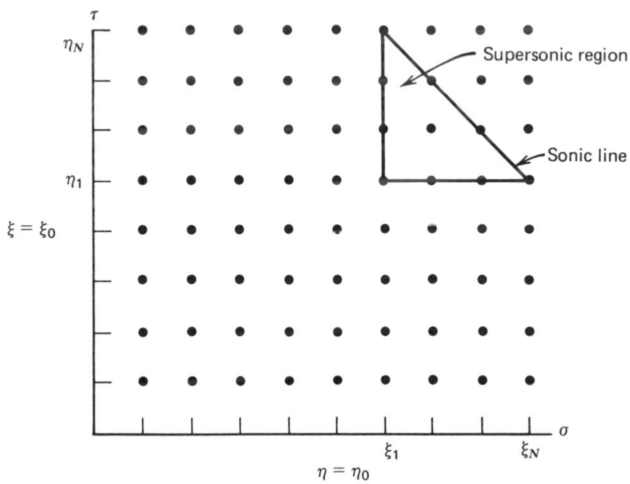

Figure 211 Supersonic region in the τ, σ plane [78].

Notice that when we enter the supersonic region, ξ is no longer equal to $\bar{\eta}$ when q and θ are real. Recall

$$s = Ce^{-i[f(q)+\theta]}, \qquad t = Ce^{-i[f(q)-\theta]} \tag{1016}$$

But for q real and greater than c^*, $f(q)$ is real and hence s and t are no longer complex conjugates. On the other hand, for each ξ on the upper sonic locus, there exists a point η on the lower sonic locus such that $\eta = \bar{\xi}$. We now choose $\xi_N = \bar{\eta}_1$ and $\eta_N = \bar{\xi}_1$. The number of grid points between ξ_1 and ξ_N, η_1 and η_N is essentially arbitrary. For convenience we may redefine $f(q)$ to be

$$f(q) = \frac{1}{\mu}\tan^{-1}\sqrt{\frac{M^{*2}-1}{\mu^{-2}-M^{*2}}} - \tan^{-1}\sqrt{\frac{M^{*2}-1}{1-M^{*2}\mu^{-2}}} \tag{1017}$$

where $M^* = q/c^*$. This $f(q)$ differs from the one in equation (980) only by a constant $(\pi/4)(1 - 1/\mu)$. The advantage of this change is that $f(c^*)$ now equals zero. The points ξ_N, η_1 and ξ_1, η_N then lie on the sonic line. To find the solution in the supersonic region, we continue our numerical integration from the point (ξ_1, η_1) to the points (ξ_2, η_1), (ξ_3, η_1), ... (ξ_N, η_1) as well as the points (ξ_1, η_2), (ξ_1, η_3), ... (ξ_1, η_N), and all the other points in the supersonic region, as shown in figure 211. Notice that all points in the supersonic region yield real solutions. For, in the supersonic region, $|s| = |t| = C$. But $s = C\exp[-i(f(q) + \theta)]$, $t = C\exp[-i(f(q) - \theta)]$. This means $f(q) + \theta$ and $f(q) - \theta$ must be real. If $f_i, f_r, \theta_i, \theta_r$ represent the real and imaginary parts of f and θ, then $f_i + \theta_i = 0$ and $f_i - \theta_i = 0$. The only way that this condition can be satisfied is if $f_i = \theta_i = 0$. But $f(q)$ real will be satisfied by q real and greater than c^*. Thus, each point in the supersonic region corresponds to a real q and θ and hence to a real x and y. The result of the integration is thus the same as if we had prescribed data on the sonic line and solved for the flow in the supersonic region in the conventional manner. Referring once again to Figure 211, we see that the solution in the supersonic region corresponds to the solution in the hodograph plane between the real characteristics corresponding to ξ_1 = constant and η_1 = constant as shown in Figure 212. The shaded region identifies the region in which the solu-

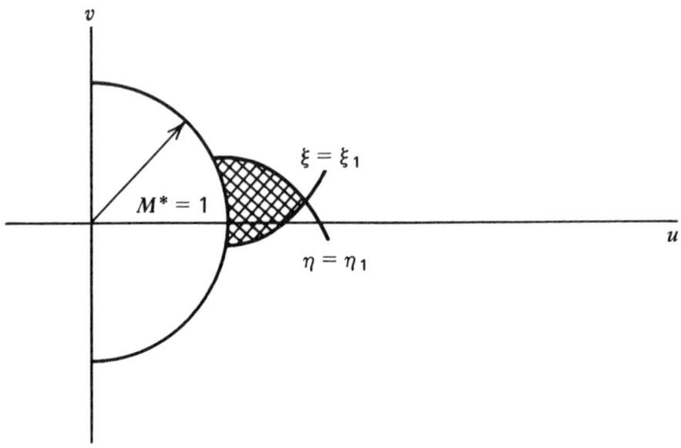

Figure 212 Area of solution in the supersonic region shown on the hodograph plane [78].

tion is found. Notice that for $\eta = \eta_1, f(q) + \theta$ = real constant. This identifies the characteristic in the real hodograph plane. Similarly, for $\xi = \xi_1, f(q) - \theta$ = real constant, which identifies the other characteristic in the real hodograph plane.

In order to find meaningful solutions in the physical plane, appropriate

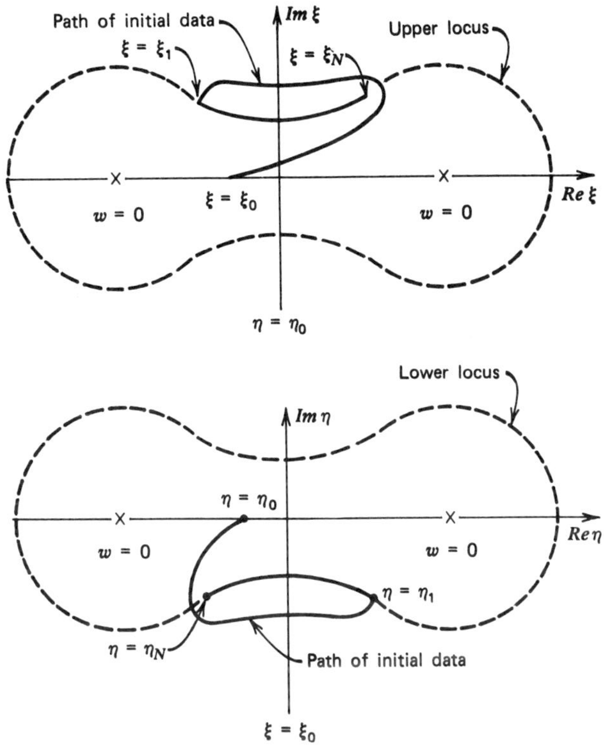

Figure 213 Typical paths of integration [240].

paths $\eta(\tau)$ and $\xi(\sigma)$ must be chosen. Figure 213 illustrates two typical paths which have been used to find the flow around an airfoil.

We now define the stream function by the equations

$$\rho_0 \psi_x = \rho v, \qquad \rho_0 \psi_y = -\rho u \tag{1018}$$

where ρ_0 is the stagnation density. Then

$$\psi = \frac{1}{\rho_0} \int (\rho v \, dx - \rho u \, dy) \tag{1019}$$

The lines $\psi = \text{constant}$ represent the streamlines. Once u and v are known as functions of x and y, equation (1019) may be integrated between any two points over any path in the flow field. ρ is known as a function of velocity from the isentropic relations. The stagnation conditions are presumed known.

The actual flow pattern will depend on the coefficients chosen in the initial condition given by equation (1009). A good starting point is to take the coefficients from a known incompressible solution and see how the profile is modified by the compressibility. The resulting profile may then be altered by changing individual coefficients by trial and error until the desired flow properties are obtained.

While there are 72 free parameters in the proposed boundary condition, certain constraints are required in order to obtain a meaningful flow. These are: $x_u = y_u = \psi = 0$ at the tail. The first two follow from the Kutta condition and the third follows from taking $\psi = 0$ on the surface of the body. This may be accomplished by taking the constant of integration in equation (1019) so that $\psi = 0$ when $q = 0$.

The criterion that the flow be around a closed body is that the integral (1019) over a closed path around the point at infinity be zero. Since $z = \infty$ when $\eta = 0$, we may take the integral around $\eta = 0$ in the real hodograph plane where $\eta = \bar{\xi}$; that is, around the point $\xi = 0, \eta = 0$. It may be shown that around this point

$$\frac{1}{\rho_0}\oint(\rho v\,dx - \rho u\,dy) = -\frac{2\pi}{\rho_0}[X_1(0,0)(\rho v)_\eta - Y_1(0,0)(\rho u)_\eta] = 4\pi\frac{\rho_\infty}{\rho_0}\mathrm{Re}\,[B\alpha] \tag{1020}$$

where ρ_∞ represents the free stream density and $\alpha = X_1(0, 0)$. The criterion that the flow be around a closed body is thus that the right-hand side of equation (1020) be zero, which is equivalent to saying that ψ must be single valued. It follows then that

$$\mathrm{Re}[B\alpha] = 0 \tag{1021}$$

for this condition to be met.

In practice it is found that the transonic flow around an airfoil is very sensitive to the boundary layer. The boundary layer in turn generates a viscous wake. The inviscid flow solution is valid only outside the boundary layer and outside the wake. In practice, then, the "body" is taken to include the boundary layer and the wake. Since the wake extends to infinity, this means that condition (1021) should not be satisfied.

The pressure may be found from the isentropic flow relations once the velocity is known and the pressure coefficient may then be evaluated in the usual manner,

$$c_p = \frac{p - p_\infty}{\frac{1}{2}\rho_\infty u_\infty^2} \tag{1022}$$

where the subscript ∞ refers to the free stream conditions.

Figure 214 shows a typical supercritical transonic airfoil designed by the present method. The supersonic region is delineated by the real physical characteristics shown in the figure. It will be observed that in spite of the large supersonic region, no shock is present (hence the name, supercritical airfoil). A surprising factor is the relatively large thickness ratio, larger than for conventional high subsonic airfoils. This has proved a boon to aircraft designers permitting more fuel to be carried in the wings and simplifying structural and design problems.

It will be observed that the body is not closed at the trailing edge. The space between the upper and lower surface at the trailing edge is occupied by the

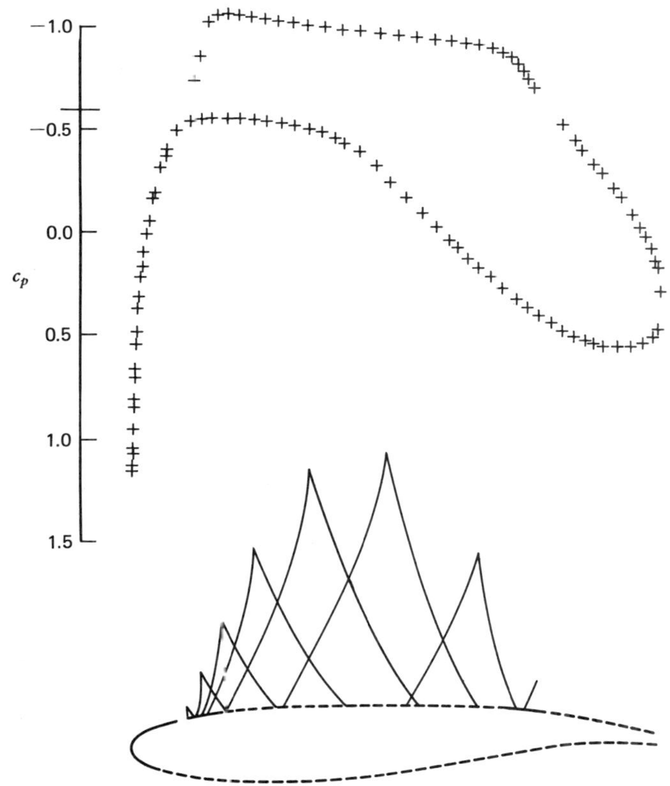

$M = 0.750$ $CL = 0.667$ $CD = 0.0334$ $T/C = 0.151$

Figure 214 Typical solution obtained by the hodograph method [78].

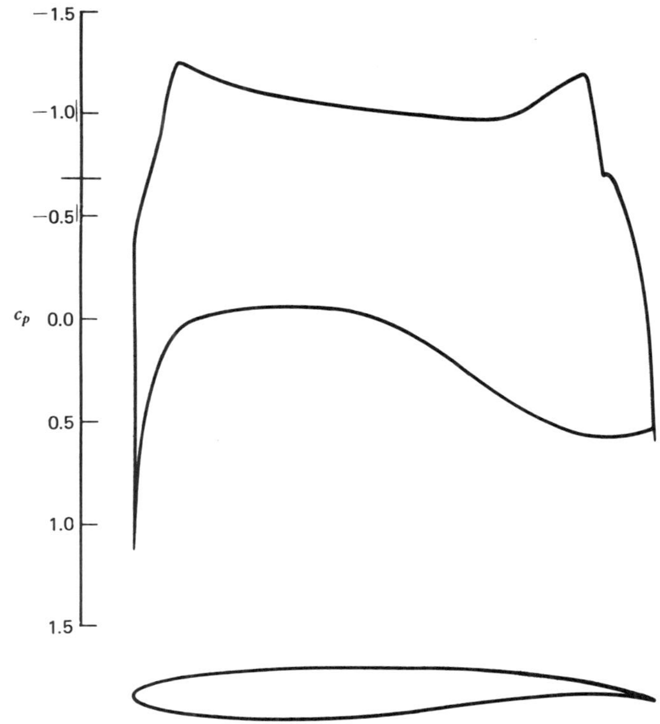

$M = 0.725$ $T/C = 0.104$ $CL = 1.187$ $CD = 0.008$ ALP = 0

$M \times N = 160 \times 30$ NCY = 800   EP = −0.50

Figure 215 Solution obtained by the hodograph method with the boundary layer neglected [78].

boundary layer over the upper surface. A typical result when the boundary layer is neglected is shown in Figure 215. In this case, however, the theoretical results do not agree well with experiment. Because the method described here is a potential flow method, it cannot be used when shocks are present. We may recall, however, the explanation of von Karman that shocks occur in transonic flow when the characteristics of the same family begin to cross. The envelope of the intersection of such characteristics is called a limiting line. An illustration

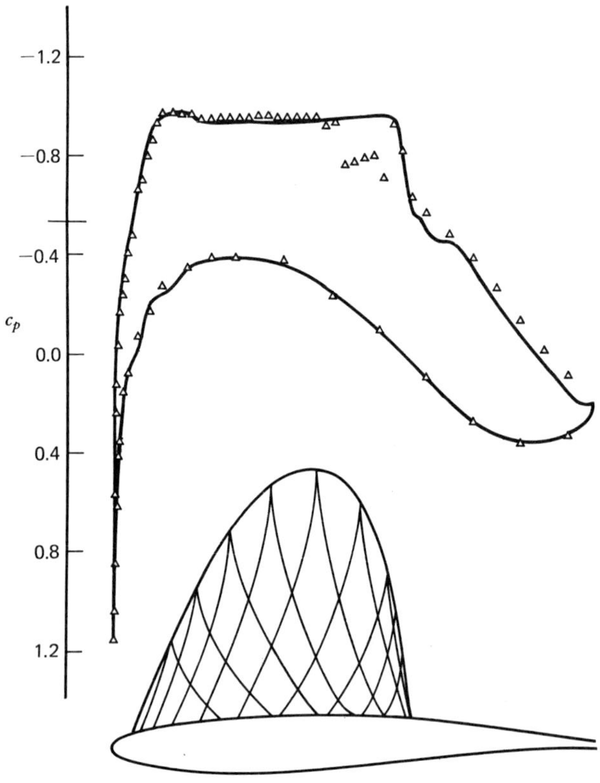

AIRFOIL 75-06-12   M∗N= 160∗30   NCY= 400   R=20 MILLION

— THEORY   M= 0.765   ALP= 0.72   CL= 0.576   CD= 0.0127

△ EXPERIMENT   M= 0.765   ALP= 0.89   CL= 0.576   CD= 0.0110

Figure 216 Hodograph solution indicating shock formation [80].

of such a case is shown in Figure 216. The potential flow method thus predicts when shocks will be present, but it cannot be relied on to give accurate solutions in such cases. On the other hand, Bauer et al. [80] have extended the present method to include the case when shocks are present. More recently, the present method has also been extended to the case of transonic cascades [81].

Internal Flow

We wish now to deal with the transonic flow in a deLaval nozzle on a two-dimensional or axisymmetric basis. It will be recalled that we first dealt with this problem on the basis of quasi-one-dimensional flow. The two-dimensional problem has been investigated by numerous authors. The first known treatment is that of Meyer [82]. Among subsequent analyses, the most noteworthy are those of Taylor [83], Görtler [84], Sauer [85], Tomatika and Tamada [86], and Oswatitsch and Rothstein [87]. An excellent historical survey of this problem has been given by Hall and Sutton [88].

Let

$$u' = \frac{u}{c^*} - 1, \qquad v' = \frac{v}{c^*} \tag{1023}$$

In the transonic range, i.e., near the throat of the nozzle, u' and v' are thus in the nature of perturbation velocities, and can be assumed small compared to one. The flow is assumed irrotational and isentropic. Hence,

$$u'_y - v'_x = 0 \tag{1024}$$

and a velocity potential ϕ exists such that

$$u' = \phi'_x, \qquad v' = \phi'_y \tag{1025}$$

Recall, now, the potential equation for two-dimensional, steady, irrotational, isentropic flow:

$$(c^2 - u^2)u_x - 2uvu_y + (c^2 - v^2)v_y = 0 \tag{1026}$$

where

$$c^2 + \frac{\gamma - 1}{2}(u^2 + v^2) = \frac{\gamma + 1}{2}c^{*2} \tag{1027}$$

In terms of the perturbation potential ϕ', these equations may be written

$$\left(-2\phi'_x - \phi'^2_x - \frac{\gamma-1}{\gamma+1}\phi'^2_y\right)\phi'_{xx} - \frac{4}{\gamma+1}(1+\phi'_x)\phi'_y\phi'_{xy}$$
$$+\left(\frac{2}{\gamma+1} - \frac{2(\gamma-1)}{\gamma+1}\phi'_x - \frac{\gamma-1}{\gamma+1}\phi'^2_x - \phi'^2_y\right)\phi'_{yy} = 0 \tag{1028}$$

The problem thus reduces to finding a solution of equation (1028) which satisfies the boundary conditions. These may be specified by the function $y = f(x)$ giving the shape of the wall, as in Figure 217. At the wall the flow direction must be parallel to the wall.

Various investigators have treated this problem in various ways, but most have assumed a solution in the form of a power series. Meyer [82] began with the simple assumption that the velocity varies linearly along the x axis; i.e.,

$$u' = Kx, \qquad \phi' = \tfrac{1}{2}Kx^2 \tag{1029}$$

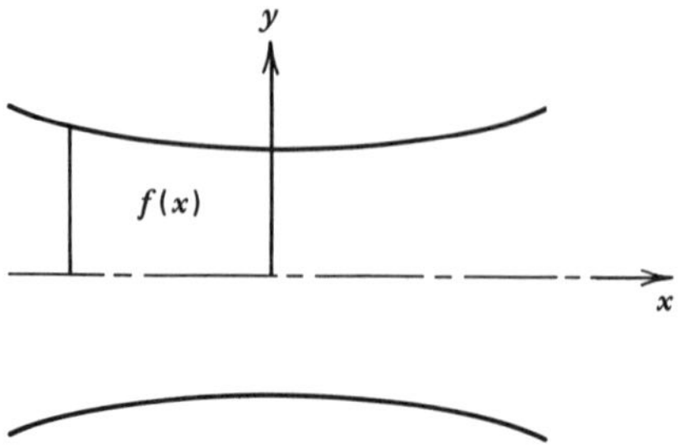

Figure 217 Nomenclature for two-dimensional flow in a de Laval nozzle.

where K is a constant, and assumed a solution ϕ' of the form

$$\phi' = \sum \sum \phi_{mn} x^m y^n \tag{1030}$$

Substituting in equation (1028) and equating power of x and y, Meyer found the coefficients up to the sixth order ($m + n \leq 6$). The result is

$$\phi' = \frac{1}{2}Kx^2 + \frac{\gamma + 1}{2}K^2xy^2 + \frac{(\gamma + 1)}{24}K^3y^4 + \frac{(\gamma + 1)(2\gamma - 1)}{4}K^3x^2y^2 + \ldots \tag{1031}$$

whence

$$u' = Kx + \frac{\gamma + 1}{2}K^2y^2 + \ldots \tag{1032}$$

$$v' = (\gamma + 1)K^2xy + \frac{(\gamma + 1)^2}{6}K^2y^3 + \ldots \tag{1033}$$

From the symmetry of the problem, it is clear that u' must be an even function of y while v' must be an odd function of y. On the sonic line v' is very small and hence to a good approximation $u' = 0$. The equation of the sonic line is, therefore, given by

$$x = -\frac{\gamma + 1}{2}Ky^2 + \ldots \tag{1034}$$

The constant K may be evaluated from the boundary conditions. Let R_s be the radius of curvature of the wall at the throat.

At the throat, $\mathbf{V} \cong c^*$. Thus, from Figure 218, we see that

$$\frac{dx}{R_s} \cong \frac{dv}{c^*} = dv' \tag{1035}$$

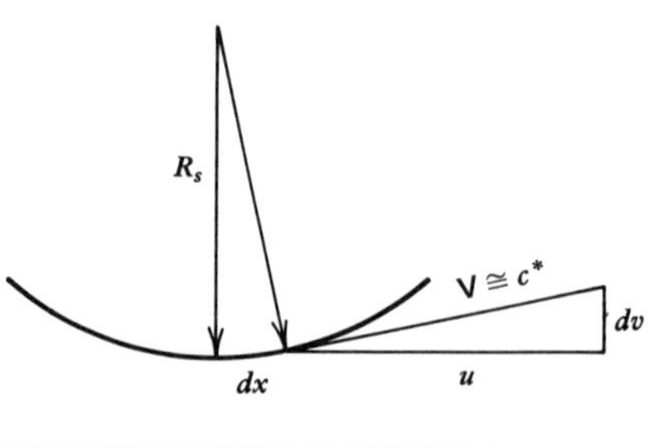

Figure 218 Radius of curvature and velocity components at the throat.

at $y = f(0)$. Hence

$$\left(\frac{\partial v'}{\partial x}\right)_{y=f(0)} = \frac{1}{R_s} = (\gamma + 1)K^2 f(0) \tag{1036}$$

from equation (1033). So

$$K = [R_s(\gamma + 1)f(0)]^{-1/2} \tag{1037}$$

An alternate approach was used by Sauer [85]. Sauer argued that in the transonic region, the higher order terms in equation (1028) could be dropped, leaving

$$-2\phi'_x\phi'_{xx} - \frac{4}{\gamma+1}\phi'_y\phi'_{xy} + \frac{2}{\gamma+1}\phi'_{yy} - \frac{2(\gamma-1)}{\gamma+1}\phi'_x\phi'_{yy} = 0 \tag{1038}$$

Since the governing equation is parabolic on the sonic line, the Mach lines there will be normal to the flow direction. This means that in the transonic region, changes in the y direction will take place very slowly. On this basis, Sauer concluded that the second and fourth terms of equation (1038) could also be dropped, leaving

$$-\phi'_x\phi'_{xx} - \frac{2}{\gamma+1}\phi'_{yy} = 0 \tag{1039}$$

The first three terms of equation (1031) for ϕ' satisfy this equation exactly.

An improved solution was obtained by Oswatitsch and Rothstein [87], who discarded the assumption that $u' = Kx$ at the centerline. We postulate a solution of the form

$$\frac{u}{c^*} = \frac{u_0}{c^*} + \frac{a}{2}y^2 + \cdots, \qquad \frac{v}{c^*} = by + \cdots \tag{1040}$$

where the coefficients u_0, a and b are functions of x. As a first approximation we will retain only the first two terms of the series for u and only the first term of the series for v.

Recall now the irrotationality condition

$$u_y - v_x = 0 \tag{1041}$$

Substituting equation (1040) in equation (1041) and equating powers of y, we get

$$a = b' \tag{1042}$$

where the prime indicates differentiation with respect to x.

Taking the wall at $y = f(x)$ as before (Figure 217), the boundary condition at the wall becomes

$$\frac{v}{c^*} = f'\frac{u}{c^*} = bf \tag{1043}$$

Now let $u_s(x)$ be the velocity obtained from quasi-one-dimensional flow theory.

Then the boundary condition may be approximated by

$$\frac{v}{c^*} = f' \frac{u_s}{c^*} \tag{1044}$$

Then

$$b = \frac{f'}{f}\frac{u_s}{c^*}, \qquad a = \frac{f''}{f}\frac{u_s}{c^*} + \frac{f'}{f}\frac{u_s'}{c^*} - \frac{f'^2}{f^2}\frac{u_s}{c^*} \tag{1045}$$

Thus, the coefficients a and b are determined from the boundary condition and the quasi-one-dimensional solution. It remains to determine the function $u_0(x)$.

We now postulate that for each value of x there exists a value of y at which $u = u_s$. The validity of this assumption may be verified by experiment. Let $y = y_s$ at $u = u_s$. Then

$$\frac{u}{c^*} = \frac{u_s}{c^*} + \frac{a}{2}(y^2 - y_s^2) \tag{1046}$$

where a is given by equation (1045). The stream density ρV may be expanded about some point 1 in the flow

$$\frac{\rho V}{\rho_1 V_1} - 1 = \beta_1^2\left(\frac{V}{V_1} - 1\right) - \frac{1}{2}\frac{\beta_1^2}{\dfrac{1}{M_1^*} - 1}\left(\frac{V}{V_1} - 1\right)^2 + \cdots \tag{1047}$$

where $\beta_1^2 = |1 - M_1^2|$. Now M and M^* are related by the relation

$$1 - M^2 = (\gamma + 1)\left(\frac{1}{M^*} - 1\right)\left[1 - (\gamma + \tfrac{1}{2})\left(\frac{1}{M^*} - 1\right) + \cdots\right] \tag{1048}$$

If we take the point 1 at the critical point, $\beta = 0$; hence, from (1047) and (1048),

$$\frac{\rho V}{\rho^* c^*} - 1 = -\frac{\gamma + 1}{2}\left(\frac{V}{c^*} - 1\right)^2 + \cdots \tag{1049}$$

Now to a good approximation, $V = u$, and u/c^* may be expressed by equation (1046). Hence

$$\frac{\rho u}{\rho^* c^*} - 1 = -\frac{\gamma + 1}{2}\left[\left(\frac{u_s}{c^*} - 1\right)^2 + \left(\frac{u_s}{c^*} - 1\right)a(y^2 - y_s^2) + \frac{a^2}{4}(y^2 - y_s^2)^2\right]$$

or

$$\frac{\rho u}{\rho^* c^*} = \frac{\rho_s u_s}{\rho^* c^*} - \frac{\gamma + 1}{2}\left[\left(\frac{u_s}{c^*} - 1\right)a(y^2 - y_s^2) + \frac{a^2}{4}(y^2 - y_s^2)^2\right] \tag{1050}$$

The mass flow rate, $\dot{m}$, is given by

$$\frac{\dot{m}}{2\rho^* c^*} = \int_0^f \frac{\rho u}{\rho^* c^*}dy = \frac{\rho_s u_s}{\rho^* c^*}f - \frac{\gamma + 1}{2}\left[\left(\frac{u_s}{c^*} - 1\right)a\left(\frac{f^3}{3} - y_s^2 f\right) + \frac{a^2}{4}\left(\frac{f^5}{5} - \frac{2}{3}f^3 y_s^2 + y_s^4 f\right)\right] \tag{1051}$$

Since we are interested only in choked flow, ρu must be a maximum at the throat. This will be true when the last term on the right-hand side of equation (1051) is a minimum; that is, when

$$\frac{d}{dy_s}\left[\frac{f^4}{5} - \frac{2}{3}f^3 y_s^2 + y_s^4 f\right] = 0 \qquad \text{or} \qquad y_s^2 = \frac{f^2}{3} \tag{1052}$$

Substituting this result for y_s in equation (1046) and with a as given by equation (1045) we get

$$u = u_s\left[1 + \frac{1}{2}\left(\frac{y^2}{f^2} - \frac{1}{3}\right)f''f\right] + \cdots \tag{1053}$$

for the velocity distribution in the plane nozzle.* Equation (1053) shows how the velocity along the center line differs from that of the quasi-one-dimensional flow analysis.

$$u_0 = u_s\left[1 - \frac{1}{6}\frac{f''}{f}\right] \tag{1054}$$

By the same method we arrive at the velocity distribution in the axially symmetric case

$$u = u_s\left[1 + \frac{1}{2}\left(\frac{y^2}{f^2} - \frac{1}{2}\right)f''f\right] \tag{1055}$$

Oswatitsch and Rothstein [87] have also given the solution when three terms are retained in the series for u and two terms are retained in the series for v. In this case we have

$$\begin{aligned} u(x, y) &= u_0 + \frac{1}{2!}a_2 y^2 + \frac{1}{4!}a_4 y^4 + \cdots \\ v(x, y) &= b_1 y + \frac{1}{3!}b_3 y^3 + \cdots \end{aligned} \tag{1056}$$

where u_0, a_i and b_i are functions of x. Recall now the irrotationality condition

$$u_y - v_x = 0 \tag{1057}$$

* The full expression for u is

$$u = u_s\left[1 + \frac{1}{2}\left(\frac{f''}{f} + \frac{f'}{f}\frac{u_s'}{u_s} - \frac{f'^2}{f^2}\right)\left(y^2 - \frac{f^2}{3}\right)\right] \tag{1053a}$$

Recall, however,

$$v = by = \frac{f'}{f}\frac{u_s}{c^*}y$$

In the transonic region, that is, in the vicinity of the throat, v is small compared to u hence justifying our assumption of $V \cong u$. This implies, however, that f'/f is small and hence the terms involving f'/f in equation (1053a) may be neglected.

Substituting the assumed form of the solution (1056) in equation (1057), we get

$$a_2 y + \frac{1}{3!} a_4 y^3 = b_1' + \frac{1}{3!} b_3' y^3 \tag{1058}$$

where the prime indicates differentiation with respect to x. Equating powers of y,

$$a_2 = b_1', \qquad a_4 = b_3' \tag{1059}$$

Instead of using equation (1027) for the speed of sound, we may use

$$c^2 = \frac{\gamma - 1}{2}[V_m^2 - (u^2 + v^2)] \tag{1060}$$

where V_m is the maximum possible value of the velocity, $\sqrt{2c_p T_0}$. Substituting equation (1060) for c in the governing equation (1026) and introducing the assumed form of the solution (1056) we may again equate powers of y. Equating terms of y to the zero power we get

$$b_1 = -\frac{V_m^2 - \dfrac{\gamma + 1}{\gamma - 1} u_0^2}{V_m^2 - u_0^2} u_0' \tag{1061}$$

If the series solutions of equation (1026) are limited to the terms shown in equation (1056), we have five unknown functions: u_0, a_2, a_4, b_1 and b_3. We thus need five equations to determine these functions. Three of these equations are (1059) and (1061). Another equation may be found from the boundary conditions.

$$\frac{v_f}{u_f} = \frac{dy}{dx} = \frac{df}{dx} = f'(x) \tag{1062}$$

Hence

$$b_1 f + \tfrac{1}{6} b_3 f'^3 = f'(u_0 + \tfrac{1}{2} a_2 + \tfrac{1}{24} b_3' f^4) \tag{1063}$$

The final equation may be found by introducing the conservation of mass.

$$\int_0^f \rho u \, dy = \text{constant} \tag{1064}$$

where

$$\frac{\rho}{\rho_0} = \left[1 - \frac{u^2 + v^2}{V_m^2}\right]^{1/(\gamma - 1)} \tag{1065}$$

As a first approximation, the velocity along the centerline from the quasi-one-dimensional flow theory may be used for $u_0(x)$. An improved approximation is given by

$$u_0 = u_s\left[1 - \frac{1}{6}\left(hh'' + 2\frac{u_s'}{u_s} hh'\right)\right] + \cdots, \qquad h(x) = \frac{f(x)}{f(0)} \tag{1066}$$

where u_s is the quasi-one-dimensional solution.* Carrying out the calculations, we find

$$v = y\frac{h'}{h} + \cdots, \qquad u = u_s + \frac{u_s}{2}\left(\frac{y^2}{h^2} - \frac{1}{3}\right)\left(hh'' + \frac{u_s'}{u_s}hh'\right) - \frac{1}{6}\frac{u_s'}{u_s}hh' + \cdots$$

A similar derivation may be carried out for the axially symmetric case. The only difference lies in the conservation of mass condition, which becomes

$$\int_0^f 2\pi r\rho u \, dr = \text{constant} \tag{1067}$$

The relation between u_0 and u_s in this case is found to be

$$u_0 = u_s\left[1 - \frac{1}{4}\left(hh'' + \frac{3}{2}\frac{u_s'}{u_s}hh'\right)\right] + \cdots \tag{1068}$$

Carrying out the calculations in the axially symmetric case we find

$$v = r\frac{h'}{h} + \cdots, \qquad u = u_s + \frac{u_s}{2}\left(\frac{r^2}{h^2} - \frac{1}{2}\right)\left(hh'' + \frac{u_s'}{u_s}hh'\right) - \frac{1}{8}\frac{u_s'}{u_s}hh' + \cdots \tag{1069}$$

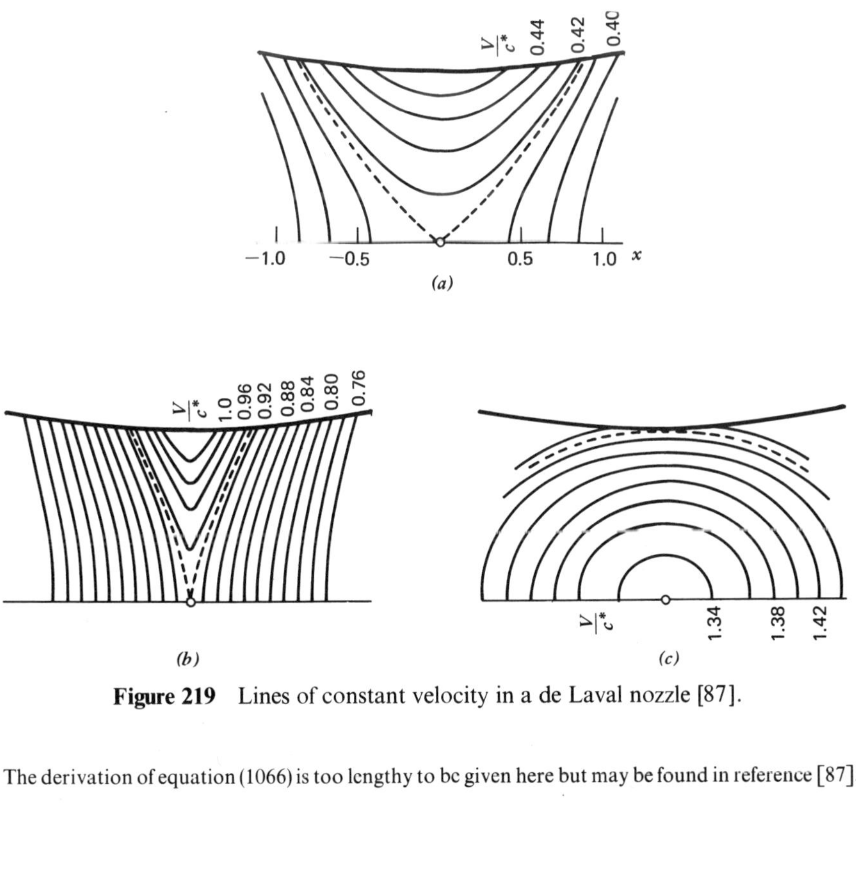

Figure 219 Lines of constant velocity in a de Laval nozzle [87].

* The derivation of equation (1066) is too lengthy to be given here but may be found in reference [87].

The method applies equally well to "symmetric" flow, that is, flow where the velocity reaches a maximum or minimum at the throat, as to asymmetric flow, where the velocity continues to increase through the throat. The former case includes flow which is subsonic throughout as well as flow which is supersonic throughout. The symmetric solution also reveals a third possibility, flow which is locally supersonic near the throat and subsonic elsewhere. The existence of such flow patterns is confirmed experimentally. Figure 219 shows three flow patterns obtained by the present method. Figure 219(a) shows symmetric flow which is wholly subsonic, Figure 219(b) shows symmetric flow which is locally supersonic, and Figure 219(c) shows flow which is wholly supersonic. The curved lines are the lines of constant velocity. Case (b) shows that supersonic flow first occurs near the walls at the throat. Such a possibility was not revealed by the quasi-one-dimensional analysis. Experiments indicate, however, that the flow pattern always passes through this intermediate stage as the back pressure is lowered, prior to going wholly supersonic in the diverging section. As the back pressure is lowered, the regions of supersonic flow grow larger until they finally join at the center line. The regions of supersonic flow will generally be terminated by a shock before joining together. After joining, the shocks terminating each region join to become a single normal shock across the nozzle as predicted by the one dimensional theory. Figure 220 shows a comparison

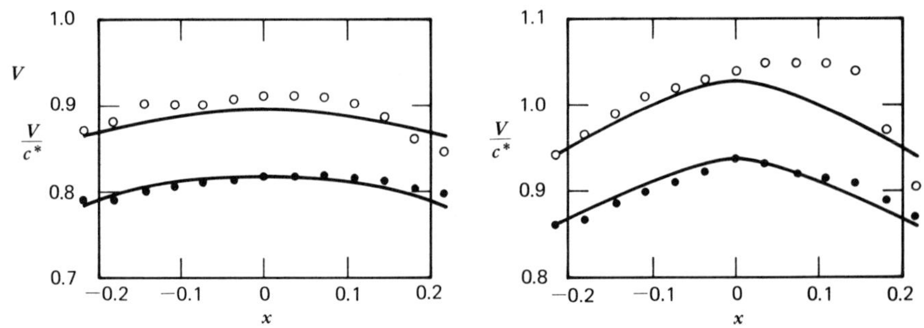

Figure 220 Velocity variation in a de Laval nozzle—theory and experiment compared[87].

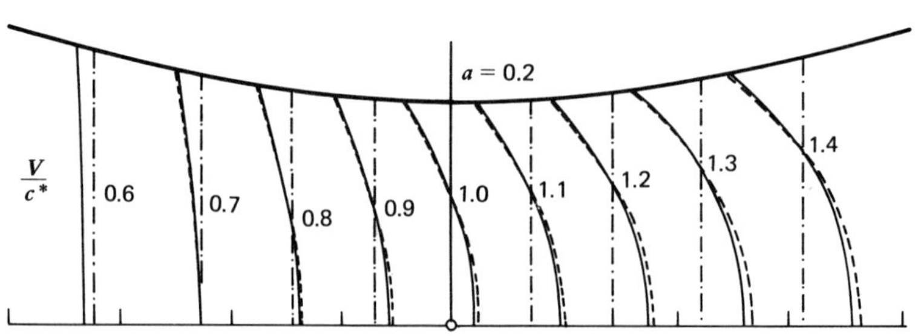

Figure 221 Lines of constant velocity in an axially symmetric de Laval nozzle from quasi-one-dimensional, first-order two-dimensional ---, and second-order two-dimensional theory [87].

between theory and experiment for the axially symmetric case with wholly subsonic flow and locally supersonic flow. The lower curve gives the velocity at the center line, the upper curve gives the velocity near the wall. Notice that theory is in error near the wall in locally supersonic flow, where the experimental data indicates the presence of a shock. The assumption of isentropic flow in the theory precludes the correct prediction of the flow pattern when shocks are present. Figure 221 shows the lines of constant velocity predicted by the theory for the asymmetric case. The shape of the nozzle here is assumed to be given by the hyperbola

$$f = \sqrt{1 + ax^2}$$

with $a = 0.2$ in Figure 221. Figure 222 shows the comparison of theory with

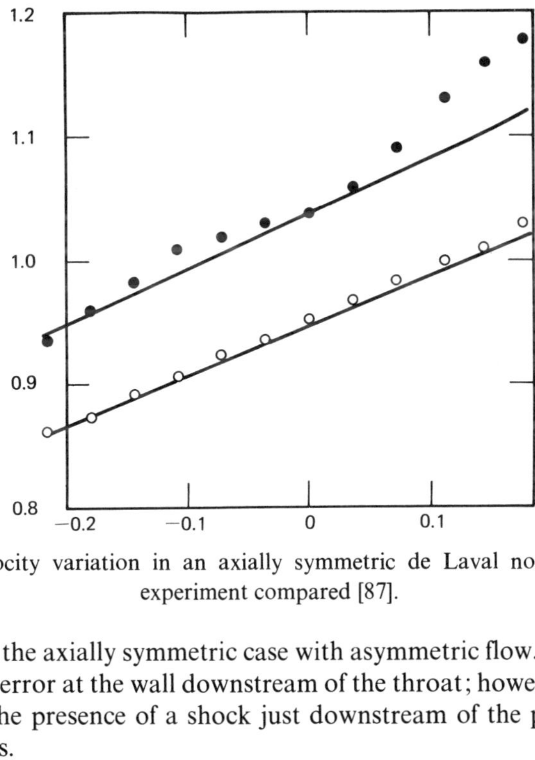

Figure 222 Velocity variation in an axially symmetric de Laval nozzle—theory and experiment compared [87].

experiment for the axially symmetric case with asymmetric flow. The prediction is seen to be in error at the wall downstream of the throat; however, this may be attributed to the presence of a shock just downstream of the point where the data terminates.

7 Laminar Boundary Layers

The Boundary Layer Equations

Since deriving the Navier–Stokes equation in Chapter 1, we have been dealing essentially with inviscid flow. We must now examine more carefully the conditions under which neglecting the viscosity is justified. As we have already observed, neglecting the viscosity in the Navier–Stokes equation changes the order of the equation and, therefore, fundamentally affects the nature of the solution.

It is well known in mathematics that the solution which one obtains by neglecting the viscosity in the fundamental equations is not necessarily the same as the solution which one obtains by solving the equations including the viscosity and then letting the viscosity go to zero in the solution. This is sometimes summarized by saying that the solution of the limit is not necessarily equal to the limit of the solution, as the viscosity goes to zero. A problem of this type is called a singular perturbation problem.

In fluid mechanics, the chief difference between the two solutions lies in the boundary condition at the wall. Except in the flow of highly rarified gases (the so-called slip flow regime), actual fluids obey the no slip condition at the wall; that is, at a solid boundary the velocity of the fluid is always equal to that of the boundary. When the viscosity is neglected in the governing equations, this condition cannot be satisfied. It will be recalled that in our analysis of inviscid flow, the boundary condition imposed at the wall was that the flow be parallel to the wall, not that the velocity be zero. In practice, this is not correct.

Experimental evidence indicates, however, that in many cases, the inviscid flow assumption gives good results. Moreover, neglecting the viscosity greatly simplifies the solution of the equations. It was Professor Ludwig Prandtl, of Göttingen, who first suggested an explanation of these results and also proposed a means whereby the boundary condition at the wall might be satisfied while retaining the simplicity of the inviscid analysis. Professor Prandtl suggested that in many cases the effect of viscosity is important only in a small region near the wall and may be neglected elsewhere. He named this region the "Grenzschichte" or boundary layer. The concept of the boundary layer made it possible not only to retain the inviscid analysis throughout most of the flow while satisfying the boundary condition at the wall, but made possible certain simplifications in the governing equations in the region where the viscosity was taken into account. The first analysis based on the boundary layer concept was published by Blasius in 1908 [89]. In all, the boundary layer concept has proved to be extremely useful.

While Prandtl was concerned chiefly with incompressible flow, we shall derive the boundary layer equations for compressible flow. In order to simplify the analysis, we shall confine ourselves at first to two-dimensional, steady flow. Recall the Navier–Stokes equations

$$\begin{aligned} \rho(uu_x + vu_y) &= -p_x + [\mu(2u_x - \tfrac{2}{3}(u_x + v_y))]_x + [u(u_y + v_x)]_y \\ \rho(uv_x + vv_y) &= -p_y + [\mu(2v_y - \tfrac{2}{3}(u_x + v_y))]_y + [\mu(u_y + v_x)]_x \end{aligned} \quad (1070)$$

For a gas, the coefficient of viscosity μ is generally small, so that the terms involving μ in the above equations will be small compared to the other terms unless the velocity gradients contained in the coefficients of μ are large. Consider now the flow over a flat or slightly curved wall, such as is encountered in many practical problems (Figure 223). Experiments show that if the free stream velocity

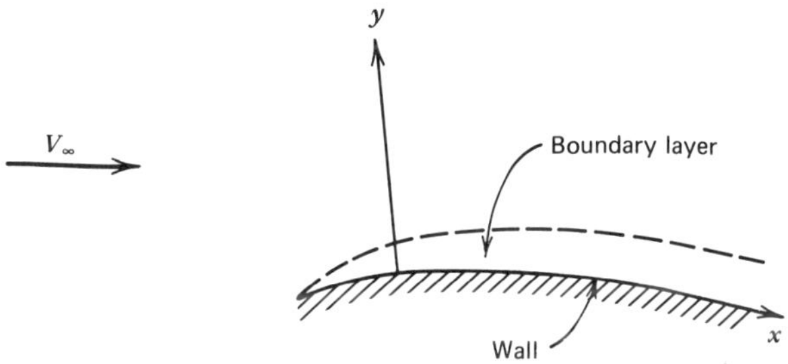

Figure 223 Boundary layer over a curved wall.

is sufficiently high, the velocity will attain essentially the value predicted by the inviscid theory a very short distance from the wall. There will thus be a thin layer adjacent to the wall in which the velocity will vary from zero at the wall to near the inviscid value a short distance from the wall. Only in this layer is the velocity gradient sufficiently large to make the viscous terms (those containing μ) in equations (1070) of comparable magnitude to the remaining terms, which are called the inertia terms. It was Prandtl's idea, then, that the viscous terms need be considered only in this small region near the wall, called the boundary layer, while the rest of the flow field could be analyzed by inviscid theory. The two solutions could then be joined together at a boundary called the outer edge of the boundary layer. The location of this outer edge is, of necessity, imprecise, but is generally taken along the line where the velocity reaches 99% of its inviscid value.

The variation of the velocity in the boundary layer is generally illustrated by a curve called a velocity profile, as shown in Figure 224. The length of the arrows at each value of y shows the magnitude of the x component of the velocity u at that value of y and x. The curve connecting the ends of these arrows is the velocity profile. The value of u is seen to change rapidly inside the boundary layer, and slowly outside it. Notice that $u = 0$ at the wall.

It will be observed that our figures show curvilinear coordinates. Equations (1070) may be applied unchanged in this coordinate system, provided the curva-

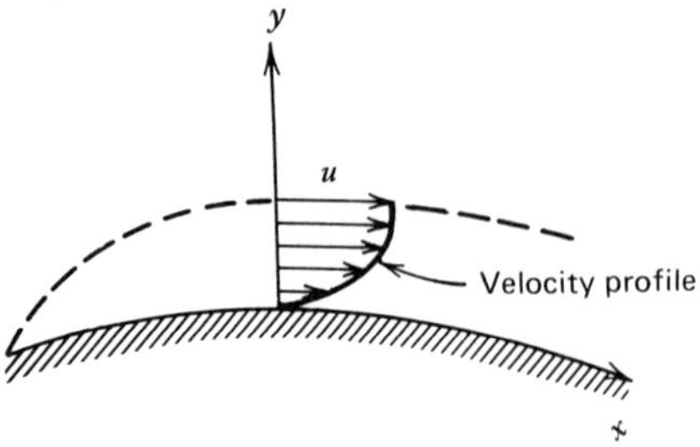

Figure 224 Typical boundary layer velocity profile.

ture of the wall is not too great. This may be seen by writing the equations in curvilinear coordinates:

$$
\begin{aligned}
\frac{R}{R+y}uu_x + vu_y + \frac{vu}{R+y} = {}& -\left(\frac{R}{R+y}\right)\frac{1}{\rho}p_x + \nu\left\{\left(\frac{R}{R+y}\right)^2 u_{xx} + u_{yy}\right. \\
& + \frac{1}{R+y}u_y - \frac{u}{(R+y)^2} + \frac{2R}{(R+y)^2}v_x \\
& \left. - \frac{R}{(R+y)^3}R_x v + \frac{Ry}{(R+y)^3}R_x u_x\right\}
\end{aligned}
\qquad (1071)
$$

$$
\begin{aligned}
\frac{R}{R+y}uv_x + vv_y - \frac{u^2}{R+y} = {}& -\frac{1}{\rho}p_y + \nu\left\{v_{yy} - \frac{2R}{(R+y)^2}u_x + \frac{1}{R+y}v_y\right. \\
& + \left(\frac{R}{R+y}\right)^2 v_{xx} - \frac{v}{(R+y)^2} + \frac{R}{(R+y)^3}R_x u \\
& \left. + \frac{Ry}{(R+y)^3}R_x v_x\right\}
\end{aligned}
$$

where R is the local radius of curvature of the wall and $\nu = \mu/\rho$. If R is large, equations (1071) reduce to equations (1070).

Although the equations are written in curvilinear coordinates, the vertical velocity component v will not be zero, even over a flat plate. This may be seen by applying the conservation of mass to a control volume $\mathcal{V}$ (Figure 225).

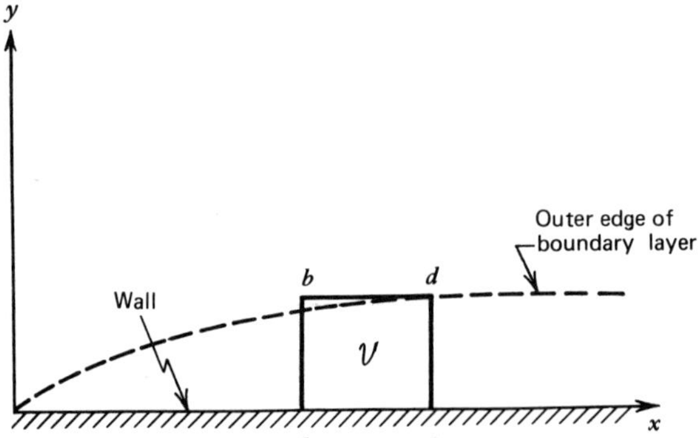

Figure 225 Control volume in boundary layer.

The mass flow in the left hand side of the control volume, *ab*, is greater than the mass flow rate out the right-hand side of the control volume, *cd*. This requires some mass to escape across the upper boundary *bd*, indicating the presence of a vertical velocity component. This component will, however, be small compared to the x component of the velocity, u. In this connection it should be noted that the outer edge of the boundary layer is not a streamline.

Referring to Figures 223, 224, and 225 it may be concluded on purely physical grounds that in addition to v being small compared to u, the x derivatives will be small compared to the y derivatives. An examination of the viscous terms in equations (1070) then shows that the term $(\mu u_y)_y$ is large compared to the other viscous terms. But because μ is small, $(\mu u_y)_y$ will itself be small compared to the inertia terms unless u_{yy} is sufficiently large. For these reasons, it is concluded that the remaining viscous terms may be neglected inside (as well as outside) the boundary layer. Furthermore, a comparison of the inertia terms shows that $uv_x \ll uu_x$ and $vv_y \ll vu_y$ inside the boundary layer. Thus, inside the boundary layer, equations (1070) reduce to

$$\begin{aligned} \rho(uu_x + vu_y) &= -p_x + (\mu u_y)_y \\ 0 &= -p_y \end{aligned} \qquad (1072)$$

Equations (1072) are called the boundary layer equations for two-dimensional, steady flow.

Of particular interest is the fact that p_y is zero in the boundary layer, i.e., the pressure does not vary across the boundary layer. This is sometimes expressed by saying that the pressure outside the boundary layer is impressed on the boundary layer. In boundary layer theory, the term *across* the boundary layer means in the y direction, and *along* the boundary layer means in the x direction.

In most practical problems, the viscosity gradients are but small and are generally neglected. Exceptions to this occur in flows with large temperature gradients.

Equations (1072) have five dependent variables: u, v, ρ, p and μ. In incompressible flow, μ and ρ would be constant and p may be taken from the inviscid solution. The addition of the continuity equation $\nabla \mathbf{V} = 0$ would thus render the solution determinate. In compressible flow, the situation is not so simple. Because ρ is a variable, the energy equation must be included in the solution.

Recall the energy equation for two-dimensional, steady flow:

$$\begin{aligned} \rho c_p(uT_x + vT_y) = k(T_{xx} + T_{yy}) + up_x + vp_y + \mu\{2(u_x^2 + v_y^2) \\ - \tfrac{2}{3}(u_x + v_y)^2 + (v_x + u_y)^2\} \end{aligned} \qquad (1073)$$

where we have made the ideal gas assumption $dh = c_p dT$ to simplify the analysis. Since the thermal conductivity k in gases is small, the terms involving k will also be small unless T_{xx} or T_{yy} are large. For the type of flow under consideration, experimental evidence indicates a temperature variation qualitatively similar to the velocity variation; that is, a rapid variation of the temperature in the y direction near the wall, followed by a much slower variation away from the wall. This suggests the existence of a thermal boundary layer, that is, a small region near the wall in which the terms involving thermal conductivity must be considered, outside of which these terms may be neglected. The thermal boundary layer may or may not be of the same thickness as the velocity boundary layer, but experimental evidence indicates that it is at least of the same order of magnitude. We may, therefore, simplify the energy equation for use near the wall, and neglect the terms involving k and μ away from the wall. The precise shape of the temperature profile in the thermal boundary layer will depend on the boundary condition at the wall, but a qualitative idea of the situation may be gained from Figure 226. The thermal boundary layer thickness may or may

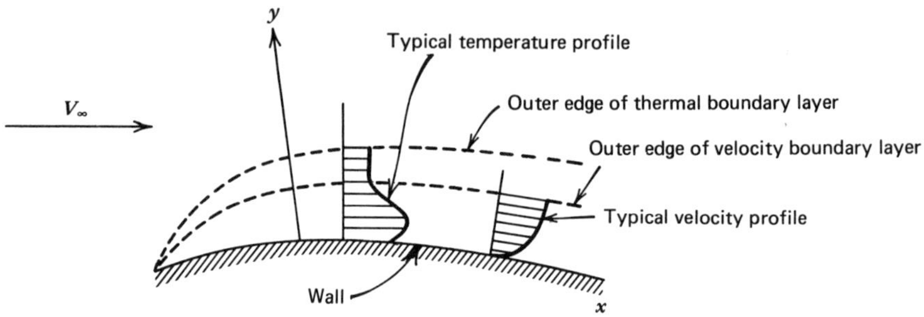

Figure 226 Typical temperature profile in thermal boundary layer.

not be greater than the velocity boundary layer thickness as shown in Figure 226, merely for example. Referring to equation (1073), then, we see that inside the thermal boundary layer, $T_{yy} \ll T_{xx}$, and $u_x, v_y \ll u_y$. In addition, we may assume $p_y \cong 0$ inside the thermal boundary layer also. Thus, near the wall, the energy equation may be written

$$\rho c_p(uT_x + vT_y) = kT_{yy} + up_x + \mu u_y^2 \tag{1074}$$

This is sometimes called the boundary layer energy equation.

The above arguments are, admittedly, not rigorous. They are physical rather than mathematical arguments. Nevertheless, the boundary layer concept has proved extremely useful in practice.

An attempt is sometimes made to put the boundary layer arguments on a more rigorous basis by nondimensionalizing the equations. If we consider the flow over a wall of length l with free stream velocity V_∞, we define the following dimensionless variables:

$$x' = \frac{x}{l}, \quad y' = \frac{y}{l}, \quad u' = \frac{u}{V_\infty}, \quad v' = \frac{v}{V_\infty}, \quad p' = \frac{p}{\rho_\infty V_\infty^2}, \quad T' = \frac{T}{T_w - T_\infty}$$

$$T' = \frac{T}{\Delta T_0}$$

where T_w is the wall temperature, $\Delta T_0 = T_w - T_\infty$, and the subscript ∞ refers to the free stream conditions. Where the wall temperature is a function of x, an average value may be taken for T_w. In terms of the new variables, the boundary layer equations become

$$u'u'_{x'} + v'u'_{y'} = -p'_{x'}\frac{\rho_\infty}{\rho} + \left(\frac{\mu}{\rho_\infty V_\infty l}\right)\frac{\rho_\infty}{\rho}u'_{y'y'} \tag{1075}$$

$$u'T'_{x'} + v'T'_{y'} = \left(\frac{k}{\rho_\infty c_p V_\infty l}\right)\frac{\rho_\infty}{\rho}T'_{y'y'} + \frac{V_\infty^2}{c_p \Delta T_0}u'p'_{x'}\frac{\rho_\infty}{\rho} + \left(\frac{\mu V_\infty}{\rho_\infty c_p l \Delta T_0}\right)\frac{\rho_\infty}{\rho}u'^2_{y'}$$

The coefficients in equations (1075) and (1076) are dimensionless groups which are usually designated by the following symbols:

$$\frac{\mu}{\rho_\infty V_\infty l} = R_l, \qquad \frac{k}{\rho_\infty c_p V_\infty l} = \frac{1}{\dfrac{c_p\mu}{k}\cdot\dfrac{\rho_\infty V_\infty l}{\mu}}, \qquad \frac{c_p\mu}{k} = Pr$$

$$\frac{V_\infty^2}{c_p\Delta T_0} = E, \qquad \frac{\mu}{\rho_\infty c_p l \Delta T_0} = \frac{V_\infty^2}{c_p \Delta T_0 \dfrac{V_\infty \rho_\infty l}{\mu}} = \frac{E}{R_l}$$

$$\frac{\rho_\infty}{\rho} = \frac{T'/T'_\infty}{\gamma_{p'}M_\infty^2 T'_\infty}, \qquad M_\infty^2 = \frac{V_\infty^2}{\gamma R T_\infty} \tag{1076}$$

R_l is called the Reynolds number based on l, Pr is called the Prandtl number, and E is called the Eckert number. These parameters are similarity parameters, since for given boundary conditions in the transformed coordinates and given values of R_l, Pr, M_∞, γ and E, the solution will be the same regardless of how the individual parameters $\rho_\infty, V_\infty, l, \mu, c_p, k, T_\infty$ and T_w vary. The same dimensionless groups R_l, Pr and E could have been obtained using Buckingham's π theorem.

We may examine the orders of magnitude of the various terms in equations (1075) and (1076) in the transformed coordinate system. Referring to Figure 227,

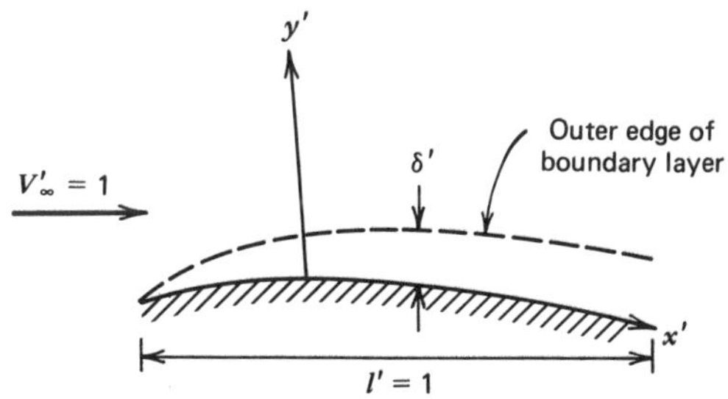

Figure 227 Boundary layer in nondimensional coordinates.

we assume that the thickness of the thermal and velocity boundary layers are of the same order of magnitude and so designate it by the single term δ'. In terms of our dimensionless groups, the boundary layer equations become

$$u'u'_{x'} + v'u'_{y'} = -p'_{x'}\frac{\rho_\infty}{\rho} + \frac{1}{R_l}\frac{\rho_\infty}{\rho}u'_{y'y'} \tag{1077}$$

$$u'T'_{x'} + v'T'_{y'} = \frac{1}{Pr\,R_l}\frac{\rho_\infty}{\rho}T'_{y'y'} + Eu'p'_{x'}\frac{\rho_\infty}{\rho} + \frac{E}{R_l}\frac{\rho_\infty}{\rho}u'^2_{y'} \tag{1078}$$

Referring to Figure 227, we may infer the following orders of magnitude:

$$u' = \mathrm{O}(1), \qquad u'_{x'} = \mathrm{O}(1), \qquad v' = \mathrm{O}(\delta'), \qquad u'_{y'} = \mathrm{O}\left(\frac{1}{\delta'}\right)$$

$$u'_{y'y'} = \mathrm{O}\left(\frac{1}{\delta'^2}\right), \qquad T'_{x'} = \mathrm{O}(1), \qquad T'_{y'} = \mathrm{O}\left(\frac{1}{\delta'}\right), \qquad T'_{y'y'} = \mathrm{O}\left(\frac{1}{\delta'^2}\right)$$

$$u'^2_{y'} = \mathrm{O}\left(\frac{1}{\delta'^2}\right)$$

In addition, $p'_{x'}$ must be of the same order as the inertia terms at the outer edge of the boundary layer and therefore can be no larger than 0(1). ρ will usually remain of the same order as ρ_∞ and hence $\rho/\rho_\infty = 0(1)$. Thus we see that all the terms in equations (1077) and (1078) will be of order 1 if

$$R_l = \mathrm{O}\left(\frac{1}{\delta'^2}\right), \qquad Pr = \mathrm{O}(1), \qquad E = \mathrm{O}(1)$$

For gases, the Prandtl number is, in fact, of order 1. For adiabatic flow, the Eckert number will also be of order 1. This may be seen by considering that in the free stream

$$\frac{V_\infty^2}{2c_p} = T_0 - T_\infty$$

while for an insulated wall, the temperature T_w is very close to the free stream stagnation temperature T_0. For an isothermal wall of prescribed temperature this may not necessarily be the case. The condition on the Reynolds number will be satisfied if the Reynolds number is sufficiently large. The boundary layer thickness tends to increase as the Reynolds number decreases. Since the entire boundary layer concept is predicated on δ' being small, the boundary layer theory will be valid only for large Reynolds numbers.

Distinction Between Laminar and Turbulent Flow

Flow in boundary layers, as in pipes, may be either laminar or turbulent. The reader will recall from elementary fluid dynamics that there are two distinct flow regimes in pipes. This was illustrated in a famous experiment by Osborne Reynolds in which a thin filament of dye was injected into the flow in a transparent pipe. As long as the flow was laminar, the dye filament was transported downstream intact. When the flow became turbulent, the dye filament was broken up and the entire flow was stained (Figure 228). The breakup of the dye is caused by the presence of random velocity fluctuations in the flow which are present when the flow is turbulent and absent when the flow is laminar. Whether the flow is laminar or turbulent in a pipe will depend on the Reynolds number

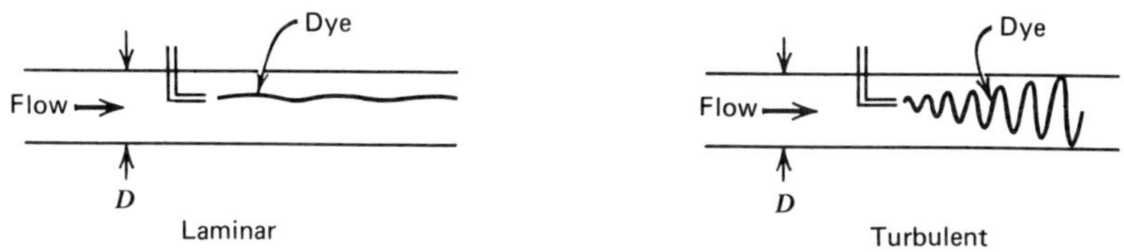

Figure 228 Laminar and turbulent flow in a pipe.

based on the diameter, $VD\rho/\mu$. If the Reynolds number is sufficiently large, the flow will be turbulent. Near the entrance of a pipe in a reservoir, the flow may be laminar for some distance and then become turbulent. This change from laminar to turbulent is called transition. The point of transition will depend on the Reynolds number based on x, $Vx\rho/\mu$, where x is the distance from the entrance of the pipe and V is the average velocity in the pipe. The change from laminar to turbulent flow usually takes place over a very short distance.

The situation in boundary layers is analogous to the situation in pipes. Near the leading edge of a surface, the flow in the boundary layer will be laminar. A certain distance from the leading edge, the flow will become turbulent, the point of transition depending on the Reynolds number based on x, $V_\infty x\rho_\infty/\mu$, where x is the distance from the leading edge. Except in the case of rarified gases, the change from laminar to turbulent will take place over a very short distance. A typical situation is shown in Figure 229.

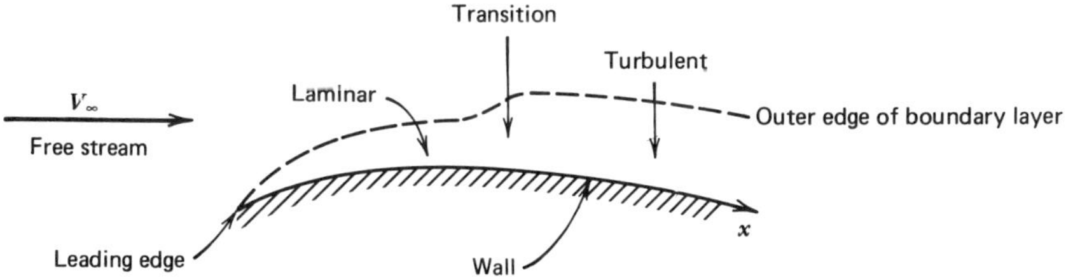

Figure 229 Boundary layer transition.

Laminar and turbulent flows are fundamentally different. In a laminar flow, transverse stresses (shear stresses) are transmitted mainly by the random fluctuations of the molecules. In turbulent flows, stresses parallel to the mean flow direction are caused by the fluctuation of macroscopic fluid particles. For this reason, the two types of flows require different mathematical analyses. In the present chapter, we shall confine ourselves to the analysis of laminar boundary layers. Turbulent boundary layers will be considered in the next chapter.

Elementary Solutions: The Crocco–Busemann Integrals

The system of equations to be solved is now the following:

$$\rho(uu_x + vu_y) = -p_x + (\mu u_y)_y \tag{1079}$$

$$\rho c_p(uT_x + vT_y) = up_x + (kT_y)_y + \mu u_y^2 \tag{1080}$$

$$(\rho u)_x + (\rho v)_y = 0 \tag{1081}$$

$$p = \rho RT \tag{1082}$$

It will be observed that the continuity equation is unchanged in the boundary layer. An equation of state is required and for simplicity we assume, for the present, the perfect gas law (1082). In order to render the problem determinate we must know how μ, c_p and k vary with p, T and ρ. In addition, we require a suitable set of boundary conditions. As a first approximation, we will suppose μ, c_p and k to be constant. At the wall, the boundary condition on the velocity is that u and v be zero. The temperature at the wall may be arbitrarily specified, but for simplicity the wall is usually assumed to be adiabatic or isothermal; that is, the wall is either assumed to be insulated (no heat transfer) or of constant and uniform temperature. In the former case we have $(T_y)_w = 0$ and in the latter case $T_w =$ constant. The condition at the outer edge of the boundary layer is more difficult, since the exact thickness of the boundary layer is not known. In practice, the following procedure has been found to give good results: the pressure and velocity distributions at the wall are calculated on the basis of inviscid flow theory and are then used as the conditions at the outer edge of the boundary layer. Since the thickness of the boundary layer is not known a priori, or even precisely defined, the conditions are applied at $y = \infty$ in the mathematical solution. In summary, the conditions at $y = 0$ for inviscid flow are assumed to prevail at $y = \infty$ in the viscous flow problem. No rigorous mathematical justification for this apparently contradictory procedure has ever been published. The pragmatic justification is that the procedure is found to give good results. Solutions found in this way show that the velocity reaches 99% of its inviscid value very quickly; that is, only a short distance from the wall, thus conforming to the boundary layer concept. Figure 230 shows a typical solution found in

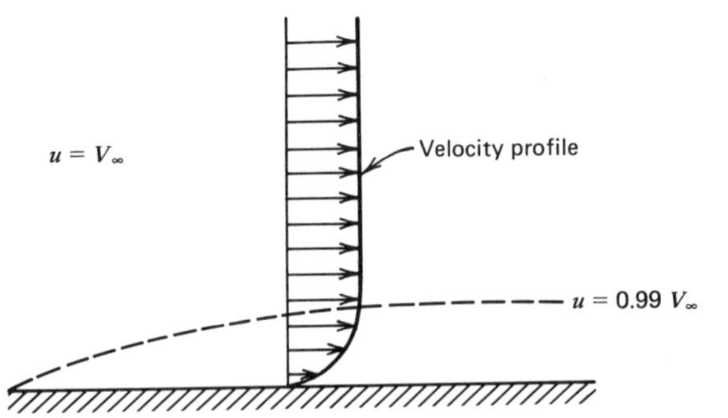

Figure 230 Typical solution of boundary layer equations.

this way for a semi-infinte flat plate. The results also agree well with experiment. An attempt to rationalize this procedure has been made by von Mises [90] in terms of the so-called inner and outer problem. The analysis of von Mises is not really very satisfying and will not be given here. Readers interested in pursuing this question further may consult reference [90].

The boundary conditions for equations (1079) to (1082) are then

$$\begin{aligned} y=0: \quad & u=v=0 \\ & T=T_w=\text{constant (isothermal wall)} \\ & \left(\frac{\partial T}{\partial y}\right)_w=0 \text{ (adiabatic wall)} \\ y=\infty: \quad & u=U(x) \\ & T=T_1(x) \end{aligned}$$

where $U(x)$ and $T_1(x)$ represent the inviscid solutions at the wall. To find $T(x)$ the inviscid flow is usually assumed to be adiabatic. In addition, the pressure in the boundary layer is found from

$$\rho U U_x = -p_x \tag{1083}$$

The problem is now determinate.

An elementary solution of the problem has been given by Crocco [91] and Busemann [92]. Suppose T is a function of u only. Then the energy equation (1080) may be written,

$$\rho c_p T_u(uu_x + vu_y) = up_x + (kT_u u_y)_y + \mu u_y^2 \tag{1084}$$

Introducing the momentum equation (1079) we have

$$\rho c_p T_u[-p_x + (\mu u_y)_y] = up_x + T_u(ku_y)_y + (T_{uu}k + \mu)u_y^2 \tag{1085}$$

Recall now the definition of the Prandtl number $Pr = c_p\mu/k$. Introducing the Prandtl number and rearranging, equation (1085) may be written

$$-p_x(c_p T_u + u) + c_p \frac{Pr-1}{Pr} T_u(\mu u_y)_y = (T_{uu}k + \mu)u_y^2 \tag{1086}$$

It is well known that the Prandtl number for most common gases is in the vicinity of 1 at standard temperature and pressure. For air it is approximately 0.75. A rough, first-order solution may, therefore, be obtained by assuming $Pr = 1$. By inspection it may be seen that such a solution of equation (1086) is

$$T_{uu} = -\frac{\mu}{k} = -\frac{1}{c_p} \tag{1087}$$

Integrating equation (1087) we get

$$T = -\frac{u^2}{2c_p} + c_1 u + c_2 \tag{1088}$$

The constants will depend on the boundary conditions. For the adiabatic wall we have

$$y = 0: \quad u = 0, \qquad T_y = T_u u_y = 0 \Rightarrow T_u = 0$$

$$y = \infty: \quad u = U(x), \qquad T = T_1(x)$$

The solution is then

$$T = T_1 + \frac{1}{2c_p}(U^2 - u^2) \tag{1089}$$

It will be noted that this solution implies that the wall temperature equals the stagnation temperature at the outer edge of the boundary layer

$$T_{aw} = T_1 + \frac{U^2}{2c_p} \tag{1090}$$

where T_{aw} is the adiabatic wall temperature. Actually, Emmons [93] has shown that

$$T_{aw} = T_1 + R\frac{U^2}{2c_p} \tag{1091}$$

where R is called the recovery factor. For subsonic laminar flow $R \cong \sqrt{Pr}$ which, for air, is approximately 0.86, while for supersonic and turbulent flow $R \cong 0.90$. From equation (1091) it may be seen that the recovery factor may be defined by

$$R = \frac{T_{aw} - T_1}{T_0 - T_1} = \frac{T_{aw} - T_1}{\dfrac{U^2}{2c_p}} = \frac{\dfrac{T_{aw}}{T_1} - 1}{\dfrac{\gamma - 1}{2}M^2} \tag{1092}$$

where T_0 is the stagnation temperature at the outer edge of the boundary layer. Thus, the recovery factor may also be written

$$T_{aw} = T_1\left(1 + R\frac{\gamma - 1}{2}M^2\right) \tag{1093}$$

These results have important consequences in the design of supersonic aircraft. At equilibrium, the skin temperature of such an aircraft will reach approximately the adiabatic wall temperature. As an example, let us consider a supersonic aircraft designed to cruise at a Mach number of 2.2. At this speed, the adiabatic wall temperature is given by

$$\frac{T_{aw}}{T_1} = 1 + 0.2(4.84)(0.9) = 1.87$$

If the aircraft is cruising at 60,000 feet, the ambient temperature will be approximately 393° R. Thus we have

$$T_{aw} = 735°\text{R} = 275°\text{F}$$

For an aircraft designed to cruise at a Mach number of 3, on the other hand, we find

$$\frac{T_{aw}}{T_1} = 2.62$$

or

$$T_{aw} = 1030°\mathrm{R} = 570°\mathrm{F}$$

Since aircraft aluminum alloy retains its structural strength at 275°F but has lost much of it at 570°F, it follows that aluminum alloy is a suitable structural material for an aircraft designed to cruise at Mach 2.2 but not for an aircraft designed to cruise at Mach 3.0. The higher cruising speed would thus require resort to higher temperature materials such as titanium and steel. Because of the many advantages of aluminum alloy as well as the more highly developed aircraft aluminum technology, high adiabatic wall temperatures represent a real barrier to higher supersonic cruising speeds in aircraft.

If the wall temperature is fixed (isothermal wall) the boundary conditions become

$$y = 0: \quad u = 0, \qquad T = T_w = \text{constant}$$
$$y = \infty: \quad u = U, \qquad T = T_1(x)$$

and the solution to equation (1087) becomes

$$\frac{T - T_w}{T_1} = \left(1 - \frac{T_w}{T_1}\right)\frac{u}{U} + \frac{U^2}{2c_p T_1}\frac{u}{U}\left(1 - \frac{u}{U}\right) \tag{1094}$$

Differentiating with respect to u we find

$$\frac{U}{T_1}\left(\frac{dT}{du}\right)_w = \frac{U^2}{2c_p T_1} - \frac{T_w - T_1}{T_1} \tag{1095}$$

According to Fourier's law, heat will flow into the wall when $(\partial T/\partial y)_w$ is positive while heat will flow out of the wall when $(\partial T/\partial y)_w$ is negative. Now

$$\left(\frac{\partial T}{\partial y}\right)_w = \left(\frac{\partial T}{du}\right)_w \left(\frac{\partial u}{\partial y}\right)_w$$

and for boundary layer flow $(\partial u/\partial y)_w$ is always positive. Thus, heat will flow into the wall when $(dT/du)_w$ is positive and heat will flow out of the wall when $(dT/du)_w$ is negative. Equation (1095) shows heat may flow into the wall even though the wall is hotter than the free stream. Not until

$$\frac{T_w - T_1}{T_1} > \frac{U^2}{2c_p T_1}$$

does heat flow out of the wall and into the fluid (assuming $Pr = 1$). Some typical temperature profiles are shown in Figure 231. Figure 231 (a) shows heat flowing

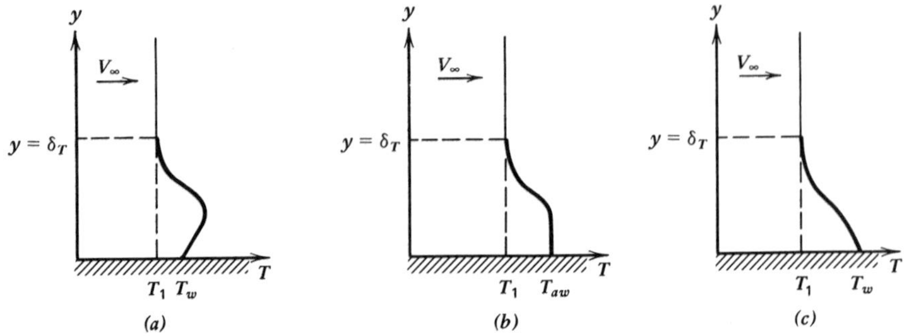

Figure 231 Variation of temperature profiles with wall temperature.

into the wall even though $T_w > T_1$. Figure 231 (b) shows the adiabatic case; Figure 231 (c) shows heat flowing into the wall from the fluid. $y = \delta_T$ shows the approximate thickness of the thermal boundary layer.

Physically, a number of different mechanisms are at work here. If the flow were adiabatic, that is, if h_0 were the same everywhere, the rise in temperature near the wall could be accounted for by the decrease in velocity. In the present case, however, the flow is not adiabatic because two other factors are present: first, the term $(kT_y)_y$ in equation (1080) shows the presence of heat transfer through conduction, and second, the term μu_y^2 in equation (1080) shows that heat is being generated by friction. Comparison with equations (186) and (187) in Chapter 2 shows that the term μu_y^2 is all that remains of the viscous dissipation function times the viscosity, $\mu\Phi$, after the two-dimensional boundary layer assumptions have been introduced. As was pointed out in Chapter 2, the term $\mu\Phi$ represents the mechanical energy being turned into heat by friction. It is, therefore, the combination of heat conduction, viscous dissipation and change in velocity which determines the local temperature.

Reynold's analogy for compressive flow

The reader will recall the Reynold's analogy for incompressible flow.

$$q_w = \frac{k}{\mu}\tau_w \frac{T_w - T_1}{U}$$

where q_w represents the heat transfer rate at the wall, T_w the sheer stress at the wall, and T_∞ the free stream temperature. An analogous expression may be developed for compressible flow. Recall equation (1094)

$$\frac{T - T_w}{T_1} = \left(1 - \frac{T_w}{T_1}\right)\frac{u}{U} + \frac{U^2}{2c_p T_1}\frac{u}{U}\left(1 - \frac{u}{U}\right) \tag{1094}$$

For a perfect gas,

$$\frac{U^2}{2c_p T_1} = \frac{\gamma - 1}{2} M_1^2 \tag{1096}$$

where M_1 represents the Mach number at the outer edge of the boundary layer. Thus,

$$\frac{T-T_w}{T_1}=\left(1-\frac{T_w}{T_1}\right)\frac{u}{U}+\frac{\gamma-1}{2}M_1^2\frac{u}{U}\left(1-\frac{u}{U}\right) \tag{1097}$$

Differentiating with respect to y,

$$\frac{U}{T_1}\frac{\partial T}{\partial y}=\left(1-\frac{T_w}{T_1}\right)\frac{\partial u}{\partial y}+\frac{\gamma-1}{2}M_1^2\left[\frac{\partial u}{\partial y}\left(1-\frac{u}{U}\right)+\frac{u}{U}\frac{\partial u}{\partial y}\right] \tag{1098}$$

At the wall, $u=0$ and hence

$$\frac{U}{T_1}\left(\frac{\partial T}{\partial y}\right)_w=\left(1-\frac{T_w}{T_1}\right)\left(\frac{\partial u}{\partial y}\right)_w+\frac{\gamma-1}{2}M_1^2\left(\frac{\partial u}{\partial y}\right)_w \tag{1099}$$

Now the heat transfer at the wall, $q_w=-k(T_y)_w$ and hence

$$Uq_w=-k(u_y)_w\left[T_1+\frac{\gamma-1}{2}M_1^2-T_w\right]=-k(u_y)_w[T_{01}-T_w] \tag{1100}$$

where T_{01} is the stagnation temperature at the outer edge of the boundary layer. Now $\tau_w=\mu(u_y)_w$ and hence

$$k(u_y)_w=\frac{k}{\mu}\tau_w=\frac{c_p}{Pr}\tau_w=c_p\tau_w \tag{1101}$$

since $Pr=1$ has been assumed. Moreover, $T_{01}=T_{aw}$ for $Pr=1$. Thus,

$$q_w=c_p\tau_w\frac{T_w-T_{aw}}{U}=\frac{k}{\mu}\tau_w\frac{T_w-T_{aw}}{U} \tag{1102}$$

Now recall for $Pr=1$

$$T_{aw}=T_1\left(1+\frac{\gamma-1}{2}M_1^2\right) \tag{1103}$$

For incompressible flow, $M_1=0$, $T_{aw}=T_1$, and equation (1102) reduces to equation (1095).

The Velocity Profile

The solution of equation (1087) gives us the temperature as a function of the velocity but does not give us the temperature or velocity profile as a function of y. The solution to this problem was first addressed by von Karman and Tsien [94]. von Karman and Tsien employed the transformation of von Mises [95]. Although von Mises derived this transformation for incompressible flow, it may be extended to compressible flow. Define a stream function ψ so that

$$\psi_y=\rho u, \qquad \psi_x=-\rho v \tag{1104}$$

Notice that in terms of ψ, the continuity equation (1081) is identically satisfied. We now change the independent variables from x and y to x and ψ in the momentum equation (1079).

$$\rho u u_x = -p_x + \rho u(\mu \rho u u_\psi)_\psi \tag{1105}$$

If we restrict ourselves to the case $p_x = 0$ (flat plate) we get

$$u_x = (\mu \rho u u_\psi)_\psi \tag{1106}$$

This equation may be solved by successive approximations. As a start, the well known solution of Blasius [89] for incompressible flow may be used for u in the right-hand side of equation (1106). If, in addition, we make the perfect gas assumption $p = \rho RT$, ρ will vary inversely with T since p is constant. T may be expressed as a function of u using the Crocco–Busemann integral. μ may be taken as constant or proportional to T to some power. Thus, equation (1106) may be integrated and the resulting $u(x,\psi)$ may then be used in the right-hand side to find a new $u(x,\psi)$, and so on. In practice it is found that the solution converges rapidly. Since the solution of Blasius is known only in tabular form, the actual integration is best carried out on a computer. Von Karman and Tsien carried out the integration in nondimensional form using the variables

$$u' = \frac{u}{U}, \qquad x' = \frac{x}{L}, \qquad \psi' = \frac{\psi}{UL}\sqrt{\frac{\rho_\infty UL}{\mu_\infty}}, \qquad \rho' = \frac{\rho}{\rho_\infty}, \qquad \mu' = \frac{\mu}{\mu_\infty}, \quad T' = \frac{T}{T_\infty} \tag{1107a}$$

$$\zeta = \frac{\psi'}{\sqrt{x'}}$$

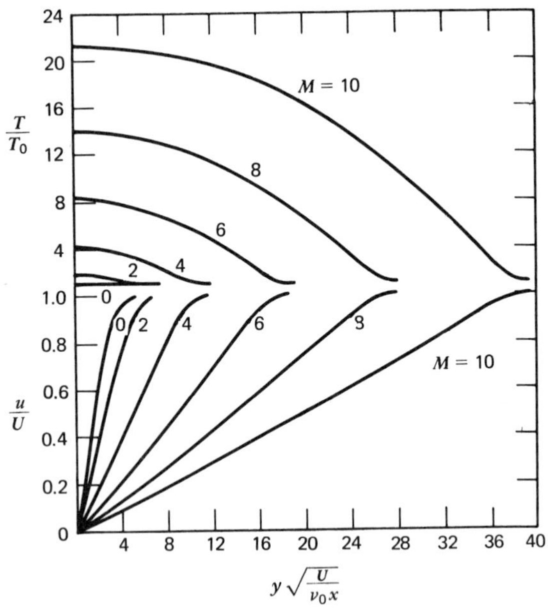

Figure 232 Temperature and velocity profiles obtained by Karman and Tsien [94].

where L is the length of the plate. ψ was defined by

$$\rho u = \rho_0 \psi_y, \qquad \rho v = -\rho_0 \psi_x$$

and μ was taken as proportional to $T^{.76}$. Equation (1106) then becomes

$$-\frac{\zeta}{2} u'_\zeta = (u' \rho' \mu' u'_\zeta)_\zeta \tag{1107}$$

If $u'\rho'\mu' = f(\zeta)$ is taken from the Blasius solution and the Crocco–Busemann integral as explained above, the solution to equation (1107) is given by

$$u' = C \int_0^\zeta \frac{F}{f} d\zeta \qquad \text{where} \qquad F = \exp\left[-\int_0^\zeta \frac{\zeta d\zeta}{\mathrm{f}}\right] \qquad \text{and} \qquad C = \int_0^\infty \frac{F}{f} d\zeta \tag{1108}$$

from the boundary condition $u = U$ at $y = \infty$. The resulting temperature and velocity profiles for an adiabatic wall are shown in Figure 232, plotted for various Mach numbers.

The Howarth Transformation

The Howarth transformation is given by the equation

$$\eta = \int_0^y \frac{\rho}{\rho_\infty} dy \tag{1109}$$

Although this transformation is commonly called the Howarth transformation in the west, it was actually first employed by Dorodnitsyn [96]. Let us define

$$\eta = \int_0^y \frac{\rho}{\rho_\infty} dy, \qquad u = \psi_\eta \tag{1110}$$

In the case of the flat plate ($p_x = 0$), the boundary layer momentum equation becomes

$$\psi_\eta \psi_{\eta x} - \psi_x \psi_{\eta\eta} = \frac{\rho\mu}{\rho_\infty^2} \psi_{\eta\eta\eta} \tag{1111}$$

If we now introduce the approximation $\mu \sim T$ (a good approximation) then, for $p =$ constant, $\rho\mu =$ constant for a perfect gas. Hence,

$$\frac{\rho\mu}{\rho_\infty^2} = \frac{\rho_\infty \mu_\infty}{\rho_\infty^2} = \frac{\mu_\omega}{\rho_\infty} = v_\infty = \text{constant}$$

and equation (1111) becomes

$$\psi_\eta \psi_{\eta x} - \psi_x \psi_{\eta\eta} = v_\infty \psi_{\eta\eta\eta} \tag{1112}$$

This will be recognized as the incompressible form of the boundary layer momentum equation. The effect is thus to uncouple the momentum and energy equations. No longer is it necessary to solve the energy equation in order to solve the momentum equation. Moreover, the solution to equation (1112) is well known. It is the solution first given by Blasius [89] for the incompressible case. If we make the change of variables

$$\zeta = \frac{\eta}{2}\sqrt{\frac{U}{\nu_\infty x}}, \qquad \psi = \sqrt{\nu_\infty x U}\, f(\zeta) \tag{1113a}$$

equation (1112) becomes

$$f f_{\zeta\zeta} + f_{\zeta\zeta\zeta} = 0 \tag{1113}$$

which is the well-known Blasius equation. The boundary conditions are

$$\zeta = 0: \quad f = 0, \qquad f_\zeta = 0; \qquad \zeta = \infty: \quad f_\zeta = 2 \tag{1114}$$

Equation (1113) must be integrated numerically. Under the assumption of constant stagnation enthalpy in the boundary layer (Crocco–Busemann solution, $Pr = 1$), the inverse transformation becomes

$$\begin{aligned} y &= \int_0^\eta \frac{T}{T_\infty}\, d\eta = \int_0^\eta \left[1 + \frac{\gamma - 1}{2c_1^2}(U^2 - u^2)\right] d\eta \\ &= \left[1 + \frac{\gamma - 1}{2} M_1^2\right]\eta - \frac{\gamma - 1}{2} M_1^2 \int_0^\eta \left(\frac{u}{U}\right)^2 d\eta \\ &= \frac{1}{2}\left(\frac{U}{\nu_\infty x}\right)^{-1/2}\left\{\zeta + \frac{\gamma - 1}{2} M_1^2\left[\zeta - \frac{1}{2}\int_0^\eta f_\zeta^2\, d\zeta\right]\right\} \end{aligned} \tag{1115}$$

Since $u/U = \frac{1}{2} f_\zeta(\zeta)$, we may plot the compressible velocity profile as a function of Mach number once the Blasius solution is known. The results are given in Figure 233. The results are observed to agree well with those obtained by von Karman and Tsien.

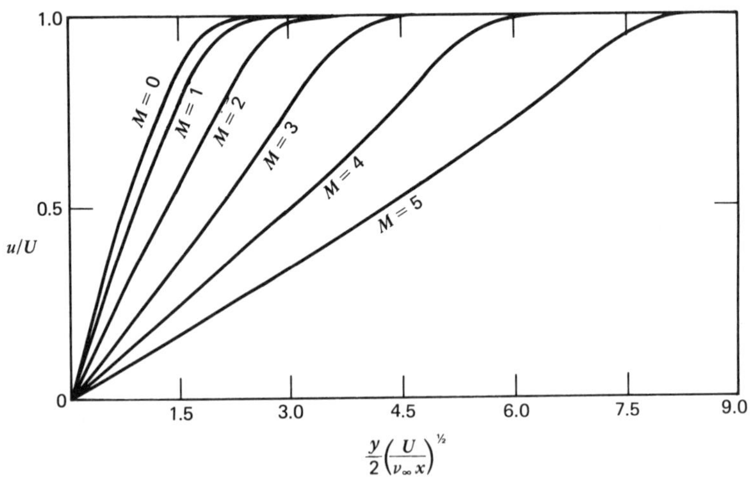

Figure 233 Boundary layer velocity profiles obtained by Howarth transformation.

Chapman and Rubesin [97] extended the solution of the compressible laminar boundary layer over a flat plate to find the temperature profile for the case of wall temperature varying with x. Under the assumption $\rho\mu =$ constant, the momentum equation is transformed to the Blasius equation (1113) as in the previous section. Under the assumption that $Pr =$ constant, the boundary layer energy equation may be written in the form

$$\rho u T_x + \rho v T_y = \frac{1}{Pr}(\mu T_y)_y + \mu u_y^2 \tag{1116}$$

We now introduce the same dimensionless variables as were used by von Karman and Tsien (1107a). Performing the von Mises transformation, the boundary layer momentum and energy equations become

$$u'_{x'} = (u' u'_{\psi'})_{\psi'} \tag{1117}$$

$$T'_{x'} = \frac{1}{Pr}(u' T'_{\psi'})_{\psi'} + (\gamma - 1) M_\infty^2 u' u'_{\psi'} \tag{1118}$$

Let ζ and $f(\zeta)$ be defined as in (1113a). Then

$$f(\zeta) = \frac{\psi'}{\sqrt{x'}} \tag{1119}$$

Equation (1117) reduces to the Blasius equation (1113), and equation (1118) becomes

$$T'_{\zeta\zeta} + Prf_\zeta T'_\zeta - 2Prf_\zeta x' T'_{x'} = -\frac{Pr}{4}(\gamma - 1) M_\infty^2 f_{\zeta\zeta}^2 \tag{1120}$$

Since $f(\zeta)$ is known, $T'(x',\zeta)$ is the only unknown in equation (1120). We may now impose the boundary condition

$$T'(x',0) = T'_{aw} + T'_2(x) \tag{1121}$$

$$T'(x',\infty) = 1$$

where T'_{aw} represents the dimensionless adiabatic wall temperature and $T'_2(x)$ represents an arbitrary temperature distribution subject to the constraint that $(\partial T/\partial x)_w$ is of the same order as $(T_w - T_\infty)/\delta$ or less, where δ is the boundary layer thickness.

The solution of equation (1120) is given by

$$T'(x',\zeta) = N(\zeta) + \sum_0^\infty a_n x'^n Y_n(\zeta) \tag{1122}$$

where

$$N(\zeta) = 1 + \frac{\gamma - 1}{2} M_\infty^2 r(\zeta) \tag{1123}$$

$$r(\zeta) = \frac{Pr}{2}\int_\zeta^\infty \left(\frac{d^2 f}{d\zeta^2}\right) d\zeta \tag{1124}$$

and $Y_n(\zeta)$ represents the solution to the equation

$$\frac{d^2 Y_n}{d\zeta} + Prf\frac{dY_n}{d\zeta} - 2n\,Pr\frac{df}{d\zeta}\frac{dY_n}{d\zeta} = 0 \tag{1125}$$

subject to the boundary conditions $Y_n(0) = 1, Y_n(\infty) = 0$. The function $r(\zeta)$ will be recognized as the recovery factor defined earlier (1093) and may be evaluated by numerical integration of equation (1124). The result is plotted in Figure 234. Equation (1125) can likewise be integrated numerically and the

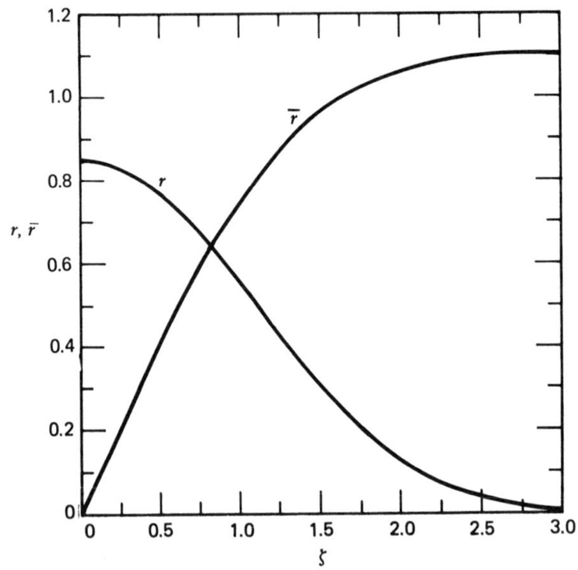

Figure 234 Recovery factor obtained by Chapman and Rubesin [97].

results are shown in Figure 235. The series $\sum_0^\infty a_n x'^n$ is given by

$$T_2'(x') = \sum_0^\infty a_n x'^n \tag{1126}$$

as may be seen by reference to the boundary conditions. The given temperature variation must thus be expressed as a power series in x'.

To determine the result in the physical plane we need the inverse transformation from ζ to y. This is given by

$$\frac{y}{2}\sqrt{\frac{U_\infty}{v_\infty x}} = \zeta + \frac{\gamma - 1}{2} M_\infty \bar{r}(\zeta) + \sum_0^\infty a_n x'^n \bar{Y}_n(\zeta) \tag{1127}$$

where

$$\bar{r}(\zeta) = \int_0^\zeta r(\zeta)\,d\zeta, \qquad \bar{Y}_n(\zeta) = \int_0^\zeta Y_n(\zeta)\,d\zeta \tag{1128}$$

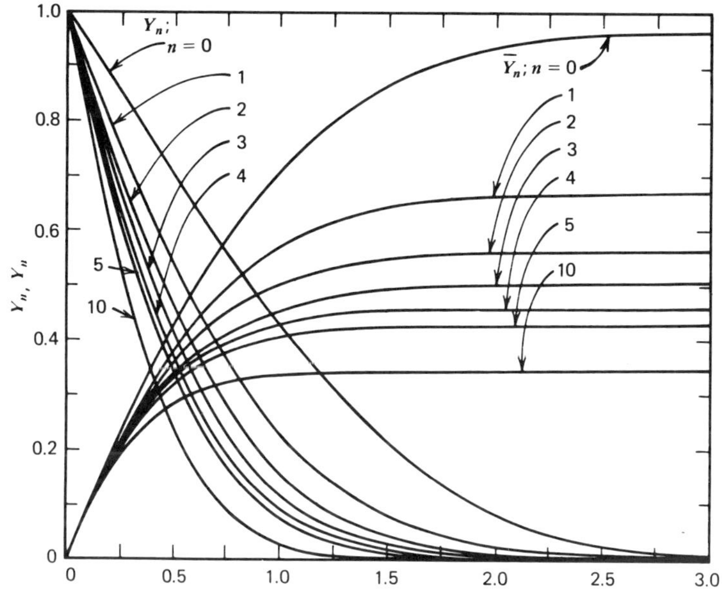

Figure 235 Functions y_n and $\bar{y}_n$ from equations (1125) and (1128) from [97].

$\bar{r}$ and $\bar{y}$ are plotted in Figures 234 and 235. It may be noted that though the momentum and energy equations have been uncoupled, the actual shape of the velocity profile depends on the temperature boundary conditions.

The heat transfer rate at the wall is given by

$$q = -k(T_y)_w = -\frac{kT_\infty}{2}\sqrt{\frac{U}{v_\infty x}}\sum_0^\infty a_n x^{*n}\left(\frac{dY_n}{d\zeta}\right)_{\zeta=0} \tag{1129}$$

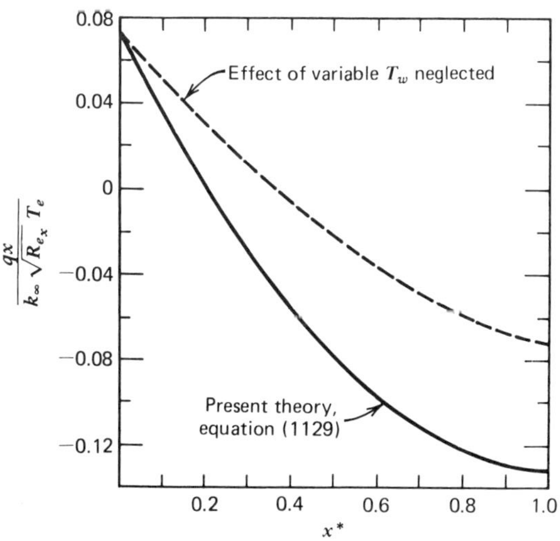

Figure 236 Heat transfer rate, as found by Chapman and Rubesin [97].

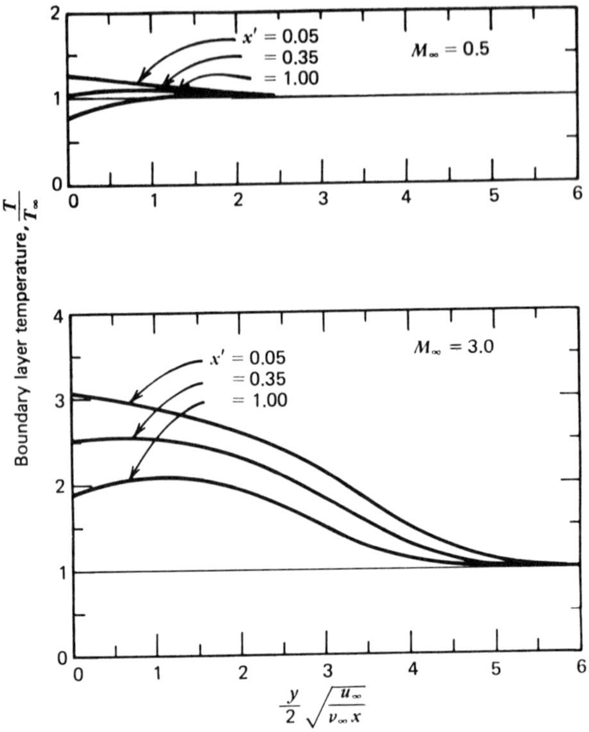

Figure 237 Temperature profiles, as found by Chapman and Rubesin [97].

As an example, consider the case

$$T_2'(x') = T_{aw}'(0.25 - 0.83x' + 0.33x'^2)$$

for $M_\infty = 0.5$ and $M_\infty = 3$. Then

$$a_0 = 0.25\,T_{aw}', \qquad a_1 = -0.83\,T_{aw}', \qquad a_2 = 0.33\,T_{aw}', \qquad a_n = 0, \qquad n > 2$$

The heat transfer rate may be found directly from equation (1129). The result is plotted in Figure 236, where $\mathrm{Re}_x = Ux/\nu_\infty$. The temperature distribution is found from equation (1122) and has been plotted in Figure 237, using equation (1127) to convert from ζ to y. Notice that $N(\zeta), r(\zeta), \bar{r}(\zeta), Y_n(\zeta)$ and $\bar{Y}_n(\zeta)$ are functions of ζ only and are independent of the temperature distribution. These functions are thus in the nature of universal functions and vary only with the Prandtl number. Once these functions have been found, all that is necessary is to substitute the appropriate a_n in equation (1122) to find the temperature distribution.

Since the universal function $f(\zeta)$ is known, it is only necessary to use equation (1127) to find the velocity distribution in the physical plane. The result is shown in Figure 238. For purposes of comparison, the incompressible Blasius velocity distribution is also shown. It will be observed that the departure from the

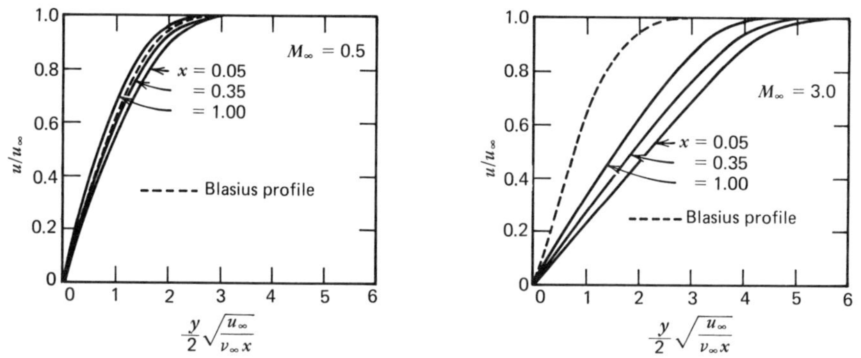

Figure 238 Velocity profiles, as found by Chapman and Rubesin [97].

incompressible solution is greater for the higher Mach number, as is to be expected.

Flow with Pressure Gradient

So far we have been dealing with flat plate solutions. Howarth [98] and Stewartson [99] extended the solution to include flow with pressure gradients. Pr is assumed to be 1 and μ is assumed proportional to T. The Crocco–Busemann integral is then valid in the boundary layer. Thus we have, for the adiabatic wall case,

$$\rho u u_x + \rho v u_y = -p_x + (\mu u_y)_y$$

$$\frac{T}{T_1} = 1 + \frac{\gamma - 1}{2c_1^2}(U^2 - u^2)$$

where the subscript 1 indicates the outer edge of the boundary layer. We now employ the following change of variables.

$$Y = \frac{c_1}{c_0}\frac{1}{\sqrt{\nu_0}}\int_0^y \frac{\rho}{\rho_0}\,dy, \qquad \rho u = \rho_0\sqrt{\nu_0}\;\psi_y, \qquad Z = \int_0^x \left(\frac{c_1}{c_0}\right)^{\frac{(3\gamma-1)}{(\gamma-1)}} dx$$

$$V_1 = c_0 M_1$$

where the subscript 0 indicates the conditions in the free stream at $x = 0$. The flow outside the boundary layer is presumed to be isentropic. The momentum equation then becomes

$$\psi_{YYY} - \psi_Y\psi_{YZ} + \psi_Z\psi_{YY} = -V_1V_{1Z} \tag{1130}$$

with boundary conditions

$$Y = 0:\quad \psi = \psi_Y = 0; \quad Y = \infty:\quad \psi_Y = V_1 \tag{1131}$$

For the isothermal wall case, the Crocco–Busemann integral may be written in the form

$$\frac{T}{T_1} = 1 + \frac{T_0}{T_1}\left(1 + \frac{\gamma - 1}{2} M_0^2\right)s + \frac{\gamma - 1}{2c_1^2}(U^2 - u^2) \tag{1132}$$

The momentum equation then becomes

$$\psi_{YYY} - \psi_Y \psi_{YZ} + \psi_Z \psi_{YY} = -V_1 V_{1Z}(1 + s) \tag{1133}$$

In terms of the new variables, the energy equation (1084) becomes

$$\psi_Y s_Z - \psi_Z s_Y = s_{YY} \tag{1134}$$

with boundary conditions

$$Y = 0: \quad s = s_w; \quad Y = \infty: \quad s = 0 \tag{1135}$$

Equations (1130) and (1133) are incompressible in form. In incompressible boundary layer theory it is shown that under certain conditions, similarity solutions may be obtained. In particular, Falkner and Skan [100] determined that similarity will exist when the free stream velocity varies as x^m. By analogy, we may postulate $V_1 = CZ^m$ and make the change of variables

$$\psi = \left(\frac{2ZV_1}{m + 1}\right)^{1/2} f(\eta), \qquad \eta = Y\left(\frac{(m + 1)V_1}{2Z}\right)^{1/2} \tag{1136}$$

Working backward, we may verify that $df/d\eta$ then equals u/U. In terms of ψ and η, equations (1133) and (1134) become,

$$f''' + ff'' = \beta(f'^2 - 1 - s) \tag{1137}$$

$$s'' + fs' = 0 \tag{1138}$$

where $\beta = 2m/(m + 1)$ and the primes indicate differentiation with respect to η. The momentum and energy equations thus reduce to ordinary differential equations which may be integrated numerically. The boundary conditions are

$$\eta = 0: \quad f = f' = 0, \quad s = s_w$$

$$\eta = \infty: \quad f' = 1, \qquad s = 0$$

The numerical integration of equations (1137) and (1138) has been performed by Cohen and Reshotko [101]. The solutions for $s_w = -1.0$ are shown in Figure 239. The asterisks indicate that the solutions for negative beta are not unique.

The derivation of Howarth and Stewartson has been generalized by Levy [102]. We begin by writing the boundary layer equations in the form

$$(\rho u)_x + (\rho v)_y = 0 \tag{1139}$$

$$\rho(uu_x + vu_y) = -p_x + (\mu u_y)_y \tag{1140}$$

$$\rho(uT_{0x} + vT_{0y}) = (\mu T_{0y})_y + \left(\frac{1}{Pr} - 1\right)(\mu T_y)_y \tag{1141}$$

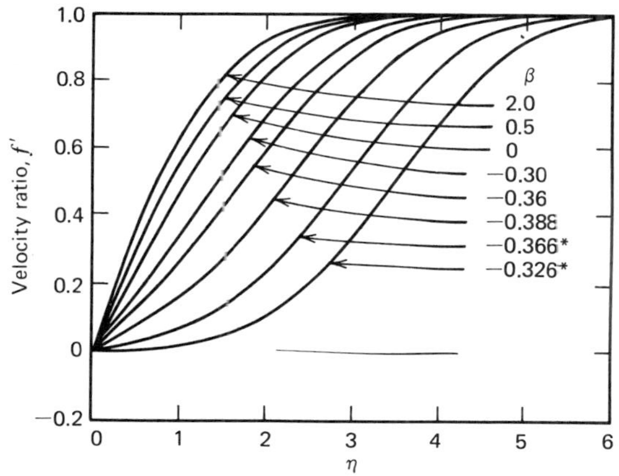

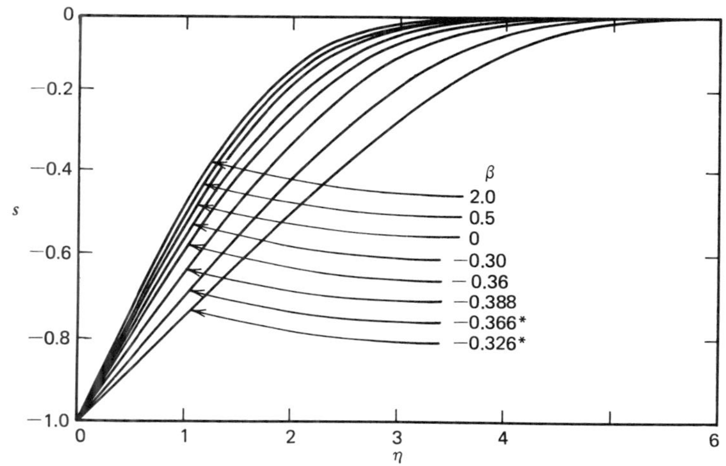

Figure 239 Velocity profile function f (1136) and temperature profile function s (1132) as found by Cohen and Reshotko [101].

where T_0 is the local stagnation temperature. The boundary conditions are

$$y=0:\quad u=v=0;\qquad y=\infty:\quad u=U,\qquad -p_x=\rho_1 UU_x,\qquad T_{0x}=0,$$
$$u_y=u_{yy}=0$$

We now introduce the following changes of variable:

$$\rho u=\psi_y,\qquad \rho v=-\psi_x,\qquad \xi=\int_0^x \rho_1\mu_1 U\,dx,\qquad \eta=\frac{U\rho_1}{\sqrt{2\xi}}\int_0^y \frac{T_1}{T}dy \tag{1142}$$

$$\frac{u}{U}=f'(\eta),\qquad \frac{T_0}{T_{0_1}}=g(\eta),\qquad \chi=\frac{\rho\mu}{\rho_1\mu_1} \tag{1143}$$

where $f'(\eta)=df/d\eta$. We note that similarity has been assumed. The conditions under which this assumption is valid may be determined by substituting the change of variables (1142) (1143) into the system of equations (1139–1141). There results

$$(\chi f'')'+ff''-\frac{2\xi}{U}U_\xi\left(f'^2-\frac{T}{T_1}\right)=0 \tag{1144}$$

$$(\chi g')'+fg'+\left(\frac{1}{Pr}-1\right)\frac{T_1}{T_{0_1}}\left(\chi\frac{T'}{T_1}\right)=0 \tag{1145}$$

where the primes indicate differentiation with respect to η. Comparison with equations (1137) and (1138) immediately brings out some special features. First we notice the effect of variable $\rho\mu$. In the previous derivation, $\rho\mu$ was assumed constant. Secondly, we see the effect of the Prandtl number when it is not equal to 1. Third, we observe the criteria for similarity: all the terms of equations (1144) and (1145) must be functions of η only. From the definition of $g(\zeta)$ and $f(\zeta)$, (1142), we have

$$\frac{T}{T_1}=g(\eta)+\frac{\gamma-1}{2}M_1^2[g-f'^2] \tag{1146}$$

Notice that this does not mean that T_0 is presumed constant in the boundary layer. Substituting equation (1146) in equations (1144) and (1145) we get, assuming $\chi=1$,

$$f'''+ff''-\frac{2\xi}{U}U_\xi\left(1+\frac{\gamma-1}{2}M_1^2\right)(f'^2-g)=0 \tag{1147}$$

$$g''+fg'+\left(\frac{1}{Pr}-1\right)\frac{T_1}{T_{0_1}}\left(1+\frac{\gamma-1}{2}M_1^2\right)g''-\frac{T_1}{T_{0_1}}\left(\frac{1}{Pr}-1\right)\left(\frac{\gamma-1}{2}M_1^2\right)(f'^2)''=0 \tag{1148}$$

Now

$$\frac{T_{0_1}}{T_1}=1+\frac{\gamma-1}{2}M_1^2,\qquad \text{and}\qquad \left(1+\frac{\gamma-1}{2}M_1^2\right)\frac{2\xi}{U}U_\xi=\frac{2\xi}{M_1}M_{1\xi}\equiv\beta\,(\text{say}) \tag{1149}$$

So that

$$f''' + ff'' + \beta(g - f'^2) = 0 \tag{1150}$$

$$g'' + Prfg' - (1 - Pr)\frac{\frac{\gamma - 1}{2}M_1^2}{1 + \frac{\gamma - 1}{2}M_1^2}(f'^2)'' = 0 \tag{1151}$$

Thus, in order for our original assumption to be valid, that is, u/U and T_0/T_{0_1} to be functions of η only, β must be a constant, Pr must be a function of η only or constant, and either Pr must equal 1 or M_1 must be constant. While these conditions may appear restrictive, they are approximately satisfied, even when M_1 is variable, provided M_1 is either small or large. If M_1 is small, we may neglect the term containing M_1 in the energy equation and take T_1 = constant. Then similarity will be satisfied if

$$U = b(2\xi)^{\beta/2}\sqrt{\gamma R T_1} \tag{1152}$$

where b and β are constants. In this case the equations reduce to their incompressible form. If M_1 is large,

$$\frac{\frac{\gamma - 1}{2}M_1^2}{1 + \frac{\gamma - 1}{2}M_1^2} \longrightarrow \frac{\frac{\gamma - 1}{2}M_1^2}{\frac{\gamma - 1}{2}M_1^2} = 1$$

and similarity is satisfied if

$$M_1 = b(2\xi)^{\beta/2} \tag{1153}$$

Levy has solved the momentum and energy equations numerically for the former case, M_1 small. In this case the governing equations reduce to

$$f''' + ff'' + \beta(g - f'^2) = 0 \tag{1150}$$

$$g'' + Prfg' = 0 \tag{1154}$$

with boundary conditions,

$$\eta = 0: \qquad f = f' = 0, \qquad g = T_w/T_{0_1}$$

$$\eta = \infty: \qquad f' = g = 1$$

Some sample results are shown in Figure 240. It is of interest to observe that if $\rho\mu$ and T are constant across the boundary layer, and if $U \sim x^m$, then equation (1144) reduces to

$$f''' + ff'' + \frac{2m}{m + 1}(1 - f'^2) = 0 \tag{1155}$$

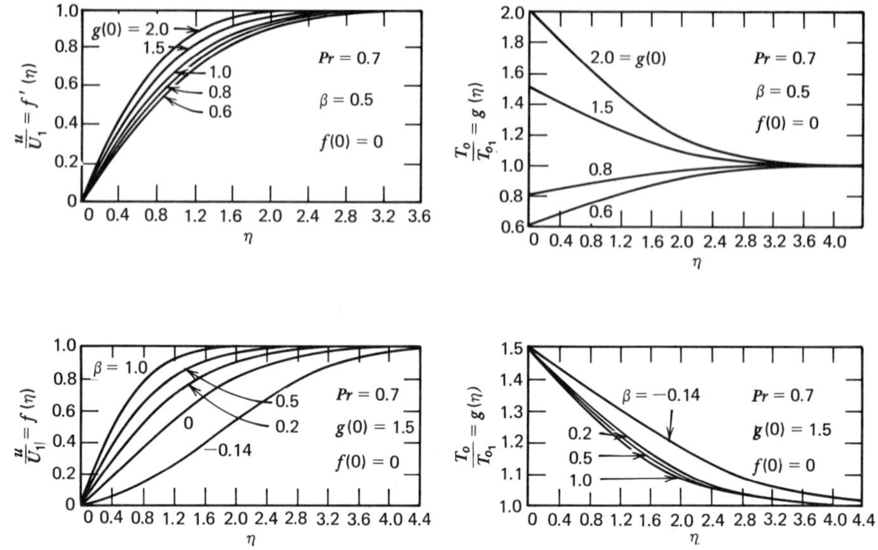

Figure 240 Velocity and temperature profiles, as obtained by Levy [102].

which will be recognized as the Falkner–Skan equation from incompressible flow. For the flat plate case, if T and $\rho\mu$ are constant, β is zero and equation (1150) reduces to

$$f''' + ff'' = 0 \tag{1156}$$

which will be recognized as the Blasius equation. Comparison of equation (1150) with equation (1137) shows that the β of Howarth, Stewartson, Cohen and Reshotko is equal to the β of Levy, and $1 + s$ in equation (1137) is equal to g in equation (1150).

If χ and Pr are to be assumed constant, the question arises as to what values should be used for these parameters. In practice, χ and Pr are essentially functions of temperature only. This has led to the so-called reference enthalpy or reference temperature method. It has been found that if one evaluates χ and Pr at the right temperature, quite accurate results can be obtained. For the flat plate, Eckert [103] has proposed the formula

$$\frac{T^*}{T_1} = 0.5 + 0.039\, M_1^2 + 0.5\frac{T_w}{T_1} \tag{1157}$$

where T^* is the reference temperature. Good results can be obtained by evaluating $\rho\mu$ and Pr at $T = T^*$. For stagnation point flow, Walker [104] has determined that good results can be obtained if the reference enthalpy h^* is taken as

$$h^* = \tfrac{1}{2}(h_{01} + h_w) \tag{1158}$$

where h_{01} is the free stream stagnation enthalpy and h_w is the enthalpy at the wall. $\rho\mu$ and Pr are then evaluated at $h = h^*$.

Method of Crocco. Crocco [105] has proposed a method particularly suited to taking variable c_p and $\rho\mu$ into account, even though Pr is still assumed constant. The energy equation is written in the form

$$\rho(uh_x + vh_y) = up_x + \mu u_y^2 + \left(\frac{k}{c_p}h_y\right)_y \tag{1159}$$

The independent variables are then changed from x and y to x and u. The momentum and continuity equations become

$$-\rho u y_x + \rho v = \tau_u - p_x y_u \tag{1160}$$

$$y_u(\rho u)_x - y_x(\rho u)_u + (\rho v)_u = 0 \tag{1161}$$

where the partials with respect to x are now taken with u held constant. Eliminating ρv between equations (1160) and (1161) we get

$$u(\rho y_u)_x + \tau_{uu} - p_x y_{uu} = 0 \tag{1162}$$

But,

$$\tau = \mu u_y = \frac{\mu}{y_u}$$

Hence

$$u\left(\frac{\rho\mu}{\tau}\right)_x + \tau_{uu} - p_x\left(\frac{\mu}{\tau}\right)_u = 0 \tag{1163}$$

Similarly, the energy equation becomes

$$\frac{1-Pr}{Pr}h_u\tau_{uu} + \frac{h_{uu}+Pr}{Pr}\tau^2 - \rho\mu u h_x + \mu(h_u + u)p_x = 0 \tag{1164}$$

The dependent variables are thus τ and h. Once τ and h have been found as functions x and u, the inverse transformation may be performed using the formula

$$y = \int_0^u \frac{\mu\, du}{\tau(x,u)} \tag{1165}$$

Using the Crocco method, van Driest [106] has performed extensive calculations for a flat plate using variable $\rho\mu$ and c_p while maintaining Pr constant. The results are of interest mainly to determine the accuracy of calculations holding $\rho\mu$ and c_p constant. It turns out that for atmospheric free stream conditions and wall to free stream temperature ratios of up to 6, the reference temperature method gives very accurate results. The effect of variable c_p turns out to be almost negligible.

Actual viscosity variation with temperature is shown in Figure 241. The exponential curve is given by

$$\frac{\mu}{\mu_\infty} = \left(\frac{T}{T_\infty}\right)^{0.76}$$

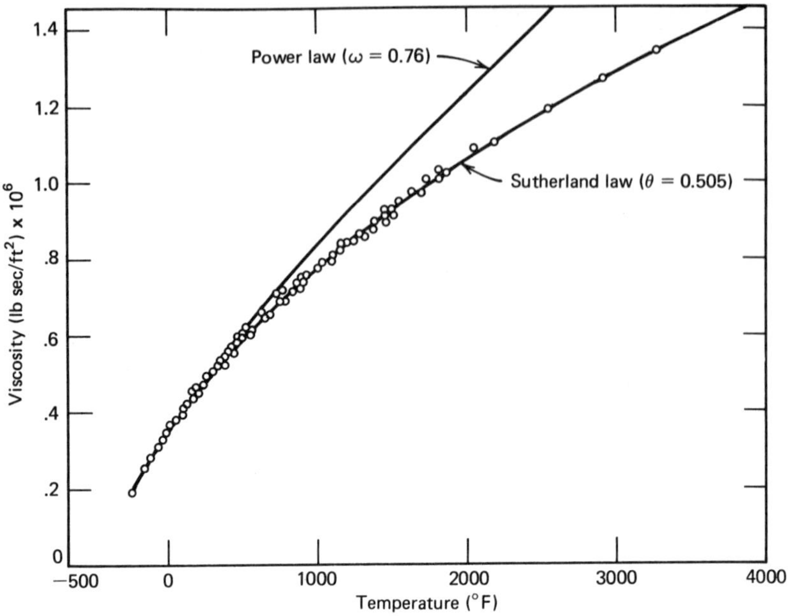

Figure 241 Variation of Viscosity with temperature [106].

The Sutherland law is given by

$$\frac{\mu}{\mu_\infty} = \left(\frac{T}{T_\infty}\right)^{1/2} \frac{1 + 0.505}{1 + 0.505\dfrac{T_\infty}{T}}$$

The Sutherland law is seen to agree closely with experiment and was used by van Driest in his calculations. *Pr* was assumed to be 0.75. Some of the results are shown in Figure 242 and 243. Once again, we observe the fact that the velocity becomes almost a straight line at hypersonic Mach number. The curves of Figure 243 are for an insulated flat plate. In this connection it should be pointed out that for hypersonic Mach numbers, the adiabatic wall temperature exceeds that for which the ideal gas law, $p = \rho RT$, is valid. Thus, while the calculations by the reference temperature method agree closely with those of van Driest, both methods assume the ideal gas law and must be expected to fail when significant departure from this law occurs. This question will be discussed further in the chapter on real gas effects.

Stagnation point flow In the earliest work on boundary layer theory, Blasius [89] already pointed out that the boundary layer equations continue to be valid near the stagnation point of a blunt body. This is because the assumptions u_{xx} and p_y small continue to be valid there. While Blasius was considering incompressible flow, the conclusion is equally valid for compressible flow.

The similarity solution near the stagnation point of a blunt body has been

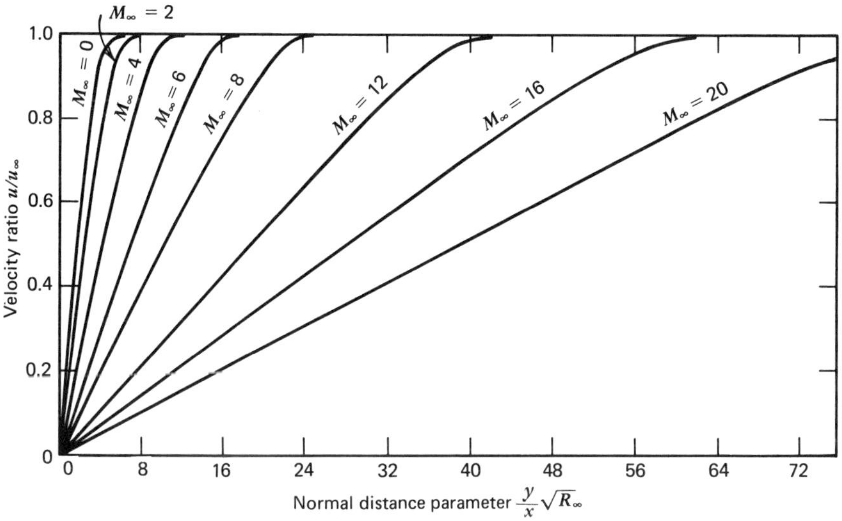

Figure 242 Velocity profiles as obtained by van Driest [106].

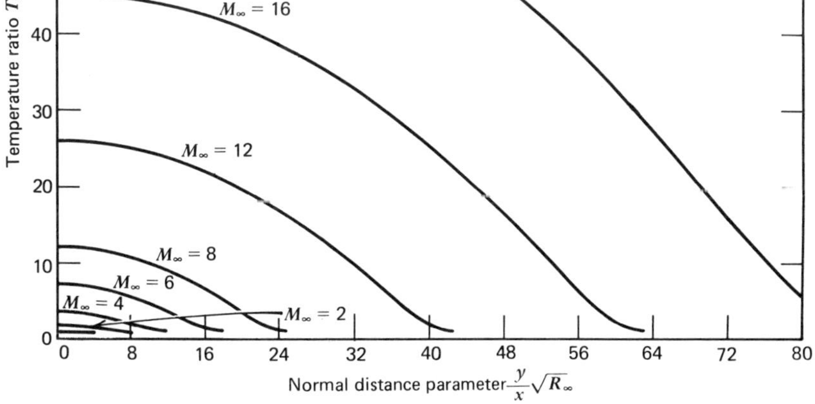

Figure 243 Temperature profiles as obtained by van Driest [106].

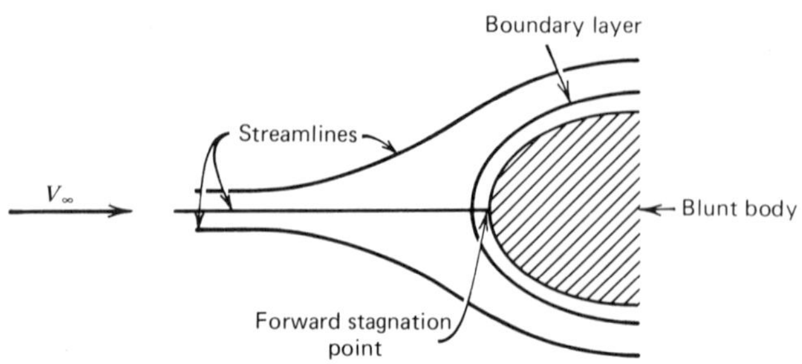

Figure 244 Flow pattern near the stagnation point of a blunt body.

discussed by Lees [107]. The physical situation is illustrated in Figure 244. Equation (1145) may be rewritten in the following way:

$$\left(\frac{\chi}{Pr}g'\right)' + fg' + \frac{U^2}{2h_{01}}\left[2\chi\left(1-\frac{1}{Pr}\right)f'f''\right]' = 0 \tag{1166}$$

Near the stagnation point, U is small and hence the last term in equation (1166) may be neglected. If, in addition, χ/Pr is taken as 1, the energy equation reduces to

$$g'' + fg' = 0 \tag{1167}$$

Recall now the momentum equation (1150) (for $\chi = 1$)

$$f''' + ff'' + \beta(g - f'^2) = 0 \tag{1150}$$

where β is given by equation (1149)

$$\beta = \frac{2\xi}{U}U_\xi\left(1 + \frac{\gamma-1}{2}M_1^2\right) \tag{1149}$$

Near the stagnation point, M_1 is small and hence

$$\beta \cong \frac{2\xi}{U}U_\xi \tag{1168}$$

We may expand U as a power series in x around the stagnation point. Near the stagnation point we may retain only the first term of the series, giving us $U = bx$, where b is a constant. Now recall

$$\xi = \int_0^x \rho_1\mu_1 U\,dx$$

Lees extended the solution to the axisymmetric case by letting

$$\xi = \int_0^x \rho_1\mu_1 Ur^2\,dx, \quad \eta = \frac{\rho_1 U}{\sqrt{2\xi}}\int_0^y \frac{r\rho}{\rho_1}dy$$

where r is the local radius of the body. The equations in the transformed coordinates are then unchanged from the plane case.

If $\rho_1 \mu_1$ and T_1 are taken as constant near the stagnation point, and $U \sim x$, we have

$$\beta = \frac{2\xi}{U} U_\xi = 1$$

and the system of equations to be solved becomes

$$g'' + fg' = 0$$
$$f''' + ff'' + g - f'^2 = 0 \tag{1169}$$

The boundary conditions are

$$\eta = 0: \quad f = f' = 0, \quad g = \text{constant (isothermal wall)}$$
$$\text{or} \quad g' = 0 \text{ (adiabatic wall)} \tag{1170}$$
$$\eta = \infty: \quad f' = g = 1$$

The solution to the system (1169) with boundary conditions (1170) is given in Figure 245 for $g(0) = 0.8$.

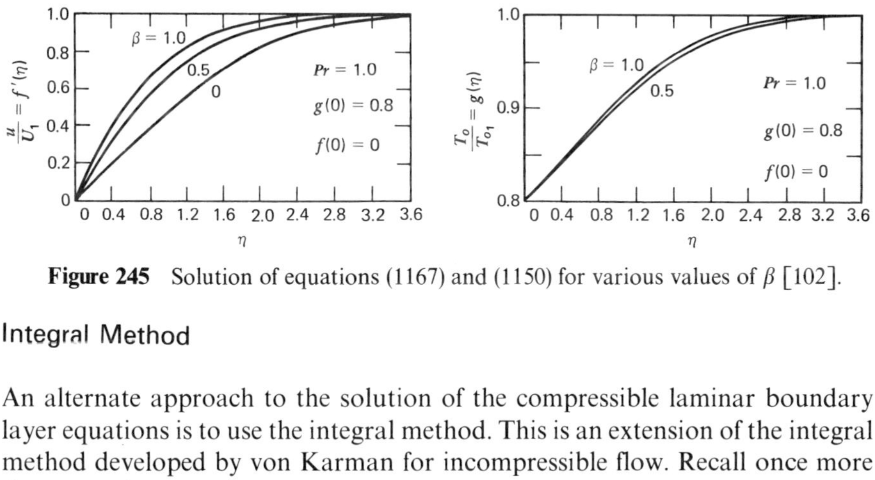

Figure 245 Solution of equations (1167) and (1150) for various values of β [102].

Integral Method

An alternate approach to the solution of the compressible laminar boundary layer equations is to use the integral method. This is an extension of the integral method developed by von Karman for incompressible flow. Recall once more the governing equations:

$$(\rho u)_x + (\rho v)_y = 0$$
$$\rho(uu_x + vu_y) = \rho_1 U U_x + (\mu u_y)_y \tag{1171}$$
$$\rho(uh_x + vh_y) = up_x + \mu u_y^2 + \left(\frac{\mu}{Pr} h_y\right)_y$$

and assume a perfect gas with constant specific heat. The boundary conditions are

$$y = 0: \quad u = v = 0, \quad h_y = 0$$
$$y = \infty: \quad u = U(x), \quad u_y = u_{yy} = 0, \quad h = h_1(x) \tag{1172}$$

where we have assumed an insulated wall. The addition of the boundary conditions $u_y = u_{yy} = 0$ at $y = \infty$ renders the problem overdeterminate, but these conditions must be satisfied by the solution and so are not inconsistent with the other boundary conditions.

Integrating the system (1171) across the boundary layer, i.e., integrating the equations from $y = 0$ to $y = \delta$ for fixed x, we get

$$\theta_x + \frac{\theta}{U} U_x \left(2 + \frac{\delta^*}{\theta} - M^2 \right) = \frac{\mu_w}{\rho_1 U^2} (u_y)_w \tag{1173}$$

$$\sigma_x + \frac{\sigma}{U} U_x (3 - (2 - \gamma) M^2) = \frac{2}{\rho_1 U^3} \int_0^\delta \mu u_y^2 \, dy \tag{1174}$$

where

$$\theta = \int_0^\delta \frac{\rho u}{\rho_1 U} \left(1 - \frac{u}{U} \right) dy, \qquad \delta^* = \int_0^\delta \left(1 - \frac{\rho u}{\rho_1 U} \right) dy$$

$$\sigma = \int_0^\delta \frac{\rho u}{\rho_1 U} \left[1 - \left(\frac{u}{U} \right)^2 \right] dy, \qquad M = \frac{U}{c_1} \tag{1175}$$

δ^*, θ and σ are called the displacement thickness, momentum thickness and energy thickness, respectively. We now define η by the equation

$$\eta = \frac{1}{\delta_1} \int_0^y \frac{\rho}{\rho_1} dy, \qquad \text{where,} \qquad \delta_1 = \int_0^\delta \frac{\rho}{\rho_1} dy \tag{1176}$$

and assume a velocity distribution in the boundary layer of the form

$$\frac{u}{U} = c_1 \eta + c_2 \eta^2 + c_3 \eta^3 + c_4 \eta^4 \tag{1177}$$

Using the boundary conditions (1172) we may write each of the c_i in terms of a single parameter:

$$c_1 = 2 + \frac{\Lambda}{6}, \qquad c_2 = -\tfrac{1}{2}\Lambda, \qquad c_3 = \tfrac{1}{2}\Lambda - 2, \qquad c_4 = 1 - \frac{\Lambda}{6} \tag{1178}$$

where

$$\Lambda = \frac{\rho_1 \delta_1^2}{\rho_w \nu_1} U_x \tag{1179}$$

Λ is called the shape factor. The value of Λ itself may be determined from the momentum integral equation. Furthermore, we assume

$$\left(1 - \frac{u}{U} \right) \frac{\rho_1}{\rho} = b_0 + b_1 \eta + b_2 \eta^2 + b_3 \eta^3 + b_4 \eta^4 + b_5 \eta^5 \tag{1180}$$

The values of five of the b_i may be determined from the boundary conditions (1172). In addition, b_0 must satisfy the relation

$$b_0 = \frac{\rho_1}{\rho_w} = \frac{T_w}{T_1} \tag{1181}$$

θ/δ_1 and σ/δ_1 may now be written in terms of Λ

$$\frac{\theta}{\delta_1} = \frac{37}{315} - \frac{1}{945}\Lambda - \frac{1}{9072}\Lambda^2 \tag{1182}$$

$$\frac{\sigma}{\delta_1} = \frac{798{,}048 - 4{,}656\,\Lambda - 758\,\Lambda^2 - 7\Lambda^3}{4{,}324{,}320}$$

Substituting these results in the momentum and energy equations (1173) and (1174) we get

$$\frac{\theta U}{v_1}\theta_x = F_1 - \frac{K}{b_0}(2 - M^2 F_2) \tag{1183}$$

$$b_0 = \left(1 + \frac{\gamma - 1}{2}M^2\right)\frac{1 + M^2 F_3}{1 + M^2 F_4} \tag{1184}$$

where

$$K = \Lambda\left(\frac{\theta}{\delta_1}\right)^2 = b_0\frac{\theta^2}{v_1}U_x \tag{1185}$$

$$F_1(K) = 2\frac{\theta}{\delta_1}\left(1 - \frac{1}{15}\Lambda + \frac{1}{240}\Lambda^2\right) \tag{1186}$$

$$F_2(K) = 1 - \frac{\gamma - 1}{2}\frac{\delta_1}{\theta}\left(\frac{\theta_E}{\delta_1} - Pr\frac{(12 + \Lambda)^2}{2160}\right) \tag{1187}$$

$$F_3(K) = \frac{\gamma - 1}{2}\left[2 - \frac{\delta_1}{\theta}\left(\frac{\theta_E}{\delta_1} - Pr\frac{(12 + \Lambda)^2}{2160}\right)\right] \tag{1188}$$

$$F_4(K) = \frac{\gamma - 1}{2}\left[2 - \frac{\delta_1}{\theta}\left(\frac{\theta_E}{\delta_1} - \frac{(12 + \Lambda)^2}{2160}\right)\right] \tag{1189}$$

and

$$\theta_E = \int_0^\delta \frac{\rho u}{\rho_1 U}\left(1 - \frac{u^2}{U^2}\right)dy$$

The functions F_i have been tabulated by Gruschwitz [108] for $-12 < \Lambda < 12$. Equations (1183) and (1184) may now be integrated to give $\theta(x)$. The method is best suited for cases where there is a forward stagnation point. Here $U = 0$ at $x = 0$ and this fixes the initial value of K from equation (1183). θ at $x = 0$ may then be evaluated from (1185). The initial value of b_0 is 1 and so

$$\theta_0 = \sqrt{\frac{K_0 v_1}{(U_x)_0}}$$

where the subscript zero indicates $x = 0$. K_0 turns out to be 0.0770. When $U = 0, d\theta/dx$ is of the indeterminate form 0/0. However, it may be shown that in the limit as $U \to 0$,

$$\frac{d\theta}{dx} \to -0.424\frac{(U_{xx})_0}{(U_x)_0}$$

With these initial conditions, equation (1183) may be integrated forward numerically. Once a θ has been found, K may be found from equation (1185). Combining equations (1185) and (1182), Λ may be found from

$$K = \Lambda\left[\frac{37}{315} - \frac{1}{945}\Lambda - \frac{1}{9025}\Lambda^2\right] \tag{1190}$$

Then δ, may be found from equation 1185. The shearing stress at the wall is given by

$$\frac{\tau_w}{\rho_1 U^2} = \frac{\nu_1}{U\delta_1}\left(2 + \frac{\Lambda}{6}\right) \tag{1191}$$

Separation of the boundary layer from the wall occurs when the velocity gradient u_y at the wall equals zero. This phenomenon is illustrated in Figure 246. Beyond

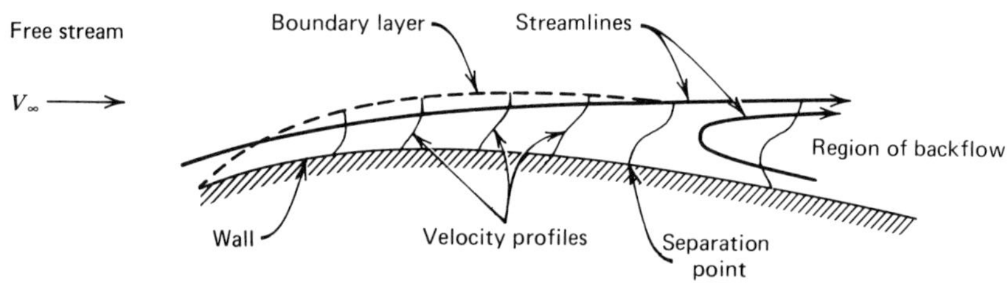

Figure 246 Boundary layer separation.

the separation point, the boundary layer concept no longer has meaning since the basic premises are no longer fulfilled. Separation is usually caused by an adverse pressure gradient and is responsible for the phenomenon known as stall on wings and compressor blades. From equation (1191) we see that the momentum integral method predicts that stall will occur when $\Lambda = -12$. We notice that this corresponds to an adverse pressure gradient since Λ is proportional to U_x. Since

$$\rho U U_x = -p_x$$

a negative Λ means a positive p_x and hence an adverse pressure gradient.

The method of Gruschwitz outlined above is found to be quite accurate where the underlying assumptions are justified; i.e. c_p = constant, perfect gas, adiabatic wall, laminar boundary layer exists. According to White [109], however, the separation point occurs nearer $\Lambda = -7$ than $\Lambda = -12$ in practice.

A similar method was developed contemporaneously by Howarth and Stewartson [110]. Kalikhman [111] extended the integral method to include wall temperatures varying with x. At the same time, Ginzel [112] applied the method to heated walls. Morris and Smith [113] extended the integral method to arbitrary Prandtl numbers and variable viscosity for both adiabatic and isothermal walls. Another adiabatic wall treatment was given by Cope and Hartree [114], but their method assumes constant stagnation temperature in the boundary layer. Simplified, approximate versions of the integral method have been developed by Fernholz [115] and Geropp [116]. A further discussion of the integral method for both adiabatic and isothermal walls has been given by Walz [117]. An iterative procedure by Yang [118] improves the accuracy of the results of the integral method but requires extensive computations.

Numerical Integration

An alternate method of solving the compressible laminar boundary layer problem is numerical integration of the equations. Such an approach was taken by Smith and Clutter [119]. For purposes of illustration, we will apply this technique to the flow over a blunt body of revolution. The coordinate system is shown in Figure 247. The boundary layer equations may be written

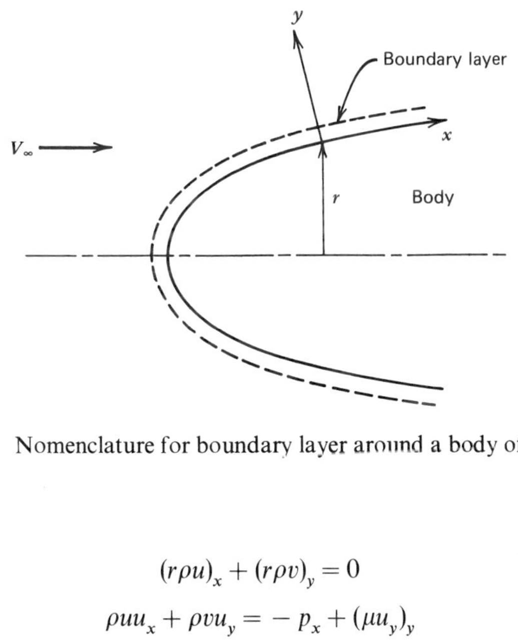

Figure 247 Nomenclature for boundary layer around a body of revolution.

in the form

$$(r\rho u)_x + (r\rho v)_y = 0 \tag{1192}$$

$$\rho u u_x + \rho v u_y = -p_x + (\mu u_y)_y \tag{1193}$$

$$\rho u h_{0x} + \rho v h_{0y} = \left[\frac{\mu}{Pr} h_{0y} - \mu\left(1 - \frac{1}{Pr}\right) u u_y\right]_y \tag{1194}$$

We now introduce the following change of variables:

$$\eta = \sqrt{\frac{u_1}{\rho_\infty \mu_\infty x}} \int_0^y \rho\, dy, \qquad \rho r u = (r\psi)_y, \qquad \rho r v = -(\psi r)_x \tag{1195}$$

$$g = \frac{h_0}{h_{0_1}}, \qquad \psi = (\rho_\infty \mu_\infty x u_1)^{1/2} f(x, \eta)$$

The continuity equation is then identically satisfied. The momentum and energy equations become

$$\frac{1}{\chi_\infty}(\chi f'')' + \frac{x}{u_1}\frac{du_1}{dx}\left(\frac{\rho_1}{\rho} - f'^2\right) + \left[\frac{1}{2}\frac{x}{u_1}\frac{du_1}{dx} + \frac{1}{2} + \frac{x}{r}\frac{dr}{dx}\right] ff''$$

$$- x\left[f'\frac{\partial f'}{\partial x} - f''\frac{\partial f}{\partial x}\right] \tag{1196}$$

$$\frac{1}{\chi_\infty}\left[\frac{\chi}{Pr}g' + \frac{u_1^2}{h_0}\chi\left(1 - \frac{1}{Pr}\right)f'f''\right] = -\left[\frac{1}{2}\frac{x}{u_1}\frac{du_1}{dx} + \frac{1}{2} + \frac{x}{r}\frac{dr}{dx}\right] fg'$$

$$+ x\left[f'\frac{\partial g}{\partial x} - g'\frac{\partial f}{\partial x}\right]$$

where the primes indicate differentiation with respect to η and the partials with respect to x are taken with η held constant. For purposes of computation it is found desirable to introduce two new functions ϕ and θ, where

$$\phi = f - \eta, \qquad \text{and} \qquad \theta = g - 1 \tag{1197}$$

Equations (1196) then become

$$(\chi\phi'')' = \chi_\infty \frac{x}{u_1}\frac{du_1}{dx}\left[\phi'^2 + 2\phi + 1 - \frac{\rho_1}{\rho}\right] - \chi_\infty N(\phi + \eta)\phi''$$

$$+ \chi_\infty x[(\phi' + 1)\phi'_x - \phi''\phi_x] \tag{1198}$$

$$\left[\frac{\chi}{Pr}\theta' + \frac{u_1^2}{h_{01}}\chi\left(1 - \frac{1}{Pr}\right)(\phi' + 1)\phi''\right]' = -\chi_\infty N(\phi + \eta)\theta' + \chi_\infty x[(\phi' + 1)\theta_x - \theta'\phi_x]$$

where

$$N = \frac{x}{r}\frac{dr}{dx} + \frac{1}{2}\frac{x}{u_1}\frac{du_1}{dx} + \frac{1}{2}, \qquad \chi_\infty = \frac{\rho_\infty \mu_\infty}{\rho_1 \mu_1}$$

The boundary conditions may be taken as

$$\eta = 0: \quad \phi = 0, \quad \phi' = -1; \quad \theta = g_w(x) - 1, \quad \text{or} \quad \theta' = g'_w(x) - 1$$

$$\eta = \infty: \quad \phi' = 0, \quad \theta = 0 \tag{1199}$$

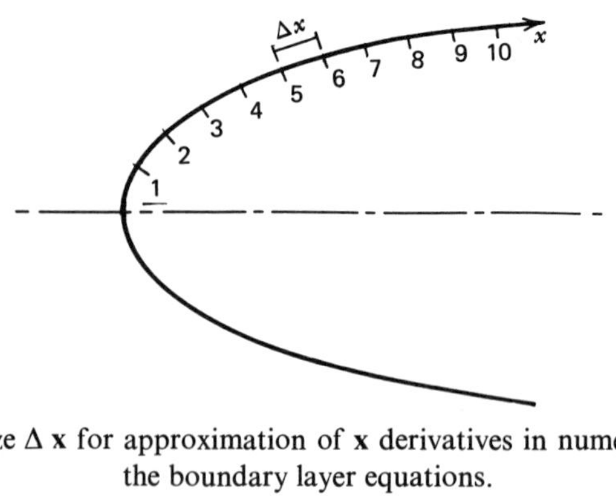

Figure 248 Step size Δ **x** for approximation of **x** derivatives in numerical integration of the boundary layer equations.

We thus allow for arbitrary thermal condition at the wall. The wall may be either adiabatic, or isothermal, or the wall temperature may vary with x. Let us choose a number of points along the wall numbered sequentially as in Figure 248. At each point $n(n=1,2,3,\ldots,\text{etc.})$ the x derivations may be approximated by the Lagrangian formula

$$\frac{\partial F_n}{\partial x}=\frac{1}{2\Delta x}(3F_n-4F_{n-1}+F_{n-2}) \tag{1200}$$

where $F_n=\phi_n,\theta_n,\phi'_n$ and Δx is the distance between points. Substituting (1199) in equations (1198) we get

$$\begin{aligned}(\chi\phi''_n)'=\chi_\infty\frac{x}{u_1}\frac{du_1}{dx}\left[\phi'^2_n+2\phi_n+1-\frac{\rho_1}{\rho}\right]-\chi_\infty N(\phi_n+\eta)\phi''_n\\+\chi_\infty x\left[(\phi'+1)\frac{1}{2\Delta x}(3\phi'_n-4\phi'_{n-1}+\phi'_{n-2})\right.\\\left.-\frac{\phi''}{2\Delta x}(3\phi_n-4\phi_{n-1}+\phi_{n-2})\right]\end{aligned} \tag{1201a}$$

$$\begin{aligned}\left[\frac{\chi}{Pr}\theta'_n+\frac{u_1}{h_{01}}\chi\left(1-\frac{1}{Pr}\right)(\phi'+1)\phi''\right]'\\=-\chi_\infty N(\phi+\eta)\theta'_n\\+\chi_\infty x\left[(\phi'+1)\frac{1}{2\Delta x}(3\theta_n-4\theta_{n-1}+\theta_{n-2})\right.\\\left.-\theta'_n\frac{1}{2\Delta x}(3\phi_n-4\phi_{n-1}+\phi_{n-2})\right]\end{aligned} \tag{1201b}$$

If ϕ and θ are known functions of η at $x=x_{n-1}$ and $x=x_{n-2}$, then equations (1201) become two coupled ordinary differential equations for $\phi_n(\eta)$ and $\theta_n(\eta)$ which may be integrated numerically. Once $\phi(\eta)$ and $\theta(\eta)$ have been found at $x=x_n$, we may proceed to find the solution at $x=x_{n+1}$, and so on. In this way, the velocity and enthalpy profiles may be found at each point along the boundary layer. It will be observed that variable Pr and χ have been allowed for. We may begin the procedure at $x=0$ where the x derivatives in equations (1198) drop out. At $x=x_1$, we may take

$$\frac{\partial F_1}{\partial x}=\frac{F_1-F_0}{\Delta x} \tag{1202}$$

instead of the formula (1200). Thereafter, equation (1200) may be used for the remaining points.

Equations (1201) may now be integrated subject to the boundary conditions (1199). Boundary conditions are given both at $\eta=0$ and $\eta=\infty$. This type of a problem is called a jury problem (they all decide together). As a result, one cannot simply integrate forward from $\eta=0$, as one could if all the boundary conditions were given at $\eta = 0$. To illustrate this point consider a simple ordinary differential equation of the third order for $F(\eta)$.

$$f(F, F', F'', F''') = 0 \tag{1203}$$

where the primes indicate differentiation with respect to η. If $F(0), F'(0)$ and $F''(0)$ are given, $F'''(0)$ may be found from the equation (1203) and the solution may be found by integrating forward numerically from $\eta = 0$. This may be done, for example, by writing a simple Taylor series:

$$F(\Delta\eta) = F(0) + F'(0)\Delta\eta + \frac{F''(0)}{2!}\Delta\eta^2 + \frac{F'''(0)}{3!}\Delta\eta^3 + \ldots$$

$$F'(\Delta\eta) = F'(0) + F''(0)\Delta\eta + \frac{F'''(0)}{2!}\Delta\eta^2 + \ldots$$

$$F''(\Delta\eta) = F''(0) + F'''(0)\Delta\eta + \ldots$$

Once F, F' and F'' are known at $\eta = \Delta\eta$, $F'''(\Delta\eta)$ may be found from equation (1103) and the process repeated for $\eta = 2\Delta\eta$. And so on. In general we have

$$F(\eta + \Delta\eta) = F(\eta) + F'(\eta)\Delta\eta + \frac{F''(\eta)}{2!}\Delta\eta^2 + \frac{F'''(\eta)}{3!}\Delta\eta^3 + \ldots$$

$$F'(\eta + \Delta\eta) = F'(\eta) + F''(\eta)\Delta\eta + \frac{F'''(\eta)}{2!}\Delta\eta^2 + \ldots$$

$$F''(\eta + \Delta\eta) = F''(\eta) + F'''(\eta)\Delta\eta + \ldots$$

If more accuracy is desired, more terms can be used in the Taylor series because $F^{(n)}(\eta), n > 3$, can be found simply by differentiating the original equation (1103). Or smaller step sizes $\Delta\eta$ may be used. Actually, more sophisticated formulas exist than simple Taylor series. These save on computer time and increase accuracy for a given step size. Such formulas will be discussed in Chapter 10. For the present, we are concerned only with discussing the effect of where the boundary conditions are given.

A problem in which all the boundary conditions are given at one point from which the solution may be found by integrating forward is called a marching problem (we march forward from the point). In the present case, and in the case of boundary layer equations in general, the boundary conditions are given at two different points: the wall and the outer edge of the boundary layer, or $y = 0$ and $y = \infty$. Boundary layer problems are thus jury problems. One way to solve jury problems is to turn them into marching problems and use a process of trial and error. Returning to our example, equation (1103), suppose now that the boundary conditions given are $F(0)$, $F'(0)$, and $F'(\infty)$. By guessing a value of $F''(0)$ and integrating forward we can turn the jury problem into a marching problem. The test of whether we guessed $F''(0)$ correctly is to see if $F'(\infty)$ comes out to the right value. If not, we guess another $F''(0)$ and repeat the process until the boundary condition at ∞ is satisfied. This is the crudest and most direct way of solving jury problems, and it is the method used by Smith and Clutter [119]. The main drawback of the method is that it is expensive in computer time. Other ways exist and will be discussed in Chapter 10.

Returning now to equations (1201), we note that the boundary conditions for ϕ are $\phi(0) = 0$, $\phi'(0) = -1$, $\phi'(\infty) = 0$. If, therefore, the coefficients in equation (1201a) are known as functions of η, the equation becomes of the type (1203) and the method of solution described above may be used. We guess a $\phi''(0)$, integrate forward, check $\phi'(\infty)$, and repeat until the boundary condition at ∞ is satisfied. Smith and Clutter also used an iterative process on the coefficients. That is, a variation of the coefficients with η is assumed, and the momentum equation is solved for $\phi(\eta)$. The result is used to solve the energy equation (1201b) and this solution is used to find the coefficients as functions of η. The procedure is then repeated until the solution converges. All this is very wasteful of computer time. An improved calculation procedure for this problem has been developed by Cebeci and Smith [120].

The boundary conditions on the energy-equation (1201b) are likewise given at 0 and ∞. The energy equation may thus be solved in the same way as the momentum equation, by turning the jury problem into a marching problem. Because the energy equation (1201b) is linear in θ_n, however, it may easily be solved by direct means for two point boundary conditions. (A two point boundary condition problem is one for which boundary conditions are given at two different points.) These techniques will be discussed in Chapter 10.

Results of a trial and error run are shown in Figure 249. The solution is seen

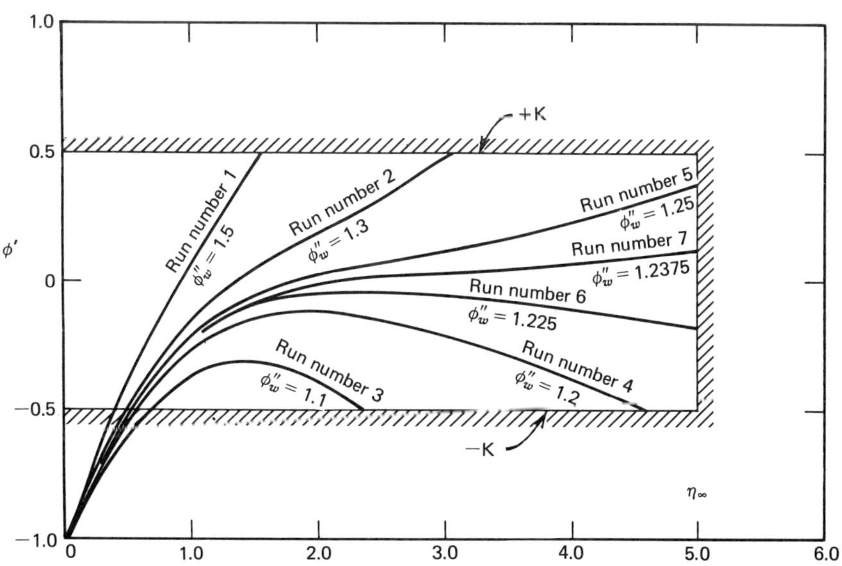

Figure 249 Effect of assumed velocity gradient at the wall on numerical solution of the boundary layer equations [119].

to be unstable but to converge rapidly. Unstable means that if the wrong value is chosen for $\phi'(0)$, the solution goes to either $+\infty$ or $-\infty$. Instability is a general characteristic of the boundary layer equations and is not restricted to the present case. As a test of the accuracy of the numerical technique, Smith and Clutter compared the results of the above method with that of Chapman and Rubesin, discussed earlier, for the case of a flat plate with

$$\frac{T_w}{T_1} = \frac{T_{ad}}{T_1}(1.25 - 0.83x + 0.33x^2)$$

where

$$\frac{T_{ad}}{T_1} = 1 + 0.169\,M_\infty^2$$

Agreement for $g_w'/(g_w)_{ad}$ was found to be within 0.4%. As another test, calculations were made for the case of a flat nosed cylinder with a corner radius equal to one-fourth the cylinder radius. The results are shown in Figure 250 for the

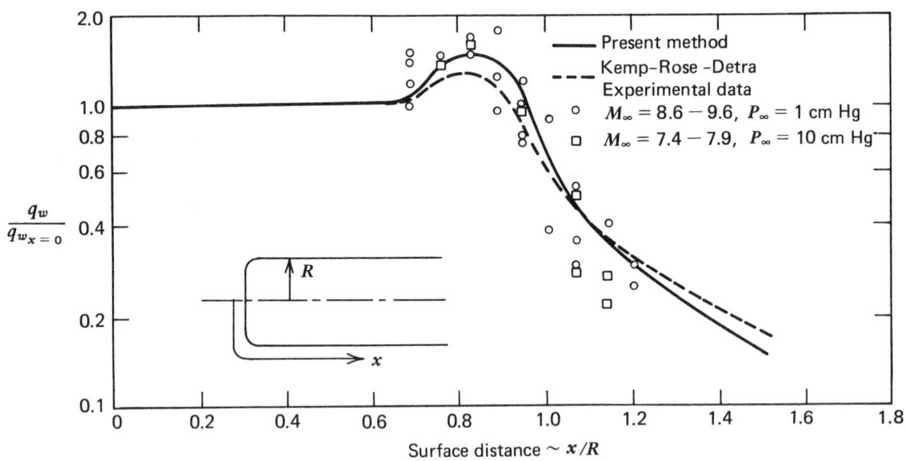

Figure 250 Heat transfer rate over a blunt body—theory [119] and experiment [121] compared.

heat transfer at the wall. Because of the high Mach number, strong variations of Pr and $\rho\mu$ with η are present. Comparison is made with experimental and theoretical results of Kemp, Rose and Detra [121]. Agreement is seen to be reasonably good, considering the uncertainties in the transport properties involved and the scatter in the experimental data.

Flow over a Cone

We now consider the laminar boundary layer over a cone at zero angle of attack when the free stream velocity is supersonic and the shock is attached. It will be recalled that the inviscid solution to this problem was given by Busemann [34] and Taylor and Macoll [35] as discussed in Chapter 4. The physical situation is shown in Figure 251, and the coordinate system to be used is shown in Figure 252. The governing equations may be written in the same form as equations

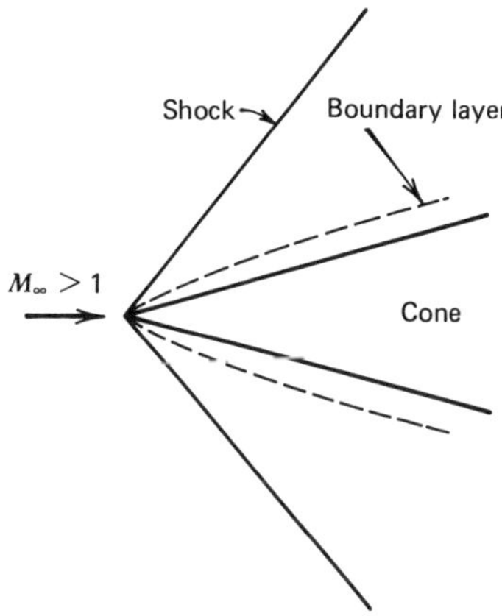

Figure 251 Cone at zero angle of attack in supersonic flow.

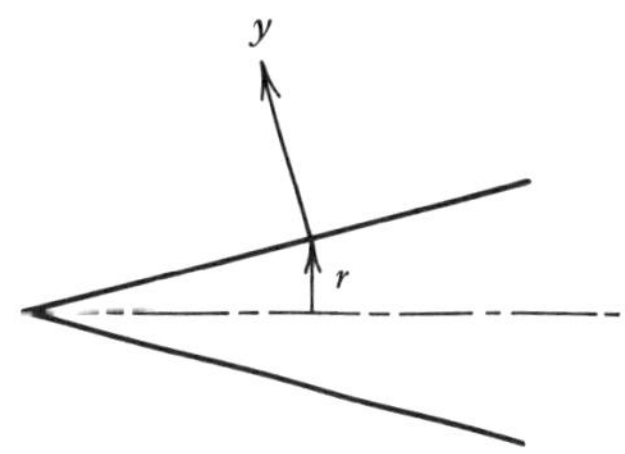

Figure 252 Coordinate system for cone in supersonic flow.

(1192–1194). Although the cone lends itself ideally to a spherical coordinate system, if the boundary layer is thin (as we assume it is), $R\,d\theta \cong dy$, where R and θ are spherical coordinates, and the variation of r with y in the boundary layer may be neglected. We now assume $Pr = 1$, $c_p =$ constant, adiabatic wall, from which it follows that the solution to equation (1194) is $h_0 =$ constant as shown earlier (Crocco–Busemann integral). Moreover, the pressure over the surface of the cone in the present instance will be constant as was shown by Busemann and Taylor and Macoll. The system of equations to be solved thus reduces to

$$(\rho ur)_x + (\rho vr)_y = 0 \tag{1204}$$

$$\rho uu_x + \rho vu_y = -(\mu u_y)_y \tag{1205}$$

We now introduce a stream function ψ defined as follows:

$$\rho ru = \psi_y, \qquad \rho rv = -\psi_x \tag{1206}$$

Notice that this is not the same definition as was used in the previous section (equation 1195). In addition, we introduce the Howarth transformation

$$Y = \int_0^y \rho\, dy \tag{1207}$$

Equation (1205) then becomes

$$\psi_Y\left(\frac{1}{r}\psi_Y\right)_X - \psi_X\left(\frac{1}{r}\psi_Y\right)_Y = \rho\mu\psi_{YYY} \tag{1208}$$

where we have assumed $\rho\mu =$ constant and $x = X$. The partial derivative with respect X is, therefore, taken with Y held constant. It will be observed that equation (1205) has now been reduced to incompressible form. Next, we introduce the transformation of Mangler [122]

$$\xi = \int_0^X r^2\, dX, \qquad \eta = rY \tag{1209}$$

Equation (1208) then becomes,

$$\psi_\eta\psi_{\eta\xi} - \psi_\xi\psi_{\eta\eta} = \rho\mu\psi_{\eta\eta\eta} \tag{1210}$$

where $u = \psi_\eta$. Because this equation is identical in form to the equation for the incompressible boundary layer over a flat plate, we may introduce the Blasius transformation to obtain a similarity solution. Let

$$\lambda = \eta\sqrt{\frac{U}{\rho\mu\xi}}, \qquad \psi = \sqrt{\rho\mu\xi U}\, f(\lambda) \tag{1211}$$

where U is the velocity at the outer edge of the boundary layer which, in the present case, is constant. Notice that under the transformation (1211) $u/U = f'(\lambda)$. Introducing the transformation (1211) into equation (1210) we get

$$ff_{\lambda\lambda} + 2f_{\lambda\lambda\lambda} = 0 \tag{1212}$$

which is the Blasius equation, the solution to which is well known, as discussed earlier. Finally, the temperature profile in the boundary layer is given by

$$c_pT + U^2f^2 = \text{constant} \tag{1213}$$

and hence the problem is solved.

The above problem with isothermal wall has been solved by Ostrach [123]. As we have observed before, the assumption $\rho\mu =$ constant uncouples the momentum and energy equations. Thus, the momentum equation is solved as before. The energy equation (1194) in the axisymmetric coordinate system is unchanged in form from the plane case and hence the Crocco–Busemann solution for isothermal walls applies. The temperature profile in this case is

$$T - T_1 = \left(T_w - T_1 + \frac{U^2}{2}\frac{u}{U}\right)\left(1 - \frac{u}{U}\right) \tag{1214}$$

where $u/U = f'(\lambda)$ as before.

The inverse transformation may readily be performed in either case to obtain the solution in the physical plane. As will be recalled,

$$\lambda = \eta\sqrt{\frac{U}{\rho\mu\xi}}, \qquad \xi = \int_0^x r^2\, dX, \qquad X = x, \qquad \eta = rY, \qquad Y = \int_0^y \rho\, dy$$

Since r is known as a function of x, ξ may easily be found as a function of x, and η may be found as a function of x and y. From the perfect gas law, $\rho = p/RT$ which is known as a function of λ, $G(\lambda)$ say. Then

$$Y = \int_0^y G(\lambda)dy = \int_0^y G(\eta, \xi)dy = \int_0^y G(\eta)dy = \int_0^y G(Y)dy$$

for a given value of x, since ξ and r are functions of x only. Thus

$$\frac{dY}{G(Y)} = dy$$

giving us the inverse transformation between Y and y.

The solution has been extended to cones at angle of attack by Moore [124] and Ferri [125].

Unsteady Boundary Layers

Thus far, we have been confining ourselves to boundary layers in steady flow. We now turn our attention to boundary layers in unsteady flow. Before proceeding, it might be well to review two classical incompressible solutions. First the case of the suddenly accelerated plane wall. An infinite wall has a velocity of zero at time $t < 0$. At time $t = 0$ it is suddenly given a velocity U_0 in its own plane. The Navier–Stokes equation for this case is given by

$$u_t - \nu u_{yy} = 0 \tag{1215}$$

The boundary conditions are

$$t < 0: \quad u = 0; \qquad t \geq 0: \quad y = 0, \quad u = U_0; \qquad y = \infty, \qquad u = 0$$

Since equation (1215) has the form of the heat conduction equation, the solution is well known and is given by

$$u = U_0 \operatorname{erfc} \eta, \qquad \eta \frac{y}{2\sqrt{\nu t}} \tag{1216}$$

where

$$\operatorname{erfc} \eta = 1 - \operatorname{erfc} \eta = 1 - \frac{2}{\sqrt{\pi}} \int_0^\eta e^{-\eta} \, d\eta$$

We have thus a similarity solution. From the form of the similarity parameter η, it is sometimes said that the vorticity diffuses upward as the square root of the time. The above problem is usually called the Rayleigh problem after the work of Lord Rayleigh [126], but was in fact, first solved by Stokes [127] about 60 years earlier.

The other incompressible problem we wish to review is that of a body in an accelerating stream as considered by Blasius [89]. The boundary layer equations now become

$$u_t + uu_x + vu_y = -\frac{1}{\rho} p_x + \nu u_{yy} \tag{1217}$$

$$u_x + v_y = 0$$

The boundary conditions are

$$t \leq 0: \quad u = v = 0$$
$$t \geq 0: \quad y = 0: \quad u = v = 0; \qquad y = \infty: \quad u = U(x, t) = tw(x)$$

In this classic solution, Blasius thus assumes a linear variation of the velocity with the time at the outer edge of the boundary layer. Introducing the stream function ψ,

$$u = \psi_y, \qquad v = -\psi_x$$

we postulate a solution of the form

$$\psi(x, y, t) = 2\sqrt{vt}\,\{tw\zeta_0(\eta) + t^3 ww_x \zeta_1(\eta) + \ldots\}$$

where $\eta = y/2\sqrt{vt}$, which will be recognized as the similarity variable from the Rayleigh problem. Substituting in the momentum equation (1217) and equating powers of t we get a series of ordinary differential equations the first two of which are

$$\begin{aligned} \zeta_0''' + 2\eta\zeta_0'' - 4\zeta_0' &= -4 \\ \zeta_1''' + 2\eta\zeta_1'' - 12\zeta_1 &= -4 + 4(\zeta_0'^2 - \zeta_0\zeta_0'') \end{aligned} \tag{1218}$$

with boundary conditions

$$\eta = 0: \quad \zeta_0 = \zeta_0' = \zeta_1 = \zeta_1' = 0$$
$$\eta = \infty: \quad \zeta_0' = 1, \qquad \zeta_1' = 0$$

For an infinite flat plate, $w(x)$ is a constant, and hence the solution is given by $u = U\zeta_0'$. The solution for ζ_0' exists in closed form and is given by

$$\zeta_0' = 1 + \frac{2}{\sqrt{\pi}}\eta e^{-\eta^2} - (1 + 2\eta^2)\,\mathrm{erfc}\,\eta \tag{1219}$$

In general, the velocity is given by

$$u = U\left(\zeta_0' + t^2\frac{dw}{dx}\zeta_1' + \cdots\right) \tag{1220}$$

While the solution for ζ_1' may also be found in closed form, the higher order terms ζ_n', $n > 1$, must be found by numerical integration. The solution may also be used to predict the time at which separation first occurs. Separation will occur when

$$\zeta_0''(0) + t^2\frac{dw}{dx}\zeta_1''(0) + \cdots = 0 \tag{1221}$$

For a circular cylinder, we have

$$U(x, t) = tw(x) = 2bt \sin\frac{x}{R} \tag{1222}$$

where R is the radius of the cylinder, b is a constant and x is the distance along the cylinder from the forward stagnation point. Thus, separation first occurs at the downstream stagnation point, where

$$\cos\frac{x}{R} = 1$$

We now turn to compressible unsteady boundary layers. Perhaps the simplest example would be that of an infinite flat plate accelerating in its own plane. This problem was considered by Illingworth [128]. The situation is shown in Figure 253.

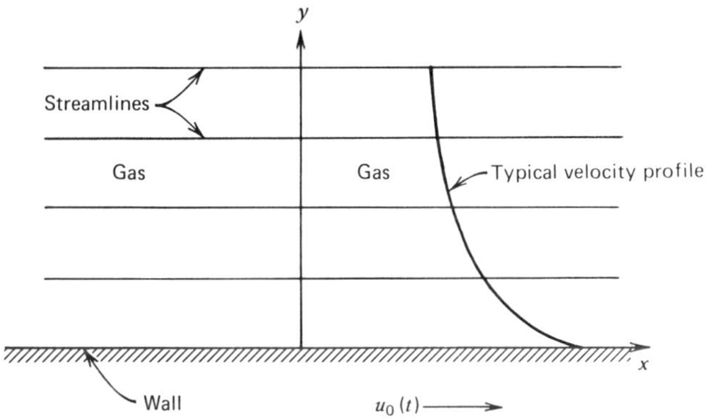

Figure 253 Velocity distribution over an accelerating, infinite plane.

The coordinate system is fixed in space (not in the plate). The governing equations may then be written

$$\rho_t + (\rho v)_y = 0 \tag{1223}$$

$$\rho(u_t + vu_y) = (\mu u_y)_y \tag{1224}$$

$$\rho(h_t + vh_y) = \frac{1}{Pr}(\mu h_y)_y + \mu u_y^2 \tag{1225}$$

Notice that there is no x dependence because the plate is infinite. We now define a function ψ by the equations

$$\psi_y = \rho, \qquad \psi_t = -\rho v \tag{1226}$$

Notice that ψ is not a stream function in the strictest sense of the term. In addition, we define the dimensionless variables

$$\bar{u} = \frac{u}{u_0}, \qquad \bar{\rho} = \frac{\rho}{\rho_1}, \qquad \bar{T} = \frac{T}{T_1}, \qquad \chi = \frac{\rho\mu}{\rho_1\mu_1}, \qquad \bar{p} = \frac{p}{p_1}, \qquad \bar{h} = \frac{h}{h_1} \tag{1227}$$

where u_0 is the velocity of the plate and the sub-script 1 refers to the outer edge of the boundary layer.

Equation (1223) is satisfied identically by ψ as defined in equation (1226). The independent variables in equations (1224) and (1225) may now be changed from t and y to t and ψ in a manner analogous to that of von Mises [95]. Equations (1224) and (1225) then become

$$\frac{\bar{u}}{u_0}\frac{du_0}{dt} + \bar{u}_t = \rho_1 \mu_1 (\chi \bar{u}_\psi)_\psi \tag{1228}$$

$$\bar{h}_t = \frac{\rho_1 \mu_1}{Pr}(\chi \bar{h}_\psi)_\psi + \rho_1 \mu_1 \frac{u_0^2}{h_1} \chi u_\psi^2 \tag{1229}$$

For subsonic flow, $u_0^2 \ll h_1$ and hence the last term on the right-hand side of equation (1229) may be neglected. We now define the variable η as

$$\eta = \frac{\psi}{2\sqrt{\rho_1 \mu_1 t}} \tag{1230}$$

Equations (1228) and (1229) then become,

$$(\chi \bar{u}_\eta)_\eta + 2\eta \bar{u}_\eta + \frac{\bar{u}}{u_0}\frac{du_0}{dt} = 0 \tag{1231}$$

$$(\chi \bar{h}_\eta)_\eta + 2Pr\eta \bar{h}_\eta = 0 \tag{1232}$$

where we have assumed $u_0^2 \ll h_1$ in equation (1232). We now assume

$$u_0 = \beta_m t^m, \qquad \chi = 1 \tag{1233}$$

Then equation (1231) becomes

$$\bar{u}_{\eta\eta} + 2\eta \bar{u}_\eta - 4m\bar{u} = 0 \tag{1234}$$

Clearly, $\bar{u}$ a function of η only is a solution to this equation, so that η becomes a similarity variable and equation (1239) may be treated as an ordinary differential equation. When m is an integer, the solution to this equation is

$$\bar{u}(\eta) = e^{-n^2}[e^{\eta^2}(c_1 + c_2 \operatorname{erf} \eta)]'' \tag{1235}$$

where the primes indicate differentiation with respect to η. c_1 and c_2 are constants of integration and are determined by the boundary conditions

$$\bar{u}(0) = 1, \qquad \bar{u}(\infty) = 0 \tag{1236}$$

For $m = 1$, the solution becomes

$$\bar{u} = (1 + 2\eta^2)(1 - \operatorname{erf} \eta) - \frac{2}{\sqrt{\pi}} \eta e^{-\eta 2} \tag{1237}$$

We observe that this solution is identical to the Blasius solution, (1219). When $m = 2$, the solution is

$$\bar{u} = \frac{1}{3}\left[(3 + 12\eta^2 + 4\eta^4)(1 - \operatorname{erf} \eta) - \frac{2}{\sqrt{\pi}} \eta(5 + 3\eta^2)e^{-\eta^2}\right] \tag{1238}$$

When m is not an integer, the solution is

$$\bar{u}(\eta) = 1 + \sum_{s=0}^{\infty} \frac{m(m-1)\ldots(m+1-s)}{(2s)!}(2\eta)^{2s} + c_2\eta\left[1 + \sum_{s=1}^{\infty} \frac{(2m-1)(2m-3)\ldots(2m+1-2s)}{(2s+1)!}2^s\eta^{2s}\right] \quad (1239)$$

where c_2 is chosen to make $\bar{u}(\infty) = 0$.

For $\chi = 1$, the solution to the energy equation (1232) for isothermal wall is

$$h(\eta) = \bar{h}(0) + [1 - \bar{h}(0)]\,\mathrm{erf}(\eta\sqrt{Pr}) \quad (1240)$$

where $\bar{h}(0)$ is the dimensionless enthalpy at the wall. For adiabatic wall, the solution to equation (1232) is, clearly,

$$\bar{h} = 1 \quad (1241)$$

This corresponds to constant stagnation enthalpy since we have neglected the term u_0^2/h_1. The solution may also be found including the last term of equation (1229), if $Pr = 1$. In this case we have,

$$\bar{h} = 1 + \frac{u_0^2}{h_1}\bar{u}(1 - \tfrac{1}{2}\bar{u}) \quad (1242)$$

Notice that this implies that the adiabatic wall enthalpy equals the free stream stagnation enthalpy relative to the wall; i.e.,

$$h_w = h_1 + \tfrac{1}{2}u_0^2 \quad (1243)$$

Since equation (1234) is linear, solutions for different integral values of m may be added when the velocity of the plate is given by a power series in t. That is, when

$$u_0 = \sum_m \beta_m t^m$$

the velocity distribution is given by

$$\bar{u} = \sum_m e^{-\eta^2}\frac{d^{2m}}{d\eta^{2m}}[e^{\eta^2}(c_1 + c_2\,\mathrm{erf}\,\eta)] \quad (1244)$$

We now proceed to the case of the semi-infinite flat plate. The physical situation is shown in Figure 254. The plate starts from rest and accelerates in

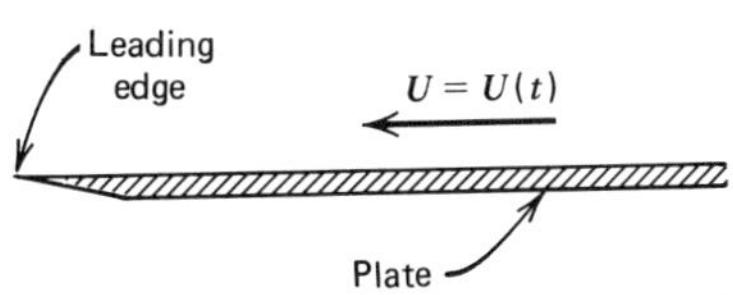

Figure 254 Accelerating semi-infinite flat plate.

its own plane with velocity $U(t)$. The plate is assumed to be of zero thickness. Only the flow over the top of the plate will be considered. Recall, once more, the two dimensional, unsteady, compressible boundary layer equations for zero pressure gradient.

$$\rho_t + (\rho u)_x + (\rho v)_y = 0$$

$$\rho[u_t + uu_x + vu_y] = (\mu u_y)_y \tag{1245}$$

$$\rho[h_t + uh_x + vh_y] = \frac{c_p}{Pr}(\mu T_y)_y + \mu u_y^2$$

The assumption of zero pressure gradient for the flat plate with leading edge is not, strictly speaking, accurate; but, as long as the velocities are not too great (hypersonic), the error involved is small. We will assume in addition that c_p, Pr and $\rho\mu$ are constant and that $p = \rho RT$.

In dealing with accelerating bodies we must bear in mind that we cannot simply fix the coordinate system in the body as we did in the case of steady flow. The equations (1245) are valid only in an inertial coordinate system; i.e., one which is not accelerating. We can, however, transform the equations by a suitable change of variables. Let x', y', t' represent a coordinate system fixed in the plate with $x' = 0$ at the leading edge (Figure 255). Let x, y, t represent a coordinate system

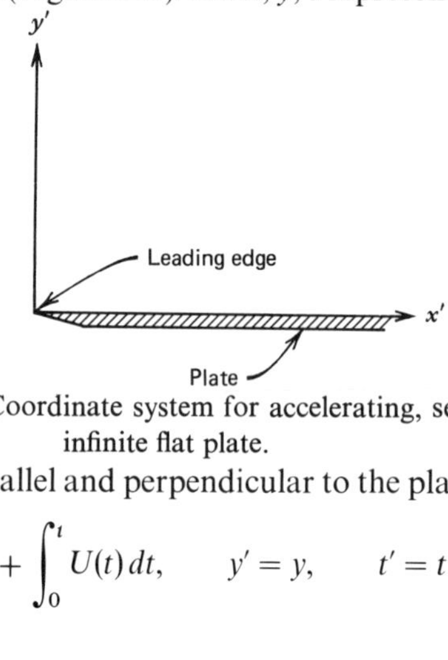

Figure 255 Coordinate system for accelerating, semi-infinite flat plate.

fixed in space, x and y parallel and perpendicular to the plate, respectively. Then

$$x' = x + \int_0^t U(t)\,dt, \qquad y' = y, \qquad t' = t$$

$$u' = u + U(t), \qquad v' = v \tag{1246}$$

In terms of the coordinates (1246), equations (1245) become

$$\rho(u'_{t'} + u'u'_{x'} + v'u'_{y'}) = \rho U_{t'} + (\mu u'_{y'})_{y'}$$

$$\rho(T_{t'} + u'T_{x'} + v'T_{y'}) = \frac{1}{Pr}(\mu T_{y'})_{y'} + \frac{\mu}{c_p}(u'_{y'})^2 \tag{1247}$$

$$\rho_{t'} + (\rho u')_{x'} + (\rho v')_{y'} = 0$$

We may note the presence of the U'_t term in the momentum equation. This takes the place of the pressure gradient in causing the flow and is sometimes called the forcing function.

We now introduce the Howarth transformation and the stream function defined by

$$\eta = \int_0^y \frac{\rho}{\rho_\infty} dy, \qquad u = \psi_\eta \tag{1248}$$

Equations (1247) then become

$$\psi_{\eta t} + \psi_\eta \psi_{\eta x'} - \psi_{x'} \psi_{\eta\eta} = U_{t'} + v\psi_{\eta\eta\eta} \tag{1249}$$

$$T_{t'} + \psi_\eta T_{x'} - \psi_{x'} T_\eta = \frac{v}{Pr} + \frac{v}{Pr}(\psi_{\eta\eta})^2 \tag{1250}$$

where $v = \mu_\infty/\rho_\infty$. The effect is twofold: 1. The momentum and energy equations are uncoupled, and 2. The equations are reduced to their incompressible form.

In order to solve equation (1249) the function $U(t)$ must be specified. We will consider the case of constant acceleration. Then $U = At$. The following change of variables now suggests itself:

$$z = \frac{\eta}{\sqrt{vt'}}, \qquad \zeta = \frac{x'}{At^2}, \qquad \psi = \sqrt{v}\, At^{3/2} g(z, \zeta) \tag{1251}$$

The definition of z is simply the similarity variable of Rayleigh used by Blasius and Illingworth (op. cit.). The distance traveled by the plate is $At^2/2$, leading to the definition of ζ. The definition of ψ comes from postulating

$$u = Atg_z(z, \zeta) \tag{1252}$$

The same dimensionless groupings could also have been obtained by using Buckingham's π theorem on the supposition that $u = u(A, v, x', \eta, t')$. Introducing the variables (1251) into equation (1249) we get

$$g_z - \tfrac{1}{2} z g_{zz} - 2\zeta g_{z\zeta} + g_z g_{z\zeta} - g_\zeta g_{z\zeta} - 1 + g_{zzz} \tag{1253}$$

The effect of the transformation (1250) is thus to reduce the number of independent variables from three to two, just as the earlier transformation reduces the number of dependent variables from four (u, v, ρ, μ) to one (ψ). The boundary conditions are

$$g_z(z, 0) = g_z(\infty, \zeta) = 1, \qquad g_z(0, \zeta) = g(0, \zeta) = 0 \tag{1254}$$

We now seek to solve equation (1253) subject to the boundary conditions (1254). One of the most common ways of solving partial differential equations is by the method of series expansion. For ζ large, we may postulate a solution of the form

$$g(z, \zeta) = \sum_{n=0}^{\infty} g_n(z)\zeta^{-n} \tag{1255}$$

Substituting into equation (1253) and equating powers of ζ we get

$n = 0$: $$g_0' - \tfrac{1}{2} z g_0'' = 1 + g_0''' \tag{1256}$$

$n > 0$: $$g_n' = 0 \tag{1257}$$

where the primes indicate differentiation with respect to z. Equation (1256) is, in fact, the same as equation (1218) for ζ_0 as found by Blasius and Illingworth (op. cit.) for a uniformly accelerating infinite flat plate. (The difference in coefficients is due to the inclusion of the factors 2 and $\frac{1}{2}$ in the definition of ψ and η by Blasius). The solution is known in closed form and may be written

$$g_0' = -\tfrac{1}{2}z^2 + \frac{1}{\sqrt{\pi}}ze^{-z^2/4} + (z^2+2)\frac{1}{\sqrt{\pi}}\int_0^{z/2} e^{-z^2}\,dz \tag{1258}$$

which is the same as equations (1219) and (1237). The results (1256) and (1257) show the futility of attempting to perturb the downstream solution to obtain the upstream solution, a fact already observed by Cheng [129], Cheng and Elliot [129a], and Stewartson [130] in the incompressible case.

As an alternative, we may try to expand the solution around the leading edge for ζ small. Recalling the Blasius solution for the steady incompressible flow over a semi-infinite flat plate,

$$\psi = \sqrt{\nu x U}\, f\left(y\sqrt{\frac{U}{\nu x}}\right) \tag{1259}$$

an expansion which suggests itself is

$$g(z,\zeta) = \sqrt{\zeta}\sum_{n=0}^{\infty} g_n\left(\frac{z}{2\sqrt{\zeta}}\right)\zeta^n \tag{1260}$$

Substituting in equation (1253) and equating powers of ζ, we get

$$g_0 g_0'' + g_0''' = 0 \tag{1261}$$

$$g_1(3g_0'') - g_1'(2g_0') + g_1''(g_0) + g_1''' = -8 + 4g_0' + 2Zg_0''' \tag{1262}$$

and so on, where the primes indicate differentiation with respect to $Z = z/2\sqrt{\zeta}$. Equation (1261) will be recognized as the Blasius equation (1156) for steady flow, but now $g_n' \neq 0$ for $n > 0$. The higher order terms thus give us the error involved in assuming the flow to be quasi-steady. The equations must be solved numerically. Fortunately, each equation involves only one new unknown so that the equations may be solved separately rather than simultaneously.

The solution may be extended to the case where U is an arbitrary function of t provided the acceleration is not too great. We postulate a solution of the form

$$\psi = U\sqrt{\frac{\nu x'}{U}}\, f(\sigma, \zeta_0, \zeta_1, \zeta_2, \ldots) \tag{1263}$$

where

$$\sigma = \frac{1}{2}\frac{\eta}{\sqrt{x'}}\sqrt{\frac{U(t)}{\nu}}, \qquad \zeta_n = \frac{x'^{n+1}}{U^{n+2}}\frac{d^{n+1}U}{dt^{n+1}} \tag{1264}$$

and

$$f(\sigma,\zeta_n) = F(\sigma) + \sum_{n=0}^{\infty}\zeta_n f_n(\sigma) + \sum_{n=0}^{\infty}\zeta_0\zeta_n f_{0n}(\sigma) + \cdots \tag{1265}$$

Substituting (1263) in (1249) and equating coefficients of the various powers and products of ζ_n we get

$$F''' + FF'' = 0 \tag{1266}$$

$$f_0''' + Ff_0'' - 2F'f_0' + 3F''f_0 = -4(2 - F') + 2\sigma F'' \tag{1277}$$

and so on, where the primes indicate differentiation with respect to σ. The boundary conditions are

$$\begin{aligned} F'(\infty) &= 2, \qquad F'(0) = F(0) = 0 \\ f_n'(\infty) &= f_n'(0) = f_n(0) = 0 \end{aligned} \tag{1278}$$

Equation (1266) will be recognized as the Blasius equation. The remaining terms in equation (1265), therefore, give us the departure of the solution from the quasi-steady-state condition.

The wall shear stress is given by

$$\tau_w = \left(\mu \frac{\partial u}{\partial y}\right)_w = \left(\frac{1}{2}\rho_\infty U^2\right)\left(\frac{1}{2}\sqrt{\frac{v}{Ux'}}\right)\left[F''(0) + \frac{x'}{U^2}\frac{dU}{dt}f_0''(0) + \frac{x'^2}{U^3}\frac{d^2U}{dt^2}f_1''(0) + \cdots\right] \tag{1279}$$

From the numerical solution of equations (1266), (1277) et. seq. we find

$$\tau_w = \tfrac{1}{2}\rho_\infty U^2 \sqrt{\frac{v}{Ux'}}(0.664)\left[1 + (2.555)\frac{x'}{U^2}\frac{dU}{dt} - (1.414)\frac{x'^2}{U^3}\frac{d^2U}{dt^2} + \cdots\right] \tag{1280}$$

A steady, positive acceleration thus causes a greater friction drag than is indicated by the quasi-steady value.

For the energy equation, we postulate a solution of the form

$$\frac{T}{T_\infty} = 1 + \frac{\gamma - 1}{2}M_\infty^2(t)r(\sigma, \zeta_0, \zeta_1, \zeta_2, \ldots) \tag{1281}$$

The function r will be recognized as the recovery factor when $\sigma = 0$. The wall is presumed to be insulated. The function r is presumed to be of the form

$$r(\sigma, \zeta_n) = R(\sigma) + \sum_{n=0}^{\infty} \zeta_n r_n(\sigma) + \sum_{n=0}^{\infty} \zeta_0 \zeta_n r_{0n}(\sigma) + \cdots \tag{1282}$$

Substituting equations (1282) and (1265) in equation (1250) and equating coefficients of the various powers and products of ζ_n we get

$$R'' + PrFR' + \tfrac{1}{2}Pr(F'')^2 = 0 \tag{1283}$$

$$r_0'' + PrFr_0' - 2PrF'r_0 = Pr(-F''f_0'' + 8R + 2\sigma R' - 3R'f_0) \tag{1284}$$

and so on, where the primes indicate differentiation with respect to σ. The boundary conditions are

$$R'(0) = R(\infty) = 0, \qquad r_n'(0) = r_n(\infty) = 0 \tag{1285}$$

The recovery factor is given by

$$r(0) = R(0) + \frac{x'}{U^2}\frac{dU}{dt}r_0(0) + \frac{x^2}{U^3}\frac{d^2U}{dt^2}r_1(0) + \cdots \tag{1286}$$

For a Prandtl number of 0.72 this works out to

$$r(0) = 0.484\left[1 - 2.852\frac{x'}{U^2}\frac{dU}{dt} + 3.266\frac{x'}{U^3}\frac{d^2U}{dt^2} + \cdots\right] \tag{1287}$$

The leading term is the quasi-steady solution. A steady positive acceleration thus results in a lower surface temperature.

The solution of the energy equation for the isothermal wall case has been given by Ostrach [131]. Let

$$\theta = \frac{T - T_\infty}{T_w - T_\infty} = h(\sigma, \zeta_n) + \frac{U^2}{2c_p(T_w - T_\infty)}s(\sigma, \zeta_n) \tag{1288}$$

and let

$$h(\sigma, \zeta_n) = H(\sigma) + \sum_{n=0}^{\infty} \zeta_n h_n(\sigma) + \sum_{n=0}^{\infty} \zeta_0 \zeta_n h_{0n}(\sigma) + \cdots \tag{1289}$$

$$s(\sigma, \zeta_n) = S(\sigma) + \sum_{n=0}^{\infty} \zeta_n s_n(\sigma) + \sum_{n=0}^{\infty} \zeta_0 \zeta_n s_{0n}(\sigma) + \cdots \tag{1290}$$

Substituting in the energy equation (1250) and equating the coefficients of the various powers and products of ζ_n we get

$$H'' + PrFH' = 0 \tag{1291}$$

$$S'' + PrFS' + \frac{Pr}{2}(F'')^2 = 0 \tag{1292}$$

$$s_0'' + PrFs_0' - 2PrF's_0 = Pr(2\sigma S' + 8S' - F''f_0'' - 3S'f_0) \tag{1239}$$

$$h_0'' + PrFh_0' - 2PrF'h_0 = PrH'(2\sigma - 3f_0) \tag{1294}$$

and so on, where the primes indicate differentiation with respect to σ. The boundary conditions are,

$$\begin{aligned} &H(\infty) = 0, \qquad H(0) = 1, \qquad h_n(\infty) = h_n(0) = 0 \\ &S(\infty) = S(0) = 0, \qquad s_n(\infty) = s_n(0) = 0 \end{aligned} \tag{1295}$$

The heat-transfer rate at the wall is given by

$$q = -\left(k\frac{\partial T}{\partial y}\right)_w = -\frac{c_p}{2Pr}(T_w - T_\infty)\sqrt{\frac{U\rho_\infty\mu_\infty}{x'}}\left[\frac{\partial h}{\partial \sigma} + \frac{U^2}{2c_p(T_w - T_\infty)}\frac{\partial s}{\partial \sigma}\right]_w \tag{1296}$$

From the numerical solutions of equations (1291) et. seq., we find

$$q = 0.4106\, c_p (T_w - T_\infty) \sqrt{\frac{U\rho_\infty \mu_\infty}{x'}} \left\{ 1 - 0.4240(\gamma - 1) M_\infty^2 \frac{T_\infty}{T_w - T_\infty} \right.$$

$$\left. - 0.06923\zeta_0 \left[1 + 0.2746(\gamma - 1) M_\infty^2 \frac{T_\infty}{T_w - T_\infty} \right] - \cdots \right\} \qquad (1297)$$

The first two terms represent the quasi-steady solution.

8 Turbulent Boundary Layers

Basic Concepts

As was indicated in the previous chapter, turbulent flow is characterized by random fluctuations of the velocity around some average value. While the Navier–Stokes equation is valid for turbulent as well as laminar flow, in practice it has proved impossible to obtain exact solutions for $\mathbf{V}(x, y, z, t)$ in turbulent flow. The method approach which is used, therefore, is to try to find solutions only for the average velocity. Let

$$u = \bar{u} + u' \qquad v = \bar{v} + v' \qquad w = \bar{w} + w' \qquad p = \bar{p} + p' \qquad \rho = \bar{\rho} + \rho' \tag{1298}$$

and so on, where the bar superscript (as in $\bar{u}$) indicates the average value and the prime superscript (as in u') indicates the variation due to turbulent fluctuation. The time average of a fluid property, $\bar{f}$ say, ($f = u, v, \rho$, etc.) is defined by

$$\bar{f} = \frac{1}{\tau}\int_0^\tau f\,dt \tag{1299}$$

Because a turbulent flow may be unsteady in the mean, τ must be large compared to the average period of the random fluctuations, but small relative to the time rate of change of the average fluid properties. In practice this usually presents no problem, since the average period of a random fluctuation is less than a hundredth of a second. It follows from (1299) that

$$\bar{f}' = 0 \tag{1300}$$

In order to facilitate our analysis, it will be helpful to review some of the properties of average quantities. Let f and g be two fluid properties. Then

$$\bar{\bar{f}} = \bar{f} \qquad \bar{f} + \bar{g} = \overline{f + g} \qquad \overline{\bar{f} \cdot g} = \bar{f} \cdot \bar{g} \tag{1301}$$

In words, the average of the average equals the average, the average of the sum equals the sum of the averages, but the average of the product does *not* equal the product of the averages. Now let s be an independent variable ($s = x, y, z, t$). Then

$$\frac{\partial \bar{f}}{\partial s} = \overline{\frac{\partial f}{\partial s}}, \qquad \int \bar{f}\,ds = \overline{\int f\,ds} \tag{1302}$$

In words, the derivative of the average equals the average of the derivative and the integral of the average equals the average of the integral.

Recall now the continuity equation

$$\rho_t + (\rho u)_x + (\rho v)_y = 0 \tag{1303}$$

for two-dimensional flow. Substituting the definitions (1298) and applying the laws (1301) and (1302),

$$\bar{\rho}_t + \rho'_t + (\overline{\rho u})_x + (\rho u)'_x + (\overline{\rho v})_y + (\rho v)'_y = 0 \tag{1304}$$

where we have let $\rho u = \overline{\rho u} + (\rho u)'$ and $\rho v = \overline{\rho v} + (\rho v)'$. Notice that this is not the same as $\overline{\rho v} + \rho' v'$. We now take the time average of both sides of equation (1304). Then

$$\bar{\rho}_t + (\overline{\rho u})_x + (\overline{\rho v})_y = 0 \tag{1305}$$

since the time average of the fluctuations is zero. Now

$$\begin{aligned} \overline{\rho u} &= \overline{(\bar{\rho} + \rho')(\bar{u} + u')} = \bar{\rho}\bar{u} + \overline{\rho' u'} \\ \overline{\rho v} &= \overline{(\bar{\rho} + \rho')(\bar{v} + v')} = \bar{\rho}\bar{v} + \overline{\rho' v'} \end{aligned} \tag{1306}$$

so that

$$\rho_t + (\bar{\rho}\bar{u})_x + (\bar{\rho}\bar{v})_y + (\overline{\rho' u'})_x + (\overline{\rho' v'})_y = 0 \tag{1307}$$

Consider now the momentum equation. From the Navier–Stokes equations in Chapter 1 we have, for two-dimensional flow

$$\rho(u_t + u u_x + v u_y) = -p_x + \frac{\partial \sigma_{xx}}{\partial x} + \frac{\partial \sigma_{xy}}{\partial y} \tag{1308}$$

In accordance with the boundary layer assumptions we may neglect $\partial \sigma_{xy}/\partial x$. In order to avoid confusion, we will let $\sigma_{xy} \equiv \tau$, in keeping with the customary boundary layer notation. τ then represents the shear stress in two-dimensional boundary layer flow and all subscripts will indicate partial differentiation. Then

$$\rho(u_t + u u_x + v u_y) = -p_x + \tau_y \tag{1309}$$

Introducing the definitions (1298) and taking the time averages of both sides we get

$$\bar{\rho}\bar{u}_t + \overline{\rho u}\,\bar{u}_x + \overline{\rho v}\,\bar{u}_y = -\bar{p}_x + \bar{\tau}_y - (\overline{\rho' u'})_t - [\overline{(\rho u)' u'}]_x - [\overline{(\rho v)' u'}]_y \tag{1310}$$

If the flow is steady in the mean,

$$\bar{\rho}\bar{u}_t = (\overline{\rho' u'})_t = 0 \tag{1311}$$

Since μ is but little affected by the random fluctuations we may take $\mu = \bar{\mu}$. Then

$$\bar{\tau} = \overline{\bar{\mu}(u_y + v_x)} \cong \bar{\mu}(\bar{u}_y + \bar{v}_x) \cong \bar{\mu}\bar{u}_y \tag{1312}$$

within the boundary layer assumptions. If the fluctuations are truly random, u' and v' will be of the same order of magnitude. In keeping with the boundary layer assumption that x derivatives are small compared to y derivatives,

$$[\overline{(\rho u)' u'}]_x \ll [\overline{(\rho v)' u'}]_y \tag{1313}$$

From (1311), (1312) and (1313), equation (1310) becomes

$$\overline{\rho u}\,\bar{u}_x + \overline{\rho v}\,\bar{u}_y = -\bar{p}_x + (\mu \bar{u}_y)_y + [\overline{-(\rho v)' u'}]_y \tag{1314}$$

Viewed in terms of the mean flow properties, we see that the effect of turbulence is to add another stress term to the right-hand side of the equation. In addition to the pressure stress $\bar{p}_x$, and the viscous stress $\mu\bar{u}_y$, we now have the turbulent stress, $[-\overline{(\rho v)' u'}]_y$. This last term is called the Reynold's stress. Physically, it may be interpreted as the additional momentum flux across a line parallel to the flow direction due to the random fluctuation of the velocity. All turbulent flow problems boil down to finding the Reynold's stress, which is large compared to the viscous stress except near a wall. So far it has proved impossible to obtain exact analytic solutions for the Reynold's stress. All turbulent flow theories are thus semiempirical in nature.

In the case of compressible boundary layers, it is also necessary to consider the energy equation. Recall for $\dot{q} = 0$, equation (186)

$$\rho \frac{Dh}{Dt} = \frac{Dp}{Dt} + \mu\Phi \tag{1315}$$

where

$$\Phi = 2[u_x^2 + v_y^2] + (v_x + u_y)^2 - \tfrac{2}{3}(u_x + v_y)^2 \tag{1316}$$

for two-dimensional flow. Using (1298), (1314) and (1307), taking time averages, and introducing the boundary layer assumptions we get, for steady in the mean, two-dimensional flow of a perfect gas,

$$\overline{\rho u}(\bar{c}_p\bar{T})_x + \overline{\rho v}(\bar{c}_p\bar{T})_y = \bar{u}\bar{p}_x + (\bar{k}\bar{T}_y)_y + \bar{\mu}(\bar{u}_y)^2 - (\bar{c}_p\overline{(\rho v)'T'})_y - \overline{(\rho v)'u'}\,\bar{u}_y \tag{1317}$$

(Van Driest [132]). k and c_p are assumed to vary but little due to the turbulent fluctuations and hence $k \cong \bar{k}, c_p \cong \bar{c}_p$.

Having established the basic governing equations, (1307), (1314) and (1317), it will now be useful to review some elementary concepts regarding turbulent boundary layers. Although these concepts were first developed for incompressible flow, they form the starting point for compressible flow analysis as well.

One of the earliest ideas considered in the analysis of turbulent boundary layers was to define a turbulent viscosity, ε, in analogy with the laminar viscosity, μ. ε is defined by

$$-\overline{(\rho v)'u'} = \varepsilon\bar{u}_y \tag{1318}$$

Equation (1318) is called the Boussinesq approximation, after J. Boussinesq, the man who first published it [133]. The coefficient ε is sometimes called the eddy viscosity.

An examination of the energy equation (1317) shows that a similar analogy may be made. The term $-(\bar{c}_p\overline{(\rho v)'T'})_y$ is the contribution made by the turbulence to the heat transfer. A turbulent heat transfer coefficient k_t may thus be defined by

$$\bar{c}_p\overline{(\rho v)'T'} = k_t\bar{T}_y \tag{1317a}$$

The coefficient k_t is sometimes called the eddy thermal conductivity.

The term $-\overline{(\rho v)' u'} \bar{u}_y$ represents the kinetic energy being turned into heat due to the turbulence. It is thus analogous to the viscous dissipation function Φ, and is sometimes called the apparent dissipation function.

Prandtl's mixing length theory

One of the first attempts to evaluate ε theoretically was made by L. Prandtl [134]. Both turbulent and viscous shear are caused by momentum transfer perpendicular to the flow direction. Consider an average velocity profile as shown in Figure 256, and consider an imaginary line (dotted line) parallel to

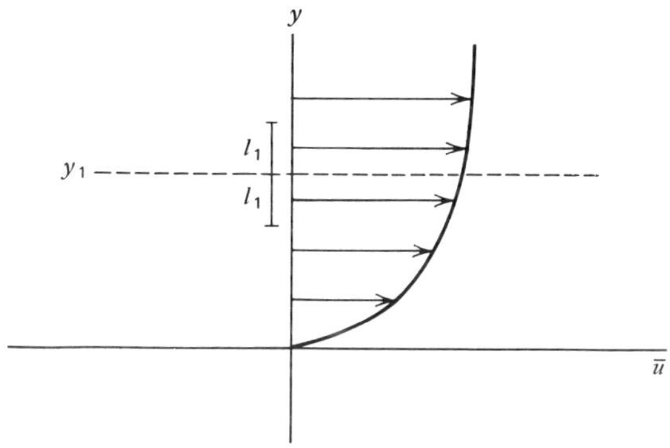

Figure 256 Prandtl's mixing length, l_1.

the flow direction. We wish to determine the force exerted by the fluid above the line on the fluid below the line due to turbulent fluctuations. Suppose that on average, each particle of fluid which crosses the line due to turbulent fluctuation originates a distance l_1 from the line, and suppose that it retains the x momentum ($\bar{u}$ being in the x direction) which it had at $y = y_1 \pm l_1$ even as it crosses the dotted line. (The $\pm$ depending on whether the particle originates above or below the line). At the line, then, the average value of the absolute perturbation velocity is given by

$$\overline{|u'|} = |\bar{u}(y_1 \pm l_1) - \bar{u}(y_1)| = \left| l_1 \left(\frac{d\bar{u}}{dy} \right)_1 \right| \tag{1318a}$$

Since a vertical displacement requires a vertical velocity component, we may postulate

$$\overline{|v'|} \sim \overline{|u'|} = c \left| l_1 \left(\frac{d\bar{u}}{dy} \right)_1 \right| \tag{1319}$$

where c is some positive constant. According to our model, a positive u' implies a negative v' and vice versa. Hence,

$$v' = -cu' \qquad \text{and} \qquad \overline{v'u'} = -\overline{cu'u'} = -\overline{cu'^2} \tag{1320}$$

While the average of the squares is not precisely equal to the square of the averages, the difference, in the present case, is not significant in light of the highly approximate nature of our theory. Hence we may say

$$-\overline{v'u'} \cong c\overline{|u'|}\,\overline{|u'|} = l_1^2\left(\frac{d\bar{u}}{dy}\right)_1^2 = l^2\left(\frac{d\bar{u}}{dy}\right)^2 \tag{1321}$$

absorbing the c into the l, and dropping the subscripts. In the incompressible case ρ is constant, and hence the Reynolds stress becomes

$$-\rho\overline{v'u'} = \rho l^2\left(\frac{d\bar{u}}{dy}\right)^2 = \varepsilon\frac{d\bar{u}}{dy} \tag{1322}$$

in which case

$$\varepsilon = \rho l^2\frac{d\bar{u}}{dy} \tag{1323}$$

This points up the weakness of the Boussinesq approximation, since ε turns out to be a function of $d\bar{u}/dy$ while l is still undetermined. It should be noted that equation (1323) is only accurate for $d\bar{u}/dy$ positive. The formula may be generalized by writing

$$\varepsilon = \rho l^2\left|\frac{d\bar{u}}{dy}\right| \tag{1324}$$

This is the mixing length theory as originally presented by Prandtl [134].

In the compressible case, the situation is not quite so simple. If we assume that in the compressible case

$$u' = l\frac{d\bar{u}}{dy} \tag{1325}$$

where l is now a variable that fluctuates with u', then

$$-\overline{(\rho v)'u'} = -\overline{(\rho v)'l\frac{d\bar{u}}{dy}} = -\overline{(\rho v)'l}\frac{d\bar{u}}{dy} = \varepsilon\frac{d\bar{u}}{dy} \tag{1326}$$

Hence,

$$\varepsilon = -\overline{(\rho v)'l} \tag{1327}$$

If the mechanism for turbulent heat transfer is the same as the mechanism for turbulent momentum transfer, then we may say

$$T' \cong l\frac{d\bar{T}}{dy} \tag{1328}$$

and

$$-\bar{c}_p\overline{(\rho v)'T'} \cong -\bar{c}_p\overline{(\rho v)'l}\frac{d\bar{T}}{dy} \cong k_t\frac{d\bar{T}}{dy} \tag{1329}$$

Hence,

$$k_t \cong -\bar{c}_p\overline{(\rho v)'l} \tag{1330}$$

Carrying our analogy with laminar flow one step further, we may define a turbulent Prandtl number, Pr_t, such that

$$Pr_t = \frac{c_p \varepsilon}{k_t} \tag{1331}$$

Substituting equation (1327) and (1330) into (1331),

$$Pr_t \cong \frac{c_p[-\overline{(\rho v)'l}]}{-\bar{c}_p\overline{(\rho v)'l}} \cong 1 \tag{1332}$$

Although the argument is crude, experiments show that the conclusion is substantially correct. The actual turbulent Prandtl number in air is around 0.80 [135].

Structure of the Turbulent Boundary Layer

Because the velocity must be zero at the wall, it is clear that as we approach the wall, the turbulent fluctuations must die out. The viscous shear, on the other hand, retains its value even at the wall. Thus, there will be a thin region of the boundary layer near the wall in which viscous effects will predominate over turbulent effects. This region is called the laminar sublayer (Figure 257).

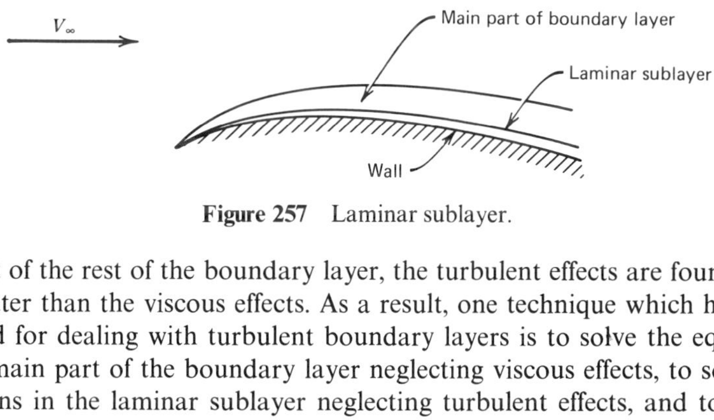

Figure 257 Laminar sublayer.

In most of the rest of the boundary layer, the turbulent effects are found to be far greater than the viscous effects. As a result, one technique which has been adopted for dealing with turbulent boundary layers is to solve the equations in the main part of the boundary layer neglecting viscous effects, to solve the equations in the laminar sublayer neglecting turbulent effects, and to match up the solutions at some intermediate point. Obviously, there will also be a region in which the turbulent effects and the laminar effects are of equal orders of magnitude. Some authors have tried to divide the boundary layer into three regions—a laminar, a turbulent and a transition. But in general, the two-layer approach has been preferred.

Elementary solution of the energy equation. In line with the above argument, the viscous effects may be neglected in the main part of the boundary layer. The governing equations then become

$$(\overline{\rho u})_x + (\overline{\rho v})_y = 0 \tag{1333}$$

$$\overline{\rho u}\,\bar{u}_x + \overline{\rho v}\,\bar{u}_y = \bar{p}_x + (\varepsilon \bar{u}_y)_y \tag{1334}$$

$$\overline{\rho u}\,\bar{c}_p \bar{T}_x + \overline{\rho v}\,\bar{c}_p \bar{T}_y = \bar{u}\bar{p}_x + (k_t \bar{T}_y)_y + \varepsilon \bar{u}_y^2 \tag{1335}$$

These equations will be observed to have the same form as the equations for the laminar boundary layer (1079–1081). It follows that if we postulate that $\bar{T}=\bar{T}(\bar{u})$,

$$-p_x(\bar{c}_p\bar{T}_{\bar{u}}+\bar{u})+\bar{c}_p\frac{Pr_t-1}{Pr_t}\bar{T}_u(\varepsilon\bar{u}_y)_y=(\bar{T}_{\bar{u}\bar{u}}k_t+\varepsilon)\bar{u}_y^2 \tag{1336}$$

in analogy with equation (1086). As we have stated, Pr_t is close to 1 for air, and hence we may assume $Pr_t=1$ as a first-order approximation. For a flat plate, $p_x=0$, and hence the solution to equation (1336) is the same as the Crocco–Busemann integral (ef. equation 1088)

$$\bar{T}=-\frac{\bar{u}^2}{2\bar{c}_p}+c_1\bar{u}+c_2 \tag{1337}$$

Since in the laminar sublayer $\bar{u}=u$, $\bar{T}=T$, and equation (1088) is valid, equation (1337) may be taken all the way across the boundary layer, and the constants evaluated as before. Thus we have for the insulated wall,

$$\bar{c}_p\bar{T}_{aw}=\bar{c}_p\bar{T}+\frac{\bar{u}^2}{2}=\text{constant} \tag{1338a}$$

and for the isothermal wall,

$$\frac{\bar{T}}{T_w}=\frac{T_w}{T_\infty}-\left(\frac{T_w}{T_\infty}-1\right)\frac{\bar{u}}{U}+\frac{\gamma-1}{2}M_\infty^2\frac{\bar{u}}{U}\left(1-\frac{\bar{u}}{U}\right) \tag{1338b}$$

where T_w is the wall temperature, T_∞ the free stream temperature, and U the velocity at the outer edge of the boundary layer, constant and equal to V_∞ in the present (flat plate) case.

Solution for the Coefficient of Friction

Van Driest's analysis [*136*]

As in the laminar case, equations (1338) give us only a relation between the velocity and the temperature, not a relation for the velocity and temperature as a function of x and y. If $\bar{u}(x, y)$ were known, however, $\bar{T}(x, y)$ could then be found from equation (1338). Recall for the turbulent shear stress

$$\tau_t=-\overline{(\rho v)'u'} \tag{1339}$$

where we have introduced the symbol τ_t for the Reynolds stress. Now from the laws of means,

$$-\overline{(\rho v)'u'}=-\bar{\rho}\overline{u'v'}-\bar{v}\overline{\rho'u'}-\overline{\rho'u'v'} \tag{1340}$$

The second and third terms of the right-hand side of equation (1340) are small compared to the first term. Hence,

$$-\overline{(\rho v)'u'}\cong-\bar{\rho}\overline{u'v'} \tag{1341}$$

Under this approximation, the entire Prandtl mixing length theory may be extended to compressible flow. Thus,

$$\tau_t = \bar{\rho} l^2 \left(\frac{d\bar{u}}{dy}\right)^2 \tag{1342}$$

Since $l \to 0$ as $y \to 0$, it is customary, in incompressible flow, to use $l = Ky$ as a first approximation, where K is some suitable constant. This has been found to yield surprisingly accurate results. We therefore try the same approximation in compressible flow. Then

$$\tau_t = \bar{\rho} K^2 y^2 \left(\frac{d\bar{u}}{dy}\right)^2 \tag{1343}$$

Now if $p_x = 0$, and the flow is steady, the shear stress cannot be a function of y. This may easily be seen by taking an element of volume and summing the forces over it. If we neglect the viscous portion of the shear stress then in the main portion of the boundary layer, the turbulent shear stress must equal the shear stress at the wall, τ_w. Then

$$\tau_w = \bar{\rho} K^2 y^2 \left(\frac{d\bar{u}}{dy}\right)^2 \qquad \text{or} \qquad \frac{d\bar{u}}{dy} = \frac{1}{K}\sqrt{\frac{\tau_w}{\bar{\rho}}}\left(\frac{1}{y}\right) \tag{1344}$$

Now for an isothermal wall, $\bar{\rho}/\rho_w = T_w/\bar{T}$, which is known as a function of $\bar{u}$ from equation (1388). Equation (1344) may thus be integrated to give us $\bar{u}$ as a function of y. There results

$$\frac{1}{A}\sin^{-1}\left[c_1\left(\frac{\bar{u}}{U}\right) - c_2\right] = C + \frac{1}{UK}\sqrt{\frac{\tau_w}{\rho_w}}\ln y \tag{1345}$$

where C is a constant of integration and

$$c_1 = \frac{2A^2}{(B^2 + 4A^2)^{1/2}}, \qquad c_2 = \frac{B}{(B^2 + 4A^2)^{1/2}}, \qquad A^2 = \frac{\gamma - 1}{2} M_\infty^2 \frac{T_\infty}{T_w}$$

$$B = \frac{T_\infty}{T_w}\left[1 + \frac{\gamma - 1}{2} M_\infty^2\right] - 1 \tag{1346}$$

The solution is not complete, however, unless we know T_w. Because the solution (1345) cannot be expected to be valid all the way to the wall, we cannot say $\tau_w = \mu \bar{u}_y$. On the other hand we know that the integral method yields good results even when the velocity profile is only approximately correct. The integral method is therefore a natural choice in the present case. From equation (1173) we have

$$\tau_w = \frac{d}{dx}\int_0^\delta \bar{\rho}\bar{u}(U - \bar{u})\, dy \tag{1347}$$

Using (1338) and (1345) we get

$$\tau_w = \frac{D\mu_w U}{K}\frac{d}{dx}\left\{Ja^2 \exp\left[\frac{a}{A}\sin^{-1}\frac{B}{(B^2 + 4A^2)^{1/2}}\right]\right\} \tag{1348}$$

where

$$D = e^{-FK} \qquad F = \text{constant of integration}$$

$$a = \frac{KU}{\sqrt{\tau_w/\rho_w}}, \qquad z = \frac{\bar{u}}{U}$$

$$J = \int_0^1 \frac{z(1-z)}{(1 + Bz - A^2 z^2)} \exp\left[\frac{a}{A}\sin^{-1}\frac{2A^2 z - B}{(B^2 + 4A^2)^{1/2}}\right] dz$$

The integral J may be evaluated by series expansion. Dropping higher order terms, we get

$$J = a^{-2}(1 + B - A^2)^{-1/2} \exp\left[\frac{a}{A}\sin^{-1}\frac{2A^2 - B^2}{(B^2 + 4A^2)^{1/2}}\right] \tag{1349}$$

Substituting (1349) into (1348) and integrating, we get

$$\frac{\sqrt{2}}{A\sqrt{c_{fw}}}(\sin^{-1}\alpha + \sin^{-1}\beta) = \text{constant} + \frac{1}{K}\ln\left[c_{fw}R_w\sqrt{\frac{T_\infty}{T_w}}\right] \tag{1350}$$

where

$$c_{fw} = \frac{\tau_w}{\frac{1}{2}\rho_w U^2}, \qquad R_w = \frac{\rho_w U x}{\mu_w}$$

$$\alpha = \frac{2A^2 - B}{(B^2 + 4A^2)^{1/2}}, \qquad \beta = \frac{B}{(B^2 + 4A^2)^{1/2}}$$

If we assume a power law for the viscosity, $\mu \sim T^\omega$, equation (1350) may be rewritten

$$\left[\frac{2T_\infty}{A^2 c_{f\infty} T_w}\right]^{1/2}(\sin^{-1}\alpha + \sin^{-1}\beta)$$

$$= \text{constant} + \frac{1}{K}\left[\ln(R_\infty c_{f\infty}) - \frac{1 + 2\omega}{2}\ln\frac{T_w}{T_\infty}\right] \tag{1351}$$

where

$$c_{f\infty} = \frac{\tau_w}{\frac{1}{2}\rho_\infty U^2}, \qquad R_\infty = \frac{\rho_\infty U x}{\mu_\infty}$$

The constant of integration and K may be evaluated by matching the solution up with an incompressible solution such as that of von Karman [137] when $M_\infty \to 0$.

$$c_{f\infty}^{-1/2} = 1.70 + 4.15 \log_{10} R_\infty c_{f\infty} \tag{1352}$$

There results the formula

$$\frac{0.242}{A\sqrt{c_{f\infty} T_w/T_\infty}}(\sin^{-1}\alpha + \sin^{-1}\beta) = 0.41 + \log_{10}(R_\infty c_{f\infty}) - \frac{1 + 2\omega}{2}\log_{10}\frac{T_w}{T_\infty} \tag{1353}$$

The total skin friction coefficient, C_{fw}, is defined by

$$\tfrac{1}{2}C_{fw}\rho_w U^2 x = \int_0^x \tau_w \, dx \tag{1354}$$

Proceeding as before, we find

$$\frac{0.242}{A\sqrt{C_{f\infty}T_w/T_\infty}}(\sin^{-1}\alpha + \sin^{-1}\beta) = \log_{10}(R_\infty C_{f\infty}) - \frac{1+2\omega}{2}\log_{10}\frac{T_w}{T_\infty} \tag{1355}$$

In the incompressible case, $M_\infty = 0$, $T_w/T_\infty = 1$, this formula reduces to the incompressible formula of von Karman [137]

$$0.242\, C_{f\infty}^{-1/2} = \log_{10}(R_\infty C_{f\infty}) \tag{1356}$$

Equation (1355) has been plotted by van Driest [136] for Mach numbers from zero to ten. The results for $M = 2$ and 4 are shown in Figure 258. Experimental evidence indicates that these results are accurate to within $\pm 10\%$ [137].

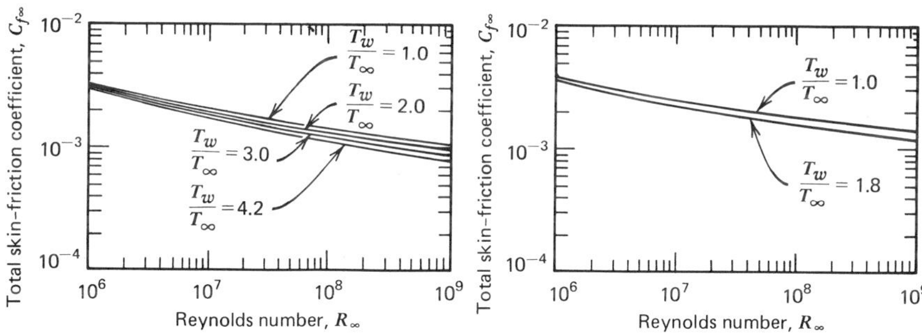

Figure 258 Turbulent boundary layer skin friction coefficient, as obtained by van Driest [136].

The heat transfer at the wall may be found by differentiating equation (1338).

$$\left(\frac{\partial T}{\partial y}\right)_w = \frac{T_\infty}{U}\left[\left(1 - \frac{T_\omega}{T_\infty}\right) + \frac{\gamma-1}{2}M_\infty^2\right]\left(\frac{\partial \bar{u}}{\partial y}\right)_w \tag{1357}$$

Hence,

$$q_w = k_w\left(\frac{\partial T}{\partial y}\right)_w = \frac{k_w}{\mu_w}\frac{T_\infty}{U}\left[\left(1 - \frac{T_w}{T_\infty}\right) + \frac{\gamma-1}{2}M_\infty^2\right]\mu_w\left(\frac{\partial \bar{u}}{\partial y}\right)_w$$

$$= c_{pw}\frac{T_\infty}{U}\left[\left(1 - \frac{T_w}{T_\infty}\right) + \frac{\gamma-1}{2}M_\infty^2\right]\tau_w \tag{1358}$$

for $Pr = 1$. If $T_w = T_{aw}$, the adiabatic wall temperature,

$$c_p T_{aw} = c_p \bar{T} + \frac{\bar{u}^2}{2} = \text{constant} \tag{1359}$$

for $c_p = \text{constant}$, and hence

$$q_w = \frac{c_p \tau_w}{U}(T_{aw} - T_w) \tag{1360}$$

If we now define a film coefficient, H, and a local heat transfer coefficient $c_{H\infty}$, by the equations

$$q_w = H(T_{aw} - T_w) \qquad c_{H\infty} = \frac{H}{c_p \rho_\infty U} \tag{1361}$$

then

$$c_{H\infty} = \tfrac{1}{2} c_{f\infty} \tag{1362}$$

$c_{H\infty}$ is called the Stanton number.

Wilson's method

An alternate approach was taken by R. E. Wilson [138], using von Karman's formula for the mixing length rather than Prandtl's. We will first briefly review von Karman's mixing length theory [139]. Recall equation (252)

$$\frac{D\zeta}{Dt} = v\nabla^2 \zeta \tag{252}$$

where $\boldsymbol{\zeta} = \nabla \times \mathbf{V}$. Where turbulent effects predominate over viscous effects, the right-hand side of equation (252) may be taken as zero. We now postulate an incompressible flow in which $\bar{u} = \bar{u}(y)$, $\bar{v} = 0$ and define a stream function ψ' such that

$$\bar{u} + u' = \bar{u} + \psi'_y, \qquad \bar{v} + v' = -\psi'_x \tag{1363}$$

Substituting in (252) and taking $v = 0$, we get

$$\nabla^2 \psi'_t + (\bar{u} + \psi'_y)\nabla^2 \psi'_x - \psi'_x(\bar{u}_{yy} + (\nabla^2 \psi')_y) = 0 \tag{1364}$$

Around any point 0 in the flow we may expand $\bar{u}$ in a Taylor series

$$\bar{u} = \bar{u}_0 + (y - y_0)(\bar{u}_y)_0 + \tfrac{1}{2}(y - y_0)^2(\bar{u}_{yy})_0 + \cdots \tag{1365}$$

In addition, we switch to a coordinate system moving with velocity $\bar{u}_0$. Define a new set of variables x', y', t':

$$\begin{aligned} y &= y_0 + ly' \qquad & t &= \text{constant} + \frac{l^2 t'}{A} \\ x &= \text{constant} + lx' \qquad & \psi' &= Af(x', y', t') \end{aligned} \tag{1366}$$

We now postulate that the eddying motions are at all points similar in the moving coordinate system and differ only in scale of length and time (and hence velocity). Thus, the function f must be independent of l, A, y_0, $(d\bar{u}/dy)_0$, and $(d^2\bar{u}/dy^2)_0$. Substituting (1365) and (1366) into (1364) we get

$$\begin{aligned} &\frac{A}{l^2}(\bar{u}_y)_0 y'(\nabla^2 f)_{x'} - \frac{A}{l}(\bar{u}_{yy})_0[f_{x'} - \tfrac{1}{2}y'^2(\nabla^2 f)_{x'}] \\ &- \frac{A^2}{l^4}[\nabla^2 f)_{t'} + f_{y'}(\nabla^2 f)_{x'} - f_{x'}(\nabla^2 f)_{y'}] = 0 \end{aligned} \tag{1367}$$

The conditions of similarity will be met if

$$l = \frac{\kappa(\bar{u}_y)}{(\bar{u}_{yy})_0} \tag{1368}$$

where κ is some suitable constant. This is von Karman's mixing length hypothesis. Substituting (1368) in (1342), we get

$$\tau_t = \frac{\bar{\rho}\kappa(\bar{u}_y)^4}{(\bar{u}_{yy})^2} \tag{1369}$$

We take this equation as our point of departure.

In the case of incompressible boundary layers, L. Prandtl [134] suggested that

$$\bar{u} = \bar{u}(\tau_w, \rho, \mu, y) \tag{1370}$$

From dimensional analysis it then follows that

$$\frac{u}{\sqrt{\tau_w/\rho}} = f\left(\frac{\rho\sqrt{\tau_w/\rho}\, y}{\mu}\right) \tag{1371}$$

The quantity τ_w/ρ is called the friction velocity and will be designated by u_τ. Following Prandtl, we may define two new variables in the compressible case, η and ϕ, such that

$$\eta = \frac{\rho_w u_\tau y}{\mu_w}, \quad \text{and} \quad \phi = \frac{\bar{u}}{u_\tau} \quad u_\tau = \frac{\sqrt{\tau_w}}{\rho_w} \tag{1372}$$

and assume that ϕ is a function of η only. Substituting in equation (1369) we get

$$-\kappa\phi_{\eta\eta}^2 = \sqrt{\frac{\rho_w}{\bar{\rho}}}\,\phi_{\eta\eta} \tag{1373}$$

For an adiabatic surface,

$$\frac{\rho_\infty}{\rho} = \frac{T}{T_\infty} = 1 + \frac{\gamma - 1}{2}M_\infty^2 - \frac{\gamma - 1}{2}M_\infty^2\left(\frac{\bar{u}}{U}\right)^2 \tag{1374}$$

whence equation (1373) becomes

$$\frac{d\ln(d\phi/d\eta)}{d\phi} = -\frac{\kappa}{\sqrt{1 - \sigma(\phi/\phi_\infty)}} \tag{1375}$$

where

$$\sigma = \frac{\dfrac{\gamma - 1}{2}M_\infty^2}{1 + \dfrac{\gamma - 1}{2}M_\infty^2} \tag{1376}$$

The assumption of ϕ as a function of η only is thus justified and (1375) is an oridinary differential equation. The boundary conditions are that the solution match up with the laminar sublayer, which is presumed to extend to $\eta = s$, where s is a constant. As a first approximation, we may use the value of s given by von Karman for the incompressible case, 11.5. The value $d\phi/d\eta$ at $\eta = s$ must likewise be a constant which we shall call c. If in the laminar sublayer we assume $\bar{u} = (\bar{u}_y)_w y$, then at $\eta = s$, $\phi = \eta$, and the boundary conditions become

$$\eta = s: \quad \phi = s \qquad \phi_\eta = c \tag{1377}$$

Integrating the differential equation (1375) subject to the boundary conditions (1376) we get, approximately,

$$\eta = \frac{\phi_\infty [\kappa\phi_\infty \sqrt{1 - \sigma(\phi/\phi_\infty)^2} + \sigma(\phi/\phi_\infty)]}{c[(\kappa\phi_\infty)^2 + \sigma]} e^{\beta} \tag{1378}$$

where

$$\beta = \frac{\kappa\phi_\infty}{\sqrt{\sigma}} \left[\sin^{-1}\left(\sqrt{\sigma} \frac{\phi}{\phi_\infty} \right) - \sin^{-1}\left(\sqrt{\sigma} \frac{s}{\phi_\infty} \right) \right] \tag{1379}$$

Recall now the momentum integral equation

$$\tau_w = \rho_\infty U^2 \theta_x \qquad \theta = \int_0^\delta \frac{\bar{\rho}\bar{u}}{\rho_\infty U} \left(1 - \frac{\bar{u}}{U} \right) dy \tag{1380}$$

Using (1372), (1374), and (1376) we get

$$\frac{\rho_w u_\tau \theta}{\mu_w} = \frac{\phi_\infty}{1 + \frac{\gamma - 1}{2} M_\infty^2} \int_0^1 \frac{\frac{\phi}{\phi_\infty}\left(1 - \frac{\phi}{\phi_\infty} \right)}{1 - \sigma(\phi/\phi_\infty)^2} \frac{d\eta}{d\phi} d\left(\frac{\phi}{\phi_\infty} \right) \tag{1381}$$

which may be integrated to give

$$\frac{\rho_w u_\tau \theta}{\mu_w} \cong \frac{\phi_\infty (\kappa\phi_\infty - 2)}{c\kappa^3 \phi_\infty^3 \left[1 + \frac{\gamma - 1}{2} M_\infty^2 \right]} e^{\Sigma} \tag{1382}$$

where

$$\Sigma = \frac{\kappa\phi_\infty}{\sqrt{\sigma}} \left[\sin^{-1}\sqrt{\sigma} - \sin^{-1}\left(\frac{\sqrt{\sigma}s}{\phi_\infty} \right) \right] \tag{1383}$$

We now define

$$c_f = \frac{\tau_w}{\frac{1}{2}\rho_w U^2}, \qquad R_w = \frac{\rho_w U x}{\mu_w} \tag{1384}$$

whence

$$\phi_\infty = \sqrt{2/c_f} \tag{1385}$$

Using equations (1374), (1382), (1384) and (1385), equation (1380) becomes

$$dR_w = \frac{1}{c\kappa^3 e^{\kappa s}} \left[-\frac{\sqrt{2}}{c_f^{3/2}} + \frac{2\kappa c_1}{c_f^2} - \frac{\sqrt{2}\kappa^2 c_1}{c_f^{5/2}} \right] \exp\left[\kappa c_1 \sqrt{2/c_f}\right] dc_f \tag{1386}$$

where

$$c_1 = \frac{\sin^{-1}\sqrt{\sigma}}{\sqrt{\sigma}}$$

Equation (1386) may be integrated to give

$$\kappa\sqrt{2}c_1 c_f^{-1/2} \cong \ln\left(\tfrac{1}{2}c\kappa^2 e^{\kappa s}\right) + \ln c_f R_w \tag{1387}$$

In order that this result might reduce to the incompressible formula of von Karman (1352) we choose our constants and rewrite the solution in the form

$$c_1 c_f^{-1/2} = 1.7 + 4.15 \log_{10}(c_f R_w) \tag{1388}$$

Comparison with experiment. Figure 259 shows a comparison between the

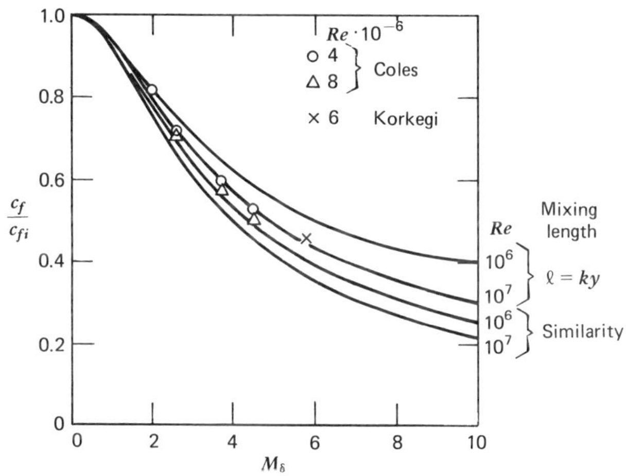

Figure 259 Turbulent skin friction coefficients—theory of van Driest and Wilson compared to experiments of Coles and Korgeri [136].

results of the theories of van Driest and Wilson and the experimental results of Coles [140] and Korkegi [141]. The van Driest theory is seen to give results that are somewhat high, while the Wilson theory gives results that are somewhat low. It should be kept in mind, however, that the experimental data itself may involve an error within $\pm$ 10%.

This fact is illustrated by Figure 260, in which different experimental data is used (Chapman and Kester [142]). Here, the theoretical data was evaluated at $R_w = 10^7$. The experimental data of Chapman and Kester seems to coincide with the theoretical prediction of Wilson, but in fact, both are subject to errors of about $\pm$ 10%.

Van Driest [136], [143], [144], was able to show that the results of using either the mixing length formula of Prandtl or the mixing length formula of

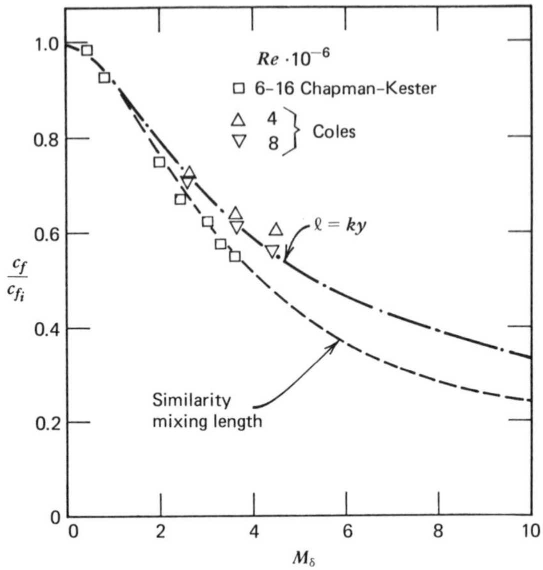

Figure 260 Turbulent skin friction coefficients—theory of van Driest and Wilson compared to experiments of Coles, Chapman and Kester [136].

von Karman may be combined into a single equation:

$$\frac{0.242}{\left[\frac{\gamma-1}{2}M_\infty^2 c_{f\infty}\right]^{1/2}}(\sin^{-1}\alpha+\sin^{-1}\beta)=0.41+\log_{10}(R_\infty c_{f\infty})$$

$$-f(\omega)\log_{10}\left(\frac{T_w}{T_\infty}\right) \quad (1353a)$$

where $f(\omega)=(1/2)+\omega$ when the Prandtl mixing length $l=\kappa y$ is used, and $f(\omega)=\omega$ when the von Karman mixing length $l=\kappa(du/dy)(d^2u/dy^2)$ is used. The other symbols in equation (1353a) are the same as those in 1353, and indeed the two formulas are identical when $f(\omega)=(1/2)+\omega$.

A similar generalization may be made for the total drag coefficient.

$$C_{f\infty}=\frac{1}{l}\int_0^1 c_{f\infty}\,dx$$

$$\frac{0.242}{\left[C_{f\infty}\frac{\gamma-1}{2}M_\infty^2\right]^{1/2}}(\sin^{-1}\alpha+\sin^{-1}\beta)=\log_{10}(R_\infty C_{f\infty})-f(\omega)\log_{10}\left(\frac{T_w}{T_\infty}\right) \quad (1355a)$$

Equation (1355a) reduces to equation (1355) for $f(\omega)=(1/2)+\omega$. A check of equation (1355a) using $f(\omega)=\omega$ has been made by comparing the results with the free flight data of Sommer and Short [145] (see below) for $T_w/T_\infty \cong 1$ (Figure 260a). ω was taken as 0.76. Agreement is good keeping in mind a possible experimental error of $\pm$ 10%.

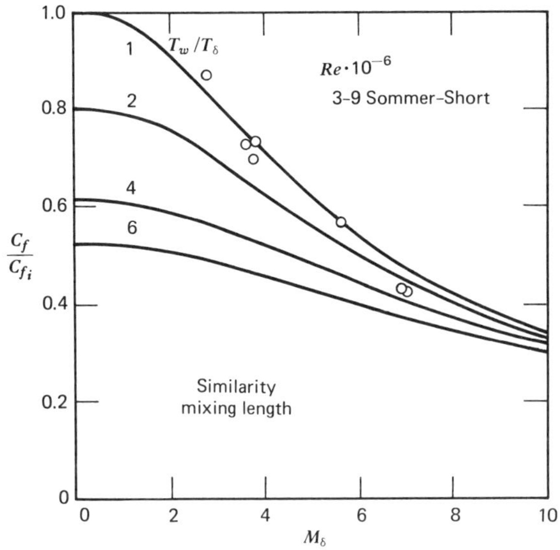

Figure 260(a) Turbulent skin friction coefficient M prediction of equation 1355 (a) compared to free flight data of Sommer and Short [145].

Method of Eckert [146]

An alternative approach was used by Eckert. Studies made on incompressible flow in pipes with fully developed turbulent flow show that the velocity distribution may be represented with considerable accuracy by the formula

$$\frac{\bar{u}}{U} = \left(\frac{y}{r}\right)^{1/n} \tag{1389}$$

where r is the radius of the pipe, y the distance from the wall, U the velocity on the centerline and n a parameter which varies from 6 to 10 as the Reynold's number based on diameter increases from roughly 10^3 to 10^6. Now for a pipe, the shearing stress at the wall is given by

$$\tau_w = \frac{\Delta p}{\Delta x}\frac{r}{2} \tag{1390}$$

We now define λ by

$$\tau_w = \tfrac{1}{8}\lambda\rho\bar{u}_m^2 \tag{1391}$$

where $\bar{u}_m$ is the mean flow velocity in the pipe. On the basis of numerous experiments, Blasius was able to show that

$$\lambda \cong 0.3164\left(\frac{\bar{u}_m d}{v}\right)^{-1/4} \tag{1392}$$

where d is the diameter of the pipe. This formula is valid for Reynold's numbers based on d of less that 10^5. Substituting (1392) into (1391) we get

$$\tau_w = 0.03955\,\rho\bar{u}_m^{7/4}\nu^{1/4}d^{-1/4} \tag{1393}$$

If we now introduce the friction velocity $u_\tau = \sqrt{\tau_w/\rho}$ and $r = d/2$, (1393) can be written

$$\frac{\bar{u}_m}{u_\tau} = 6.99\left(\frac{u_\tau r}{\nu}\right)^{1/7} \tag{1394}$$

Now the mean velocity is related to the maximum velocity by the formula

$$\frac{\bar{u}_m}{U} = \frac{2n^2}{(n+1)(2n+1)} \tag{1395}$$

where n is the exponent in (1389). If we assume that $n = 1/7$, $\bar{u}_m/U \cong 0.8$, and (1394) can be written

$$\frac{U}{u_\tau} = 8.74\left(\frac{u_\tau r}{\nu}\right)^{1/7} \tag{1396}$$

But if equation (1389) is valid, this means

$$\frac{U}{u_\tau} = 8.74\left(\frac{u_\tau y}{\nu}\right)^{1/7} \tag{1397}$$

This is the famous one seventh power law.

In compressible flow it is found that good results can be obtained if the one seventh power law is applied to plates with a leading edge in the form

$$\frac{u}{U} = \left(\frac{y}{\delta}\right)^{1/7} \tag{1398}$$

This formula is found to be valid for Reynold's numbers based on l (the length of the plate) up to 10^7.

Eckert now assumes that the one-seventh power law is equally valid for compressible flow when suitable average values are used for ρ and μ. The shear stress then becomes

$$\tau_w = 0.0395\,\bar{\bar{\rho}}\bar{u}_m^2\left(\frac{\bar{\bar{\mu}}}{\bar{u}_m\bar{\bar{\rho}}d}\right)^{1/4} \tag{1399}$$

where $\bar{\bar{\rho}} = \dfrac{1}{u_m A}\displaystyle\int_A \rho\bar{u}\,dA$, $\bar{\bar{\mu}} =$ viscosity taken at $\bar{\bar{T}} = \dfrac{P}{\bar{\bar{\rho}}R}$

and A is the cross-sectional area of the pipe. Now for the one seventh power law, $\bar{u}_m/U = 0.817$. In addition, the $\mu \sim T^\omega$. In terms of the values at the center of the pipe

$$\tau_w = 0.0277\,\rho_c U^2\left(\frac{\mu_c}{U\rho_c d}\right)^{1/4}\left(\frac{\bar{\bar{T}}}{T_c}\right)^{(\omega-3)/4} \tag{1400}$$

where the subscript c indicates the center of the pipe where the velocity is U.

We now suppose that equation (1400) is equally valid for the turbulent compressible boundary layer over a flat plate if we replace d by 2δ, subscript c is replaced by subscript ∞, and U is the velocity at the outer edge of the boundary layer. Recall the momentum integral equation (1380):

$$\tau_w = \rho_\infty U^2 \theta_x = \rho_\infty U^2 \frac{d}{dx} \int_0^\delta \frac{\bar{\rho}\bar{u}}{\rho_\infty U}\left(1 - \frac{\bar{u}}{U}\right) dy \tag{1380}$$

It then follows that for a plate

$$\theta_x = 0.0233\left(\frac{\mu_\infty}{U\rho_\infty \delta}\right)\left(\frac{\bar{\bar{T}}}{T_\infty}\right)^{(\omega-3)/4} \tag{1401}$$

On the basis of the assumption of similarity of the velocity profiles, θ/δ is not a function of x. It follows that

$$\frac{\theta}{\delta}\frac{d\delta}{dx} = 0.0233\left(\frac{\mu_\infty}{U\rho_\infty \delta}\right)^{1/4}\left(\frac{\bar{\bar{T}}}{T_\infty}\right)^{(\omega-3)/4} \tag{1402}$$

This may be integrated to give

$$\frac{\delta}{x} = 0.059\left(\frac{\theta}{\delta}\right)^{-4/5}\left(\frac{\bar{\bar{T}}}{T_\infty}\right)^{(\omega-3)/5} Re^{-1/5}, \qquad Re = \frac{U\rho_\infty x}{\mu_\infty} \tag{1403}$$

It follows by comparison with (1402) that $d\delta/dx = (4/5)\delta/x$. The local coefficient of friction from the momentum theorem is, then,

$$c_f = \frac{\tau_w}{\frac{1}{2}\rho_\infty U^2} = 2\frac{\theta}{\delta}\frac{d\delta}{dx} = 0.0944\left(\frac{\theta}{\delta}\right)^{1/5}\left(\frac{\bar{\bar{T}}}{T_\infty}\right)^{(\omega-3)/5} Re^{-1/5} \tag{1404}$$

where

$$\frac{\theta}{\delta} = \int_0^1 \frac{\rho}{\rho_\infty}\frac{\bar{u}}{U}\left(1 - \frac{\bar{u}}{U}\right) d\left(\frac{y}{\delta}\right) \tag{1405}$$

If we presume

$$T_\infty + \frac{1}{2}\frac{U^2}{c_p} = \bar{T} + \frac{1}{2}\frac{\bar{u}^2}{c_p}, \qquad \bar{p} = \bar{P}R\bar{T}$$

and p = constant, and if we let $\bar{u}/U = (y/\delta)^{1/n}$, then

$$\frac{\theta}{\delta} = \int_0^1 \frac{T_\infty}{\bar{T}} \eta^{1/n}(1-\eta)\, d\eta \qquad \text{and} \qquad \frac{T_\infty}{\bar{T}} = \left[1 - \frac{\gamma - 1}{2} M_\infty^2 (\eta^{2/n} - 1)\right]^{-1} \tag{1406}$$

Equation (1406) is plotted in Figure 261 for various Mach numbers. For equation (1404), $n = 7$ is appropriate. In a similar way, $\bar{\bar{T}}/T_\infty$ may be evaluated, and the result is shown in Figure 262.

For incompressible flow, with $n = 7$, $\theta/\delta = 0.0972$, and T = constant. Thus, the ratio of the local coefficient of friction for compressible flow to the local

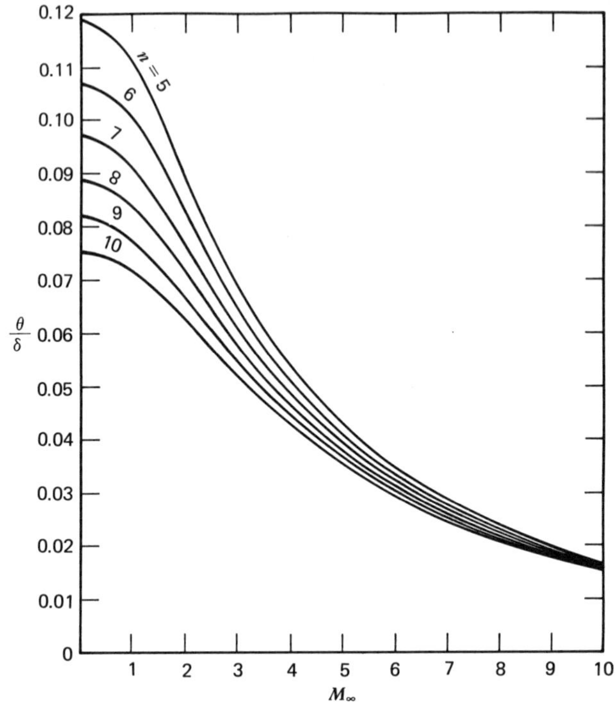

Figure 261 Momentum thickness over boundary layer thickness for various power law profiles [146].

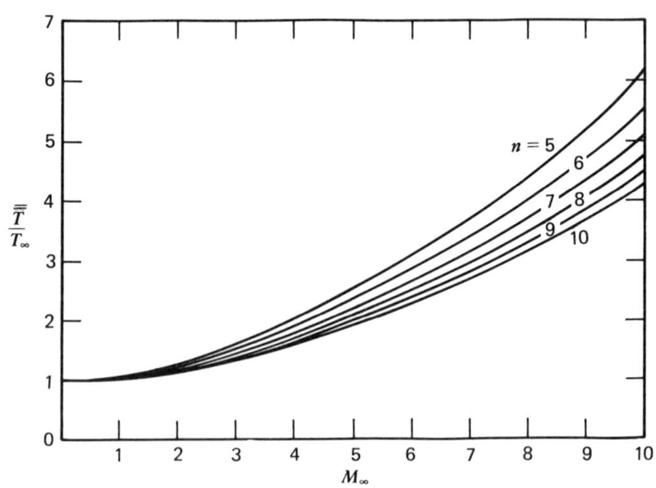

Figure 262 Reference temperature versus Mach number for various power law profiles [146].

coefficient of friction for incompressible flow becomes

$$\frac{c_f}{c_{f_i}} = \left(\frac{\theta/\delta}{0.0972}\right)^{1/5} \left(\frac{T}{T_\infty}\right)^{(\omega-3)/5} \qquad (1407)$$

at the same Reynolds number, Re. This result is plotted in Figure 263.

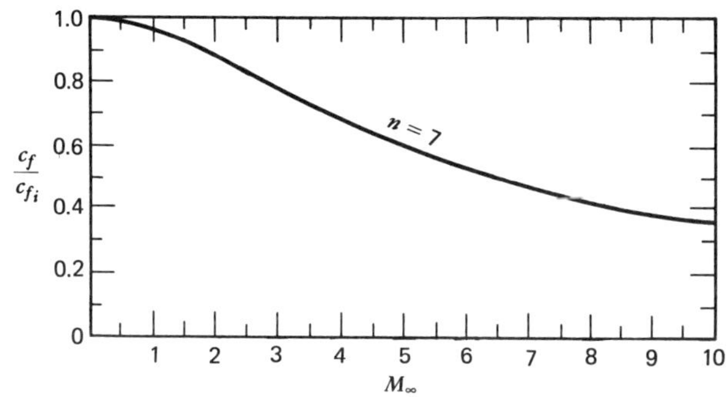

Figure 263 Skin friction coefficient from reference temperature method [146].

Unlike the theories of van Driest and Wilson, there is no Reynolds number dependence here. The curve is close to that predicted by van Driest for $R_w = 10^6$, but flatter. Experimental data confirms the trend predicted by Walker and van Driest of a drop in c_f/c_{fi} with an increase in Reynolds number at a given Mach number. The theory of Eckert must therefore be counted as inferior to the other two theories. It does have the virtue of providing a simple formula for quick rough estimates, however, if the Reynolds number is not too high (i.e., $R_w < 10^7$).

The method of Sommer and Short [145]

Sommer and Short developed an empirical method based on the reference enthalpy or reference temperature concept mentioned earlier. First a suitable formula is chosen for the total friction drag of an incompressible turbulent boundary layer over a flat plate of length l. The formula of von Karman [137] is selected (equation (1356)).

$$\frac{0.242}{\sqrt{C_{f_i}}} = \log(C_{f_i} R_\infty) \qquad (1356)$$

where $R_\infty = Ul\rho_\infty/\mu$ and we have written C_{fi} instead of $C_{f\infty}$ to emphasize that C_{fi} is for an incompressible flow.

$$C_{f_i} = \frac{1}{l}\int_0^l \frac{\tau_{wi}}{\frac{1}{2}\rho_\infty U^2} dx \qquad (1408)$$

ρ in an incompressible flow is, of course, constant and equal to ρ_∞ everywhere.

We now define C'_f by

$$\frac{0.242}{\sqrt{C'_f}} = \log_{10}(C'_f R') \tag{1409}$$

where $R' = Ul\rho'/\mu'$ is the Reynolds number based on some reference condition $T'\rho', \mu'$. T' is found from the empirical formula

$$\frac{T'}{T_\infty} = 1 + 0.035 M_\infty^2 + 0.45\left(\frac{T_w}{T_\infty} - 1\right) \tag{1410}$$

μ' is found from the Sutherland formula

$$\frac{\mu'}{\mu_\infty} = \left(\frac{T'}{T_\infty}\right)^{1.5} \left(\frac{T_\infty + 216^0 R}{T' + 216^0 R}\right) \tag{1411}$$

and $\rho'/\rho_\infty = T_\infty/T'$. C'_f is presumed equal to

$$C'_f = \frac{1}{l}\int_0^l \frac{\tau_w}{\frac{1}{2}\rho' U^2} dx = \frac{\int_0^l \tau_w dx}{\frac{1}{2}\rho' U^2 l} \tag{1412}$$

where $\tau_w(x)$ is the actual shear stress at the wall. Finally, C_f is defined by

$$C_f = \frac{1}{l}\int_0^l \frac{\tau_w}{\frac{1}{2}\rho_\infty U^2} dx \tag{1413}$$

so that

$$C_f = C'_f \rho'/\rho_\infty \tag{1414}$$

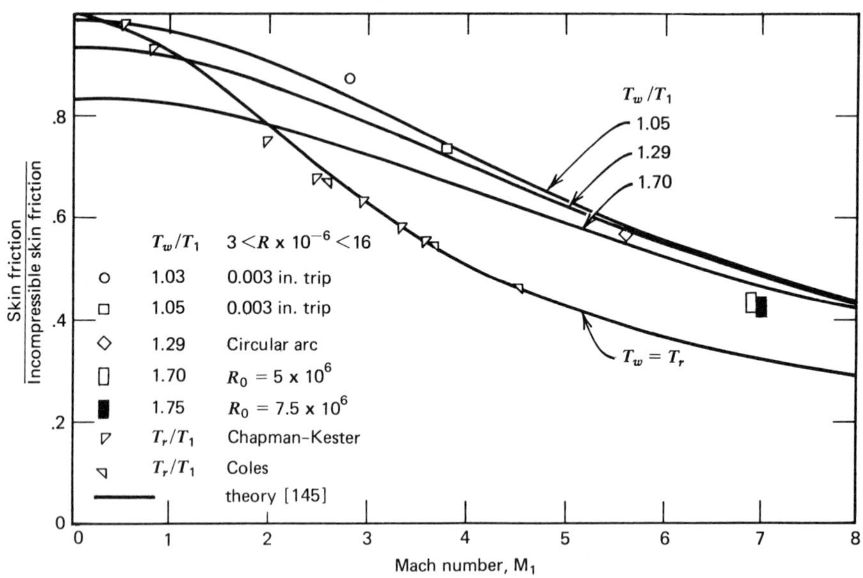

Figure 264 Turbulent skin friction coefficient—theory of Sommer and Short [145] compared to experiment.

C_f is thus presumed to be the actual, total coefficient of friction for the compressible flow over a flat plate of length l, based on the free stream conditions.

Figure 264 shows a comparison of C_f/C_{fi} obtained by this method with experimental results obtained by Sommer and Short [145], as well as Chapman and Kester [142], and Coles [140], cited earlier. $R_\infty = 10^7$ was used for all theoretical curves in the figure. Results are seen to be very good for insulated walls, but somewhat less accurate for the isothermal wall (heat transfer) cases.

The Recovery Factor

For Prandtl number unity, the adiabatic wall temperature is equal to the free stream stagnation temperature both in laminar and turbulent flow. In actuality, the Prandtl number is somewhat less than one and the adiabatic wall temperature is somewhat less than the free stream stagnation temperature. The difference is measured by the recovery factor r, which is defined by

$$r = \frac{T_{aw} - T_\infty}{T_0 - T_\infty}$$

where T_0 is the free stream stagnation temperature. It follows that for a perfect

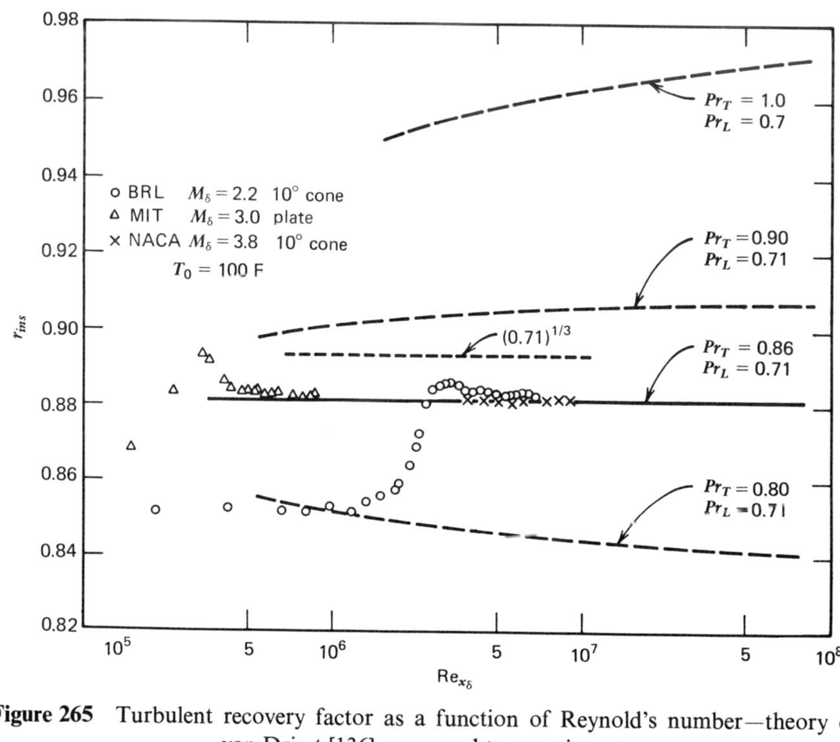

Figure 265 Turbulent recovery factor as a function of Reynold's number—theory of van Driest [136] compared to experiment.

gas,

$$T_{aw} = T\left(1 + r\frac{\gamma - 1}{2}M_\infty^2\right)$$

For laminar flow we already mentioned that $r \cong \sqrt{Pr}$, as demonstrated by Emmons [93]. For turbulent flow, van Driest [136] has demonstrated that

$$r = Pr_t\left[1 + \frac{2}{\kappa}\sqrt{\frac{\lambda c_f}{2}}(1 - Pr_t)\left[\frac{\pi^2}{6} + \tfrac{3}{2}(1 - Pr_t)\right] + 12.5\lambda c_f\left\{\left(\frac{Pr_l}{Pr_t} - 1\right)\right.\right.$$
$$+ 2\ln\left[1 + \frac{5}{6}\left(\frac{Pr_l}{Pr_t} - 1\right)\right] + (\ln 6)\ln\left[1 + \frac{7}{8}\left(\frac{Pr_l}{Pr_t} - 1\right)\right]$$
$$\left.\left. - (\ln 6)\ln\left[1 + \frac{1}{4}\left(\frac{Pr_l}{Pr_t} - 1\right)\right]\right\}\right] \quad (1415)$$

where

$$\lambda = 1 + r\frac{\gamma - 1}{2}M_\infty^2 \quad 1416)$$

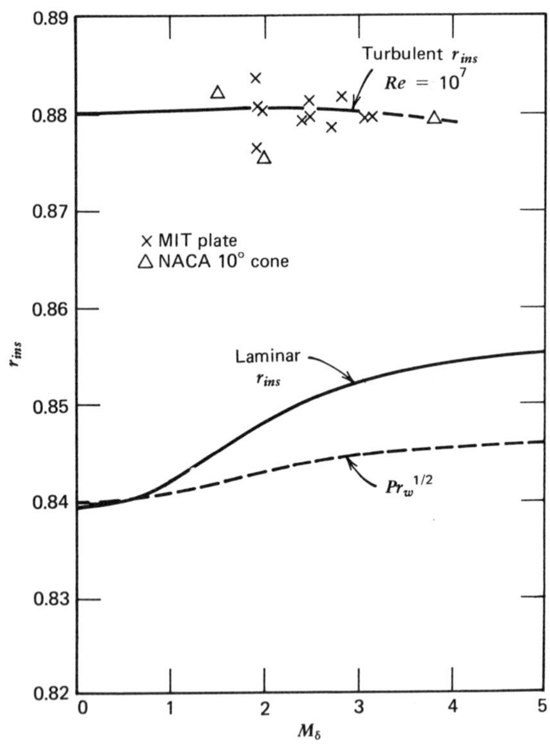

Figure 266 Turbulent recovery factor as a function of Mach number—theory of van Driest compared to experiment [136].

Pr_l is the (laminar) Prandtl number, Pr_t is the turbulent Prandtl number, κ is the proportionality constant in von Karman's mixing length theory, usually taken as 0.4. Figure 266 shows equation (1415) plotted versus Mach number for $Pr_l = 0.71$, $Pr_t = 0.86$. Also shown are measured recovery factors from wind tunnel data. Agreement is seen to be quite good, especially considering the scatter of the experimental data. A noteworthy feature of Figure 266 is the relative absence of Mach number dependence of the turbulent recovery factor, justifying the approximation $r \cong 0.88$, at least to a Mach number of 3. The laminar recovery factor is shown for purposes of comparison, along with Emmon's approximation $r \cong \sqrt{Pr_w}$, where Pr_w is the laminar Prandtl number at the wall. The recovery factor is equally independent of the Reynolds number, as shown in Figure 265 where equation (1393) is plotted for various turbulent Prandtl numbers with $M_\infty = 0$ and compared with experimental data.

Heat transfer

In the method of van Driest, we demonstrated that $c_{H\infty} = (1/2)c_{f\infty}$ for a Prandtl number of 1. When $Pr \neq 1$, a correction may be introduced using the Reynolds analogy factor, s, defined by

$$c_{H\infty} = \frac{1}{s}\frac{c_{f\infty}}{2} \tag{1417}$$

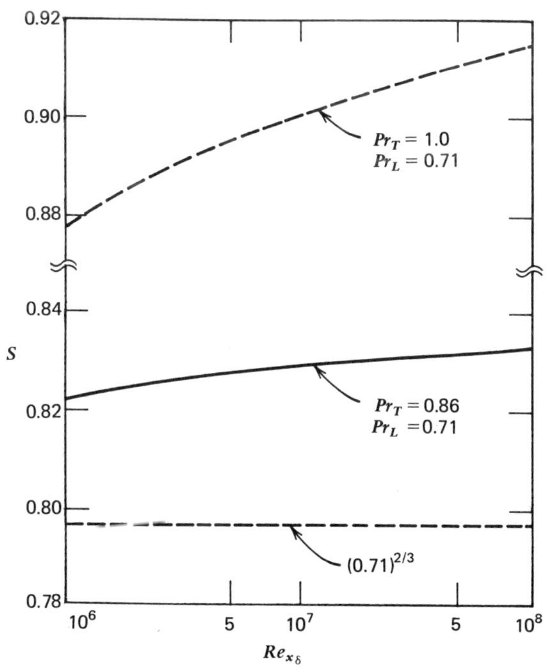

Figure 267 Turbulent Reynold's analogy factor versus Reynold's number—theory of van Driest (1418) from [136].

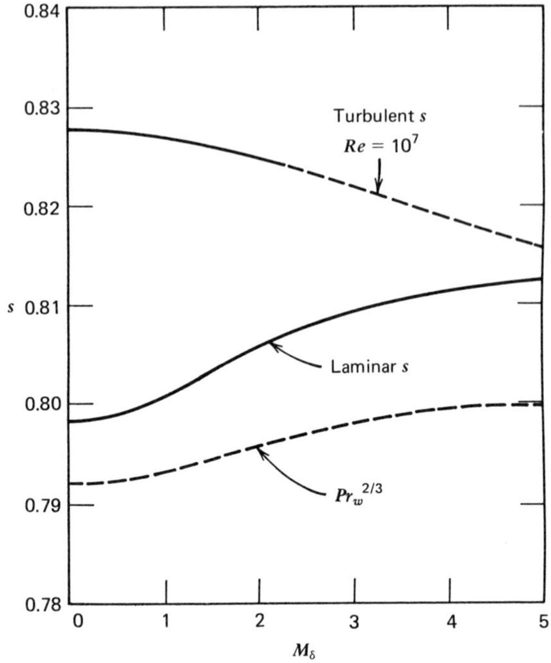

Figure 268 Experimentally obtained turbulent Reynold's analogy factor [136].

Van Driest has shown that

$$s = Pr_t\left[1 + 5\sqrt{\frac{c_f}{2}}\left\{\frac{1}{\quad}(1 - Pr_t)\left[\frac{\pi^2}{6} + \frac{3}{2}(1 - Pr_t)\right] + \left[\frac{Pr_l}{Pr_t} - 1\right]\right.\right.$$
$$\left.\left. + \ln\left[1 + \frac{5}{6}\left(\frac{Pr_l}{Pr_t} - 1\right)\right]\right\}\right] \tag{1418}$$

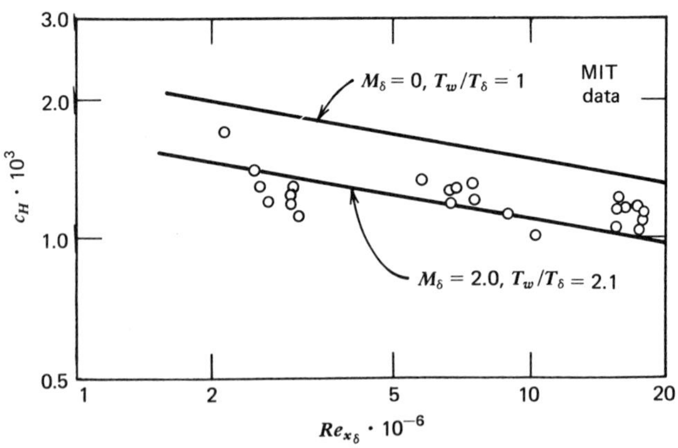

Figure 269 Stanton number versus Reynold's number—equations (1417) and (1355a) compared to experiment [136].

Figure 267 shows a plot of s from equation (1396) versus Reynold's number for various laminar and turbulent Prandtl numbers. When $Pr_t = 1$, equation (1396) reduces to that derived earlier by von Karman [146a]. Colburn [147] has proposed an approximation $s = (Pr_l)^{2/3}$. This approximation is shown for $Pr_l = 0.71$. Experimental results indicate that the curve for $Pr_t = 0.86$, $Pr_l = 0.71$ yields the best results. Figure 268 shows values of s obtained experimentally in a wind tunnel. The laminar factor is shown for purposes of comparison as is Colburn's factor $(Pr_l)^{2/3}$, with $Pr_w = Pr_l$ at the wall. Figure 269 and 270

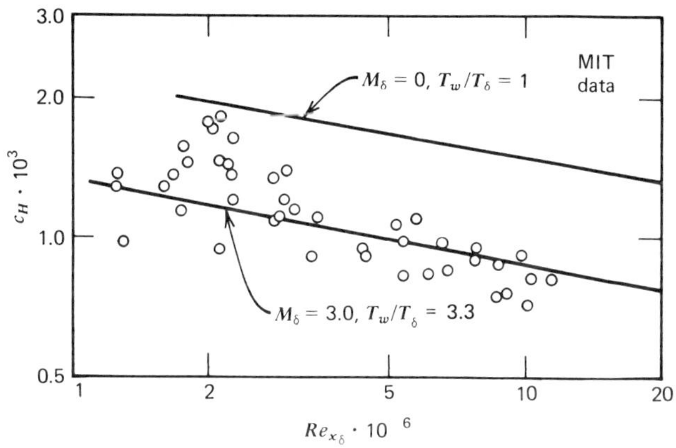

Figure 270 Stanton number versus Reynold's number—equations (1417) and (1355a) compared to experiment [136].

show actual heat-transfer coefficients (Stanton numbers) obtained experimentally. The solid lines are the theoretical results obtained by using c_f from equation (1355a) with $f(\omega) = \omega$ (von Karman mixing length), and finding $c_{H\infty}$ from equation (1417). using 0.825 for s. Agreement is seen to be fairly good, considering the scatter of the experimental data.

Turbulent Boundary Layers with Pressure Gradient

Method of Reshotko and Tucker [148]

Recall the momentum and continuity equations

$$(\overline{\rho u})_x + (\overline{\rho v})_y + \frac{\overline{\rho u}}{R} R_x = 0 \tag{1419}$$

$$\overline{\rho u}\,\bar{u}_x + \overline{\rho v}\,\bar{u}_y = -\bar{p}_x + \bar{\tau}_y \tag{1420}$$

where τ is the total shear stress (laminar and turbulent) R is the body radius for axisymmetric flow, and $R = 1$ $(dR/dx = 0)$ for plane flow. We now introduce the transformation

$$\psi_y = \frac{\overline{\rho u}\,R}{\rho_0}, \qquad \psi_x = -\frac{\overline{\rho v}\,R}{\rho_0}, \qquad X = \int_0^x \frac{p_e c_e}{p_0 c_0}\,dx, \qquad Y = \frac{c_e}{c_0}\int_0^y \frac{\rho}{\rho_0}\,dy$$

$$UR = \psi_Y, \qquad VR = -\psi_X, \qquad U = \bar{u}\frac{c_0}{c_e} \tag{1421}$$

where the subscripts e and 0 indicate the outer edge of the boundary layer and the stagnation conditions in the free stream, respectively. All state properties are understood to be mean values even though the bar has been omitted. In the flat plate case, the free stream conditions and the conditions at the outer edge of the boundary layer were the same. In the present case, however, the free stream refers only to the undisturbed flow ahead of the body. Substituting the transformation (1421) into equations (1419) and (1420) we get

$$U_X + V_Y + \frac{U}{R}R_X = 0 \tag{1422}$$

$$U U_X + V U_Y = U_e U_{eX}\frac{h_s}{h_0} + \frac{\tau_Y}{\rho_0 \dfrac{p_e}{p_0}\left(\dfrac{c_e}{c_0}\right)^2} \tag{1423}$$

where h is the enthalpy and the subscript s designates the local stagnation value. Perfect gas with constant c_p has been assumed. We now multiply equation (1400) by $U_e - U$ and subtract equation (1423). There results

$$\begin{aligned}&[U(U_e - U)]_X + [V(U_e - U)]_Y + U_{eY}\left[U_e - U + U_e\left(\frac{h_s}{h_0} - 1\right)\right] \\ &+ (U_e - U)\frac{U}{R}R_X = -\left[\rho_0\frac{p_e}{p_0}\left(\frac{c_e}{c_0}\right)^2\right]^{-1}\tau_Y\end{aligned} \tag{1424}$$

Integrating this equation cover Y from $Y = 0$ to $Y = \Delta$, where Δ is sufficiently large so that $U = U_e$ and $h_s = h_0$ there, yields a momentum integral equation,

$$\tilde{\theta}_X + \frac{\tilde{\theta}}{U_e}U_{ex}\left[2 + H_i + \frac{1}{\tilde{\theta}}\int_0^\Delta\left(\frac{h_s}{h_0} - 1\right)dy\right] + \frac{\tilde{\theta}}{R}R_X = \left[\rho_0\frac{p_e}{p_0}\left(\frac{c_e}{c_0}\right)^2 U_e^2\right]^{-1}\tau_w \tag{1425}$$

where

$$\tilde{\theta} = \int_0^\Delta \frac{U}{U_e}\left[1 - \frac{U}{U_e}\right]dY, \quad H_i = \frac{1}{\tilde{\theta}}\int_0^\Delta\left(1 - \frac{U}{U_e}\right)dY \tag{1426}$$

and τ_w is the shear stress at the wall. H_i is called the incompressible form factor. Another integral equation, the moment of momentum equation, is obtained by multiplying equation (1424) by Y and integrating over Y.

$$\begin{aligned}&\frac{\partial}{\partial X}\int_0^\Delta YU_e^2\frac{U}{U_e}\left(1 - \frac{U}{U_e}\right)dY + U_e\int_0^\Delta Y\frac{\partial}{\partial Y}\left[V\left(1 - \frac{U}{U_e}\right)\right]dY \\ &+ \frac{U_e^2}{R}\frac{dR}{dX}\int_0^\Delta Y\frac{U}{U_e}\left(1 - \frac{U}{U_e}\right)dY + U_e\frac{dU_e}{dX}\int_0^\Delta\left[Y\left(1 - \frac{U}{U_e}\right) + Y\left(\frac{h_s}{h_0} - 1\right)\right]dY \\ &= \left[\rho_0\frac{p_e}{p_0}\left(\frac{c_e}{c_0}\right)^2\right]^{-1}\int_0^\Delta Y\tau_Y\,dY\end{aligned} \tag{1427}$$

In order to evaluate the above integrals, a knowledge of the velocity profile is necessary; however, as pointed out earlier, the integral method is not sensitive to the exact shape of the velocity profile, so that an approximate shape will do. We therefore select a power law relation

$$\frac{U}{U_e} = \left(\frac{Y}{\Delta}\right)^N \tag{1428}$$

Then H_i becomes $H = 1 + 2N$ (from equation (1404)). Equation (1405) may be written

$$\begin{aligned}\frac{dH_i}{dX} = -\frac{1}{2U_e}\frac{dU_e}{dX}\left[H_i(H_i+1)^2(H_i-1)\right]\Bigg\{1 + 2\left[(H_i+1)\tilde{\theta}\right]^{-1}\int_0^\Delta\left(\frac{h_s}{h_0}-1\right)dY \\ - \frac{2(H_i-1)}{H_i^2(H_i+1)\tilde{\theta}^2}\int_0^\Delta\left(\frac{h_s}{h_0}-1\right)Y\,dY \\ + \frac{\tau_w(H_i^2-1)}{\rho_0\frac{p_e}{p_0}\left(\frac{c_e}{c_0}\right)^2 U_e\tilde{\theta}}\left[H_i - (H_i+1)\int_0^1\frac{\tau}{\tau_w}d\left(\frac{Y}{\Delta}\right)\right]\Bigg\}\end{aligned} \tag{1429}$$

In order to integrate equations (1425) and (1429), c_f must be known in terms of $\tilde{\theta}$ and H_i. For this purpose, the expression of Ludwig and Tillmann [149] for incomressible flow is adapted to compressible flow by the reference enthalpy method of Eckert [146]. The expression of Ludwig and Tillmann is

$$c_{fi} = \frac{\tau_w}{\frac{1}{2}\rho u_e^2} = 0.246\, e^{-1.561\,H}\, Re_\theta^{-0.268} \tag{1430}$$

where H is the form factor δ^*/θ, δ^* being the displacement thickness and θ the momentum thickness. Re_θ is the Reynold's number based on θ, $u_e\theta/\nu$. For a flat plate, the incompressible momentum integral equation is

$$\frac{d\theta}{dx} = \frac{c_{fi}}{2} = 0.123\, e^{-1.561\,H}\left(\frac{u_e\theta}{v}\right)^{-0.268} \tag{1431}$$

Integrating equation (1409) for θ assuming H to be constant we get

$$\theta = \left[(1.268)(0.123)\, e^{-1.56\,H}\left(\frac{u_e}{\nu}\right)^{-0.268} x\right]^{1/1.268} \tag{1432}$$

whence

$$c_{fi} = \frac{\tau_w}{\frac{1}{2}\rho u_e^2} = 0.364\, e^{-1.23\,H}\left(\frac{u_e x}{\nu}\right)^{-0.211} \tag{1433}$$

We now define a reference enthalpy, h_{ref}, by the empirical equation

$$\frac{h_{\text{ref}}}{h_e} = \frac{1}{2}\left(\frac{h_w}{h_e}+1\right) + 0.22\left(\frac{h_{aw}}{h_e}-1\right) \tag{1434}$$

with T_{ref} being the corresponding reference temperature and assume

$$\frac{\tau_w}{\frac{1}{2}\rho_{\text{ref}} u_e^2} = 0.364\, e^{-1.23\,H_i}\left(\frac{u_e x}{\nu_{\text{ref}}}\right)^{-0.211} \tag{1435}$$

or

$$c_f = \frac{\tau_w}{\frac{1}{2}\rho_e u_e^2} = 0.364\, e^{-1.23 H_i}\left(\frac{u_e x}{\nu_{\text{ref}}}\right)^{-0.211} \frac{T_e}{T_{\text{ref}}} \tag{1436}$$

In terms of momentum thickness, this becomes

$$c_f = \frac{\tau_w}{\frac{1}{2}\rho_e u_e^2} = 0.246\, e^{-1.56 H_i}\left(\frac{u_e \theta}{\nu_{\text{ref}}}\right)^{-0.268} \left(\frac{T_e}{T_{\text{ref}}}\right)^{1.268} \tag{1437}$$

Finally, in terms of our transformed momentum thickness, $\tilde{\theta}$,

$$c_f = \frac{\tau_w}{\frac{1}{2}\rho_e u_e^2} = 0.246\, e^{-1.561 H_i}\left(\frac{M_e c_0 \tilde{\theta}}{\nu_0}\right)^{-0.268} \left(\frac{T_e}{T_{\text{ref}}}\right)^{0.732} \left(\frac{T_e}{T_0}\right)^{0.268} \tag{1438}$$

Experimental evidence indicates that this is a good approximation in the case of moderate pressure gradients as well as in the flat plate case.

Parenthetically, it may be observed that for the present approximation,

$$\frac{c_f}{c_{fi}} = (1 + 0.144\, M_e^2)^{-0.578} \tag{1439}$$

Thus, the variation with Reynolds number is not accounted for.

In order to integrate equation (1429) we also need an estimate for

$$\int_0^1 \frac{\tau}{\tau_w}\, d\left(\frac{Y}{\Delta}\right) \tag{1440}$$

Based on an analysis made for incompressible flow by Maskell [150], the integral (1440) is approximated by

$$\int_0^1 \frac{\tau}{\tau_w}\, d\left(\frac{Y}{\Delta}\right) = 1.03\, \frac{H_i}{H_i + 1} \tag{1441}$$

This approximation is not considered critical since the integral term (1418) is small compared to the pressure gradient term.

In analogy with the laminar flow analysis of Cohen and Reshotko [101], it is presumed that the enthalpy integrals in equation (1429) may be approximated by

$$\frac{1}{\tilde{\theta}} \int_0^\Delta \left(\frac{h_s}{h_0} - 1\right) dY = \left(\frac{h_w}{h_0} - 1\right) H_{if}, \quad \int_0^\Delta \left(\frac{h_s}{h_0} - 1\right) Y\, dY = \Delta^2 \left(\frac{h_w}{h_0} - 1\right) \frac{H_{if} - 1}{2(H_{if} + 3)} \tag{1442}$$

where H_{if} represents the incompressible form factor for a flat plate which may, to a good approximation, be taken as 1.3.

Using equations (1438), (1441), (1442), and (1421), equations (1425) and (1407) become

$$\tilde{\theta}_x + \frac{\tilde{\theta}}{M_e} M_{e_x}\left[2 + H_i + \left(\frac{h_w}{h_0} - 1\right) H_{if}\right] + \frac{\tilde{\theta}}{R} R_x = A \tag{1443}$$

where

$$A = 0.123\, e^{-1.561 H_i}\left(\frac{M_e c_0 \tilde{\theta}}{\nu_0}\right)^{-0.268}\left(\frac{T_e}{T_{\mathrm{ref}}}\right)^{0.732}\left(\frac{T_e}{T_0}\right)^{3.268} \tag{1444}$$

$$H_{i_x} = \frac{-M_{e_x}}{2M_e}\left[H_i\left(H_i+1\right)^2\left(H_i-1\right)\right]\left[1+\left(\frac{h_w}{h_0}-1\right)\left(\frac{2H_{if}}{H_i+1}\right.\right.$$
$$\left.\left.-\frac{1}{2}\left(\frac{H_i+1}{H_i-1}\right)\left(\frac{H_{if}-1}{H_{if}+3}\right)\right)\right] - 0.03\, H_i(H_i^2-1)\frac{A}{\tilde{\theta}} \tag{1445}$$

For purposes of calculation, equation (1443) may be approximated by

$$\hat{\theta}_x + 1.2155\frac{\hat{\theta}}{R}R_x = 0.01173\left(\frac{T_e}{T_{\mathrm{ref}}}\right)^{0.732}\left(\frac{T_e}{T_0}\right)^{3.268} - B\frac{\hat{\theta}}{M_e}M_{e_x} \tag{1446}$$

where

$$\hat{\theta} = \tilde{\theta}\left(\frac{M_e c_0 \tilde{\theta}}{\nu_0}\right)^{0.2155} \qquad B = 4.2 + 1.2155\, H_{if}\left(\frac{h_w}{h_0}-1\right) \tag{1447}$$

The numbers are obtained from known results for incompressible turbulent flow, notably those of Maskell [150], and Skopets [151]. Equation (1424) is a linear differential equation for $\hat{\theta}$ and may be integrated forward from either the point of transition or the front of the body, since the conditions at the outer edge of the boundary layer and $R(x)$ are presumed known. Once $\hat{\theta}$ has been obtained, H_i may be obtained by integrating equation (1445) numerically. The coefficient of friction is then obtained by substituting back in equation (1438).

Heat transfer For compressible, turbulent boundary layers with pressure gradient, Reshotko and Tucker (op. cit.) have found an appropriate Reynolds analogy factor to be

$$s = Pr^{2/3} e^{-1.561(H_i - H_{if})} \tag{1448}$$

where Pr is the laminar Prandtl number. The local Stanton number, c_{He}, is now defined by

$$c_{He} = \frac{q}{\rho_e u_e (h_{aw} - h_w)} \tag{1449}$$

where q is the local heat transfer rate. It follows that

$$c_{He} = Pr^{-2/3} e^{1.561(H_i - H_{if})} c_f/2 \tag{1450}$$

h_{aw} may be evaluated by

$$h_{aw} = h_e\left[1 + Pr_{ref}^{1/3}\left(\frac{\gamma-1}{2}\right)M_e^2\right] \tag{1451}$$

The method of Reshotko and Tucker has proved successful for flows with favorable pressure gradients, zero pressure gradients and mildly adverse pressure gradients. In the case of moderate or highly adverse pressure gradients, however, it has proved to be unreliable.

Method of Susman and Cresci

An attempt to improve the accuracy of the method in the case of adverse gradients has been made by Susman and Cresci [152]. The method is identical to that of Reshotko and Tucker through equation (1429). However, the shear stress integral in equation (1429), instead of being evaluated by equation (1441), is presumed given by

$$\int_0^1 \frac{\tau}{\tau_w} d\left(\frac{Y}{\Delta}\right) = 0.011\left[\frac{H_i - 1}{H_i}\right]^2 \left(\frac{c_{f_i}}{2}\right)^{-1} \tag{1452}$$

In the adiabatic case with Prandtl number unity,

$$\frac{h_s}{h_0} - 1 = 0 \tag{1453}$$

and hence the integrals involving this term in equations (1425) and (1429) are zero. For constant surface temperature, Susman and Cresci assume that the Crocco solution is valid and assume a power law for the velocity profile. Thus

$$\frac{h_s - h_w}{h_0 - h_w} = \frac{u}{u_e} = \frac{U}{U_e} = \left(\frac{Y}{\Delta}\right)^{N_i} \tag{1454}$$

and hence

$$\int_0^\Delta \left(\frac{h_s}{h_0} - 1\right) dY = \left(\frac{h_w}{h_0} - 1\right) H_i \tilde{\theta}, \qquad H_i = 2N_i + 1 \tag{1455}$$

$$\int_0^\Delta \left(\frac{h_s}{h_0} - 1\right) Y\, dY = \left(\frac{h_w}{h_0} - 1\right) \frac{H_i^2 (H_i + 1)^2}{2(H_i - 1)(H_i + 3)} \tilde{\theta}^2 \tag{1456}$$

These approximations may be compared with those used by Reshotko and Tucker in equation (1442). Including all approximations, equations (1425) and (1429) then become

$$\tilde{\theta}_X + \frac{\tilde{\theta}}{U_e} U_{e_X} [2 + g_w H_i] + \frac{\tilde{\theta}}{R} R_X = 0.123 e^{-1.561 H_i} \left[\frac{U_e \tilde{\theta}}{\nu_0}\right]^{-0.263} \left(\frac{T_e}{T_0}\right)\left(\frac{\bar{\mu}}{\mu_0}\right)^{0.268} \tag{1457}$$

$$H_{iX} = -\frac{1}{2U_e} U_{e_X} \left[H_i (H_i + 1)^2 (H_i - 1)\right] \left[1 + (g_w - 1)\frac{H_i^2 + 4H_i - 1}{(H_i + 1)(H_i + 3)}\right]$$

$$+ \frac{H_i^2 - 1}{\tilde{\theta}} \left(\frac{\bar{T}}{T_e}\right) \left[\frac{H_i \tau_w}{\rho_e U_e^2} - \frac{(H_i + 1)\tau_w}{\rho_e U_e^2} \int_0^1 \frac{\tau}{\tau_w} d\left(\frac{Y}{\Delta}\right)\right], \quad g_w = \frac{h_w}{h_0} \tag{1458}$$

Unlike Reshotko and Tucker, who uncouple the equations for θ and H_i through

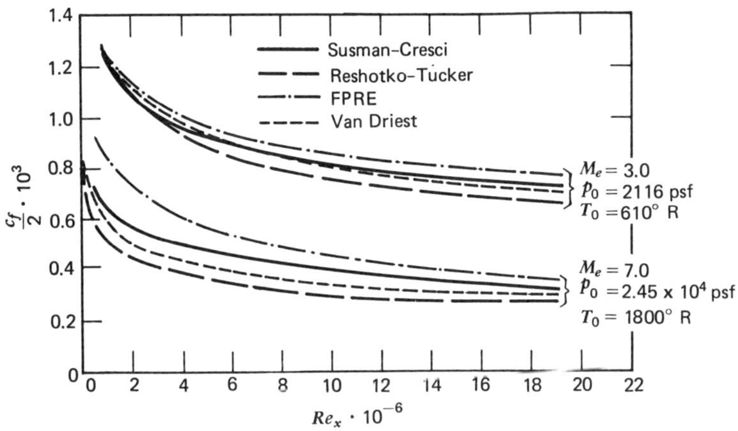

Figure 271 Adiabatic flat plate turbulent skin friction coefficients fron various theories compared [152].

further approximations, Susman and Cresci solved equations (1457) and (1458) simultaneously by computer. Figure 271 shows a comparison of the c_f obtained for an adiabatic flat plate by the methods of van Driest, flat plate reference enthalpy (FPRE), Reshotko and Tucker, and Susman and Cresci. Figure 272

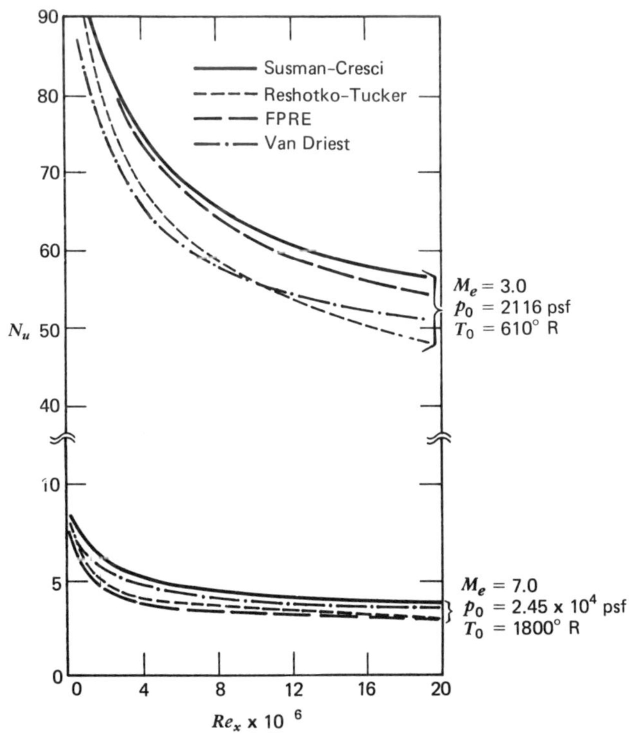

Figure 272 Nusselt number for isothermal flat plate with turbulent boundary layer obtained by various methods [152].

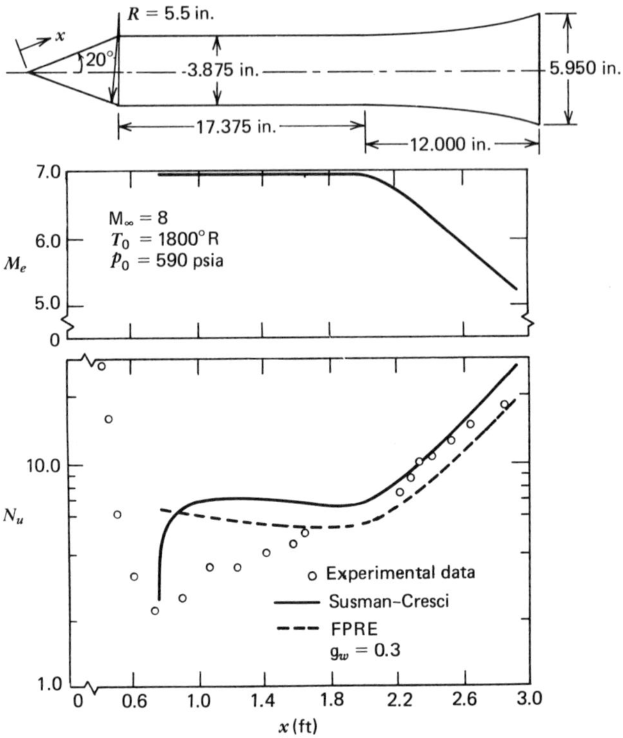

Figure 273 Nusselt number for turbulent boundary layer with pressure gradient—theory and experiment compared [152].

shows a comparison of the Nusselt number obtained by the same four methods over an isothermal flat plate with $g_w = 3$.

Figure 273 shows the results for the Nusselt numbers for the case of hypersonic flow over a body obtained by Susman and Cresci, reference enthalpy method, and experiment. The Reynolds analogy,

$$Nu = Pr^{1/3} \frac{h_{aw} - h_w}{h_0 - h_w} \frac{\rho_e u_e L}{\mu_0} \frac{c_f}{2} \tag{1459}$$

has been employed. L is a reference length and is taken as one inch. The adiabatic wall temperature is found from the relation

$$\frac{h_{aw} - h_e}{h_0 - h_e} = Pr^{1/3} \tag{1460}$$

Figure 273 shows that the method of Susman and Cresci works well in the region of mildly adverse pressure gradient near the tail of the body. Near the nose of the body, the flow is, of course, laminar.

Figures 274 through 276 show the results of the method of Susman and Cresci compared to the results of Reshotko and Tucker for the case of a strongly

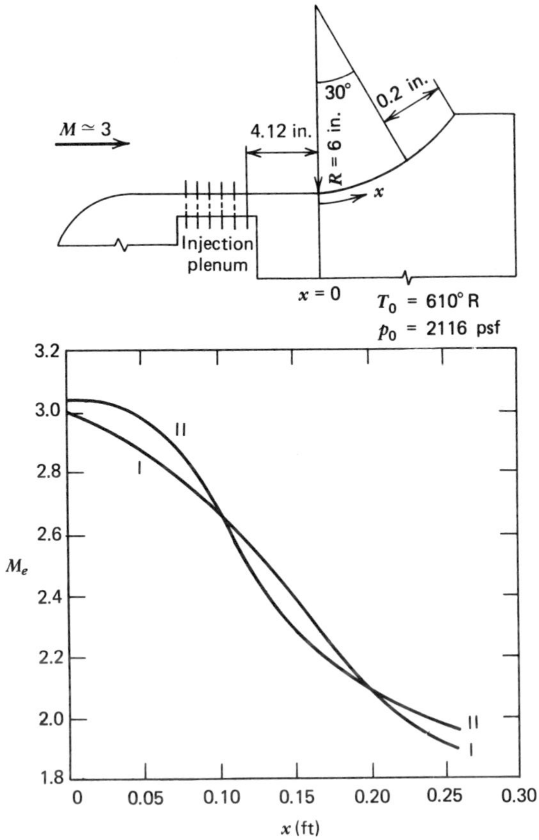

Figure 274 Mach number variation for two flows with pressure gradient [152].

adverse pressure gradient. The method of Reshotko and Tucker is seen to fail badly in this case, while the method of Susman and Cresci gives good results, so far as can be determined. (No direct measurement of c_f was made). Figure 277 shows a comparison of the results of Susman and Cresci with the experimental data of Winter et al. [153] for supersonic flow over a wasp-waisted body. Agreement is seen to be reasonably good. This is a severe test, containing, as it does, both favorable and unfavorable pressure gradients.

A number of methods for turbulent boundary layers have been developed at the United Aircraft Research Laboratories. Unfortunately, these methods have not been published in sufficient detail for presentation here. Flaherty [154] has given a method quite similar to that of Susman and Cresci. Camarata and McDonald [155] have developed a method which takes into account the upstream history of the turbulence. Fish and McDonald [241] have developed a method suitable for the transition range from laminar to turbulent, as wall as turbulent flow. In this method, the momentum and energy equations are reduced to ordinary differential equations in $\eta = y/\delta^*$ by writing the x derivatives in

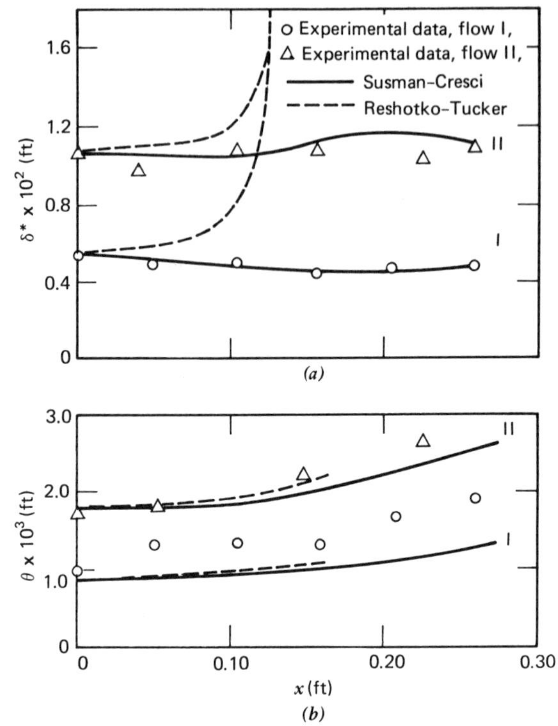

Figure 275 Turbulent displacement and momentum thickness for the two flows of Figure 274 obtained by the method of Susman and Cresci [152] and Reshotko and Tucker [148], compared to experiment. (*a*) Displacement thickness. (*b*) Momentum thickness.

finite difference form. The equations may then be integrated numerically at each value of x.

McDonald [156] has presented a comparison (Figure 278) of the results of Flaherty, Camarata and McDonald, and Fish and McDonald, as well as the results of the Reshotko–Tucker method compared to the experimental results of Winter et al. (op. cit.). The Reshotko–Tucker method is seen to be considerably in error in this case.

Method of Economos and Boccio [156a]

This method is an example of the compressibility transform type. The idea is to try to find a transformation which will transform the compressible boundary layer equations with ρ and μ variable into the incompressible boundary layer equations with ρ and μ constant.

All quantities are taken to be mean quantities with respect to the turbulent fluctuations, even though the bar has been omitted. The governing equations may then be written in the form

$$(\rho u)_x + (\rho v)_y = 0 \tag{1461}$$

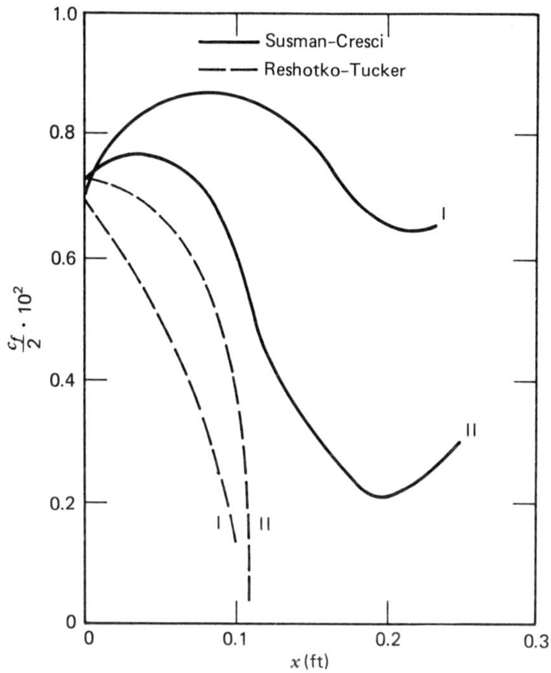

Figure 276 Turbulent skin friction coefficients for the two flows of Figure 274 obtained by the method of Susman and Creci [152] and Reshotko and Tucker [148].

$$\rho u u_x + \rho v u_y = -p_{e_x} + [(\rho\varepsilon + \mu)u_y]_y \tag{1462}$$

$$\rho u g_x + \rho v g_y = \left[\frac{\rho\varepsilon + \mu}{Pr_e} g_y + \frac{u_e}{2h_{se}}(\rho\varepsilon + \mu)\left(1 - \frac{1}{Pr_e}\right)\left(\frac{u^2}{u_e^2}\right)_y\right]_y \tag{1463}$$

where $\rho\varepsilon$ is the eddy viscosity, $g = h_s/h_{se}$, h_s being the local stagnation enthalpy and h_{se} being the stagnation enthalpy at the outer edge of the boundary layer. Pr_e is called the effective Prandtl number and is defined by

$$Pr_e = \left\{\left[Pr_L\left(1 + \frac{\rho\varepsilon}{\mu}\right)\right]^{-1} + \left[Pr_T\left(1 + \frac{\mu}{\rho\varepsilon}\right)\right]^{-1}\right\}^{-1} \tag{1464}$$

where Pr_L and Pr_T are the laminar and turbulent Prandtl numbers, respectively.

The boundary conditions are,

$$y = \infty: \quad u = u_e(x), \qquad g = 1; \quad y = 0: \quad u = v = 0, \qquad g = g_w, \qquad Pr_e = Pr_{Lw} \tag{1465}$$

We now seek transformations which will transform equations (1461) and (1462) into

$$U_X + V_Y = 0 \tag{1466}$$

$$\hat{\rho} U U_X + \hat{\rho} V V_Y = p_{ex} + [(\hat{\rho}\hat{\varepsilon} + \hat{\mu})U_Y]_Y \tag{1467}$$

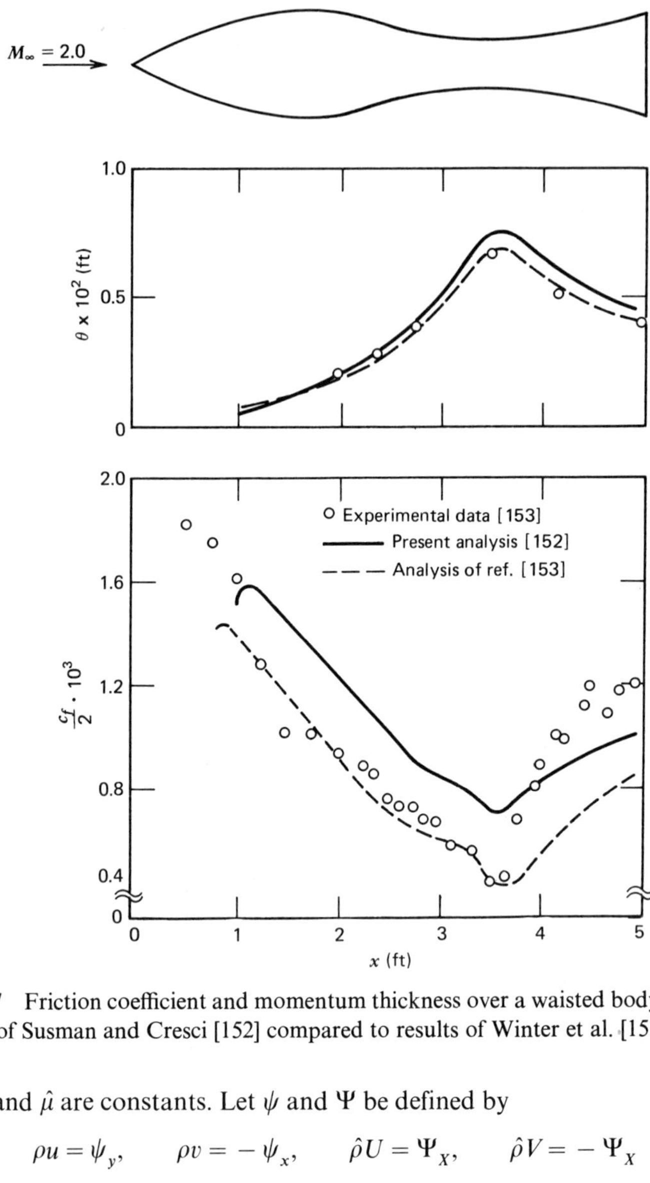

Figure 277 Friction coefficient and momentum thickness over a waisted body by method of Susman and Cresci [152] compared to results of Winter et al. [153].

where $\hat{\rho}$ and $\hat{\mu}$ are constants. Let ψ and Ψ be defined by

$$\rho u = \psi_y, \qquad \rho v = -\psi_x, \qquad \hat{\rho} U = \Psi_X, \qquad \hat{\rho} V = -\Psi_X \tag{1468}$$

and assume

$$\Psi(X, Y) = \sigma(x)\psi(x, y), \qquad dY = \left(\frac{\rho}{\hat{\rho}}\right)\eta(x)\,dy$$
$$dX = \xi(x)\,dx \tag{1469}$$

It follows that

$$\frac{u}{U} = \frac{\eta}{\sigma}, \qquad \frac{u}{u_e} = \frac{U}{U_e} \tag{1470}$$

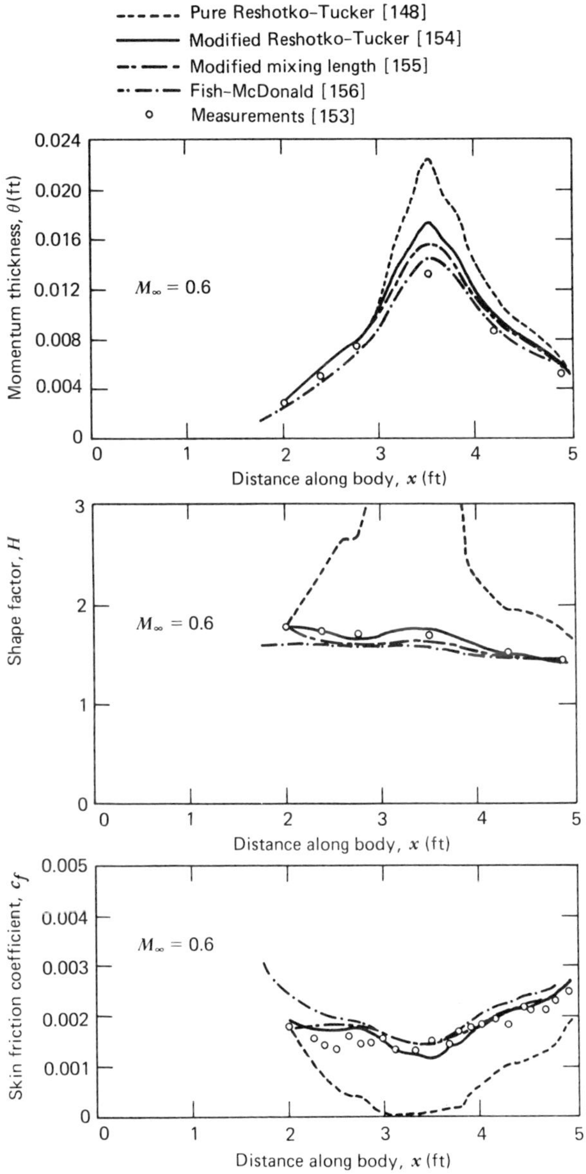

Figure 278 Momentum thickness, shape factor and skin friction for the waisted body of Figure 277 computed by different methods [156].

and hence,

$$(\ln \tilde{\sigma})_{\hat{\chi}} - (\ln \tilde{\eta})_{\hat{\chi}} - [\ln(U_e/U_{eo})]_{\hat{\chi}} + \left(\ln \frac{u_e}{v_e}\right)_{\chi} \chi_{\hat{\chi}} = 0 \tag{1471}$$

where

$$\tilde{\sigma} = \frac{\sigma \mu_e}{\hat{\mu}}, \qquad \tilde{\eta} = \frac{\eta \rho_e}{\hat{\rho}}, \qquad \hat{\chi} = \frac{U_{eo}}{\hat{v}}(X - X_0), \qquad \chi = \frac{u_{eo}}{v}(x - x_0) \tag{1472}$$

and the subscript $_0$ designates an initial value of x. The shear stress in the actual and transformed flows is given by

$$\tau_w = \mu(u_y)_w, \qquad \hat{\tau}_w = \hat{\mu}(U_y)_w \tag{1473}$$

and hence the coefficients of friction are related by

$$\frac{c_f}{\hat{c}_f} = \tilde{\sigma}(\tilde{\rho}\tilde{\mu})_w \tag{1474}$$

where

$$\tilde{\rho} = \rho/\rho_e, \qquad \tilde{\mu} = \mu/\mu_e \tag{1475}$$

For steady flow we have

$$\rho_e u_e u_{e_x} = -p_{e_x} = (\tau_y)_w \tag{1476}$$

Hence

$$(\ln U_e)_{\hat{\chi}} - \left(\frac{\tilde{\eta}}{\tilde{\sigma}}\right)_0 \left(\frac{1}{\tilde{\rho}}\right)_w \left(\frac{1}{\tilde{\rho}\tilde{\mu}}\right)_w \left(\frac{1}{\tilde{\sigma}\tilde{\eta}}\right) \frac{d}{d\chi}(\ln u_e)$$
$$- \left(\frac{\tilde{\eta}}{\tilde{\sigma}}\right)_0 \phi^{-2} \left(\frac{u_e}{v_e}\right)_0^{-1} \left[\frac{\partial}{\partial Y}(\ln \tilde{\rho}\tilde{\mu})\right]_w = 0 \tag{1477}$$

where

$$\phi = \frac{U_e}{U_\tau}, \qquad U_\tau = \left(\frac{\hat{\tau}_w}{\hat{\rho}}\right)^{1/2}$$

and hence

$$\phi = \left(\frac{2}{\hat{c}_f}\right)^{1/2} \tag{1478}$$

Also, the initial values, designated by the subscript $_0$, are related by

$$\left[\frac{U_e/\hat{v}}{u_e/v}\right]_0 = \left(\frac{\tilde{\sigma}}{\tilde{\eta}}\right)_0 \tag{1479}$$

In addition, we may impose the requirement on the transformation that the momentum integral equation for the compressible flow transform into the momentum integral equation for incompressible flow. For compressible flow the momentum integral equation may be written in the form

$$\frac{d\theta}{dx} + \theta\left\{2\frac{d\ln u_e}{dx} + \frac{d\ln \rho_e}{dx}\right\} + \delta^* \frac{d\ln u_e}{dx} = \frac{\tau_w}{\rho_e u_e^2} = \frac{c_f}{2} \tag{1480}$$

where θ and δ^* are the momentum and displacement thickness, respectively. According to our change of variables,

$$\theta = \frac{\hat{\theta}}{\tilde{\eta}}, \qquad \delta^* = \frac{1}{\tilde{\eta}}\left\{\int_0^{\hat{\delta}} \left(\frac{1}{\tilde{\rho}} - 1\right) dY + \hat{\delta}^*\right\} \tag{1481}$$

where $\hat{\theta}, \hat{\delta}$ and $\hat{\delta}^*$ represent the momentum thickness, boundary layer thickness, and displacement thickness in the transformed flow. The boundary layer thickness in the transformed flow may be eliminated in the traditional way by defining

$$\Sigma = \frac{\hat{\theta}}{\hat{\delta}}, \qquad \Omega = \frac{\hat{\delta}^*}{\hat{\delta}}, \qquad \hat{\eta} = \frac{Y}{\hat{\delta}} \tag{1482}$$

Then the form factor δ^*/θ becomes

$$\frac{\delta^*}{\theta} = \frac{1}{\Sigma}\left\{\int_0^1 \left(\frac{1}{\tilde{\rho}} - 1\right) d\hat{\eta} + \Omega\right\} \tag{1483}$$

Defining now a Reynolds number

$$\hat{R} = \frac{U_{e0}\hat{\delta}}{\hat{v}} \tag{1484}$$

we find

$$\frac{d \ln \theta}{d\chi} = \frac{1}{\chi'}\{(\ln \Sigma)' + (\ln \hat{R})' - (\ln \tilde{\eta})'\} \tag{1485}$$

where the primes indicate differentiation with respect to $\hat{\chi}$. The momentum integral equation in the transformed plane may be written

$$(\ln \hat{R})' + (\ln \Sigma)' + (\ln U_e)'\left(2 + \frac{\Omega}{\Sigma}\right) = (\phi^2 \hat{R}\Sigma)^{-1} \tag{1486}$$

Combining (1486) with (1471) and (1485)

$$\frac{d \ln \theta}{d\chi} = \frac{1}{\chi'}\left\{(\phi^2 \hat{R}\Sigma)^{-1} - \left(2 + \frac{\Omega}{\Sigma}\right)\left[(\ln \tilde{\sigma})' + \chi' \frac{d \ln u_e/v_e}{d\chi}\right] + (\ln \tilde{\eta})'\left(1 + \frac{\Omega}{\Sigma}\right)\right\} \tag{1487}$$

Substituting this equation and (1474) into equation (1480) yields

$$\begin{aligned} &-(\ln \tilde{\sigma})'\left(2 + \frac{\Omega}{\Sigma}\right) + (\ln \tilde{\eta})'\left(1 + \frac{\Omega}{\Sigma}\right) \\ &+ \chi'\left\{\frac{d \ln u_e}{d\chi}\left[\frac{\delta^*}{\theta} - \frac{\Omega}{\Sigma}\right] + \frac{d \ln \mu_e}{d\chi}\left[2 + \frac{\Omega}{\Sigma}\right]\right\} \\ &+ \chi'\left\{\frac{d \ln \rho_e}{d\chi}\left[1 + \frac{\Omega}{\Sigma}\right] - (\phi^2 \hat{R}\Sigma)^{-1}\left(\frac{\tilde{\sigma}\tilde{\sigma}_0\tilde{\eta}}{\eta}\right)(\tilde{\rho}\tilde{\mu})_w\right\} = -(\phi^2 \hat{R}\Sigma)^{-1} \end{aligned} \tag{1488}$$

This is the condition which must be satisfied in order that the momentum integral equation for the compressible flow, (1458), transform into the momentum integral equation for the incompressible flow.

In summary, then, the governing equations for the compressible turbulent boundary layer, (1461), (1462), can be transformed into the corresponding incompressible equations (1466) and (1467) by means of the transformations (1469), provided that equations (1470), (1471), (1474), (1475), (1481), (1483), (1486) and (1488) are satisfied.

The energy equation has not been transformed. This is because in solving the incompressible flow equations, the momentum and continuity equations may be solved without reference to the energy equation. In order to determine suitable functions $\sigma(x)$, $\eta(x)$ and $\xi(x)$, however, knowledge of

$$\left[\frac{\partial}{\partial Y}[\ln \rho\mu]\right]_w \quad \text{and} \quad \int_0^{\eta}\left(\frac{1}{\tilde{\rho}} - 1\right)dY \tag{1489}$$

is required, since these quantities occur in equations (1477) and (1481). The easiest solution to this problem is to assume the validity of the Crocco integral. Then

$$g = g_w + (1 - g_w)\tilde{u}, \qquad \tilde{u} = \frac{u}{u_e} \tag{1490}$$

It follows that

$$\frac{1}{\tilde{\rho}} = H_e[g_w + (1 - g_w)\tilde{u}] + (1 - H_e)\tilde{u}^2, \qquad H_e = \frac{h_{se}}{h_e} \tag{1491}$$

$$\left[\frac{\partial}{\partial Y}(\ln \tilde{\rho}\tilde{\mu})\right]_w = (n - 1)\frac{1 - g_w}{g_w}\frac{U_e}{\phi^2}\frac{\tilde{\sigma}_0}{\tilde{\eta}_0}\frac{u_{e0}}{v_{e0}} \tag{1492}$$

where n is the exponent in the assumed viscosity power law relation

$$\tilde{\mu} = \left(\frac{1}{\tilde{\rho}}\right)^n \tag{1493}$$

and

$$\int_0^{\hat{\eta}}\left(\frac{1}{\tilde{\rho}} - 1\right)dY = (H_e g_w - 1)\hat{\eta} + H_e(1 - g_w)I_1(\hat{\eta}) + (1 - H_e)I_2(\hat{\eta}) \tag{1494}$$

where

$$I_1(\hat{\eta}) = \int_0^{\hat{\eta}} \tilde{u}\, d\hat{\eta} \qquad I_2(\hat{\eta}) = \int_0^{\hat{\eta}} \tilde{u}^2\, d\hat{\eta} \tag{1495}$$

An alternate approach, that of numerical integration of the energy equation (1441) has also been presented by Economos and Boccio, but will not be discussed here.

Once the transformations which transform the equation from compressible to incompressible form have been rendered determinate, a solution to the incompressible equations must be found. For this purpose, a two-layer model is assumed. Based on the work of Coles [157], let

$$\tilde{u} = \begin{cases} \hat{R}U_e\hat{\eta}/\phi^2 & 0 < \hat{\eta} < \hat{\eta}_1 \\ \left(\dfrac{A}{\phi}\right)\left[\ln\left(\dfrac{b\hat{R}U_e\hat{\eta}}{\phi}\right) + 2\pi(3\hat{\eta}^2 - 2\hat{\eta}^3)\right] & \hat{\eta}_1 < \hat{\eta} < 1 \end{cases} \tag{1496}$$

where π is a function of $\hat{\chi}$, which remains to be determined, and $\tilde{\eta}_1$ represents the outer edge of the laminar sublayer. A and b are constants. Based on comparison with experimental results, $A = 2.43$ and $b = 7.5$. It follows from equation (1474) that

$$\frac{A}{\phi}\left[\ln\left(\frac{\hat{R}U_e}{\phi}\right) + 2\pi\right] = 1 \tag{1497}$$

$$\Omega = \frac{A}{\phi(1+\pi)} \tag{1498}$$

$$\Sigma = \frac{A}{\phi}\left\{1 + \pi - \frac{A}{\phi}\left[1 + \frac{19}{6}\pi + \frac{52}{35}\pi^2\right]\right\} \tag{1499}$$

The three dependent variables, $\hat{R}$, ϕ, and π, are linked to the independent variable X by the momentum integral equation (1486), with Ω and Σ as given by (1498) and (1499), and the skin friction law (1497). A third equation is obtained by integrating the momentum equations:

$$\begin{aligned}\frac{\hat{\tau} - \hat{\tau}_w}{\hat{\rho}_e U_e^2 \hat{R}} &= \frac{d}{d\hat{\chi}}\int_0^{\hat{\eta}} \tilde{u}^2\, d\hat{\eta} - \tilde{u}\frac{d}{d\hat{\chi}}\int_0^{\hat{\eta}} \tilde{u}\, d\hat{\eta} \\ &+ \frac{d \ln U_e}{d\hat{\chi}}\left[\int_0^{\hat{\eta}} 2\tilde{u}^2\, d\hat{\eta} - \tilde{u}\int_0^{\hat{\eta}} \tilde{u}\, d\hat{\eta} - \hat{\eta}\right] \\ &+ \frac{d \ln \hat{R}}{d\hat{\chi}}\left[\int_0^{\hat{\eta}} \tilde{u}^2\, d\hat{\eta} - \tilde{u}\int_0^{\hat{\eta}} \tilde{u}\, d\hat{\eta}\right]\end{aligned} \tag{1500}$$

The shear stress $\hat{\tau}$ is given by

$$\hat{\tau} = (\hat{\rho}\hat{\varepsilon} + \hat{\mu})\frac{\partial U}{\partial Y} \tag{1501}$$

Introducing the model of Clauser [158]

$$\hat{\varepsilon} + \hat{\nu} = \beta U_e \hat{\theta} \tag{1502}$$

where β is a proportionality constant. Based on experimental evidence, β is taken as 0.021). Substituting equations (1496) and (1502) into equation (1500) and integrating, we get

$$\begin{aligned}&\frac{d \ln \phi}{d\hat{\chi}}\left\{\hat{\eta}\left(\frac{A}{\phi}\right)\left[-\frac{4A}{\phi}(q_2 + 2\pi q_3 + 2\pi^2 q_4) - (\tilde{u} - 2)(q_5 + 2\pi q_6)\right\}\right. \\ &+ \frac{d \ln \hat{R}}{d\hat{\chi}}\left[\int_0^{\hat{\eta}} \tilde{u}^2\, d\hat{\eta} - \int_0^{\hat{\eta}} \tilde{u}\, d\hat{\eta}\right] + \frac{d\pi}{d\hat{\chi}}\left\{\frac{\hat{\eta}A}{\phi}\left[4\frac{A}{\phi}(q_3 + 2\pi q_4) - 2q_6(\tilde{u} - 2)\right]\right\} \\ &+ \frac{d \ln U_e}{d\hat{\chi}}\left[2\int_0^{\hat{\eta}} \tilde{u}\, d\hat{\eta} - \tilde{u}\int_0^{\hat{\eta}} \tilde{u}\, d\hat{\eta} - \hat{\eta}\right] \\ &+ \frac{1}{\hat{R}}\left\{\beta(1+\pi)\left(\frac{A}{\phi}\right)^2 \hat{\eta}[\hat{\eta}^{-2} + 12\pi(1 - \hat{\eta})] - \phi^{-2}\right\} = 0\end{aligned} \tag{1503}$$

where

$$q_2 = \tfrac{1}{2}(\ln \hat{\eta})^2 - \ln \hat{\eta} - 1$$
$$q_3 = (\ln \hat{\eta})\left(\hat{\eta}^2 - \frac{\hat{\eta}^3}{3} - 1\right) + \frac{\hat{\eta}^3}{8} - \frac{\hat{\eta}^2}{2} + 1$$
$$q_4 = \tfrac{4}{7}\hat{\eta}^6 - 2\hat{\eta}^5 + \tfrac{9}{5}\hat{\eta}^4 + \hat{\eta}^3 - 2\hat{\eta}^2 + 1$$
$$q_5 = \ln \hat{\eta} - 1$$
$$q_6 = -\tfrac{1}{2}\hat{\eta}^3 + \hat{\eta}^2 - 1$$

In order to render the solution determinate, one more assumption is needed. The assumption which will be made is that the Reynolds number based on y at the outer edge of the laminar sublayer is invariant under our transformation; that is

$$\frac{\rho_1 \mu_1 y_1}{\mu_1} = \frac{\hat{\rho} U_1 Y_1}{\hat{\mu}} \tag{1504}$$

Considerable evidence exists to support this contention [159], [160]. It follows that

$$\tilde{\sigma} = \frac{\tilde{\rho}_1}{\tilde{\mu}_1} \frac{1}{Y_1} \int_0^{Y_1} \frac{dY}{\tilde{\rho}} = \frac{\tilde{\rho}_1}{\tilde{\mu}_1} \frac{1}{Y_1^+} \int_0^{Y_1^+} \frac{dY^+}{\tilde{\rho}} \equiv \frac{\tilde{\rho}_1}{\tilde{\mu}_1} \Gamma \tag{1505}$$

where Y^+ represents the "inner variable" $U_\tau Y/\hat{v}$. From experiment, we find $Y_1^+ \cong 10.6$. Then from (1469)

$$\frac{1}{\tilde{\rho}_1} = H_e\left[g_w + 10.6\frac{1-g_w}{\phi}\right] + 112.36\frac{1-H_e}{\phi^2} \tag{1506}$$

and

$$\Gamma = \frac{1}{Y_1^+}\int_0^{Y_1^+} \frac{1}{\tilde{\rho}}\, dY^+ = H_e\left[g_w + 5.3\frac{1-g_w}{\phi}\right] + 37.45\frac{1-H_e}{\phi^2} \tag{1507}$$

If we define Λ as $\tilde{\rho}_1/\tilde{\mu}_1 = \tilde{\rho}_1^{1+n}$ equation (1505) may be written

$$\ln \tilde{\sigma} = \ln \Gamma + \ln \Lambda \tag{1508}$$

In summary, then, we have developed a system of seven equations and seven unknowns, which renders the solution determinate. The seven unknowns (dependent variables) are,

$$\phi, \pi, \hat{R}, U_e, \tilde{\sigma}, \hat{\eta}, \chi \tag{1509}$$

The seven equations are, (1471), (1477), (1486), (1488), (1503), (1508) and (1497). The only independent variable is χ.

The initial conditions and conditions at the outer edge of the boundary layer are presumed known. At the wall, the wall temperature, and hence g_w, is presumed given. For the adiabatic wall cases, $g_w = 1$.

The system of seven equations in seven unknowns has been solved numerically by computer for a number of cases. As an example, Figure 279 shows the

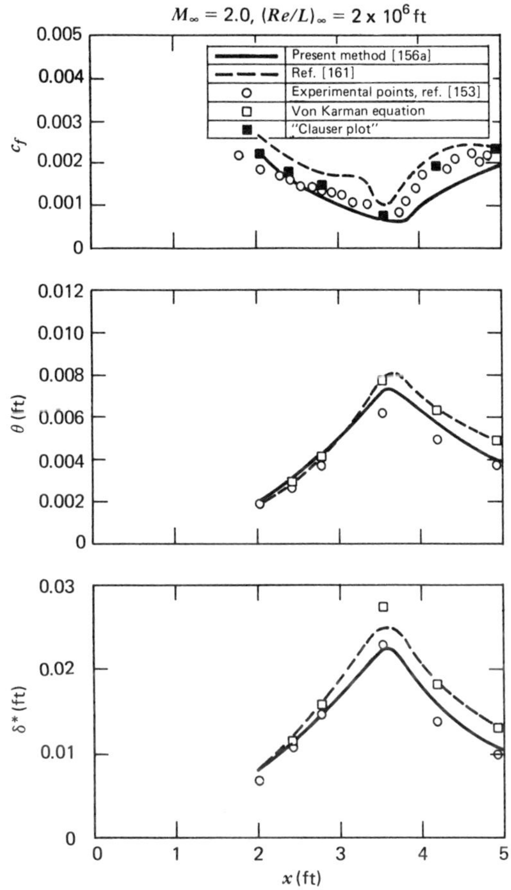

Figure 279 Results of Economos and Boccio [156a] for the waisted body of Figure 277.

results for c_f for a waisted axially symmetric body. Conversion to the axially symmetric case was made by applying the Mangler transformation. For purposes of comparison, the experimental results of Winter Smith and Rotta [153] are also shown. The results of the numerical method of Herring and Mellor [161] is also shown. This method will be discussed next.

The method of Herring and Mellor [161].

This method is an example of the direct numerical integration type. On the basis of a rather lengthy order of magnitude analysis which will not be reproduced here, Herring and Mellor conclude that the governing equations should be written in the form

$$(\bar{\rho}\bar{u})_x + (\bar{\rho}\bar{v} + \overline{\rho' v'})_y = 0 \tag{1510}$$

$$\bar{\rho}\bar{u}\bar{u}_x + (\bar{\rho}\bar{v} + \overline{\rho' v'})\bar{u}_y = -\bar{p}_x + \bar{\tau}_y \tag{1511}$$

$$\bar{\rho}\bar{u}\bar{h}_{s_x} + (\bar{\rho}\bar{v} + \overline{\rho' v'})\bar{h}_{s_y} = (\bar{q} + \bar{u}\bar{\tau})_y \tag{1512}$$

$$\bar{h}_s = \bar{h} + \frac{\bar{u}^2}{2}, \qquad \bar{p}_e = \frac{\gamma - 1}{2}\bar{\rho}\bar{h} \tag{1513}$$

$$\bar{\tau} = \bar{\mu}\bar{u}_y - \bar{\rho}\,\overline{u'v'}, \qquad \bar{q} = \bar{k}\bar{h}_y - \bar{\rho}\overline{v'h'} \tag{1514}$$

where the bars indicate the mean quantities and the primes the turbulent fluctuations. The reader may wish to compare this form of the equations with those derived at the beginning of this chapter. As in every turbulent flow solution, the average of the turbulent fluctuations must in some way be related to the mean flow variables. The method chosen here begins with the Boussinesq formulation,

$$\bar{\tau} = \bar{\rho}\nu_E\bar{u}_y \tag{1515}$$

where ν_E is an effective kinematic viscosity. We now divide the boundary layer into three separate layers: an outer layer, an inner layer and an overlap layer. In the outer layer, we assume that

$$\nu_E = \nu_E(\nabla, y, \bar{u}_y), \qquad \nabla = \int_0^\infty (u_e - \bar{u})\,dy \tag{1516}$$

In the inner layer it is assumed that

$$\nu_E = \nu_E(\bar{\nu}, y, \bar{u}_y) \tag{1517}$$

where $\bar{\nu}$ is conventional kinematic viscosity. In the overlap layer, both conditions, (1516) and (1517), must be satisfied. By dimensional analysis, it follows that in the outer layer,

$$\frac{\nu_E}{\nabla} = \Phi\left(\frac{y^2}{\nabla}\bar{u}_y\right) \tag{1518}$$

while in the inner layer,

$$\frac{\nu_E}{\bar{\nu}} = \phi\left(\frac{y^2}{\bar{\nu}}\bar{u}_y\right) \tag{1519}$$

In the overlap region,

$$\nu_E = \nabla\Phi = \bar{\nu}\phi \tag{1520}$$

It follows that in the overlap region

$$\nu_E = k^2 y^2 \bar{u}_y \tag{1521}$$

where k^2 is a proportionality constant. This result will be observed to be the same as that obtained by Prandtl in his mixing length theory (op. cit.). The hypothesis indicates that the inner layer is thicker than the laminar sublayer discussed earlier. Thus, as $y \to 0$, $\phi \to 1$. The division of turbulent boundary layers into outer and inner layers, rather than turbulent main part and laminar sublayer, has become increasingly fashionable in recent years. The inner layer is also sometimes called the wall layer. It will be observed that in the outer layer, the (molecular) viscosity is not considered, implying that turbulent momentum transfer is large compared to molecular momentum transfer. In the inner layer, viscous forces are important, but turbulent forces are not neglected. We now postulate a form of v_E valid for the entire boundary layer.

$$v_E = \bar{v}\phi(XR) + \nabla[\Phi(X) - X] \tag{1522}$$

where

$$X = \frac{ky}{\nabla}\sqrt{\frac{\bar{\tau}}{\bar{\rho}}}, \qquad R = \frac{\nabla}{\bar{v}} \tag{1523}$$

For the approximation of the heat transfer rate, $\bar{q}$, we define an effective heat diffusivity, α_E

$$\bar{q} = \bar{\rho}\alpha_E \bar{h}_y \tag{1524}$$

Both v_E and α_E may be expressed as the sum of a turbulent and a laminar contribution. Thus

$$\alpha_E = \alpha_t + \bar{\alpha}, \qquad v_E = v_t + \bar{v} \tag{1525}$$

where α_t and v_t are the turbulent contribution and $\bar{\alpha}$ and $\bar{v}$ are the laminar contributions. The laminar and turbulent Prandtl numbers may be written

$$Pr = \frac{\bar{v}}{\bar{\alpha}}, \qquad Pr_t = \frac{v_t}{\alpha_t} \tag{1526}$$

so that

$$\alpha_E = \frac{\bar{v}}{Pr} + \frac{v_t}{Pr_t} \tag{1527}$$

In numerical solutions, account may be taken of the variation of the Prandtl numbers if that variation is known as a function of the other dependent variables. Since the variation of the turbulent Prandtl number is not known exactly, Pr_t (and Pr) were assumed constant in the present example. The variation of $\bar{v}$ may be expressed by the Sutherland formula, which may be written in the form

$$\frac{\bar{v}}{v_\infty} = \left(\frac{\bar{h}}{h_e}\right)^{5/2} \frac{(h_e/c_p) + 110^0}{(\bar{h}/c_p) + 110^0} \tag{1528}$$

where h_e/c_p is in degrees Kelvin. The implication here, as in the definition of α_E, is that c_p is constant.

Boundary Conditions. The numerical integration is begun at some value of $x = x_0$, where the solution is presumed known. Thus

$$\bar{u}(y, x_0) = u_0(y), \qquad \bar{h}_s(y, x_0) = h_{s0}(y) \tag{1529}$$

$u_0(y)$ and $h_{s0}(y)$ are the given initial conditions. At $y = 0$, we have the usual boundary conditions

$$\bar{u}(0, x) = 0, \qquad \bar{v}(0, x) = 0, \qquad \bar{h}(0, x) = h_w(x) \tag{1530}$$

where $h_w(x)$, the wall enthalpy, is presumed given as a function of x. Notice again that the numerical technique allows a greater flexibility than more analytical methods in allowing the wall temperature to vary with x. Alternatively, the heat transfer at the wall may be specified

$$\bar{q}(0, x) = q_w(x) \tag{1531}$$

The adiabatic case, $q_w = 0$, becomes a special case of this form of the boundary condition. At the outer edge of the boundary layer we have

$$\lim_{y\to\infty} \bar{u}(x, y) = u_e(x) \tag{1532}$$

$$\lim_{y\to\infty} \bar{h}_s(x, y) = h_{se}(x) \tag{1533}$$

The solution is now determinate. Summarizing, we solve the system of equation (1510)–(1512), with $\bar{\tau}$ and $\bar{q}$ given by equations (1515) and (1524), where v_E and α_E are given by equations (1522) and (1527). The system is subject to the boundary conditions (1529)–(1533).

In practice, it is found more convenient to work in the following dimensionless variables:

$$\eta = \frac{y}{\delta^*} \qquad \frac{df}{d\eta} = \frac{\rho_e u_e - \bar{\rho}\bar{u}}{\rho_e u_e} = f'(\eta, x), \qquad j(\eta, x) = \frac{\rho_e}{\bar{\rho}}$$

$$\frac{dg}{d\eta} = \frac{h_{se} - \bar{h}_s}{\bar{h}_{se} - h_r} = g'(\eta, x), \qquad \delta^* = \frac{1}{\rho_e u_e}\int_0^\infty (\rho_e u_e - \bar{\rho}\bar{u})\,dy \tag{1534}$$

where the prime indicates differentiation with respect to η, and h_r is an arbitrary reference value. The boundary conditions then become

$$f'(\eta, x_0) = f'_0(\eta), \qquad f(0, x) = 0, \qquad f'(0, x) = 1, \qquad f(\infty, x) = 1$$
$$g'(\eta, x_0) = g'_0(\eta), \qquad g'(\infty, x) = 0 \tag{1535}$$

and

$$g'(0, x) = \frac{h_{se} - h_w(x)}{h_{se} - h_r} \qquad \text{or} \qquad g''(0, x) = -\frac{\delta^* q_w(x)}{\rho_w v_w (h_{se} - h_r)} \tag{1536}$$

Substituting the change of variables (1534) into equations (1520) and (1512), we get

$$\left\{\frac{\delta_k}{\delta^*}\frac{\Upsilon}{j}[j(1-f')]'\right\}' = [Qj(\eta - f) - \delta^* j f_x]f'' + [(V_j - \delta^* j_x)(f' - 2)$$
$$+ \delta^*(jf'_x - j'f_x) + Qj'(\eta - f)]f' + Qj'f + V(j-1) + \delta^*(j_x - jf'_x + j'f'_x) - Qj'\eta \tag{1537}$$

$$\left\{\frac{\delta_k}{\delta^*}\frac{\Upsilon}{j}\left[-g'' + \frac{\frac{\gamma-1}{2}M_e^2}{H\left(1 + \frac{\gamma-1}{2}M_e^2\right)}\left(\frac{\Upsilon}{\Upsilon_h} - 1\right)(j^2(1-f')^2)'\right]\right\}' = \tag{1538}$$

$$[Q(\eta - f) - \delta^* f_x]g'' - \delta^*(1 - f')g'_x$$

where the primes indicate differentiation with respect to η, the subscript x indicates differentiation with respect to x, and

$$\delta_k = \frac{\nabla}{u_e}, \qquad Q = \frac{(\rho_e u_e \delta^*)_x}{\rho_e u_e}, \qquad V = \frac{u_{e_x} \delta^*}{u_e}$$

$$H = \frac{h_{se} - h_r}{h_{se}}, \qquad \Upsilon = \frac{v_E}{\delta_k u_e} = \frac{1}{R}\phi(R, X) + \Phi(X) - X \tag{1539}$$

$$\Upsilon_h = \frac{\alpha_E}{\nabla} = \frac{1}{RPr} + \frac{1}{Pr_t}\left(\Upsilon - \frac{1}{R}\right)$$

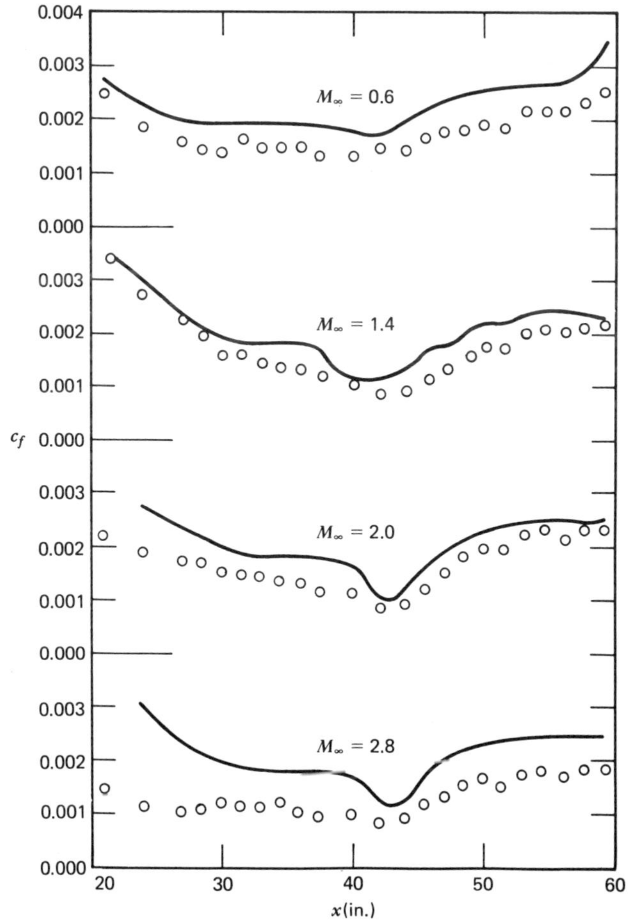

Figure 280 Results of Herring and Mellor [161] for the waisted body of figure 277 (——) compared to the experimental results of Winter et al. (ooo).

From equations (1513)

$$j = \frac{2\left(1 + \frac{\gamma - 1}{2} M_e^2\right)(1 - Hg')}{1 + \left[1 + \frac{\gamma - 1}{2} M_e^2 (1 - f')^2 \left(1 + \frac{\gamma - 1}{2} M_e^2\right)(1 - Hg')\right]^{1/2}} \tag{1540}$$

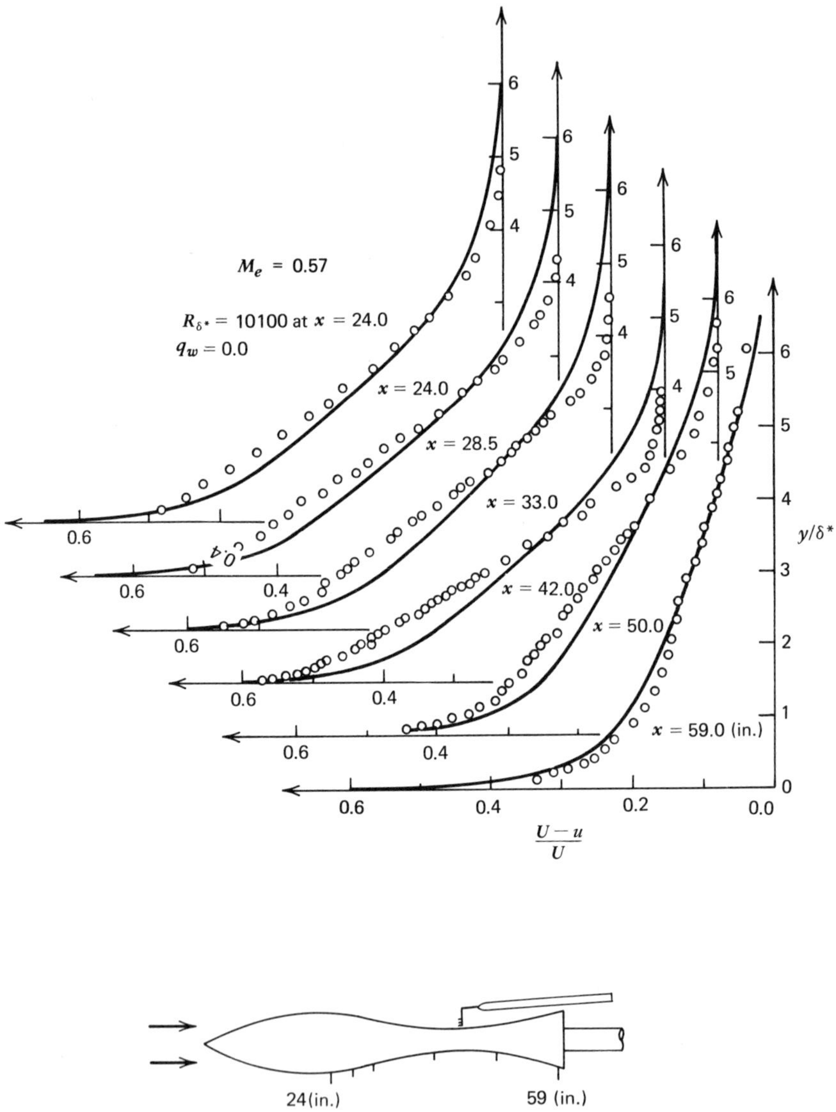

Figure 281 Velocity profiles found by Herring and Mellor [161] over a waisted body(——), compared to experimental results (ooo).

ϕ and Φ are based on known solutions for constant pressure incompressible profiles. Φ is approximated by

$$\Phi = X, \qquad X \le 0.016; \qquad \Phi = k, \qquad X > 0.016, \qquad k = 0.41 \qquad (1541)$$

ϕ is approximated by

$$\phi = 1 + \frac{\chi^4}{\chi^3 + (6.9)^3}, \qquad \chi = \frac{ky}{\bar{\nu}}\sqrt{\frac{\bar{\tau}}{\bar{\rho}}} \qquad (1542)$$

The integration of equations (1537) and (1538) may now proceed. The following technique is used. All x derivatives are first written in finite difference form:

$$f'_x = \frac{f' - f'_b}{\Delta x}, \qquad g'_x = \frac{g' - g'_b}{\Delta x}, \qquad \delta^* - \frac{\delta^* - \delta^*_b}{\Delta x}, \qquad \Delta x = x - x_b \quad (1543)$$

where the subscript b indicates known solutions at $x = x_b$. Once the relations (1543) have been substituted in equations (1537) and (1538), these become ordinary differential equations in η. They may then be integrated numerically by any of the well-known standard techniques. (Herring and Mellor used a fourth degree Runge–Kutta method.)

Sample results. The method was applied to the case investigated experimentally by Winter Smith and Rotta [153] and one of the results of this test have already been given in Figure 279. Further illustrations of the results are shown in Figure 280 and 281. Where the boundary-layer thickness was no longer small compared to the radius of the body, the equations were rewritten in axially symmetric form to obtain the results shown in these figures.

Method of White and Christoph. [162]

This is an integral method which differs in certain important respects from other integral methods. Recall the definition of the so-called inner variables, u^+ and y^+ :

$$u^+ = \frac{u}{u_\tau}, \qquad y^+ = \frac{yu_\tau}{\nu_w}, \qquad u_\tau = \sqrt{\frac{\tau_w}{\rho_w}} \qquad (1544)$$

where u_τ is the friction velocity. We begin making the assumption that

$$u^+ = f(y^+, \alpha, \beta, \gamma) \qquad (1545)$$

where α, β, and γ are three parameters containing pressure gradient, heat transfer and compressibility. On the basis of dimensional analysis it may be concluded that

$$\alpha = \frac{\nu_w}{\tau_w u_\tau}\frac{dp}{dx}, \qquad \beta = \frac{q_w \nu_w}{T_w, k_w u_\tau}, \qquad \gamma = \frac{ru_\tau^2}{2c_p T_w} \qquad (1546)$$

where r is the turbulent recovery factor.

In the absence of theoretical justification, the assumption (1545) can only be verified a posteriori. We know of course that in incompressible flow with α, β and

γ equal to zero, $u^+ = f(y^+)$, at least to a good approximation. In addition, White [163] has demonstrated that the functional relation (1545) is valid for incompressible flow, at least in certain cases. Thus, the present method is an extension of White's incompressible theory to the case of compressible flow.

The temperature and density variation across the boundary layer is presumed given by the Crocco–Busemann integral,

$$T = a + bu + cu^2$$

In terms of the variables given above, this becomes

$$\frac{\rho_w}{\rho} = \frac{T}{T_w} = 1 + \beta u^+ - \gamma u^{+2} \tag{1547}$$

In the present case, all the variables in equation (1547) are functions of x. The shear stress is approximated by

$$\tau = \tau_w + \frac{dp}{dx} y + \cdots = \rho \kappa y^2 \left(\frac{du}{dy}\right)^2 \tag{1548}$$

where κ is the von Karman proportionality constant, taken as 0.4. Substituting for ρ from equation (1547) we get a differential equation for u^+ in terms of y^+ with α, β and γ as parameters.

$$\frac{du^+}{dy^+} = \frac{1}{ky^+}(1 + \alpha y^+)^{1/2}(1 + \beta u^+ - \gamma u^{+2})^{1/2} \tag{1549}$$

This may be integrated to give

$$u^+ = \frac{1}{2\gamma}\left[\beta + Q \sin\left\{\phi + \frac{\sqrt{\gamma}}{k}\left[2(s - s_0) + \ln\left(\frac{s-1}{s+1}\cdot\frac{s_0+1}{s_0-1}\right)\right]\right\}\right] \tag{1550}$$

where

$$\phi = \sin^{-1}\left(\frac{2\gamma u_0^+ - \beta}{Q}\right), \qquad Q = (\beta^2 + 4\gamma)^{1/2} \qquad s = (1 + \alpha y^+)^{1/2} \tag{1551}$$

The subscript $_0$ refers to the value of y^+ at which u^+ is taken as zero. White and Christoph have taken $u^+ = 0$ at $y^+ = 0.1108$. This is based on the assumption that for $\alpha = \beta = \gamma = 0$, u^+ may be approximated by the standard log formula

$$u^+ = -\frac{1}{\kappa}\ln y^+ + B \tag{1552}$$

for which Nikuradse [164] has determined that for flat plates a good approximation is given by $\kappa = 0.40$ and $B = 5.5$. Thus u^+ will equal zero when

$$y^+ = e^{-\kappa B} = 0.1108 \tag{1553}$$

All this is very approximate, but may be justified, as it is proposed to use the resulting velocity profile to integrate the boundary layer momentum equation across the boundary layer in the manner of the von Karman integral method in which the solution for the shear stress is not sensitive to the exact shape of the velocity profile used. We will return to this point below.

Accordingly, we now propose to integrate the boundary layer equations

$$(\rho u)_x + (\rho v)_y = 0 \tag{1554}$$

$$\rho u u_x + \rho v u_y = -p_y + \tau_y \tag{1555}$$

across the boundary layer, using the velocity profile of equation (1550). First, however, we eliminate v by introducing the compressible stream function ψ, where

$$\rho u = \psi_y, \qquad \rho v = -\psi_x \tag{1556}$$

Next, we rewrite the equations in terms of our new variables. The change of variables is

$$x, y \rightarrow y^+, \alpha, \beta, \gamma \tag{1557}$$

Then, in light of equations (1545) and (1547),

$$\psi = \mu_\infty \int_0^{y^+} \frac{u^+ \, dy^+}{1 + \beta u^+ - \gamma u^{+2}} \tag{1558}$$

The momentum equation becomes

$$\rho u_\tau u^+ (u_\tau u^+)_x - \psi_x \frac{u_\tau}{v_w} (u_\tau u^+)_{y^+} = \rho_e u_e u_{e_x} + \frac{u_\tau}{v_w} \tau_{y^+} \tag{1559}$$

where

$$\frac{\partial}{\partial x} = \frac{\partial y^+}{\partial x}\frac{\partial}{\partial y^+} + \frac{\partial \alpha}{\partial x}\frac{\partial}{\partial \alpha} + \frac{\partial \beta}{\partial x}\frac{\partial}{\partial \beta} + \frac{\partial \gamma}{\partial x}\frac{\partial}{\partial \gamma} \tag{1560}$$

Equation (1559) is now integrated over y^+ from 0 to δ^+, where δ^+ is defined as the value of y^+ where $u = u_e$. The result is

$$\rho_w u_\tau G \frac{du_\tau}{dx} + \rho_w u_\tau^2 H \frac{d\alpha}{dx} = \rho_e u_e \delta^+ \frac{du_e}{dx} - \frac{u_\tau}{v_w} \tau_w \tag{1561}$$

where

$$G = \int_0^{\delta^+} \frac{\rho}{\rho_w} \left(u^{+2} - \beta u^+ u_\beta^+ + 2\gamma u^+ u_\gamma^+ + \frac{\beta}{\rho v_w} \psi_\beta u_{y^+}^+ - \frac{\gamma}{\rho v_w} \psi_\gamma u_{y^+}^+ \right) dy^+ \tag{1562}$$

$$H = \int_0^{\delta^+} \left(\frac{\rho}{\rho_w} u^+ u_\alpha^+ - \frac{1}{\mu_w} \psi_\alpha u_{y^+}^+ \right) dy^+ \tag{1563}$$

In equation (1561), the only independent variable is x, since u_τ, α and u_e are functions of x only. G and H may be evaluated since α, β and γ are presumed given as function of x. u^+ is taken from equation (1550) and ψ is evaluated from equation (1558). Equation (1561) is thus an ordinary differential equation for u_τ. It should be pointed out that in deriving equation (1561), the assumption has been made that $\tau = 0$ when $u = u_e$, an assumption which is not fulfilled by the assumed velocity profile (1550). Also, terms containing $d\beta/dx$ have been neglected, thus assuming only slow changes in β with x.

White and Christoph have given two simplified ways of solving equation (1561). We begin by defining the following nondimensional variables:

$$x^* = \frac{x}{L}, \qquad V = \frac{u_e}{u_0}, \qquad \lambda = \sqrt{\frac{2}{c_f}}, \qquad c_f = \frac{\tau_w}{\frac{1}{2}\rho_e u_e^2} \tag{1564}$$

where L and u_0 are reference values, usually the length of the body and the free stream velocity. Equation (1561) then becomes

$$(G - 3\alpha H)\lambda' + \frac{V'}{V}\lambda(\lambda^2\delta^+ - G) - \lambda^4 \frac{1}{V''}\frac{H}{R_L} - R_L V = 0 \tag{1565}$$

where

$$R_L = \frac{u_0 L}{\nu_e}\frac{\mu_e}{\mu_w}\left(\frac{T_e}{T_w}\right)^{1/2} \tag{1566}$$

and the primes indicate differentiation with respect to x^*. α, β and γ may be expressed in terms of λ:

$$\alpha = \left(\frac{1}{V}\right)' \frac{\lambda^3}{R_L}, \qquad \beta = \frac{\dfrac{T_{aw}}{T_w} - 1}{\lambda\sqrt{\dfrac{T_e}{T_w}}}, \qquad \gamma = \frac{\dfrac{T_{aw}}{T_e} - 1}{\lambda^2} \tag{1567}$$

where T_{aw} is the adiabatic wall temperature using the recovery factor r as given in the definition of γ (equation (1546)). δ^+ may be related to λ by noting that $u^+ = u_0^+$ when $y^+ = \delta^+$ and

$$u_e^+ = \lambda\sqrt{\frac{T_e}{T_w}} \tag{1568}$$

u^+ and y^+ being related by equation (1550).

The calculations may be simplified by use of the following curve fits:

$$G \cong 8.0\left(\frac{\delta^+}{0.1108}\right)^{1.2} + 1.2\alpha H, \qquad H = 0.5\delta^{+2}\left[\lambda\left(\frac{T_e}{T_w}\right)^{1/2} - 0.4\alpha\delta^+\right] \tag{1569}$$

Equation (1565) may then be integrated forward given an initial value of λ. The curve fits (1569) are considered only approximate and not accurate at high Mach numbers. An alternate graphical method is, therefore, proposed. Equation (1565) may be rewritten in the form

$$0.16 f^* A^3 \frac{d\lambda}{dx^*} = \frac{(1/V)'}{(1/V)}[1 + 9A^{-2} g^* R^{*0.07}] + \frac{(1/V)''}{(1/V)'}[3A^2 g^* R^{*0.07}] \tag{1570}$$

where

$$f^* = (2.434Z + 1.443Z^2)\exp(-44Z^6)$$

$$g^* = 1 - 2.3Z + 1.76Z^3, \qquad Z = 1 - \frac{\lambda}{\lambda_{max}} \tag{1571}$$

$$R^* = \frac{R_L}{(1/V)'}, \quad A = \frac{\sqrt{(T_{aw}/T_e) - 1}}{\sin^{-1}\frac{a}{c} + \sin^{-1}\frac{b}{c}}, \quad \lambda_{max} = 8.7A \log_{10} R^* \tag{1572}$$

$$a = \frac{T_{aw} + T_w}{T_e} - 2, \quad b = \frac{T_{aw} - T_w}{T_e}, \quad c = \left[\left(\frac{T_{aw} + T_w}{T_e}\right)^2 - 4\frac{T_w}{T_e}\right]^{1/2} \tag{1573}$$

λ_{max} refers to the value for which $(G - 3\alpha H) = 0$. At the separation point, $c_f = 0, \lambda = \infty$, and hence $d\lambda/dx^* = \infty$. Since the other terms in equation (1565) remain bounded, this means $G - 3\alpha H$ must be zero there. Since equation (1550) is not exact, however, λ_{max} is finite. Equations (1571) and (1572) are curve fits, which are accurate in the range $0.36 \leq \lambda/\lambda_{max} \leq 1.0$. Equation (1570) is restricted to adverse pressure gradients since R^* is required to be positive. The functions A, f^* and g^* are plotted in Figures 282 and 283 to facilitate the integration of

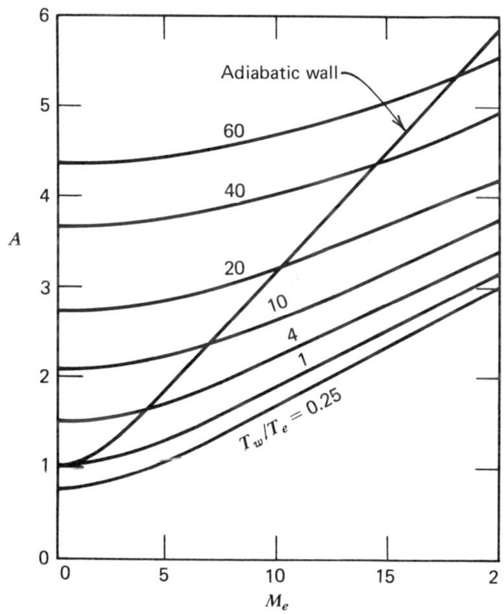

Figure 282 Parameter A for use in the method of Christoph and White [162].

equation (1570). For $\lambda/\lambda_{max} < 0.36$ or $R^* < 0$ (favorable pressure gradient), equation (1565) may be approximated by

$$\frac{d\lambda}{dx^*} = \frac{1}{8} R_L V \exp\left[\frac{-0.48\lambda}{A}\right] - 5.5\frac{V'}{V} \tag{1574}$$

Equations (1570) and (1574) may be integrated numerically by hand with the aid of Figure 282 and 283, or by machine.

Flat Plate Case. As a test of the theory we may first examine the flat plate case for which experimental data is available. In this case $\alpha = 0$ and equation (1565)

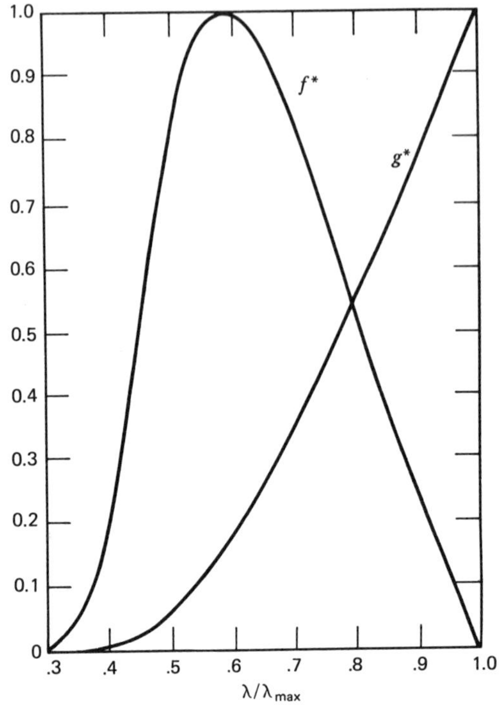

Figure 283 Functions f^* and g^* for use in the method of Christoph and White [162].

reduces to

$$G\frac{d\lambda}{dx^*} = R_L \qquad \text{where} \qquad G \cong 8.0\exp\left[\frac{0.48\lambda}{A}\right] \tag{1575}$$

Thus

$$R_x = \frac{u_e x}{\nu_e} = \frac{\mu_w}{\mu_e}\left(\frac{T_w}{T_e}\right)^{1/2}\int_0^\lambda G\,d\lambda \tag{1576}$$

Substituting for G, integrating and rearranging we get

$$c_f = 0.455\left\{A^2 \ln^2\left[0.06\,R_x\left(\frac{\mu_e}{\mu_w}\right)\left(\frac{T_e}{T_w}\right)^{1/2}\Big/ A\right]\right\}^{-1} \tag{1577}$$

This formula is found to compare favorably with experimental data over a wide range of Reynolds numbers, Mach numbers and wall temperatures. (The reader may wish to compare this formula with some of the flat plate formulas derived earlier in this chapter.) The accuracy of the present formula is about the same as that given by van Driest (cf. equation (1353)). In this connection it might be pointed out that the accuracy of both formulas ((1353) and (1577)) is considerably less for the case of flow with heat transfer than for the case of adiabatic wall. In a comparison with a large body of experimental data, White and Christoph [162], [165] compared the average error of prediction of the flat plate formulas of a number of theories, including those discussed in this chapter. Some of the results are presented in Table 3. The method of van Driest and the present method is seen to give the best results. However, the average error in the heat-transfer cases is roughly twice that encountered in the adiabatic cases. Considering that experimental errors run as high as 10%, the predictions of van Driest and the present formula are excellent in the adiabatic cases but less accurate in the heat-transfer cases. Overall, the formulas of van Driest and of White and Christoph are seen to give the best results for flat plates.

Table 3

	Average Error (%)	
Method	Adiabatic	Heat Transfer
Eckert	9.06	25.56
Sommer and Short	7.77	20.14
Van Driest	5.46	13.81
White and Christoph	5.26	11.28

In the case of favorable pressure gradient, comparison has been made with the results of Brott et al. [166] (Figure 284). Agreement is seen to be good

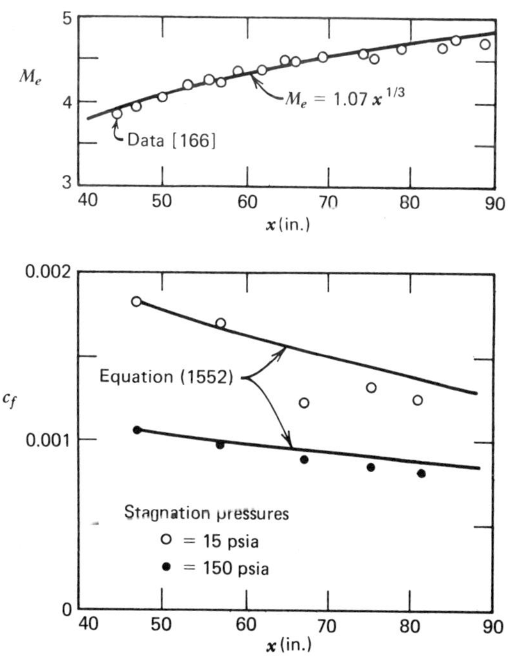

Figure 284 Results of the method of Christoph and White [162] for flow with a favorable pressure gradient.

at the higher pressure but only fair at the lower pressure. For the case of unfavorable pressure gradient, comparison has been made with the experimental results of Winter et al. (op. cit.) (Figure 285). These are the same results which

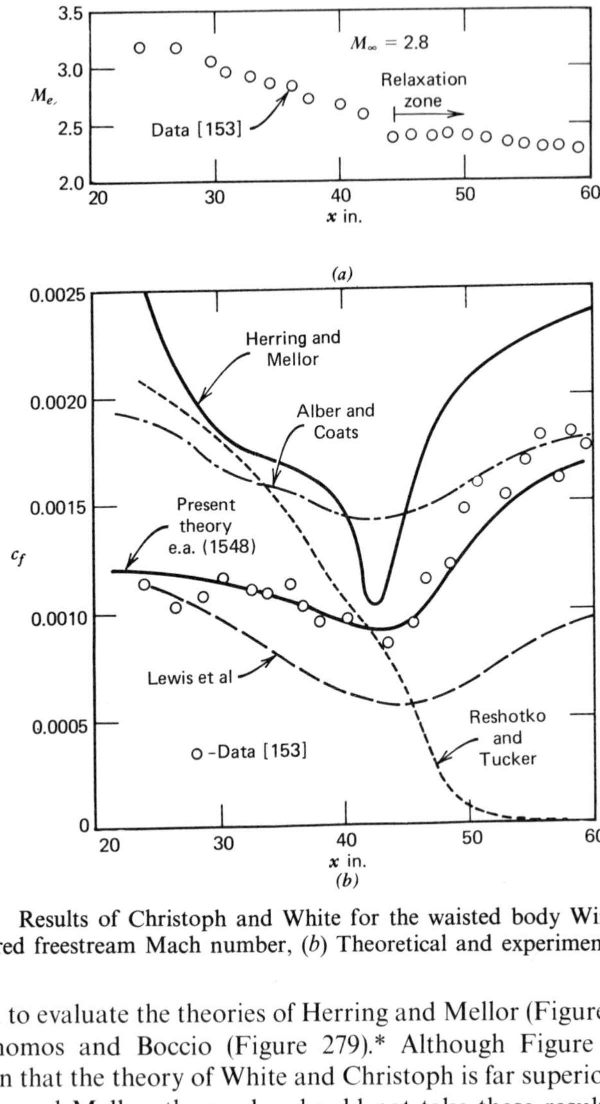

Figure 285 Results of Christoph and White for the waisted body Winter et al. [153]. (*a*) Measured freestream Mach number, (*b*) Theoretical and experimental skin friction.

were used to evaluate the theories of Herring and Mellor (Figures 280 and 281) and Economos and Boccio (Figure 279).* Although Figure 285 gives the impression that the theory of White and Christoph is far superior to the theory of Herring and Mellor, the reader should not take these results too literally. The same caveat applies to other comparisons of theory and experiment in compressible turbulent boundary layers. An example will serve to illustrate

* In applying their theory to the axially symmetric, waisted body considered by Winter et al., White and Christoph retained the assumption of plane flow. Lateral and longitudinal curvature were neglected.

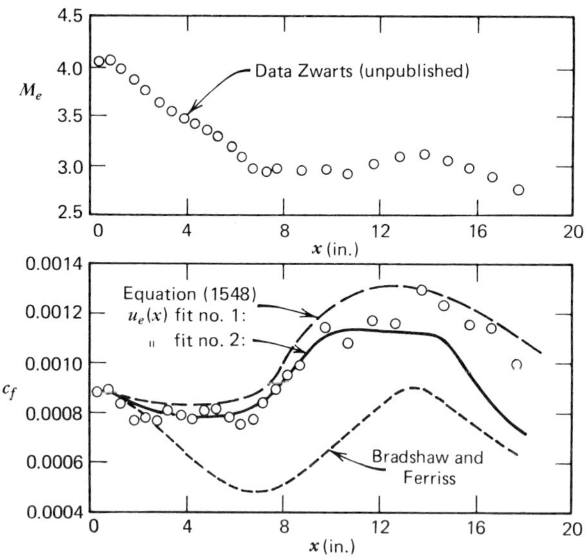

Figure 286 Results of Christoph and White for a flow with an unfavorable pressure gradient [162].

this point. Figure 286 shows a comparison of the theoretical results of White and Christoph with unpublished experimental results of Zwarts. It is seen that two theoretical curves are presented. In making their calculations, White and Christoph used two different curve fits to approximate the given external velocity, $u_e(x)$. The results are seen to be quite sensitive to which curve fit is used. In general compressible turbulent boundary layer theories contain so many free parameters, matchings to experimental data at some point or points, curve fit approximations, etc., that there is considerable room for fudging. Comparisons of theoretical and experimental results should, therefore, be taken as approximate, not exact, indications of the accuracy of the theory.

9 Real Gas Effects

Properties of Gases at High Temperatures

Throughout most of this book we have been assuming that gases, notably air, behave like perfect gases. By this we mean that the internal energy is a function of temperature only and that

$$p = \rho R T$$

where R is a constant. In much of the work we have also ignored viscosity. The perfect gas assumption is, in fact, a good one in most practical applications. Viscosity, as we have seen, may usually be ignored except in a thin layer near the wall. In some cases we have gone even further and assumed constant c_p and c_v. These assumptions are more restrictive and apply in a less broad range of cases. Other assumptions which have been made are constant γ and constant Prandtl number. Thermal conductivity and viscosity have sometimes been assumed constant, sometimes functions of temperature.

While the assumptions which we have made are often valid, they are not always valid. An intelligent understanding of the theory, therefore, requires a knowledge of the limitations of our assumptions and an ability to deal with problems in those cases in which the assumptions are not valid.

A breakdown of the assumptions occurs mainly in those cases where high temperatures and high temperature gradients are present. This would include flow of very hot gases or plasmas, particularly over colder walls, very high velocity flow where high temperatures are attained in the boundary layer even though the free stream, and even the wall, may be cool, and shock waves, particularly strong shock waves, where high temperatures and high temperature gradients are present.

In order to deal with the problem, we must first develop an understanding of how the properties of gases, in particular air, vary with temperature. The properties of gases may be divided into two groups. Thermodynamic properties and transport properties. Thermodynamic properties are those which depend only on the state of the gas at a particular point. Examples of thermodynamic properties are internal energy, enthalpy and specific heat. Transport properties, on the other hand, are those which describe the movement of mass, momentum, and energy through the gas as a result of the motion of individual particles such as molecules, atoms, ions, and electrons. Examples of transport properties are viscosity and thermal conductivity.

Thermodynamic Properties

In order to understand the variation of thermodynamic properties with temperature, a brief review of elementary kinetic theory may be in order. The temperature of a gas, according to kinetic theory, is a measure of the average translational kinetic energy of the random motion of the molecules. When heat is added to a gas, in the absence of work, the energy of the gas rises. At low temperatures, most of this energy rise takes place in the form of an increase in translational kinetic energy—that is, temperature. As the temperature rises, however, an addition of heat adds to the energy of the molecules in other ways: an increase in rotational kinetic energy, vibrational kinetic energy and electron energy. In classical kinetic theory, this is explained by the familiar dumbell model for molecules in diatomic gases (Figure 287). Each ball represents an

Figure 287 Different forms of molecular energy.

atom and the bar connecting them the force holding them together. Rotation and vibration are thus seen to be forms of kinetic energy. Electrons absorb energy in certain discreet increments called quanta. If the temperature is sufficiently high, some of the molecules will split into separate atoms absorbing an amount of energy in the process called the bond energy. For still higher temperatures, electrons will split off from the atoms leaving a mixture of ions and free electrons. A fully ionized gas is called a plasma.

As a consequence of these factors, the specific heat of a gas varies as the temperature goes up, since only a fraction of the heat goes into raising the temperature while the rest goes into the other forms of energy. At each temperature and pressure, the distribution of the energy between the different forms is fixed and known if the gas is in equilibrium. When the temperature of a gas is changed suddenly, it takes a short time for the gas to reach its new equilibrium state. This time is known as the relaxation time. The relaxation time is quite short so that in most practical applications we are dealing with gases in equilibrium. However, there are cases where the gas may not be in equilibrium. If the relaxation time is long compared to time in which the temperature changes, the gas may sometimes be assumed to retain its original energy distribution even after the temperature has changed. Such a gas is said to be frozen.

The proportion of the energy in each form in a gas is determined by the partition functions. The partition function for a pure gas may be expressed as the product of the partition functions for each mode of energy. Thus

$$Q = Q_t Q_r Q_v Q_e \tag{1578}$$

where Q is the total partition function, and Q_t, Q_r, Q_v and Q_e are the translational, rotational, vibrational, and electronic partition functions, respectively. In statistical mechanics, it is shown that for a diatomic gas,

$$Q_t = \left(\frac{2\pi mkT}{h^2}\right)^{3/2}\frac{RT}{p}, \qquad Q_r = \sum_{j=0}^{\infty}(2j+1)\exp\left\{-\frac{h^2 j(j+1)}{8\pi^2 IkT}\right\}$$

$$Q_v = \sum_{n=0}^{\infty} e^{-nh\nu/kT} = (1 - e^{-h\nu/kT})^{-1}; \qquad Q_e = \sum_{n=0}^{\infty} g_n e^{-\varepsilon_n/kT} \tag{1579}$$

where m is the mass of the molecule, k is the Boltzmann constant, h is Planck's constant, I is the molecular moment of inertia, ν the vibrational frequency, g_n the degeneracy of the nth electronic state (number of states having the same energy level) and ε_n the energy of the nth electronic state. The energy and enthalpy of the gas are related to the partition function by the formulas

$$\frac{E - E_0}{RT} = T\left(\frac{\partial \ln Q}{\partial T}\right)_\rho = T\frac{d \ln Q_c}{dT} \tag{1580}$$

$$\frac{H - E_0}{RT} = T\left(\frac{\partial \ln Q}{\partial T}\right)_p = T\frac{d \ln Q_p}{dT} \tag{1581}$$

where R is the gas constant per mole, E and H are the energy and enthalpy per mole, E_0 is an arbitrary constant representing the energy of the gas at absolute zero, and

$$Q_c = \frac{p}{RT}Q, \qquad Q_p = pQ \tag{1582}$$

The specific heats per mole are then given by

$$c_v = \left(\frac{\partial E}{\partial T}\right)_\rho, \qquad c_p = \left(\frac{\partial H}{\partial T}\right)_p \tag{1583}$$

Substituting equations (1578) and (1579) into equations (1580), (1581) and (1583), we find

$$\frac{E - E_0}{RT} = \frac{5}{2} + \frac{h\nu}{kT}[e^{h\nu/kT} - 1]^{-1} + A \tag{1584}$$

$$\frac{H - E_0}{RT} = \frac{7}{2} + \frac{h\nu}{kT}[e^{h\nu/kT} - 1]^{-1} + A \tag{1585}$$

$$\frac{c_v}{R} = \frac{5}{2} + \left(\frac{h\nu}{RT}\right)^2\left(\sinh\frac{h\nu}{2kT}\right)^{-2} + B - A^2 \tag{1586}$$

$$\frac{c_p}{R} = \frac{7}{2} + \left(\frac{h\nu}{RT}\right)^2\left(\sinh\frac{h\nu}{2kT}\right)^{-2} + B - A^2 \tag{1587}$$

where

$$A = \frac{\sum \frac{\varepsilon_n}{kT} g_n \exp\left[-\frac{\varepsilon_n}{kT}\right]}{\sum g_n \exp\left[-\frac{\varepsilon_n}{kT}\right]} \tag{1588}$$

and

$$B = \frac{\sum \left(\frac{\varepsilon_n}{kT}\right)^2 g_n \exp\left[-\frac{\varepsilon_n}{kT}\right]}{\sum g_n \exp\left[-\frac{\varepsilon_n}{kT}\right]} \tag{1589}$$

The entropy of the gas is given by

$$\frac{S}{R} = \ln Q + T\left(\frac{\partial \ln Q}{\partial T}\right)_p \tag{1590}$$

The partition functions for nitrogen and oxygen are well known and will be given here as these gases constitute the principal components of air, which is the most common compressible fluid

$$N_2: \quad \ln Q = \frac{7}{2}\ln T + 0.42 - \ln\left[1 - \exp\left(-\frac{3390}{T}\right)\right] - \ln p \tag{1591}$$

$$O_2: \quad \ln Q = \frac{7}{2}\ln T + 0.11 - \ln\left[1 - \exp\left(-\frac{2270}{T}\right)\right] - \ln p + \ln\left[3 + 2\exp\left(-\frac{11390}{T}\right) + \exp\left(-\frac{18990}{T}\right)\right] \tag{1592}$$

where T is in degrees Kelvin and p is in atmospheres. By convention, E_0 is taken as zero for the molecules of nitrogen and oxygen. It follows that E_0 for atoms will be higher, since energy input is required to split molecules into atoms. E_0 for atoms will thus be half the energy required to dissociate the (diatomic) molecule.

Dissociation

When a gas reaches a certain temperature, the molecules will begin to break up into atoms. This is called dissociation. The fraction of molecules dissociated at each temperature and pressure is a function of the temperature and pressure and is known, provided the gas in equilibrium. When the temperature and pressure change suddenly, a short time is required for the relative number of atoms and molecules to adjust to the new equilibrium state. Thus, in a dissociated gas, relaxation involves not only the redistribution of energy among the different forms, but adjustment of the degree of dissociation. A frozen dissociated gas retains the same degree of dissociation even though its state has changed. In most practical problems the dissociated gas may be considered in equilibrium. Cases do occur, however, where the assumption of frozen flow may be more accurate.

The fraction of a gas dissociated is given by its equilibrium constant. For example, if nitrogen dissociates, we have

$$N_2 \rightarrow aN_2 + bN \tag{1593}$$

and

$$N_2 \rightleftharpoons 2N \tag{1594}$$

The fraction dissociated is then given by the equilibrium constant K_p

$$K_p = \frac{p_N^2}{p_{N_2}} \tag{1595}$$

where p_N and p_{N_2} are the partial pressures of N and N_2, respectively. Caution must be exercised in using tables of equilibrium constants for in some tables the reciprocal of equation (1595) is given. In general, the equilibrium constant for the dissociation reaction

$$\sum a_i A_i \rightleftharpoons \sum b_i B_i \tag{1596}$$

is given by

$$K_p = \frac{\Pi p^{b_i}(B_i)}{\Pi p^{a_i}(A_i)} \tag{1597}$$

where Π means the product, $p(B_i), p(A_i)$ represent the partial pressure of each component of the mixture and the exponents a_i and b_i represent the stoichiometric coefficients. The equilibrium constants may be found from the partition functions. Thus

$$\ln K_p = -\frac{\Delta E_0}{RT} + \sum b_i \ln Q_p(B_i) - \sum a_i \ln Q_p(A_i) \tag{1598}$$

where

$$\Delta E_0 = \sum b_i E_0(B_i) - \sum a_i E_0(A_i) \tag{1599}$$

In order to find the concentration of each species present, we define the concentration equilibrium constant K_c,

$$K_c = \frac{\Pi n^{b_i}(B_i)}{\Pi n^{a_i}(A_i)} \tag{1600}$$

where $n(B_i)$ and $n(A_i)$ are the moles per unit volume of each species present. It follows that

$$K_c = K_p(RT)^{\Sigma a_i - \Sigma b_i} \tag{1601}$$

and

$$T\frac{d \ln K_c}{dT} = \frac{\Delta E_0}{RT} + \sum b_i\left(\frac{E - E_0}{RT}\right)_{B_i} - \sum a_i\left(\frac{E - E_0}{RT}\right)_{A_i} \tag{1602}$$

or

$$T\frac{d \ln K_p}{dT} = \frac{d \ln K_c}{dT} + \sum b_i - \sum a_i \tag{1603}$$

The partition functions for 0 and N are given by

$$0: \quad \ln Q = \frac{5}{2}\ln T + 0.5 + \ln\left[5 + 3\exp\left(-\frac{228}{T}\right) + \exp\left(-\frac{326}{T}\right) + 5\exp\left(-\frac{22800}{T}\right) + \exp\left(-\frac{48600}{T}\right)\right] - \ln p \tag{1604}$$

$$\text{N}: \ln Q = \frac{5}{2}\ln T + 0.3 + \ln\left[4 + 10\exp\left(-\frac{27700}{T}\right) + 6\exp\left(-\frac{41500}{T}\right)\right] - \ln p$$

where, as before, T is in degrees Kelvin and p is in atmospheres. It follows that for $O_2 \rightleftharpoons 2O$,

$$\ln K_p = -\frac{59000}{T} + 2\ln Q_p(\text{O}) - \ln Q_p(\text{O}_2) \tag{1605}$$

and for $N_2 \rightleftharpoons N$

$$\ln K_p = -\frac{113200}{T} + 2 \ln Q_p(\mathrm{N}) - 2 \ln Q_p(\mathrm{N}_2) \tag{1606}$$

The relation between Q and Q_p was given in equation (1582).

At still higher temperatures, atoms will ionize, giving off an electron and absorbing a corresponding amount of energy in the process. Thus, for example,

$$\mathrm{O} \rightarrow a\mathrm{O} + b\mathrm{O}^+ + be^- \tag{1607}$$

leading to the equilibrium reaction

$$\mathrm{O} \rightleftharpoons \mathrm{O}^+ + e^- \tag{1608}$$

where O^+ represent an oxygen ion and e^- a free electron. The degree of ionization, like the degree of dissociation, may be determined by equilibrium constants and partition functions. However, ionization does not begin to be significant until very high temperatures ($\sim$ 10,000°K for air at 1 atmosphere) are attained. The equilibrium constant for equations (1607) and (1608) is given by

$$K_p = \frac{p_{\mathrm{O}^+} p_{e^-}}{p_{\mathrm{O}}} \tag{1609}$$

and

$$\ln K_p = -\frac{158000}{T} + \ln Q_p(\mathrm{O}^+) + \ln Q_p(e^-) - \ln Q_p(\mathrm{O}) \tag{1610}$$

where

$$\ln Q(\mathrm{O}^+) = \frac{5}{2} \ln T + .5 + \ln\left(4 + 10 \exp\left(-\frac{38600}{T}\right)\right.$$
$$\left. + 6 \exp\left(-\frac{58200}{T}\right)\right) - \ln p \tag{1611}$$

$$\ln Q(e^-) = \frac{5}{2} \ln T - 14.24 - \ln p \tag{1612}$$

The concentration of each species may then be found in the same manner as before. Tables of equilibrium constants may be found in numerous references including Hansen [167], Lewis and von Elbe [168] and National Bureau of Standards Tables [169]. The equilibrium composition of air has been calculated for a wide range of densities and temperatures by Moeckel and Western [170]. As an example, we show their results for atmospheric density (Figure 288).

Correction of the perfect gas equation of state

At high temperatures and low pressures, the equation of state $p = \rho R_0 T$ becomes inaccurate. The error in the equation may be corrected by introduction of the compressibility Z, defined by the equation

$$p = \rho Z R_0 T \tag{1613}$$

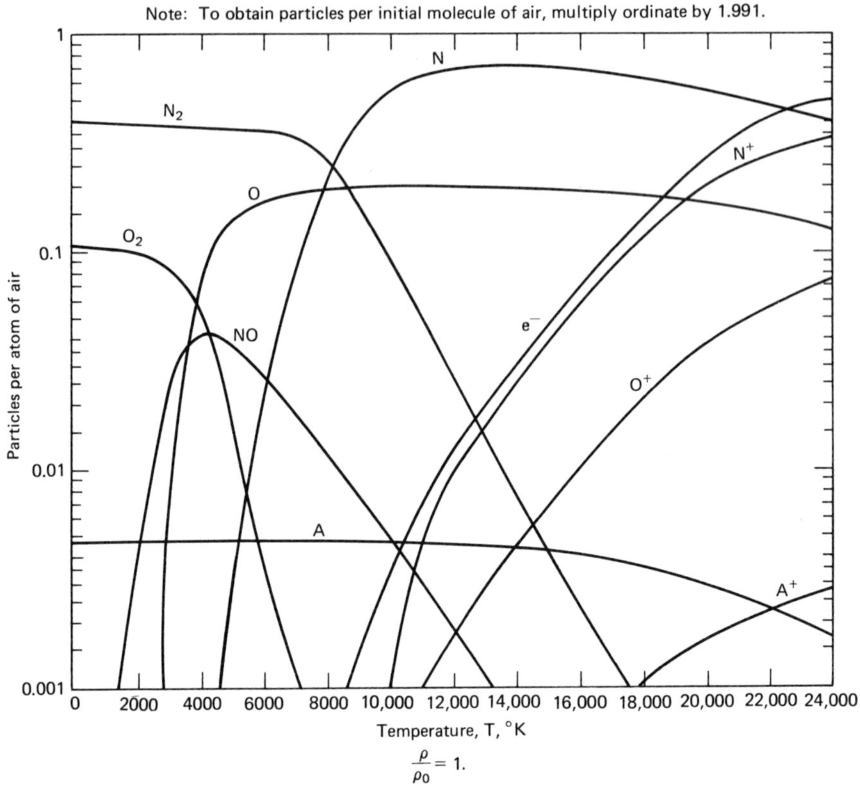

Figure 288 Equilibrium composition of air at atmospheric density [170].

The gas constant R_0 here is per unit mass of gas in its undissociated state. At high temperatures and low pressures a good approximation is that each species present in a gas continues to satisfy the ideal gas equation of state so that the error in the equation for the gas as a whole is caused mainly by the change in the number of particles present due to dissociation and ionization. In this case Z is simply given by

$$Z = 1 + \varepsilon \qquad (1614)$$

where ε is the fractional increase in the number of particles due to dissociation and ionization. Since ε may be calculated from the equilibrium constants, the compressibility Z is readily determined. Figure 289 shows a plot of the compressibility of air versus temperature for various pressures.

When a gas is a mixture of different species (molecules, atoms, ions and electrons) the energy of the gas is given by the sum of the energies of each species present. Thus

$$E = \sum x_i E_i \qquad (1615)$$

where E is the energy per mole of the gas, x_i the mole fraction of each species

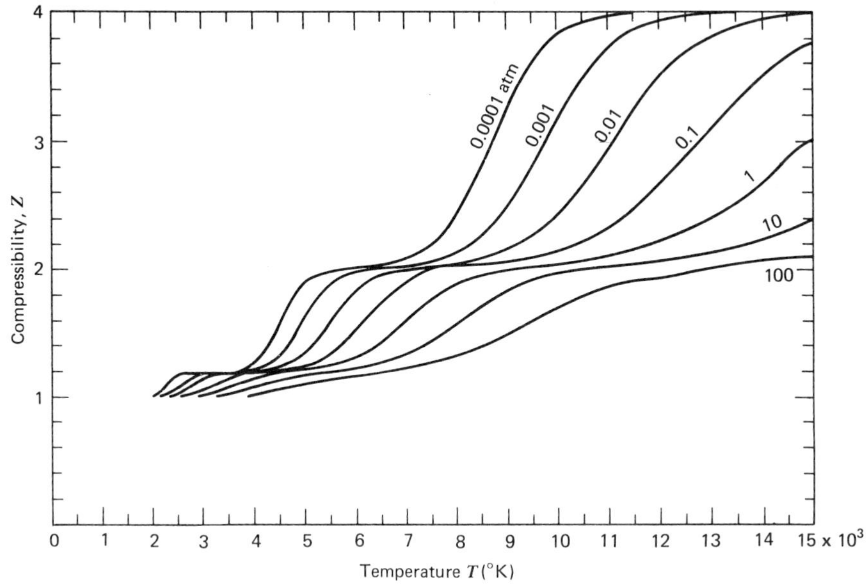

Figure 289 The compressibility of air [167].

present, and E_i the energy per mole of each species present. For each state the mole fractions may be determined from the equilibrium constants and the energy of each species from the partition functions as outlined above. Figure 290 shows the energy of air as a function of temperature for various pressures, expressed in terms of the dimensionless parameter ZE/RT. The enthalpy is then given by

$$H = E + RT \tag{1616}$$

The entropy for a gas containing several species is given by

$$\frac{S}{R} = \sum x_i \frac{S_i}{R} - \sum x_i \ln x_i - \ln \frac{p}{p_0} \tag{1617}$$

where p_0 is an arbitrary reference value, usually taken as 1 atmosphere, and S_i represents the entropy per mole of each species in the gas. S_i for each species may be found from equation (1590). Figure 291 shows the entropy of air as a function of temperature for various pressures expressed in terms of the dimensionless parameter ZS/R.

In finding the specific heats we must keep in mind that the composition of a gas at higher temperatures changes with temperature due to changes in the amount dissociated and ionized. This also changes the average molecular weight of the gas. As a result, the mass of one mole of gas at low temperature will not be the same as the mass of one mole of the gas at higher temperature. If M_0 is the average molecular weight of an undissociated gas and $\bar{M}$ is the average molecular weight at higher temperatures, then

$$Z = \frac{M_0}{\bar{M}} \tag{1618}$$

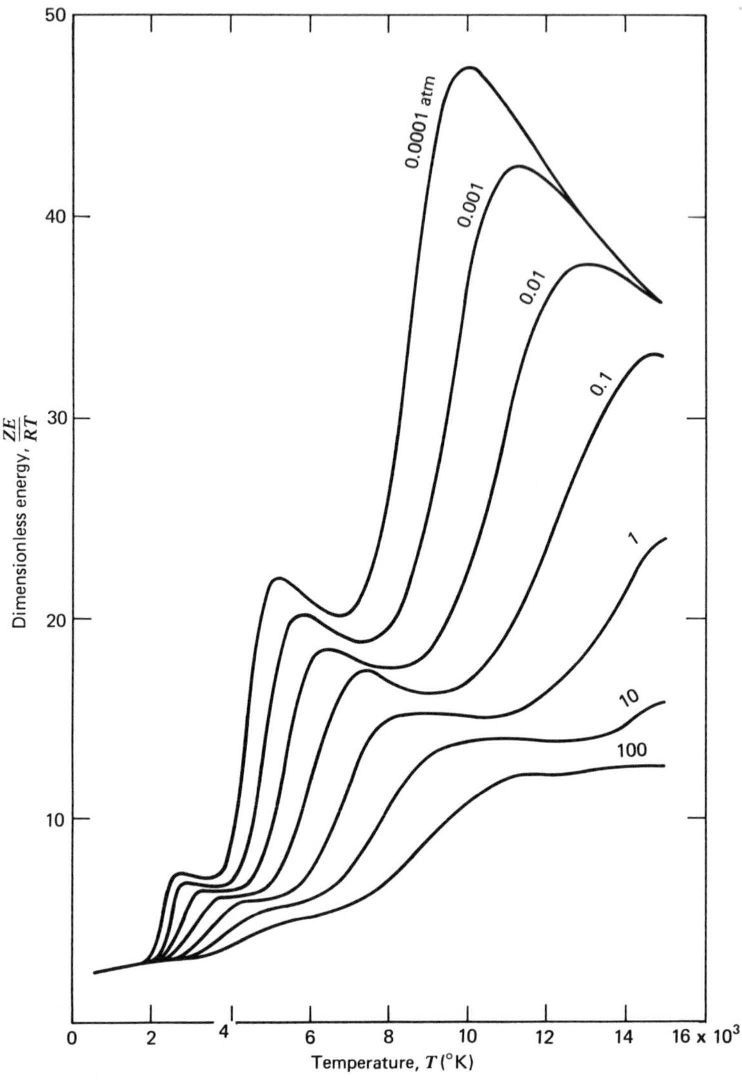

Figure 290 Internal energy of air [167].

The energy per mole of undissociated gas is then given by

$$ZE = Z\sum x_i E_i \tag{1619}$$

Hence, the specific heat per mole of undissociated gas is given by

$$ZC_v = \left(\frac{\partial ZE}{\partial T}\right)_v = \frac{\partial}{\partial T}(Z\sum x_i E_i)_v \tag{1620}$$

$$= Z\sum x_i \left(\frac{\partial E_i}{\partial T}\right)_v + \sum E_i \left(\frac{\partial}{\partial T}(x_i Z)\right)_v \tag{1621}$$

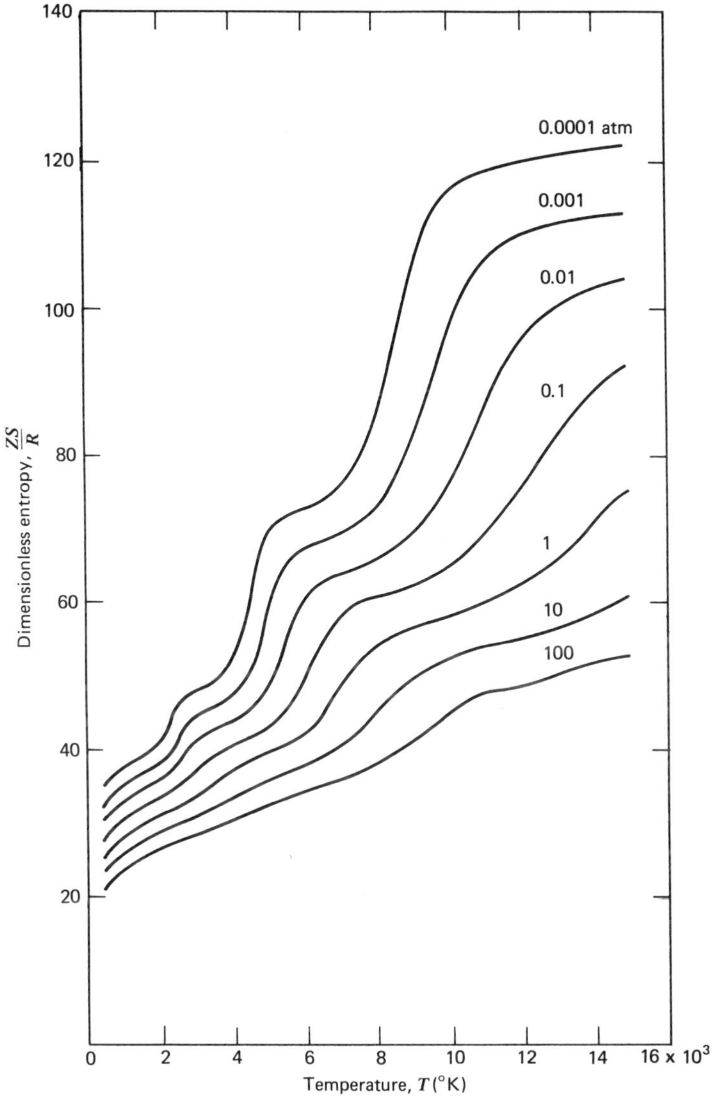

Figure 291 Entropy of air [167].

$$= Z\sum x_i Cv_i + \sum E_i\left(\frac{\partial}{\partial T}(x_i Z)\right)_v \tag{1622}$$

where C_v is the specific heat of the gas per mole at the given temperature and pressure, E_i the energy of each species present per mole of that species, x_i the mole fraction of the ith species per mole of that species. The second term of equation (1622) thus takes into account the change in composition of the gas when the temperature changes. Sometimes, as when dealing with frozen flow, it is more convenient to deal with the partial, or frozen specific heat, C'_v, defined by

$$C'_v = \sum x_i C_{v_i} \tag{1623}$$

In contrast, we may refer to the specific heat of equation (1622) as the equilibrium specific heat. The equilibrium specific heat is shown in Figure 292 plotted versus temperature for various pressures.

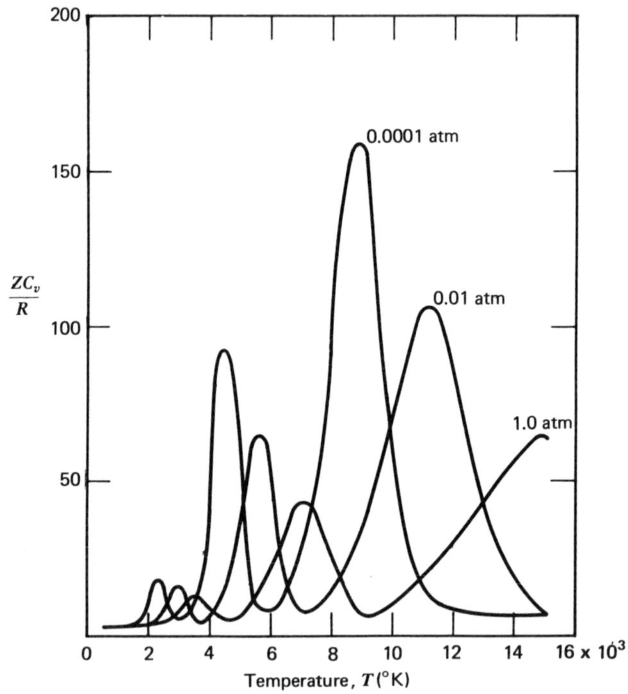

Figure 292 Specific heat at constant volume of air [167].

Similarly, the equilibrium specific heat at constant pressure per mole of undissociated gas is given by

$$ZC_p = Z\sum x_i(C_{v_i} + R) + \sum(E_i + RT)\left(\frac{\partial}{\partial T}(x_i Z)\right)_p \tag{1624}$$

The ratio of specific heats, γ, may then be obtained by dividing equation (1622) into equation (1624). Figure 293 shows a plot of γ for air versus temperature for various pressures.

It should be emphasized that the above theoretical treatment assumes that each species of a gas mixture behaves like a perfect gas. That is

$$p_i = \frac{\rho_i RT}{M_i}, \qquad E_i = E_i(T) \tag{1625}$$

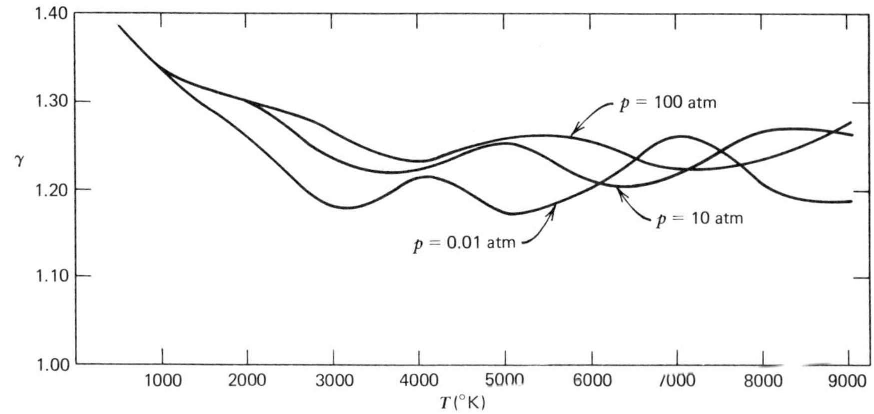

Figure 293 Ratio of specific heats of air [167].

This assumption greatly simplifies the calculations, both in determining the properties of a gas and in solving problems in high temperature gas dynamics. The error involved is small, on the order of a few percent over a range of temperatures from 150°K to 15,000°K and densities from 10^{-6} to 10^2 Amagat for air.

Viscosity

In kinetic theory it is shown that for a gas mixture such as air, the viscosity may be approximated by the formula

$$\mu = \sum \mu_i \qquad \text{where} \qquad \mu_i = \frac{5\pi}{32}\rho_i u_i \lambda_i \qquad \text{so that} \qquad \mu = \frac{5\pi}{32}\sum \rho_i u_i \lambda_i \tag{1626}$$

where ρ_i is the density, u_i the average random velocity, and λ_i the mean free path of the ith species. Let μ_0 be the viscosity of undissociated air at the same temperature. Then

$$\mu_0 = \frac{5}{32}\pi \rho_0 u_0 \lambda_0 \tag{1627}$$

and to a good approximation

$$\mu_0 = 1.462 \times 10^{-5} \frac{\sqrt{T}}{1 + \dfrac{112}{T}} \quad \frac{\text{gm}}{\text{cm sec}} \tag{1628}$$

where T is in degrees Kelvin. Also, assuming the average kinetic energy of each kind of particle present is the same,

$$\frac{u_i}{u_0} = \sqrt{\frac{M_0}{M_i}} \tag{1629}$$

The ratio of the density of the ith species present to the density of the undissociated gas at the same temperature and pressure is given by

$$\frac{\rho_i}{\rho_0} = \frac{M_i}{M_0} x_i \tag{1630}$$

It may also be shown (see, for example, Kennard [171] that the ratio of mean free paths is given by

$$\frac{\lambda_0}{\lambda_i} = \sum_j x_j \frac{S_{ij}}{S_0} \left[\frac{1}{2} \left(1 + \frac{M_i}{M_j} \right) \right]^{1/2} \tag{1631}$$

where S_{ij} represents the collision cross section of particle i when colliding with particle j, and S_0 represents the collision cross section of the undissociated molecules. The viscosity ratio thus becomes

$$\frac{\mu_i}{\mu} = \sqrt{\frac{M_i}{M_0}} x_i \frac{\lambda_i}{\lambda_0} \quad \text{and hence} \quad \frac{\mu}{\mu_0} = \sum_i \sqrt{\frac{M_i}{M_0}} x_i \frac{\lambda_i}{\lambda_0} \tag{1632}$$

The viscosity ratio for air as a function of temperature and pressure is given in Figure 294.

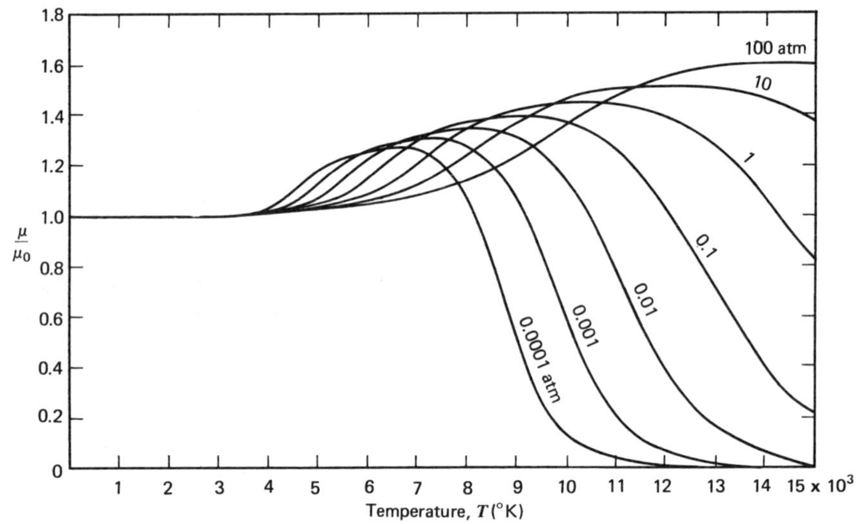

Figure 294 Viscosity of air [167].

Thermal Conductivity

In a dissociating gas, heat conduction occurs by two different mechanisms. The first is the usual mechanism of energy transfer by collision of the molecules and atoms present. The second results from the diffusion of atoms and molecules and their dissociation and recombination. When there is a temperature gradient in a dissociating gas, there will also be a concentration gradient, unless the gas is frozen. The concentration of atoms will increase in the direction of increasing temperature and the concentration of molecules will increase in the direction of decreasing temperature. Thus, atoms will diffuse in the direction of decreasing temperature and molecules will diffuse in the direction of increasing temperature. When the atoms arrive in a region of lower temperature, some of them will recombine, giving off heat, while when the molecules arrive at a region of higher temperature some of them will dissociate, absorbing heat. The net effect is to provide a powerful mechanism of heat transfer, which may exceed that due to mere collision. Notice, however, that the direction of heat transfer due to both mechanisms is the same—from higher to lower temperatures. Also, the rate of heat transfer due to both mechanisms will increase with the temperature gradient. We therefore define the thermal conductivity k as

$$k = k_n + k_r \tag{1633}$$

where k_n is the thermal conductivity due to collision and k_r is the thermal conductivity due to dissociation. In kinetic theory, it is shown that

$$k_n = \frac{5\pi}{32} \sum_i \rho_i u_i \frac{\lambda_i}{M_i}\left(\frac{5}{2}C_t + C_{int}\right)_i \tag{1634}$$

where

$$C_t = \frac{dE_t}{dT}, \qquad C_{int} = \frac{dE_{int}}{dT}, \qquad E = E_t + E_{int} \tag{1635}$$

E_t refers to the translational kinetic energy of the species and E_{int} refers to the other types of energy of the species (rotational, vibrational and electron). Total derivatives are used in (1635) because the formula (1634) is based on the assumption that each species behaves like an ideal gas. In this case

$$C_i = (C_t + C_{int})_i \qquad \text{and} \qquad C_t = \tfrac{3}{2}R \tag{1636}$$

from which it follows that

$$k_n = \frac{5}{32}\pi \sum_i \rho_i u_i \lambda_i \left(\frac{C_i}{M_i} + \frac{9}{4}\frac{R}{M_i}\right) \tag{1637}$$

The terms outside the parenthesis in (1637) will be observed to coincide with the expression for the viscosity, (1626). We define k_0 to be the thermal conductivity of a gas with translational and rotational energy only. Then from kinetic theory

$$C_{int} = C_r = R$$

and

$$k_0 = \frac{\mu_0}{M_0}\left(\frac{5}{2}C_t + C_r\right) = \frac{\mu_0}{M_0}\left(\frac{5}{2}\frac{3}{2}R + R\right) \tag{1638}$$

or

$$k_0 = \mu_0 \frac{19}{4}\frac{R}{M_0} \tag{1639}$$

Parenthetically, it might be observed that experimental values for oxygen and nitrogen are

$$\left(\frac{k}{\mu c_v}\right)_{N_2} = 1.91 \qquad \left(\frac{k}{\mu c_v}\right)_{O_2} = 1.92 \tag{1640}$$

at zero degrees C, as compared to a value of 1.90 for equation (1639). From equations (1637) and (1639) it follows that

$$\frac{k_n}{k_0} = \sum_i \left[\sqrt{\frac{M_i}{M_0}}\, x_i \frac{\lambda_i}{\lambda_0}\right] \frac{M_0}{M_i}\left[\frac{4}{19}\frac{C_i}{R} + \frac{9}{19}\right] \tag{1641}$$

From equation (1632), this is equivalent to

$$\frac{k_n}{k_0} = \sum_i \frac{\mu_i}{\mu_0}\frac{M_0}{M_i}\left[\frac{4}{19}\frac{C_i}{R} + \frac{9}{19}\right] \tag{1642}$$

The heat transfer coefficient due to diffusion, dissociation and recombination, k_r, has been worked out, for an arbitrary mixture of species with an arbitrary number of chemical reactions, by Hirschfelder [172] and Butler and Brokaw [173]. The mole flux due to diffusion of species i, g_i, in a mixture, is given by

$$g_i = -\frac{n^2}{\rho}\sum_j D_{ij} M_j \frac{dx_j}{dz} \tag{1643}$$

where

n = number of moles per unit volume
ρ = density of the mixture
D_{ij} = multicomponent diffusion coefficient
M_j = molecular weight of jth species
x_i = mole fraction of ith species
dx_j/dz = concentration gradient of jth species

and g_i is in moles per second per unit area. Now

$$\frac{dx_i}{dz} = \frac{dx_i}{dT}\frac{dT}{dz} \tag{1644}$$

where dx_i/dT is known if equilibrium is assumed and dT/dz is the temperature gradient. The thermal conductivity due to reaction then becomes

$$k_r = \frac{n^2}{\rho}\sum_{i,j} h_i D_{ij} M_j \frac{dx_j}{dT} \tag{1645}$$

where h_i is the enthalpy of the ith species. Evaluating k_r from equation (1645) requires quite lengthy computations if more than one reaction is going on at the same time. Even if only one reaction is going on, Butler and Brokaw have shown that

$$k_r = \frac{\Delta H^2}{RT^2}\left[\sum_{j=1}^{\mu-1}\sum_{m=j+1}^{\mu} \Delta_{mj}\frac{(n_j x_m - n_m x_j)^2}{x_m x_j}\right]^{-1} \tag{1646}$$

where ΔH is the heat of reaction, μ the number of species present, x_m, x_j the mole fractions. The n_j, n_m are the stoichiometric coefficients in the ongoing reaction written in the form

$$X' = \sum_{j=2}^{\mu} n_j X^j \tag{1647}$$

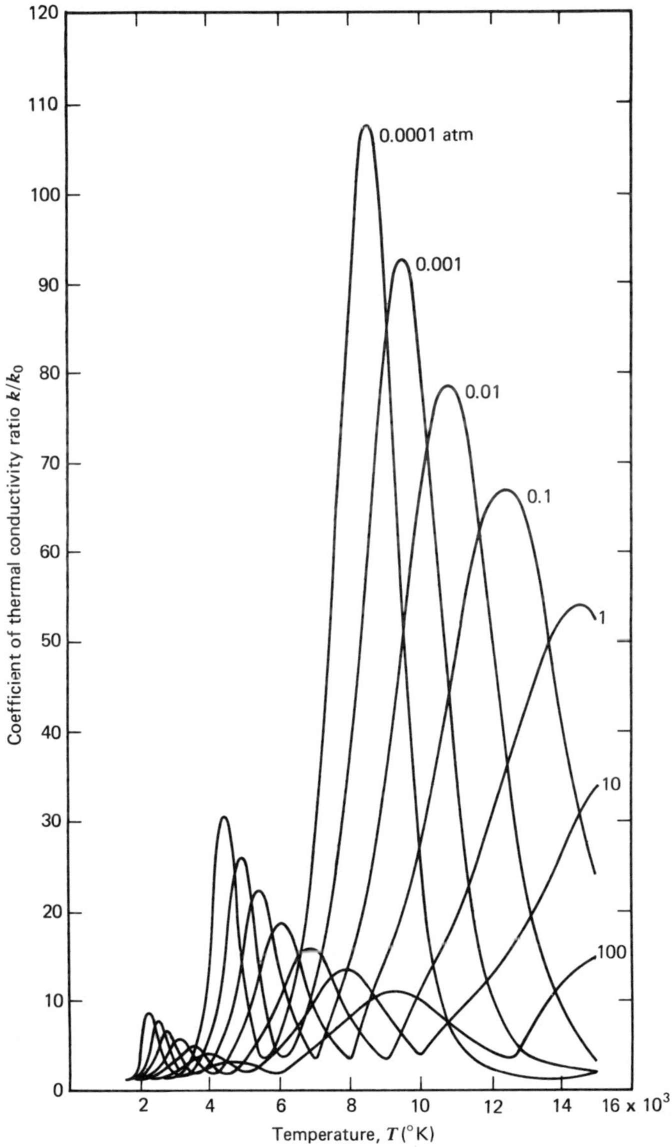

Figure 295 Thermal conductivity of air [167].

Δ_{mj} is defined by

$$\Delta_{m_j} = \frac{RT}{D_{m_j} p} \tag{1648}$$

where D_{mj} is the binary diffusion coefficient. For a simple dissociation reaction, $A \rightleftharpoons nB$, equation (1646) reduces to

$$k_r = \frac{D_{AB} p}{RT} \frac{\Delta H^2}{RT^2} \frac{x_A x_B}{n x_A + x_B} \tag{1649}$$

In dissociating air there are generally two reactions going on simultaneously,

$$N_2 \rightleftharpoons 2N \qquad \text{and} \qquad O_2 \rightleftharpoons 2O \tag{1650}$$

Because oxygen and nitrogen are similar in size and weight, air may be approximated by "air molecules" representing an average of N_2 and O_2 and "air atoms" representing an average of N and O, and an approximate reaction thermal conductivity may be found on this basis. The binary diffusion coefficient is given by

$$\frac{1}{nD_{ij}} = \frac{8}{3}\sqrt{\frac{2}{\pi}}\left(\frac{M_i M_j}{M_i + M_j}\right)^{1/2} \frac{N_0}{\sqrt{RT}} S_{ij} \tag{1651}$$

where n is the number of moles per unit volume, N_0 is Avagadro's number and S_{ij} the collision cross section of particles i when colliding with particles j. Equation (1651) shows once again the importance of knowing the collision cross section when calculating transport properties. Collision cross sections are known with some confidence up to about 10,000°K but are still subject to some uncertainty above that level.

The total thermal conductivity of air has been calculated by Hansen [167] (Figure 295). Hansen neglected the difference between mass and size of oxygen and nitrogen atoms and molecules, as outlined above, but included the effects of ionization at higher temperatures. The results of Hansen are considered reliable to 10,000°K, but open to some question above that level due to uncertainties in the collision cross sections.

Dissociating Laminar Boundary Layers

We consider now the problem of a boundary layer of air in which the temperatures and temperature gradients are sufficiently high so that appreciable dissociation and concentration gradients across the boundary layer are present. If only two species are present, the mass flux due to diffusion is given by

$$m_i = \rho_i V_i = \rho c_i V_i = -\rho c_i \frac{D_{12}}{c_i} \frac{\partial c_i}{\partial y}, \qquad i = 1, 2 \tag{1652}$$

where c_i is the concentration (mass fraction) of the ith species, V_i the diffusion velocity, ρ the total density of the gas, ρ_i is the density of the ith species, D_{12}

the binary diffusion coefficient, y the direction across the boundary layer and n_i the mass flux of the ith species in the y direction in mass per unit area per unit time. Equation (1652) is sometimes called Fick's law and simply says that the mass flux due to diffusion is proportional to the concentration gradient. If more than two species are present the situation is not so simple and we must turn to equation (1643) to get the mass flux. The multicomponent diffusion coefficients, D_{ij}, are, in general, quite sensitive to the composition of the mixture. As Penner and Lees [174] have pointed out, however, in a mixture such as air consisting chiefly of molecules and atoms roughly similar in size and weight (oxygen and nitrogen) the multicomponent diffusion equation may be simplified by using an equivalent or average binary diffusion coefficient, D_{12}, which may be found, for example, by putting average values in equation (1651). One thus defines D_{12} on the basis of "air molecules" diffusing into "air atoms" and vice versa. The mass flow rate per unit area of the ith species due to diffusion then becomes

$$\dot{m}_i = \rho c_i V_i = -\rho c_i \frac{D_{12}}{c_i} \frac{\partial c_i}{\partial y} \tag{1653}$$

where c_i is the concentration (mass fraction) of the ith species, and $i = 1, 2$ (1 for air molecules, 2 for air atoms). The heat transfer rate in the y direction then becomes

$$\dot{q} = -k_n \frac{\partial T}{\partial y} + \sum_i \rho c_i V_i h_i = -\left[k_n \frac{\partial T}{\partial y} + \rho D_{12} \sum_i h_i \frac{\partial c_i}{\partial y} \right] \tag{1654}$$

k_n is sometimes called the partial, or frozen, thermal conductivity. h_i is the enthalpy of the ith species, given by

$$h_i = \int_0^T c'_{p_i} + h_i^0 \tag{1655}$$

where h_i^0 is the enthalpy of formation of the ith species at 0°K. By definition, h_i^0 is zero for atoms and hence has a nonzero value only for the air molecules. For the enthalpy h of the mixture we may write

$$dh = \bar{c}'_p \, dT + \sum_i h_i \, dc_i \tag{1656}$$

where $\bar{c}'_p = \sum c_i c'_{p_i}$. Notice that the perfect gas assumption $c'_{p_i} = c'_{p_i}(T)$ is made for each species. As pointed out earlier, the error involved is small in most practical cases. The heat transfer rate becomes

$$\dot{q} = \frac{k_n}{c'_p} \left[\left(\frac{\partial h}{\partial y} - \sum_i h_i \frac{\partial c_i}{\partial y} \right) + \frac{\rho D_{12} \bar{c}'_p}{k_n} \sum_i h_i \frac{\partial c_i}{\partial y} \right] \tag{1657}$$

The quantity $\rho D_{12} \bar{c}'_p / k_n$ is known as the Lewis number (sometimes called the Lewis-Semenov number) and will be designated by Le. Thus

$$Le = \frac{\rho D_{12} \bar{c}'_p}{k_n}$$

where the $\bar{c}'_p$ is the partial, or frozen, specific heat as defined earlier. Equation (1657) may thus be rewritten

$$\dot{q} = \frac{k_n}{\bar{c}'_p}\left[\frac{\partial h}{\partial y} + (Le - 1)\sum_i h_i \frac{\partial c_i}{\partial y}\right] \tag{1658}$$

It is seen that when the Lewis number equals 1, the heat transfer is proportional to the enthalpy gradient.

In terms of heat transfer to the wall, we may now envisage three limiting cases.

1. The boundary layer is everywhere in equilibrium.
2. The boundary layer is everywhere frozen.
3. The boundary layer is frozen everywhere except at the wall where all atoms striking the wall recombine to form molecules, giving up the heat thus released to the wall.

This last case is known as catalytic recombination and in this case the wall is said to be catalytic. In case 2 on the other hand, the wall is said to be noncatalytic. Whether a wall is catalytic or not depends on the material of which the wall is made. In practice, most materials are catalytic. There are, however, some known materials which are noncatalytic or at least not entirely catalytic. Among these are certain glassy materials. Whether the boundary layer is frozen or in equilibrium will depend on the net rate at which atoms recombine to form molecules relative to the rate at which they diffuse through the boundary layer. Of course in practice, boundary layers will be neither frozen nor in equilibrium but somewhere in between. Because of the complexity of analyzing the "in between" cases, the assumption of frozen or equilibrium boundary layer is usually made. This is not as restrictive as it sounds because in practice most boundary layers are essentially in equilibrium except at low densities. Moreover, it turns out that when the Lewis number is close to one the heat transfer is about the same whether cases 1 or 3 are assumed. In other words, the heat transfer to the wall will be about the same whether atoms recombine in the boundary layer, giving off heat which is then transferred through the boundary layer to the wall by conduction, or whether the atoms drift all the way to the wall by diffusion and then give up their heat to the wall by catalytic recombination at the wall. Furthermore, the Lewis number in air is, in fact, close to 1, varying roughly between 1.4 and 0.6 over a wide range of temperatures and pressures (Figure 296).

Derivation of the Governing Equations

We now proceed to derive the boundary layer equations for the case of a dissociating gas. Consider the general case in which the gas may not be in equilibrium. In this case an additional equation will be needed to describe the composition of the gas at each point. This equation is the continuity of species equation. For each species we have

$$\nabla(\rho(\mathbf{V} + \mathbf{V}_i)c_i) = w_i \tag{1659}$$

where $\mathbf{V}$ is the average velocity of the gas, $\mathbf{V}_i$ the diffusion velocity, c_i the concentration (mass fraction) and w_i the rate of formation of the ith species. As

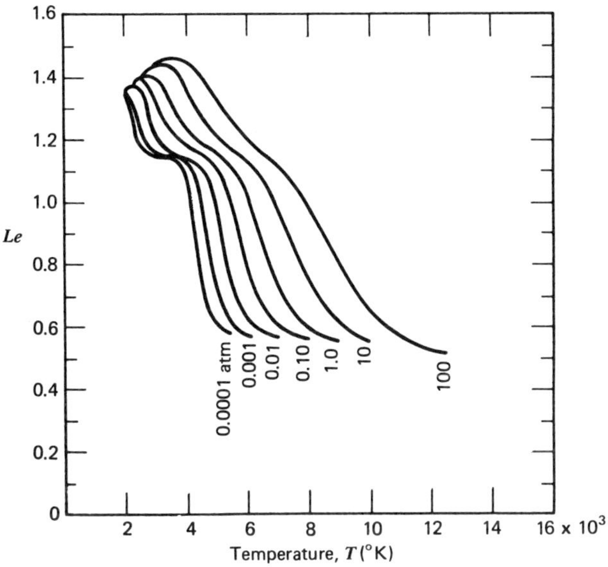

Figure 296 Lewis number of air [167].

a rough approximation we may say

$$\mathbf{V}_i = -\frac{D_i}{c_i}\nabla c_i \tag{1660}$$

For a multicomponent mixture, D_i will be a function of the other concentration gradients, but for a binary mixture, $D_i = D_{12}$, the binary diffusion coefficient, and equation (1660) is Fick's law. Conservation of mass requires that

$$\sum_i w_i = 0 \tag{1661}$$

and

$$\sum_i \mathbf{V}_i c_i = 0 \tag{1662}$$

It follows that for steady average flow,

$$\nabla(\rho\mathbf{V}) = 0 \tag{1663}$$

The overall continuity equation is thus unchanged. In the boundary layer, changes in the y direction are large compared to changes in the x direction. From (1659), (1660) and (1663) we have then,

$$\rho u\frac{\partial c_i}{\partial x} + \rho v\frac{\partial c_i}{\partial y} - \frac{\partial}{\partial y}\left(\rho D_i\frac{\partial c_i}{\partial y}\right) = w_i \tag{1664}$$

This is the continuity of species equation for the boundary layer. The momentum equation is unchanged:

$$\rho u\frac{\partial u}{\partial x} + \rho v\frac{\partial u}{\partial y} = -\frac{dp}{dx} + \frac{\partial}{\partial y}\left(\mu\frac{\partial u}{\partial y}\right) \tag{1665}$$

The energy equation may be found by adding the diffusive heat transfer effects from equation (1635) to the energy equation in the form (1194). There results

$$\rho\left(u\frac{\partial h_s}{\partial x}+v\frac{\partial h_s}{\partial y}\right)=\frac{\partial}{\partial y}\left(\frac{\mu}{Pr}\frac{\partial h_s}{\partial y}\right)+\frac{\partial}{\partial y}\left\{\mu\left(1-\frac{1}{Pr}\right)\frac{\partial}{\partial y}\frac{u^2}{2}\right\}$$
$$+\frac{\partial}{\partial y}\left\{\rho D_{12}\left(1-\frac{1}{Le}\right)\sum_i h_i\frac{\partial c_i}{\partial y}\right\} \quad (1666)$$

This form of the equation brings out clearly the effect of assuming that the Prandtl and Lewis numbers are 1. In air, both the Prandtl and the Lewis number differ from one by about the same order of magnitude, and the errors involved in assuming the Lewis number equal to one are about the same order of magnitude as those involved in assuming the Prandtl number equal to one.

It remains only to write the equations of state. As we did in determining the thermodynamic and transport properties of air we will assume the perfect gas equation for each species. Then

$$p_i=\rho_i R_i T \quad (1667)$$

where p_i is the partial pressure, ρ_i the density and R_i the gas constant per unit mass of the ith species. Assuming Dalton's law we have

$$p=\rho R_m T \qquad R_m=\sum_i R_i c_i \quad (1668)$$

The system of equations (1663)–(1668) may be applied equally to plane, two-dimensional, or axially symmetric flow, provided the boundary layer thickness is small compared to the radius of curvature. The only difference occurs in writing out the overall, sometimes called global, continuity equation

$$\frac{\partial(\rho u r^\delta)}{\partial x}+\frac{\partial(\rho v r^\delta)}{\partial y}=0 \quad (1669)$$

where r represents the local radius of the body in axially symmetric flow, $\delta=1$ for axially symmetric flow and $\delta=0$ for plane flow.

Solution of the Governing Equations

We now seek a solution to the system of equations (1663)–(1669). Our strategy will be to seek similarity solutions and to determine under what conditions these solutions will be valid. We begin by introducing the following transformations:

$$\eta=\frac{r^\delta u_e}{\sqrt{2\xi}}\int_0^y \rho\,dy, \qquad \xi=\int_0^x \rho_w\mu_w u_e r^{2\delta}\,dx \quad (1670)$$

These transformations will be observed to include the transformations of Howarth, Illingworth and Mangler and were first proposed by Lees [175]. In addition, let

$$\frac{u}{u_e}=\frac{\partial f}{\partial \eta}, \qquad g=\frac{h_s}{h_{se}}, \qquad s_i=\frac{c_i}{c_{ie}} \quad (1671)$$

Then ρv may be found from the continuity equation

$$\rho v = \frac{1}{r}[(2\xi f_\xi + f)\xi_x/\sqrt{2\xi} + \sqrt{2\xi} f_\eta \eta_x] \tag{1672}$$

where the subscripts indicate partial differentiation. The continuity of species equation then becomes

$$\frac{\chi}{Pr}(Le\, s_{i\eta})_\eta + f\, s_{i\eta} + \frac{2\xi}{\rho u_e}\frac{dx}{d\xi}\frac{w_i}{c_{ie}} = 2\xi(f_\eta s_{i\xi} - f_\xi s_{i\eta}) + 2 f_\eta s_i \frac{d \ln c_{ie}}{d \ln \xi} \tag{1673}$$

where Pr is the frozen, or partial, Prandtl number

$$Pr = \frac{\bar{c}'_p \mu}{k_n} \tag{1674}$$

and Le is the frozen Lewis number

$$Le = \frac{D_{12}\rho \bar{c}'_p}{k_n} \tag{1675}$$

For a multicomponent mixture we would have Le_i instead of Le where

$$Le_i = \frac{D_i \rho \bar{c}'_p}{k_n} \tag{1676}$$

The use of a single diffusion coefficient D_{12} anticipates our approximation of air as a binary mixture of air atoms and air molecules. Also,

$$\chi = \frac{\rho\mu}{\rho_w \mu_w} \tag{1677}$$

The momentum equation becomes

$$(\chi f_{\eta\eta})_\eta + f f_{\eta\eta} + 2\frac{d \ln u_e}{d \ln \xi}\left[\frac{\rho_e}{\rho} - f_\eta^2\right] = 2\xi(f_\eta f_{\eta\xi} - f_\xi f_{\eta\eta}) \tag{1678}$$

The energy equation becomes

$$\left(\frac{\chi}{Pr} g_\eta\right)_\eta + f g_\eta + \frac{u_e^2}{h_{se}}\left[\left(1 - \frac{1}{Pr}\right)\chi f_\eta f_{\eta\eta}\right]_\eta + \left[\frac{\chi}{Pr}\sum_i \frac{c_{ie}}{h_{se}} h_i (Le - 1) s_{i\eta}\right]_\eta$$

$$= 2\xi(f_\eta g_\xi - f_\xi g_\eta) \tag{1679}$$

where

$$h_i = \int_0^T c'_{p_i}\, dT + h_i^0 \tag{1680}$$

For similarity in η to exist, then,

$$\frac{\chi}{pr}, \quad Le, \quad \chi, \quad \frac{\rho_e}{\rho}, \quad h_i, \quad \frac{w_i}{\rho}$$

must be functions of η only, and

$$\frac{2\xi dx}{c_{ie} d\xi}, \quad \frac{d \ln c_{ie}}{d \ln \xi}, \quad \frac{d \ln u_e}{d \ln \xi}, \quad \frac{u_e^2}{h_{se}} \quad \frac{c_{ie}}{h_{se}}$$

must be constants. If the boundary layer is frozen or in equilibrium, the continuity of species equation becomes unnecessary. The requirement for similarity then reduces to

$$\chi, \quad \frac{\rho_e}{\rho}, \quad pr, \quad h_i, \quad Le$$

function of η only and

$$\frac{d \ln u_e}{d \ln \xi}, \quad \frac{u_e^2}{h_{se}}, \quad \frac{c_{ie}}{h_{se}}$$

constant. These conditions may be satisfied in four practical cases.

1. Flow near the forward stagnation point of a blunt nosed body.
2. Constant conditions at the outer edge of the boundary layer (flat plate).
3. Hypersonic solutions ($u_e^2/h_{se} \cong 2$).
4. Slowly varying conditions at the outer edge of the boundary layer (local similarity).

Near the forward stagnation point, u_e may be expanded as a power series in x and

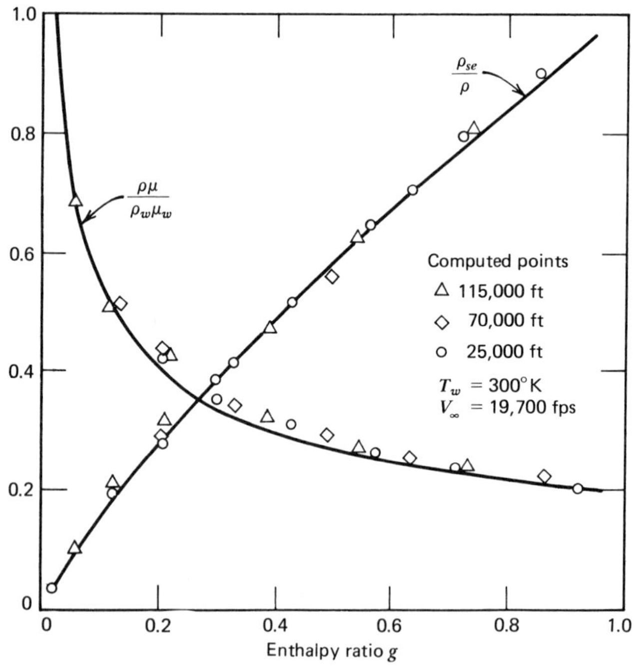

Figure 297 $\rho\mu$ and density variation with enthalpy in the boundary layer [176].

only the first term retained. The state variables T_e, p_e, ρ_e vary quite slowly near the stagnation point and may be taken as constant. u_e^2 will be much less than h_{se} and h will be close to h_s. Under these conditions, Pr, Le, χ and ρ_e/ρ may all be taken as functions of g only which in turn is a function of η only for similarity. A plot of χ and ρ_e/ρ versus g near the stagnation point is shown in Figure 297 for equilibrium conditions. Under these conditions, and assuming equilibrium, h_i and c_{ie}/h_{se} may also be taken approximately as functions of η only. The conditions for similarity are thus satisfied. The governing equations then reduce to

$$(\chi f_{\eta\eta})_\eta + f_\eta f_{\eta\eta} + \beta\left(\frac{\rho_e}{\rho} - f_\eta^2\right) = 0 \tag{1681}$$

$$\left(\frac{\chi}{Pr} g_\eta\right)_\eta + f g_\eta + \left[\frac{\chi}{Pr}\sum_i \frac{c_{ie}}{h_{ie}} h_i (Le - 1) s_{i\eta}\right]_\eta = 0 \tag{1682}$$

where

$$\beta = 2\frac{d \ln u_e}{d \ln \xi} \tag{1683}$$

Near the stagnation point we may say

$$u_e = x\left(\frac{du_e}{dx}\right)_{x=0} \qquad r \cong x \qquad \xi \sim \frac{x^{2(1+\delta)}}{2(1+\delta)}$$

so that $\beta = 1/(1 + \delta)$. Equation (1682) may be rewritten

$$\left[\frac{\chi}{Pr}(1 + d) g_\eta\right]_\eta + f g_\eta = 0 \tag{1684}$$

where

$$d = (Le - 1)\sum_i h_i \left(\frac{\partial c_i}{\partial h}\right)_p$$

the subscript p denoting the partial derivative at constant pressure. The boundary conditions are

$$\begin{aligned} f(0) = f_\eta(0) = 0, &\qquad f_\eta(\infty) = 1 \\ g(0) = g_w, &\qquad g(\infty) = 1 \end{aligned} \tag{1685}$$

Equations (1681) and (1684), subject to the boundary conditions (1685) were solved numerically by Fay and Riddell [176]. Notice that χ, Pr, Le, ρ_e/ρ, and d may be evaluated at each point once g is known there by using a table of the equilibrium properties of air such as that of Hansen [167]. Fay and Riddell assumed Le and Pr constant across the boundary layer. Their solution for $g(\eta)$ is shown in Figure 298. The rate of heat transfer at the wall is given by

$$q_w = \left(k_n \frac{\partial T}{\partial y}\right)_{y=0} + \left[\sum_i \rho h_i D_{12} \frac{\partial c_i}{\partial y}\right]_{y=0} \tag{1686}$$

The second term is the heat released by the atoms recombining upon hitting

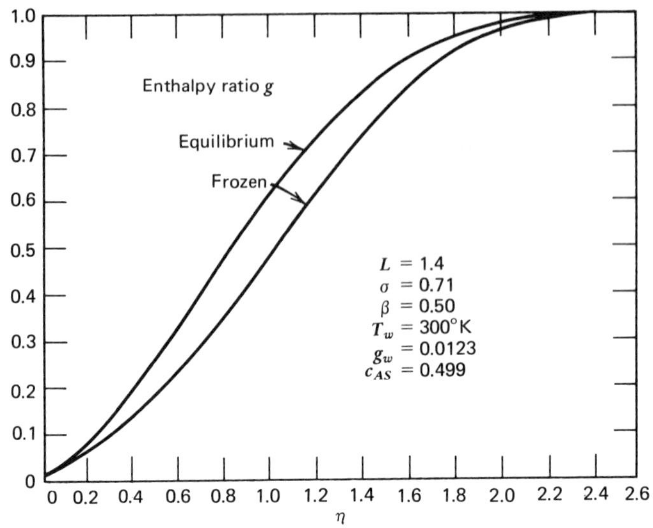

Figure 298 Enthalpy variation in the boundary layer—solution of Fay and Riddell[176].

the wall. This term is thus relevant only for the case of a frozen boundary layer with a cold catalytic wall. Generally speaking, even at a catalytic wall, only the equilibrium condition will be reached. Thus, not all the atoms hitting the wall will recombine. Material considerations dictate, however, that a wall will generally be cool enough to make equilibrium dissociation negligible there. For the same reason, hardly any atoms will reach the wall in an equilibrium boundary layer. In terms of the transformed variables,

$$q_w = \frac{r^\delta k_{nw} \rho_w u_e h_{se}}{\sqrt{2\xi \bar{c}'_p}} \left[g_\eta + \sum c_{ie} \frac{h_i}{h_{se}} (Le - 1) s_{i\eta} \right]_{\eta = 0} \tag{1687}$$

Substituting for $g(\eta)$ it is found that a good curve fit for the heat transfer is given by

$$q_w = \frac{0.76}{Pr^{0.6}} (\rho\mu)_w^{0.1} (\rho\mu)_{0e}^{0.4} \left[1 + (Le^{0.52} - 1) \frac{h_{De}}{h_{se}} \right] (h_{se} - h_w) \sqrt{\left(\frac{du_e}{dx} \right)_{x=0}} \tag{1688}$$

Let Δh_j be the difference in enthalpy between element j molecules and element j atoms. For air, $j = 1, 2$, where 1 refers to oxygen and 2 refers to nitrogen. Take Δh_j positive. Let c_j be the mass fraction of atoms of element j present. Then

$$h_D = \sum_i \Delta h_j c_j \tag{1689}$$

The accuracy of the formula (1688) has been tested by comparison with the experimental shock tube results of Rose and Stark [177]. (Figure 299). Considering the scatter of the experimental data, the agreement can only be described as remarkable.

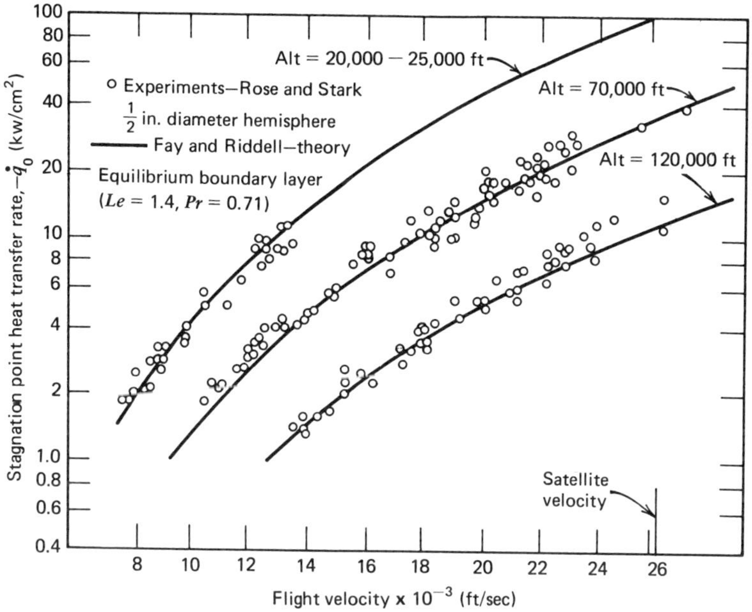

Figure 299 Stagnation point heat transfer—theory of Fay and Riddell compared to experiments of Rose and Stark [177].

Frozen Case. For the frozen case with catalytic recombination at the wall, the continuity of species equation must be solved along with the other equations. Since the boundary layer is frozen, $w_i = 0$. The boundary condition at the wall is taken as the equilibrium condition at the wall temperature and pressure and the boundary condition at the outer edge of the boundary layer is taken as the equilibrium condition there. The continuity of species equation in this case is

$$\frac{\chi}{Pr}(Le\, s_{i\eta})_\eta + f s_{i\eta} = 0 \tag{1690}$$

Numerical solutions for the frozen case with catalytic recombination at the wall were also found by Fay and Riddell. The results for the heat transfer rate in this case may be correlated by the formula

$$q_w = \frac{0.763}{Pr^{0.6}}(\rho\mu)_e^{0.4}(\rho\mu)_w^{0.1}\sqrt{\left(\frac{du_e}{dx}\right)_{x=0}}\,(h_{se} - h_w)\left[1 + (Le^{0.63} - 1)\frac{h_{De}}{h_{se}}\right] \tag{1691}$$

Comparison with equation (1688) shows the difference to be small.

For the case of frozen boundary layer with noncatalytic wall, on the other hand, it is found that the heat transfer rate is much reduced. In this case no recombination occurs anywhere in the boundary layer and hence no concentration gradients exist. The concentration of each species is everywhere the same as at the outer edge of the boundary layer. The solution thus reduces to that of a nonreacting gas having the given composition. As an example, for the case $Le = 1.4, Pr = 0.71, \beta = 0.5, T_w = 300°K$ it turns out that

$$\frac{[Nu/\sqrt{Re}]_{\text{catalytic wall}}}{[Nu/\sqrt{Re}]_{\text{noncatalytic wall}}} = 2.19 \tag{1692}$$

where Nu is the Nusselt number, defined by

$$Nu = \frac{q_w x \bar{c}'_{pw}}{k_{nw}(h_{se} - h_w)} \tag{1693}$$

and Re is the Reynolds number defined by

$$Re = \frac{u_e x}{\nu_w} \tag{1694}$$

x being the distance from the stagnation point along the wall.

Finite recombination rates

The case of finite recombination rates may also be solved. In this case, an expression must be developed for w_i, the mass rate of formation of the ith species per unit volume, for use in integrating the continuity of species equation (1664). Evidence indicates that recombination takes place as a result of three body rather than two body collisions. Thus,

atom + atom + particle → molecule + particle

where the particle may be either another atom or a molecule. If N_A is the number of moles of atoms per unit volume and N is the total number of moles per unit volume present, the recombination rate may be written

$$\frac{dN_A}{dt} = -K_1 N_A^2 N T^{-1.5} \tag{1695}$$

The value of K_1 is somewhat uncertain but has been estimated as $T^{1.5} \times 5 \times 10^{14}$ cc^2/mole^2 sec for oxygen. The dissociation rate may be written

$$\frac{dN_A}{dt} = K_2(T) N_M N \tag{1696}$$

where N_m is the number of moles of molecules per unit volume. The net rate of change of the concentration of atoms is, therefore, given by

$$\left(\frac{dN_A}{dt}\right)_{net} = K_2(T) N_M N - K_1 N_A^2 N T^{-1.5} \tag{1697}$$

Since $(dN_A/dt)_{net} = 0$ at equilibrium, it follows that

$$K_2(T) = \frac{K_1 N_{AE}^2 N T^{-1.5}}{N_{ME} N} \tag{1698}$$

where the subscript E indicates the equilibrium concentration of atoms and molecules at the given temperature. Thus,

$$\left(\frac{dN_A}{dt}\right)_{net} = -K_1 N_A^2 N T^{-1.5}\left[1 - \frac{N_M}{N_{ME}} \cdot \frac{N_{AE}^2}{N_A^2}\right] \tag{1699}$$

For a mixture of molecules of molecular weight MW and atoms of weight $MW/2$, the concentration of atoms is given by

$$c_A = \frac{N_A}{N_A + 2N_M} \tag{1700}$$

and the density is given by

$$\rho = MW\left(N_M + \frac{N_A}{2}\right) \tag{1701}$$

The net mass rate of formation of atoms per unit volume is, therefore, given by

$$w_A = \frac{MW}{2}\frac{dN_A}{dt} = -\frac{2K_1\rho^3 T^{-1.5}}{MW^2}\left[\frac{(1+c_A)(c_A^2 - c_{AE}^2)}{1 - c_{AE}^2}\right] \tag{1702}$$

For similarity with finite recombination rates, the continuity of species equation becomes

$$\left[\frac{\chi}{Pr} Le\, s_{i\eta}\right]_\eta + f s_{i\eta} + \frac{w_i}{\rho c_{ise} 2\left(\frac{du_e}{dx}\right)_{se}} = 0 \tag{1703}$$

Since $\sum w_i = 0 = w_A + w_M$, it follows that once w_A is known, the solution is again rendered determinate. From (1702) it follows that

$$\frac{w_A}{2\rho\left(\frac{du_e}{dx}\right)_{se}} = -\frac{K_1 p_{se}^2}{T_{se}^{3.5}\,\mathscr{R}^2\left(\frac{du_e}{dx}\right)_s}\left[\left(\frac{T_e}{T}\right)^{3.5}\frac{(c_A^2 - c_{AE}^2)}{(1 + c_A)}\right] \tag{1704}$$

where $\mathscr{R}$ is the universal gas constant and the term $1 = c_{AE}^2$ has been neglected since its variation is small compared to that of the other terms.

We define

$$C_1 \equiv \frac{K_1 p_{se}^2}{T_{se}^{3.5}\,\mathscr{R}^2\left(\frac{du_e}{dx}\right)_s} \tag{1705}$$

and call C_1 the recombination rate parameter. The continuity of species equation (1703), along with the momentum and energy equations, may then be solved for various values of C_1. For $C_1 = 0$ the flow is frozen, while for $C_1 = \infty$ the flow is in equilibrium. For nonequilibrium flow it is more convenient to write the energy equation in terms of the temperature. Let $\theta = T/T_e$ and

$\bar{c} = \bar{c}'_p/\bar{c}'_{pw}$. Then the energy equation becomes

$$\left(\frac{\bar{c}\chi}{Pr}\theta_\eta\right)_\eta + \bar{c}f\theta_\eta + \frac{\chi}{Pr}\theta_\eta \sum_i \frac{c'_{pi}}{\bar{c}'_{pw}} Le\, c_{ise} s_{i\eta} - \frac{1}{2\left(\dfrac{du_e}{dx}\right)_{se}} \sum_i \frac{w_i}{\rho} \frac{h_i}{\bar{c}'_{pw} T_{se}} = 0$$

(1706)

Assuming the gas consists of a mixture of air atoms and air molecules, the third term in equation (1706) may be written

$$\frac{Le\chi}{Pr}\theta_\eta c_{A\eta}\left[\frac{c'_{pA} - c'_{pM}}{\bar{c}'_{pw}}\right] \tag{1707}$$

Using equation (1704), (1705), and (1689), the fourth term may be written

$$C_1 \frac{h_D}{\bar{c}'_{pw} T_{se}}\left[\frac{c_A^2 - c_{AE}^2}{\theta^{3.5}(1 + c_{Ase})}\right] \tag{1708}$$

The frozen specific heats for air atoms and air molecules may be approximated by

$$c'_{pA} = \frac{5\mathcal{R}}{M}, \qquad c'_{pM} = \frac{\mathcal{R}}{M}\left[\frac{7}{2} + e^{-\tau}\right], \qquad \tau = \left(\frac{T_v}{T}\right)^2 \tag{1709}$$

$e^{-\tau}$ represents the vibrational contribution to the specific heat and $T_v \cong 800°$ K for air. Thus

$$\bar{c} = \frac{\bar{c}'_p}{\bar{c}_{pw}} = \frac{10}{7}c_A + \left[1 + \frac{2}{7}e^{-\tau}\right][1 - c_A] \tag{1710}$$

and

$$\frac{c'_{pA} - c'_{pM}}{\bar{c}'_{pw}} = \frac{3}{7} - \frac{2}{7}e^{-\tau} \tag{1711}$$

χ may be approximated by

$$\chi = (1 + c_A)^{-3/2}\frac{\theta_w}{\theta}F(\theta) \tag{1712}$$

where $F(\theta)$ is a curve fit obtained by assuming that the atoms and molecules have the same collision cross section. The equilibrium mass fraction of atoms may be obtained from any suitable table of the equilibrium properties of air, such as Hilsenrath and Beckett [178].

The momentum equation in the form (1681) may be used for computational purposes in the nonequilibrium case with

$$\frac{\rho_{se}}{\rho} = \theta\left[\frac{1 + c_A}{1 + c_{Ase}}\right] \tag{1713}$$

and χ as given by equation (1712).

The continuity of species equation may be written in the form

$$\left(\frac{\chi Le}{Pr} c_{A\eta}\right)_\eta + f c_{A\eta} - C_1 m = 0 \tag{1714}$$

where

$$m = \frac{c_A^2 - c_{AE}^2}{\theta^{3.5}(1 + c_A)}$$

Equations (1681), (1706) and (1714) may now be solved simultaneously subject to the boundary conditions

$\eta = 0$: $\quad f = f_\eta = 0, \quad \theta = \theta_w \quad c_A - 0$ (catalytic wall)

$c_{A\eta} = 0$ (noncatalytic wall)

$\eta = \infty$: $\quad f_\eta = 1, \quad c_A = c_{Ase}$

In terms of the Nusselt number (1693) and the Reynolds number (1694), the heat-transfer rate is given by

$$\frac{Nu}{\sqrt{Re}} = \frac{\sqrt{2}\bar{c}'_{pw} T_{se}}{h_{se} - h_w}\left[\theta_\eta + Le \sum_i \frac{c_{ise} h_i s_{i\eta}}{\bar{c}'_p T_{se}}\right]_{\eta = 0} \tag{1715}$$

Figure 300 shows a plot of $Nu/\sqrt{Re}$ versus C_1 for the specific case $Le = 1.4$, $Pr = 0.71$, $\beta = \frac{1}{2}$, $T_w = 300°K$.

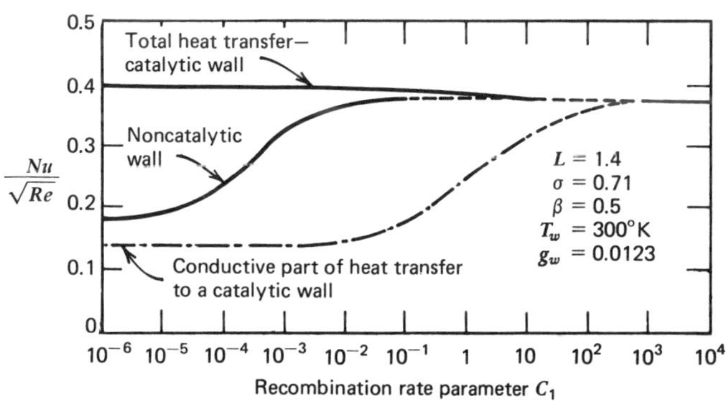

Figure 300 Effect of recombination rate on heat transfer [176].

The question of finite recombination rate at the wall with frozen boundary layer has been considered by Goulard [179]. By finite recombination rate at the wall we mean a wall which is neither catalytic nor noncatalytic but somewhere in between. In a frozen boundary layer only a portion of the atoms hitting such a wall would recombine. Goulard has shown that in such a case

$$\rho_w D_{12}\left(\frac{\partial c_A}{dy}\right)_w = (k_{RW} \rho c_A)_w \tag{1716}$$

where k_{RW} is a constant depending on the surface material and is called the surface recombination rate coefficient. Dorrance [180] has shown that for a frozen boundary layer, a least to a rough approximation,

$$s'_A(0) \cong 0.47\,Sc^{1/3}[1 - s_A(0)] \tag{1717}$$

where the prime indicates differentiation with respect to η and Sc is the Schmidt number, defined by

$$Sc = \frac{\mu}{\rho D_{12}} \tag{1718}$$

In terms of our transformed variables, equation (1716) becomes

$$s'_A(0) = \sqrt{\frac{\rho_e \mu_e}{2\left(\dfrac{du_e}{dx}\right)_s}}\,\frac{k_{RW}}{\rho_w D_{12}}\,s_A(0) \tag{1719}$$

Using equation (1717) for $s'_A(0)$,

$$s_A(0) = \left[1 + \sqrt{\frac{\rho_e \mu_e}{2\left(\dfrac{du_e}{dx}\right)_s}}\,\frac{k_{RW}}{0.47 Sc^{1/3}\rho_w D_{12}}\right]^{-1} \tag{1720}$$

For a noncatalytic wall, $k_{RW} = 0$ and $s_A(0) = 1$. For catalytic wall, $k_{RW} = \infty$ and $s_A(0) = 0$ if the wall is sufficiently cool. Also

$$s'_A(0) = 0.47\,Sc^{1/3}\,\Phi, \qquad \Phi = \left[1 + \frac{0.47\,Sc^{-2/3}\sqrt{2\left(\dfrac{du_e}{dx}\right)_s \rho_e \mu_e}}{\rho_w k_{RW}}\right]^{-1} \tag{1721}$$

Goulard has shown that a suitable curve fit for the heat transfer rate for arbitrary Φ is given by

$$q_w = 0.66\,Pr^{-2/3}\sqrt{\rho_e \mu_e}\left(\frac{du_e}{dx}\right)_s^{1/2}(h_{se} - h_w)\left[1 + (Le^{2/3}\Phi - 1)\frac{h_e}{h_{se} - h_w}\right] \tag{1722}$$

where $h_e = \sum h_{iw}(c_{ie} - c_{iw})$. Notice that for the case $\Phi = 1$ (catalytic wall) this equation is similar, but not identical, to equation (1691). A plot of $q_w/[q_w(\Phi = 1)]$ is shown in Figure 301. It will be observed that metallic surfaces tend to be most catalytic while Pyrex glasses are least catalytic, with oxides in between.

The chief application of the theory of stagnation point heat transfer in dissociating flow is in the calculation of the maximum heat transfer rate on the nose of space vehicles reentering the earth's atmosphere. Earth satellites reenter the earth's atmosphere at approximately 25,000 fps, while lunar vehicles reenter at about 30,000 fps. For equilibrium flow at 25,000 fts, the stagnation temperature to free stream temperature ratio is about 35 at 100,000 feet altitude. (Martin [181]). At a free stream temperature of 400° R (222° K), this translates into a

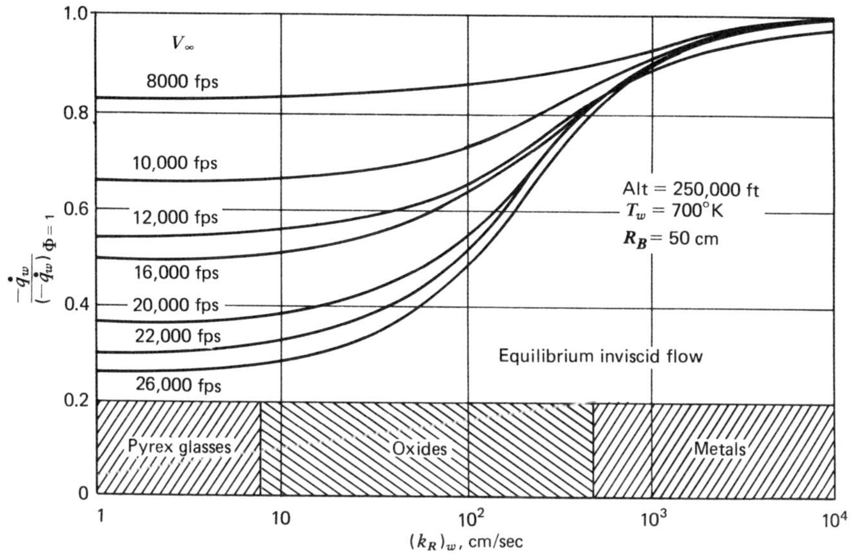

Figure 301 Effect of wall catalyticity on heat transfer rate [179].

stagnation temperature of 7770° K which is well within the dissociation range. For lunar and planetary reentry, stagnation temperatures may exceed 10,000° K, causing ionization in the flow. The question of calculating stagnation point heat transfer with ionization present has been discussed by a number of authors, including Cohen [182], Adams [183], Schreier [184], and Martin [181]. Interestingly enough, it appears that a purely formal extension of the correlation formulas of Fay and Riddell to the higher velocity range yields good results.

In the case of the boundary layer over the nose of a reentry vehicle, the boundary layer is essentially frozen only during the initial part of the reentry trajectory, which may be considered to begin at about 400,000 feet. Here, however, heat transfer rates are relatively low. Maximum heat transfer rates occur at about 100,000 feet. Here the boundary layer will be essentially in equilibrium. Thus in this particular application, the catalyticity of the surface material is not critical since wall temperatures will generally be sufficiently cool so that little dissociation will be present at or near the wall in an equilibrium boundary layer.

Constant Pressure Solutions

When the pressure at the outer edge of the boundary layer is constant, the term

$$\beta\left(\frac{\rho_e}{\rho} - f_\eta^2\right)$$

in the momentum equation (1681) vanishes. In theory, the pressure is constant in flow over flat plates. In practice, this is not strictly correct due to viscous interaction (sometimes called viscid-inviscid interaction). That is, the boundary layer distorts the external flow inducing a pressure gradient. The pressure gradient thus induced is, however, generally small.

Again we assume air to be a mixture of atoms and molecules and define

$$h_A^0 = h_A - h_M \tag{1723}$$

where h_A is the specific enthalpy of the atoms and h_M the specific enthalpy of the molecules. h_A^0 is thus positive since heat is given up on the formation of molecules from atoms. Now let

$$h_f = h_s - c_A h_A^0 \qquad \text{and} \qquad g_f = \frac{h_f}{h_{fe}} \tag{1724}$$

the subscript e referring as usual to the outer edge of the boundary layer. Then

$$g(\eta) \equiv \frac{h_s}{h_{se}} = \frac{h_{fe}}{h_e} g_f + \frac{c_{Ae} h_A^0}{h_{se}} s_A \tag{1725}$$

The energy equation (1679) in the case of similarity may be rewritten

$$\left(\frac{\chi}{Pr} g'\right)' + fg' = \frac{u_e^2}{h_{se}}\left[\left(1 - \frac{1}{Pr}\right)\chi f' f''\right]' + \left[\frac{\chi}{Sc}\left(\frac{1}{Le} - 1\right)\frac{h_A^0}{h_{se}} c_{Ae} s_A'\right] \tag{1726}$$

where the primes indicate differentiation with respect to η.

We now consider the case of frozen flow. Then $w_i = 0$ and the continuity of species equation (1703) can be written in the form

$$\left(\frac{\chi}{Sc} s_A'\right)' + f s_A' = 0 \tag{1727}$$

where Sc is the Schmidt number, defined by $Sc = Pr/Le$. If we now substitute equation (1725) into the energy equation (1703) and subtract (1704) we get

$$\left(\frac{\chi}{Pr} g_f'\right)' + f g_f' = \frac{u_e^2}{h_{se}}\left[\left(1 - \frac{1}{Pr}\right)\chi f' f''\right]' \tag{1728}$$

For $Pr = 1$ or $u_e^2 \ll h_{se}$ this reduces to

$$\left(\frac{\chi}{Pr} g_f'\right)' + f g_f' = 0 \tag{1729}$$

The momentum equation (1681) for similarity with constant pressure becomes

$$(\chi f'')' + ff'' = 0 \tag{1730}$$

The three equations—(1727), (1729), and (1730)—thus determine the composition, enthalpy and velocity distributions in the present case. The boundary conditions are:

$$\begin{aligned} \eta = 0: &\qquad s_A = s_w, \quad f = f' = 0, \quad g_f = g_{fw} \\ \eta = \infty: &\qquad s_A = f' = g_f = 1 \end{aligned} \tag{1731}$$

We observe that the form of the equations (1727) and (1729), as well as the form of their boundary conditions, is identical. Moreover, the system (1729) and (1730) is identical to that which was already solved for undissociated boundary layers by Cohen and Reshotko [101], Levy [102], and Li and Nagamatsu [102a], as discussed in Chapter 7. These authors assumed constant χ and Pr, with Li and Nagamatsu taking $\chi = Pr = 1$, Levy taking $\chi = 1$, $Pr = 0.7$, and Cohen and Reshotko taking $\chi = Pr = 1$.

The case of the flow of a laminar dissociating boundary layer in equilibrium over a flat plate was first considered by Moore [185]. Moore wrote the energy equation in the form

$$\rho u h_x + \rho v h_y = (k_n T_y)_y + \mu u_y^2 \tag{1732}$$

The implication is thus that the Lewis–Semenov number is equal to one. The momentum equation in the present case is

$$\rho u u_x + \rho v u_y = (\mu u_y)_y \tag{1733}$$

and conservation of mass (global continuity) is

$$(\rho u)_x + (\rho v)_y = 0 \tag{1734}$$

The continuity of species equation is not needed since equilibrium is assumed. Moore assumes all dependent variables are functions only of $z = y/\sqrt{x}$. Equations (1732), (1733) and (1734) then reduce to ordinary differential equations in z. The independent variable may then be changed from z to u. Combining the momentum and energy equations, we get

$$G'h' = \left(\frac{1}{Pr} Gh'\right)' + G \tag{1735}$$

where $G = \mu\, du/dz$ and the primes indicate differentiation with respect to u. From momentum and continuity we get

$$GG'' = -\tfrac{1}{2}\rho u \mu \tag{1736}$$

This form of the equations was first derived by Hantzsche and Wendt [186]. Define now

$$\eta = \frac{u}{\sqrt{h_B}}, \qquad g(\eta) = \frac{G(u)}{\sqrt{\frac{1}{2}\rho_B \mu_B h_B^{3/2}}}, \qquad j = \frac{h}{h_B} \tag{1737}$$

where B indicates a reference value taken at 400° R. Then equations (1735) and (1736) become

$$g'j' = \left(\frac{g}{Pr} j'\right)' + g \tag{1738}$$

$$gg'' = -\chi_B \eta \tag{1739}$$

where $\chi_B = \rho\mu/\rho_B\mu_B$. These equations may be solved numerically. A difficulty is caused by the fact that $1/g \to \infty$ as $y \to \infty$. It is convenient, therefore, to define a new independent variable t and a new dependent variable w, such that

$$g \equiv \frac{d\eta}{dt}, \qquad \frac{dw}{dt} \equiv \chi_B \eta \tag{1740}$$

Then equations (1715) and (1716) become

$$-w\frac{dj}{dt} = \frac{d}{dt}\left(\frac{1}{Pr}\frac{dj}{dt}\right) + g^2 \tag{1741}$$

$$\frac{dg}{d\eta} = -w \tag{1742}$$

These equations were solved numerically by Moore subject to the boundary conditions at $\eta = 0$:

$$u = v = 0; \qquad w = 0; \qquad j = j_w; \qquad g = g_w; \qquad \frac{dj}{dt} = \left[\frac{dj}{dt}\right]_w \tag{1743}$$

By specifying all the boundary conditions at the wall, the problem is turned into a marching problem and the equations may be integrated forward from the wall. The free stream conditions are thus not specified but found from the solution.

The shear stress at the wall is given by

$$\tau_w = \left[\frac{\rho_B \mu_B (h_B)^{3/2}}{2x}\right]^{1/2} g_w \tag{1744}$$

The local skin friction coefficient may be expressed as

$$c_{fe}\sqrt{Re_e} = \frac{\tau_w}{\frac{1}{2}\rho_e u_e^2}\left[\frac{\rho_e u_e x}{\mu_e}\right]^{1/2} = \sqrt{2}\frac{g_w}{\eta_e^{3/2}}\left[\frac{\rho_e \mu_e}{\rho_B \mu_B}\right]^{1/2} \tag{1745}$$

The free stream Mach number is given by

$$M_e = \frac{u_e}{c_e} = \frac{\sqrt{gJh_B\eta_e}}{\sqrt{\gamma g R T_B j_e}} = 1.5858\frac{\eta_e}{\sqrt{j_e}} \tag{1746}$$

where g and J are the usual conversion factors and $h_B = 96.8$ BTU/lb, $T_B = 400°R$. These are arbitrary reference values. A plot of $c_{fe}\sqrt{Re_e}$ versus M_e is shown in Figure 302. Unfortunately, Moore used values of the Prandtl number which are off by factors varying from 2 to 6 from the correct values for equilibrium air. The results presented in Figure 302 are, therefore, only qualitatively correct. Roming and Dore [187] recalculated the results of Moore using a corrected Prandtl number but inaccurate values for the equilibrium properties of air. Nevertheless, their results are considered accurate enough to be of interest. Roming and Dore found that their results for skin friction and heat transfer varied less than 15% from those calculated by the method of van Driest (discussed in Chapter 7) under the assumption of a perfect gas for the same free stream stagnation enthalpy. As Hayes and Probstein [188] have pointed out,

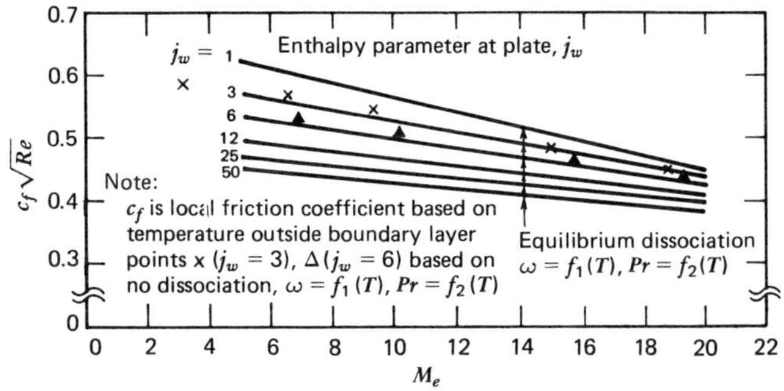

Figure 302 Friction coefficient in a laminar, dissociating boundary layer, according to Moore [185].

this indicates that for $Le = 1$, it is the total enthalpy difference across the boundary layer which determines the heat-transfer rate, regardless of what form the energy takes inside the boundary layer. The results of Roming and Dore also correlate well with the results of the reference enthalpy method of Eckert [103], discussed in Chapter 7. In the present case, the reference enthalpy method leads to the formulas

$$c_f\sqrt{Re_x} = 0.664\sqrt{\frac{\rho^*\mu^*}{\rho_e\mu_e}}, \qquad \frac{Nu}{\sqrt{Re_x}} = 0.332\sqrt{\frac{\rho^*\mu^*}{\rho_e\mu_e}}(Pr^*)^{1/3} \tag{1747}$$

where the superscript * indicates that the quantity was evaluated at the reference enthalpy, given by

$$h^* = 0.5(h_e - h_w) + 0.22(h_r - h_e) \tag{1748}$$

where

$$h_r = h_e + \tfrac{1}{2}\sqrt{Pr^*}\,u_e^2$$

Both c_f and Re_x are based on free stream values. That is

$$c_f = \frac{\mu\left(\dfrac{\partial u}{\partial y}\right)_w}{\frac{1}{2}\rho_e u_e^2}, \quad Re_x = \frac{u_e x \rho_e}{\mu_e} \tag{1749}$$

and the Nusselt number is defined by

$$Nu = \frac{\dot{q} c_{pe} x}{k_e(h_r - h_w)} \tag{1750}$$

Local Similarity Concept

Constant pressure solutions may be extended to the case of free stream conditions varying along the boundary layer if the variation is slow enough so that the derivatives with respect to x may be neglected. An exception to this is the term

$$2\frac{d \ln u_e}{d \ln \xi}$$

in equation (1678), which may be taken as some constant β, equal to its local value. If, in addition, we assume that the Lewis number is equal to one, equations (1678) and (1679) become

$$(\chi f_{\eta\eta})_\eta + f f_{\eta\eta} + \beta\left(\frac{\rho_e}{\rho} - f_\eta^{\,2}\right) = 0 \tag{1751}$$

$$\left(\frac{\chi}{Pr} g_\eta\right)_\eta + f g_\eta + \frac{u_e^2}{h_{se}}\left[\left(1 - \frac{1}{Pr}\right)\chi f_\eta f_{\eta\eta}\right]_\eta = 0 \tag{1752}$$

These equations may then be integrated at a series of points over the wall or body with the boundary conditions at the wall and in the free stream taken at their values at each point. In particular, g_w equals its local value at the wall, and ρ_e, u_e, h_{se} and β are taken at their local values in the free stream. The boundary conditions are

$$\eta = 0: \qquad f = f_\eta = 0; \qquad g = g_w$$

$$\eta = \infty: \qquad f_\eta = g = 1 \tag{1753}$$

Numerical solutions of these equations have been given by Kemp, Rose and Detra [189]. Kemp, Rose and Detra assumed the boundary layer to be in equilibrium and used the equilibrium thermodynamic properties of air as given by Hilsenrath and Beckett [190] which are considered reasonably accurate. The Prandtl number was assumed constant at 0.71 and the Sutherland formula was used for the viscosity. The equations were integrated for four values of β: 0, $\frac{1}{2}$, 1.0 and 2.0. From the discussion of the work by Fay and Riddell (op. cit.) it will be recalled that $\beta = \frac{1}{2}$ near the stagnation point.

In terms of the notation of Fay and Riddell, and for a Lewis number of one,

$$q = \left(k_n \frac{\partial T}{\partial y}\right)_{y=0} = \frac{r\rho_w \mu_w u_e h_{se}}{\sqrt{2\xi}\, Pr} g_\eta(0) \tag{1754}$$

Near the stagnation point,

$$r \cong x, \qquad u_e \cong \left(\frac{du_e}{dx}\right)_0 x, \qquad \xi = \rho_{wo}\mu_{wo}\left(\frac{du_e}{dx}\right)_0 \cdot \frac{1}{4}$$

so that

$$\frac{r\rho_w \mu_w u_e}{\sqrt{2\xi}} = \sqrt{2\rho_{wo}\mu_{wo}\left(\frac{du_e}{dx}\right)_0} \tag{1755}$$

where the subscript 0 indicates the forward stagnation point. It follows that the ratio of the heat transfer at the forward stagnation point of a body to the heat transfer at a point downstream on the body is given by

$$\frac{q}{q_0} = \frac{r\rho_w \mu_w u_e/\sqrt{\xi}}{2\sqrt{\rho_{wo}\mu_{wo}\left(\frac{du_e}{dx}\right)_0}} \frac{g_\eta(0)}{g_{\eta o}(0)} \tag{1756}$$

If we define the Nusselt number and the Reynolds number by

$$Nu = \frac{qxc'_{pw}}{k_{nw}(h_{se} - h_w)}, \qquad Re = \frac{u_e x}{\nu_w} \tag{1757}$$

then near the forward stagnation point,

$$\frac{Nu}{\sqrt{Re}} = \frac{\sqrt{2}g_\eta(0)}{1 - g_w} \tag{1758}$$

for a Lewis number of one. As a result of their numerical solution, Fay and Riddell found that $Nu/\sqrt{Re}$ varied linearly with $(\rho\mu)_{e0}/(\rho\mu)_w$ as shown in Figure 303. Kemp, Rose and Detra, from their numerical solutions, found that a

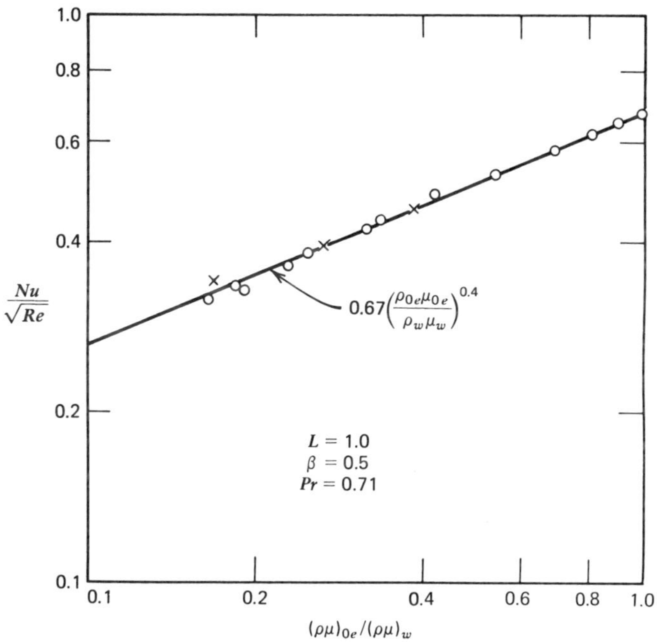

Figure 303 Correlation of $Nu/\sqrt{Re}$ versus $(\rho\mu)_{0e}/(\rho\mu)_w$ as found by Fay and Riddell [176].

similar correlation could be established for variable β (Figure 304).

$$\frac{\sqrt{2}g_{\eta w}}{1 - g_w} = 0.648(1 + 0.096\sqrt{\beta})\left(\frac{\rho_e\mu_e}{\rho_w\mu_w}\right)^{0.438} \tag{1759}$$

It must be emphasized, however, that away from the stagnation point, the left-hand side of equation (1759) is no longer equal to $Nu/\sqrt{Re}$. If the wall enthalpy distribution is known, $g_\eta(0)/g_{\eta 0}(0)$ may be found from equation (1759). Substitution in equation (1756) then yields q/q_0.

As an illustration of the use of their method, Kemp, Rose and Detra calculated q/q_0 for a cylinder with a hemispherical nose. Free stream conditions were

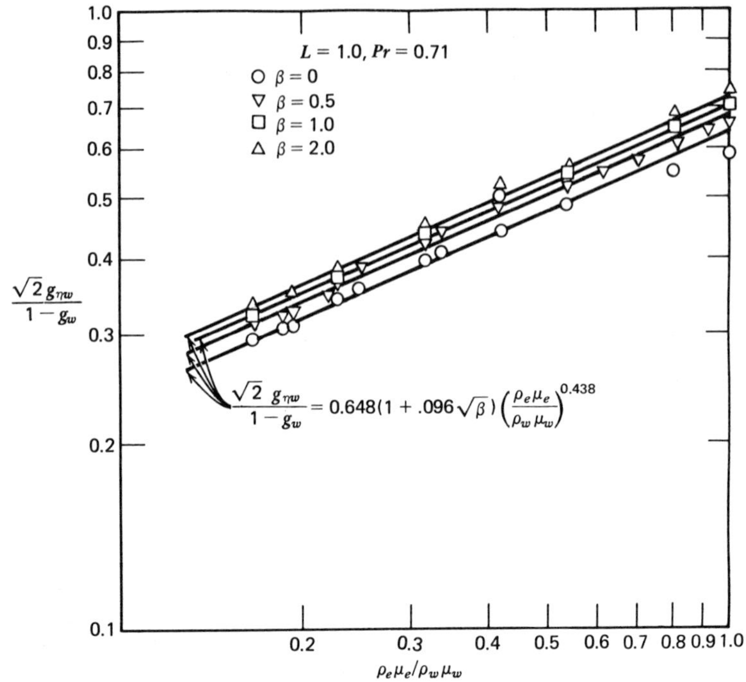

Figure 304 Correlation of $Nu/\sqrt{Re}$ versus $\rho_e \mu_e / \rho_w \mu$, as found by Kemp, Rose and Detra [189].

based on a flight velocity of 18,000 fps at 70,000 feet in order to make the results comparable to available experimental data. The pressure and velocity at the outer edge of the boundary layer were calculated on the basis of Newtonian flow. Newtonian flow assumes that when a fluid strikes a surface it loses all its momentum normal to that surface. Consider the point "*a*" on the body (Figure 305). The tangent to the surface at point "*a*" makes an angle α with the

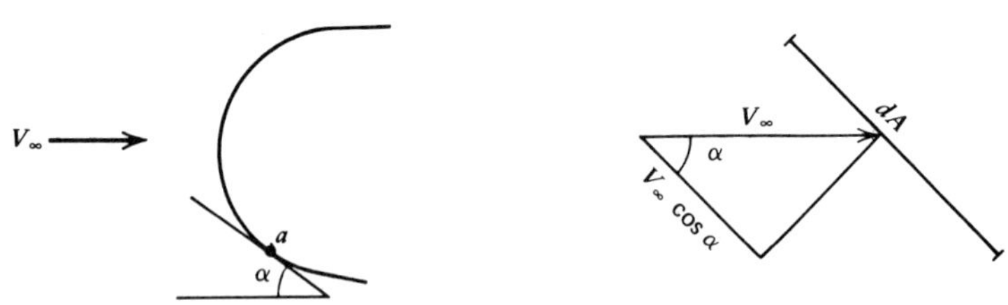

Figure 305 Notation for Newtonian flow.

horizontal. Under the assumption of Newtonian flow, then, the velocity along the surface at point "*a*" will be $V_\infty \cos \alpha$. The mass flow rate hitting an area dA of the surface is $\rho_\infty V_\infty \, dA \sin \alpha$. The rate of change of momentum normal to the surface is then

$$(\rho_\infty V_\infty dA \sin\alpha)(V_\infty \sin\alpha) = \rho_\infty V_\infty^2 \sin^2\alpha\, dA \qquad (1760)$$

causing a force on the surface dF_n,

$$dF_n = \frac{d(mV_n)}{dt} = \rho_\infty V_\infty^2 \sin^2\alpha\, dA \qquad (1761)$$

and hence the pressure due to this force is

$$\frac{dF_n}{dA} = \rho_\infty V_\infty^2 \sin^2\alpha \qquad (1762)$$

It follows that the pressure distribution over a hemispherical nose is given by

$$\frac{p}{p_0} = 1 - \left(1 - \frac{p_\infty}{p_0}\right)\sin^2\frac{x}{R} \qquad (1763)$$

where R is the nose radius and x the distance along the surface from the forward stagnation point. At the Mach numbers under consideration, this primitive technique yields surprisingly accurate results (Figure 306). One may explain

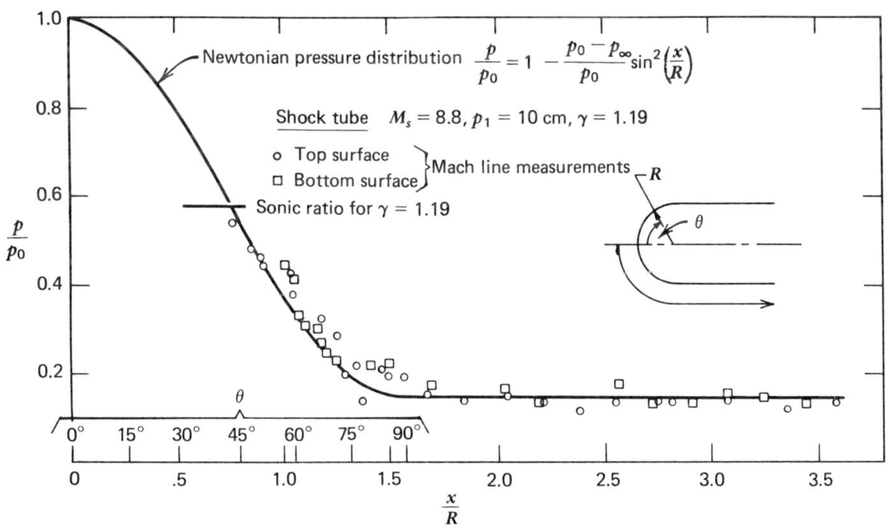

Figure 306 Newtonian pressure distribution over blunt nosed body [187].

this on physical grounds by noting that at hypersonic speeds the shock wave tends to wrap itself around the body causing the flow inside the shock wave to follow the contour of the body.

Returning to the local similarity solution for the boundary layer in the present case, the results are shown in Figure 307. Considering the scatter of the experimental data, the results are seen to be quite good, though possibly a little on the high side. Also shown for purposes of comparison is the solution of Lees

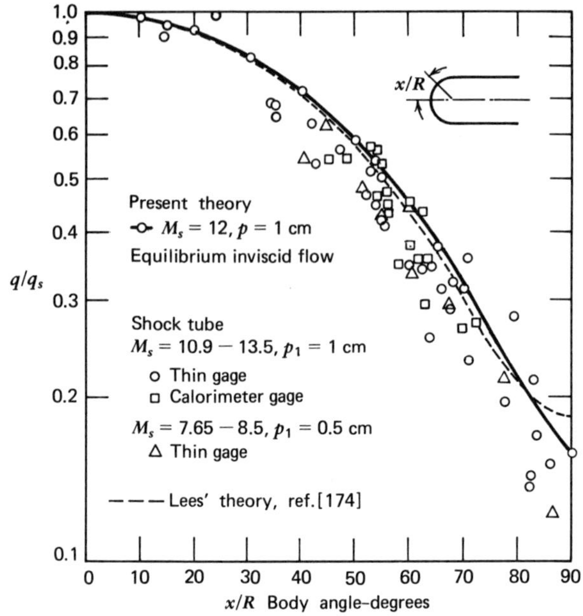

Figure 307 Heat-transfer rate over blunt nosed body [189].

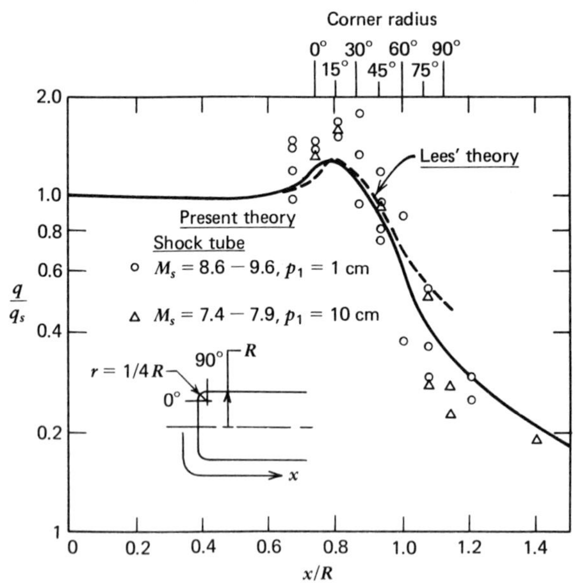

Figure 308 Heat transfer rate over blunt end of cylinder, from Kemp, Rose and Detra [189].

[174] (op. cit.) On the basis of numerical solutions by Cohen and Reshotko [101], Levy [102], and Chapman and Rubesin [97] (all of which we have discussed earlier), Lees concludes that over the surface of a blunt nosed body, $g_n(0)$ may be approximated "quite accurately" by 0.47 $Pr^{1/3}$ so long as $\rho_w/\rho_e \gg 1$. If Pr is assumed constant, $g_n(0)$ is constant over the surface and hence cancels out in equation (1756). This assumption is seen to lead to good results up to a value of x/R equal to about 80 degrees but, not surprisingly, breaks down beyond that point. Figure 308 shows the results of the theory of Kemp, Rose and Detra and the theory of Lees applied to a flat nosed cylinder. The results of both theories are seen to be quite good, though the applicability of Lees' theory is limited to the front of the body. The heat-transfer rate at the forward stagnation point may, of course, be calculated by the method of Fay and Riddell [176] (op. cit.).

Hypersonic Similarity Solution

By hypersonic similarity solution is meant the case where $\frac{1}{2}u_e^2 \gg h_e$. Then

$$\frac{u_e^2}{h_{se}} = \frac{u_e^2}{h_e + \frac{1}{2}u_e^2} \cong \frac{u_e^2}{\frac{1}{2}u_e^2} = 2 \tag{1764}$$

and equations (1678) and (1679) become

$$(\chi f'')' + ff'' + 2\frac{d \ln u_e}{d \ln \xi}\left(\frac{\rho_e}{\rho} - f'^2\right) = 0 \tag{1765}$$

$$\left(\frac{\chi}{Pr}g'\right)' + fg' + 2\left[\chi\left(1 - \frac{1}{Pr}\right)f'f''\right]' = 0 \tag{1766}$$

for a Lewis number of 1. Similarity will then exist if $d \ln u_e/d \ln \xi =$ constant and if equilibrium is assumed. For a perfect gas, the former condition will be satisfied if $p \sim \bar{x}^n$, where

$$\bar{x} = \int_0^x r^{2\delta}\, dx \tag{1767}$$

Then (1742) becomes

$$(\chi f'')' + ff'' + \beta(g - f'^2) = 0 \tag{1768}$$

where

$$\beta = -\left(\frac{\gamma - 1}{\gamma}\right)\left(\frac{n}{n+1}\right) \tag{1769}$$

Solutions to equations (1766) and (1768) have been calculated by Reshotko and Cohen [101] for the case $Pr = 0.7$, $\beta = 1$. In the case of real gases, no such simple similarity criterion as $p \sim x^n$ exists. The only obvious one is $du_e/d\xi = 0$ which we have already discussed under constant pressure solutions. Another alternative is to use local similarity to solve equation (1766) with equation (1765) as given.

Turbulent dissociating boundary Layers

In order to analyze this problem we begin with the equations stated earlier for the laminar dissociating boundary layer. For convenience we restate these equations here:
The global continuity equation:

$$\frac{\partial(\rho u)}{\partial x} + \frac{\partial(\rho v)}{\partial y} = 0 \tag{1770}$$

The continuity of species equation:

$$\rho u \frac{\partial c_i}{\partial x} + \rho v \frac{\partial c_i}{\partial y} - \frac{\partial}{\partial y}\left(\rho D_i \frac{\partial c_i}{\partial y}\right) = w_i \tag{1771}$$

The momentum equation:

$$\rho u \frac{\partial u}{\partial x} + \rho v \frac{\partial u}{\partial y} = -\frac{\partial p}{\partial x} + \frac{\partial}{\partial y}\left(\mu \frac{\partial u}{\partial y}\right) \tag{1772}$$

The energy equation:

$$\rho\left(u \frac{\partial h_s}{\partial x} + v \frac{\partial h_s}{\partial y}\right) = \frac{\partial}{\partial y}\left(\frac{\mu}{Pr}\frac{\partial h_s}{\partial y} + \mu\left(1 - \frac{1}{Pr}\right)\frac{\partial u^2}{\partial y}\right)$$
$$+ \frac{\partial}{\partial y}\left(\rho D_{12}\left(1 - \frac{1}{Le}\right)\sum_i h_i \frac{\partial c_i}{\partial y}\right) \tag{1773}$$

In addition, recall

$$h = \sum_i h_i c_i, \qquad h_i = \int_0^T c_{pi}\, dT + h_i^0, \qquad V_i = \frac{D_{12}}{c_i}\frac{\partial c_i}{\partial y} \tag{1774}$$

As in our study of undissociated turbulent boundary layers we now express each of the dependent variables as the sum of a mean value, designated by a bar, and a perturbation value, designated by a prime. Thus,

$$\begin{aligned} u &= \bar{u} + u', \qquad & \rho u &= \overline{\rho u} + (\rho u)', \qquad & h &= \bar{h} + h' \\ h_i &= \bar{h}_i + h'_i \qquad & \rho v &= \overline{\rho v} + (\rho v)', \qquad & c_i &= \bar{c}_i + c'_i \\ & & \rho_i V_i &= \overline{\rho_i V_i} + (\rho_i V_i)' & & \end{aligned} \tag{1775}$$

Introducing these expressions into the governing equations (1770) through (1773) and, as before, taking time averages and omitting higher order terms, we get

$$\frac{\partial \overline{\rho u}}{\partial x} + \frac{\partial \overline{\rho v}}{\partial y} = 0 \tag{1776}$$

$$\overline{\rho u}\frac{\partial \bar{c}_i}{\partial x} + \overline{\rho v}\frac{\partial \bar{c}_i}{\partial y} = \frac{\partial}{\partial y}\left[\bar{\rho} D_{12}\frac{\partial \bar{c}_i}{\partial y} - \overline{(\rho v)' c'_i}\right] + w_i \tag{1777}$$

$$\overline{\rho u}\frac{\partial \bar{u}}{\partial x} + \overline{\rho v}\frac{\partial \bar{u}}{\partial y} = -\frac{\partial \bar{p}}{\partial x} + \frac{\partial}{\partial y}\left[\mu\frac{\partial \bar{u}}{\partial y} - \overline{(\rho v)'u'}\right] \tag{1778}$$

$$\overline{\rho u}\frac{\partial \bar{h}}{\partial x} + \overline{\rho v}\frac{\partial \bar{h}}{\partial y} = \mu\left(\frac{\partial \bar{u}}{\partial y}\right)^2 - \overline{(\rho v)'u'}\left(\frac{\partial \bar{u}}{\partial y}\right)$$
$$+ \frac{\partial}{\partial y}\left\{k\frac{\partial \bar{T}}{\partial y} - \sum_i \bar{c}_i\overline{(\rho v)'h_i'} - \sum_i h_i\left[\overline{\rho_i V_i} + \overline{(\rho v)'c_i'}\right]\right\} \tag{1779}$$

We now define ε, K and D_T by the equations

$$\varepsilon = -\frac{\overline{(\rho v)'u'}}{\dfrac{\partial \bar{u}}{\partial y}}, \qquad \sum_i \bar{c}_i\overline{(\rho v)'h_i'} = -\sum_i c_i\frac{K}{c_p}\frac{\partial \bar{h}_i}{\partial y} \tag{1780}$$

$$\sum_i \rho D_T\frac{\partial \bar{c}_i}{\partial y} = -\sum_i \overline{(\rho v)'c_i'} \tag{1781}$$

where c_p is the frozen average c_p of the mixture, $\sum_i \bar{c}_i c_{p_i}$. We then define the turbulent Lewis number, L_T, and the turbulent Prandtl number, P_T, by

$$L_T = \frac{c_p\bar{\rho}D_T}{K}, \qquad P_T = \frac{c_p\varepsilon}{K} \tag{1782}$$

The system of governing equations then becomes

$$\rho u\frac{\partial c_i}{\partial x} + \rho v\frac{\partial c_i}{\partial y} = \frac{\partial}{\partial y}\left[\rho\frac{\partial c_i}{\partial y}(D_{12} + D_T)\right] + w_i \tag{1783}$$

$$\rho u\frac{\partial u}{\partial x} + \rho v\frac{\partial u}{\partial y} = -\frac{\partial p}{\partial x} + \frac{\partial}{\partial y}\left[(\mu + \varepsilon)\frac{\partial u}{\partial y}\right] \tag{1784}$$

$$\rho u\frac{\partial h}{\partial x} + \rho\frac{\partial h}{\partial y} = (\mu + \varepsilon)\left(\frac{\partial u}{\partial y}\right)^2 + \frac{\partial}{\partial y}\left[\left(\frac{\mu}{Pr} + \frac{\varepsilon}{Pr}\right)\frac{\partial h}{\partial y}\right]$$
$$+ \frac{\partial}{\partial y}\left\{\left[\frac{k}{c_p}(Le - 1) + \frac{K}{c_p}(L_T - 1)\right]\sum_i h_i\frac{\partial c_i}{\partial y}\right\} \tag{1785}$$

where all dependent variables are time average quantities and hence the bars have been omitted. Alternatively, the energy equation may be written

$$\rho u\frac{\partial h_s}{\partial x} + \rho v\frac{\partial h_s}{\partial y} = \frac{\partial}{\partial y}\left[\left(\frac{\mu}{Pr} + \frac{\varepsilon}{P_T}\right)\frac{\partial h_s}{\partial y}\right] + \frac{\partial}{\partial y}\left\{\left[\frac{k}{c_p}(Le - 1)\right.\right.$$
$$\left.\left.+ \frac{K}{c_p}(L_T - 1)\right]\sum_i h_i\frac{\partial c_i}{\partial y}\right\} + \frac{\partial}{\partial y}\left\{\left[\mu\left(1 - \frac{1}{Pr}\right) + \varepsilon\left(1 - \frac{1}{Pr}\right)\right]\frac{1}{2}\frac{\partial u^2}{\partial y}\right\} \tag{1786}$$

Forstall and Shapiro [191], by making measurements of temperature and velocity profiles in heated jets, have concluded that

$$P_T = \frac{c_p \varepsilon}{K} \cong 0.70 \pm 10\%, \qquad L_T = \frac{\rho D_T c_p}{K} = 1.0 \pm 20\% \tag{1787}$$

These results have encouraged investigators to set P_T and L_T equal to one. For air,

$$Pr \cong 0.7, \qquad Le \cong 1.4 \tag{1788}$$

Thus, as a first approximation, these parameters may also be set equal to one. Equations (1783) through (1786) then reduce to

$$\frac{\partial(\rho u)}{\partial x} + \frac{\partial(\rho v)}{\partial y} = 0 \tag{1789}$$

$$\rho u \frac{\partial u}{\partial x} + \rho v \frac{\partial u}{\partial y} = -\frac{\partial p}{\partial x} + \frac{\partial}{\partial y}\left[(\mu + \varepsilon)\frac{\partial u}{\partial y}\right] \tag{1790}$$

$$\rho u \frac{\partial h_s}{\partial x} + \rho v \frac{\partial h_s}{\partial y} = \frac{\partial}{\partial y}\left[(\mu + \varepsilon)\frac{\partial h_s}{\partial y}\right] \tag{1791}$$

Since these equations are of the same form as those for the undissociated case, it follows that for the case $\partial p/\partial x = 0$, the Crocco integral (cf. Chapter 8) must still be a valid solution. For constant wall temperature, the Crocco integral may be written in the form

$$h_s = h_w + (h_{se} - h_w)\frac{u}{u_e} \tag{1792}$$

Then Reynolds' analogy follows directly and we have

$$St = \frac{q}{\rho_e u_e (h_{se} - h_w)} = \frac{\tau}{\rho_e u_e^2} = \tfrac{1}{2} c_f \tag{1793}$$

As will be recalled, the left-hand side of equation (1793) is the Stanton number, defined by

$$St = \frac{q}{\rho_e u_e (h_{se} - h_w)} \tag{1794}$$

The solution of the system of equations (1789)–(1791) for heat-transfer coefficient and friction coefficient over the surface of a blunt body in hypersonic flow has been considered by Rose, Probstein and Adams [192]. These authors, on the basis of rather convoluted and highly speculative philosophical arguments, arrive at the conclusion that the heat-transfer and friction coefficient data for this case may be correlated by the same formulas as laminar undissociated and even incompressible boundary layers. Nevertheless, shock tube experiments performed by the same authors appear to bear out their contentions.

If the Nusselt number and the Reynolds number based on x are defined by

$$Nu = \frac{q x c_{pe}}{k_e (h_{se} - h_w)}, \qquad Re_x = \frac{\rho_e u_e x}{\mu_e} \tag{1795}$$

we have the relation

$$Nu = St\,Pr\,Re_x \tag{1796}$$

For laminar boundary layers and for turbulent boundary layers with low velocities and small temperature differences, Colburn has shown that

$$St = \tfrac{1}{2}c_f Pr^{-2/3} \tag{1797}$$

Rose, Probstein and Adams conclude that this formula is equally applicable to the case of blunt bodies in hypersonic flow, thus representing a correction for the case $Pr \neq 1$. Equation (1797) may be rewritten

$$St = \frac{c_f}{c_{f_i}} \cdot \frac{c_{f_i}}{2} Pr^{-2/3} \tag{1798}$$

where c_{fi} is the value for incompressible turbulent boundary layers, which has been found to be

$$\tfrac{1}{2}c_{f_i} = 0.0296\,Re_x^{-1/5} \tag{1799}$$

for Reynolds numbers between 5×10^5 and 1×10^7 (see, for example, Schlichting [193]).

Correction for the case of $Le \neq 1$. In dissociated turbulent boundary layers, the degree of dissociation appears to change but little between the outer edge of the boundary layer and the outer edge of the laminar sublayer. The heat transfer may thus be expressed in terms of the heat transfer across the laminar sublayer. Recall for a laminar boundary layer we showed that (equation (1654))

$$q = k\frac{\partial T}{\partial y} + D_{12}\rho h_A^0 \frac{\partial c_A}{\partial y} \tag{1800}$$

where h_A^0 represents the difference in enthalpy between the atoms and molecules and hence is positive. If h_A^0 is taken at 0°R, (heat of formation), equation (1800) assumes that the specific heat of atoms and molecules is the same. The error introduced by this assumption is not great for high temperature air. Alternatively, we may define h_A^0 to be the local enthalpy difference between atoms and molecules, in which case the error is eliminated. From equation (1723), equation (1658) is equivalent to

$$q = \frac{k}{c_p}\frac{\partial h}{\partial y} + (Le - 1)\frac{k}{c_p} h_A^0 \frac{\partial c_A}{\partial y} \tag{1801}$$

where c_p, as pointed out above, is the average frozen c_p of the mixture. The ratio of the heat transfer to the heat transfer for $Le = 1$ is then

$$\frac{q_L}{q_1} = 1 + (Le - 1)\frac{\partial(h_A^0 c_A)/\partial y}{\partial h/\partial y} \tag{1802}$$

Assuming a linear distribution of concentration and enthalpy across the sublayer, equation (1802) may be approximated by

$$\frac{q_L}{q_1} = 1 + (Le - 1)\frac{h_{De}}{h_{se}} \tag{1803}$$

where $h_{De} = (h_A^0 c_A)_e$. The justification here lies in the shape of the enthalpy and concentration profiles which are almost vertical until the laminar sublayer is reached. Thus, the enthalpy and concentration at the outer edge of the sublayer varies but little from the enthalpy and concentration at the outer edge of the boundary layer. Moreover, near the nose of a blunt body, $h_{se} \cong h_e$. Equation (1803) may be compared to the correlation of Fay and Riddell [176] given earlier,

$$\frac{q_L}{q_1} = 1 + (Le^\beta - 1)\frac{h_{De}}{h_{se}} \tag{1804}$$

with $\beta = 0.52$ for equilibrium flow and $\beta = 0.63$ for frozen flow.

If equation (1797) for the Stanton number is taken as valid for $Le = 1$, it follows from equation (1796) for the Nusselt number that

$$Nu = \frac{qPrx}{\mu_e(h_{se} - h_w)} = \frac{c_f}{2} Pr^{1/3} Re_x \tag{1805}$$

Introducing equation (1804) to correct for the case of Lewis number not equal to one, with β left open, we get

$$Nu = \frac{qPrx}{\mu_e(h_{se} - h_w)} = \frac{c_f}{2}\left[1 + (Le^\beta - 1)\frac{h_{De}}{h_{se}}\right] Pr^{1/3} Re_x \tag{1806}$$

Recall now

$$Nu = St\, Pr\, Re_x = \frac{c_f}{2} Pr^{-2/3} Pr\, Re_x = \frac{c_f}{2} Pr^{1/3} Re_x = \frac{c_f}{c_{f_i}} \frac{c_{f_i}}{2} Pr^{1/3} Re_x \tag{1807}$$

If, then, the incompressible value for c_f is used in equation (1806), that is, if

$$Nu = \frac{c_{f_i}}{2}\left[1 + (Le^\beta - 1)\frac{h_{De}}{h_{se}}\right] Pr^{1/3} Re_x \tag{1808}$$

then

$$\frac{c_f}{c_{f_i}} = 1 + (Le^\beta - 1)\frac{h_{De}}{h_{se}} \tag{1809}$$

and c_{fi} is given by equation (1799). The Nusselt number then becomes

$$\begin{aligned} Nu &= 0.029\, Re_x^{-0.2}\left[1 + (Le^\beta - 1)\frac{h_{De}}{h_{se}}\right] Pr^{1/3} Re_x \\ &= 0.029\, Re_x^{0.8}\left[1 + (Le^\beta - 1)\frac{h_{De}}{h_{se}}\right] Pr^{1/3} \end{aligned} \tag{1810}$$

For a Lewis number of one, a similar correlation formula for laminar flow is given by (cf. eq. 1747)

$$Nu = 0.38\, Pr^{1/3}\, Re_x^{1/2} \tag{1811}$$

Comparison with Experiment

Rose, Probstein and Adams (op. cit.) performed shock tube experiments to test the above correlation formulas. In Figure 309 heat-transfer data for the flow

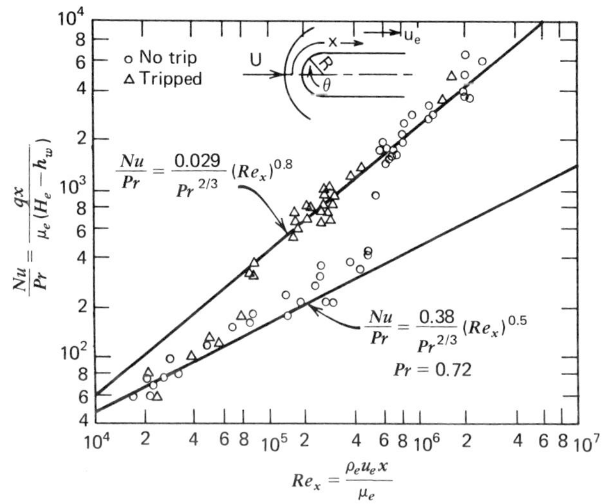

Figure 309 Effect of boundary layer transition on the heat transfer over a blunt nosed body, from Rose, Probstein and Adams [192].

over a hemisphere cylinder combination is shown correlated by equation (1810) and (1811) under the assumption of a Lewis number of one. Shock Mach numbers from 7 to 17 and downstream pressures of 0.01 to 10 cm Hg were used. Mach numbers at the edge of the boundary layer ranged from a maximum of 1.7 to 2.2. The degree of dissociation encountered is shown in Figure 310. Figure 309 has a number of interesting features. We observe first the natural transition occurring over the body, showing the importance of both the laminar and turbulent regions of the body. Tests were also made with a trip wire to move transition forward. The correlation formulas are seen to correlate well with the experimental results, both in the laminar and in the turbulent ranges. Of further interest is the correlation of the heat transfer rate parameter $q/Re_x^{0.8}$ with the fraction of energy in dissociation according to equation (1810). For purposes of comparison, results for Lewis numbers of 1.0 and 1.4 are shown in Figure 311 along with the results of the reference enthalpy method of Eckert (cf. Chapter 7). The present correlation with a Lewis number of 1.4 is seen to give the best results.

An additional discussion of turbulent dissociating boundary layers has been

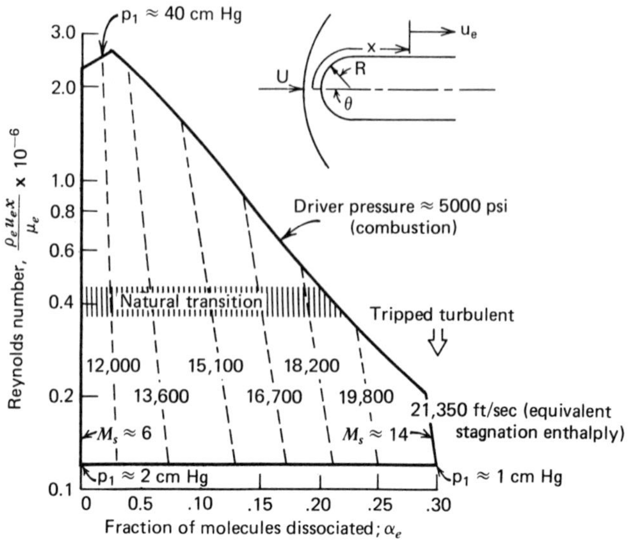

Figure 310 Degree of dissociation versus equivalent flight velocity [192].

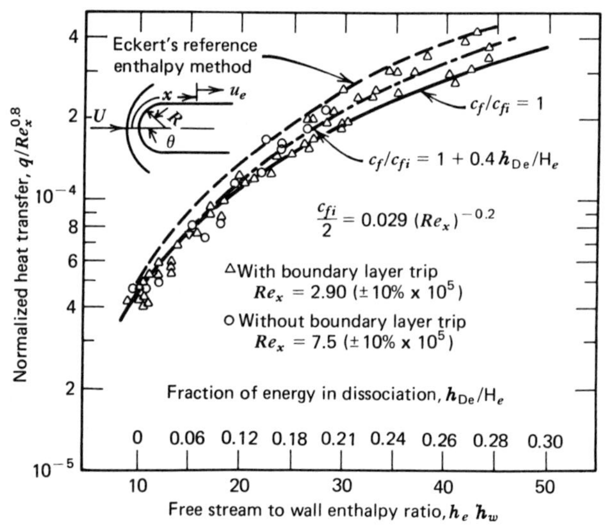

Figure 311 Modified Reynolds analogy for dissociating flow, from Rose, Probstein and Adams (op. cit.).

given by Dorrance [180], [194]. The derivations of Dorrance are quite lengthy and will not be given here. However, Dorrance finds

$$q = \rho_e u_e Pr^{-2/3}(h_r - h_w)\frac{c_{f0}}{2}\left[2\left(\frac{1+c_{Aw}}{1+c_{Ae}}\right)^{1/7} - 1\right]\left[1 + (Le^{2/3} - 1)\frac{h_D}{h_r - h_w}\right] \tag{1812}$$

where $h_r = h_e + r^2 u_e^2/2$, r = recovery factor = $Pr^{1/3}$

$h_D = (c_{Ae} - c_{Aw})(h_1^0 - h_2^0)$

h_1^0 = heat of formation of atoms

h_2^0 = heat of formation of molecules

c_{f0} = local skin friction coefficient for no dissociation

$$\frac{c_f}{c_{fo}} = 2.0\left[\frac{1 + c_{Ae}}{1 + c_{Aw}}\right]^{1/7} - 1 \tag{1813}$$

Equation (1789) may be rewritten

$$Nu = Re_x \frac{c_{fo}}{2} Pr^{1/3}\left[2\left(\frac{1 + c_{Ae}}{1 + c_{Aw}}\right)^{1/7} - 1\right]\left[1 + (Le^{2/3} - 1)\frac{h_D}{h_r - h_w}\right] \tag{1814}$$

which may be compared to equation (1808) of Rose, Probstein and Adams. Dorrance made fewer assumptions in the derivation of his equations and finds good agreement with the experimental results of Rose, Probstein and Adams, as shown in Figure 312.

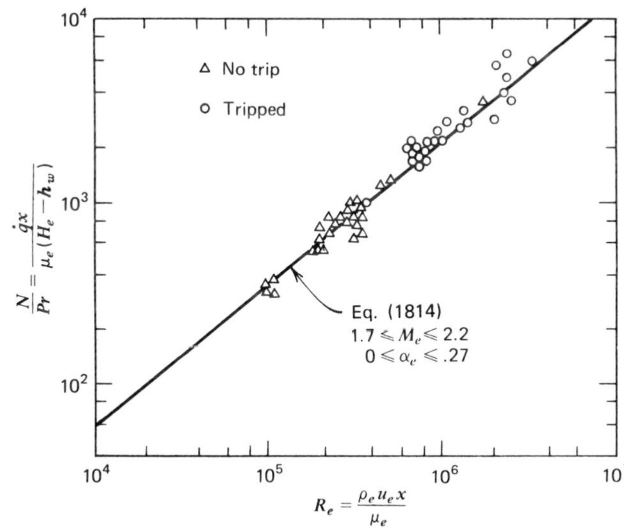

Figure 312 Correlation of Nusselt number versus Reynolds number, according to Dorrance [180, 194].

Shock waves:

In Chapter 2 we treated shock waves from the standpoint of ideal gas theory. In the case of strong shocks and/or high temperatures, this assumption will not be justified. The final conditions across a shock wave in a real gas may be found with the help of Mollier diagrams such as those given by Feldman [195] (Figures 313, 314). Using the subscripts 1 and 2 to indicate the conditions before

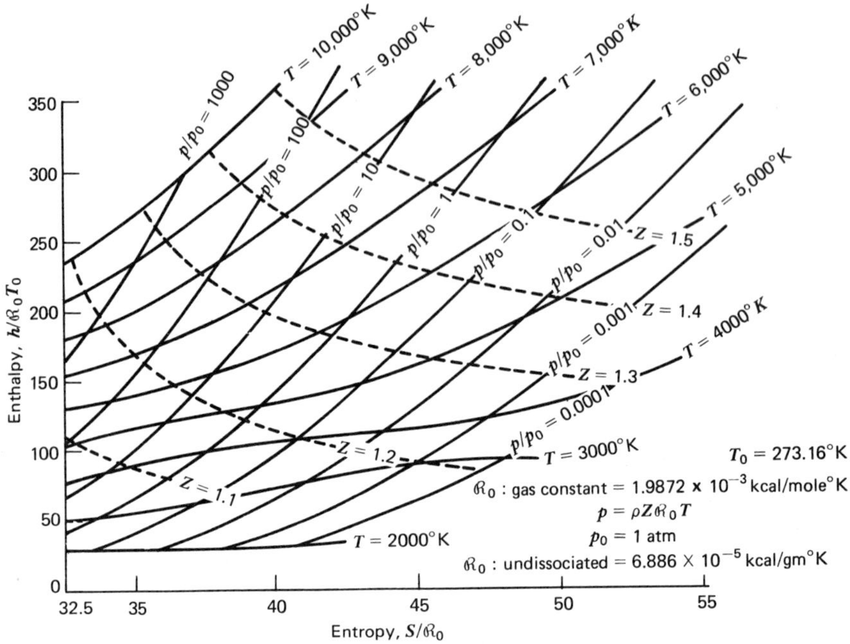

Figure 313 Mollier diagram for air, from Feldman [195].

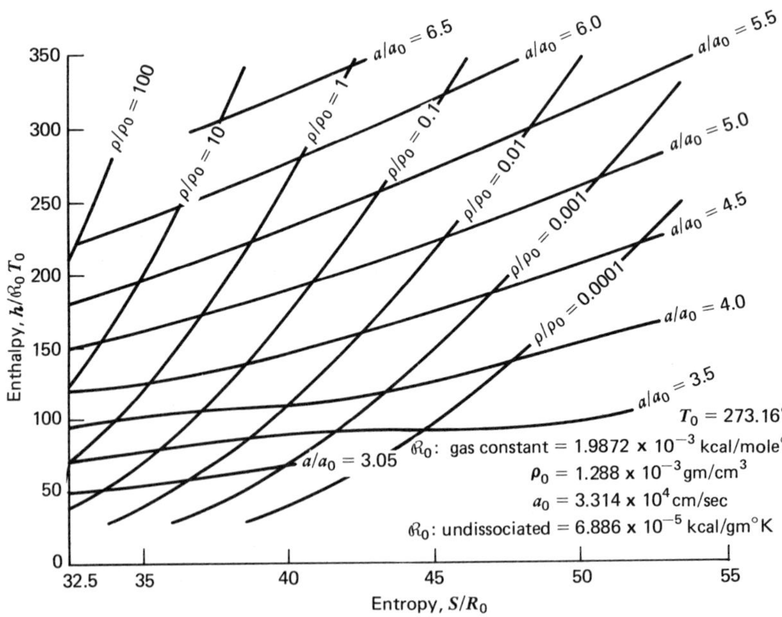

Figure 314 Mollier diagram for air, from Feldman [195].

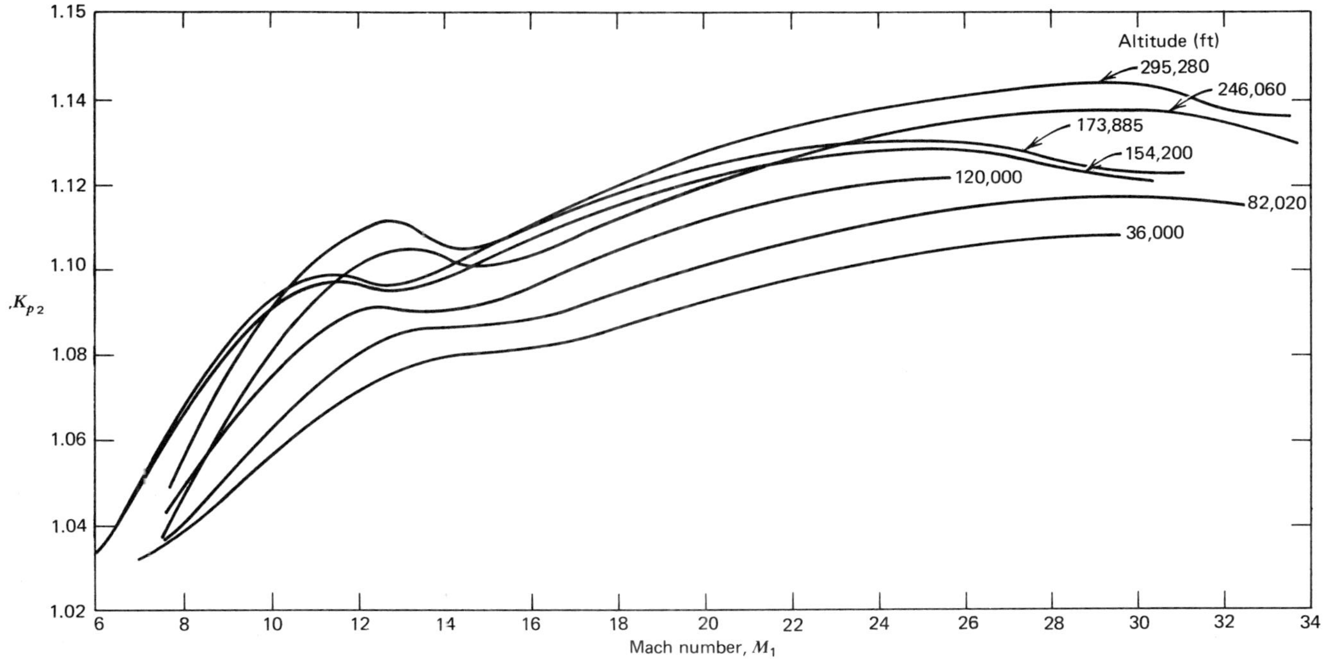

Figure 315 Actual versus ideal shock pressure ratios, from Huber [196].

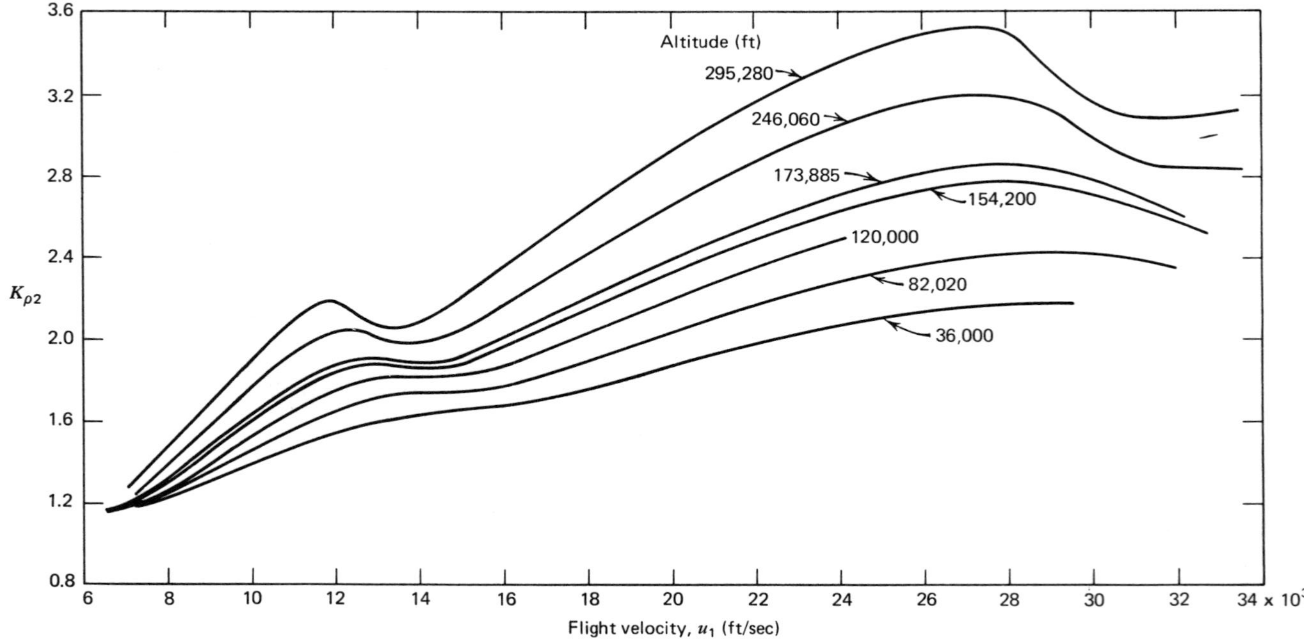

Figure 316 Actual versus ideal shock density ratios, from Huber [op. cit.].

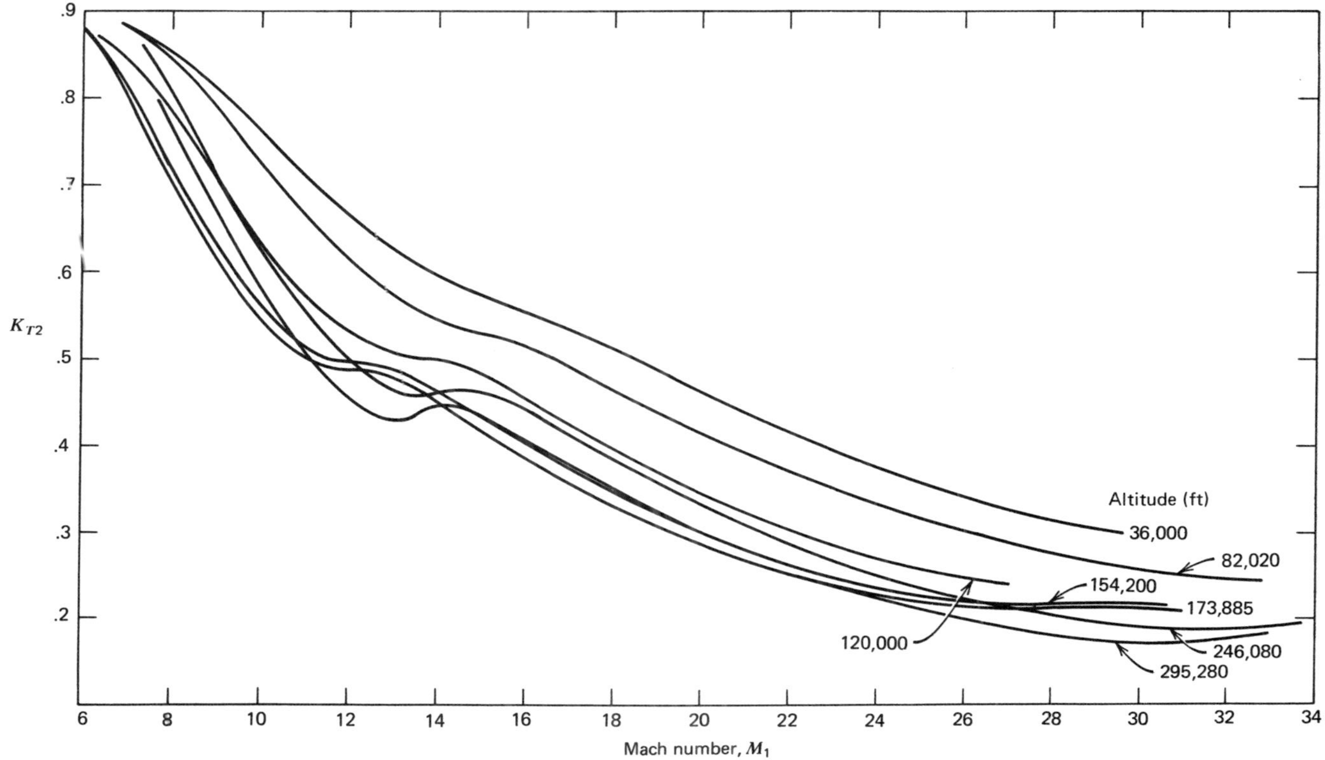

Figure 317 Actual versus ideal shock temperature ratios, from Huber [op. cit.].

and after the shock, the momentum equation may be written

$$p_2 - p_1 = \rho_1 V_1^2 \left(1 - \frac{\rho_2}{\rho_1}\right) \tag{1815}$$

The energy equation may be written

$$h_2 - h_1 = \tfrac{1}{2} V_1^2 \left(1 - \frac{\rho_2}{\rho_1}\right) \tag{1816}$$

Given the conditions ahead of the shock, the procedure is as follows:

1. Guess ρ_2/ρ_1.
2. Calculate p_2 and h_2 from equations (1815) and (1816).
3. Using these values of p_2 and h_2, find ρ_2 from Figures 313 and 314.
4. Calculate the corresponding ρ_2/ρ_1 and compare to the original estimate.
5. Repeat until the two values of ρ_2/ρ_1 agree.

Tables of property ratios across shock waves in air treated as a real gas have been given by Huber [196] and Rozhdestvenski [197]. Figures 315, 316, and 317 show the actual versus ideal temperature, pressure and density ratios across a normal shock in air. K_{p2}, $K_{\rho 2}$ and K_{T2} represent the ratios of the actual to the ideal pressure, density and temperature behind (downstream of) the shock. Of particular interest is the actual temperature downstream of the shock which may be 50% or less of the ideal value, showing the activation of the higher degrees of freedom.

10 Computational Methods

General Comments

Throughout this text we have concluded derivations by saying "the resulting equations must be solved numerically." The advent of modern computers with large capacity, high computing speed and low operating cost has made the use of digital computers standard procedure at some stage of nearly all compressible flow problems. The stage at which a computer enters the proceedings depends on the extent to which the problem has first been solved analytically. In the case of the method of characteristics, for example, the computer may be confined to doing simple arithmetic, though refinements may be introduced to plot the characteristic curves more accurately. In the case of boundary layer solutions where the governing partial differential equations are reduced to one or more ordinary differential equations, the computer will be employed to solve the ordinary differential equations numerically where no closed form solutions exist. Where it is not possible to reduce the governing partial differential equations to ordinary differential equations, the computer may be used to solve the partial differential equations subject to the given boundary conditions. An example of such a case would be solution of the equation for the velocity potential in the case of isentropic, subsonic flow over a body of some arbitrary shape.

The most extreme example of the use of the computer would be the direct solution of the original equations without any preliminary analytical work. This is sometimes referred to as the brute force approach. An example of such an application would be the direct solution of the Navier–Stokes equations for a given set of boundary conditions.

Generally speaking, the more analytical work is done before the computer is resorted to, the better. This is because the more analytical work has been done, the broader the application of the solution is likely to be and the smaller the cumulative error due to the numerical integration. In the case where universal functions are generated, as in certain boundary layer solutions, the applicability of the solution will be very broad. In the case of brute force solutions, on the other hand, as in the direct solution of the Navier–Stokes equations for a given set of boundary conditions, the applicability of the solution will be restricted to that one particular case.

Ordinary Differential Equations

We consider the numerical integration of an ordinary differential equation of the form

$$F\left(\frac{d^n y}{dx^n}, \frac{d^{n-1} y}{dx^{n-1}}, \ldots, \frac{dy}{dx}, y, x\right) = 0 \tag{1817}$$

If we are given the value of the function and the first $n-1$ derivatives at some point x, we may find the nth derivative from the equation and integrate forward from the point x by methods which we shall discuss below. A boundary condition of this type is called a one point boundary condition (all the boundary conditions are given at the same point). The problem of solving the equation subject to the given boundary conditions is then called an initial value problem or marching problem, after the suggestion of Richardson [198]. If, on the other hand, the boundary conditions are given at two or more different points, the problem is no longer so simple. In this case we speak of a two (or multi) point boundary condition and the problem is called a jury problem (they all decide together) again after Richardson.

The simplest way of integrating forward in a marching problem is by means of a Taylor series. Given the solution at x, the solution at $x + \Delta x$ may be found from the Taylor series

$$\begin{aligned} f(x+\Delta x) &= f(x) + f'(x)(\Delta x) + \frac{1}{2!} f''(x)(\Delta x)^2 + \frac{1}{3!} f'''(x)(\Delta x)^3 + \ldots \\ &\quad + \frac{1}{(m-1)!} f^{m-1}(x)(\Delta x)^{m-1} + \frac{1}{m!} f^m(x + \theta \Delta x)(\Delta x)^m \end{aligned} \tag{1818}$$

where θ is between zero and one. The last term is known as the remainder term. Since the value of θ is unknown unless the solution is known, the last term may not be evaluated exactly. It does, however, enable us to estimate the bounds on the error involved in terminating the series after a finite number of terms. The accuracy of the solution may be increased without limit by decreasing the size of Δx and increasing the number of terms employed.* The size of Δx is known as the step size. The number of terms may be increased by differentiating the original equation as many times as desired. Once the first $n-1$ derivatives are known at a point x, all other derivatives may thus be found there, assuming they exist. The derivatives at $x + \Delta x$ may also be found by Taylor series

$$f^m(x+\Delta x) = f^m(x) + f^{m+1}(x)(\Delta x) + \frac{1}{2!} f^{m+2}(x)(\Delta x)^2 + \ldots \tag{1819}$$

to any desired accuracy.* Thus solution may be carried forward until the entire range of x has been covered.

Runge–Kutta Method

The name Runge–Kutta is used to denote approximation formulas of a type

*Provided the function $f(x)$ has derivatives of all orders in the interval Δx and provided that the absolute value of the last term tends to zero uniformly as $m \to \infty$.

first derived by Runge [199] and Heun [200] and later modified by Kutta [201]. It is Kutta's modification of Runge's formulas which is generally used today under the heading of Runge–Kutta method.

Consider a differential equation of the type

$$\frac{dy}{dx}=f(x,y) \tag{1820}$$

Although y, and therefore dy/dx, is a function of x only, we may treat dy/dx as a function of x and y. Taking the differential of $f(x,y)$ we get

$$df=f_x dx+f_y dy \tag{1821}$$

where the subscripts denote the partial derivative. Dividing by dx we get

$$\frac{df}{dx}=f_x+f_y\frac{dy}{dx}=f_x+f_yf=\frac{d^2y}{dx^2} \tag{1822}$$

Treating d^2y/dx^2 as a function of x and y, we take the differential

$$d(f_x+f_yf)=\frac{\partial}{\partial x}(f_x+f_yf)dx+\frac{\partial}{\partial y}(f_x+f_yf)dy \tag{1823}$$

and dividing by dx we get

$$\frac{d^3y}{dx^3}=f_{xx}+2f_{xy}f+f_{yy}f^2+(f_x+ff_y)f_y \tag{1824}$$

and so on for the higher derivatives.

We now expand y in a Taylor series about x_0. Then

$$y=y_0+\left(\frac{dy}{dx}\right)_0(x-x_0)+\frac{1}{2!}\frac{d^2y}{dx^2}(x-x_0)^2+\cdots \tag{1825}$$

Letting $x-x_0=h$ and taking $dy/dx, d^2y/dx^2$ from equations (1797) and (1799) with similar substitutions for the higher order derivatives, we get

$$y-y_0=\Delta y=f_0h+\tfrac{1}{2}(f_{\dot{x}}+f_yf)_0h^2+\tfrac{1}{6}(f_{xx}+2f_{xy}f+f_{yy}f^2$$
$$+f_x+ff_y)_0f_yh^3+\cdots \tag{1826}$$

where the subscript 0 indicates that the functions are to be evaluated at $x=x_0$, $y=y_0$.

Alternatively, we may expand $f(x,y)$ in a Taylor series for two independent variables. Then

$$f(x_0+p,y_0+q)=f_0+(f_x)_0p+(f_y)_0q+\tfrac{1}{2}[(f_{xx})_0p^2+(f_{xy})_0pq+(f_{yy})_0q^2] \tag{1827}$$

We now define

$$\Delta' y=f(x_0,y_0)\Delta x \tag{1828}$$

$$\Delta'' y=f(x_0+mh,y_0+m\Delta' y)h \tag{1829}$$

$$\Delta'''y = f(x_0 + \lambda h, y_0 + \rho\Delta''y + (\lambda - \rho)\Delta'y)h \tag{1830}$$

Thus, we have selected two different values for p and q. In (1829), $p = mh$, $q = m\Delta'y$; in (1830), $p = \lambda h$,

$$q = \rho\Delta''y + (\lambda - \rho)\Delta'y$$

where m, λ, and ρ have been left as free parameters. Equations (1829) and (1830) may now be expressed in a Taylor series for a function of two independent variables. Moreover, the $\Delta'y$ and $\Delta''y$ from the right-hand side of each equation may be eliminated by use of the previous equation. There results

$$\begin{aligned}
\Delta'y &= f_0 h \\
\Delta''y &= f_0 h + m(f_x + f_y f)_0 h^2 + \tfrac{1}{2}m^2(f_{xx} + 2f_{xy}f + f_{yy}f^2)_0 h^3 \\
\Delta'''y &= f_0 h + \lambda(f_x + f_y f)_0 h^2 + \tfrac{1}{2}[\lambda(f_{xx} + 2f_{xy}f + f_{yy}f^2) \\
&\quad + 2m\rho(f_x + f_y f)f_y]_0 h^3
\end{aligned} \tag{1831}$$

We now wish to express Δy in the form

$$\Delta y = a\Delta'y + b\Delta''y + c\Delta'''y \tag{1832}$$

Substituting for $\Delta'y, \Delta''y$ and $\Delta'''y$ from equations (1831) and comparison with the Taylor series expansion of equation (1826) shows that the free parameters must satisfy the following relations:

$$\begin{aligned}
a + b + c &= 1 \\
bm + cy &= \tfrac{1}{2} \\
bm^2 + c\lambda^2 &= \tfrac{1}{3} \\
c\rho m &= \tfrac{1}{6}
\end{aligned} \tag{1833}$$

Since there are six free parameters and only four constraining equations, two of the parameters may be chosen at random. We shall give only one solution here, Kutta's third-order rule. Kutta used

$$\begin{aligned}
\lambda &= 1 \qquad & a &= \tfrac{1}{6} \\
\rho &= 2 & b &= \tfrac{2}{3} \\
m &= \tfrac{1}{2} & c &= \tfrac{1}{6}
\end{aligned} \tag{1834}$$

Then

$$\Delta y = \tfrac{1}{6}(\Delta'y + 4\Delta''y + \Delta'''y) \tag{1835}$$

where

$$\begin{aligned}
\Delta'y &= f(x, y)h \\
\Delta''y &= f(x + \tfrac{1}{2}h, y + \tfrac{1}{2}\Delta'y)h \\
\Delta'''y &= f(x + h, y + 2\Delta''y - \Delta'y)h
\end{aligned} \tag{1836}$$

The accuracy of the solution is equivalent to that of a Taylor series of equal step size, $\Delta x = h$, keeping only the first three terms. The advantage lies in the saving of computational time.

The above method may obviously be carried out to any order desired except that after the fourth order the algebra tends to become intractable. We have derived the third-order Runge–Kutta method above. Actually, the most common Runge–Kutta method used in practice is the fourth order. In this case

$$\Delta y = a\Delta' y + b\Delta'' y + c\Delta''' y + d\Delta'''' y \tag{1837}$$

where

$$\Delta'''' y = f(x + \mu h, y + \sigma\Delta''' y + \tau\Delta'' y + (\mu - \sigma - \tau)\Delta' y)h$$

The method of evaluating the constants is the same as before. We shall give only one of the possible results, called the Kutta–Simpson one-third rule. (The terminology here is somewhat vague. In practice, all the formulas given here are usually referred to as Runge–Kutta).

$$\Delta y = \tfrac{1}{6}(\Delta' y + 2\Delta'' y + 2\Delta''' y + \Delta'''' y) \tag{1838}$$

where

$$\begin{aligned} \Delta' y &= f(x, y)h \\ \Delta'' y &= f(x + \tfrac{1}{2}h, y + \tfrac{1}{2}\Delta' y)h \\ \Delta''' y &= f(x + \tfrac{1}{2}h, y + \tfrac{1}{2}\Delta'' y)h \\ \Delta'''' y &= f(x + h, y + \Delta''' y)h \end{aligned} \tag{1839}$$

We notice the elegant simplicity of these formulas.

Integral Formulas

There are numerous formulas for the numerical evaluation of integrals. These formulas are usually derived by expressing the integral as a polynomial and integrating term by term. Such formulas go under the general name of quadratures. The best known of the quadratures is Simpson's rule.

We wish to evaluate the integral

$$I = \int_{x_1}^{x_1 + \Delta x} f(x)\, dx \tag{1840}$$

where Δx is a small interval. Define a new variable ξ such that

$$\xi = x - (x_1 + \tfrac{1}{2}\Delta x) \tag{1841}$$

and let $\Delta x/2 = h$. Then

$$I = \int_{-h}^{h} f(\xi)\, d\xi \tag{1842}$$

We now assume that the Taylor series expansion of $f(x)$ exists and converges. Then

$$f(\xi) = f(0) + f'(0)\xi + \frac{1}{2!}f''(0)\xi^2 + \frac{1}{3!}f'''(0)\xi^3 + \ldots$$

Retaining only the first four terms and substituting in the integral,

$$I = \int_{-h}^{h} \left[f(0) + f'(0)\xi + \frac{1}{2!} f''(0)\xi^2 + \frac{1}{3!} f'''(0)\xi^3 \right] d\xi \tag{1843}$$

$$I = \frac{h}{3}[2f''(0)h^2 + 6f(0)] \tag{1844}$$

Now let y_1, y_2 and y_3 equal the values of $f(\xi)$ at $\xi = -h, 0$, and $+h$, respectively. Then

$$\begin{aligned} y_1 &= -f'''(0)h^3 + f''(0)h^2 - f'(0)h + f(0) \\ y_2 &= f(0) \\ y_3 &= f'''(0)h^3 + f''(0)h^2 + f'(0)h + f(0) \end{aligned} \tag{1845}$$

from which it follows that

$$2f''(0)h^2 + 6f(0) = y_1 + 4y_2 + y_3 \tag{1846}$$

and hence

$$\int_{x_1}^{x_1+\Delta x} f(x)\,dx = \frac{h}{3}(y_1 + 4y_2 + y_3) \tag{1847}$$

This is called Simpson's one-third rule.

Other integration formulas may be found writing the derivatives in Taylor's series in finite difference form. That is, divide the x axis into a series of increments of width h (Figure 318). Then approximately,

$$f'(x_0) = \frac{y_m - y_{m-1}}{h}, \qquad y_m = f(x_m) \tag{1848}$$

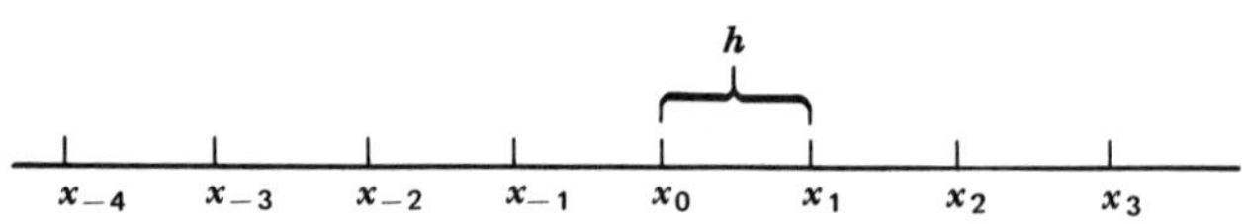

Figure 318 Step size on the x axis.

where $m = 0, \frac{1}{2}$, or 1. For $m = 1$, the formula (1848) is called the forward difference or descending difference, for $m = \frac{1}{2}$, the formula (1848) is called the central difference, and for $m = 0$, the formula (1848) is called the backward difference or ascending difference. Similarly, the second derivative may be approximated by

$$f''(x_0) = \frac{\dfrac{y_{n+1} - y_n}{h} - \dfrac{y_n - y_{n-1}}{h}}{h} = \frac{y_{n+1} - 2y_n + y_{n-1}}{h^2} \tag{1849}$$

We wish now to evaluate the integral

$$I = \int_{x_0}^{x_0+\Delta x} f(x)\,dx$$

where Δx is a small interval Presuming the Taylor series of $f(x)$ about x_0 to exist and converge,

$$I \cong \int_{x_0}^{x_0+\Delta x} \left[f(x_0) + \frac{1}{2!} f'(x_0)(x - x_0) + \frac{1}{3!} f''(x_0)(x - x_0)^2 \right] dx \quad (1850)$$

We now define a new variable, ξ, such that $\xi = x - x_0$, and express f as a function of ξ. Then

$$I = \int_0^{\Delta x} (f(0) + f'(0)\xi + \frac{1}{2!} f''(0)\xi^2)\, d\xi \quad (1851)$$

$$= f(0)\Delta x + f'(0)\frac{(\Delta x)^2}{2} + \frac{1}{2!} f''(0)\frac{(\Delta x)^3}{3}$$

We now define a small interval h such that $\Delta x = jh$ where j is an integer. Then

$$I \cong f_0 jh + (f_m - f_{m-1})\frac{(jh)^2}{2h} + (f_{n+1} - 2f_n + f_{n-1})\frac{(jh)^3}{6h^2} \quad (1852)$$

$$= h\left[f_0 j + (f_m - f_{m-1})\frac{j^2}{2} + (f_{n+1} - 2f_n + f_{n-1})\frac{j^3}{6} \right] \quad (1853)$$

where $m, n = 0$ refer to the function evaluated at $\xi = 0$. Clearly, a number of different formulas may now be written for the integral I, depending on the values of m, n and j which are used. For example, taking $m = n = 1, j = 2$, we get

$$I = \frac{2h}{3}[2f_0 - f_1 + 2f_2] \quad (1854)$$

Even more formulas may be obtained if more terms are retained in the Taylor series. Kunz [202] gives a list of 28 such formulas. The smaller h and j are taken, and the larger the number of terms retained in the Taylor series, the more accurate the formula.

Integration formulas may also be derived in terms of operators. Define Δ as the forward difference operator. Thus,

$$\Delta y_i = y_{i+1} - y_i \quad (1855)$$

Define E as the incremental operator; that is,

$$Ey_i = y_{i+1} \quad (1856)$$

Define D as the differential operator:

$$D = \frac{d}{dx} \quad (1857)$$

It can be shown that these operators may be treated as variables in accordance with the rules of algebra. For example,

$$Ey_i = 1 + \Delta y_i \quad (1858)$$

Thus we may say

$$E = 1 + \Delta \quad (1859)$$

Also, if $Ef(x) = f(x + h)$,

$$Ef(x) = \left[1 + hD + \frac{h^2}{2!} D^2 + \cdots \right] f(x) \quad (1860)$$

or

$$E = 1 + hD + \frac{h^2}{2!}D^2 + \cdots$$

If E and D are treated as algebraic quantities we may say

$$E = e^{hD} \qquad \text{or} \qquad D = \frac{\ln E}{h} = \frac{1}{h}\ln(1 + \Delta) \tag{1861}$$

Expanding the log on the right-hand side we have

$$D = \frac{1}{h}\left(\Delta - \frac{\Delta^2}{2} + \frac{\Delta^3}{3} + \cdots\right) \tag{1862}$$

Now

$$D\int_a^{a+h} f(x)\,dx = f(x)\Big|_a^{a+h} \tag{1863}$$

Again, D may be treated like an algebraic quantity, so that

$$\int_a^{a+h} f(x)dx = \frac{f(x)}{D}\Big|_a^{a+h} = \frac{f(a+h) - f(a)}{D} = \frac{E-1}{D}f(a) \tag{1864}$$

Now define E^j so that

$$E^j f(x) = f(x + jh) - f(x) \tag{1865}$$

and Δ^j so that

$$\Delta^j f(x) = \Delta[\Delta^{j-1} f(x)] \tag{1866}$$

Then

$$E^j = (1 + \Delta)^j = 1 + j\Delta + \frac{j}{2}\Delta^2 + \cdots$$

and

$$\int_a^{a+jh} f(x)dx = \frac{E^j - 1}{D}f(a) = \frac{(1+\Delta)^j - 1}{D}f(a) \tag{1867}$$

Expressing numerator and denominator as power series in Δ, and dividing the denominator into the numerator, we get

$$\int_a^{a+jh} f(x)dx = jh\left[1 + \frac{j}{2}\Delta + \frac{j(2j-3)}{12}\Delta^2 + \frac{j(j-2)^2}{24}\Delta^3 + \cdots\right]f(a) \tag{1868}$$

Again, numerous integration formulas may be generated from this result. For example, if $j = 1$ and all higher order terms are neglected, we get

$$\int_a^{a+h} f(x)\,dx = \frac{h}{2}[f(a) + f(a+h)] \tag{1869}$$

Similarly, taking $j = 2$ and retaining terms up to Δ^3, we get

$$\int_a^{a+2h} f(x)\,dx = \frac{h}{3}[f(a) + 4f(a+h) + f(a+2h)] \tag{1870}$$

which will be recognized as Simpson's rule.

Predictor–Corrector Methods

While entire solutions may be generated by the use of Taylor series or Runge–Kutta methods, solutions of equal accuracy using less computer time may be generated by the predictor corrector technique. Suppose, for example, we wish to integrate the equation

$$y' = f(x, y)$$

Applying the formula (1854) over the interval x_{n-3} to x_{n+1}, we have

$$y_{n+1} - y_{n-3} = \int_{x_{n-3}}^{x_{n+1}} y'\,dx = \frac{4h}{3}[2y'_{n-2} - y'_{n-1} + 2y'_n] \tag{1871}$$

It may be shown that the error here is

$$\frac{15}{45}h^5 y^{(5)}(\xi_1), \qquad x_{n-3} < \xi_1 < x_{n+1} \tag{1872}$$

Although the value of ξ_1 is not known, the error bounds may be controlled by the size of h.

The formula (1871) gives us y_{n+1} provided y is known up to y_n. The formula (1871) is known as the predictor. An improved value of y_{n+1} may be obtained by using an integration formula which contains y'_{n+1} on the right-hand side. For example, Simpson's rule:

$$y_{n+1} - y_{n-1} = \int_{x_{n-1}}^{x_{n+1}} y'\,dx = \frac{h}{3}(y'_{n-1} + 4y'_n + y'_{n+1}) \tag{1873}$$

Using the value of y'_{n+1} obtained by the predictor in the right-hand side of equation (1873), we get an improved value of y_{n+1}. The formula (1873) is known as the corrector. Obviously, there are as many predictor-corrector techniques as there are integrations formulas. When the formulas given above are used, this particular algorithm is called Milne's method.

Equation (1873) may be iterated for improved accuracy, but this is poor use of computer time. Accuracy may be improved more cheaply by decreasing h.

The predictor-corrector technique is not self-starting. In order to obtain y_{n+1}, the preceding three values had to be known. It has become customary to obtain starting values by the Runge–Kutta method and then to generate the rest of the solution by a predictor corrector technique.

Adams method

Like the terms Runge–Kutta and predictor-corrector, the term Adams method tends to be a generic term applied to a number of variations. Often, the variations are designated by a hyphen, such as Adams-Bashforth, Adams-Moulton, Adams-Stormer, and so on.

Adams method may be derived in terms of the backward difference operator ∇:

$$\nabla y_i = y_i - y_{i-1} \tag{1874}$$

Thus

$$\nabla = 1 - \frac{1}{E} = 1 - \left[1 - hD + \frac{h^2D^2}{2!} - \cdots\right] = 1 - e^{-hD} \tag{1875}$$

Now

$$\nabla^2 y_i = \nabla(\nabla y_i) = \nabla(y_i - y_{i-1}) = (y_i - y_{i-1}) - (y_{i-1} - y_{i-2}) \tag{1876}$$

$$\begin{aligned}\nabla^3 y_i &= \nabla(\nabla^2 y_i) = \nabla(y_i - 2y_{i-1} + y_{i-2}) \\ &= (y_i - y_{i-1}) - 2(y_{i-1} - y_{i-2}) + (y_{i-2} - y_{i-3})\end{aligned} \tag{1877}$$

and so on. Now

$$\nabla^2 = (1 - e^{-hD})^2 = h^2D^2 - h^3D^3 + \tfrac{7}{12}h^4D^4 + \cdots \tag{1878}$$

$$\nabla^3 = (1 - e^{-hD})^3 = h^3D^3 - \tfrac{3}{2}h^4D^4 + \tfrac{5}{4}h^5D^5 + \cdots \tag{1879}$$

and so on. Conversely,

$$D = -\frac{1}{h}\ln(1 - \nabla) = \frac{1}{h}\left(\nabla + \frac{\nabla^2}{2} + \frac{\nabla^3}{3} + \cdots\right) \tag{1880}$$

$$D^2 = \left[-\frac{1}{h}ln(1 - \nabla)\right]^2 = \frac{1}{h^2}\left(\nabla^2 + \nabla^3 + \frac{11}{12}\nabla^4 \cdots\right) \tag{1881}$$

From equation (1875) it follows that

$$\nabla y_{i+1} = \left[hD - \frac{h^2D^2}{2} + \frac{h^3D^3}{6} - \cdots\right] y_i \tag{1882}$$

$$= \left[1 - \frac{hD}{2} + \frac{h^2D^2}{6} - \cdots\right] hDy_i \tag{1883}$$

Substituting for D in terms of ∇ in accordance with equations (1880)–(1881) and noting that $Dy_i = y_i'$ we get

$$y_{i+1} - y_i = h\left[1 + \frac{\nabla}{2} + \frac{5}{2}\nabla^2 + \frac{3}{8}\nabla^3 + \cdots\right] y_i' \tag{1884}$$

This is Adams formula. We note that this is an open formula. An integration formula, or quadrature, is called open if it does not contain the upper limit, and closed if it does contain the upper limit. Thus, predictor formulas are open, while corrector formulas are closed. In the case of Adams formula, it is considered of sufficient accuracy so that the corrector may be dispensed with in many cases. However, a number of modifications of Adams method use one or another corrector formula to improve accuracy. Perusal of Adams formula and equations (1874)–(1877) shows that four starting values are needed. To get y_{i+1} we need y_i, y_{i-1}, y_{i-2} and y_{i-3}. Once these values are known, y_i', y_{i-1}', y_{i-2}' and y_{i-3}' may be obtained from the given differential equation

$$y' = f(x, y)$$

which we are trying to solve. In practice, starting values are usually obtained by the Runge–Kutta method. This combination of Adams and Runge–Kutta has proved very successful and is widely used.

If the fourth-order term is retained, Adams formula becomes

$$y_{i+1} - y_i = h\left[1 + \frac{1}{2}\nabla + \frac{5}{12}\nabla^2 + \frac{3}{8}\nabla^3 + \frac{251}{720}\nabla^4\right]y_i' \qquad (1885)$$

The method is then called Adams–Bashforth. The accuracy of Adams method may be improved by the use of a corrector formula. Kunz [202] recommends

$$y_{i+1} - y_i = h\left[1 - \frac{1}{2}\nabla - \frac{1}{12}\nabla^2 - \frac{1}{24}\nabla^3 - \frac{19}{720}\nabla^4\right]y_{i+1}' - \frac{3h^6}{160}y^{(6)}(\xi) \qquad (1886)$$

where $x_{i-3} < \xi < x_{i+1}$, and the last term indicates the error bounds. In general, if a corrector of the type

$$y_{i+1} - y_i = \sum_{k=-1}^{n} b_k y_{i-k}' \qquad (1887)$$

is used, the method is known as Adams–Moulton.

Higher Order Equations

So far we have been dealing mainly with first-order equations. This is not as restrictive as it sounds. All higher order equations can be reduced to a system of first-order equations. It is for this reason that numerical analysis has concentrated on first-order equations.

Consider the equation

$$f(x, y, y', y'', \ldots, y^{(n)}) = 0 \qquad (1888)$$

where y is a function of x and the primes indicate differentiation with respect to x. By treating the derivatives of y as independent variables, equation (1865) may be changed to a system of n simultaneous first-order equations. Let

$$y^{(i)} = y_i \qquad (1889)$$

Then the system is

$$\begin{aligned} y' &= y_1 \\ y_1' &= y_2 \\ y_2' &= y_3 \\ &\vdots \\ y_{n-2}' &= y_{n-1} \\ y_n = y_{n-1}' &= g(x, y, y_1, y_2, \ldots, y_{n-1}) \end{aligned} \qquad (1890)$$

To solve this system we require that the initial conditions be given. If x_0 is the initial value of x, we require

$$y(x_0)$$
$$y_i(x_0) = y^{(i)}(x_0), \qquad i = 1 \text{ to } n-1$$

$y_n(x_0)$ is then determined by substituting in the differential equation. The simplest method of integrating this system is the so-called method of Euler.

$$\left(\frac{dy_{n-1}}{dx}\right)_0 = (g)_0 \tag{1891}$$

where the subscript 0 indicates the initial value. Then write

$$\left(\frac{\Delta y_{n-1}}{\Delta x}\right)_0 = (g)_0, \qquad \Delta y_{n-1} = (g)_0 \Delta x, \qquad y_{n-1}(x_0 + \Delta x) = y_{n-1}(x_0) + (g)_0 \Delta x$$

Similarly

$$\left(\frac{dy_{n-2}}{dx}\right)_0 = (y_{n-1})_0$$

$$\Delta y_{n-2} = (y_{n-1})_0 \Delta x, \qquad y_{n-2}(x_0 + \Delta x) = y_{n-2}(x_0) + (y_{n-1})_0 \Delta x$$

and so on. This gives us a new set of values for $x, y, y_i, i = 1, 2, \ldots, n-1$. The process may then be repeated. This method is presented for purposes of illustration only and is not recommended for use in practice.

An improvement in the method of Euler may be obtained by use of a primitive corrector formula. When the first step has been completed, calculate

$$\frac{dx_i}{dy} = \frac{1}{2}\left[\left(\frac{dx_i}{dy}\right)_0 + \left(\frac{dx_i}{dy}\right)_1\right] \tag{1892}$$

where the subscript 1 indicates the value for $x_0 + \Delta x$. Use this new value of dx_i/dy in place of $(dx_i/dy)_0$ and repeat the first step. Equation (1892) may then be applied again and the process repeated until no further change is observed. Although this method is not recommended to find a complete solution, it may be used to find starting values.

The method of Runge–Kutta may also be extended to solve systems of simultaneous equations. The appropriate formulas are derived in a manner analogous to that used for solution of a single equation. Consider the system of two simultaneous first-order equations.

$$\frac{dy}{dx} = f(x, y, z), \qquad \frac{dz}{dx} = g(x, y, z) \tag{1893}$$

The Kutta–Simpson one-third rule in this case becomes

$$\begin{aligned} \Delta y &= \tfrac{1}{6}(k_0 + 2k_1 + 2k_2 + k_3) \\ \Delta z &= \tfrac{1}{6}(l_0 + 2l_1 + 2l_2 + l_3) \end{aligned} \tag{1894}$$

where we have employed the notation

$$k_{i-1} = \Delta^{(i)}y, \qquad l_{i-1} = \Delta^{(i)}z \tag{1895}$$

It may be shown that in the present case

$$\begin{aligned} k_0 &= f(x_0, y_0, z_0)h \\ k_1 &= f(x_0 + \tfrac{1}{2}h, y_0 + \tfrac{1}{2}k_0, z_0 + \tfrac{1}{2}l_0)h \\ k_2 &= f(x_0 + \tfrac{1}{2}h, y_0 + \tfrac{1}{2}k_1, z_0 + \tfrac{1}{2}l_1)h \\ k_3 &= f(x_0 + h, y_0 + k_2, z_0 + l_2)h \end{aligned} \tag{1896}$$

Similarly,

$$\begin{aligned} l_0 &= g(x_0, y_0, z_0)h \\ l_1 &= g(x_0 + \tfrac{1}{2}h, y_0 + \tfrac{1}{2}k_0, z_0 + \tfrac{1}{2}l_0)h \\ l_2 &= g(x_0 + \tfrac{1}{2}h, y_0 + \tfrac{1}{2}k_1, z_0 + \tfrac{1}{2}l_1)h \\ l_3 &= g(x_0 + h, y_0 + k_2, z_0 + l_2)h \end{aligned} \tag{1897}$$

Again the initial values x_0, y_0, z_0 are presumed given. This result may be applied directly to the solution of a second-order equation. If

$$\frac{d^2y}{dx^2} = g\left(x, y, \frac{dy}{dx}\right) \tag{1898}$$

we have the system of first-order equations

$$\frac{dz}{dx} = g(x, y, z) \tag{1899}$$

$$\frac{dy}{dx} = z \tag{1900}$$

The l_i are then as given by equations (1897). The k_i reduce to

$$k_0 = z_0 h, \qquad k_1 = (z_0 + \tfrac{1}{2}l_0)h, \qquad k_2 = (z_0 + \tfrac{1}{2}l_1)h, \qquad k_3 = (z_0 + l_2) \tag{1901}$$

In a similar way, the method may be extended to third-order equations (see, for example, Weeg and Reed [203]).

While an entire solution may be generated in this way, in practice the Runge–Kutta method is preferred to obtain starting values, while a continuation method, such as Adams method, is used to obtain the rest of the solution. Adams method may be extended to higher order equations by rewriting the equation as a system of first-order equations. For example, given the second-order equation

$$y'' = f(x, y, y') \tag{1902}$$

where the primes indicate differentiation with respect to x. We write

$$y' = g(x, y) \tag{1903}$$

$$g' = f(x, y, g) \tag{1904}$$

Then from equation (1885) we have

$$y_{i+1} - y_i = h\left[1 + \frac{1}{2}\nabla + \frac{5}{12}\nabla^2 + \frac{3}{8}\nabla^3 + \frac{251}{720}\nabla^4\right]y_i' \tag{1905}$$

$$g_{i+1} - g_i = h\left[1 + \frac{1}{2}\nabla + \frac{5}{12}\nabla^2 + \frac{3}{8}\nabla^3 + \frac{251}{720}\nabla^4\right]g_i' \tag{1906}$$

We thus require that the starting values $y_i, y_{i-1}, y_{i-2}, y_{i-3}$ and $g_i, g_{i-1}, g_{i-2}, g_{i-3}$ be given. (The reader should take care not to confuse the subscripts here, which refer the value of x at which the function is evaluated, with the subscripts in equations (1890), which refer to the order of the derivative.)

The method may clearly be extended to equations of any order simply by writing each equation as a system of equations of the first order. When Adams method is used as outlined above, it is called the Adams–Störmer method. For additional accuracy, a corrector formula, such as equation (1886), may be employed at each step.

Not all numerical methods for solving higher order equations require rewriting the equation as a series of first-order equations. As we have already seen, the method of Taylor series expansion may be applied to equations of any order. The drawback of this method is that it is not considered cost effective in comparison with other methods, such as Adams and Runge–Kutta.

Milne's method may also be extended to second-order equations. Consider, for example, the equation

$$y'' = f(x, y, y') \tag{1907}$$

The initial values, x_0, y_0 and y_0' are presumed given. In addition, four starting values are presumed to have been found by some other method (Euler, Taylor series, Runge–Kutta, etc.). Then from equation (1871) we may write

$$y_{n+1}' - y_{n-3} = \frac{4h}{3}(2y_n'' - y_{n-1}'' + 2y_{n-2}'') \tag{1908}$$

where the primes indicate differentiation with respect to x and the subscripts indicate the value of x at which the functions are evaluated. y_i, y_i', y_i'' are presumed known for $i = n, n-1, n-2, n-3$. With y_{n+1}' known from equation (1885) we may find y_{n+1} from Simpson's rule:

$$y_{n+1} - y_{n-1} = \frac{h}{3}(y_{n+1}' - 4y_n' + y_{n-1}') \tag{1909}$$

The differential equation (1907) may then be used to find y_{n+1}''. Simpson's rule may then be used as a corrector for y_{n+1}':

$$y_{n+1}' - y_{n-1}' = \frac{h}{3}(y_{n+1}'' + 4y_n'' + y_{n-1}'') \tag{1910}$$

Another method which may be used for the integration of second order equations is Numerov's method. This method is suitable for linear equations of the type

$$\frac{d^2y}{dx^2}+f(x)y=F(x) \tag{1911}$$

Define δ as the central difference operator. Thus,

$$\delta y_i = y_{i+1/2} - y_{i-1/2} \tag{1912}$$

Recall the operator E, defined by

$$E^j y_i = y_{i+j} - y_i \tag{1913}$$

Then

$$\delta = E^{1/2} - E^{-1/2} \tag{1914}$$

Recall also

$$E = e^{hD} \tag{1915}$$

It follows that

$$\delta^2 = 2\cosh(hD) - 2 = h^2D^2 + \frac{h^4D^4}{12} + \frac{h^6D^6}{360} + \cdots \tag{1916}$$

or

$$D^2 = \frac{1}{h^2}\left[\delta^2 - \frac{\delta^4}{12} + \frac{\delta^6}{90} - \cdots\right] \tag{1917}$$

Substituting (1917) in the differential equation (1911)

$$\left(\delta^2 - \frac{\delta^4}{12} + \frac{\delta^6}{90} - \cdots\right)y_i + h^2 f_i y_i = h^2 F_i \tag{1918}$$

Multiplying both sides by $1 + \delta^2/12$ we get

$$\left(\delta^2 + \frac{\delta^6}{240} - \frac{13\delta^8}{15{,}120} + \cdots\right)y_i + h^2 f_i y_i + \frac{h^2\delta^2}{12}(f_i y_i) = h^2 F_i + \frac{h^2\delta^2}{12}F_i \tag{1919}$$

or

$$y_{i+1} - 2y_i + y_{i-1} + h^2 f_i y_i + \frac{h}{12}(f_{i+1}y_{i+1} - 2f_i y_i + f_{i-1}y_{i-1})$$
$$= h^2F_i + \frac{h^2}{12}(F_{i+1} - 2F_i + F_{i-1}) - \left(\frac{\delta^6}{240} - \frac{13\delta^8}{15{,}120} + \cdots\right)y_i \tag{1920}$$

Neglecting the last term of solving for y_{i+1},

$$y_{i+1} = \frac{1}{1+\frac{h^2}{12}f_{i+1}}\left[\left(2 - \frac{5}{6}h^2 f_i\right)y_i - \left(1 + \frac{h^2}{12}f_{i-1}\right)y_{i-1} + \frac{h^2}{12}(F_{i-1} + 10F_i + F_{i+1})\right] \tag{1921}$$

This is Numerov's recursion formula. Two starting values are needed: y_i and y_{i-1}. The error involved at each step is of order h^6.

Boundary conditions

In any practical solution of a differential equation, certain boundary conditions must be met. We have already mentioned the two principal types of boundary value problems. Initial value problems, also called marching problems, and two, or multipoint boundary value problems, also called jury problems. Most numerical integration schemes assume that all the boundary conditions are given at one point. Only initial value problems can be integrated forward directly. It sometimes happens, however, that boundary conditions are given at two or more different points. The solution of boundary layer problems is a notable example of this. Boundary conditions are given at the wall and at the outer edge of the boundary layer. Except in the simplest cases, boundary layer equations must be integrated numerically. Techniques are, therefore, necessary for dealing with problems of this type.

The simplest and most obvious way of dealing with two point boundary value problems is trial and error. Say we have a second-order equation

$$f(x, y, y', y'') = 0 \tag{1922}$$

with boundary conditions given at $x = a$ and $x = b$:

$$y(a) = A, \qquad y(b) = B \tag{1923}$$

We begin by guessing a value for $y'(a)$, A' say, and solve the problem for the boundary conditions

$$y(a) = A, \qquad y'(a) = A' \tag{1924}$$

We then compare the resulting value of $y(b)$ with B and, if it is different, we try another value of A' and repeat the procedure. Sometimes convergence is quite rapid and computers can easily be programmed to do the hunting automatically. $y(b)$ may be brought as close to B as desired. This method is called the shooting method. A disadvantage of the shooting method is that it tends to be expensive in computer time. On an IBM system 370, solution of a typical boundary layer equation by shooting method will take about 30 seconds.

Results of a typical shooting method run have already been shown in Figure 249. This figure illustrates another problem associated with initial value problems in general and boundary layer equations in particular: the problem of instability. It will be observed that when the assumed initial value of the slope the velocity profile, dy/du, is too low, the velocity goes to plus infinity while if the initial value of the slope of the velocity profile is too high, the velocity goes to minus infinity. Only for the correct value of the initial slope does the velocity remain bounded. This phenomenon is called instability. Sometimes the term is used to mean merely that a small change in the initial conditions causes a large change in the solution. In the case of initial value problems, where all the boundary conditions are given at one point, instability can cause real problems in numerical solutions, since the cumulative error due to the numerical approximation can cause the solution to go to plus or minus infinity, when in fact, it should be bounded. In the case of two-point boundary conditions, on the other hand, instability need not cause any problems and can, in fact be an advantage. In boundary layer problems, for example, one is typically given the free stream velocity and seeks the slope of the velocity profile at the wall. If the equation is unstable, a large change in the free stream value of the solution will cause only a small change in the velocity profile at the wall. Thus, if the solution satisfies the free stream conditions, the chances are the solution near the wall will be very accurate. In this connection it might be pointed out that in boundary layer problems, one boundary condition is usually given at $y = \infty$. In practice, it is therefore necessary to impose this boundary condition at some value $y < \infty$. Usually, a value three or four times the expected boundary layer thickness will prove adequate. The further out the value is chosen, the more accurate the solution and the more expensive the computer time. And vice versa. A little trial and error will show the effect of moving the point taken as "$y = \infty$." As a rule, no significant change will be observed once y exceeds 4δ.

In the case of linear equations, a much better way exists of solving two-point boundary value problems. Consider, for example, the equation

$$y'' = p(x)y + q(x)y' + r(x) \tag{1925}$$

with the boundary conditions

$$y(0) = 0, \qquad y(a) = A \tag{1926}$$

We first solve the problem

$$Y'' = p(x)Y + q(x)Y' + r(x) \tag{1927}$$

subject to the boundary conditions

$$Y(0) = 0, \qquad Y'(0) = 1 \tag{1928}$$

Next we solve the problem

$$\eta'' = p(x)\eta + q(x)\eta' \tag{1929}$$

subject to the boundary conditions

$$\eta(0) = 0, \qquad \eta'(0) = 1$$

Clearly,

$$y(x) = Y(x) + \alpha\eta(x) \tag{1930}$$

is a solution of equation (1902), where α is a free parameter. Now set

$$y(a) = A = Y(a) + \alpha\eta(a) \tag{1931}$$

Thus

$$\alpha = \frac{A - Y(a)}{\eta(a)} \tag{1932}$$

Inserting this value of α into equation (1930) gives a solution of equation (1925) subject to boundary conditions (1926). Notice that no trial and error is involved.

The method can clearly be extended to other boundary conditions. For example, if $y'(a) = A'$ is given instead of $y(a)$, we merely rewrite equation (1931) in the form

$$y'(a) = A' = Y'(a) + \alpha\eta'(a) \tag{1933}$$

and use

$$\alpha = \frac{A' - Y'(a)}{\eta'(a)} \tag{1934}$$

If the boundary conditions given are

$$y(a) = A, \qquad y(b) = B \tag{1935}$$

we solve equation (1927) subject to the boundary conditions

$$Y(a) = A \qquad Y'(a) = 1 \tag{1936}$$

and equation (1929) subject to the boundary conditions

$$\eta(a) = 0, \qquad \eta'(a) = 1 \tag{1937}$$

The solution is then given by

$$y = Y + \alpha\eta \tag{1938}$$

where

$$\alpha = \frac{B' - Y'(b)}{\eta'(b)} \tag{1939}$$

If the boundary condition at b is changed to $y'(b) = B'$, then

$$\alpha = \frac{B' - Y'(b)}{\eta'(b)} \tag{1940}$$

and so on.

Method of Successive Integrations

This is a relaxation method which may be used for nonlinear second-order equations. Consider first the case

$$y'' = f(x, y) \tag{1941}$$

with

$$y = y_0 = y_a \quad \text{at} \quad x = a \quad \text{and} \quad y = y_n = y_b \quad \text{at} \quad x = b \tag{1942}$$

We divide x between a and b into n equal increments of width h so that

$$x_i - x_{i-1} = h, \qquad y_i = y(x_i) \tag{1943}$$

We then guess at set of values y_i, taking care that $y_0 = y_a$ and $y_n = y_b$. Using the differential equation (1941), we then get a set of values y_i''. We then find

$$y_i' - y_{i-1}' = \int_{x_{i-1}}^{x_i} y_i''\,dx \qquad \text{and} \qquad y_i' - y_0' = \int_{x_0}^{x_i} y_i''\,dx \tag{1944}$$

Now

$$\int_{x_0}^{x_i} (y_i' - y_0')\,dx = \int_{x_0}^{x_i} y_i'\,dx - \int_{x_0}^{x_i} y_0'\,dx = \int_{x_0}^{x_i} y_i'\,dx - ihy_0' \tag{1945}$$

and

$$y_i - y_0 = \int_{x_0}^{x_i} y_i'\,dx = \int_{x_0}^{x_i} (y_i' - y_0')\,dx + ihy_0' \tag{1946}$$

or

$$y_i = \int_{x_0}^{x_i} (y_i' - y_0')\,dx + ihy_0' + y_0 \tag{1947}$$

where the integrals must be evaluated numerically. Since y_0' is a free parameter here, we may choose it so that $y_n = y_b$ in equation (1947). Equation (1947) thus gives us a new set of y_i, which we may then use to repeat the procedure until changes in successive integrations are sufficiently small. The procedure is generally found to converge.

The method may be extended to the more general case

$$y'' = f(x, y, y') \tag{1948}$$

In this case the y_i' must also be guessed, but this may be done by letting

$$y_i' = \frac{y_{i+1} - y_{i-1}}{2h} \tag{1949}$$

Method of simultaneous equations

An alternate method of handling two-point boundary value problems with second-order equations is to replace the equation (1948) by a set of finite difference equations. Using the finite difference forms

$$y_i' = \frac{y_{i+1} - y_{i-1}}{2h}, \qquad y_i'' = \frac{1}{h^2}(y_{i+1} - 2y_i + y_{i-1}) \tag{1950}$$

we get

$$y_{i+1} - 2y_i + y_{i-1} = h^2 f\left(x_i, y_i, \frac{y_{i+1} - y_{i-1}}{2h}\right) \tag{1951}$$

with boundary conditions $y_0 = y_a, y_n = y_b$. Equation (1951) must be satisfied for all values of $i = 1, 2, \ldots, n-1$. Since y_0 and y_n are known, equation (1951) then represents a set of $n-1$ equations for $n-1$ unknowns y_i, which may be solved simultaneously.

Third-Order Linear Equations

The method outlined above for linear second-order equations may be extended to linear third-order equations. Consider a third-order equation of the form

$$m_1 g + m_2 g' + m_3 g'' + m_4 g''' = R \tag{1952}$$

with boundary conditions

$$g(0) = g'(0), \qquad g'(b) = 1 \tag{1953}$$

Consider now the homogeneous equation

$$m_1 g_h + m_2 g_h' + m_3 g_h'' + m_4 g_h''' = 0 \tag{1954}$$

Let g_h be the solution to this equation subject to the boundary conditions

$$g_h(0) = g_h'(0) = 0, \qquad g_h''(0) = 1$$

Let g_p be the solution of equation (1929) subject to the boundary conditions

$$g_p(0) = g_p'(0) = g_p''(0) = 0 \tag{1955}$$

The complete solution will then be given by

$$g = c_1 g_h + g_p \tag{1956}$$

The boundary conditions at $x = 0$ are met automatically. The boundary condition at $x = b$ will be met if

$$g'(b) = c_1 g_h'(b) + g_p'(b) = 1$$

which will be true if

$$c_1 = \frac{1 - g_p'(b)}{g_h'(b)} \tag{1957}$$

and hence the problem is solved.

Simultaneous Linear Equations of the Third Order This is a problem which arises in the field of boundary layer theory (see, for example, Schreier [204]). Given two equations

$$k_1 f + k_2 f' + k_3 f'' + k_4 f''' + k_5 g = R_f \tag{1958}$$

$$m_1 g + m_2 g' + m_3 g'' + m_4 g''' + m_5 f = R_g \tag{1959}$$

where k_i, m_i, R_f and R_g are known functions of x, with boundary conditions

$$\begin{aligned} f(0) = f'(0) = 0, &\qquad f'(\infty) = 1 \\ g(0) = g'(0) = 0, &\qquad g'(\infty) = 1 \end{aligned} \tag{1960}$$

Let f_{h_1} and g_{h_1} be the solutions to the equations

$$k_1 f_h + k_2 f_h' + k_3 f_h'' + k_4 f_h''' + k_5 g_h = 0 \tag{1961}$$

$$m_1 g_h + m_2 g_h' + m_3 g_h'' + m_4 g_h''' + m_5 f_h = 0 \tag{1962}$$

subject to the boundary conditions

$$\begin{aligned} f_{h_1}(0) = f_{h_1}'(0) = 0, &\qquad f_{h_1}''(0) = 1 \\ g_{h_1}(0) = g_{h_1}'(0) = 0, &\qquad g_{h_1}''(0) = 0 \end{aligned} \tag{1963}$$

and let f_{h_2} and g_{h_2} be the solutions of equations (1961) and (1962) subject to the boundary conditions

$$\begin{aligned} f_{h_2}(0) &= f'_{h_2}(0) = 0\,; \qquad f''_{h_2}(0) = 0 \\ g_{h_2}(0) &= g'_{h_2}(0) = 0\,; \qquad g''_{h_2}(0) = 1 \end{aligned} \tag{1964}$$

Furthermore, let f_p and g_p be the solutions of equations (1958) and (1959) subject to the boundary conditions

$$\begin{aligned} f_p(0) &= f'_p(0) = f''_p(0) = 0 \\ g_p(0) &= g'_p(0) = g''_p(0) = 0 \end{aligned} \tag{1965}$$

The solution of the original problem will then be given by

$$\begin{aligned} f &= c_1 f_{h_1} + c_2 f_{h_2} + f_p \\ g &= c_1 g_{h_1} + c_2 g_{h_2} + g_p \end{aligned} \tag{1966}$$

We observe that the boundary conditions

$$f(0) = f'(0) = g(0) = g'(0) = 0 \tag{1967}$$

are automatically met. To meet the third boundary condition we must first choose a value of $x < \infty$ at which to take the upper limit, as discussed earlier. In the case of a boundary layer problem, if the boundary layer thickness is known approximately, a value three or four times as high will serve well. Even twice the boundary layer thickness will give an accurate answer. Otherwise, a process of trial and error must be used, with $x = b$ taken in place of ∞, where b is sufficiently large so that taking $x > b$ produces no significant change in the result. The upper boundary condition

$$f'(\infty) = g'(\infty) = 1 \tag{1968}$$

will then be satisfied if

$$c_1 f'_{h_1}(b) + c_2 f'_{h_2}(b) + f'_p(b) = 1 \tag{1969}$$

$$c_1 g'_{h_1}(b) + c_2 g'_{h_2}(b) + g'_p(b) = 1 \tag{1970}$$

Solving equations (1969) simultaneously for c_1 and c_2 and substituting these values into equations (1966) then gives the complete solution.

In order to perform the actual numerical integration of the equations, we may rewrite the third-order equations as eighteen first-order equations. Let

$$\begin{aligned} x_1 &= f_{h_1} & x_7 &= g_{h_1} & x_{13} &= f_p \\ x_2 &= f'_{h_1} & x_8 &= g'_{h_1} & x_{14} &= f'_p \\ x_3 &= f''_{h_1} & x_9 &= g''_{h_1} & x_{15} &= f''_p \\ x_4 &= f_{h_2} & x_{10} &= g_{h_2} & x_{16} &= g_p \\ x_5 &= f'_{h_2} & x_{11} &= g'_{h_2} & x_{17} &= g'_p \\ x_6 &= f''_{h_2} & x_{12} &= g''_{h_2} & x_{18} &= g''_p \end{aligned}$$

The eighteen equations and their boundary conditions then become

1. $x_1' = x_2, \quad x_1(0) = 0$
2. $x_2' = x_3, \quad x_2(0) = 0$
3. $-k_y x_3' = k_1 x_1 + k_2 x_2 + k_3 x_3 + k_5 x_7, \quad x_3(0) = 1$
4. $x_4' = x_5, \quad x_4(0) = 0$
5. $x_5' = x_6, \quad x_5(0) = 0$
6. $-k_4 x_6' = k_1 x_4 + k_2 x_5 + k_3 x_6 + k_5 x_{10}, \quad x_6(0) = 0$
7. $x_{13}' = x_{14}, \quad x_{13}(0) = 0$
8. $x_{14}' = x_{15}, \quad x_{14}(0) = 0$
9. $-k_4 x_{15} = k_1 x_{13} + k_2 x_{14} + k_3 x_{15} + k_5 x - R_f, \quad x_{15}(0) = 0$

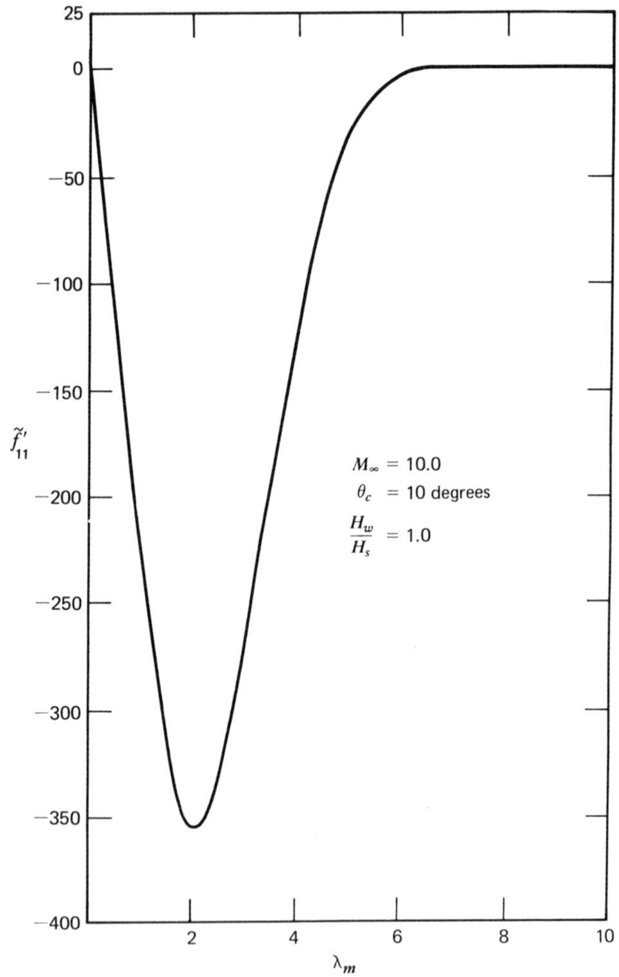

Figure 319 Numerical solution of a third-order equation of the type (1958) from [205].

10. $x_7' = x_8, \quad x_7(0) = 0$
11. $x_8' = x_9, \quad x_8(0) = 0$
12. $-m_4 x_9' = m_1 x_7 + m_2 x_8 + m_3 x_9 + m_5 x_1, \quad x_9(0) = 0$
13. $x_{10}' = x_{11}, \quad x_{10}(0) = 0$
14. $x_{11}' = x_{12}, \quad x_{11}(0) = 0$
15. $-m_4 x_{12}' = m_1 x_{10} + m_2 x_{11} + m_3 x_{12} + m_5 x_4, \quad x_{12}(0) = 1$
16. $x_{16}' = x_{17}, \quad x_{16}(0) = 0$
17. $x_{17}' = x_{18}, \quad x_{17}(0) = 0$
18. $-m_4 x_{18}' = m_1 x_{16} + m_2 x_{17} + m_3 x_{18} + m_5 x_{13} - R_g, \quad x_{18}(0) = 0$

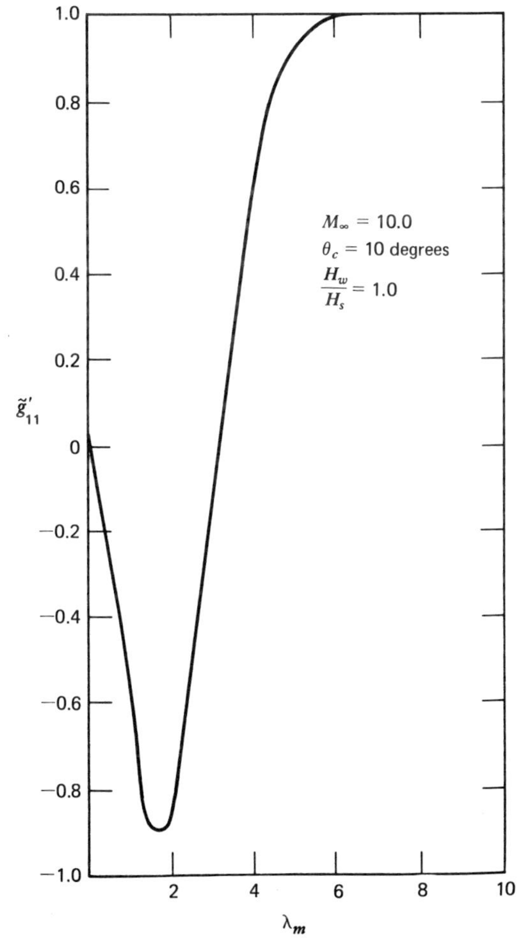

Figure 320 Numerical solution of a third order equation of the type (1959) from [205].

The constants c_1 and c_2 are now given by

$$c_1 x_2(b) + c_2 x_5(b) + x_{14}(b) = 1$$
$$c_1 x_8(b) + c_2 x_{11}(b) + x_{17}(b) = 1$$

There are now three sets of six first-order equations, each set of which may be solved separately. With the given initial values we find the first increment in the solution of equations 1, 2, and 3, and the first increment in the solution of equations 10, 11 and 12. The new value of x_7 is then used to find the second increment in equations 1, 2, and 3, and the new value of x_1 is used to find the next increment in equations 10, 11 and 12, and so on. Similarly for equations 4, 5 and 6, which must be solved together with equations 13, 14 and 15 and equations 7, 8 and 9 which are solved together with equations 16, 17 and 18. Starting values may be obtained by Taylor series, after which any suitable method for first-order equations may be used to carry the solution forward, such as Adams method or predictor-corrector.

Figures 319 and 320 show the results of one such calculation [205]. The particular functions refer to the velocity profiles in the boundary layer about an oscillating cone in supersonic flight. The longitudinal and circumferential velocity components are given by

$$u = f'_{00}(\lambda) + c_{01}\alpha\cos\phi f'_{01}(\lambda) + c_{11}\dot{\alpha}\cos\phi f'_{11}(\lambda)x + \cdots$$
$$w = k_{01}\sin\phi\,\alpha g'_{01}(\lambda) + k_{11}\sin\phi\,\dot{\alpha} g'_{11}(\lambda)x + \cdots$$

α being the angle of attack and ϕ the circumferential angle. Figures 319 and 320 show the solutions for f'_{11} and g'_{11}. λ_m, $\tilde{f}_{11}$ and $\tilde{g}_{11}$ differ only by constant factors from λ, f_{11}, and g_{11}. Infinity was here taken at $\lambda_m = 9.2$. We observe how quickly the solutions approach their asymptotic values as $\lambda_m \to \infty$. All the higher order derivatives were found to be essentially zero at $\lambda_m = 9.2$.

Errors and Error Bounds

Numerical methods are, by nature, approximate. The exact error cannot, however, be determined unless the exact solution is known, in which case there is no need for numerical methods. The best that can be hoped for is to determine error bounds; that is, the maximum and minimum possible values of the error.

There are two principal types of error associated with numerical analysis: round off error and truncation error. Round off error comes from approximating fractions by a finite number of decimal places. Such errors may be minimized by resorting to more decimal places, such as going to double precision in a computer, which increases the number of decimal places retained at each step from 8 to 16. Round off error at each step is generally small but may accumulate if a large number of steps is involved. Again, the trade-off must be recognized since going to double precision increases the computer time. Truncation error is the result of approximating the solution by an infinite series but retaining only a finite number of terms at each step. Truncation error can be decreased by using formulas which retain more terms, with a corresponding increase in computer time. To make such trade-offs intelligently, it is necessary to have some idea of the error bounds involved.

Error Bounds

The formulas for continuing the solution which we have derived for the integration of the first-order equation

$$\frac{dy}{dx} = F(x, y) \tag{1971}$$

such as Adam's method, predictor-corrector, etc., may be written in the general form

$$y_{n+1} = y_{n-p} + h \sum_{k=-1}^{r} (\alpha_k y'_{n-k}) = y_{n-p} + h \sum_{k=-1}^{r} \alpha_k F(x_{n-k}, y_{n-k}) \tag{1972}$$

The y_n resulting from these formulas are approximate due to round off and truncation error. Let Y_n be the true solution and E_n be the error involved in the nth step. Then

$$Y_{n+1} = Y_{n-p} + h \sum_{k=-1}^{r} \alpha_k F(x_{n-k}, Y_{n-k}) + E_n \tag{1973}$$

Now define

$$\varepsilon_n = Y_n - y_n \tag{1974}$$

Then subtracting (1972) from (1973) we get

$$\varepsilon_{n+1} = \varepsilon_{n-p} + h \sum_{k=-1}^{r} \alpha_k [F(x_{n-k}, Y_{n-k}) - F(x_{n-k}, y_{n-k})] + E_n \tag{1975}$$

We now seek a bound on the magnitude of the ε_i.

From the mean value theorem we have

$$F(x_i, Y_i) - F(x_i, y_i) = (Y_i - y_i) F_y(x_i, \eta_i) = \varepsilon_i F_y(x_i, \eta_i) \tag{1976}$$

where η_i lies between y_i and Y_i. Equation (1975) may then be rewritten in the form

$$[1 - h\alpha_{-1} F_y(x_{n+1}, \eta_{n+1})]\varepsilon_{n+1} = \varepsilon_{n-p} + h \sum_{k=0}^{r} \alpha_k \varepsilon_{n-k} F_y(x_{n-k}, \eta_{n-k}) + E_n \tag{1977}$$

Now define K such that

$$|F_y(x, y)| \leq K \tag{1978}$$

for all values of x and y involved, and define E such that

$$|E_n| \leq E \tag{1979}$$

for all n. Then

$$|1 - h\alpha_{-1} F_y(x_{n+1}, \eta_{n+1})||\varepsilon_{n+1}| \leq |\varepsilon_{n-p}| + Kh \sum_{k=0}^{r} |\alpha_k||\varepsilon_{n-k}| + E \tag{1980}$$

We now define e_n by the relation

$$(1 - Kh|\alpha_{-1}|)e_{n+1} = e_{n-p} + Kh \sum_{k=0}^{r} |\alpha_k| e_{n-k} + E \tag{1981}$$

Notice that if $\alpha_{-1} = 0$, the formula (1972) is of the open type. If $\alpha_{-1} \neq 0$, the formula is of the closed type and we postulate h sufficiently small so that

$$Kh|\alpha_{-1}| < 1 \tag{1982}$$

Then if

$$|\varepsilon_{n-k}| \le e_{n-k}, \qquad k = 1, \ldots, r$$

it follows that

$$|1 - h\alpha_{-1} F_y(x_{n+1}, \eta_{n+1})|\,|\varepsilon_{n+1}| \le (1 - Kh|\alpha_{-1}|)e_{n+1} \tag{1983}$$

and hence

$$|\varepsilon_{n+1}| \le e_{n+1} \tag{1984}$$

for all n. The initial value $y_0 = Y_0$ is presumed prescribed. Hence ε_0 is limited to round off error. The starting values $y_1, \ldots, y_r$ are also presumed given, and the errors $\varepsilon_1, \ldots, \varepsilon_r$ associated with these starting values are presumed known. Let $\bar{e}$ be a positive number such that

$$|\varepsilon_i| < \bar{e}, \qquad i = 1, \ldots, r \tag{1985}$$

Let e_i be a set of positive numbers which satisfies equation (1981). If

$$e_i \ge \bar{e}, \qquad i = 0, 1, \ldots, r \tag{1986}$$

then

$$|\varepsilon_i| \le e_i \tag{1987}$$

over the entire range of the solution. Equation (1981) may be thought of as a set of equations with e_i as the unknowns. One possible "solution" of this equation is $e_i =$ constant for all i. Let this constant be called $-\lambda$. Then

$$\lambda = \frac{E}{Kh\sigma}, \qquad \sigma \equiv \sum_{k=-1}^{r} |\alpha_k| \tag{1988}$$

It may be shown that $\sigma \ge p + 1$. Consider now the "homogeneous" equation

$$(1 - Kh|\alpha_{-1}|)e_{n+1} = e_{n-p} + Kh \sum_{k=0}^{r} |\alpha_k| e_{n-k} \tag{1989}$$

Any solution of this equation added to $-\lambda$ will also be a solution of equation (1981). Let β^n be such a solution. Then

$$(1 - Kh|\alpha_{-1})\beta^{r+1} = - Kh \sum_{k=0}^{r} |\alpha_k| \beta^{r-k} - \beta^{r-p} = 0 \tag{1990}$$

For h small it may be shown that a positive real root of this equation is

$$\beta_0 = 1 + \frac{Kh\sigma}{p+1} + 0(h^2) \tag{1991}$$

A solution to equation (1981) is, then

$$e_i = c\beta_0^i - \lambda \tag{1992}$$

where c is an arbitrary constant. Since e_i increases monotomically with i, choose c so that $e_0 = \bar{e}$. Then

$$e_i = \bar{e}\beta_0^i + \lambda(\beta_0^i - 1) \tag{1993}$$

Since this $e_i \geq \bar{e}$ for all i, it follows that

$$|\varepsilon_i| \leq \bar{e}\beta_0^i + \lambda(\beta_0^i - 1) \tag{1994}$$

Provided that λ and β_0 can be determined, equation (1994) provides us with a bound on the error involved in the numerical integration.

When the α_k are all positive, it may be shown that

$$\lambda = \frac{E}{Kh(p+1)}, \qquad \beta_0^n \cong \exp\left[K(x_n - x_0)\right] \tag{1995}$$

The value of $\bar{e}$ can generally be estimated since it involves only the error associated with the starting values. Each numerical approximation formula has an error term associated with it, usually of the form

$$ah^m y^{(m)}(\xi) \tag{1996}$$

where a is some fraction and $x_n \leq \xi \leq x_{n+1}$. The term (1996) represents the truncation error at each step although the exact value of ξ is, of course, unknown. Some of these terms have been given above. For example, in Milne's method we have

$$y_{n+1} = y_{n-3} + \frac{4h}{3}(2y_n' - y_{n-1}' + 2y_{n-2}') + \frac{14}{45}h^5 y^{(5)}(\xi)$$

$$y_{n+1} = y_{n-1} + \frac{h}{3}(y_{n+1}' + 4y_n' + y_{n-1}') - \frac{1}{90}h^5 y^{(5)}(\xi)$$

The magnitude of this term may be obtained at each step of the calculation by setting

$$h^m y^{(m)}(\xi) = h\Delta^{m-1} y' \tag{1997}$$

where Δ is the forward difference operator. The value of E may then be taken as larger than the maximum value of (1997) in the integration. K can similarly be estimated a posteriori from the numerical solution for $\partial F/\partial y$.

Round off error cannot be estimated with any precision at this time. Because it fluctuates randomly between positive and negative, however, it is generally considered less important than truncation error.

Partial Differential Equations

The numerical integration of partial differential equations represents the least desirable method of solving compressible flow problems. The solutions obtained in this way are generally restricted to the particular problem at hand and tell us little about the general nature of the flow. The amount of computer time involved is usually large and the potential for error greater than in the case of ordinary differential equations. Stability problems tend also to be more severe. Moreover, the mathematical difficulties involved are greater than in the case of ordinary differential equations. On the other hand, there are cases when numerical solutions of partial differential equations will give us the only possible method of obtaining an accurate solution to a particular problem. This can be the case, for example, when very complicated physical boundary conditions are involved, as in the case of wing-body interaction in the design of aircraft.

A mathematical difficulty in the numerical solution of partial differential equations arises from the change in the nature of the solution when the equation goes from elliptic to parabolic to hyperbolic. Thus, it is not possible in partial differential equations, as it was in ordinary differential equations, to simply postulate $n-1$ boundary conditions at some point and integrate forward. The boundary conditions required depend on the nature of the equation. When the nature of the equation changes in the flow field, the problem becomes particularly difficult, as we have seen in the chapter on transonic flow.

Consider the general, quasilinear, second-order partial differential equation

$$a\phi_{xx} + b\phi_{xy} + c\phi_{yy} + f = 0 \tag{1998}$$

where a, b, c, and f are functions of x, y, ϕ_x and ϕ_y. As we have already mentioned in our discussion of the method of characteristics, such an equation will be elliptic, hyperbolic, or parabolic, depending on whether $b^2 - 4ac$ is less than, equal to, or greater than zero. The nomenclature comes from the geometric shape of the curve

$$ay^2 + bxy + cx^2 + ky + mx + n = 0 \tag{1999}$$

in the $x - y$ plane, where k, m and n are constants.

We have already shown that a change in the boundary conditions at one point changes solution of a hyperbolic equation only in a part of the $x - y$ plane, called the domain of dependence, whereas a change in the boundary condition at one point changes the solution of an elliptic equation everywhere. On the other hand, if the boundary conditions are given along a line in the $x - y$ plane, the solution of the hyperbolic equation is rendered determinate throughout a finite region. For an elliptic equation, such a boundary condition is insufficient. In mathematical terms, such a problem is not "well posed" (the expression is due to Hadamard). That is, the given boundary condition does not uniquely determine the solution. To determine the solution of an elliptic equation in a given region of the $x - y$ plane, the boundary conditions must be given over the entire closed curve surrounding the region. The problem of finding the solution of an elliptic equation in a simply connected region when the solution is given on the boundary of the region is called the Dirichlet problem. The problem of finding the solution of a hyperbolic equation in a finite region when suitable boundary conditions are given along some curve in the region is called the Cauchy problem.

Cononical Forms

The discussion of the numerical integration of equations of the type (1998) is greatly simplified by noting that such an equation can always be put in the form

$$\phi_{\xi\xi} \pm \phi_{\eta\eta} = g(\xi, \eta, \phi_\xi, \phi_\eta) \tag{2000}$$

where the plus sign applies for elliptic equations and the minus sign applies for hyperbolic equations, or in the form

$$\phi_{\eta\eta} = g(\xi, \eta, \phi, \phi_\xi, \phi_\eta) \tag{2001}$$

if the equation is parabolic. (2000) and (2001) are called the canonical forms of the equation.

In compressible flow, the equation for the velocity potential is of the form (1998). In the case of incompressible flow, this equation reduces to the Laplace equation. As was shown in Chapter 3, the equation for the velocity potential can also be reduced to Laplace's equation in the case of subsonic compressible flow with small perturbations, through the use of suitable approximations and changes of variables.

When the equation takes the form

$$\phi_{xx} + \phi_{yy} = f(x, y) \tag{2002}$$

it is called Poisson's equation. While the equation

$$\phi_{xx} + \phi_{yy} = f(x, y, \phi_x, \phi_y) \tag{2003}$$

is not, strictly speaking, Poisson's equation, it can be treated as such in numerical integration through the use of open integration formulas if the solution is known up to a point.

Numerical Solution of Laplace's Equation

Divide the $x - y$ plane into mesh, or lattice, as shown in Figure 321. Each point of the mesh may then be designated by two numbers, m and n. For convenience, we take all points a distance h apart. This is not really necessary. We could take points a distance h apart in the x direction and k apart in the y direction. We could even let the distance between points vary over the plane. This is known as variable mesh size and is particularly useful in regions of a flow where the solution is changing rapidly with x and/or y. But for purposes of illustration, let h be constant. Consider now the point n, m (Figure 322). The derivatives at this point may be approximated by

$$\frac{\partial \phi}{\partial x} = \frac{\phi_{n+1,m} - \phi_{n-1,m}}{2h} \qquad \frac{\partial \phi}{\partial y} = \frac{\phi_{n,m+1} - \phi_{n,m-1}}{2h}$$

$$\frac{\partial^2 \phi}{\partial x^2} = \frac{\phi_{n+1,m} - 2\phi_{n,m} + \phi_{n-1,m}}{h^2}; \qquad \frac{\partial^2 \phi}{\partial y^2} = \frac{\phi_{n,m+1} - 2\phi_{n,m} + \phi_{n,m-1}}{h^2}$$

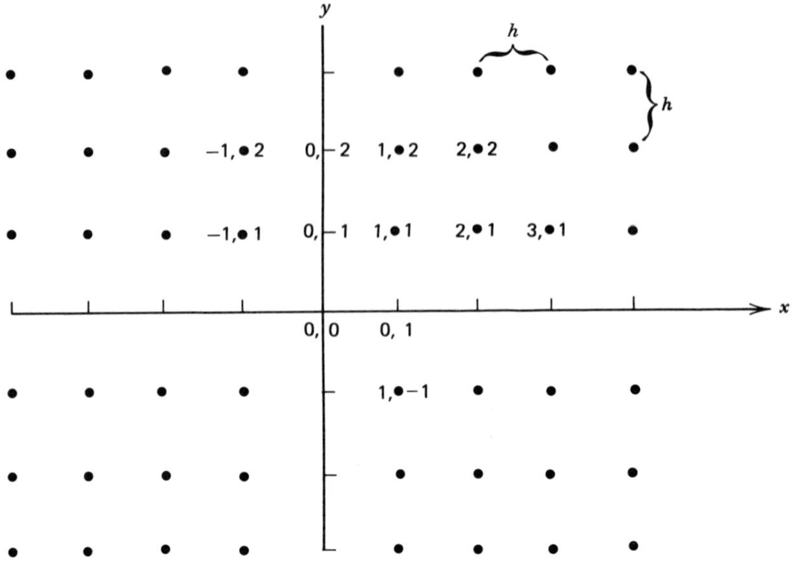

Figure 321 Sample mesh for solution of Laplace's equation.

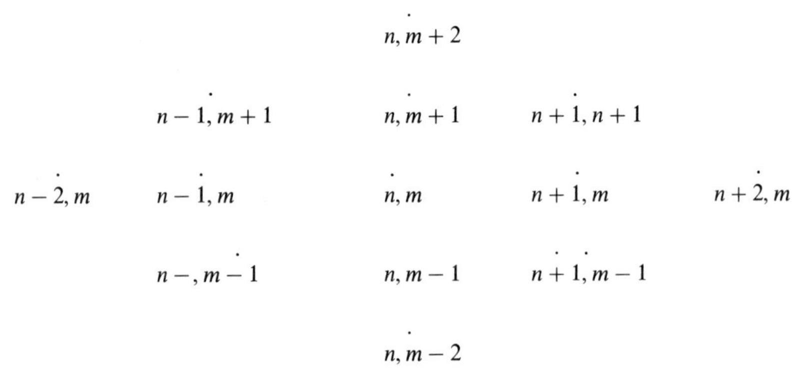

Figure 322 Mesh points for obtaining second derivative.

Substitution in Laplace's equation yields

$$\frac{\phi_{n+1,m} - 2\phi_{n,m} + \phi_{n-1,m}}{h^2} + \frac{\phi_{n,m+1} - 2\phi_{n,m} + \phi_{n,m-1}}{h^2} = 0 \qquad (2004)$$

or

$$\phi_{n+1,m} + \phi_{n-1,m} + \phi_{n,m+1} + \phi_{n,m-1} - 4\phi_{n,m} = 0 \qquad (2005)$$

Suppose now that ϕ is given on the boundary of a region in the $x - y$ plane, say the rectangle a, b, c, d, and we wish to find the solution inside the rectangle. Again, the rectangle is chosen merely for convenience. Any simple closed curve

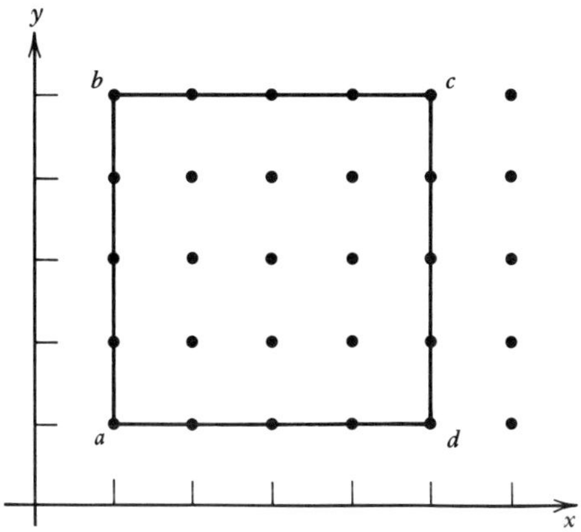

Figure 323 Sample mesh and boundary points for solution of Laplace's equation.

will do. In the closed region bounded by a, b, c, d in Figure 323, 25 mesh points are shown. The solution is given at all mesh points on the boundary. This leaves nine points where the solution is unknown. At each of these points we write equation (2005). This gives us a set of nine simultaneous equations for nine unknowns and the solution is thus rendered determinate.

Liebmann Method An alternative to the method of simultaneous equations is the method of successive iterations. This is a relaxation method. Observe that according to equation (2005), the value of ϕ at each point is the average of the values of ϕ at the four surrounding points. We begin by postulating a value of the solution at each point. Such values may be selected by observing the boundary conditions and taking values of the same order of magnitude. One may even postulate the same value at each interior point as a first guess. We then replace the value at each point by the average value of the four surrounding points. This will give us a new set of interior points. Notice, however, that the points on the boundary remain unchanged. A repetition of the procedure will thus give us still another set of interior points, and so on. Depending on the choice of initial values and the accuracy desired, ten or twenty iterations may be required for the solution to converge. As the number of the iterations increases, the change from successive iterations will decrease. Thus the user, weighing the accuracy desired and the expense of the computer time, can decide when the point of diminishing returns has been reached. For a given mesh size, the solution by the Liebmann method will converge to the same result as the solution by simultaneous equations.

Relaxation Method of Southwell (Simple Relaxation) As in the previous method, we begin by postulating a value of the solution at each point. Using these values we calculate at each point the value of $R_{n,m}$ given by

$$R_{n,m} = \phi_{n+1,m} + \phi_{n-1,m} + \phi_{n,m+1} + \phi_{n,m-1} - 4\phi_{n,m} \tag{2006}$$

$R_{n,m}$ is called the residue at the point n, m. From equation (1985) we see that for the correct solution, the residue at each point is zero. Next we select the point with the largest residue and change the value of the solution there so that the residue at this point will be zero. That is, we select $\Delta\phi_{n,m}$ so that

$$\phi_{n+1,m} + \phi_{n-1,m} + \phi_{n,m+1} + \phi_{n,m-1} - 4(\phi_{n,m} + \Delta\phi_{n,m}) = 0$$

and take as the new value of ϕ at the point n, m, the value $\phi_{n,m} + \Delta\phi_{n,m}$. This change of ϕ at n, m will change the residues at the four points surrounding n, m; namely, at n, $m+1$; n, $m-1$; $n+1$, m; $n-1$, m. We next compute these new residues. Now we select the point with the largest remaining residue and repeat the procedure. And so on. As we proceed, the size of the largest remaining residue will tend towards zero. The number of iterations required at each point will depend on how well the initial values are chosen and on the accuracy desired, but ten to twenty interactions at each point should give good results.

The experienced programmer may hasten convergence by taking $\Delta\phi_{n,m}$ larger or smaller than required to make $R_{n,m}$ zero, thereby reducing the residues at the surrounding points as well. This is called under or over relaxation, depending on whether $\Delta\phi_{n,m}$ is taken small or larger than required to make $R_{n,m}$ zero.

Curved Boundaries When the value of the function is given on a curved boundary, no mesh points may be on the boundary.

Let the curve C in Figure 324 be part of the bounbary where the solution

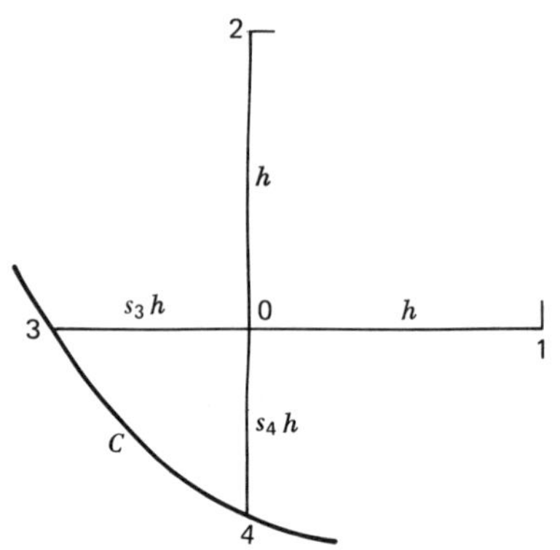

Figure 324 Sample of boundary not falling on mesh points.

is given and let points 3 and 4 lie on C. Let 0, 1 and 2 be interior mesh points where the solution is sought. Let h be the mesh size and let $s_3 h$ and $s_4 h$ be the distance between points 0 and 3, and 0 and 4, respectively. The first derivatives at 0 may then be approximated by

$$\left(\frac{\partial\phi}{\partial x}\right)_0 \cong \frac{\phi_1 - \phi_3}{(s_3+1)h}, \quad \left(\frac{\partial\phi}{\partial y}\right)_0 \cong \frac{\phi_2 - \phi_4}{(s_4+1)h} \tag{2007}$$

The second derivatives may be approximated by

$$\left(\frac{\partial^2\phi}{\partial x^2}\right)_0 \cong \frac{\dfrac{\phi_1-\phi_0}{h}-\dfrac{\phi_0-\phi_3}{s_3h}}{(s_3+1)h}, \qquad \left(\frac{\partial^2\phi}{\partial y^2}\right)_0 = \frac{\dfrac{\phi_2-\phi_0}{h}-\dfrac{\phi_0-\phi_4}{s_4h}}{(s_4+1)h}$$

Laplace's equation then becomes

$$\frac{1}{h^2}\left[\frac{\phi_1-\phi_0}{s_3+1}-\frac{\phi_0-\phi_3}{s_3(s_3+1)}+\frac{\phi_2-\phi_0}{s_4+1}-\frac{\phi_0-\phi_4}{s_4(s_4+1)}\right]=0 \qquad (2008)$$

or

$$\frac{\phi_1}{s_3+1}+\frac{\phi_2}{s_4+1}+\frac{\phi_3}{s_3(s_3+1)}+\frac{\phi_4}{s_4(s_4+1)}-\frac{s_3+s_4}{s_3s_4}\phi_0=0 \qquad (2009)$$

Equations of the type (2009) thus replace equation (2005) for points near the boundary. The method of simultaneous equations, Liebmann method, or simple relaxation method may then be applied.

Poisson's Equation

Recall Poisson's equation

$$\phi_{xx}+\phi_{yy}=f(x,y) \qquad (2010)$$

Poisson's equation is an elliptic equation and thus requires boundary conditions of the Dirichlet type. The methods discussed above for Laplace's equation may also be applied to Poisson's equation.

Method of Simultaneous Equations The equation to be satisfied at each point is now

$$\phi_{n+1,m}+\phi_{n-1,m}+\phi_{n,m+1}+\phi_{n,m-1}-4\phi_{n,m}=f(x_n,y_m)h^2 \qquad (2011)$$

Referring again to Figure 323, the boundary conditions are given as before on the rectangle a, b, c, d. Since $f(x_n, y_m)$ is presumed known at every point, the number of equations and the number of unknowns is the same as before and hence the equations may be solved simultaneously for the value of ϕ at each interior point. Notice that if the function on the right-hand side of equation (2010) contains ϕ_x and ϕ_y, i.e., has the form

$$f(x,y,\phi_x,\phi_y) \qquad (2012)$$

the method of simultaneous equations may still be employed if ϕ_x and ϕ_y are replaced by suitable finite difference formulas such as

$$\phi_x=\frac{\phi_{n+1,m}-\phi_{n-1,m}}{2h}, \qquad \phi_y=\frac{\phi_{n,m+1}-\phi_{n,m-1}}{2h} \qquad (2013)$$

since neither the number of unknowns nor the number of equations is increased thereby.

Liebmann's Method The procedure is the same as before. We postulate a value of the solution at each point of the mesh. Now, however, we employ equation (2011) to obtain a new value at each point. The new value at each point is now given by

$$\phi_{n,m} = \tfrac{1}{4}[\phi_{n+1,m} + \phi_{n-1,m} + \phi_{n,m+1} + \phi_{n,m-1} - f(x_n, y_m)h^2] \qquad (2014)$$

Again, the procedure is applied as many times as required to obtain the desired degree of accuracy.

Relaxation Method The simple relaxation method described for Laplace's equation may also be applied to Poisson's equation. The residue is now given by

$$R_s = \phi_{n+1,m} + \phi_{n-1,m} + \phi_{n,m+1} + \phi_{n,m-1} - 4\phi_{n,m} - f(x_n, y_m)h^2 \qquad (2015)$$

As before, a value of the solution is postulated at every point. The corresponding residues are then calculated. The value of the solution at the point having the largest residue is then changed to eliminate that residue and the procedure is carried forward in the same way as for the Laplace equation.

Generalized Form of Poisson's Equation. Gauss–Seidel Method

Poisson's equation may be written in the noncanonical form

$$\alpha\phi_{xx} + \beta\phi_{yy} = f(x, y) \qquad (2016)$$

where α and β are functions of x and y such that the equation is elliptic. Introducing the finite difference form,

$$\frac{\alpha}{h^2}[\phi_{n+1,m} - 2\phi_{n,m} + \phi_{n-1,m}] + \frac{\beta}{h^2}[\phi_{n,m+1} - 2\phi_{n,m} + \phi_{n,m-1}] = f(x, y) \qquad (2017)$$

This may be rewritten in the form

$$\gamma_1\phi(x+h, y) + \gamma_2\phi(x-h, y) + \gamma_3\phi(x, y+h) + \gamma_4\phi(x, y-h) + \gamma_5\phi(x, y) = f(x, y) \qquad (2018)$$

or

$$\phi(x, y) = v_1\phi(x+h, y) + v_2\phi(x-h, y) + v_3\phi(x, y+h) + v_4\phi(x, y-h) + g(x, y) \qquad (2019)$$

Again one begins by postulating a value of the solution at each point. One then applies the formula (2019) to recompute the value using the updated values as they become available. Thus, if during the jth iteration, new values have been computed at $x-h$, y and x, $y-h$, giving us

$$\phi_{j+1}(x-h, y), \qquad \phi_{j+1}(x, y-h)$$

but have not as yet been computed at $x+h$, y and x, $y+h$, then the new value of $\phi(x, y)$ is computed by

$$\phi_{j+1}(x, y) = v_1\phi_j(x+h, y) + v_2\phi_{j+1}(x-h, y) + v_3\phi_j(x, y+h) + v_4\phi_{j+1}(x, y-h) + g(x, y) \quad (2020)$$

This method is also called the method of successive displacements. The reader should be cautioned that the method of successive displacement does not always converge.

Parabolic Equations

For purposes of illustration we take as an example a flat plate moving in its own plane in incompressible flow. Then

$$\frac{\partial u}{\partial t} = v\frac{\partial^2 u}{\partial y^2} \quad (2021)$$

In finite difference form, the derivatives may be written

$$\frac{\partial u}{\partial t} \simeq \frac{u_{n,m+1} - u_{n,m}}{h}, \qquad \frac{\partial^2 u}{\partial y^2} = \frac{u_{n+1,m} - 2u_{n,m} + u_{n-1,m}}{h^2} \quad (2022)$$

where

$$u_{n,m} = u(y_n, t_m) \quad (2023)$$

and h is the step size. The differential equation then becomes

$$\frac{u_{n,m+1} - u_{n,m}}{h} = v\frac{u_{n+1,m} - 2u_{n,m} + u_{n-1,m}}{h^2} \quad (2024)$$

Solving for $u_{n,m+1}$,

$$u_{n,m+1} = \left(1 - 2\frac{v}{h}\right)u_{n,m} + \frac{v}{h}\left(u_{n+1,m} + u_{n-1,m}\right) \quad (2025)$$

The boundary conditions are

$$u(0, t) = f_1(t), \qquad u(\infty, t) = f_2(t), \qquad u(y, 0) = g(y) \quad (2026)$$

f_2 is usually zero. In practice, it is necessary to substitute some finite value, $y = L$ say, for $y = \infty$ in the second boundary condition. We begin by finding the solution at time $t = t_1$ for all values of y. The formula is then

$$u_{n,1} = \left(1 - 2\frac{v}{h}\right)u_{n,0} + \frac{v}{h}(u_{n+1,0} + u_{n-1,0}) \quad (2027)$$

$n = 1, 2, \ldots$. This gives us the solution at time $t = t_1 = h$ for all values of y. We then find the solution at $t = t_2 = 2h$ for all y.

$$u_{n,2} = \left(1 - 2\frac{v}{h}\right)u_{n,1} + \frac{v}{h}(u_{n+1,1} + u_{n-1,1}) \quad (2028)$$

and so on.

It may be shown that the solution of the difference equation converges to the solution of the differential equation as $h \to 0$ only so long as $0 < \nu/h < \frac{1}{2}$; [206], [207]. Thus one cannot improve the accuracy of the solution by taking h smaller beyond a certain point. If ν is small, as it usually is, this is not necessarily fatal. The accuracy of the solution may be improved, however, by taking different step sizes for y and t. If the step size for y is taken as h and the step size for t is taken as l, we have

$$\frac{\partial u}{\partial t} = \frac{u_{n,m+1} - u_{n,m}}{l}, \qquad \frac{\partial^2 u}{\partial y^2} = \frac{u_{n+1,m} - 2u_{n,m} + u_{n-1,m}}{h^2} \tag{2029}$$

The differential equation then becomes

$$\frac{u_{n,m+1} - u_{n,m}}{l} = \nu \frac{u_{n+1,m} - 2u_{n,m} + u_{n-1,m}}{h^2} \tag{2030}$$

or

$$u_{n,m+1} = (1 - 2\delta)u_{n,m} + \delta(u_{n+1,m} + u_{n-1,m}) \tag{2031}$$

where

$$\delta = \frac{\nu l}{h^2} \tag{2032}$$

Thus the solution may be made progressively more accurate by taking l and h smaller so long as $0 < \delta < \frac{1}{2}$.

Two-Dimensional Case Consider the equation

$$\frac{\partial u}{\partial t} = A\frac{\partial^2 u}{\partial x^2} + C\frac{\partial^2 u}{\partial y^2} \tag{2033}$$

where A and C may be known functions of x and y. It is desired to find the

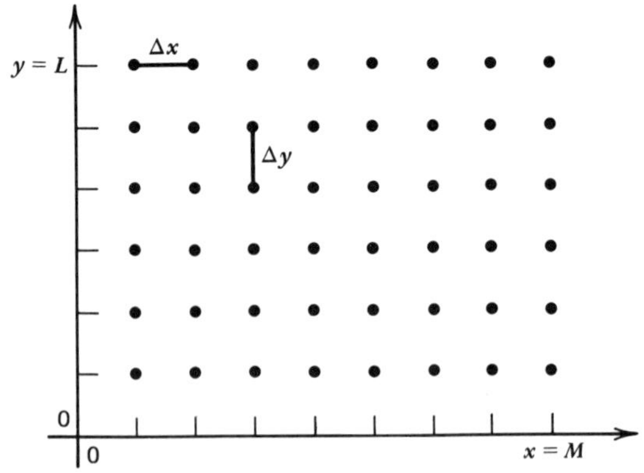

Figure 325 Sample grid for solution of the two-dimensional unsteady heat conduction equation (2033).

solution over a finite region $0 \leq x \leq M$, $0 \leq y \leq L$. The boundary conditions are

$$u(x, y, 0) = g(x, y) \qquad u(0, y, t) = f_1(t)$$
$$u(M, y, t) = f_2(t) \qquad u(x, 0, t) = f_3(t)$$
$$u = (x, L, t) = f_4(t)$$

We consider the grid shown in Figure 325.

The boundary conditions give the solution on the boundaries at all times and at every point for $t = 0$. Let

$$u(x_i, y_j, t_k) = u_{i,j,k} \tag{2034}$$

The finite difference representation of equation (2033) then becomes

$$\begin{aligned}\frac{u_{i,j,k+1} - u_{i,j,k}}{\Delta t} &= A\frac{u_{i+1,j,k} - 2u_{i,j,k} + u_{i-1,j,k}}{(\Delta x)^2} \\ &\quad + C\frac{u_{i,j+1,k} - 2u_{i,j,k} + u_{i,j-1,k}}{(\Delta y)^2}\end{aligned} \tag{2035}$$

We begin by setting $k = 0$ in the right-hand side of the equation. $k = 0$ corresponds to $t = 0$; that is, $t_0 = 0$. This gives us

$$\begin{aligned}\frac{u_{i,j,1} - u_{i,j,0}}{\Delta t} &= A\frac{u_{i+1,j,0} - 2u_{i,j,0} + u_{i-1,j,0}}{(\Delta x)^2} \\ &\quad + C\frac{u_{i,j+1,0} - 2u_{i,j,0} + u_{i,j-1,0}}{(\Delta y)^2}\end{aligned} \tag{2036}$$

Since $u_{i,j,0}$ is known for all values of i and j, this equation may be solved for $u_{i,j,1}$ for all values of i and j. We then set $k = 1$ on the right-hand side of equation (2015), solve for $u_{i,j,2}$, and so on. The condition that the solution converge to the exact solution as Δx, $\Delta y \to 0$ in this case becomes

$$\frac{A\Delta t}{(\Delta x)^2} + \frac{C\Delta t}{(\Delta y)^2} < \frac{1}{2} \tag{2037}$$

Hyperbolic Equations

Although the method of characteristics may be used to turn the solution of hyperbolic equations into what is essentially an arithmetic problem, hyperbolic equations may also be solved directly by the method of finite differences. As an example, consider the one-dimensional wave equation

$$\frac{\partial^2 \phi}{\partial t^2} = c^2 \frac{\partial^2 \phi}{\partial x^2} \tag{2038}$$

given the boundary conditions

$$\phi(0, t) = f_1(t) \qquad \phi(L, t) = f_2(t)$$
$$\phi(x, 0) = g_1(x) \qquad \frac{\partial \phi}{\partial t}(x, 0) = g_2(x) \tag{2039}$$

The difference equation corresponding to equation (2039) may be written

$$\frac{c^2}{h^2}(\phi_{i+1,j} - 2\phi_{i,j} + \phi_{i-1,j}) = (\phi_{i,j+1} - 2\phi_{i,j} + \phi_{i,j-1})\frac{1}{h^2} \tag{2040}$$

where

$$\phi_{i,j} = \phi(x_i, t_j) \tag{2041}$$

If c is a constant, equation (2040) may be simplified by taking different step sizes for x and t. Let h be the step size for x and l be the step size for t. Then equation (2040) becomes

$$\frac{c^2}{h^2}(\phi_{i+1,j} - 2\phi_{i,j} + \phi_{i-1,j}) = \frac{1}{l^2}(\phi_{i,j+1} - 2\phi_{i,j} + \phi_{i,j-1}) \tag{2042}$$

Now take h and l so that

$$\frac{c^2 l^2}{h} = 1 \tag{2043}$$

Then equation (2042) reduces to

$$\phi_{i,j+1} = \phi_{i+1,j} + \phi_{i-1,j} - \phi_{i,j-1} \tag{2044}$$

The last boundary condition in (2039) may be written

$$\frac{\phi_{i,1} - \phi_{i,0}}{l} = g_2(x_i) \tag{2045}$$

The boundary conditions may also be written in the form

$$\begin{aligned}
&\phi(x_0, t_j) = f_1(t_j), \qquad \phi(x_L, t_j) = f_2(t_j) \\
&\phi(x_i, t_0) = g_1(x_i), \qquad x_0 = 0, \qquad t_0 = 0 \\
&\phi(x_i, t_1) = \phi(x_i, t_0) + l g_2(x_i) = g_1(x_i) + l g_2(x_i)
\end{aligned} \tag{2046}$$

Thus the solution is known for all x for $t = t_0$ and $t = t_1$. If, therefore, we let $j = 1$ in equation (2044) we get

$$\phi_{i,2} = \phi_{i+1,1} + \phi_{i-1,0} - \phi_{i,0} \tag{2047}$$

giving us the solution at time $t = t_2$ for all x. We then set $j = 2$ in the right-hand side of equation (2044) solve for $\phi_{i,3}$, and so on. The solution may thus be considered a marching problem in time.

If

$$\delta = \frac{c^2 l^2}{h^2} \neq 1$$

is used, the problem may still be solved using equation (2042) instead of equation (2040). For $\delta > 1$, however, the solution will not usually converge. Thus δ should always be taken ≤ 1.

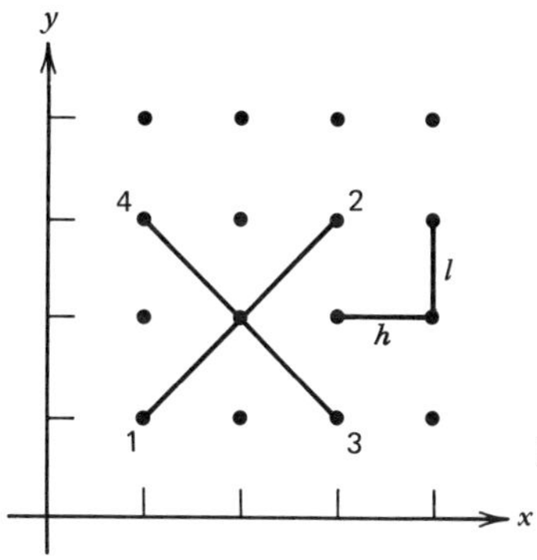

Figure 326 Sample grid for the numerical solution of the wave equation (2038).

Consider the net shown in Figure 326. For the diagonals 1,2 and 3,4 we have

$$\frac{dx}{dt} = \pm \frac{h}{l}, \qquad x = \pm \frac{h}{l} + \text{constant} \tag{2048}$$

But for $\delta = 1$, $h/l = c$. The equations of the diagonals in this case are

$$x \pm ct = \text{constant} \tag{2049}$$

The diagonals are thus the characteristics of equation (2038).

A More Complicated Example

We conclude by presenting an example of how the elementary concepts discussed above may be extended to more complicated cases, in particular cases of practical interest for which analytic solutions are not available.

Supersonic flow against the blunt end of a slab [208] We wish to determine the flow field around the forward end of the slab as shown in Figure 327, includ-

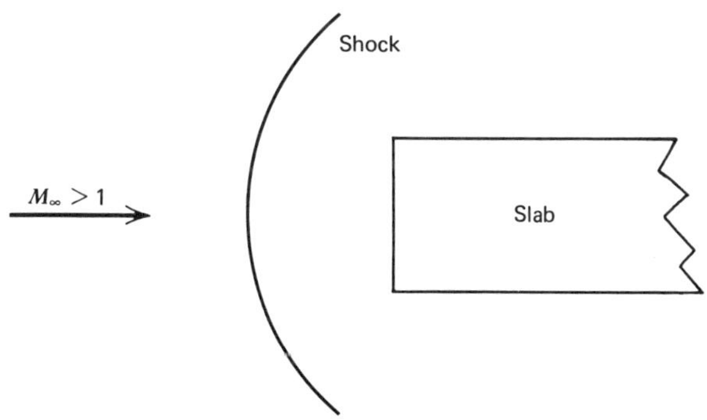

Figure 327 Supersonic flow against the blunt end of a slab.

ing the shock location. This problem is particularly complicated because of the mixed subsonic supersonic flow behind the shock. The problem may be solved by a scheme which is also applicable to other mixed flow problems. Instead of treating the problem as a steady flow problem with the given boundary conditions, we treat the problem as an unsteady flow problem. The initial conditions are taken as $V = V_\infty$ everywhere except on the forward end of the slab where the velocity is zero. This corresponds to starting the flow impulsively from rest. The flow will then approach its steady-state condition asymptotically as $t \to \infty$. The advantage of this method is that the unsteady equations are hyperbolic in t, making it possible to treat the problem as a marching problem, integrating forward in t from $t = 0$. Since the asymptotic solution as $t \to \infty$ is approached rather quickly, one soon arrives at a good approximation to the steady-state solution.

We shall use the coordinate system shown in Figure 328. Since the flow is

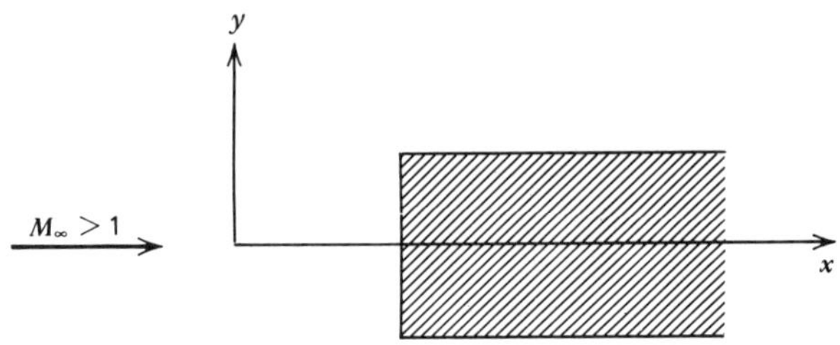

Figure 328 Coordinate system for numerical solution of the supersonic flow against the blunt end of a slab.

symmetric about the x axis, this may be considered equivalent to flow over a step. The governing equations may be written in the form

$$\rho_t + (\rho u)_x + (\rho v)_y = 0 \tag{2050}$$

$$(\rho u)_t + (\rho u^2 + p)_x + (\rho uv)_y = 0 \tag{2051}$$

$$(\rho v)_t + (\rho uv)_x + (\rho v^2 + p)_y = 0 \tag{2052}$$

$$\left[\tfrac{1}{2}\rho(u^2 + v^2) + \rho h - p\right]_t + \left\{\rho u\left[\tfrac{1}{2}(u^2 + v^2) + h\right]\right\}_x + \left\{\rho v\left[\tfrac{1}{2}(u^2 + v^2) + h\right]\right\}_y = 0 \tag{2053}$$

There are five unknowns here – ρ, u, v, p and h, and four equations. An equation of state supplies the fifth equation. The simplest example is

$$p = \rho RT, \qquad h = c_p T \tag{2054}$$

It will be observed that the equations (2050)–(2053) are of the form

$$\xi_t + \eta_x + \zeta_y = 0 \tag{2055}$$

For purposes of illustration, we may write this equation in a very crude finite difference form

$$\frac{\xi(x,y,t+\Delta t) - \xi(x,y,t)}{\Delta t} + \frac{\eta(x+\Delta x,y,t) - \eta(x,y,t)}{\Delta x} + \frac{\zeta(x,y+\Delta y,t) - \zeta(x,y,t)}{\Delta y} = 0 \tag{2056}$$

We are given

$$\xi(x,y,0), \qquad \eta(x,y,0), \qquad \zeta(x,y,0) \tag{2057}$$

for all x and y. Starting at $t=0$, equation (2056) gives us a value of $\xi(x,y,\Delta t)$. Similarly, the other equations give us values of $\eta(x,y,\Delta t)$ and $\zeta(x,y,\Delta t)$. Using all five equations we may thus integrate forward in time to obtain the solution for ξ, η and ζ for all values of x and y and as many values of t as desired. In practice, the time derivatives are written in the form

$$\frac{\partial \omega}{\partial t} = \frac{1}{\Delta t}\{\omega(x,y,t+\Delta t) - \tfrac{1}{4}[\omega(x+\Delta x,y,t) + \omega(x-\Delta x,y,t) + \omega(x,y+\Delta y,t) + \omega(x,y-\Delta y,t)]\} \tag{2058}$$

where ω is any dependent variable. The boundary conditions are shown in Figure 329. The shock location is not known a priori and is sketched merely for

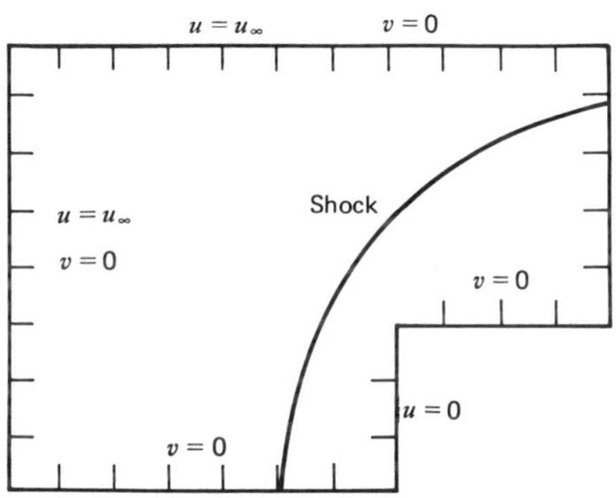

Figure 329 Boundary conditions for the numerical solution of the supersonic flow against the blunt end of a slab.

orientation purposes. The solution is to be found inside the boundaries shown. Since no boundary conditions are given on the upper right-hand boundary, the formula (2038) may not be applied there. It may, however, be replaced by

$$\frac{\partial \omega}{\partial t} = \frac{1}{\Delta t}\{\omega(x,y,t+\Delta t) - \tfrac{1}{2}[\omega(x-\Delta x,y,t) + \omega(x,y+\Delta y,t)]\} \tag{2059}$$

Similarly, the x derivative may be replaced by

$$\frac{\partial \eta}{\partial x} = \frac{\eta(x,y,t) - \eta(x-\Delta x,y,t)}{\Delta x} \tag{2060}$$

The solution turns out not to be sensitive to the particular approximation used on the upper right-hand boundary.

On the solid boundaries, the mesh is carried one step inside the solid boundaries. The values of the dependent variables assigned at these points are taken as the same as those one step outside the solid boundary except for the velocity components, which are zero on the boundary, for which the negative of the value one step outside the boundary is assigned to the points one step inside the boundary. This makes it possible to use of the four point formula (2038) at points on the boundary.

The step size must be chosen in such a way that the stability conditions are met. In the present case this requires

$$\frac{\Delta t}{\Delta x} \leq \frac{1}{c+|u|}, \qquad \frac{\Delta t}{\Delta y} \leq \frac{1}{|v|+c} \tag{2061}$$

where c is the speed of sound. Otherwise, the solution may not converge to the exact solution as the step size is decreased.

Figure 330 shows the results for the density variation at one value of y.

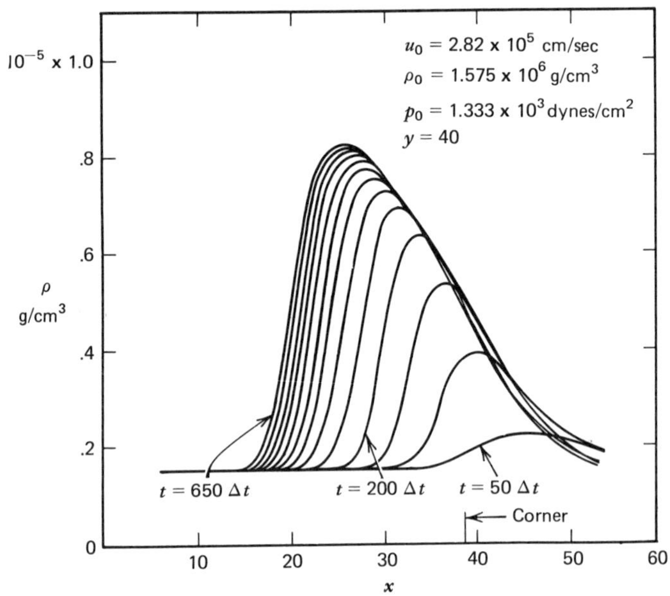

Figure 330 Density variation for the supersonic flow against the blunt end of a slab, from [208].

The shock location may be taken at the midpoint of the density rise. In this way the shock may be plotted for each value of the time. The results of such a calculation are shown in Figure 331. The shock is seen to approach its limiting value after 650 iterations.

In the axially symmetric case, the governing equations become

$$(r\rho)_t + (r\rho u)_x + (r\rho v)_r = 0 \tag{2062}$$

$$(r\rho u)_t + [r(\rho u^2 + p)]_x + (r\rho uv)_r = 0 \tag{2063}$$

$$(r\rho v)_t + (r\rho uv)_x + [r(\rho v^2 + p)]_r = p \tag{2064}$$

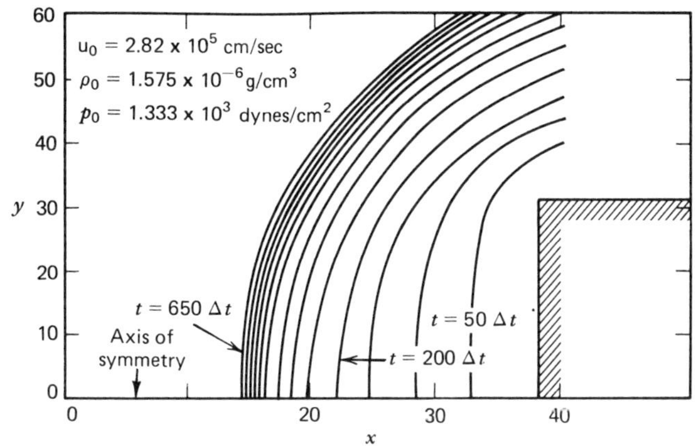

Figure 331 Shock wave location for supersonic flow against the blunt end of a slab, from [208].

$$\{[\tfrac{1}{2}\rho(u^2+v^2)+\rho h-p]r\}_t+\{r\rho u[\tfrac{1}{2}(u^2+v^2)+h]\}_x + \{r\rho v[\tfrac{1}{2}(u^2+v^2)+h]\}_r=0 \tag{2065}$$

where r is the radial coordinate. A difficulty arises where $r=0$. This may be eliminated by dividing all the equations by r and applying L'Hospital's rule where $r=0$. The boundary conditions and the computational procedure is otherwise analogous to the plane case. The result for the final shock shape in the flow over the blunt end of a cylinder is shown in Figure 332. An interesting

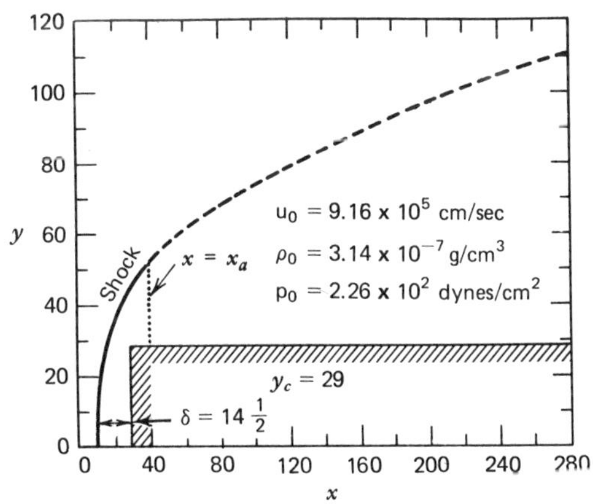

Figure 332 Calculated shock wave shape for flow over the blunt end of a cylinder, from [208].

result in the cylindrical case which was not observed in the plane case is that after 300 iterations the shock tends to oscillate about its equilibrium position.

The procedure has been extended by Bohachevsky and Rubin to the case of

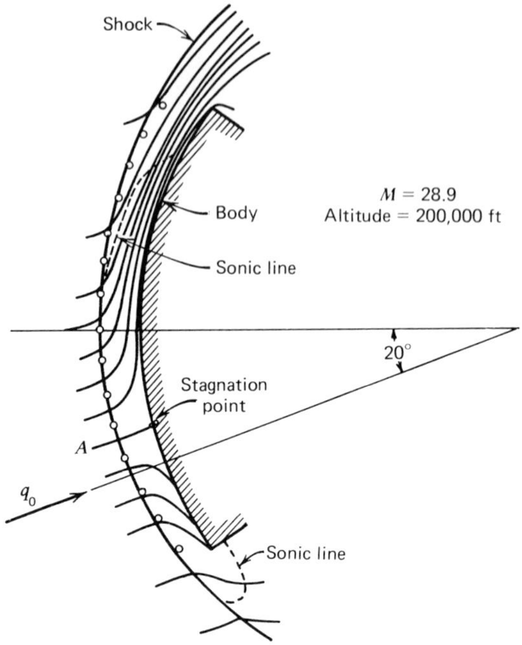

Figure 333 Numerical solution of the flow over the front of a reentry vehicle, from [210].

dissociating flow [209]. Use of this method is not restricted to flows over a step or cylinder. The flow over any reasonably shaped obstacle may be calculated. Bohachevsky and Mates [210] have employed the method calculate the flow over an actual re-entry vehicle. Figures 333 and 334 show the streamline pattern obtained in this manner.

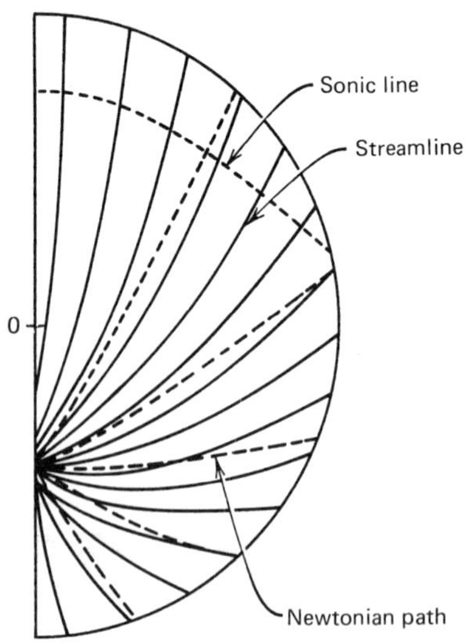

Figure 334 Numerical solution of actual streamline paths over the front of reentry vehicle, from [210].

Literature in Computational Gas Dynamics

During the last fifteen years, a considerable body of literature has grown up in computational fluid dynamics. We conclude with a brief survey of this literature, particularly of compendia of outstanding papers.

The compendium of C. K. Chu, "Computational Fluid Dynamics" [211] features outstanding papers written prior to 1968. Compressible flow problems addressed in this work include unsteady one-dimensional shocks, by Neumann and Richtmeyer [212], chemically reacting flows by Moretti [213], supersonic wake flows by Amsden and Harlow [214], hypersonic flat plate flow, by Butler [215], solution of the Navier–Stokes equations for supersonic flow about a cylinder by Scala and Gordon [216], kinetic theory calculation of the formation of shock waves by Chu [217], and low-density supersonic flow about an obstacle by Vogenitz et al. [218].

The proceedings of the second international conference on numerical methods in fluid dynamics [219] contain a number of papers of interest to the student of compressible flow. Among the problems addressed are those of wing body interaction, by Kutler and Lomax [220], unsteady transonic flow through nozzles by Laval [221], high-speed subsonic flow over bodies by Holt and Masson [222] and conical wings in hypersonic flow by Gonor et al. [223].

The proceedings of the conference on computational methods and problems in aeronautical fluid dynamics at the University of Manchester [224] are of particular interest to the aircraft and engine designer. Problems discussed include subsonic flows in turbomachines by Marsh [225], unsteady flow about aircraft by Hancock [226], subsonic flows about aircraft, by Lock [227], supersonic flow about wings, by Purshouse and Nangia [228], and unsteady transonic flow about airfoils, by Nixon [229].

An excellent discussion of Russian numerical work in compressible flow has been given by Holt [230]. Included are discussions of the BVLR method for calculation of compressible flow about bodies of arbitrary shape, including mixed subsonic-supersonic flow and three-dimensional unsteady flow, integral method for transonic flow, supersonic flow about blunt bodies, boundary layers with suction or injection, separated flows, wakes and base flows.

A highly theoretical treatment of numerical methods has been given by Chung in "Finite Element Analysis in Fluid Dynamics" [231]. While this work contains a section on compressible flow, it is of interest chiefly to mathematicians rather than engineers.

Mention must also be made of the compendium of Winz and Smolderen, "Numerical Methods in Fluid Dynamics" [232]. This work contains extensive discussions of transonic flow calculations by Jameson, Ballhaus and Schmidt as well as shorter discussions of compressible viscous flow calculations and rarefied gas flows by Belotserkovskii.

Finally, we must mention the work "Computational Gas Dynamics" by Napotitano and Belotserkovskii [237]. The second half of this work, "Methods of Computational Gas Dynamics", by O. M. Belotserkovskii, is of particular interest to student of numerical methods in compressible flow.

References

1 Euler, L. Principes généreaux du movement des fluides. *Mé. Acad. Sci. Berlin* **11** (1755): 274–315.

2 Lagrange, J. L. Mémoire sur la théorie du mouvement des fluides. *Nouv. Mém. Acad. Sci. Berlin* (1781): 151–198.

3 Newton, I. *Philosophiae Naturalis Principia Mathematica.* Royal Society, London, 1687.

4 Navier, M. Memoire sur les lois du mouvement des fluides. *Mém. Acad. Sci. Inst. France* **6**, (1822): 389–440.

5 Stokes, G. G. On the theories of the internal friction of fluids in motion and of the equilibrium and motion of elastic solids. *Trans. Camb. Phil. Soc.* **8** (1844–1849): 287–319.

6 Becker R. Stosswelle und Detonation. *Z. f. Physik* **8** (1922): 321–362.

7 Thomas, L. H. Note on Becker's theory of the shock front. *J. Chem. Phys.* **12** (1944): 449–453.

8 Morduchow, M. and Libby, P. A. On a complete solution of the one-dimensional flow equations of a viscous, heat conducting, compressible gas. *J. Aero Sci.* **16** (1949): 674–684, 704.

9 Ballard, H. N. and Venable, D. Shock front thickness measurements by an electron beam technique. *Phys. Fluids* **1** (1958): 225ff.

10 Bradley, J. N. *Shock Waves in Chemistry and Physics.* John Wiley and Sons, New York, 1962.

11 Apostal, T. M. *Mathematical Analysis.* Addison-Wesley, Reading, MA, 1957.

12 Courant, R. and Hilbert, D. *Methoden der Mathematischen Physik.* Julius Springer Verlag, Berlin, 1937.

13 Sneddon, I. A. *Elements of Partial Differential Equations.* McGraw-Hill, New York, 1957.

14 Petrovsky, I. G. *Lectures on Partial Differential Equations.* Interscience, New York, 1954.

15 Czarnecki, K. R. and Jackson, M. W. Theoretical pressure distributions over arbitrarily shaped periodic waves in subsonic compressible flow and comparison with experiments. NASA TN D-5984 (1970).

16 Hildebrand, F. B. *Advanced Calculus for Applications.* Prentice-Hall, Inc., Englewood Cliffs, NJ, 1962.

17 Bateman, H. Notes on a differential equation which occurs in the two-dimensional motion of a compressible fluid and the associated variational problems. *Proc. Roy. Soc. London* **A125** (1929): 598–618.

18 Strutt, J. W., Lord Rayleigh: On the flow of a compressible fluid past an obstacle. *Phil. Mag.* 32 (1916): 1–6.

19 Ritz, W. Über eine neue Methode zur Lösung gewisser Variations-probleme der mathematischen Physik. *Journ. f. d. Reine u. Angew. Math.* **135** (1909): 1–61.

20 Wang, C. T. Variational method in the theory of compressible fluids. *J. Aero. Sci.* **15** (1948): 675–685.

21 Taylor, G. I. and Sharman, C. F. A mechanical method for solving problems of flow in compressible fluids. *Proc. Roy. Soc. London* **A121** (1928): 194–217.

22 See reference 20.

23 Southwell, R. V. *Relaxation Methods in Theoretical Physics.* Oxford University Press, New York, 1946.

24 Shortley, G. H. and Weller, R. The numerical solution of Laplace's equation. *J. Appl. Phys.* **9** (1938): 334–348.

25 Kormes, M. Numerical solution of the boundary value problem for the potential equation by means of punched cards. *Rev. Sci. Inst.* **14** (1943): 248–250.

26 Emmons, H. W. The numerical solution of compressible fluid flow problems. NACA TN 932 (1944).

27 Emmons, H. W. Flow of a compressible fluid past a symmetrical airfoil in a wind tunnel and in free air. NACA TN (1948).

28 Amick, J. L. Comparison of the experimental pressure distribution on an NACA 0012 profile at high speeds with that calculated by relaxation method. NACA TN 2174 (1950).

29 Ludford, G. S. S. The behavior at infinity of the potential function of a two dimensional subsonic compressible flow. *J. Math. Phys.* **30** (1952): 117–130.

30 Whalen, P. J. and Mulholland, G. P. Expansion solution for subsonic compressible flow. *Appl. Sci. Res.* **25** (1972): 445–473.

31 Thornhill, C. K. The numerical method of characteristics for hyperbolic problems in three independent variables. British ARC R and M No. 2615 (1952).

32 Ferri, A. *Elements of Aerodynamics of Supersonic Flows*. MacMillan Co., New York, 1949.

33 Ferri, A. Application of the method of characteristics to supersonic rotational flow. NACA TN 1135 (1946).

34 Busemann, A. Drücke auf kegelförmige Spitzen bei Bewegung mit Überschallgeschwindigkeit. *ZAMM* **9** (1929): 496–498.

35 Taylor, G. I. and Macoll, J. W. The air pressure on a cone moving at high speed. *Proc. Roy. Soc. London*, **A139** (1933): 278ff.

36 Moeckel, W. E. Approximate method for predicting forms and location of detached shock waves ahead of plane or axially symmetric bodies. NACA TN 1921 (1949).

37 Ward, L. C. and Pugh, P. G. Shock stand off distances of blunt and sharp cones. *AIAA J.* **6** (1968): 2018–2019.

38 Kaatari, G. E. Shock envelopes of blunt bodies at large angles of attack. NASA TN D-1980 (1963).

39 Owen, R. V. Aerodynamic characteristics of spherically blunted cones at Mach numbers from 0.5 to 5.0. NASA TN D-3088, (1965).

40 Guy, T. B. A note on the shock stand off distance for a spherical body in supersonic flow. *AIAA. J.* **12** (1974): 380–381.

41 Witham, G. B. A new approach to problems of shock dynamics, part I, two dimensional problems. *J. Fluid Mech.* **2** (1957): 145–171.

42 Schwartz, R. N. and Eckerman, J. Shock location in front of a sphere as a function of real gas effects. *J. Appl. Phys.* **27** (1956): 169–174.

43 Heberle, J. W., Wood, G. P., and Gooderum, P. B. Data on shape and location of detached shock waves on cones and spheres. NASA TM 2000, (1950).

44 Van Dyke, M. D. and Milton, D. The supersonic blunt body problem-review and extension. *J. Aero. Sci.* **25** (1958): 485–496.

45 Van Dyke, M. D. and Gordon, H. D. Supersonic flow past a family of blunt axisymmetric bodies. NASA TR R-1, (1959).

46 Radvogin, Y. B. Dependence of the shock standoff distance on the free stream Mach number. *Sov. Phys. Dokl.* **19** (1974): 179–180.

47 Von Karman, T. and Moore, N. B. The resistance of slender bodies moving with supersonic velocities with special reference to projectiles. *Trans. ASME*, **54** (1932): 303–310.

48 Thompson, J. R. A rapid graphical method for computing the pressure distribution at supersonic speeds on a slender body of revolution. NACA TN 1768 (1949).

49 Taylor, Sir G. I. The formation of a blast wave by a very intense explosion. *Proc. Roy. Soc. London ser* **A201** (1950): 159–186.

50 Brinkley, S. R. and Kirkwood, J. G. Theory of the propagation of shock waves. *Phys. Rev.* **71** (1974): 606–611.

51 Sachdev, P. L. Propagation of a blast wave in uniform or non-uniform media: a uniformly valid analytic solution. *J. Fluid Mech.* **52** (1972): 369–378.

52 Landau, L. D. On shock waves at a large distance from the location of explosion. *Prikl. Mat. Mekh* **20** (1945): 286ff.

53 Witham, G. B. The propagation of spherical blast. *Proc. Roy. Soc. London*, **A 203** (1950): 571–581.

54 Sedov, L. I. *Similarity and Dimensional Methods in Mechanics.* Academic Press, New York, 1959.

55 Plooster, M. N. Shock waves from line sources. Numerical solutions and experimental measurements. *Phys. Fluids* **13** (1970): 2665–2675.

56 Lutzky, M. and Lehto, D. L. Shock propagation in spherically symmetric exponential atmospheres. *Phys. Fluids* **11** (1968): 1466–1472.

57 Tricomi, F. Sulle equazioni lineari alle derivate parziali di 2° ordine di tipo misto. *R. C. Accad. Lincei* **14** (1923): 134ff.

58 Guderley, K. G. *The Theory of Transonic Flow.* Pergamon Press, Oxford, 1962.

59 Guderley, K. G. On the transition from a transonic potential flow to a flow with shocks. Report F-TR-2160 ND, Headquarters, Air Material Command, Wright Field, OH (1947).

60 Oswatitsch, K. Die Geschwindigkeitsverteilung bei lokalen Überschallgebieten an flachen Profilen. *ZAMM* **30** (1950): 228–271.

61 Von Karman, T. Similarity law of transonic flow. *J. Math. Phys.* **26** (1947): 182ff.

62 Ferrari, C. and Tricomi, F. *Transonic Aerodynamics.* Academic Press, New York, 1968.

63 Guderley, K. G. and Yoshihara, H. The flow over a wedge profile at Mach number one. *J. Aero. Sci* **17** (1950): 723ff.

64 Von Karman, T. Compressibility effects in aerodynamics. *J. Aero. Sci.* **8** (1941): 337–356.

65 Nikolskii, A. A. and Taganov, G. I., Dvizhenie gaza v mestnoi sverkhzvukovoi zone i nekotorye uslovia razrushenia potentsialnogo techenia. *Prikl. Mat. Mek.* **10** (1946): 481–502. (English translation, NACA TM 1213 (1949)).

66 Pearcey, H. H. The aerodynamic design of section shapes for swept wings. *Adv. Aero. Sci.* **3** (1962): 277–322

67 Oswatitsch, K. Die Geschwindigkeitsverteilung an symmetrischen Profilen beim Auftreten lokaler Überschallgebiete. *Acta Physica Austriaca* **4** (1950): 230–271.

68 Spreiter, J. R. and Alksne, A. Theoretical predictions of pressure distributions on non lifting airfoils at high subsonic speeds. NACA TN 3096 (1954)

69 Nixon, D. The high subsonic flow around a two dimensional airfoil with a trailing edge control surface. *Aero. Quart.* **24** (1973): 273–283.

70 Nixon, D. Extended integral equation method for transonic flows. *AIAA* J. **13** (1975): 934–935

71 Nixon, D. A comparison of two integral equation methods for high subsonic lifting flows. *Aero. Quart.* **26** (1975): 56–58.

72 Nixon, D. and Patel, J. The evaluation of an integral equation method for two dimensional shock free flows. *Aero. Quart.* **26** (1975): 59–70.

73 Nixon, D. and Hancock, G. J. High subsonic flow past a steady two-dimensional airfoil. ARC CP 1280 (1974).

74 Hansen, H. Airfoil design for a prescribed velocity distribution in transonic flow by an integral method. In *Symposium Transsonicum II*, Springer-Verlag, Berlin, 1976, pp. 183–196.

75 Norstrud, H. High speed flow past wings. NASA CR-2246 (1973).

76 Frohn, A. Lösung der nichtlinearen Integralgleichung der schallaher Strömug, *ZAMM* **54** (1974): T116-T117.

77 Boerstoel, J. W. Review of the application of hodograph theory to transonic airfoil design and theoretical and experimental analysis of shock free airfoils. In *Symposium Transsonicum II*, Springer Verlag, Berlin, 1976, pp. 109–133.

78 Bauer, F., Garabedian, P., and Korn, D. *Supercritical Wing Sections (monograph). Lecture Notes in Economics and Mathematical Systems*, No. 66, Springer Verlag, Berlin, 1972.

79 See reference 78.

80 Bauer, F., Garabedian, P., Korn, D., and Jameson, A. *Supercritical Wing Sections II (monograph). Lecture Notes in Economics and Mathematical Systems*, No. 108, Springer Verlag, Berlin, (1975).

81 Garabedian, P. and Korn, D. A systematic method for computer design of supercritical airfoils in cascade. *Commun Pure Appl. Math.* **29** (1976): 369–382.

82 Meyer, T. Über zweidimensional Bewegungsvorgänge in einem Gas das mit Überschallgeschwindigkeit strömt. DVI Forschung a.d. Gebiet d. Ing., *Forschungsheft* **62**, (1908).

83 Taylor, G. I. The flow of air at high speeds past curved surfaces. British ARC R and M, No. 1381 and 1382 (1930).

84 Görtler, H. Zum Übergang von Unterschall zu Überschallgeschwindigkeiten in Düsen. *ZAMM* **19** (1939); 325–337.

85 Sauer, R. General characteristics of the flow through nozzles at near critical speeds. NACA TM 1147 (1947).

86 Tomotika, S. and Tameda, K. Studies on two-dimensional transonic flows of compressible fluids. *Quart. Appl. Math.* **7** (1950): 381ff.

87 Oswatitsch, K. and Rothstein, W. Das Strömungsfeld in einer Laval Düse. *Jahrbuch der Deutschen Luftfahrtforschung* **1**, (1942) (English translation, NACA TM 1215).

88 Hall, I. M. and Sutton, E. P. Transonic flow in ducts and nozzles; A survey. In *Symposium Transsonicum*, Springer-Verlag, Berlin, 1964 pp. 325–344.

89 Blasius, H. Grenzschichten in Flüssigkeiten mit kleiner Reibung. Dissertation, Göttingen, (1907); *ZAMP* **56** (1908): 1–37; (English translation, NACA TM 1256).

90 Von Mises, R. *Mathematical Theory of Compressible Fluid Flow*. Academic Press, New York, 1958.

91 Crocco, L. Sulla transmissione del calore da una lamina piana a un fluido scorrente ad alta velocità. L' Aerotecnica, **12** (1932) pp. 181–197.

92 Busemann, A. Gasströmung mit laminarer Grenzschicht entlang einer Platte. *ZAMM* **15** (1935): 23–25.

93 Emmons, H. W. and Brainerd, J. G. Temperature effects in a laminar compressible fluid boundary layer along a flat plate. *J. Appl. Mech.* **8**, A105, (1941) and **9**, 1 (1942).

94 Von Karman, T. and Tsien, H. S. Boundary layer in compressible fluid. *J. Aero. Sci.* **5** (1938): 227ff.

95 Von Mises, R. Bemerkungen zur Hydrodynamik. *ZAMM* **7** (1927): 425–431.

96 Dorodnitsyn, A. A. Laminar boundary layer in compressible flow. *Dokl. Akad. Nauk. SSSR* **34** (1942): 213–219; see also *Prikl. Mat. Mekh.* **6**, pt. 6 (1942).

97 Chapman, D. R. and Rubesin, M. W. Temperature and velocity profiles in compressible laminar boundary layers with arbitrary distribution of surface temperature. *J. Aero. Sci.* **16** (1949): 547–565.

98 Howarth, L. Concerning the effect of compressibility on laminar boundary layers and their separation. *Proc. Roy. Soc. London.* **A194** (1948): 16–42.

99 Stewartson, K. Correlated compressible and incompressible boundary layers. *Proc. Roy. Soc. London* **A200** (1949): 84–100.

100 Falkner, V. M. and Skan, S. W. Some approximate solutions of the boundary layer equations. *Phil. Mag.* **12** (1931): 865–896.

101 Cohen, C. B. and Reshotko, E. Similarity solutions for the compressible laminar boundary layer with heat transfer and arbitrary pressure gradient. NACA Report 1293 (1956).

102 Levy, S. Effect of large temperature changes (including viscous heating) upon laminar boundary layers with variable free stream velocity. *J. Aero. Sci.* **21** (1954): 459–474.

102a Li, T. Y. and Nagamatsu, H. T. Similar solutions of compressible boundary layer equations. *J. Aero. Sci.* **22** (1955): 607–616.

103 Eckert, E. R. G. Engineering relations for friction and heat transfer to surfaces in high velocity flow. *J. Aero. Sci.* **22** (1955): 585–587.

104 Walker, G. K. Some comments on laminar and turbulent heat transfer equations. General Electric Missile and Space Vehicle Division Aerophysics Engineering Technical Memorandum 147 (1959).

105 Crocco, L. Lo strato limite laminare nei gas. Monografie Scientifiche di Aeronautica No. 3, Min. del Difesa-Aero, Roma, (1946); English translation: North American Aviation Aerophysics Lab. Report AL-684 (1948).

106 Van Driest, E. R. Investigation of laminar boundary layer in compressible fluids using the Crocco method. NACA TN 2597 (1952).

107 Lees, L. Laminar heat transfer over blunt nosed bodies at hypersonic flight speeds. *Jet Propulsion* **26** (1956): 259–269.

108 Gruschwitz, E. Calcul approche de la couche limite laminaire en ecoulement compressible sur une paroi non-conductrice de la chaleur. ONERA, Paris, Publication No. 47 (1950).

109 White, F. M. *Viscous Fluid Flow*. McGraw-Hill, New York, 1974, p. 611.

110 See references 98 and 99.

111 Kalikhman, B. L. Heat transmission in the boundary layer. NACA TM 1229 (1949).

112 Ginzel, J. Ein Pohlhausen Verfahren zur Berechnung laminarer kompressibler Grenzschichten an einer geheitzten Wand. *ZAMM* **29** (1949): 6, 321.

113 Morris, D. N. and Smith, J. W. The compressible laminar boundary layer with arbitrary pressure and surface temperature gradients. *J. Aero. Sci.* **20** (1953): 805–818.

114 Cope, W. F. and Hartree, D. R. The laminar boundary layer in compressible flow. *Phil. Trans. Roy. Soc. London* **A241** (1948): 1ff.

115 Fernholz, H. Halbempirische Gesetze zur Berechnung turbulenter Grenzschichten nach der Methode der Integralbedingungen. *Ing. Arch.* **33** (1961): 384–395.

116 Geropp, D. Näherungstheorie fur kompressible laminare Grenzschichten mit zwei Formparametern für das Geschwindigkeits profil. Ph.D. thesis, Technische Hochschule, Karlsruhe (1963).

117 Walz, A. *Boundary Layers of Flow and Temperature*. MIT Press, Cambridge, MA 1969.

118 Yang, K. T. An improved integral procedure for compressible laminar boundary layer analysis. *J. Appl. Mech.* **28** (1961): 9–20.

119 Smith, A. M. O. and Clutter, D. W. Machine calculation of compressible laminar boundary layers. *AIAA* J. **3** (1965): 639–647.

120 Cebeci, T. and Smith, A. M. O. A finite difference method for calculating compressible laminar and turbulent boundary layers. ASME paper 70-FE-A 1970.

121 Kemp, N. H., Rose, P. H., and Detra, R. W. Laminar heat transfer around blunt bodies in dissociated air. *J. Aero. Sci.* **26** (1959): 421–430.

122 Mangler, W. Zusammenhang zwischen ebenen und rotations-symmetrischen Grenzschichten in kompressiblen Flüssigkeiten. *ZAMM* **28** (1948): 97–103.

123 Ostrach, S. Unpublished.

124 Moore, F. K. Laminar boundary layer on a circular cone in supersonic flow at a small angle of attack. NACA TN 2521 (1951).

125 Ferri, A. Supersonic flow around circular cones at angle of attack. NACA TN 2236 (1950).

126 Strutt, J. W., Lord Rayleigh. On the motion of solid bodies through viscous liquid. *Phil. Mag.* **21** (1911): 697ff.

127 Stokes, G. G. On the effect of the internal friction of fluids on the motion of pendulums. *Trans. Camb. Phil. Soc.* **9** (1851): 8–106.

128 Illingworth, C. R. Unsteady laminar flow of gas near an infinite flat plate. *Proc. Camb. Phil. Soc.* **46** (1950): 603–613.

129 Cheng, S. I. Some aspects of unsteady boundary layer flow. Princeton University, Dept. Aero. Eng., Rep. No. 311 (1955).

129a Cheng, S. I. and Elliott, D. The unsteady laminar boundary layer on a flat plate. Princeton University, Dept. Aero. Eng., Rep. No. 318 (1955).

130 Stewartson, K. On the impulsive motion of a flat plate in a viscous fluid. *Quart. J. Mech. Appl. Math.* **4** (1951): 182–198. See also Stewartson, K. On the motion of a flat plate at high speed in a viscous compressible fluid. I. Impulsive motion. *Proc. Camb. Phil. Soc.* **51** (1955): 202–219.

131 Ostrach, S. Compressible laminar boundary layers and heat transfer for unsteady motions of a flat plate. NACA TN 3569 (1955).

132 Van Driest, E. R. Turbulent boundary layer in compressible fluids. *J. Aero. Aci.* **18** (1951): 145–160.

133 Boussinesq, J. Theorie de l'ecoulement tourbillant. Mem. Pre. par div. Sav. Paris, **23** (1877).

134 Prandtl, L. Über die ausgebildete Turbulenz. *ZAMM* **5** (1925): 136. (See also: *Proc. 2nd Int. Congr. Appl. Mech.* Zürich 1926.

135 Ludwig, H. Bestimmung des Verhältnisses der Austauschkoeffizienten für Wärme und Impuls bei turbulenten Grenzschichten. *ZFW* **4** (1956): 73–81. (See also: Kestin, J. and Richardson, P. D. Wärmeübergang in turbulenten Grenzschichten. Forschg. Ing.-Wes. **29** (1963): 93–104.)

136 See reference 132. See also Van Driest, E. R. The problem of aerodynamic heating. *Aero. Eng. Rev.* (1956): 26–41; Van Driest, E. R. The turbulent boundary layer with variable Prandtl number. In *Fünfzig Jahre Grenzschicht Forschung*, Tollimen and Görtler, ed's, Vieweg Verlag, Braunschweig, 1955.

137 Von Karman, T. Turbulence and skin friction. *J. Aero. Sci.* **1** (1934): 1ff.

138 Wilson, R. E. Turbulent boundary layer characteristics at supersonic speeds. *J. Aero. Sci.* **17** (1950): 585–594.

139 V. Karman, T. Mechanische Ähnlichkeit und Turbulenz. Ges. Wiss. Göttingen, Math. Phys. Klasse, 1930, p. 58 See also Proc. 3rd Int. Congr. Appl. Mech., Stockholm, (1930), p. 85, Part I, also, NACA TM 611 (1931).

140 Coles, D. Measurements of turbulent friction on a smooth flat plate in supersonic flow. *J. Aero. Sci.* **21** (1954): 433–438.

141 Korkegi, R. H. Transition studies and skin friction measurements on an insulated flat plate at a Mach number of 5.8. *J. Aero. Sci.* **23** (1956): 97–107, 192.

142 Chapman, D. R. and Kester, R. H. Measurement of turbulent skin friction on cylinders in axial flow at subsonic and supersonic velocities. *J. Aero. Sci.* **20** (1953): 441–448.

143 van Driest, E. R. On the boundary layer with variable Prandtl number. *Jahrbuch der Wissenschaftlichen Gesellschaft für Luftfahrt e.V.* (*WGL*) (1954): 66–75. Published by Friedr. Vieweg u. Sohn, Brauschweig.

144 See reference 136.

145 Sommer, S. C. and Short, B. J. Free flight measurements of turbulent boundary layer skin friction in the presence of severe aerodynamic heating at Mach numbers from 2.8 to 7.0. NACA TN 3391 (1955).

146 Eckert, H. U. Characteristics of the turbulent boundary layer on a flat plate in compressible flow from measurements of friction in pipes. *J. Aero. Sci.* **17** (1950): 573–584.

146a von Karman, T. The analogy between fluid friction and heat transfer. *Trans. ASME*, **61** (1939): 705–710.

147 Colburn, A. P. A method of correlating forced convection heat transfer data and a comparison with fluid friction. *Amer. Inst. Chem. Eng.* **29** (1933): 174–209.

148 Reshotko, E. and Tucker, M. Approximate calculation of the compressible turbulent boundary layer with heat transfer and arbitrary pressure gradient. NACA TN 4154 (1957).

149 Ludwig, H. and Tillmann, W. Investigation of the wall shearing stress in turbulent boundary layers. NACA TM 1285 (1950).

150 Maskell, E. C. Approximate calculation of the turbulent boundary layer in two-dimensional incompressible flow. British RAE, report no. AERO 2443 (1951).

151 Skopets, M. B. An approximate method of calculating turbulent boundary layer in the presence of heat exchange *Zh. Tekhn. Fiz.* **25** (1955): 864–876.

152 Sasman, P. K. and Cresci, R. J. Compressible turbulent boundary layer with pressure gradient and heat transfer. *AIAA J.* **4** (1966): 19–25.

153 Winter, K. G. Smith, K. G., and Rotta, J. C. Turbulent boundary layer studies on a waisted body of revolution in subsonic and supersonic flow. AGARDograph 97 (1965).

154 Flaherty, R. J. Arbitrary pressure gradient integral technique for predicting boundary layer and thermal parameters. *J. Aircraft* **11** (1974): 293–301.

155 Camarata, F. J. and McDonald, H. A procedure for predicting characteristics of compressible turbulent boundary layers which includes the treatment of upstream history. United Aircraft Research Laboratories Report No. G 212238–3 (1968).

156 McDonald, H. An assessment of certain procedures for computing the compressible turbulent boundary layer development. NASA SP-216 (1969): 181–229.

156a Economos, C. and Boccio, J. An investigation of the high speed turbulent boundary layer with heat transfer and arbitrary pressure gradient. NASA CR-1679, 1680, 1681 (1970).

157 Coles, D. The turbulent boundary layer in a compressible fluid. *Phys. Fluids.* **7** (1964): 1403–1423.

158 Clauser, F. The turbulent boundary layer. *Advances in Applied Mechanics*, **4** (1956).

159 Baronti, P. O. and Libby, P. A. Velocity profiles in turbulent compressible boundary layers. *AIAA J.* **4**, (1966): 193–202.

160 Libby, P. A. and Baronti, P. O. A transformation theory of the turbulent compressible boundary layer with pressure gradient and heat transfer. GASL TR-455 (1964).

161 Herring, H. J. and Mellor, G. L. A method of calculating compressible turbulent boundary layers. NASA SP-216 (1969): 27–133.

162 White, F. M. and Christoph, G. H. A simple theory for the two dimensional compressible turbulent boundary layer. *J. Basic Eng.* **94** (1972) : 636–642.

163 White, F. M. A new integral method for analyzing the turbulent boundary layer with arbitrary pressure gradient. *J. Basic Eng.* **91** (1969): 371–378.

164 Nikuradse, J. Turbulente Strömung in nicht kreisförmigen Rohren. *Ing-Archiv.* **1** (1930): 306–332.

165 White, F. M. *Viscous Fluid Flow.* McGraw-Hill, New York (1974): 645.

166 Brott, D. L., Yanta, W. J., Voisinet, R. L., and Lee, R. E. An experimental investigation of the compressible turbulent boundary layer with a favorable pressure gradient. AIAA Paper no. 69–685 (1969).

167 Hansen, C. F. Approximations for the thermodynamic and transport properties of high temperature air. NASA TR R-50 (1959).

168 Lewis, B. and von Elbe, G. *Combustion, Flames and Explosions of Gases.* Academic Press, New York, 1951.

169 Anon. Selected values of chemical thermodynamic properties. Series III. National Bureau of Standards, Washington, D.C.

170 Moeckel, W. E. and Weston, K. C. Composition and thermodynamic properties of air in chemical equilibrium. NACA TN 4265 (1958).

171 Kennard, E. H. *Kinetic Theory of Gases.* McGraw-Hill, New York, 1938.

172 Hirschfelder, J. O. Heat transfer in chemically reacting mixtures. *Int. J. Chem. Phys.* **26** (1957): 274–281.

173 Butler, J. N. and Brokaw, R. S. Thermal conductivity of gas mixtures in chemical equilibrium. *J. Chem. Phys.* **26** (1957): 1636–1643.

174 Lees, L. Laminar heat transfer over blunt nosed bodies at hypersonic flight speeds. *Jet Propulsion* **26** (1956): 259–269, 274 (see footnote, p. 260 of this paper).

175 See reference 174.

176 Fay, J. A. and Riddell, F. R. Theory of stagnation point heat transfer in dissociated air. *J. Aero. Sci.* **25** (1958): 73–85, 121.

177 Rose, P. H. and Stark, W. I. Stagnation point heat transfer measurements in dissociated air. *J. Aero. Sci.* **25** (1958): 86–97.

178 Hilsenrath, J. and Beckett, C. Tables of thermodynamic properties of argon free air to 15,000°K. Arnold Engineering Development Center, USAF, Rep. no. AEDC-TN-56-12 (1956).

179 Goulard, R. On catalytic recombination rates in hypersonic stagnation heat transfer. *Jet Propulsion* **28** (1958): 737–745.

180 Dorrance, W. H. *Viscous Hypersonic Flow*. McGraw-Hill, New York, 1962.

181 Martin, J. J. *Atmospheric Reentry*. Prentice Hall, Englewood Cliffs, NJ (1966).

182 Cohen, N. B. Boundary layer similar solutions and correlation equations for laminar heat transfer distribution in equilibrium air at velocities up to 41,000 feet per second. NASA TR R-118 (1961).

183 Adams, M. C. A look at the heat transfer problem at super satellite speeds. AVCO Report AMP 53 (1960).

184 Schreier, S. A. comparison of five methods of calculating aerodynamic stagnation point heat transfer at velocities of 25,000 to 40,000 feet per second. *J. Appl. Mech.* **30** (1963): 430–434.

185 Moore, L. L. A solution of the laminar boundary layer equations for a compressible fluid with variable properties, including dissociation. *J. Aero. Sci.* **19** (1952): 505–518.

186 Hantzsche, W. and Wendt, H. The laminar boundary layer of the flat plate with and without heat transfer considering compressibility. *Jahrbuch der Deutschen Luftfahrtforschung* (1942): Part I, 40–50.

187 Romig, M. F. and Dore, F. J. Solutions of the compressible laminar boundary layer including the case of a dissociated free stream. Convair Rept. No. ZA-7-012, San Diego, Calif. (1954).

188 Hayes, W. D. and Probstein, R. F. *Hypersonic Flow Theory*. Academic Press, New York, 1959.

189 Kemp, N. H., Rose, P. H., and Detra, R. W. Laminar heat transfer around blunt bodies in dissociated air. *J. Aero. Sci.* **26** (1959): 421–430.

190 See reference 178.

191 Forstall, W. and Shapiro, A. H. Momentum and mass transfer in coaxial jets. *J. Appl. Mech.* **17** (1950): 399ff.

192 Rose, P. H., Probstein, R. F., and Adams, M. C. Turbulent heat transfer through a highly cooled partially dissociated boundary layer. *J. Aero. Sci.* **25** (1958): 751–760.

193 Schlichting, H. *Boundary Layer Theory*. McGraw-Hill, New York, 1979.

194 Dorrance, W. H. Dissociation effects upon compressible turbulent boundary layer skin friction and heat transfer. *ARS J.* **31** (1961): 61–70.

195 Feldman, S. Hypersonic gas dynamic charts for equilibrium air. Published by Avco Everett Research Lab., Everett, Mass. (1957).

196 Huber, P. W. Tables and graphs of normal shock parameters at hypersonic Mach numbers and selected altitudes. NACA TN 4352 (1958).

197 Rozhdestvenski, I. B. Thermodynamic and gas dynamic properties of a flow of air behind primary densification shock wave considering dissociation and ionization of air. In *Physical Gas Dynamics*, Akad. Nauk. Moscow 1959.

198 Richardson, L. F. and Gaunt, J. A. The deferred approach to the limit. *Phil. Trans. Roy. Soc. London* **A226** (1927): 300ff.

199 Runge, C. Über die numerische Auflösung von Differential-Gleichungen *Math. Ann.* **46** (1895): 167–178.

200 Heun, K. Neue Methode zur approximativen Integration der Differentialgleichungen einer unabhängingen Veränderlichen *Z. Math. Phys.* **45** (1900): 23–38.

201 Kutta, W. Beitrag zur näherungsweisen Integration totaler Differentialgleichungen. *Z. Math. Phys.* **46** (1901): 435–453.

202 Kunz, K. S. *Numerical Analysis*. McGraw-Hill, New York, 1957.

203 Weeg, G. P. and Reed, G. B. Introduction to numerical analysis. Blaisdell Publishing, Waltham, MA 1966.

204 Schreier, S. Solution of the boundary layer equations for a cone flying at supersonic speed with varying angle of attack and non-zero pitch rate. AVCO Systems Division, Wilmington, Mass. Report AFFDL-TR-68-37 (1969).

205 See reference 204.

206 Hildebrand, F. B. On the convergence of numerical solutions of the heat flow equation. *J. Math. Phys.* **31** (1952): 35–41.

207 Forsythe, G. E. and Wasow, W. R. Finite Difference Methods for Partial Differential Equations. *John Wiley*, New York, 1960.

208 Bohachevsky, I. O. and Rubin, E. L. A direct method for computation of non equilibrium flows with detached shock waves. *AIAA J.* **4** (1966): 600–606.

209 See reference 208.

210 Bohachevsky, I. O. and Mates, R. E. A direct method for calculation of the flow about an axisymmetric blunt body at angle of attack. *AIAA J.* **4** (1966): 776–782.

211 Chu, C. K. *Computational Fluid Dynamics*. AIAA Selected Reprint Series, vol. 4, (1968).

212 von Neumann, J. and Richtmeyer, R. D. A method for the numerical calculation of hydrodynamic shocks. *J. Appl. Phys.* **21** (1950): 232–237.

213 Moretti, G. A new technique for the numerical analysis of non-equilibrium flows. *AIAA J.*, **3** (1965): 223–229.

214 Amsden, A. A. and Harlow, F. H. Numerical calculation of supersonic wake flow. *AIAA J.* **3** (1965): 2081–2086.

215 Butler, T. D. Numerical solutions of hypersonic sharp leading edge flows. *Phys. Fluids* **10** (1967): 1205–1215.

216 Scala, S. M. and Gordon, P. Solution of the time dependent Navier Stokes equations for the flow around a circular cylinder. *AIAA J.* **6** (1968): 815–822.

217 Chu, C. K. Kinetic theoretic description of shock wave formation. *Phys. Fluids* **8** (1965): 1450–1455.

218 Vogenitz, F. W., Bird, G. A., Broadwell, J. E. and Rungaldier, H. Theoretical and experimental study of low density supersonic flows about several simple shapes. AIAA paper 68–6 (1968).

219 Holt, M., Editor. *Proceedings of the Second International Conference on Numerical Methods in Fluid Dynamics*. Springer Verlag, Berlin (1970).

220 Kutler, P. and Lomax, H. The computation of supersonic flow fields about wing-body combinations by "shock capturing" finite difference techniques. In reference 219, pp. 24–29.

221 Laval, P. Time dependent calculation method for transonic nozzle flows. In reference 219, pp. 187–192.

222 Holt, M. and Masson, B. S. The calculation of high subsonic flow past bodies by the method of integral relations. In reference 219, pp. 207–214.

223 Gonor, A. L., Lapygin, V. I. and Ostapenko, N. A. The conical wing in hypersonic flow. In reference 219.

224 Hewitt, B. L., Illingworth, C. R., Lock, R. C., Mangler, K. W., McDonnell, J. H., Richards, C. and Walkden, F., editors *Computational Methods and Problems in Aeronautical Fluid Dynamics*. Academic Press, London, 1976.

225 Marsh, H. Finite element and difference methods for cascades. In reference 224, pp. 117–139.

226 Hancock, G. J. Some problems of unsteady flow about aircraft. In reference 224 pp. 214–241.

227 Lock, R. C. Methods for elliptic problems in external aerodynamics. In reference 224, pp. 53–99.

228 Purshouse, M. and Nangia, R. K. Applications of linearized supersonic wing theory to the calculation of some aircraft interference flows. In reference 224, pp. 383–423.

229 Nixon, D. An extended integral equation method for unsteady transonic flow past a two-dimensional airfoil. In reference 224.

230 Holt, M. *Numerical Methods in Fluid Dynamics*. Springer Verlag, Berlin, 1977.

231 Chung, T. J. *Finite Element Analysis in Fluid Dynamics*. McGraw-Hill, New York, 1978.

232 Wirz, H. J. and Smolderen, J. J. *Numerical Methods in Fluid Dynamics*, McGraw-Hill, New York, 1978.

233 Ackeret, J. Über Luftkräfte bei sehr grossen Geschwindigkeit insbesondere bei ebenen Strömungen. *Helvetia Phys. Acta* **1** (1928): 301–322.

234 Shapiro, A. H. Dynamics and Thermodynamics of Compressible Fluid Flow. Ronald Press, New York, (1953).

235 Hantzsche, W. and Wendt, H. Cones in supersonic flow. NACA TM 1157 (1947).

236 Dailey, C. L. and Wood, F. C. Computation Curves for Compressible Flow Problems. John Wiley and Sons, New York (1949).

237 Napolitano, L. G. and Belotserkovsky, O. M. Computational Gasdynamics. International Centre for Mechanical Sciences, Courses and Lectures: Vol. 40 Springer-Verlag, Wien, New York (1978).

238 Busemann, A. A review of analytical methods for the treatment of flows with detached shocks. NACA TM 1858 (1949).

239 Ames Research Staff: Equations, tables and charts for compressible flow. NACA Report 1135 (1953).

240 Korn, D. G. Computation of shock free flows for airfoil design, New York University, Report NYO–1480–125 (1968).

241 Fish, R. W. and McDonald, H. Practical Calculation of Transitional Boundary Layers. United Aircraft Research Labs Report UAR H48 (1969).

Problems

Chapter 1

1 Demonstrate by expansion into components that equation (80) is indeed equivalent to equations (66).

2 In deriving the Navier–Stokes equations, the displacement due to the average rate of rotation of a particle is subtracted from the total displacement so as to leave only pure strain. This is equivalent to taking all displacements relative to a set of moving axes. Could a different set of moving axes be used?

3 Equations (61) state that the normal stresses consist of an average stress, called the pressure, and a departure from the average caused by velocity gradients. Look up the viscosity for air and water and determine the ratio of the total normal stress to the average normal stress as a function of velocity gradient in incompressible flow for standard atmospheric pressure. How might this ratio be affected by compressibility?

4 Obtain equations (63) by substituting equations (60) and (61) into equations (62).

5 In incompressible flow, there are a number of well-known exact solutions of the Navier–Stokes equations. One of these is parallel flow in a long channel as a result of a pressure gradient in the flow direction. Given a compressible viscous parallel flow in a long channel of width $2h$, say, due to a pressure gradient in the flow direction x, determine the velocity profile in the channel. What assumptions must one make to solve this problem? How does the solution compare to the incompressible solution?

6 Another well-known incompressible solution of the Navier–Stokes equation is that of Couette flow. A fluid is contained between two long parallel walls. The lower wall is at rest and the upper wall moves in its own plane at some constant speed. There may or may not be a pressure gradient in the flow direction. Find the velocity profile for this flow when the fluid is compressible. What assumptions must one make to solve this problem? How does the solution compare to the incompressible solution?

7 In elementary thermodynamics, the first law is sometimes written

$$\delta q = de_t + p\, d\mathscr{V}$$

It has been argued that the first law for a moving system may be obtained simply by taking the substantial derivative of this equation, yielding

$$\dot{q} = \frac{De_t}{Dt} - \frac{p}{\rho}\frac{D\rho}{Dt}$$

which is equivalent to equation (24). Is this a valid way of deriving the first law for a moving system?

8 Another well-known exact solution of the incompressible Navier–Stokes equation is that of the Rayleigh (more properly, Stokes) problem. An infinite wall is suddenly accelerated in its own plane. At time $t < 0$, the velocity of the wall is zero. At time $t \geq 0$, the velocity of the wall is $u_0 =$ constant. Suppose now that the fluid over the wall is a perfect gas whose viscosity varies linearly with the temperature. Find the velocity distribution over the plate as a function of time.

9 Two concentric cylinders are rotating, one inside the other, at different speeds. The space between the cylinders is filled with a compressible fluid. Find the velocity distribution is the fluid, making suitable assumptions.

10 Sometimes it is more convenient to work in polar coordinates. Show that in polar coordinates, the Navier–Stokes equations may be written

$$\rho\left(u_t + uu_r + \frac{v}{r}u_\theta - \frac{v^2}{r} + wu_z\right) = -p_r + \mu\left(u_{rr} + \frac{1}{r}u_r - \frac{u}{r^2} + \frac{1}{r^2}u_{\theta\theta} - \frac{2}{r^2}v_\theta + u_{zz}\right)$$

$$\rho\left(v_t + uv_r + \frac{v}{r}v_\theta + \frac{uv}{r} + wv_z\right) = -\frac{1}{r}p_\theta + \mu\left(v_{rr} + \frac{1}{r}v_r - \frac{v}{r^2} + \frac{1}{r^2}v_{\theta\theta} + \frac{2}{r^2}u_\theta + v_{zz}\right)$$

$$\rho\left(w_t + uw_r + \frac{v}{r}w_\theta + ww_z\right) = -p_z + \mu\left(w_{rr} + \frac{1}{r}w_r + \frac{1}{r^2}w_{\theta\theta} + w_{zz}\right)$$

Chapter 2

1 At one point in a quasi-one-dimensional flow of a perfect gas the temperature is 70°F, the pressure is 14.7 psi, and the velocity is 2000 fps. Find the stagnation pressure, the stagnation temperature, the Mach number, the density, and the stagnation density at the point. Assume $\gamma = 1.4$, MW $= 29$.

2 An isentropic flow of air has a mass flow rate of 1 slug per second, a stagnation temperature of 1000°R and a stagnation pressure of 100 psi. Find

a. The minimum area which will pass this flow and the velocity, temperature and pressure at this minimum area.

b. The maximum velocity which this flow may attain and the pressure, temperature, cross sectional area and Mach number at this point.

Which of these results would continue to be valid if the flow were adiabatic but not isentropic?

3 Air is flowing in a duct with a velocity of 3000 fps. The stagnation temperature is 2000°R, the stagnation pressure is 200 psi. A normal shock now occurs. Find the velocity, Mach number, temperature, pressure and stagnation pressure after the shock and the entropy change across the shock.

4 A de Laval nozzle has a ratio of exit area to throat area of 2. The upstream stagnation pressure and temperature are 1500°R and 150 psi, respectively. At a point in the diverging section of the nozzle where the area ratio is 1.5, a normal shock occurs. Find the pressure, temperature, velocity and Mach number at the exit of the nozzle. Also, find the change in the critical quantities ($p^*, \rho^*, T^*, c^*, A^*$) across the shock.

5 Suppose in Problem 4 the exit pressure is increased by 10%. Find the new shock location.

6 Plot equation (205) for air at a Mach number of 2. Take p_1, ρ_1 and μ at standard temperature and pressure, $\gamma = 1.4$, and determine the shock wave thickness predicted by Becker's solution in accordance with the definition of Figure 36. Then, instead of $\mu = \mu_1$, take an average value of μ across the wave and see what effect this has on the shock wave thickness as predicted by Becker.

7 A closed loop supersonic wind tunnel employing a Kantrowitz–Donaldson diffuser is designed for a Mach number of 2. The conditions in the test section are $T = 70°\text{F}, p = 14.7$ psi. The cross-sectional area of the test section is 1 ft^2. Suppose a shock stands downstream of the test section where the Mach number is 1.2. Determine the minimum power required to run this tunnel.

8 For flow in a constant area duct with energy transfer, demonstrate that
a. T is a maximum when $\text{M} = 1/\sqrt{\gamma}$.
b. T_0 is a maximum when $\text{M} = 1$.

9 A ram jet engine combustion chamber, which is shaped like a constant area duct, has the following entrance conditions before combustion: $V_1 = 500$ fps, $M_1 = 0.5$, $T_{0^1} = 500°\text{R}$, $P_1 = 20$ psia, $A = 1$ ft^2. Combustion is now initiated and the stagnation temperature at the exit of the chamber rises to 2000°R. Find the change in the mass flow rate through the engine assuming the gas properties remain those of air at 80°F.

10 It is desired to maintain the Mach number constant while adding heat to a flow. This may be done by suitable change in the cross-sectional area. Determine the change in the cross-sectional area required as a function of the change in the stagnation temperature. How does the loss in stagnation pressure in this case compare to the loss if the area is maintained constant?

Chapter 3

1 Demonstrate by direct expansion that equation (272) is equivalent to equation (270) in the two-dimensional case.

2 Consider a Couette flow of air. The distance between the bottom wall, which is at rest, and the top wall, which is moving at 1000 fps, is 100 ft. V^2 varies linearly across the flow from 0 to 10^6 ft^2/sec^2. T varies linearly from 100°R to 1000°R and s varies linearly from 2 BTU/lb°R to 20 BTU/lb°R. what is the vorticity half way between the top and bottom walls?

3 A two-dimensional, radial flow has velocity $1/r$ at every point r being the distance from the origin. Demonstrate that this flow is a potential flow and find the velocity potential as a function of x and y.

4 At one point on an airfoil in an isentropic flow, the Mach number reaches 1. Determine the pressure coefficient at this point in terms of the free stream Mach number.

5 Consider the vortex motion of a perfect gas. The streamlines are circles around the origin. The velocity is inversely proportional to the radius. The motion is assumed to have constant entropy and stagnation enthalpy.
a. Demonstrate that this is a potential flow
b. Give the equation for the velocity potential
c. Determine the minimum radius for which this motion may exist
d. Determine the radius for which the velocity equals the speed of sound

6 Consider the equation

$$(c^2 - \phi_x^2)\phi_{xx} - 2\phi_x\phi_{xt} - \phi_{tt} = 0$$

Determine if this equation is elliptic, hyperbolic, or parabolic.

7 Using the Rayleigh Janzen method, determine the velocity potential for the flow around a circular cylinder. Find the ϕ_n in equation (335) up to $n = 2$.

8 Consider the incompressible flow about a circular cylinder. Determine the shape of the corresponding profile in the compressible plane according to the theory of von Karman and Tsien.

9 Consider the flow between two wavy shaped walls which are the mirror image of each other. Determine the velocity potential for this flow.

10 Starting with equation (272), determine the equation for the velocity potential in cylindrical and spherical coordinates.

Chapter 4

1 Consider the nozzle shown in the figure. Along the curve *ab*, the Mach number is equal to 1, the pressure is 100 psi, and the temperature is 2000°R. Downstream of the line *ab* the flow is supersonic. Assume that the flow is isentropic, the gas is air. Using the method of characteristics, find the velocity, temperature, pressure, and Mach number distribution downstream of the line *ab* when the flow is (a) plane two-dimensional, (b) axially symmetric, (c) redesign this nozzle so as to obtain uniform parallel flow at the exit.

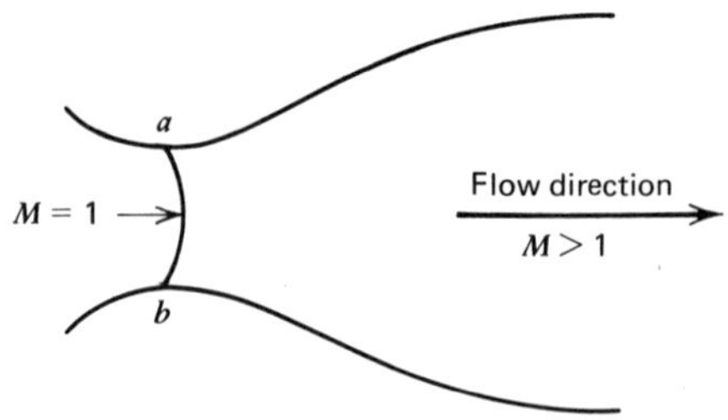

2 Consider the flow in a supersonic jet as shown in the figure. $p_1 = 7$ psia, $M_1 = 3, T_0 = 4000°$R. The gas is air. The back pressure is 14 psia. Find the pressure, temperature and velocity in regions 2, 3, 4, and 5, using (a) isentropic flow theory, (b) oblique shock theory.

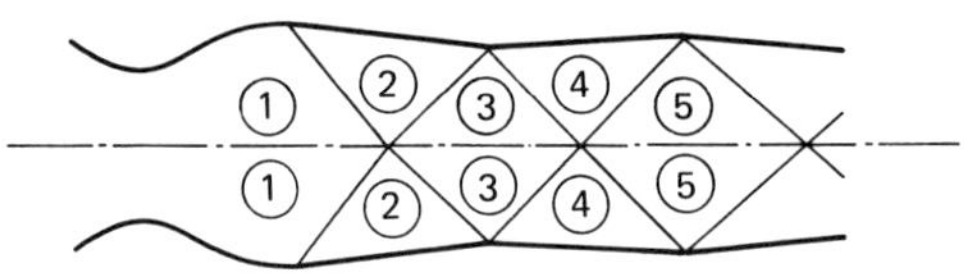

3 Consider the supersonic flow over a cone at zero angle of attack, given the following data: at the cone surface, $M^* = 1.406, M = 1.5677$. Immediately behind the shock, $M = 1.693$, $\theta = 8.55°$, $M^* = 1.48$. The shock wave angle is 37.8°. Using the graphical method of Busemann, find the velocity distribution between the wave and the cone, and the cone angle.

4 Consider a curved shock wave as shown in the figure.

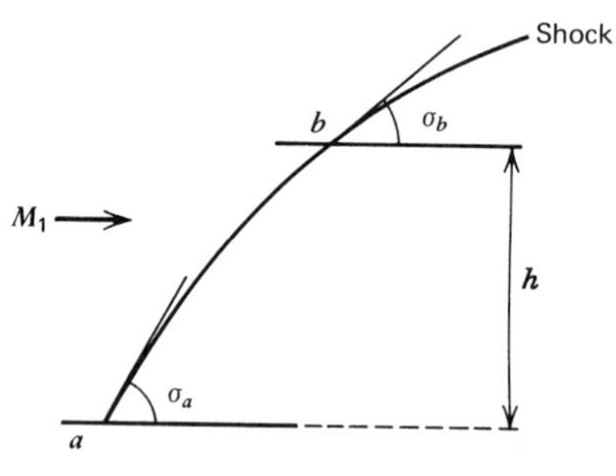

$M_1 = 2, h = 0.1$ft, $\sigma_a = 35°$, $\sigma_b = 55°$, $T_0 = 530$T° R. The gas is air. Determine the vorticity just downstream of the shock at a point midway between a and b.

5 For the wedge type airfoil shown in Figure 123, assume $M_1 = 2, T_1 = 500°$R, with the airfoil as shown in the figure.

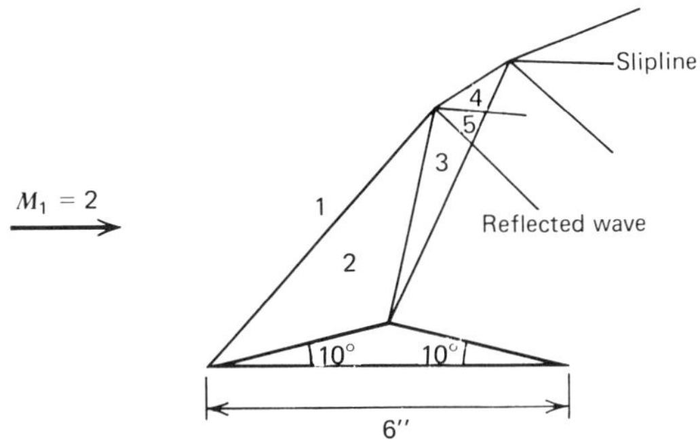

Find the shock wave shape and the velocity and flow directions in regions 2, 3, 4, and 5. Assume air. Assume 1° change in flow direction across each fan wave shown.

6 For the airfoil shown in Figure 126, the following data is given:

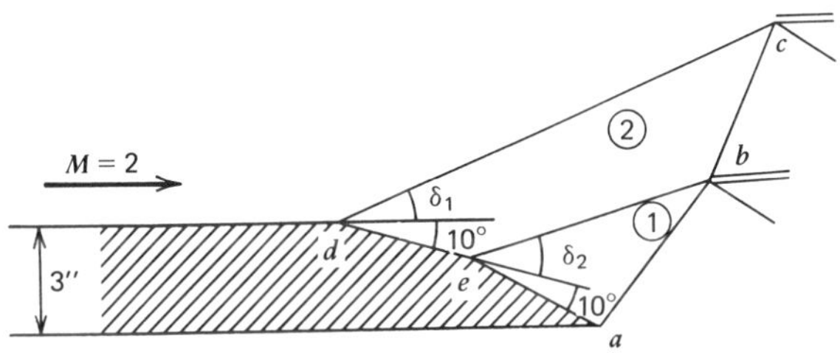

Determine the shock wave shape *abc*, the flow properties in regions 1 and 2, the angles δ_1 and δ_2 and the angles the slip lines make with the horizontal.

7 Consider the airfoil shown in Figure 143. Assume the airfoil consists of

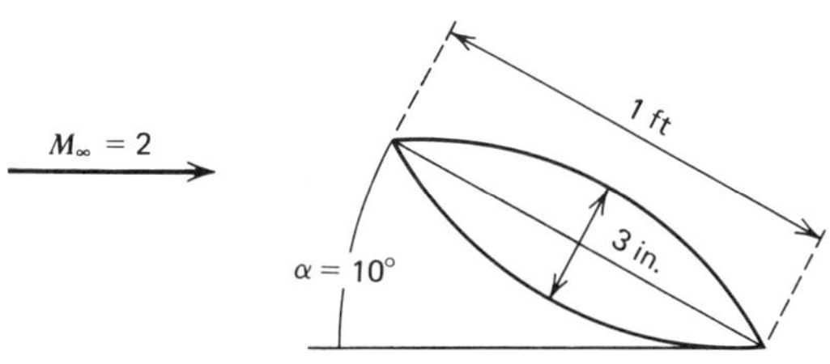

two intersecting circular arcs. Using supersonic small disturbance theory, find the lift and drag of this airfoil assuming (a) isentropic flow throughout, (b) leading edge shock followed by isentropic flow. Repeat the calculation assuming Prandtl–Meyer flow (c) with leading edge shock, (d) without leading edge shock.

8 Consider the flow of air in a duct with a ramp. The conditions in region 1 are: $p = 14.7$ psia, $T = 530°R$, $M = 3$. Find the velocity, pressure and temperature in regions 2 and 3, and the angles which the waves between

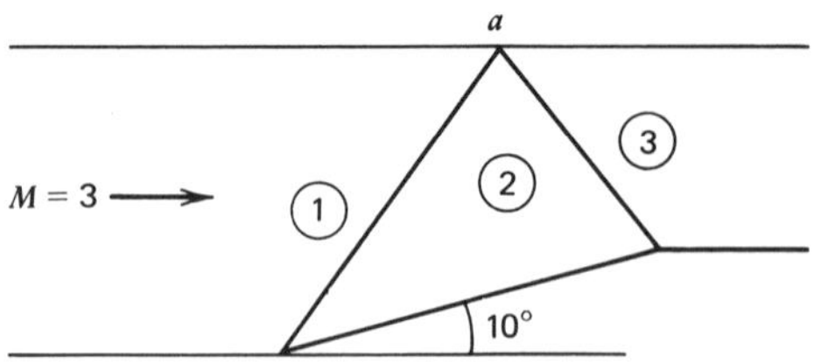

1 and 2, and 2 and 3 make with the horizontal. What is the largest ramp angle for which regular reflection is possible at point a?

9 Starting with equation (563), derive the equation for the circle in Figure 103.

10 Demonstrate that for supersonic flow over a cone at zero angle of attack,

$$-U_{\sigma\sigma}\left[\frac{\gamma+1}{2}U_\sigma^2-\frac{\gamma-1}{2}(W_{\max}^2-U^2)\right]-\frac{\gamma-1}{2}U_\sigma^3\cot\sigma-\gamma UU_\sigma^2$$

$$+\frac{\gamma-1}{2}(W_{\max}^2-U^2)U_\sigma\cot\sigma+(\gamma-1)U(W_{\max}^2-U^2)=0$$

where U is the velocity component in the radial direction in spherical coordinates and $W_{\max}=\sqrt{2c_pT_0}$. (This is the equation originally derived by Taylor and Macoll.)

Chapter 5

1 Consider the situation shown in Figure 160. At time $t<0$, the piston is at rest 1000 ft from the closed end of the tube. At time $t\geq 0$ the piston accelerates to the left with velocity At, where A equals 100 ft/sec^2. Taking time intervals of one half second between point 1, 2, 3 and 4, draw the physical characteristics to scale and determine the pressure, temperature, and velocity in regions a, b, c, d, e, f, g, i, and h. Assume air, isentropic flow, with an initial pressure of 100 psi and an initial temperature of 2000°R.

2 A piston is describing simple harmonic motion between a and b, as shown in the figure. The distance between a and b is 2 ft and the distance between b and c is 0.2 ft. The average speed of the piston is 100 fps. The gas in the

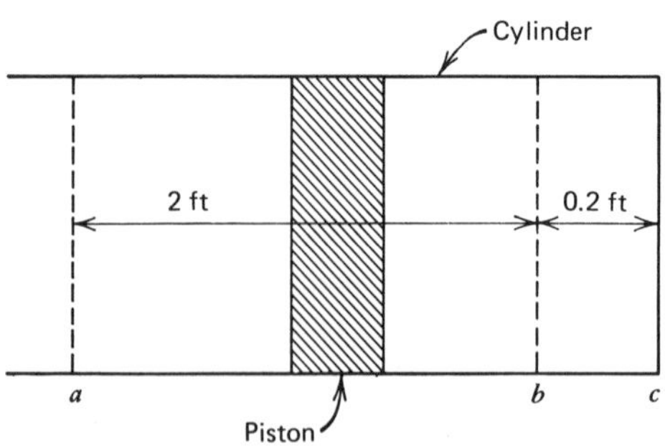

cylinder is air. When the piston is at a, the air in the cylinder is at rest with a pressure of 14 psi and a temperature of 100°F. Plot to scale the physical characteristics as shown in Figure 164, using only a single wave going back and forth. Assume isentropic flow.

3 Consider the situation shown in Figure 172. A shock wave moving at 3000 fps strikes a piston moving in the same direction at 1000 fps. The temperature and pressure in region 1 are 70°F and 20 psi, respectively. Determine the velocity, temperature and pressure in regions 2 and 3. Repeat the problem assuming the piston is moving to the left, i.e., opposite in direction to the wave.

4 Consider the situation shown in Figure 174. A shock wave moves toward the open end of a tube with a velocity of 1500 fps. The temperature and pressure in region 1 are 0°F and 10 psia, respectively. Assuming the pressure at the exit remains constant, compute the temperature, pressure and velocity in regions 2 and 3. Repeat the problem assuming the compression wave to be isentropic.

5 Consider the situation shown in Figure 175. Shock a is moving with a velocity of 1300 fps and shock b is moving with a velocity of 1800 fps. The pressure and temperature in region 2 is 30 psi and 10°F. Determine the pressure, temperature velocity and entropy in regions 1, 3, 4, and 5. Repeat the problem assuming the waves to be isentropic.

6 Consider the situation shown in Figure 162. The wave between regions 1 and 2 is moving at 1000 fps and the wave between regions 2 and 3 is moving at 800 fps. In region 1, the velocity is zero and the pressure and temperature are 14.7 psia and 70°F. Assuming constant pressure at the open end, find the pressure, temperature and velocity in regions 2, 3, 4 and 5. Assume air, $\gamma = 1.4$.

7 Consider the shock tube shown in Figure 184. Assume first that both regions 1 and 3 are filled with air. Take $p_3 = 2000$ psi and $p_1 = 1$ psi, $T_1 = T_3 = 70$°F. The diaphragm is now broken and the flow is as shown in Figure 185. Determine the pressure, temperature, entropy and velocity in regions 2 and 4. Draw to scale the diagram as shown in Figure 186. Repeat the problem if the driver is helium instead of air.

8 Consider the situation shown in Figure 187. Assuming air throughout, take the same initial conditions as in Problem 7, with the ratio of cross-sectional areas $A_4/A_3 = 2$. Determine the velocity, pressure, temperature and entropy in regions 2 and 4. Draw this process to scale on the state plane as illustrated in Figure 189.

9 Continue Problem 8 by considering the situation shown in Figure 190. The initial shock wave now reflects off a wall with a small diameter tube leading from it as shown in Figure 190(b). The area ratio $A_5/A_6 = 10$.

Determine the velocity, pressure, temperature and entropy in regions 5, 6 and 7. Draw this process to scale on the state plane as shown in Figure 191.

10 Consider the unsteady flow in a duct of varying cross-sectional area. The cross-sectional area at 1 is one square foot. The duct is diverging at an angle of 20 degrees. The velocity varies linearly along the line between the points 1 and 2 in the $x-t$ plane from 1500 fps at 1 to 1400 fps at 2. The

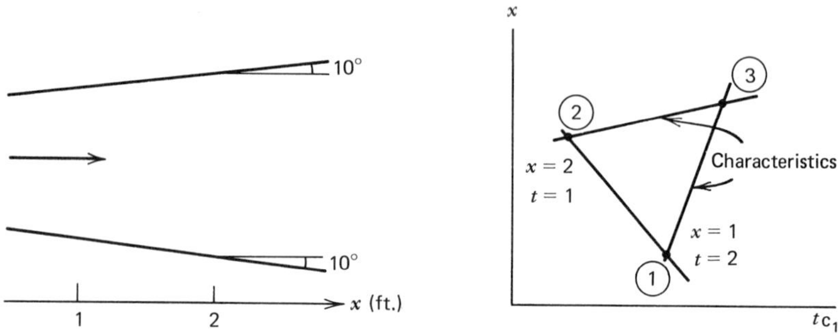

entropy also varies linearly from 0.60 Btu/1b° R to 0.50 Btu/1b°R going from 1 to 2. The temperature likewise varies linearly from 1000°R at 1 to 900°R at 2. Assuming air, $\gamma = 1.4$, find the location of the point 3 in the physical plane and the velocity and entropy there.

Chapter 6

1 Demonstrate that for a normal shock, approximately,

$$u'_1 - \frac{u'^2_1}{2} = u'_2 - \frac{u'^2_2}{2}$$

(equation (858)).

2 Verify equation (864) by combining equations (861)–(863) and omitting higher order terms (linearized potential equation).

3 Derive equation (889) by substituting equation (886) into equation (421) (Tricomi's equation).

4 Express the transonic similarity parameter of Oswatitsch, equation (857), in terms of M_∞, τ and γ, and compare it to the transonic similarity parameter of von Kármán in equation (879). Are the two equivalent?

5 Consider a circular arc airfoil, as shown in the figure. Take $t = 1$ ft., $l = 10\,ft$, $u_\infty = 1000$ fps, $T_\infty = 520°$R, $p_\infty = 14.7$ psi. Assuming that equations (896) and (897) are valid, determine the flow field around this airfoil using equations (899) and (900). Using the solution, plot the region of supersonic flow over the airfoil. Also using the solution, determine the magnitude of the term $u'u'_{x'}$, which was neglected, and compare the magnitude of this

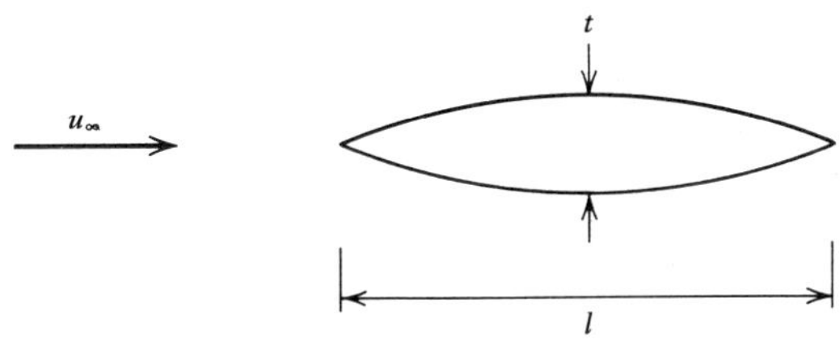

term with that of the terms in equation (896). Was neglect of this term justified in the present case? Note: you may wish to evaluate the integrals numerically, using a computer. Care must be exercised in evaluating the integrals around the singularities $x = \xi$. The Cauchy principal value is to be used.

6 Find the velocity distribution at the surface of the airfoil given in problem 5 with the same free stream conditions, using the integral method of Oswatitsch. Compare the results using equation (927) and Figure 204 with those obtained using equation (926) and Figure 203.

7 In incompressible flow, the complex potential for the flow over a source and sink is given by

$$F = -m \ln(z-a) + m \ln(z+a) + V_\infty z$$

Using the method of characteristics in the complex domain, determine the characteristic coordinates ξ and η, and find the appropriate function $g(\eta)$ for this case.

8 Demonstrate that

$$\oint(\rho v\, dx - \rho u\, dy) = -2\pi[X_1(0{,}0)(\rho v)_\eta - Y_1(0{,}0)(\rho u)_\eta]$$

$$= 4\pi \rho_\infty Re[BX_1(0{,}0)]$$

where the integral is taken over a closed path around the point $\xi = 0$, $\eta = 0$ (equation (1020)). Show that if this integral is zero, the stream function represents the flow around a closed body.

9 Consider a deLaval nozzle. The walls near the throat are circular arcs with a radius of 10 ft. The width of the nozzle at the throat is 1 ft. Assume air is flowing through the nozzle with $\gamma = 1.4$. The stagnation pressure and temperature are 100 psi and 1000°R, respectively. Consider the cross section at the throat, $x = 0$. Using the theory of Oswatitsch and Rothstein, determine (a) the Mach number on the centerline at $x = 0$ for which the velocity at the wall first attains sonic speed; and (b) the Mach number at the wall for $x = 0$ at which the velocity first exceeds or equals the speed of sound across an entire cross section of the nozzle.

10 In developing his theory, Meyer assumed that u varies linearly with x

along the centerline. Using the nozzle shape and stagnation conditions of Problem 9, plot the velocity along the centerline for asymmetric flow, i.e. for flow accelerating isentropically from subsonic to supersonic, using the theory of Oswatitsch and Rothstein and the theory of Meyer, to show the extent to which Meyer's assumption is valid.

Chapter 7

1 Demonstrate that the von Mises transformation may be applied to compressible flow; i.e., using equation (1104), derive equation (1105) from equation (1079).

2 (a) Using the Howarth transformation and the stream function as defined in equation (1110), derive equation (1111) from equation (1079). What would be the form of this equation for $p_x \neq 0$? (b) Using the transformation (1113a), derive equation (1113) from equation (1112) and show that $2u/U = f_\zeta$.

3 Using the von Mises transformation, derive the energy equation (1118) from equation (1116). What assumptions must one make in order to obtain this equation?

4 Consider an insulated flat plate at zero angle of attack in a stream of air. The plate is 2 ft wide and 10 ft long. The free stream conditions are $T_\infty =$ 520°R and $p_\infty = 14.7$ psia. Using the results of the method of Karman and Tsien (Figure 232), calculate and plot the drag and the drag coefficient of the plate as a function of Mach number.

5 Consider the flow over a flat plate at zero angle of attack whose surface temperature is constant. The plate is 1 ft wide and 5 ft long. The free stream temperature, pressure and velocity are 500° R, 10 psi, and 2000 fps, respectively. The surface temperature of the plate is one fourth of the adiabatic wall temperature. Using the method of Chapman and Rubesin, calculate the heat transfer rate to (or from) the plate.

6 Using the transformation of Levy, (1142), derive the momentum equation, (1144), and the energy equation, (1145), from equations (1139–1141).

7 In the momentum integral method, assuming that the velocity distribution is given by the fourth-order polynomial, (1177), verify the values of the constants given in (1178) by employing the boundary conditions (1172).

8 In the method of Howarth and Stewartson, consider a flow for which $V_1 = CZ^m$, with the flow outside the boundary layer isentropic. Find the corresponding relationship between U and x, and between Z and x, assuming a perfect gas with constant specific heat. Note: In the case of $m = \frac{1}{2}$, a closed form solution exists. For $m \neq \frac{1}{2}$, integration by series expansion may be necessary.

9 Consider a blunt body in a stream of air with a Mach number of 2. The free stream pressure and temperature are 460°R and 7 psia respectively. The ratio of the stagnation temperature at the outer edge of the boundary layer to the stagnation temperature at the surface is 0.8. Using the method of Lees and Figure 245, calculate the heat transfer rate per unit area at the stagnation point in Btu/ft^2.

10 A semi-infinite flate (flat plate) is accelerating in air with velocity $U = At^2$, where $A = 10$ ft/sec^3. The temperature and pressure at $y = \infty$ are 520°R and 14.7 psi, respectively.

a. Find the shear stress at the surface of the plate. Is this shear stress greater or less than that predicted by the quasisteady solution?
b. Find the recovery factor for the case when the plate is insulated. Is the recovery factor greater or less than the quasisteady value?
c. Suppose the surface temperature is 600°R. What is the heat-transfer rate in this case? Is this greater or less than the quasisteady value?
d. Repeat part (c) with a surface temperature of 400°R.
e. Can you give a physical explanation of why the unsteady solution should differ from the quasisteady solution in the way that it does?

Chapter 8

1 Starting with the momentum equation (1309), introduce the change of variables (1298), take the time average of both sides of the equation, and, using the laws of means (1300)–(1302), derive equation (1310).

2 Starting with equations (1315) and (1316), introduce the change of variables (1298), take the time averages of both sides of the energy equation, and verify that in the case of steady in the mean, two-dimensional flow, this equation reduces to the form (1317).

3 Consider the turbulent boundary layer over a flat plate whose surface temperature equals the free stream temperature. The plate is 1 ft wide and 5 ft long. The free stream temperature, pressure and velocity are 500° R, 10 psi and 2000 fps, respectively. Using the method of van Driest, calculate the shear stress at the wall as a function of x, the total drag of the plate and the heat transfer at the wall.

4 Consider the turbulent boundary layer over an insulated flat plate. Assume that the free stream conditions and the dimensions of the plate are the same as in Problem 3. Using the theory of Wilson, find the shear stress at the wall as a function of x, and the total drag of the plate. Repeat the calculation, using the method of van Driest and compare the results.

5 (a) Repeat Problem 4 using the reference temperature method of Eckert. Compare your results to those found by using the methods of van Driest and Wilson. (b) Find the drag of the plate using the method of Sommer and Short and compare the result with those found by the methods of Wilson, van Driest and Eckert.

6 In the method of Reshotko and Tucker, demonstrate that using the change of variables (1421), the continuity and momentum equations (1419) and (1420) may be written in the form (1422) and (1423).

7 It has been suggested that, in the method Reshotko and Tucker, equation (1436) might be employed directly to find the coefficient of friction, thereby avoiding the need to integrate equations (1443) and (1445). In order to evaluate H_i, use equation (1428) in equation (1426) with $N = 1/7$. Take $M_e = 3.0$, $p_0 = 2116\ \mathrm{lb/ft^2}$, $T_0 = 610°R$ where T_0 and p_0 are the stagnation conditions outside the boundary layer. Assume the wall to be adiabatic. Compare the result obtained in this way with that shown in Figure 271. Discuss the pros and cons of this approach.

8 Starting with the energy equation (1074), derive the energy equation (1141). Use this result to demonstrate that when the effective Prandtl number, Pr_e, as defined by Economos and Boccio is used (1464), the energy equation may be written in the form (1463).

9 Herring and Mellor assume that in the outer layer, $v_E = v_E(\nabla, y, \bar{u}_y)$, eq. (1516), and in the inner layer $v_E = v_E(\bar{v}, y, \bar{u}_y)$, eq. (1517). Using Buckingham's π theorem, demonstrate that in this case

$$\frac{v_E}{\nabla} = \Phi\left(\frac{y^2}{\nabla}\bar{u}_y\right) \tag{1496}$$

and

$$\frac{v_E}{\bar{v}} = \phi\left(\frac{y^2}{\bar{v}}, \bar{u}_y\right) \tag{1497}$$

Using these results, show that in the overlap layer, v_E must have the form $v_E = k^2 y^2 \bar{u}_y$ where k^2 is a constant.

10 Consider the flow over an airfoil of length L. The velocity outside the boundary layer is given by

$$U_e = U_\infty + 0.2\,U_\infty\left(1 - \frac{x}{L}\right)x$$

In addition, the following data are given

$U_\infty = 5000$ fps	$T_w = 1600°R$
$T_\infty = 400°$R	$U_0 = U_\infty$
$p_\infty = 7$ psi	$L = 10$ ft

Using the method of Christoph and White, find c_f as a function of x/L.

Chapter 9

1 Consider the equilibrium reaction

$$O_2 \rightleftharpoons 2O$$

at a temperature of 2000°K and a pressure of 1 atm. Find the equilibrium constant K_p for this reaction, using the partition functions for O_2 and O. Assuming ideal gas behavior, determine the mass fraction of O present.

2 Starting with the energy equation in the form (1194), and adding the diffusive heat transfer from equation (1657), derive the energy equation in the form (1666).

3 Using the change of variables (1670) and (1671), verify from the global continuity equation that ρv may be written in the form (1672). Using this result, derive the continuity of species equation in the form (1673).

4 Consider a blunt body with hemisphere nose flying at zero angle of attack at a Mach number of 12, at an altitude of 120,000 ft. The free stream pressure and temperature are 9.026 psfa and − 7.2°F, respectively. The surface temperature of the nose of the body is 9000°R. The Lewis number and Prandtl number are presumed constant through the boundary layer at 1.4 and 0.71, respectively. The velocity distribution at the nose is presumed to be Newtonian. The nose radius is 0.25 in. Calculate the heat transfer at the stagnation point using the formula of Fay and Ridell (1688) and compare your result with the data presented in Figure 299.

5 Consider a hemisphere cylinder combination flying through the air at zero angle of attack. The nose radius is 1 ft, the velocity is 1000 fps, the atmospheric pressure is 1 psi and the atmospheric temperature is 400°R. The pressure over the base of the body is zero. Calculate the pressure drag of this body assuming Newtonian flow.

6 Consider a normal shock in air. The upstream Mach number relative to the shock is 12. The static pressure and temperature ahead of the shock are 474.8 psfa and − 67.6°F, respectively. Using Figures 313 and 314, find the density, pressure and temperature ratios across the shock. Compare your results with those shown in Figures 315, 316, and 317. The altitude is 36,000 ft.

7 Taking time averages and applying the laws of means, demonstrate that for steady, turbulent, dissociating boundary layers, the continuity of species equation assumes the form (1777).

8 Consider the hemisphere cylinder combination shown in Figure 309. The nose radius, R, is 1 ft. The free stream velocity, U, is 20,000 fps, the free stream pressure and temperature are 474.8 psfa and − 67.6°F, respectively, corresponding to an altitude of 120,000 ft. Suppose the boundary layer over the nose becomes turbulent at $\theta = 45°$. Take $Le = 1.4$, $Pr = 0.72$, $\beta = 1.0$. Assume that the velocity distribution over the nose at the outer edge of the boundary layer is Newtonian. The wall temperature, T_w is 2000° R. Using equation (1810) and Figures 311 and 313, determine the heat transfer rate in Btu/ft^2/sec for $\theta = 50°$.

9 For the conditions of Problem 8, determine the shear stress at the wall where $\theta = 50°$.

10 Scala* has proposed the following formula for equilibrium stagnation point heat transfer:

$$q\sqrt{R_N} = (1.10)(10.0)^a(0.100)^b \frac{BTU}{\text{ft}^2\,\text{sec}}\sqrt{\text{ft}}$$

where R_N = nose radius in feet
$a = -(0.9689 + 3.888 \times 10^{-5}\,T_w)(5.626 + 9.84 \times 10^{-6}\,\text{A})$
$b = (0.9793 + 2.593 \times 10^{-5}\,T_w)(2.830 + 3.00 \times 10^{-7}\,\text{A})$
T_w = wall temperature in °R
A = altitude in feet

Check the validity of this formula by using it to calculate the stagnation point heat transfer for the conditions of Problem 4 and comparing the result with that obtained by the method of Fay and Riddell.

Chapter 10

1 Using a method analogous to that used for deriving the third order Runge–Kutta formulas (1835–1836), derive the fourth-order Runge–Kutta formula (1838–1839).

2 Consider the integral

$$\int_1^{1+\Delta x} \frac{dx}{x\sqrt{x^2+1}}$$

Taking $\Delta x = 0.001, 0.01$ and 0.1, evaluate this integral by (a) Simpson's rule (1847), (b) equation (1854) and (c) equation (1869). Compare these results with the exact value of the integral.

3 Consider the equation

$$\frac{dy}{dx} = \frac{xy}{1+x^2}$$

subject to the initial condition $y_1 = \sqrt{2}$ when $x_1 = 1$. Using step sizes $x_{n+1} - x_n = 0.001$, 0.01 and 0.1, determine y_2, y_3 and y_4 using fourth-order Runge–Kutta. Using these values as starting values, apply the predictor formula (1871) to obtain y_5. Compare your results to the exact solution.

4 Repeat Problem 4, this time using the exact solution to provide the starting values. Find the error bounds from equation (1872) and compare these

*Scala, S. M. The thermal protection of a re-entry satellite. General Electric MSVD Aerophysics Technical Memorandum 115, March, 1959.

to the actual error. Now apply the corrector formula (1873) and determine the change in the error by comparison with the exact solution.

5 Consider the equation

$$\frac{dy}{dx} = \frac{y}{x} + \sqrt{1 + \frac{y^2}{x^2}}$$

with initial values $y_1 = 0$ when $x_1 = 1$. Using step sizes $x_i - x_{i-1} = 0.001$, 0.01 and 0.1, determine y_2, y_3 and y_4 using third-order Runge–Kutta. Now find y_5 using Adams formula (1884). Compare your result with the exact solution.

6 Consider the second-order equation

$$\frac{d^2y}{dx^2} = \frac{1}{x}\frac{dy}{dx} - \frac{y}{x^2} - 1$$

with initial values $y_1 = 2$, $(dy/dx)_1 = 4$ when $x_1 = 1$.

a. Rewrite this equation as two first-order equations
b. Using the Runge–Kutta method with a step size of 0.01, find y_2, y_3, y_4, y_2', y_3', y_4', where $y' = dy/dx$.
c. Using Adams method for second order equations, formulas (1905–1906), find y_5.
d. Compare your result to the exact solution.

7 Repeat Problem 6, using Milne's method for second-order equations instead of Adams method. Which of the two methods yields the more accurate result?

8 Consider the equation

$$x^2\frac{d^2y}{dx^2} - x\frac{dy}{dx} + y = x^2$$

with boundary conditions $x = 1, y = 2; x = 10, y = 133.026$. Using Runge–Kutta to obtain starting values and Adams method to continue the solution, write a computer program to find $y(x), 1 \leq x \leq 10$. If you have access to a computer, experiment by varying the step size from 0.001 to 0.1, and comparing your result with the exact solution. Also, notice the change in computing time as the step size varies.

9 Given the parabolic equation

$$\frac{\partial u}{\partial t} = \nu\frac{\partial^2 u}{\partial y^2}$$

with boundary conditions $u(0, t) = At, u(\infty, t) = 0, u(y, 0) = 0$. Take $A = 10$, $\nu = 2 \times 10^{-4}$. Write a computer program to solve this equation numerically subject to the given boundary conditions. If you have access to a computer, experiment by taking different step sizes and different values $y = L$ for

$y = \infty$. Compare your solution to the analytical solution of this problem (see for example, Becker, *ZAMP* **11** (1960): 146–152 or Watson, *Quart. J. Mech.* **11** (1958): 302–325).

10 Given the wave equation

$$\frac{\partial^2 \phi}{\partial t^2} = c^2 \frac{\partial^2 \phi}{\partial x^2}$$

with boundary conditions $\phi(0,t) = \phi(2,t) = 0$; $\phi(x,0) = 0.1 \sin[(\pi/2)x]$, $\dfrac{\partial \phi}{\partial t}(x,0) = 0$. Take $c = 1$. Write a computer program to solve this equation numerically subject to the given boundary conditions. If you have access to a computer, experiment by taking different step sizes. Compare your solution to the analytical solution of this problem.

Appendix

Tables for compressible flow calculations*

*Sources of Tables.

Tables Al through A4: Shapiro, A. H., Hawthorne, W. R. and Edelman, G. M. The mechanics of steady one-dimensional gas flow with tables for numerical solutions. Meteor Report No. 14, Massachusetts Institute of Technology Guided Missile Program, Cambridge, Mass. (1947)

Table A5: Shapiro, A. H. and Edelman, G. M. Method of characteristics for two-dimensional supersonic flow—graphical and numerical procedures. *J. of Applied Mechanics*, **14**, (1947): A 154–A167.

Table A6: Forsythe, W. E., Ed. Smithsonian Physical Tables, Smithsonian Institution, Washington, D. C. (1969).

Table A.1 One-Dimensional Isentropic Compressible-Flow Functions
For a perfect gas with constant specific heat and molecular weight
$\gamma = 1.4$

M	M*	$\frac{A}{A^*}$	$\frac{p}{p_0}$	$\frac{\rho}{\rho_0}$	$\frac{T}{T_0}$	$\frac{F}{F^*}$	$\left(\frac{A}{A^*}\right)\left(\frac{p}{p_0}\right)$	M	M*	$\frac{A}{A^*}$	$\frac{p}{p_0}$	$\frac{\rho}{\rho_0}$	$\frac{T}{T_0}$	$\frac{F}{F^*}$	$\left(\frac{A}{A^*}\right)\left(\frac{p}{p_0}\right)$
0	0	∞	1.00000	1.00000	1.00000	∞	∞	**0.40**	.43133	1.5901	.89562	.92428	.96899	1.3749	1.4241
0.01	0.01096	57.874	.99993	.99995	.99998	45.650	57.870	**.41**	.44177	1.5587	.89071	.92066	.96747	1.3527	1.3883
.02	.02191	28.942	.99972	.99980	.99992	22.834	28.934	**.42**	.45218	1.5289	.88572	.91697	.96592	1.3318	1.3542
.03	.03286	19.300	.99937	.99955	.99982	15.232	19.288	**.43**	.46256	1.5007	.88065	.91322	.96434	1.3122	1.3216
.04	.04381	14.482	.99888	.99920	.99968	11.435	14.465	**.44**	.47292	1.4740	.87550	.90940	.96272	1.2937	1.2905
.05	.05476	11.592	.99825	.99875	.99950	9.1584	11.571	**.45**	.48326	1.4487	.87027	.90552	.96108	1.2763	1.2607
.06	.06570	9.6659	.99748	.99820	.99928	7.6428	9.6415	**.46**	.49357	1.4246	.86496	.90157	.95940	1.2598	1.2322
.07	.07664	8.2915	.99658	.99755	.99902	6.5620	8.2631	**.47**	.50385	1.4018	.85958	.89756	.95769	1.2443	1.2050
.08	.08758	7.2616	.99553	.99680	.99872	5.7529	7.2291	**.48**	.51410	1.3801	.85413	.89349	.95595	1.2296	1.1788
.09	.09851	6.4613	.99435	.99596	.99838	5.1249	6.4248	**.49**	.52432	1.3594	.84861	.88936	.95418	1.2158	1.1537
.10	.10943	5.8218	.99303	.99502	.99800	4.6236	5.7812	**.50**	.53452	1.3398	.84302	.88517	.95238	1.2027	1.12951
.11	.12035	5.2992	.99157	.99398	.99758	4.2146	5.2546	**.51**	.54469	1.3212	.83737	.88092	.95055	1.1903	1.10631
.12	.13126	4.8643	.98998	.99284	.99714	3.8747	4.8157	**.52**	.55482	1.3034	.83166	.87662	.94869	1.1786	1.08397
.13	.14216	4.4968	.98826	.99160	.99664	3.5880	4.4440	**.53**	.56493	1.2864	.82589	.87227	.94681	1.1675	1.06245
.14	.15306	4.1824	.98640	.99027	.99610	3.3432	4.1255	**.54**	.57501	1.2703	.82005	.86788	.94489	1.1571	1.04173

.15	.16395	3.9103	.98441	.98884	.99552	3.1317	3.8493	**.55**	.58506	1.2550	.81416	.86342	.94295	1.1472	1.02174
.16	.17483	3.6727	.98228	.98731	.99490	2.9474	3.6076	**.56**	.59508	1.2403	.80822	.85892	.94098	1.1378	1.00244
.17	.18569	3.4635	.98003	.98569	.99425	2.7855	3.3943	**.57**	.60506	1.2263	.80224	.85437	.93898	1.1289	.98381
.18	.19654	3.2779	.97765	.98398	.99356	2.6422	3.2046	**.58**	.61500	1.2130	.79621	.84977	.93696	1.1205	.96581
.19	.20738	3.1122	.97514	.98217	.99283	2.5146	3.0348	**.59**	.62491	1.2003	.79012	.84513	.93491	1.1126	.94839
.20	.21822	2.9635	.97250	.98027	.99206	2.4004	2.8820	**.60**	.63480	1.1882	.78400	.84045	.93284	1.10504	.93155
.21	.22904	2.8293	.96973	.97828	.99125	2.2976	2.7437	**.61**	.64466	1.1766	.77784	.83573	.93074	1.09793	.91525
.22	.23984	2.7076	.96685	.97621	.99041	2.2046	2.6178	**.62**	.65448	1.1656	.77164	.83096	.92861	1.09120	.89946
.23	.25063	2.5968	.96383	.97403	.98953	2.1203	2.5029	**.63**	.66427	1.1551	.76540	.82616	.92646	1.08485	.88416
.24	.26141	2.4956	.96070	.97177	.98861	2.0434	2.3975	**.64**	.67402	1.1451	.75913	.82132	.92428	1.07883	.86932
.25	.27216	2.4027	.95745	.96942	.98765	1.9732	2.3005	**.65**	.68374	1.1356	.75283	.81644	.92208	1.07314	.85493
.26	.28291	2.3173	.95408	.96699	.98666	1.9088	2.2109	**.66**	.69342	1.1265	.74650	.81153	.91986	1.06777	.84096
.27	.29364	2.2385	.95060	.96446	.98563	1.8496	2.1279	**.67**	.70307	1.1178	.74014	.80659	.91762	1.06271	.82740
.28	.30435	2.1656	.94700	.96185	.98455	1.7950	2.0508	**.68**	.71268	1.1096	.73376	.80162	.91535	1.05792	.81421
.29	.31504	2.0979	.94329	.95916	.98346	1.7446	1.9789	**.69**	.72225	1.1018	.72735	.79662	.91306	1.05340	.80141
.30	.32572	2.0351	.93947	.95638	.98232	1.6979	1.9119	**.70**	.73179	1.09437	.72092	.79158	.91075	1.04915	.78896
.31	.33638	1.9765	.93554	.95352	.98114	1.6546	1.8491	**.71**	.74129	1.08729	.71448	.78652	.90842	1.04514	.77685
.32	.34701	1.9218	.93150	.95058	.97993	1.6144	1.7902	**.72**	.75076	1.08057	.70802	.78143	.90606	1.04137	.76507
.33	.35762	1.8707	.92736	.94756	.97868	1.5769	1.7348	**.73**	.76019	1.07419	.70155	.77632	.90368	1.03783	.75360
.34	.36821	1.8229	.92312	.94446	.97740	1.5420	1.6828	**.74**	.76958	1.06814	.69507	.77119	.90129	1.03450	.74243
.35	.37879	1.7780	.91877	.94128	.97608	1.5094	1.6336	**.75**	.77893	1.06242	.68857	.76603	.89888	1.03137	.73155
.36	.38935	1.7358	.91433	.93803	.97473	1.4789	1.5871	**.76**	.78825	1.05700	.68207	.76086	.89644	1.02844	.72095
.37	.39988	1.6961	.90979	.93470	.97335	1.4503	1.5431	**.77**	.79753	1.05188	.67556	.75567	.89399	1.02570	.71062
.38	.41039	1.6587	.90516	.93129	.97193	1.4236	1.5014	**.78**	.80677	1.04705	.66905	.75046	.89152	1.02314	.70054
.39	.42087	1.6234	.90044	.92782	.97048	1.3985	1.4618	**.79**	.81597	1.04250	.66254	.74524	.88903	1.02075	.69070

Table A.1 (*Continued*) One-Dimensional Isentropic Compressible-Flow Functions

For a perfect gas with constant specific heat and molecular weight

$\gamma = 1.4$

M	M*	$\frac{A}{A^*}$	$\frac{p}{p_0}$	$\frac{\rho}{\rho_0}$	$\frac{T}{T_0}$	$\frac{F}{F^*}$	$\left(\frac{A}{A^*}\right)\left(\frac{p}{p_0}\right)$	M	M*	$\frac{A}{A^*}$	$\frac{p}{p_0}$	$\frac{\rho}{\rho_0}$	$\frac{T}{T_0}$	$\frac{F}{F^*}$	$\left(\frac{A}{A^*}\right)\left(\frac{p}{p_0}\right)$
0.80	.82514	1.03823	.65602	.74000	.88652	1.01853	.68110	**1.20**	1.1583	1.03044	.41238	.53114	.77640	1.01082	.42493
.81	.83426	1.03422	.64951	.73474	.88400	1.01646	.67173	**1.21**	1.1658	1.03344	.40702	.52620	.77350	1.01178	.42063
.82	.84334	1.03046	.64300	.72947	.88146	1.01455	.66259	**1.22**	1.1732	1.03657	.40171	.52129	.77061	1.01278	.41640
.83	.85239	1.02696	.63650	.72419	.87890	1.01278	.65366	**1.23**	1.1806	1.03983	.39645	.51640	.76771	1.01381	.41224
.84	.86140	1.02370	.63000	.71890	.87633	1.01115	.64493	**1.24**	1.1879	1.04323	.39123	.51154	.76481	1.01486	.40814
.85	.87037	1.02067	.62351	.71361	.87374	1.00966	.63640	**1.25**	1.1952	1.04676	.38606	.50670	.76190	1.01594	.40411
.86	.87929	1.01787	.61703	.70831	.87114	1.00829	.62806	**1.26**	1.2025	1.05041	.38094	.50189	.75900	1.01705	.40014
.87	.88817	1.01530	.61057	.70300	.86852	1.00704	.61991	**1.27**	1.2097	1.05419	.37586	.49710	.75610	1.01818	.39622
.88	.89702	1.01294	.60412	.69769	.86589	1.00591	.61193	**1.28**	1.2169	1.05810	.37083	.49234	.75319	1.01933	.39237
.89	.90583	1.01080	.59768	.69237	.86324	1.00490	.60413	**1.29**	1.2240	1.06214	.36585	.48761	.75029	1.02050	.38858
.90	.91460	1.00886	.59126	.68704	.86058	1.00399	.59650	**1.30**	1.2311	1.06631	.36092	.48291	.74738	1.02170	.38484
.91	.92333	1.00713	.58486	.68171	.85791	1.00318	.58903	**1.31**	1.2382	1.07060	.35603	.47823	.74448	1.02292	.38116
.92	.93201	1.00560	.57848	.67639	.85523	1.00248	.58171	**1.32**	1.2452	1.07502	.35119	.47358	.74158	1.02415	.37754
.93	.94065	1.00426	.57212	.67107	.85253	1.00188	.57455	**1.33**	1.2522	1.07957	.34640	.46895	.73867	1.02540	.37397
.94	.94925	1.00311	.56578	.66575	.84982	1.00136	.56754	**1.34**	1.2591	1.08424	.34166	.46436	.73577	1.02666	.37044

.95	.95781	1.00214	.55946	.66044	.84710	1.00093	.56066
.96	.96633	1.00136	.55317	.65513	.84437	1.00059	.55392
.97	.97481	1.00076	.54691	.64982	.84162	1.00033	.54732
.98	.98325	1.00033	.54067	.64452	.83887	1.00014	.54085
.99	.99165	1.00008	.53446	.63923	.83611	1.00003	.53450
1.00	1.00000	1.00000	.52828	.63394	.83333	1.00000	.52828
1.01	1.00831	1.00008	.52213	.62866	.83055	1.00003	.52218
1.02	1.01658	1.00033	.51602	.62339	.82776	1.00013	.51619
1.03	1.02481	1.00074	.50994	.61813	.82496	1.00030	.51031
1.04	1.03300	1.00130	.50389	.61288	.82215	1.00053	.50454
1.05	1.04114	1.00202	.49787	.60765	.81933	1.00082	.49888
1.06	1.04924	1.00290	.49189	.60243	.81651	1.00116	.49332
1.07	1.05730	1.00394	.48595	.59722	.81368	1.00155	.48787
1.08	1.06532	1.00512	.48005	.59203	.81084	1.00200	.48251
1.09	1.07330	1.00645	.47418	.58685	.80800	1.00250	.47724
1.10	1.08124	1.00793	.46835	.58169	.80515	1.00305	.47206
1.11	1.08914	1.00955	.46256	.57655	.80230	1.00365	.46698
1.12	1.09699	1.01131	.45682	.57143	.79944	1.00429	.46199
1.13	1.10480	1.01322	.45112	.56632	.79657	1.00497	.45708
1.14	1.11256	1.01527	.44545	.56123	.79370	1.00569	.45225
1.15	1.1203	1.01746	.43983	.55616	.79083	1.00646	.44751
1.16	1.1280	1.01978	.43425	.55112	.78795	1.00726	.44284
1.17	1.1356	1.02224	.42872	.54609	.78507	1.00810	.43825
1.18	1.1432	1.02484	.42323	.54108	.78218	1.00897	.43374
1.19	1.1508	1.02757	.41778	.53610	.77929	1.00988	.42930

1.35	1.2660	1.08904	.33697	.45980	.73287	1.02794	36697
1.36	1.2729	1.09397	.33233	.45527	.72997	1.02924	.36355
1.37	1.2797	1.09902	.32774	.45076	.72707	1.03056	.36018
1.38	1.2865	1.10420	.32319	.44628	.72418	1.03189	.35686
1.39	1.2932	1.10950	.31869	.44183	.72128	1.03323	.35359
1.40	1.2999	1.1149	.31424	.43742	.71839	1.03458	.35036
1.41	1.3065	1.1205	.30984	.43304	.71550	1.03595	.34717
1.42	1.3131	1.1262	.30549	.42869	.71261	1.03733	.34403
1.43	1.3197	1.1320	.30119	.42436	.70973	1.03872	.34093
1.44	1.3262	1.1379	.29693	.42007	.70685	1.04012	.33787
1.45	1.3327	1.1440	.29272	.41581	.70397	1.04153	.33486
1.46	1.3392	1.1502	.28856	.41158	.70110	1.04295	.33189
1.47	1.3456	1.1565	.28445	.40738	.69823	1.04438	.32896
1.48	1.3520	1.1629	.28039	.40322	.69537	1.04581	.32607
1.49	1.3583	1.1695	.27637	.39909	.69251	1.04725	.32321
1.50	1.3646	1.1762	.27240	.39498	.68965	1.04870	.32039
1.51	1.3708	1.1830	.26848	.39091	.68680	1.05016	.31761
1.52	1.3770	1.1899	.26461	.38687	.68396	1.05162	.31487
1.53	1.3832	1.1970	.26078	.38287	.68112	1.05309	.31216
1.54	1.3894	1.2042	.25700	.37890	.67828	1.05456	.30948
1.55	1.3955	1.2115	.25326	.37496	.67545	1.05604	.30685
1.56	1.4016	1.2190	.24957	.37105	.67262	1.05752	.30424
1.57	1.4076	1.2266	.24593	.36717	.66980	1.05900	.30167
1.58	1.4135	1.2343	.24233	.36332	.66699	1.06049	.29913
1.59	1.4195	1.2422	.23878	.35951	.66418	1.06198	.29662

Table A.1 (*Continued*) One-Dimensional Isentropic Compressible-Flow Functions
For a perfect gas with constant specific heat and molecular weight
$\gamma = 1.4$

M	M^*	$\frac{A}{A^*}$	$\frac{p}{p_0}$	$\frac{\rho}{\rho_0}$	$\frac{T}{T_0}$	$\frac{F}{F^*}$	$\left(\frac{A}{A^*}\right)\left(\frac{p}{p_0}\right)$
1.60	1.4254	1.2502	.23527	.35573	.66138	1.06348	.29414
1.61	1.4313	1.2583	.23181	.35198	.65858	1.06498	.29169
1.62	1.4371	1.2666	.22839	.34826	.65579	1.06648	.28928
1.63	1.4429	1.2750	.22501	.34458	.65301	1.06798	.28690
1.64	1.4487	1.2835	.22168	.34093	.65023	1.06948	.28454
1.65	1.4544	1.2922	.21839	.33731	.64746	1.07098	.28221
1.66	1.4601	1.3010	.21515	.33372	.64470	1.07249	.27991
1.67	1.4657	1.3099	.21195	.33016	.64194	1.07399	.27764
1.68	1.4713	1.3190	.20879	.32664	.63919	1.07550	.27540
1.69	1.4769	1.3282	.20567	.32315	.63645	1.07701	.27318
1.70	1.4825	1.3376	.20259	.31969	.63372	1.07851	.27099
1.71	1.4880	1.3471	.19955	.31626	.63099	1.08002	.26882
1.72	1.4935	1.3567	.19656	.31286	.62827	1.08152	.26668
1.73	1.4989	1.3665	.19361	.30950	.62556	1.08302	.26457
1.74	1.5043	1.3764	.19070	.30617	.62286	1.08453	.26248

M	M^*	$\frac{A}{A^*}$	$\frac{p}{p_0}$	$\frac{\rho}{\rho_0}$	$\frac{T}{T_0}$	$\frac{F}{F^*}$	$\left(\frac{A}{A^*}\right)\left(\frac{p}{p_0}\right)$
2.00	1.6330	1.6875	.12780	23005	.55556	1.1227	.21567
2.01	1.6375	1.7017	.12583	.22751	.55310	1.1241	.21412
2.02	1.6420	1.7160	.12389	.22499	.55064	1.1255	.21259
2.03	1.6465	1.7305	.12198	.22250	.54819	1.1269	.21107
2.04	1.6509	1.7452	.12009	.22004	.54576	1.1283	.20957
2.05	1.6553	1.7600	.11823	.21760	.54333	1.1297	.20808
2.06	1.6597	1.7750	.11640	.21519	.54091	1.1311	.20661
2.07	1.6640	1.7902	.11460	.21281	.53850	1.1325	.20515
2.08	1.6683	1.8056	.11282	.21045	.53611	1.1339	.20371
2.09	1.6726	1.8212	.11107	.20811	.53373	1.1352	.20228
2.10	1.6769	1.8369	.10935	.20580	.53135	1.1366	.20087
2.11	1.6811	1.8529	.10766	.20352	.52898	1.1380	.19947
2.12	1.6853	1.8690	.10599	.20126	.52663	1.1393	.19809
2.13	1.6895	1.8853	.10434	.19902	.52428	1.1407	.19672
2.14	1.6936	1.9018	.10272	.19681	.52194	1.1420	.19537

1.75	1.5097	1.3865	.18782	.30287	.62016	1.08603	.26042	**2.15**	1.6977	1.9185	.10113	.19463	.51962	1.1434	.19403
1.76	1.5150	1.3967	.18499	.29959	.61747	1.08753	.25838	**2.16**	1.7018	1.9354	.09956	.19247	.51730	1.1447	.19270
1.77	1.5203	1.4071	.18220	.29635	.61479	1.08903	.25636	**2.17**	1.7059	1.9525	.09802	.19033	.51499	1.1460	.19138
1.78	1.5256	1.4176	.17944	.29314	.61211	1.09053	.25436	**2.18**	1.7099	1.9698	.09650	.18821	.51269	1.1474	.19008
1.79	1.5308	1.4282	.17672	.28997	.60945	1.09202	.25239	**2.19**	1.7139	1.9873	.09500	.18612	.51041	1.1487	.18879
1.80	1.5360	1.4390	.17404	.28682	.60680	1.09352	.25044	**2.20**	1.7179	2.0050	.09352	.18405	.50813	1.1500	.18751
1.81	1.5412	1.4499	.17140	.28370	.60415	1.09500	.24851	**2.21**	1.7219	2.0229	.09207	.18200	.50586	1.1513	.18624
1.82	1.5463	1.4610	.16879	.28061	.60151	1.09649	.24660	**2.22**	1.7258	2.0409	.09064	.17998	.50361	1.1526	.18499
1.83	1.5514	1.4723	.16622	.27756	.59883	1.09798	.24472	**2.23**	1.7297	2.0592	.08923	.17798	.50136	1.1539	.18375
1.84	1.5564	1.4837	.16369	.27453	.59626	1.09946	.24286	**2.24**	1.7336	2.0777	.08784	.17600	.49912	1.1552	.18252
1.85	1.5614	1.4952	.16120	.27153	.59365	1.1009	.24102	**2.25**	1.7374	2.0964	.08648	.17404	.49689	1.1565	.18130
1.86	1.5664	1.5069	.15874	.26857	.59105	1.1024	.23919	**2.26**	1.7412	2.1154	.08514	.17211	.49468	1.1578	.18009
1.87	1.5714	1.5188	.15631	.26563	.58845	1.1039	.23739	**2.27**	1.7450	2.1345	.08382	.17020	.49247	1.1590	.17890
1.88	1.5763	1.5308	.15392	.26272	.58536	1.1054	.23561	**2.28**	1.7488	2.1538	.08252	.16830	.49027	1.1603	.17772
1.89	1.5812	1.5429	.15156	.25984	.58329	1.1068	.23385	**2.29**	1.7526	2.1734	.08123	.16643	.48809	1.1616	.17655
1.90	1.5361	1.5552	.14924	.25699	.58072	1.1083	.23211	**2.30**	1.7563	2.1931	.07997	.16458	.48591	1.1629	.17539
1.91	1.5909	1.5677	.14695	.25417	.57816	1.1097	.23039	**2.31**	1.7600	2.2131	.07873	.16275	.48374	1.1641	.17424
1.92	1.5957	1.5804	.14469	.25138	.57561	1.1112	.22868	**2.32**	1.7637	2.2333	.07751	.16095	.48158	1.1653	.17310
1.93	1.6005	1.5932	.14247	.24862	.57307	1.1126	.22699	**2.33**	1.7673	2.2537	.07631	.15916	.47944	1.1666	.17197
1.94	1.6052	1.6062	.14028	.24588	.57054	1.1141	.22532	**2.34**	1.7709	2.2744	.07513	.15739	.47730	1.1678	.17085
1.95	1.5099	1.6193	.13813	.24317	.56802	1.1155	.22367	**2.35**	1.7745	2.2953	.07396	.15564	.47517	1.1690	.16975
1.96	1.6146	1.6326	.13600	.24049	.56551	1.1170	.22204	**2.36**	1.7781	2.3164	.07281	.15391	.47305	1.1703	.16866
1.97	1.6193	1.6461	.13390	.23784	.56301	1.1184	.22042	**2.37**	1.7817	2.3377	.07168	.15220	.47095	1.1715	.16757
1.98	1.6239	1.6597	.13184	.23522	.56051	1.1198	.21882	**2.38**	1.7852	2.3593	.07057	.15052	.46885	1.1727	.16649
1.99	1.6285	1.6735	.12981	.23262	.55803	1.1213	.21724	**2.39**	1.7887	2.3811	.06948	.14885	.46676	1.1739	.16543

Table A.1 (*Concluded*) One-Dimensional Isentropic Compressible-Flow Functions
For a perfect gas with constant specific heat and molecular weight
$\gamma = 1.4$

M	M*	$\frac{A}{A^*}$	$\frac{p}{p_0}$	$\frac{\rho}{\rho_0}$	$\frac{T}{T_0}$	$\frac{F}{F^*}$	$\left(\frac{A}{A^*}\right)\left(\frac{p}{p_0}\right)$	M	M*	$\frac{A}{A^*}$	$\frac{p}{p_0}$	$\frac{\rho}{\rho_0}$	$\frac{T}{T_0}$	$\frac{F}{F^*}$	$\left(\frac{A}{A^*}\right)\left(\frac{p}{p_0}\right)$
2.40	1.7922	2.4031	.06840	.14720	.46468	1.1751	.16437	**2.80**	1.9140	3.5001	.03685	.09462	.38941	1.2182	.12897
2.41	1.7957	2.4254	.06734	.14557	.46262	1.1763	.16332	**2.81**	1.9167	3.5336	.03629	.09360	.38771	1.2192	.12823
2.42	1.7991	2.4479	.06630	.14395	.46056	1.1775	.16229	**2.82**	1.9193	3.5674	.03574	.09259	.38603	1.2202	.12750
2.43	1.8025	2.4706	.06527	.14235	.45851	1.1786	.16126	**2.83**	1.9220	3.6015	.03520	.09158	.38435	1.2211	.12678
2.44	1.8059	2.4936	.06426	.14078	.45647	1.1798	.16024	**2.84**	1.9246	3.6359	.03467	.09059	.38268	1.2221	.12605
2.45	1.8093	2.5168	.06327	.13922	.45444	1.1810	.15923	**2.85**	1.9271	3.6707	.03415	.08962	.38102	1.2230	.12534
2.46	1.8126	2.5403	.06229	.13768	.45242	1.1821	.15823	**2.86**	1.9297	3.7058	.03363	.08865	.37937	1.2240	.12463
2.47	1.8159	2.5640	.06133	.13616	.45041	1.1833	.15724	**2.87**	1.9322	3.7413	.03312	.08769	.37773	1.2249	.12393
2.48	1.8192	2.5880	.06038	.13465	.44841	1.1844	.15626	**2.88**	1.9348	3.7771	.03262	.08674	.37610	1.2258	.12323
2.49	1.8225	2.6122	.05945	.13316	.44642	1.1856	.15528	**2.89**	1.9373	3.8133	.03213	.08581	.37448	1.2268	.12254
2.50	1.8258	2.6367	.05853	.13169	.44444	1.1867	.15432	**2.90**	1.9398	3.8498	.03165	.08489	.37286	1.2277	.12185
2.51	1.8290	2.6615	.05763	.13023	.44247	1.1879	.15337	**2.91**	1.9423	3.8866	.03118	.08398	.37125	1.2286	.12117
2.52	1.8322	2.6865	.05674	.12879	.44051	1.1890	.15242	**2.92**	1.9448	3.9238	.03071	.08308	.36965	1.2295	.12049
2.53	1.8354	2.7117	.05586	.12737	.43856	1.1901	.15148	**2.93**	1.9472	3.9614	.03025	.08218	.36806	1.2304	.11982
2.54	1.8386	2.7372	.05500	.12597	.43662	1.1912	.15055	**2.94**	1.9497	3.9993	.02980	.08130	.36648	1.2313	.11916

2.55	1.8417	2.7630	.05415	.12458	.43469	1.1923	.14963
2.56	1.8448	2.7891	.05332	.12321	.43277	1.1934	.14871
2.57	1.8479	2.8154	.05250	.12185	.43085	1.1945	.14780
2.58	1.8510	2.8420	.05169	.12051	.42894	1.1956	.14691
2.59	1.8541	2.8689	.05090	.11418	.42705	1.1967	.14601
2.60	1.8572	2.8960	.05012	.11787	.42517	1.1978	.14513
2.61	1.8602	2.9234	.04935	.11658	.42330	1.1989	.14426
2.62	1.8632	2.9511	.04859	.11530	.42143	1.2000	.14339
2.63	1.8662	2.9791	.04784	.11403	.41957	1.2011	.14253
2.64	1.8692	3.0074	.04711	.11278	.41772	1.2021	.14168
2.65	1.8721	3.0359	.04639	.11154	.41589	1.2031	.14083
2.66	1.8750	3.0647	.04568	.11032	.41406	1.2042	.13999
2.67	1.8779	3.0938	.04498	.10911	.41224	1.2052	.13916
2.68	1.8808	3.1233	.04429	.10792	.41043	1.2062	.13834
2.69	1.8837	3.1530	.04361	.10674	.40863	1.2073	.13752
2.70	1.8865	3.1830	.04295	.10557	.40684	1.2083	.13671
2.71	1.8894	3.2133	.04230	.10442	.40505	1.2093	.13591
2.72	1.8922	3.2440	.04166	.10328	.40327	1.2103	.13511
2.73	1.8950	3.2749	.04102	.10215	.40151	1.2113	.13432
2.74	1.8978	3.3061	.04039	.10104	.39976	1.2123	.13354
2.75	1.9005	3.3376	.03977	.09994	.39801	1.2133	.13276
2.76	1.9032	3.3695	.03917	.09885	.39627	1.2143	.13199
2.77	1.9060	3.4017	.03858	.09777	.39454	1.2153	.13123
2.78	1.9087	3.4342	.03800	.09671	.39282	1.2163	.13047
2.79	1.9114	3.4670	.03742	.09566	.39111	1.2173	.12972

2.95	1.9521	4.0376	.02935	.08043	.36490	1.2322	.11850
2.96	1.9545	4.0763	.02891	.07957	.36333	1.2331	.11785
2.97	1.9569	4.1153	.02848	.07872	.36177	1.2340	.11720
2.98	1.9593	4.1547	.02805	.07788	.36022	1.2348	.11656
2.99	1.9616	4.1944	.02764	.07705	.35868	1.2357	.11591
3.00	1.9640	4.2346	.02722	.07623	.35714	1.2366	.11528
3.10	1.9866	4.6573	.02345	.06852	.34223	1.2450	.10921
3.20	2.0079	5.1210	.02023	.06165	.32808	1.2530	.10359
3.30	2.0279	5.6287	.01743	.05554	.31466	1.2605	.09837
3.40	2.0466	6.1837	.01512	.05009	.30193	1.2676	.09353
3.50	2.0642	6.7896	.01311	.04523	.28986	1.2743	.08902
3.60	2.0808	7.4501	.01138	.04089	.27840	1.2807	.08482
3.70	2.0964	8.1691	.00990	.03702	.26752	1.2867	.08090
3.80	2.1111	8.9506	.00863	.03355	.25720	1.2924	.07723
3.90	2.1250	9.7990	.00753	.03044	.24740	1.2978	.07380
4.00	2.1381	10.719	.00658	.02766	.23810	1.3029	.07059
4.10	2.1505	11.715	.00577	.02516	.22925	1.3077	.06758
4.20	2.1622	12.792	.00506	.02292	.22085	1.3123	.06475
4.30	2.1732	13.955	.00445	.02090	.21286	1.3167	.06209
4.40	2.1837	15.210	.00392	.01909	.20525	1.3208	.05959
4.50	2.1936	16.562	.00346	.01745	.19802	1.3247	.05723
4.60	2.2030	18.018	.00305	.01597	.19113	1.3284	.05500
4.70	2.2119	19.583	.00270	.01463	.18457	1.3320	.05289
4.80	2.2204	21.264	.00240	.01343	.17832	1.3354	.05091
4.90	2.2284	23.067	.00213	.01233	.17235	1.3386	.04904
5.00	2.2361	25.000	$189(10)^{-5}$	.01134	.16667	1.3416	.04725
6.00	2.2953	53.180	$633(10)^{-6}$	.00519	.12195	1.3655	.03368
7.00	2.3333	104.143	$242(10)^{-6}$	.00261	.09259	1.3810	.02516
8.00	2.3591	190.109	$102(10)^{-6}$	.00141	.07246	1.3915	.01947
9.00	2.3772	327.189	$474(10)^{-7}$	.000815	.05814	1.3989	.01550
10.00	2.3904	535.938	$236(10)^{-7}$	.000495	.04762	1.4044	.01263
∞	2.4495	∞	0	0	0	1.4289	0

Table A.2 RAYLEIGH LINE—One-Dimensional Compressible-Flow Functions for Stagnation-Temperature Change in the Absence of Friction and Area Change
For a perfect gas with constant specific heat and molecular weight
$\gamma = 1.4$

M	T_0/T_0^*	T/T^*	p/p^*	p_0/p_0^*	V/V^*	M	T_0/T_0^*	T/T^*	p/p^*	p_0/p_0^*	V/V^*
0	0	0	2.4000	1.2679	0	**0.40**	.52903	.61515	1.9608	1.1566	.31372
0.01	.000480	.000576	2.3997	1.2678	.000240	**.41**	.54651	.63448	1.9428	1.1523	.32658
.02	.00192	.00230	2.3987	1.2675	.000959	**.42**	.56376	.65345	1.9247	1.1480	.33951
.03	.00431	.00516	2.3970	1.2671	.00216	**.43**	.58075	.67205	1.9065	1.1437	.35251
.04	.00765	.00917	2.3946	1.2665	.00383	**.44**	.59748	.69025	1.8882	1.1394	.36556
.05	.01192	.01430	2.3916	1.2657	.00598	**.45**	.61393	.70803	1.8699	1.1351	.37865
.06	.01712	.02053	2.3880	1.2647	.00860	**.46**	.63007	.72538	1.8515	1.1308	.39178
.07	.02322	.02784	2.3837	1.2636	.01168	**.47**	.64589	.74228	1.8331	1.1266	.40493
.08	.03021	.03621	2.3787	1.2623	.01522	**.48**	.66139	.75871	1.8147	1.1224	.41810
.09	.03807	.04562	2.3731	1.2608	.01922	**.49**	.67655	.77466	1.7962	1.1182	.43127
.10	.04678	.05602	2.3669	1.2591	.02367	**.50**	.69136	.79012	1.7778	1.1140	.44445
.11	.05630	.06739	2.3600	1.2573	.02856	**.51**	.70581	.80509	1.7594	1.1099	.45761
.12	.06661	.07970	2.3526	1.2554	.03388	**.52**	.71990	.81955	1.7410	1.1059	.47075
.13	.07768	.09290	2.3445	1.2533	.03962	**.53**	.73361	.83351	1.7226	1.1019	.48387
.14	.08947	.10695	2.3359	1.2510	.04578	**.54**	.74695	.84695	1.7043	1.0979	.49696

.15	.10196	.12181	2.3267	1.2486	.05235	**.55**	.75991	.85987	1.6860	1.09397	.51001
.16	.11511	.13743	2.3170	1.2461	.05931	**.56**	.77248	.87227	1.6678	1.09010	.52302
.17	.12888	.15377	2.3067	1.2434	.06666	**.57**	.78467	.88415	1.6496	1.08630	.53597
.18	.14324	.17078	2.2959	1.2406	.07438	**.58**	.79647	.89552	1.6316	1.08255	.54887
.19	.15814	.18841	2.2845	1.2377	.08247	**.59**	.80789	.90637	1.6136	1.07887	.56170
.20	.17355	.20661	2.2727	1.2346	.09091	**.60**	.81892	.91670	1.5957	1.07525	.57447
.21	.18943	.22533	2.2604	1.2314	.09969	**.61**	.82956	.92653	1.5780	1.07170	.58716
.22	.20574	.24452	2.2477	1.2281	.10879	**.62**	.83982	.93585	1.5603	1.06821	.59978
.23	.22244	.26413	2.2345	1.2248	.11820	**.63**	.84970	.94466	1.5427	1.06480	.61232
.24	.23948	.28411	2.2209	1.2213	.12792	**.64**	.85920	.95298	1.5253	1.06146	.62477
.25	.25684	.30440	2.2069	1.2177	.13793	**.65**	.86833	.96081	1.5080	1.05820	.63713
.26	.27446	.32496	2.1925	1.2140	.14821	**.66**	.87709	.96816	1.4908	1.05502	.64941
.27	.29231	.34573	2.1777	1.2102	.15876	**.67**	.88548	.97503	1.4738	1.05192	.66159
.28	.31035	.36667	2.1626	1.2064	.16955	**.68**	.89350	.98144	1.4569	1.04890	.67367
.29	.32855	.38773	2.1472	1.2025	.18058	**.69**	.90117	.98739	1.4401	1.04596	.68564
.30	.34686	.40887	2.1314	1.1985	.19183	**.70**	.90850	.99289	1.4235	1.04310	.69751
.31	.36525	.43004	2.1154	1.1945	.20329	**.71**	.91548	.99796	1.4070	1.04033	.70927
.32	.38369	.45119	2.0991	1.1904	.21494	**.72**	.92212	1.00260	1.3907	1.03764	.72093
.33	.40214	.47228	2.0825	1.1863	.22678	**.73**	.92843	1.00682	1.3745	1.03504	.73248
.34	.42057	.49327	2.0657	1.1821	.23879	**.74**	.93442	1.01062	1.3585	1.03253	.74392
.35	.43894	.51413	2.0487	1.1779	.25096	**.75**	.94009	1.01403	1.3427	1.03010	.75525
.36	.45723	.53482	2.0314	1.1737	.26327	**.76**	.94546	1.01706	1.3270	1.02776	.76646
.37	.47541	.55530	2.0140	1.1695	.27572	**.77**	.95052	1.01971	1.3115	1.02552	.77755
.38	.49346	.57553	1.9964	1.1652	.28828	**.78**	.95528	1.02198	1.2961	1.02337	.78852
.39	.51134	.59549	1.9787	1.1609	.30095	**.79**	.95975	1.02390	1.2809	1.02131	.79938

Table A.2 (*Continued*) RAYLEIGH LINE—One-Dimensional Compressible-Flow Functions for Stagnation-Temperature Change in the Absence of Friction and Area Change
For a perfect gas with constant specific heat and molecular weight
$\gamma = 1.4$

M	$\frac{T_0}{T_0^*}$	$\frac{T}{T^*}$	$\frac{p}{p^*}$	$\frac{p_0}{p_0^*}$	$\frac{V}{V^*}$	M	$\frac{T_0}{T_0^*}$	$\frac{T}{T^*}$	$\frac{p}{p^*}$	$\frac{p_0}{p_0^*}$	$\frac{V}{V^*}$
0.80	.96394	1.02548	1.2658	1.01934	.81012	**1.20**	.97872	.91185	.79576	1.01941	1.1459
.81	.96786	1.02672	1.2509	1.01746	.82075	**1.21**	.97685	.90671	.78695	1.02140	1.1522
.82	.97152	1.02763	1.2362	1.01569	.83126	**1.22**	.97492	.90153	.77827	1.02348	1.1584
.83	.97492	1.02823	1.2217	1.01399	.84164	**1.23**	.97294	.89632	.76971	1.02566	1.1645
.84	.97807	1.02853	1.2073	1.01240	.85190	**1.24**	.97092	.89108	.76127	1.02794	1.1705
.85	.98097	1.02854	1.1931	1.01091	.86204	**1.25**	.96886	.88581	.75294	1.03032	1.1764
.86	.98363	1.02826	1.1791	1.00951	.87206	**1.26**	.96675	.88052	.74473	1.03280	1.1823
.87	.98607	1.02771	1.1652	1.00819	.88196	**1.27**	.96461	.87521	.73663	1.03536	1.1881
.88	.98828	1.02690	1.1515	1.00698	.89175	**1.28**	.96243	.86988	.72865	1.03803	1.1938
.89	.99028	1.02583	1.1380	1.00587	.90142	**1.29**	.96022	.86453	.72078	1.04080	1.1994
.90	.99207	1.02451	1.1246	1.04485	.91097	**1.30**	.95798	.85917	.71301	1.04365	1.2050
.91	.99366	1.02297	1.1114	1.00393	.92039	**1.31**	.95571	.85380	.70535	1.04661	1.2105
.92	.99506	1.02120	1.09842	1.00310	.92970	**1.32**	.95341	.84843	.69780	1.04967	1.2159
.93	.99627	1.01921	1.08555	1.00237	.93889	**1.33**	.95108	.84305	.69035	1.05283	1.2212
.94	.99729	1.01702	1.07285	1.00174	.94796	**1.34**	.94873	.83766	.68301	1.05608	1.2264

.95	.99814	1.01463	1.06030	1.00121	.95692	**1.35**	.94636	.83227	.67577	1.05943	1.2316
.96	.99883	1.01205	1.04792	1.00077	.96576	**1.36**	.94397	.82698	.66863	1.06288	1.2367
.97	.99935	1.00929	1.03570	1.00043	.97449	**1.37**	.94157	.82151	.66159	1.06642	1.2417
.98	.99972	1.00636	1.02364	1.00019	.98311	**1.38**	.93915	.81613	.65464	1.07006	1.2467
.99	.99993	1.00326	1.01174	1.00004	.99161	**1.39**	.93671	.81076	.64778	1.07380	1.2516
1.00	1.00000	1.00000	1.00000	1.00000	1.00000	**1.40**	.93425	.80540	.64102	1.07765	1.2564
1.01	.99993	.99659	.98841	1.00004	1.00828	**1.41**	.93178	.80004	.63436	1.08159	1.2612
1.02	.99973	.99304	.97697	1.00019	1.01644	**1.42**	.92931	.79469	.62779	1.08563	1.2659
1.03	.99940	.98936	.96569	1.00043	1.02450	**1.43**	.92683	.78936	.62131	1.08977	1.2705
1.04	.99895	.98553	.95456	1.00077	1.03246	**1.44**	.92434	.78405	.61491	1.09400	1.2751
1.05	.99838	.98161	.94358	1.00121	1.04030	**1.45**	.92184	.77875	.60860	1.0983	1.2796
1.06	.99769	.97755	.93275	1.00175	1.04804	**1.46**	.91933	.77346	.60237	1.1028	1.2840
1.07	.99590	.97339	.92206	1.00238	1.05567	**1.47**	.91682	.76819	.59623	1.1073	1.2884
1.08	.99600	.96913	.91152	1.00311	1.06320	**1.48**	.91431	.76294	.59018	1.1120	1.2927
1.09	.99501	.96477	.90112	1.00394	1.07062	**1.49**	.91179	.75771	.58421	1.1167	1.2970
1.10	.99392	.96031	.89086	1.00486	1.07795	**1.50**	.90928	.75250	.57831	1.1215	1.3012
1.11	.99274	.95577	.88075	1.00588	1.08518	**1.51**	.90676	.74731	.57250	1.1264	1.3054
1.12	.99148	.95115	.87078	1.00699	1.09230	**1.52**	.90424	.74215	.56677	1.1315	1.3095
1.13	.99013	.94646	.86094	1.00820	1.09933	**1.53**	.90172	.73701	.56111	1.1367	1.3135
1.14	.98871	.94169	.85123	1.00951	1.10626	**1.54**	.89920	.73189	.55553	1.1420	1.3175
1.15	.98721	.93685	.84166	1.01092	1.1131	**1.55**	.89669	.72680	.55002	1.1473	1.3214
1.16	.98564	.93195	.83222	1.01243	1.1198	**1.56**	.89418	.72173	.54458	1.1527	1.3253
1.17	.98400	.92700	.82292	1.01403	1.1264	**1.57**	.89167	.71669	.53922	1.1582	1.3291
1.18	.98230	.92200	.81374	1.01572	1.1330	**1.58**	.88917	.71168	.53393	1.1639	1.3329
1.19	.98054	.91695	.80468	1.01752	1.1395	**1.59**	.88668	.70669	.52871	1.1697	1.3366

Table A.2 (*Continued*) RAYLEIGH LINE—One-Dimensional Compressible-Flow Functions for Stagnation-Temperature Change in the Absence of Friction and Area Change

For a perfect gas with constant specific heat and molecular weight

$\gamma = 1.4$

M	$\frac{T_0}{T_0^*}$	$\frac{T}{T^*}$	$\frac{p}{p^*}$	$\frac{p_0}{p_0^*}$	$\frac{V}{V^*}$	M	$\frac{T_0}{T_0^*}$	$\frac{T}{T^*}$	$\frac{p}{p^*}$	$\frac{p_0}{p_0^*}$	$\frac{V}{V^*}$
1.60	.88419	.70173	.52356	1.1756	1.3403	**2.00**	.79339	.52893	.36364	1.5031	1.4545
1.61	.88170	.69680	.51848	1.1816	1.3439	**2.01**	.79139	.52526	.36057	1.5138	1.4567
1.62	.87922	.69190	.51346	1.1877	1.3475	**2.02**	.78941	.52161	.35754	1.5246	1.4589
1.63	.87675	.68703	.50851	1.1939	1.3511	**2.03**	.78744	.51800	.35454	1.5356	1.4610
1.64	.87429	.68219	.50363	1.2002	1.3546	**2.04**	.78549	.51442	.35158	1.5467	1.4631
1.65	.87184	.67738	.49881	1.2066	1.3580	**2.05**	.78355	.51087	.34866	1.5579	1.4652
1.66	.86940	.67259	.49405	1.2131	1.3614	**2.06**	.78162	.50735	.34577	1.5693	1.4673
1.67	.86696	.66784	.48935	1.2197	1.3648	**2.07**	.77971	.50386	.34291	1.5808	1.4694
1.68	.86453	.66312	.48471	1.2264	1.3681	**2.08**	.77781	.50040	.34009	1.5924	1.4714
1.69	.86211	.65843	.48014	1.2332	1.3713	**2.09**	.77593	.49697	.33730	1.6042	1.4734
1.70	.85970	.65377	.47563	1.2402	1.3745	**2.10**	.77406	.49356	.33454	1.6161	1.4753
1.71	.85731	.64914	.47117	1.2473	1.3777	**2.11**	.77221	.49018	.33181	1.6282	1.4773
1.72	.85493	.64455	.46677	1.2545	1.3809	**2.12**	.77037	.48683	.32912	1.6404	1.4792
1.73	.85256	.63999	.46242	1.2618	1.3840	**2.13**	.76854	.48351	.32646	1.6528	1.4811
1.74	.85020	.63546	.45813	1.2692	1.3871	**2.14**	.76673	.48022	.32383	1.6653	1.4830

1.75	.84785	.63096	.45390	1.2767	1.3901	**2.15**	.76493	.47696	.32122	1.6780	1.4849
1.76	.84551	.62649	.44972	1.2843	1.3931	**2.16**	.76314	.47373	.31864	1.6908	1.4867
1.77	.84318	.62205	.44559	1.2920	1.3960	**2.17**	.76137	.47052	.31610	1.7037	1.4885
1.78	.84087	.61765	.44152	1.2998	1.3989	**2.18**	.75961	.46734	.31359	1.7168	1.4903
1.79	.83857	.61328	.43750	1.3078	1.4018	**2.19**	.75787	.46419	.31110	1.7300	1.4921
1.80	.83628	.60894	.43353	1.3159	1.4046	**2.20**	.75614	.46106	.30864	1.7434	1.4939
1.81	.83400	.60463	.42960	1.3241	1.4074	**2.21**	.75442	.45796	.30621	1.7570	1.4956
1.82	.83174	.60036	.42573	1.3324	1.4102	**2.22**	.75271	.45489	.30381	1.7707	1.4973
1.83	.82949	.59612	.42191	1.3408	1.4129	**2.23**	.75102	.45184	.30143	1.7846	1.4990
1.84	.82726	.59191	.41813	1.3494	1.4156	**2.24**	.74934	.44882	.29908	1.7986	1.5007
1.85	.82504	.58773	.41440	1.3581	1.4183	**2.25**	.74767	.44582	.29675	1.8128	1.5024
1.86	.82283	.58359	.41072	1.3669	1.4209	**2.26**	.74602	.44285	.29445	1.8271	1.5040
1.87	.82064	.57948	.40708	1.3758	1.4235	**2.27**	.74438	.43990	.29218	1.8416	1.5056
1.88	.81846	.57540	.40349	1.3848	1.4261	**2.28**	.74275	.43698	.28993	1.8562	1.5072
1.89	.81629	.57135	.39994	1.3940	1.4286	**2.29**	.74114	.43409	.28771	1.8710	1.5088
1.90	.81414	.56734	.39643	1.4033	1.4311	**2.30**	.73954	.43122	.28551	1.8860	1.5104
1.91	.81200	.56336	.39297	1.4127	1.4336	**2.31**	.73795	.42837	.28333	1.9012	1.5119
1.92	.80987	.55941	.38955	1.4222	1.4360	**2.32**	.73638	.42555	.28118	1.9165	1.5134
1.93	.80776	.55549	.38617	1.4319	1.4384	**2.33**	.73482	.42276	.27905	1.9320	1.5150
1.94	.80567	.55160	.38283	1.4417	1.4408	**2.34**	.73327	.41999	.27695	1.9476	1.5165
1.95	.80359	.54774	.37954	1.4516	1.4432	**2.35**	.73173	.41724	.27487	1.9634	1.5180
1.96	.80152	.54391	.37628	1.4616	1.4455	**2.36**	.73020	.41451	.27281	1.9794	1.5195
1.97	.79946	.54012	.37306	1.4718	1.4478	**2.37**	.72868	.41181	.27077	1.9955	1.5209
1.98	.79742	.53636	.36988	1.4821	1.4501	**2.38**	.72718	.40913	.26875	2.0118	1.5223
1.99	.79540	.53263	.36674	1.4925	1.4523	**2.39**	.72569	.40647	.26675	2.0283	1.5237

Table A.2 (*Concluded*) RAYLEIGH LINE—One-Dimensional Compressible-Flow Functions for Stagnation-Temperature Change in the Absence of Friction and Area Change
For a perfect gas with constant specific heat and molecular weight
$\gamma = 1.4$

M	$\frac{T_0}{T_0^*}$	$\frac{T}{T^*}$	$\frac{p}{p^*}$	$\frac{p_0}{p_0^*}$	$\frac{V}{V^*}$	M	$\frac{T_0}{T_0^*}$	$\frac{T}{T^*}$	$\frac{p}{p^*}$	$\frac{p_0}{p_0^*}$	$\frac{V}{V^*}$
2.40	.72421	.40383	.26478	2.0450	1.5252	**2.80**	.67380	.31486	.20040	2.8731	1.5711
2.41	.72274	.40122	.26283	2.0619	1.5266	**2.81**	.67273	.31299	.19909	2.8982	1.5721
2.42	.72129	.39863	.26090	2.0789	1.5279	**2.82**	.67167	.31114	.19780	2.9236	1.5730
2.43	.71985	.39606	.25899	2.0961	1.5293	**2.83**	.67062	.30931	.19652	2.9493	1.5739
2.44	.71842	.39352	.25710	2.1135	1.5306	**2.84**	.66958	.30749	.19525	2.9752	1.5748
2.45	.71700	.39100	.25523	2.1311	1.5320	**2.85**	.66855	.30568	.19399	3.0013	1.5757
2.46	.71559	.38850	.25337	2.1489	1.5333	**2.86**	.66752	.30389	.19274	3.0277	1.5766
2.47	.71419	.38602	.25153	2.1669	1.5346	**2.87**	.66650	.30211	.19151	3.0544	1.5775
2.48	.71280	.38356	.24972	2.1850	1.5359	**2.88**	.66549	.30035	.19029	3.0813	1.5784
2.49	.71142	.38112	.24793	2.2033	1.5372	**2.89**	.66449	.29860	.18908	3.1084	1.5792
2.50	.71005	.37870	.24616	2.2218	1.5385	**2.90**	.66350	.29687	.18788	3.1358	1.5801
2.51	.70870	.37630	.24440	2.2405	1.5398	**2.91**	.66252	.29515	.18669	3.1635	1.5809
2.52	.70736	.37392	.24266	2.2594	1.5410	**2.92**	.66154	.29344	.18551	3.1914	1.5818
2.53	.70603	.37157	.24094	2.2785	1.5422	**2.93**	.66057	.29175	.18435	3.2196	1.5826
2.54	.70471	.36923	.23923	2.2978	1.5434	**2.94**	.65961	.29007	.18320	3.2481	1.5834

2.55	.70340	.36691	.23754	2.3173	1.5446
2.56	.70210	.36461	.23587	2.3370	1.5458
2.57	.70081	.36233	.23422	2.3569	1.5470
2.58	.69953	.36007	.23258	2.3770	1.5482
2.59	.69825	.35783	.23096	2.3972	1.5494
2.60	.69699	.35561	.22936	2.4177	1.5505
2.61	.69574	.35341	.22777	2.4384	1.5516
2.62	.69450	.35123	.22620	2.4593	1.5527
2.63	.69327	.34906	.22464	2.4804	1.5538
2.64	.69205	.34691	.22310	2.5017	1.5549
2.65	.69084	.34478	.22158	2.5233	1.5560
2.66	.68964	.34267	.22007	2.5451	1.5571
2.67	.68845	.34057	.21857	2.5671	1.5582
2.68	.68727	.33849	.21709	2.5892	1.5593
2.69	.68610	.33643	.21562	2.6116	1.5603
2.70	.68494	.33439	.21417	2.6342	1.5613
2.71	.68378	.33236	.21273	2.6571	1.5623
2.72	.68263	.33035	.21131	2.6802	1.5633
2.73	.68150	.32836	.20990	2.7035	1.5644
2.74	.68038	.32638	.20850	2.7270	1.5654
2.75	.67926	.32442	.20712	2.7508	1.5663
2.76	.67815	.32248	.20575	2.7748	1.5673
2.77	.67704	.32055	.20439	2.7990	1.5683
2.78	.67595	.31864	.20305	2.8235	1.5692
2.79	.67487	.31674	.20172	2.8482	1.5702

2.95	.65865	.28841	.18205	3.2768	1.5843
2.96	.65770	.28676	.18091	3.3058	1.5851
2.97	.65676	.28512	.17978	3.3351	1.5859
2.98	.65583	.28349	.17867	3.3646	1.5867
2.99	.65490	.28188	.17757	3.3944	1.5875
3.00	.65398	.28028	.17647	3.4244	1.5882
3.50	.61580	.21419	.13223	5.3280	1.6198
4.00	.58909	.16831	.10256	8.2268	1.6410
4.50	.56983	.13540	.08177	12.502	1.6559
5.00	.55555	.11111	.06667	18.634	1.6667
6.00	.53633	.07849	.04669	38.946	1.6809
7.00	.52437	.05826	.03448	75.414	1.6897
8.00	.51646	.04491	.02649	136.62	1.6954
9.00	.51098	.03565	.02098	233.88	1.6993
10.00	.50702	.02897	.01702	381.61	1.7021
∞	.48980	0	0	∞	1.7143

Table A.3 FANNO LINE—One-Dimensional Compressible-Flow Functions for Adiabatic Flow at Constant Area with Friction

For a perfect gas with constant specific heat and molecular weight

$\gamma = 1.4$

M	$\frac{T}{T^*}$	$\frac{p}{p^*}$	$\frac{p_0}{p_0^*}$	$\frac{V}{V^*}$	$\frac{F}{F^*}$	$4\frac{fL_{max}}{D}$	M	$\frac{T}{T^*}$	$\frac{p}{p^*}$	$\frac{p_0}{p_0^*}$	$\frac{V}{V^*}$	$\frac{F}{F^*}$	$4\frac{fL_{max}}{D}$
0	1.2000	∞	∞	0	∞	∞	**0.40**	1.1628	2.6958	1.5901	.43133	1.3749	2.3085
0.01	1.2000	109.544	57.874	.01095	45.650	7134.40	**.41**	1.1610	2.6280	1.5587	.44177	1.3527	2.1344
.02	1.1999	54.770	28.942	.02191	22.834	1778.45	**.42**	1.1591	2.5634	1.5289	.45218	1.3318	1.9744
.03	1.1998	36.511	19.300	.03286	15.232	787.08	**.43**	1.1572	2.5017	1.5007	.46257	1.3122	1.8272
.04	1.1996	27.382	14.482	.04381	11.435	440.35	**.44**	1.1553	2.4428	1.4739	.47293	1.2937	1.6915
.05	1.1994	21.903	11.5914	.05476	9.1584	280.02	**.45**	1.1533	2.3865	1.4486	.48326	1.2763	1.5664
.06	1.1991	18.251	9.6659	.06570	7.6428	193.03	**.46**	1.1513	2.3326	1.4246	.49357	1.2598	1.4509
.07	1.1988	15.642	8.2915	.07664	6.5620	140.66	**.47**	1.1492	2.2809	1.4018	.50385	1.2443	1.3442
.08	1.1985	13.684	7.2616	.08758	5.7529	106.72	**.48**	1.1471	2.2314	1.3801	.51410	1.2296	1.2453
.09	1.1981	12.162	6.4614	.09851	5.1249	83.496	**.49**	1.1450	2.1838	1.3595	.52433	1.2158	1.1539
.10	1.1976	10.9435	5.8218	.10943	4.6236	66.922	**.50**	1.1429	2.1381	1.3399	.53453	1.2027	1.06908
.11	1.1971	9.9465	5.2992	.12035	4.2146	54.688	**.51**	1.1407	2.0942	1.3212	.54469	1.1903	.99042
.12	1.1966	9.1156	4.8643	.13126	3.8747	45.408	**.52**	1.1384	2.0519	1.3034	.55482	1.1786	.91741
.13	1.1960	8.4123	4.4968	.14216	3.5880	38.207	**.53**	1.1362	2.0112	1.2864	.56493	1.1675	.84963
.14	1.1953	7.8093	4.1824	.15306	3.3432	32.511	**.54**	1.1339	1.9719	1.2702	.57501	1.1571	.78662

.15	1.1946	7.2866	3.9103	.16395	3.1317	27.932	**.55**	1.1315	1.9341	1.2549	.58506	1.1472	.72805
.16	1.1939	6.8291	3.6727	.17482	2.9474	24.198	**.56**	1.1292	1.8976	1.2403	.59507	1.1378	.67357
.17	1.1931	6.4252	3.4635	.18568	2.7855	21.115	**.57**	1.1268	1.8623	1.2263	.60505	1.1289	.62286
.18	1.1923	6.0662	3.2779	.19654	2.6422	18.543	**.58**	1.1244	1.8282	1.2130	.61500	1.1205	.57568
.19	1.1914	5.7448	3.1123	.20739	2.5146	16.375	**.59**	1.1219	1.7952	1.2003	.62492	1.1126	.53174
.20	1.1905	5.4555	2.9635	.21822	2.4004	14.533	**.60**	1.1194	1.7634	1.1882	.63481	1.10504	.49081
.21	1.1895	5.1936	2.8293	.22904	2.2976	12.956	**.61**	1.1169	1.7325	1.1766	.64467	1.09793	.45270
.22	1.1885	4.9554	2.7076	.23984	2.2046	11.596	**.62**	1.1144	1.7026	1.1656	.65449	1.09120	.41720
.23	1.1874	4.7378	2.5968	.25063	2.1203	10.416	**.63**	1.1118	1.6737	1.1551	.66427	1.08485	.38411
.24	1.1863	4.5383	2.4956	.26141	2.0434	9.3865	**.64**	1.1091	1.6456	1.1451	.67402	1.07883	.35330
.25	1.1852	4.3546	2.4027	.27217	1.9732	8.4834	**.65**	1.10650	1.6183	1.1356	.68374	1.07314	.32460
.26	1.1840	4.1850	2.3173	.28291	1.9088	7.6876	**.66**	1.10383	1.5919	1.1265	.69342	1.06777	.29785
.27	1.1828	4.0280	2.2385	.29364	1.8496	6.9832	**.67**	1.10114	1.5662	1.1179	.70306	1.06271	.27295
.28	1.1815	3.8820	2.1656	.30435	1.7950	6.3572	**.68**	1.09842	1.5413	1.1097	.71267	1.05792	.24978
.29	1.1802	3.7460	2.0979	.31504	1.7446	5.7989	**.69**	1.09567	1.5170	1.1018	.72225	1.05340	.22821
.30	1.1788	3.6190	2.0351	.32572	1.6979	5.2992	**.70**	1.09290	1.4934	1.09436	.73179	1.04915	.20814
.31	1.1774	3.5002	1.9765	.33637	1.6546	4.8507	**.71**	1.09010	1.4705	1.08729	.74129	1.04514	.18949
.32	1.1759	3.3888	1.9219	.34700	1.6144	4.4468	**.72**	1.08727	1.4482	1.08057	.75076	1.04137	.17215
.33	1.1744	3.2840	1.8708	.35762	1.5769	4.0821	**.73**	1.08442	1.4265	1.07419	.76019	1.03783	.15606
.34	1.1729	3.1853	1.8229	.36822	1.5420	3.7520	**.74**	1.08155	1.4054	1.06815	.76958	1.03450	.14113
.35	1.1713	3.0922	1.7780	.37880	1.5094	3.4525	**.75**	1.07865	1.3848	1.06242	.77893	1.03137	.12728
.36	1.1697	3.0042	1.7358	.38935	1.4789	3.1801	**.76**	1.07573	1.3647	1.05700	.78825	1.02844	.11446
.37	1.1680	2.9209	1.6961	.39988	1.4503	2.9320	**.77**	1.07279	1.3451	1.05188	.79753	1.02570	.10262
.38	1.1663	2.8420	1.6587	.41039	1.4236	2.7055	**.78**	1.06982	1.3260	1.04705	.80677	1.02314	.09167
.39	1.1646	2.7671	1.6234	.42087	1.3985	2.4983	**.79**	1.06684	1.3074	1.04250	.81598	1.02075	.08159

Table A.3 (*Continued*) FANNO LINE—One-Dimensional Compressible-Flow Functions for Adiabatic Flow at Constant Area with Friction
For a perfect gas with constant specific heat and molecular weight
$\gamma = 1.4$

M	$\frac{T}{T^*}$	$\frac{p}{p^*}$	$\frac{p_0}{p_0^*}$	$\frac{V}{V^*}$	$\frac{F}{F^*}$	$4\frac{fL_{max}}{D}$	M	$\frac{T}{T^*}$	$\frac{p}{p^*}$	$\frac{p_0}{p_0^*}$	$\frac{V}{V^*}$	$\frac{F}{F^*}$	$4\frac{fL_{max}}{D}$
0.80	1.06383	1.2892	1.03823	.82514	1.01853	.07229	**1.20**	.93168	.80436	1.03044	1.1583	1.01082	.03364
.81	1.06080	1.2715	1.03422	.83426	1.01646	.06375	**1.21**	.92820	.79623	1.03344	1.1658	1.01178	.03650
.82	1.05775	1.2542	1.03047	.84334	1.01455	.05593	**1.22**	.92473	.78822	1.03657	1.1732	1.01278	.03942
.83	1.05468	1.2373	1.02696	.85239	1.01278	.04878	**1.23**	.92125	.78034	1.03983	1.1806	1.01381	.04241
.84	1.05160	1.2208	1.02370	.86140	1.01115	.04226	**1.24**	.91777	.77258	1.04323	1.1879	1.01486	.04547
.85	1.04849	1.2047	1.02067	.87037	1.00966	.03632	**1.25**	.91429	.76495	1.04676	1.1952	1.01594	.04858
.86	1.04537	1.1889	1.01787	.87929	1.00829	.03097	**1.26**	.91080	.75743	1.05041	1.2025	1.01705	.05174
.87	1.04223	1.1735	1.01529	.88818	1.00704	.02613	**1.27**	.90732	.75003	1.05419	1.2097	1.01818	.05494
.88	1.03907	1.1584	1.01294	.89703	1.00591	.02180	**1.28**	.90383	.74274	1.05809	1.2169	1.01933	.05820
.89	1.03589	1.1436	1.01080	.90583	1.00490	.01793	**1.29**	.90035	.73556	1.06213	1.2240	1.02050	.06150
.90	1.03270	1.12913	1.00887	.91459	1.00399	.014513	**1.30**	.89686	.72848	1.06630	1.2311	1.02169	.06483
.91	1.02950	1.11500	1.00714	.92332	1.00318	.011519	**1.31**	.89338	.72152	1.07060	1.2382	1.02291	.06820
.92	1.02627	1.10114	1.00560	.93201	1.00248	.008916	**1.32**	.88989	.71465	1.07502	1.2452	1.02415	.07161
.93	1.02304	1.08758	1.00426	.94065	1.00188	.006694	**1.33**	.88641	.70789	1.07957	1.2522	1.02540	.07504
.94	1.01978	1.07430	1.00311	.94925	1.00136	.004815	**1.34**	.88292	.70123	1.08424	1.2591	1.02666	.07850

.95	1.01652	1.06129	1.00215	.95782	1.00093	.003280	**1.35**	.87944	.69466	1.08904	1.2660	1.02794	.08199
.96	1.01324	1.04854	1.00137	.96634	1.00059	.002056	**1.36**	.87596	.68818	1.09397	1.2729	1.02924	.08550
.97	1.00995	1.03605	1.00076	.97481	1.00033	.001135	**1.37**	.87249	.68180	1.09902	1.2797	1.03056	.08904
.98	1.00664	1.02379	1.00033	.98324	1.00014	.000493	**1.38**	.86901	.67551	1.10419	1.2864	1.03189	.09259
.99	1.00333	1.01178	1.00008	.99164	1.00003	.000120	**1.39**	.86554	.66931	1.10948	1.2932	1.03323	.09616
1.00	1.00000	1.00000	1.00000	1.00000	1.00000	0	**1.40**	.86207	.66320	1.1149	1.2999	1.03458	.09974
1.01	.99666	.98844	1.00008	1.00831	1.00003	.000114	**1.41**	.85860	.65717	1.1205	1.3065	1.03595	.10333
1.02	.99331	.97711	1.00033	1.01658	1.00013	.000458	**1.42**	.85514	.65122	1.1262	1.3131	1.03733	.10694
1.03	.98995	.96598	1.00073	1.02481	1.00030	.001013	**1.43**	.85168	.64536	1.1320	1.3197	1.03872	.11056
1.04	.98658	.95506	1.00130	1.03300	1.00053	.001771	**1.44**	.84822	.63958	1.1379	1.3262	1.04012	.11419
1.05	.98320	.94435	1.00203	1.04115	1.00082	.002712	**1.45**	.84477	63387	1.1440	1.3327	1.04153	.11782
1.06	.97982	.93383	1.00291	1.04925	1.00116	.003837	**1.46**	.84133	.62824	1.1502	1.3392	1.04295	.12146
1.07	.97642	.92350	1.00394	1.05731	1.00155	.005129	**1.47**	.83788	.62269	1.1565	1.3456	1.04438	.12510
1.08	.97302	.91335	1.00512	1.06533	1.00200	.006582	**1.48**	.83445	.61722	1.1629	1.3520	1.04581	.12875
1.09	.96960	.90338	1.00645	1.07331	1.00250	.008185	**1.49**	.83101	.61181	1.1695	1.3583	1.04725	.13240
1.10	.96618	.89359	1.00793	1.08124	1.00305	.009933	**1.50**	.82759	.60648	1.1762	1.3646	1.04870	.13605
1.11	.96276	.88397	1.00955	1.08913	1.00365	.011813	**1.51**	.82416	.60122	1.1830	1.3708	1.05016	.13970
1.12	.95933	.87451	1.01131	1.09698	1.00429	.013824	**1.52**	.82075	.59602	1.1899	1.3770	1.05162	.14335
1.13	.95589	.86522	1.01322	1.10479	1.00497	.015949	**1.53**	.81734	.59089	1.1970	1.3832	1.05309	.14699
1.14	.95244	.85608	1.01527	1.11256	1.00569	.018187	**1.54**	.81394	.58583	1.2043	1.3894	1.05456	.15063
1.15	.94899	.84710	1.01746	1.1203	1.00646	.02053	**1.55**	.81054	.58084	1.2116	1.3955	1.05604	.15427
1.16	.94554	.83827	1.01978	1.1280	1.00726	.02298	**1.56**	.80715	.57591	1.2190	1.4015	1.05752	.15790
1.17	.94208	.82958	1.02224	1.1356	1.00810	.02552	**1.57**	.80376	.57104	1.2266	1.4075	1.05900	.16152
1.18	.93862	.82104	1.02484	1.1432	1.00897	.02814	**1.58**	.80038	.56623	1.2343	1.4135	1.06049	.16514
1.19	.93515	.81263	1.02757	1.1508	1.00988	.03085	**1.59**	.79701	.56148	1.2422	1.4195	1.06198	.16876

Table A.3 (*Continued*) FANNO LINE—One-Dimensional Compressible-Flow Functions for Adiabatic Flow at Constant Area with Friction
For a perfect gas with constant specific heat and molecular weight
$\gamma = 1.4$

M	$\frac{T}{T^*}$	$\frac{p}{p^*}$	$\frac{p_0}{p_0^*}$	$\frac{V}{V^*}$	$\frac{F}{F^*}$	$4\frac{fL_{max}}{D}$	M	$\frac{T}{T^*}$	$\frac{p}{p^*}$	$\frac{p_0}{p_0^*}$	$\frac{V}{V^*}$	$\frac{F}{F^*}$	$4\frac{fL_{max}}{D}$
1.60	.79365	.55679	1.2502	1.4254	1.06348	.17236	**2.00**	.66667	.40825	1.6875	1.6330	1.1227	.30499
1.61	.79030	.55216	1.2583	1.4313	1.36498	.17595	**2.01**	.66371	.40532	1.7017	1.6375	1.1241	.30796
1.62	.78695	.54759	1.2666	1.4371	1.06648	.17953	**2.02**	.66076	.40241	1.7160	1.6420	1.1255	.31091
1.63	.78361	.54308	1.2750	1.4429	1.06798	.18311	**2.03**	.65783	.39954	1.7305	1.6465	1.1269	.31384
1.64	.78028	.53862	1.2835	1.4487	1.06948	.18667	**2.04**	.65491	.39670	1.7452	1.6509	1.1283	.31675
1.65	.77695	.53421	1.2922	1.4544	1.07098	.19022	**2.05**	.65200	.39389	1.7600	1.6553	1.1297	.31965
1.66	.77363	.52986	1.3010	1.4601	1.07249	.19376	**2.06**	.64910	.39110	1.7750	1.6597	1.1311	.32253
1.67	.77033	.52556	1.3099	1.4657	1.07399	.19729	**2.07**	.64621	.38834	1.7902	1.6640	1.1325	.32538
1.68	.76703	.52131	1.3190	1.4713	1.07550	.20081	**2.08**	.64333	.38562	1.8056	1.6683	1.1339	.32822
1.69	.76374	.51711	1.3282	1.4769	1.07701	.20431	**2.09**	.64047	.38292	1.8212	1.6726	1.1352	.33104
1.70	.76046	.51297	1.3376	1.4825	1.07851	.20780	**2.10**	.63762	.38024	1.8369	1.6769	1.1366	.33385
1.71	.75718	.50887	1.3471	1.4880	1.08002	.21128	**2.11**	.63478	.37760	1.8528	1.6811	1.1380	.33664
1.72	.75392	.50482	1.3567	1.4935	1.08152	.21474	**2.12**	.63195	.37498	1.8690	1.6853	1.1393	.33940
1.73	.75067	.50082	1.3665	1.4989	1.08302	.21819	**2.13**	.62914	.37239	1.8853	1.6895	1.1407	.34215
1.74	.74742	.49686	1.3764	1.5043	1.08453	.22162	**2.14**	.62633	.36982	1.9018	1.6936	1.1420	.34488

1.75	.74419	.49295	1.3865	1.5097	1.08603	.22504	**2.15**	.62354	.36728	1.9185	1.6977	1.1434	.34760
1.76	.74096	.48909	1.3967	1.5150	1.08753	.22844	**2.16**	.62076	.36476	1.9354	1.7018	1.1447	.35030
1.77	.73774	.48527	1.4070	1.5203	1.08903	.23183	**2.17**	.61799	.36227	1.9525	1.7059	1.1460	.35298
1.78	.73453	.48149	1.4175	1.5256	1.09053	.23520	**2.18**	.61523	.35980	1.9698	1.7099	1.1474	.35564
1.79	.73134	.47776	1.4282	1.5308	1.09202	.23855	**2.19**	.61249	.35736	1.9873	1.7139	1.1487	.35828
1.80	.72816	.47407	1.4390	1.5360	1.09352	.24189	**2.20**	.60976	.35494	2.0050	1.7179	1.1500	.36091
1.81	.72498	.47042	1.4499	1.5412	1.09500	.24521	**2.21**	.60704	.35254	2.0228	1.7219	1.1513	.36352
1.82	.72181	.46681	1.4610	1.5463	1.09649	.24851	**2.22**	.60433	.35017	2.0409	1.7258	1.1526	.36611
1.83	.71865	.46324	1.4723	1.5514	1.09798	.25180	**2.23**	.60163	.34782	2.0592	1.7297	1.1539	.36868
1.84	.71551	.45972	1.4837	1.5564	1.09946	.25507	**2.24**	.59895	.34550	2.0777	1.7336	1.1552	.37124
1.85	.71238	.45623	1.4952	1.5614	1.1009	.25832	**2.25**	.59627	.34319	2.0964	1.7374	1.1565	.37378
1.86	.70925	.45278	1.5069	1.5664	1.1024	.26156	**2.26**	.59361	.34091	2.1154	1.7412	1.1578	.37630
1.87	.70614	.44937	1.5188	1.5714	1.1039	.26478	**2.27**	.59096	.33865	2.1345	1.7450	1.1590	.37881
1.88	.70304	.44600	1.5308	1.5763	1.1054	.26798	**2.28**	.58833	.33641	2.1538	1.7488	1.1603	.38130
1.89	.69995	.44266	1.5429	1.5812	1.1068	.27116	**2.29**	.58570	.33420	2.1733	1.7526	1.1616	.38377
1.90	.69686	.43936	1.5552	1.5861	1.1083	.27433	**2.30**	.58309	.33200	2.1931	1.7563	1.1629	.38623
1.91	.69379	.43610	1.5677	1.5909	1.1097	.27748	**2.31**	.58049	.32983	2.2131	1.7600	1.1641	.38867
1.92	.69074	.43287	1.5804	1.5957	1.1112	.28061	**2.32**	.57790	.32767	2.2333	1.7637	1.1653	.39109
1.93	.68769	.42967	1.5932	1.6005	1.1126	.28372	**2.33**	.57532	.32554	2.2537	1.7673	1.1666	.39350
1.94	.68465	.42651	1.6062	1.6052	1.1141	.28681	**2.34**	.57276	.32342	2.2744	1.7709	1.1678	.39589
1.95	.68162	.42339	1.6193	1.6099	1.1155	.28989	**2.35**	.57021	.32133	2.2953	1.7745	1.1690	.39826
1.96	.67861	.42030	1.6326	1.6146	1.1170	.29295	**2.36**	.56767	.31925	2.3164	1.7781	1.1703	.40062
1.97	.67561	.41724	1.6461	1.6193	1.1184	.29599	**2.37**	.56514	.31720	2.3377	1.7817	1.1715	.40296
1.98	.67252	.41421	1.6597	1.6239	1.1198	.29901	**2.38**	.56262	.31516	2.3593	1.7852	1.1727	.40528
1.99	.66964	.41121	1.6735	1.6284	1.1213	.30201	**2.39**	.56011	.31314	2.3811	1.7887	1.1739	.40760

Table A.3 (*Concluded*) FANNO LINE—One-Dimensional Compressible-Flow Functions for Adiabatic Flow at Constant Area with Friction

For a perfect gas with constant specific heat and molecular weight

$\gamma = 1.4$

M	$\frac{T}{T^*}$	$\frac{p}{p^*}$	$\frac{p_0}{p_0^*}$	$\frac{V}{V^*}$	$\frac{F}{F^*}$	$4\frac{fL_{max}}{D}$	M	$\frac{T}{T^*}$	$\frac{p}{p^*}$	$\frac{p_0}{p_0^*}$	$\frac{V}{V^*}$	$\frac{F}{F^*}$	$4\frac{fL_{max}}{D}$
2.40	.55762	.31114	2.4031	1.7922	1.1751	.40989	**2.80**	.46729	.24414	3.5001	1.9140	1.2182	.48976
2.41	.55514	.30916	2.4254	1.7956	1.1763	.41216	**2.81**	.46526	.24274	3.5336	1.9167	1.2192	.49148
2.42	.55267	.30720	2.4479	1.7991	1.1775	.41442	**2.82**	.46324	.24135	3.5674	1.9193	1.2202	.49321
2.43	.55021	.30525	2.4706	1.8025	1.1786	.41667	**2.83**	.46122	.23997	3.6015	1.9220	1.2211	.49491
2.44	.54776	.30332	2.4936	1.8059	1.1798	.41891	**2.84**	.45922	.23861	3.6359	1.9246	1.2221	.49660
2.45	.54533	.30141	2.5168	1.8092	1.1810	.42113	**2.85**	.45723	.23726	3.6707	1.9271	1.2230	.49828
2.46	.54291	.29952	2.5403	1.8126	1.1821	.42333	**2.86**	.45525	.23592	3.7058	1.9297	1.2240	.49995
2.47	.54050	.29765	2.5640	1.8159	1.1833	.42551	**2.87**	.45328	.23458	3.7413	1.9322	1.2249	.50161
2.48	.53810	.29579	2.5880	1.8192	1.1844	.42768	**2.88**	.45132	.23326	3.7771	1.9348	1.2258	.50326
2.49	.53571	.29395	2.6122	1.8225	1.1856	.42983	**2.89**	.44937	.23196	3.8133	1.9373	1.2268	.50489
2.50	.53333	.29212	2.6367	1.8257	1.1867	.43197	**2.90**	.44743	.23066	3.8498	1.9398	1.2277	.50651
2.51	.53097	.29031	2.6615	1.8290	1.1879	.43410	**2.91**	.44550	.22937	3.8866	1.9423	1.2286	.50812
2.52	.52862	.28852	2.6865	1.8322	1.1890	.43621	**2.92**	.44358	.22809	3.9238	1.9448	1.2295	.50973
2.53	.52627	.28674	2.7117	1.8354	1.1910	.43831	**2.93**	.44167	.22682	3.9614	1.9472	1.2304	.51133
2.54	.52394	.28498	2.7372	1.8386	1.1912	.44040	**2.94**	.43977	.22556	3.9993	1.9497	1.2313	.51291

2.55	.52163	.28323	2.7630	1.8417	1.1923	.44247
2.56	.51932	.28150	2.7891	1.8448	1.1934	.44452
2.57	.51702	.27978	2.8154	1.8479	1.1945	.44655
2.58	.51474	.27808	2.8420	1.8510	1.1956	.44857
2.59	.51247	.27640	2.8689	1.8541	1.1967	.45059
2.60	.51020	.27473	2.8960	1.8571	1.1978	.45259
2.61	.50795	.27307	2.9234	1.8602	1.1989	.45457
2.62	.50571	.27143	2.9511	1.8632	1.2000	.45654
2.63	.50349	.26980	2.9791	1.8662	1.2011	.45850
2.64	.50127	.26818	3.0074	1.8691	1.2021	.46044
2.65	.49906	.26658	3.0359	1.8721	1.2031	.46237
2.66	.49687	.26499	3.0647	1.8750	1.2042	.46429
2.67	.49469	.26342	3.0938	1.8779	1.2052	.46619
2.68	.49251	.26186	3.1234	1.8808	1.2062	.46807
2.69	.49035	.26032	3.1530	1.8837	1.2073	.46996
2.70	.48820	.25878	3.1830	1.8865	1.2083	.47182
2.71	.48606	.25726	3.2133	1.8894	1.2093	.47367
2.72	.48393	.25575	3.2440	1.8922	1.2103	.47551
2.73	.48182	.25426	3.2749	1.8950	1.2113	.47734
2.74	.47971	.25278	3.3061	1.8978	1.2123	.47915
2.75	.47761	.25131	3.3376	1.9005	1.2133	.48095
2.76	.47553	.24985	3.3695	1.9032	1.2143	.48274
2.77	.47346	.24840	3.4017	1.9060	1.2153	.48452
2.78	.47139	.24697	3.4342	1.9087	1.2163	.48628
2.79	.46933	.24555	3.4670	1.9114	1.2173	.48803

2.95	.43788	.22431	4.0376	1.9521	1.2322	.51447
2.96	.43600	.22307	4.0763	1.9545	1.2331	.51603
2.97	.43413	.22185	4.1153	1.9569	1.2340	.51758
2.98	.43226	.22063	4.1547	1.9592	1.2348	.51912
2.99	.43041	.21942	4.1944	1.9616	1.2357	.52064
3.0	.42857	.21822	4.2346	1.9640	1.2366	.52216
3.5	.34783	.16850	6.7896	2.0642	1.2743	.58643
4.0	.28571	.13363	10.719	2.1381	1.3029	.63306
4.5	.23762	.10833	16.562	2.1936	1.3247	.66764
5.0	.20000	.08944	25.000	2.2361	1.3416	.69380
6.0	.14634	.06376	53.180	2.2953	1.3655	.72987
7.0	.11111	.04762	104.14	2.3333	1.3810	.75280
8.0	.08696	.03686	190.11	2.3591	1.3915	.76819
9.0	.06977	.02935	327.19	2.3772	1.3989	.77898
10.0	.05714	.02390	535.94	2.3905	1.4044	.78683
∞	0	0	∞	2.4495	1.4289	.82153

Table A.4 One-Dimensional Normal-Shock Functions
For a perfect gas with constant specific heat and molecular weight
$\gamma = 1.4$

M_x	M_y	$\frac{p_y}{p_x}$	$\frac{\rho_y}{\rho_x}$	$\frac{T_y}{T_x}$	$\frac{p_{0y}}{p_{0x}}$	$\frac{p_{0y}}{p_x}$	M_x	M_y	$\frac{p_y}{p_x}$	$\frac{\rho_y}{\rho_x}$	$\frac{T_y}{T_x}$	$\frac{p_{0y}}{p_{0x}}$	$\frac{p_{0y}}{p_x}$
1.00	1.00000	1.00000	1.00000	1.00000	1.00000	1.8929	**1.40**	.73971	2.1200	1.6896	1.2547	.95819	3.0493
1.01	.99013	1.02345	1.01669	1.00665	.99999	1.9152	**1.41**	.73554	2.1528	1.7070	1.2612	.95566	3.0844
1.02	.98052	1.04713	1.03344	1.01325	.99998	1.9379	**1.42**	.73144	2.1858	1.7243	1.2676	.95306	3.1198
1.03	.97115	1.07105	1.05024	1.01981	.99997	1.9610	**1.43**	.72741	2.2190	1.7416	1.2742	.95039	3.1555
1.04	.96202	1.09520	1.06709	1.02634	.99994	1.9845	**1.44**	.72345	2.2525	1.7589	1.2807	.94765	3.1915
1.05	.95312	1.1196	1.08398	1.03284	.99987	2.0083	**1.45**	.71956	2.2862	1.7761	1.2872	.94483	3.2278
1.06	.94444	1.1442	1.10092	1.03931	.99976	2.0325	**1.46**	.71574	2.3202	1.7934	1.2938	.94196	3.2643
1.07	.93598	1.1690	1.11790	1.04575	.99962	2.0570	**1.47**	.71198	2.3544	1.8106	1.3004	.93901	3.3011
1.08	.92772	1.1941	1.13492	1.05217	.99944	2.0819	**1.48**	.70829	2.3888	1.8278	1.3070	.93600	3.3382
1.09	.91965	1.2194	1.15199	1.05856	.99921	2.1072	**1.49**	.70466	2.4234	1.8449	1.3136	.93292	3.3756
1.10	.91177	1.2450	1.1691	1.06494	.99892	2.1328	**1.50**	.70109	2.4583	1.8621	1.3202	.92978	3.4133
1.11	.90408	1.2708	1.1862	1.07130	.99858	2.1588	**1.51**	.69758	2.4934	1.8792	1.3269	.92658	3.4512
1.12	.89656	1.2968	1.2034	1.07764	.99820	2.1851	**1.52**	.69413	2.5288	1.8962	1.3336	.92331	3.4894
1.13	.88922	1.3230	1.2206	1.08396	.99776	2.2118	**1.53**	.69073	2.5644	1.9133	1.3403	.91999	3.5279
1.14	.88204	1.3495	1.2378	1.09027	.99726	2.2388	**1.54**	.68739	2.6003	1.9303	1.3470	.91662	3.5667

1.15	.87502	1.3762	1.2550	1.09657	.99669	2.2661	**1.55**	.68410	2.6363	1.9473	1.3538	.91319	3.6058
1.16	.86816	1.4032	1.2723	1.10287	.99605	2.2937	**1.56**	.68086	2.6725	1.9643	1.3606	.90970	3.6451
1.17	.86145	1.4304	1.2896	1.10916	.99534	2.3217	**1.57**	.67768	2.7090	1.9812	1.3674	.90615	3.6847
1.18	.85488	1.4578	1.3069	1.11544	.99455	2.3499	**1.58**	.67455	2.7458	1.9981	1.3742	.90255	3.7245
1.19	.84846	1.4854	1.3243	1.12172	.99371	2.3786	**1.59**	.67147	2.7828	2.0149	1.3811	.89889	3.7645
1.20	.84217	1.5133	1.3416	1.1280	.99280	2.4075	**1.60**	.66844	2.8201	2.0317	1.3880	.89520	3.8049
1.21	.83601	1.5414	1.3590	1.1343	.99180	2.4367	**1.61**	.66545	2.8575	2.0485	1.3949	.89144	3.8456
1.22	.82998	1.5698	1.3764	1.1405	.99073	2.4662	**1.62**	.66251	2.8951	2.0652	1.4018	.88764	3.8866
1.23	.82408	1.5984	1.3938	1.1468	.98957	2.4961	**1.63**	.65962	2.9330	2.0820	1.4088	.88380	3.9278
1.24	.81830	1.6272	1.4112	1.1531	.98835	2.5263	**1.64**	.65677	2.9712	2.0986	1.4158	.87992	3.9693
1.25	.81264	1.6562	1.4286	1.1594	.98706	2.5568	**1.65**	.65396	3.0096	2.1152	1.4228	.87598	4.0111
1.26	.80709	1.6855	1.4460	1.1657	.98568	2.5876	**1.66**	.65119	3.0482	2.1318	1.4298	.87201	4.0531
1.27	.80165	1.7150	1.4634	1.1720	.98422	2.6187	**1.67**	.64847	3.0870	2.1484	1.4369	.86800	4.0954
1.28	.79631	1.7448	1.4808	1.1782	.98268	2.6500	**1.68**	.64579	3.1261	2.1649	1.4440	.86396	4.1379
1.29	.79108	1.7748	1.4983	1.1846	.98106	2.6816	**1.69**	.64315	3.1654	2.1813	1.4512	.85987	4.1807
1.30	.78596	1.8050	1.5157	1.1909	.97935	2.7135	**1.70**	.64055	3.2050	2.1977	1.4583	.85573	4.2238
1.31	.78093	1.8354	1.5331	1.1972	.97758	2.7457	**1.71**	.63798	3.2448	2.2141	1.4655	.85155	4.2672
1.32	.77600	1.8661	1.5505	1.2035	.97574	2.7783	**1.72**	.63545	3.2848	2.2304	1.4727	.84735	4.3108
1.33	.77116	1.8970	1.5680	1.2099	.97382	2.8112	**1.73**	.63296	3.3250	2.2467	1.4800	.84312	4.3547
1.34	.76641	1.9282	1.5854	1.2162	.97181	2.8444	**1.74**	.63051	3.3655	2.2629	1.4873	.83886	4.3989
1.35	.76175	1.9596	1.6028	1.2226	.96972	2.8778	**1.75**	.62809	3.4062	2.2791	1.4946	.83456	4.4433
1.36	.75718	1.9912	1.6202	1.2290	.96756	2.9115	**1.76**	.62570	3.4472	2.2952	1.5019	.83024	4.4880
1.37	.75269	2.0230	1.6376	1.2354	.96534	2.9455	**1.77**	.62335	3.4884	2.3113	1.5093	.82589	4.5330
1.38	.74828	2.0551	1.6550	1.2418	.96304	2.9798	**1.78**	.62104	3.5298	2.3273	1.5167	.82152	4.5783
1.39	.74396	2.0874	1.6723	1.2482	.96065	3.0144	**1.79**	.61875	3.5714	2.3433	1.5241	.81711	4.6238

Table A.4 (*Continued*) One-Dimensional Normal-Shock Functions
For a perfect gas with constant specific heat and molecular weight
$\gamma = 1.4$

M_x	M_y	$\frac{p_y}{p_x}$	$\frac{\rho_y}{\rho_x}$	$\frac{T_y}{T_x}$	$\frac{p_{0y}}{p_{0x}}$	$\frac{p_{0y}}{p_x}$
1.80	.61650	3.6133	2.3592	1.5316	.81268	4.6695
1.81	.61428	3.6554	2.3751	1.5391	.80823	4.7155
1.82	.61209	3.6978	2.3909	1.5466	.80376	4.7618
1.83	.60993	3.7404	2.4067	1.5542	.79926	4.8083
1.84	.60780	3.7832	2.4224	1.5617	.79474	4.8551
1.85	.60570	3.8262	2.4381	1.5694	.79021	4.9022
1.86	.60363	3.8695	2.4537	1.5770	.78567	4.9498
1.87	.60159	3.9130	2.4693	1.5847	.78112	4.9974
1.88	.59957	3.9568	2.4848	1.5924	.77656	5.0453
1.89	.59758	4.0008	2.5003	1.6001	.77197	5.0934
1.90	.59562	4.0450	2.5157	1.6079	.76735	5.1417
1.91	.59368	4.0894	2.5310	1.6157	.76273	5.1904
1.92	.59177	4.1341	2.5463	1.6236	.75812	5.2394
1.93	.58988	4.1790	2.5615	1.6314	.75347	5.2886
1.94	.58802	4.2242	2.5767	1.6394	.74883	5.3381

M_x	M_y	$\frac{p_y}{p_x}$	$\frac{\rho_y}{\rho_x}$	$\frac{T_y}{T_x}$	$\frac{p_{0y}}{p_{0x}}$	$\frac{p_{0y}}{p_x}$
2.20	.54706	5.4800	2.9512	1.8569	.62812	6.7163
2.21	.54572	5.5314	2.9648	1.8657	.62358	6.7730
2.22	.54440	5.5831	2.9783	1.8746	.61905	6.8299
2.23	.54310	5.6350	2.9918	1.8835	.61453	6.8869
2.24	.54182	5.6872	3.0052	1.8924	.61002	6.9442
2.25	.54055	5.7396	3.0186	1.9014	.60554	7.0018
2.26	.53929	5.7922	3.0319	1.9104	.60106	7.0597
2.27	.53805	5.8451	3.0452	1.9194	.59659	7.1178
2.28	.53683	5.8982	3.0584	1.9285	.59214	7.1762
2.29	.53561	5.9515	3.0715	1.9376	.58772	7.2348
2.30	.53441	6.0050	3.0846	1.9468	.58331	7.2937
2.31	.53322	6.0588	3.0976	1.9560	.57891	7.3529
2.32	.53205	6.1128	3.1105	1.9652	.57452	7.4123
2.33	.53089	6.1670	3.1234	1.9745	.57015	7.4720
2.34	.52974	6.2215	3.1362	1.9838	.56580	7.5319

1.95	.58618	4.2696	2.5919	1.6473	.74418	5.3878	**2.35**	.52861	6.2762	3.1490	1.9931	.56148	7.5920
1.96	.58437	4.3152	2.6070	1.6553	.73954	5.4378	**2.36**	.52749	6.3312	3.1617	2.0025	.55717	7.6524
1.97	.58258	4.3610	2.6220	1.6633	.73487	5.4880	**2.37**	.52638	6.3864	3.1743	2.0119	.55288	7.7131
1.98	.58081	4.4071	2.6369	1.6713	.73021	5.5385	**2.38**	.52528	6.4418	3.1869	2.0213	.54862	7.7741
1.99	.57907	4.4534	2.6518	1.6794	.72554	5.5894	**2.39**	.52419	6.4974	3.1994	2.0308	.54438	7.8354
2.00	.57735	4.5000	2.6666	1.6875	.72088	5.6405	**2.40**	.52312	6.5533	3.2119	2.0403	.54015	7.8969
2.01	.57565	4.5468	2.6814	1.6956	.71619	5.6918	**2.41**	.52206	6.6094	3.2243	2.0499	.53594	7.9587
2.02	.57397	4.5938	2.6962	1.7038	.71152	5.7434	**2.42**	.52100	6.6658	3.2366	2.0595	.53175	8.0207
2.03	.57231	4.6411	2.7109	1.7120	.70686	5.7952	**2.43**	.51996	6.7224	3.2489	2.0691	.52758	8.0830
2.04	.57068	4.6886	2.7255	1.7203	.70218	5.8473	**2.44**	.51894	6.7792	3.2611	2.0788	.52344	8.1455
2.05	.56907	4.7363	2.7400	1.7286	.69752	5.8997	**2.45**	.51792	6.8362	3.2733	2.0885	.51932	8.2083
2.06	.56747	4.7842	2.7545	1.7369	.69284	5.9523	**2.46**	.51691	6.8935	3.2854	2.0982	.51521	8.2714
2.07	.56589	4.8324	2.7690	1.7452	.68817	6.0052	**2.47**	.51592	6.9510	3.2975	2.1080	.51112	8.3347
2.08	.56433	4.8808	2.7834	1.7536	.68351	6.0584	**2.48**	.51493	7.0088	3.3095	2.1178	.50706	8.3983
2.09	.56280	4.9295	2.7977	1.7620	.67886	6.1118	**2.49**	.51395	7.0668	3.3214	2.1276	.50303	8.4622
2.10	.56128	4.9784	2.8119	1.7704	.67422	6.1655	**2.50**	.51299	7.1250	3.3333	2.1375	.49902	8.5262
2.11	.55978	5.0275	2.8216	1.7789	.66957	6.2194	**2.51**	.51204	7.1834	3.3451	2.1474	.49502	8.5904
2.12	.55830	5.0768	2.8402	1.7874	.66492	6.2736	**2.52**	.51109	7.2421	3.3569	2.1574	.49104	8.6549
2.13	.55633	5.1264	2.8543	1.7960	.66029	6.3280	**2.53**	.51015	7.3010	3.3686	2.1674	.48709	8.7198
2.14	.55538	5.1762	2.8683	1.8046	.65567	6.3827	**2.54**	.50923	7.3602	3.3802	2.1774	.48317	8.7850
2.15	.55395	5.2262	2.8823	1.8132	.65105	6.4377	**2.55**	.50831	7.4196	3.3918	2.1875	.47927	8.8505
2.16	.55254	5.2765	2.8962	1.8219	.64644	6.4929	**2.56**	.50740	7.4792	3.4034	2.1976	.47540	8.9162
2.17	.55114	5.3270	2.9100	1.8306	.64185	6.5484	**2.57**	.50651	7.5391	3.4149	2.2077	.47155	8.9821
2.18	.54976	5.3778	2.9238	1.8393	.63728	6.6042	**2.58**	.50562	7.5992	3.4263	2.2179	.46772	9.0482
2.19	.54841	5.4288	2.9376	1.8481	.63270	6.6602	**2.59**	.50474	7.6595	3.4376	2.2281	.46391	9.1146

Table A.4 (*Concluded*) One-Dimensional Normal-Shock Functions
For a perfect gas with constant specific heat and molecular weight
$\gamma = 1.4$

M_x	M_y	$\frac{p_y}{p_x}$	$\frac{\rho_y}{\rho_x}$	$\frac{T_y}{T_x}$	$\frac{p_{0y}}{p_{0x}}$	$\frac{p_{0y}}{p_x}$	M_x	M_y	$\frac{p_y}{p_x}$	$\frac{\rho_y}{\rho_x}$	$\frac{T_y}{T_x}$	$\frac{p_{0y}}{p_{0x}}$	$\frac{p_{0y}}{p_x}$
2.60	.50387	7.7200	3.4489	2.2383	.46012	9.1813	**2.85**	.48470	9.3096	3.7139	2.5067	.37330	10.933
2.61	.50301	7.7808	3.4602	2.2486	.45636	9.2481	**2.86**	.48402	9.3762	3.7238	2.5179	.37013	11.006
2.62	.50216	7.8418	3.4714	2.2589	.45262	9.3154	**2.87**	.48334	9.4431	3.7336	2.5292	.36700	11.080
2.63	.50132	7.9030	3.4825	2.2693	.44891	9.3829	**2.88**	.48268	9.5102	3.7434	2.5405	.36389	11.154
2.64	.50048	7.9645	3.4936	2.2797	.44522	9.4507	**2.89**	.48203	9.5775	3.7532	2.5518	.36080	11.228
2.65	.49965	8.0262	3.5047	2.2901	.44155	9.5187	**2.90**	.48138	9.6450	3.7629	2.5632	.35773	11.302
2.66	.49883	8.0882	3.5157	2.3006	.43791	9.5869	**2.91**	.48074	9.7127	3.7725	2.5746	.35469	11.377
2.67	.49802	8.1504	3.5266	2.3111	.43429	9.6553	**2.92**	.48010	9.7808	3.7821	2.5860	.35167	11.452
2.68	.49722	8.2128	3.5374	2.3217	.43070	9.7241	**2.93**	.47946	9.8491	3.7917	2.5975	.34867	11.527
2.69	.49642	8.2754	3.5482	2.3323	.42713	9.7932	**2.94**	.47883	9.9176	3.8012	2.6090	.34570	11.603
2.70	.49563	8.3383	3.5590	2.3429	.42359	9.8625	**2.95**	.47821	9.986	3.8106	2.6206	.34275	11.679
2.71	.49485	8.4014	3.5697	2.3536	.42007	9.9320	**2.96**	.47760	10.055	3.8200	2.6322	.33982	11.755
2.72	.49408	8.4648	3.5803	2.3643	.41657	10.002	**2.97**	.47699	10.124	3.8294	2.6438	.33692	11.831
2.73	.49332	8.5284	3.5909	2.3750	.41310	10.072	**2.98**	.47638	10.194	3.8387	2.6555	.33404	11.907
2.74	.49256	8.5922	3.6014	2.3858	.40965	10.142	**2.99**	.47578	10.263	3.8479	2.6672	.33118	11.984
2.75	.49181	8.6562	3.6119	2.3966	.40622	10.212	**3.00**	.47519	10.333	3.8571	2.6790	.32834	12.061
2.76	.49107	8.7205	3.6224	2.4074	.40282	10.283	**3.50**	.45115	14.125	4.2608	3.3150	.21295	16.242
2.77	.49033	8.7850	3.6328	2.4183	.39945	10.354	**4.00**	.43496	18.500	4.5714	4.0469	.13876	21.068
2.78	.48960	8.8497	3.6431	2.4292	.39610	10.426	**4.50**	.42355	23.458	4.8119	4.8751	.09170	26.539
2.79	.48888	8.9147	3.6533	2.4402	.39276	10.498	**5.00**	.41523	29.000	5.0000	5.8000	.06172	32.654
2.80	.48817	8.9800	3.6635	2.4512	.38946	10.569	**6.00**	.40416	41.833	5.2683	7.941	.02965	46.815
2.81	.48746	9.0454	3.6737	2.4622	.38618	10.641	**7.00**	.39736	57.000	5.4444	10.469	.01535	63.552
2.82	.48676	9.1111	3.6838	2.4733	.38293	10.714	**8.00**	.39289	74.500	5.5652	13.387	.00849	82.865
2.83	.48607	9.1770	3.6939	2.4844	.37970	10.787	**9.00**	.38980	94.333	5.6512	16.693	.00496	104.753
2.84	.48538	9.2432	3.7039	2.4955	.37649	10.860	**10.00**	.38757	116.50	5.7143	20.388	.00304	129.217
							∞	.37796	∞	6.000	∞	0	∞

Table A.5 Two-Dimensional Isentropic Compressible Flow Functions for Method of Characteristics
For a perfect gas with constant specific heat and molecular weight
$\gamma = 1.4$

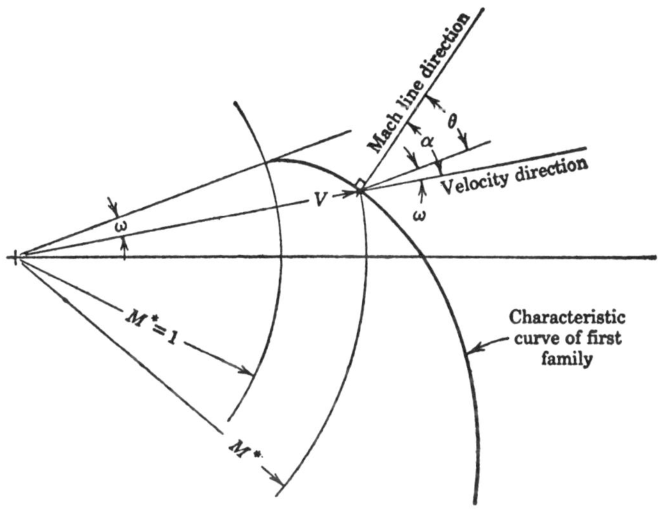

ω	α	θ	M	M*	$\frac{A}{A^*}$	$\frac{p}{p_0}$	$\frac{\rho}{\rho_0}$
0.0	90.0000	90.0000	1.0000	1.0000	1.0000	.52828	.63394
0.5	72.0988	71.5988	1.0509	1.0418	1.0021	.49735	.60720
1.0	67.5741	66.5741	1.0818	1.0668	1.0053	.47898	.59110
1.5	64.4505	62.9505	1.1084	1.0879	1.0093	.46350	.57738
2.0	61.9969	59.9969	1.1326	1.1068	1.0137	.44964	.56500
2.5	59.9500	57.4500	1.1553	1.1244	1.0187	.43688	.55350
3.0	58.1805	55.1805	1.1769	1.1408	1.0240	.42494	.54265
3.5	56.6139	53.1139	1.1976	1.1565	1.0297	.41365	.53231
4.0	55.2048	51.2048	1.2177	1.1715	1.0358	.40291	.52240
4.5	53.9204	49.4204	1.2373	1.1859	1.0423	.39263	.51284
5.0	52.7383	47.7383	1.2565	1.1999	1.0491	.38274	.50358
5.5	51.6419	46.1419	1.2753	1.2135	1.0562	.37320	.49459
6.0	50.6186	44.6186	1.2938	1.2267	1.0637	.36398	.48583
6.5	49.6583	43.1583	1.3120	1.2396	1.0715	.35506	.47729
7.0	48.7528	41.7528	1.3300	1.2522	1.0796	.34640	.46895
7.5	47.8957	40.3957	1.3478	1.2645	1.0880	.33798	.46078
8.0	47.0818	39.0818	1.3655	1.2766	1.0967	.32979	.45278
8.5	46.3065	37.8065	1.3830	1.2885	1.1058	.32182	.44493
9.0	45.5660	36.5660	1.4004	1.3002	1.1152	.31404	.43723
9.5	44.8570	35.3570	1.4177	1.3117	1.1249	.30646	.42966

Table A.5 (*Continued*) Method of Characteristics
For a perfect gas with constant specific heat and molecular weight
$\gamma = 1.4$

ω	α	θ	M	M*	$\frac{A}{A^*}$	$\frac{p}{p_0}$	$\frac{\rho}{\rho_0}$
10.0	44.1770	34.1770	1.4349	1.3230	1.1349	.29906	.42222
10.5	43.5233	33.0233	1.4521	1.3341	1.1453	.29184	.41491
11.0	42.8940	31.8940	1.4692	1.3451	1.1560	.28478	.40772
11.5	42.2869	30.7869	1.4862	1.3559	1.1670	.27788	.40064
12.0	41.7007	29.7007	1.5032	1.3666	1.1783	.27114	.39367
12.5	41.1338	28.6338	1.5202	1.3772	1.1900	.26454	.38680
13.0	40.5849	27.5849	1.5371	1.3876	1.2021	.25809	.38004
13.5	40.0529	26.5529	1.5540	1.3979	1.2145	.25178	.37338
14.0	39.5366	25.5366	1.5709	1.4081	1.2273	.24560	.36681
14.5	39.0350	24.5350	1.5878	1.4182	1.2405	.23955	.36034
15.0	38.5474	23.5474	1.6047	1.4282	1.2541	.23363	.35396
15.5	38.0730	22.5730	1.6216	1.4380	1.2680	.22783	.34766
16.0	37.6108	21.6108	1.6385	1.4478	1.2823	.22216	.34145
16.5	37.1605	20.6605	1.6555	1.4575	1.2970	.21661	.33533
17.0	36.7212	19.7212	1.6725	1.4671	1.3121	.21117	.32929
17.5	36.2925	18.7925	1.6895	1.4766	1.3277	.20584	.32334
18.0	35.8739	17.8739	1.7065	1.4860	1.3437	.20062	.31747
18.5	35.4648	16.9648	1.7235	1.4953	1.3602	.19551	.31168
19.0	35.0648	16.0648	1.7406	1.5046	1.3771	.19051	.30596
19.5	34.6735	15.1735	1.7578	1.5138	1.3945	.18562	.30032
20.0	34.2904	14.2904	1.7750	1.5229	1.4123	.18082	.29475
20.5	33.9153	13.4153	1.7922	1.5319	1.4306	.17612	.28926
21.0	33.5479	12.5479	1.8095	1.5409	1.4494	.17152	.28385
21.5	33.1877	11.6877	1.8269	1.5498	1.4687	.16702	.27851
22.0	32.8344	10.8344	1.8443	1.5586	1.4886	.16261	.27324
22.5	32.4879	9.9879	1.8618	1.5673	1.5090	.15830	.26804
23.0	32.1478	9.1478	1.8793	1.5760	1.5300	.15408	.26291
23.5	31.8138	8.3138	1.8969	1.5846	1.5515	.14995	.25786
24.0	31.4859	7.4859	1.9146	1.5932	1.5736	.14590	.25287
24.5	31.1637	6.6637	1.9324	1.6017	1.5963	.14194	.24795
25.0	30.8469	5.8469	1.9503	1.6101	1.6197	.13806	.24310
25.5	30.5355	5.0355	1.9682	1.6184	1.6437	.13427	.23831
26.0	30.2293	4.2293	1.9862	1.6267	1.6683	.13057	.23359
26.5	29.9281	3.4281	2.0044	1.6350	1.6936	.12694	.22894
27.0	29.6316	2.6316	2.0226	1.6432	1.7196	.12339	.22435
27.5	29.3397	1.8397	2.0409	1.6513	1.7464	.11992	.21982
28.0	29.0524	1.0524	2.0593	1.6594	1.7739	.11653	.21536
28.5	28.7694	0.2694	2.0778	1.6674	1.8022	.11321	.21097
29.0	28.4906	−0.5094	2.0964	1.6753	1.8312	.10997	.20664
29.5	28.2158	−1.2842	2.1151	1.6832	1.8611	.10680	.20237

Table A.5 (*Continued*) Method of Characteristics
For a perfect gas with constant specific heat and molecular weight
$\gamma = 1.4$

ω	α	θ	M	M*	$\frac{A}{A^*}$	$\frac{p}{p_0}$	$\frac{\rho}{\rho_0}$
30.0	27.9451	−2.0549	2.1339	1.6911	1.8918	.10370	.19816
30.5	27.6782	−2.8218	2.1528	1.6989	1.9233	.10068	.19401
31.0	27.4149	−3.5851	2.1718	1.7066	1.9557	.09773	.18992
31.5	27.1552	−4.3448	2.1910	1.7143	1.9891	.09484	.18590
32.0	26.8991	−5.1009	2.2103	1.7220	2.0235	.09202	.18194
32.5	26.6464	−5.8536	2.2297	1.7296	2.0588	.08927	.17804
33.0	26.3970	−6.6030	2.2492	1.7371	2.0951	.08658	.17419
33.5	26.1507	−7.3493	2.2689	1.7446	2.1324	.08395	.17040
34.0	25.9076	−8.0924	2.2887	1.7521	2.1709	.08139	.16667
34.5	25.6675	−8.8325	2.3086	1.7595	2.2105	.07889	.16300
35.0	25.4304	−9.5696	2.3287	1.7669	2.2512	.07646	.15938
35.5	25.1962	−10.3038	2.3489	1.7742	2.2931	.07408	.15582
36.0	24.9648	−11.0352	2.3693	1.7814	2.3363	.07176	.15232
36.5	24.7361	−11.7639	2.3898	1.7886	2.3807	.06950	.14887
37.0	24.5101	−12.4899	2.4105	1.7958	2.4264	.06729	.14548
37.5	24.2866	−13.2134	2.4313	1.8029	2.4736	.06514	.14215
38.0	24.0657	−13.9343	2.4523	1.8100	2.5222	.06304	.13886
38.5	23.8473	−14.6527	2.4734	1.8170	2.5722	.06100	.13563
39.0	23.6313	−15.3687	2.4947	1.8240	2.6237	.05901	.13246
39.5	23.4176	−16.0824	2.5162	1.8310	2.6768	.05707	.12934
40.0	23.2061	−16.7939	2.5378	1.8379	2.7316	.05519	.12627
40.5	22.9969	−17.5031	2.5596	1.8447	2.7881	.05335	.12325
41.0	22.7900	−18.2100	2.5816	1.8515	2.8463	.05156	.12029
41.5	22.5852	−18.9148	2.6038	1.8583	2.9063	.04982	.11738
42.0	22.3824	−19.6176	2.6261	1.8650	2.9682	.04813	.11452
42.5	22.1816	−20.3184	2.6487	1.8717	3.0321	.04648	.11170
43.0	21.9828	−21.0172	2.6714	1.8783	3.0981	.04488	.10894
43.5	21.7860	−21.7140	2.6944	1.8849	3.1662	.04332	.10622
44.0	21.5911	−22.4089	2.7176	1.8915	3.2364	.04181	.10356
44.5	21.3980	−23.1020	2.7409	1.8980	3.3089	.04034	.10094
45.0	21.2068	−23.7932	2.7644	1.9045	3.3838	.03891	.09837
45.5	21.0174	−24.4826	2.7882	1.9109	3.4611	.03752	.09585
46.0	20.8297	−25.1703	2.8122	1.9173	3.5410	.03617	.09338
46.5	20.6437	−25.8563	2.8364	1.9236	3.6236	.03486	.09095
47.0	20.4594	−26.5406	2.8609	1.9299	3.7089	.03359	.08856
47.5	20.2767	−27.2233	2.8856	1.9362	3.7971	.03235	.08622
48.0	20.0956	−27.9044	2.9105	1.9424	3.8883	.03115	.08393
48.5	19.9160	−28.5840	2.9356	1.9486	3.9827	.02999	.08168
49.0	19.7380	−29.2620	2.9610	1.9547	4.0803	.02886	.07948
49.5	19.5615	−29.9385	2.9867	1.9608	4.1812	.02777	.07732

Table A.5 (*Continued*) Method of Characteristics
For a perfect gas with constant specific heat and molecular weight
$\gamma = 1.4$

ω	α	θ	M	M*	$\frac{A}{A^*}$	$\frac{p}{p_0}$	$\frac{\rho}{\rho_0}$
50.0	19.3865	−30.6135	3.0126	1.9669	4.2857	.02671	.07520
50.5	19.2129	−31.2871	3.0388	1.9729	4.3938	.02568	.07313
51.0	19.0408	−31.9592	3.0652	1.9789	4.5058	.02469	.07110
51.5	18.8700	−32.6300	3.0919	1.9848	4.6218	.02373	.06911
52.0	18.7005	−33.2995	3.1189	1.9907	4.7419	.02280	.06716
52.5	18.5324	−33.9676	3.1462	1.9966	4.8663	.02190	.06525
53.0	18.3657	−34.6343	3.1738	2.0024	4.9953	.02103	.06338
53.5	18.2002	−35.2998	3.2016	2.0082	5.1290	.02018	.06155
54.0	18.0360	−35.9640	3.2298	2.0139	5.2676	.01936	.05976
54.5	17.8730	−36.6270	3.2583	2.0196	5.4114	.01857	.05801
55.0	17.7112	−37.2888	3.2871	2.0253	5.5606	.01781	.05629
55.5	17.5506	−37.9494	3.3162	2.0309	5.7154	.01707	.05461
56.0	17.3911	−38.6089	3.3457	2.0365	5.8761	.01636	.05297
56.5	17.2328	−39.2672	3.3755	2.0421	6.0429	.01567	.05137
57.0	17.0757	−39.9243	3.4056	2.0476	6.2162	.01500	.04981
57.5	16.9196	−40.5804	3.4361	2.0531	6.3963	.01436	.04828
58.0	16.7646	−41.2354	3.4669	2.0585	6.5834	.01374	.04678
58.5	16.6107	−41.8893	3.4981	2.0639	6.7778	.01314	.04532
59.0	16.4579	−42.5421	3.5297	2.0692	6.9799	.01257	.04389
59.5	16.3061	−43.1939	3.5616	2.0745	7.1902	.01202	.04250
60.0	16.1552	−43.8448	3.5940	2.0798	7.4090	.01148	.04114
60.5	16.0053	−44.4947	3.6268	2.0850	7.6368	.01096	.03981
61.0	15.8564	−45.1436	3.6600	2.0902	7.8739	.01047	.03851
61.5	15.7085	−45.7915	3.6936	2.0954	8.1208	.00999	.03725
62.0	15.5615	−46.4385	3.7276	2.1005	8.3780	.00953	.03602
62.5	15.4154	−47.0846	3.7620	2.1056	8.6460	.00909	.03482
63.0	15.2703	−47.7297	3.7969	2.1107	8.9254	.00867	.03365
63.5	15.1260	−48.3740	3.8323	2.1157	9.2168	.00826	.03251
64.0	14.9826	−49.0174	3.8681	2.1207	9.5208	.00786	.03140
64.5	14.8400	−49.6600	3.9044	2.1256	9.8380	.00749	.03032
65.0	14.6983	−50.3017	3.9412	2.1305	10.169	.00712	.02926
65.5	14.5574	−50.9426	3.9785	2.1353	10.515	.00678	.02823
66.0	14.4174	−51.5826	4.0163	2.1401	10.876	.00644	.02723
66.5	14.2781	−52.2219	4.0547	2.1449	11.253	.00612	.02626
67.0	14.1396	−52.8604	4.0936	2.1497	11.648	.00582	.02532
67.5	14.0019	−53.4981	4.1330	2.1544	12.061	.00552	.02440
68.0	13.8650	−54.1350	4.1730	2.1591	12.493	.00524	.02350
68.5	13.7288	−54.7712	4.2136	2.1637	12.945	.00497	.02263
69.0	13.5934	−55.4066	4.2548	2.1683	13.418	.00472	.02179
69.5	13.4587	−56.0413	4.2966	2.1728	13.913	.00447	.02097

Table A.5 (*Concluded*) Method of Characteristics
For a perfect gas with constant specific heat and molecular weight
$\gamma = 1.4$

ω	α	θ	M	M^*	$\frac{A}{A^*}$	$\frac{p}{p_0}$	$\frac{\rho}{\rho_0}$
70.0	13.3247	−56.6753	4.3390	2.1773	14.433	.00423	.02017
70.5	13.1913	−57.3087	4.3821	2.1818	14.978	.00401	.01940
71.0	13.0587	−57.9413	4.4258	2.1863	15.549	.00379	.01865
71.5	12.9268	−58.5732	4.4702	2.1907	16.148	.00359	.01792
72.0	12.7955	−59.2045	4.5152	2.1951	16.777	.00339	.01721
72.5	12.6649	−59.8351	4.5610	2.1994	17.438	.00320	.01652
73.0	12.5349	−60.4651	4.6076	2.2037	18.132	.00302	.01586
73.5	12.4055	−61.0945	4.6549	2.2080	18.862	.00285	.01522
74.0	12.2768	−61.7232	4.7029	2.2122	19.630	.00269	.01460
74.5	12.1487	−62.3513	4.7517	2.2164	20.438	.00254	.01400
75.0	12.0212	−62.9788	4.8014	2.2205	21.288	.00239	.01342
75.5	11.8943	−63.6057	4.8519	2.2246	22.183	.00225	.01285
76.0	11.7680	−64.2320	4.9032	2.2287	23.126	.00212	.01230
76.5	11.6422	−64.8578	4.9554	2.2327	24.121	.00199	.01177
77.0	11.5170	−65.4830	5.0085	2.2367	25.171	.00187	.01126
77.5	11.3924	−66.1076	5.0626	2.2407	26.279	.00176	.01077
78.0	11.2683	−66.7317	5.1176	2.2446	27.449	.00165	.01029
78.5	11.1447	−67.3553	5.1736	2.2485	28.685	.00155	.00983
79.0	11.0217	−67.9783	5.2306	2.2523	29.992	.00145	.00939
79.5	10.8992	−68.6008	5.2887	2.2561	31.374	.00136	.00896
80.0	10.7772	−69.2228	5.3479	2.2599	32.837	.00127	.00854
80.5	10.6558	−69.8442	5.4081	2.2636	34.387	.00119	.00814
81.0	10.5348	−70.4652	5.4694	2.2673	36.029	.00111	.00776
81.5	10.4143	−71.0857	5.5320	2.2710	37.770	.00104	.00739
82.0	10.2942	−71.7058	5.5959	2.2746	39.618	.00097	.00703
82.5	10.1746	−72.3254	5.6610	2.2782	41.580	.00090	.00669
83.0	10.0555	−72.9445	5.7274	2.2818	43.665	.00084	.00636
83.5	9.9369	−73.5631	5.7950	2.2853	45.882	.00078	.00604
84.0	9.8187	−74.1813	5.8640	2.2888	48.240	.00073	.00574
84.5	9.7010	−74.7990	5.9345	2.2922	50.750	.00068	.00545
85.0	9.5837	−75.4163	6.0064	2.2956	53.424	.00063	.00517
85.5	9.4668	−76.0332	6.0799	2.2990	56.276	.00058	.00490
86.0	9.3503	−76.6497	6.1550	2.3023	59.320	.00054	.00464
86.5	9.2342	−77.2658	6.2317	2.3056	62.570	.00050	.00439
87.0	9.1185	−77.8815	6.3101	2.3088	66.043	.00046	.00416
87.5	9.0032	−78.4968	6.3902	2.3120	69.758	.00043	.00393
88.0	8.8884	−79.1116	6.4720	2.3152	73.734	.00040	.00371
88.5	8.7740	−79.7260	6.5558	2.3183	77.994	.00037	.00351
89.0	8.6599	−80.3401	6.6415	2.3214	82.561	.00034	.00331
89.5	8.5462	−80.9538	6.7292	2.3245	87.463	.00031	.00312
90.0	8.4328	−81.5672	6.8190	2.3275	92.730	.00029	.00294

Table A.6 Standard Atmosphere

Altitude	Temperature		Pressure		Density		Speed of Sound	Absolute Viscosity	Kinematic Viscosity
ft	°F	R	in. Hg	lb/ft^2	lb/ft^3	ratio, ρ/ρ_0	mi/hr	lbm/ft sec	ft^2/sec
0	58.7	518.4	29.92	2116.	.002378	1.0000	760.9	3.725×10^{-7}	1.566×10^{-4}
2000	51.5	511.2	27.82	1968.	.002242	.9428	755.7	3.685	1.644
4000	44.4	504.1	25.84	1828.	.002112	.8881	750.4	3.644	1.725
6000	37.3	497.0	23.98	1696.	.001988	.8358	745.1	3.602	1.812
8000	30.2	489.9	22.22	1572.	.001869	.7859	739.7	3.561	1.905
10000	23.0	482.7	20.58	1455.	.001756	.7384	734.3	3.519	2.004
12000	15.9	475.6	19.03	1346.	.001648	.6931	728.8	3.476	2.109
14000	8.8	468.5	17.57	1243.	.001545	.6499	723.4	3.434	2.223
16000	1.6	461.3	16.21	1146.	.001448	.6088	718.7	3.391	2.342
18000	−5.5	454.2	14.94	1056.	.001355	.5698	712.2	3.348	2.471
20000	−12.6	447.1	13.75	972.1	.001267	.5327	706.6	3.305	2.608
22000	−19.8	439.9	12.63	893.3	.001183	.4974	701.1	3.261	2.756
24000	−26.9	432.8	11.59	819.8	.001103	.4640	695.3	3.217	2.916
26000	−34.0	425.7	10.62	751.2	.001028	.4323	689.5	3.173	3.086
28000	−41.2	418.5	9.720	687.4	.000957	.4023	683.7	3.128	3.268
30000	−48.3	411.4	8.880	628.0	.000889	.3740	677.9	3.083	3.468

32000	−55.4	404.3	8.101	572.9	.000826	.3472	672.0	3.038	3.678
34000	−62.5	397.2	7.377	521.7	.000765	.3218	666.0	2.992	3.911
35332	−67.3	392.4	6.926	489.8	.000727	.3058	662.0	2.962	4.073
36000	−67.3	392.4	6.711	474.4	.000705	.2963	662.0	2.962	4.204
38000	−67.3	392.4	6.098	431.1	.000640	.2692	662.0	2.962	4.625
40000	−67.3	392.4	5.544	391.9	.000582	.2448	662.0	2.962	5.089
42000	−67.3	392.4	5.038	356.2	.000529	.2225	662.0	2.962	5.599
44000	−67.3	392.4	4.578	323.7	.000480	.2021	662.0	2.962	6.161
46000	−67.3	392.4	4.162	294.2	.000437	.1838	662.0	2.962	6.778
48000	−67.3	392.4	3.782	267.4	.000397	.1670	662.0	2.962	7.459
50000	−67.3	392.4	3.438	243.1	.000361	.1518	662.0	2.962	8.206
52000	−67.3	392.4	3.124	220.9	.000328	.1379	662.0	2.962	9.028
54000	−67.3	392.4	2.840	200.8	.000298	.1254	662.0	2.962	9.933
56000	−67.3	392.4	2.581	182.5	.000271	.1140	662.0	2.962	10.93
58000	−67.3	392.4	2.346	165.9	.000246	.1036	662.0	2.962	12.02
60000	−67.3	392.4	2.132	150.8	.000224	.09415	662.0	2.962	13.23
62000	−67.3	392.4	1.938	137.1	.000203	.08557	662.0	2.962	14.56
64000	−67.3	392.4	1.761	124.6	.000185	.07777	662.0	2.962	16.02
65000	−67.3	392.4	1.679	118.7	.000176	.07414	662.0	2.962	16.80
70000	−67.3	392.4	1.322	93.53	.000139	.05839	662.0	2.962	21.33
75000	−67.3	392.4	1.042	73.66	.000109	.04599	662.0	2.962	27.09
80000	−67.3	392.4	.8202	58.01	.0000861	.03621	662.0	2.962	34.39
85000	−67.3	392.4	.6460	45.68	.0000678	.02852	662.0	2.962	43.67
90000	−67.3	392.4	.5086	35.97	.0000534	.02246	662.0	2.962	55.45
95000	−67.3	392.4	.4006	28.33	.0000421	.01769	662.0	2.962	70.41
100000	−67.3	392.4	.3156	22.31	.0000331	.01394	662.0	2.962	89.41

Index